Systems Science

Addressing Global Issues

Systems Science

Addressing Global Issues

Edited by

Frank A. Stowell
Daune West
and James G. Howell

University of Paisley
Paisley, Scotland, United Kingdom

Plenum Press • New York and London

Library of Congress Cataloging-in-Publication Data

Systems Science : addressing global issues / edited by Frank A.
Stowell, Daune West, James G. Howell.
p. cm.
"Proceedings of the United Kingdom Systems Society Conference on
Systems Science: Addressing Global Issues, held July 27-30, 1993, in
Paisley, Scotland, United Kingdom"--T.p. verso.
Includes bibliographical references and index.
ISBN 0-306-44522-0
1. Social sciences--Methodology--Congresses. 2. Systems theory-
-Congresses. 3. System analysis--Congresses. 4. Cybernetics-
-Congresses. 5. Social systems--Congresses. I. Stowell, Frank A.
II. West, Daune. III. Howell, James G. IV. United Kingdom Systems
Society Conference on Systems Science: Addressing Global Issues
(1993 : Paisley, Scotland)
H61.S995 1993
300--dc20 93-19328
CIP

Proceedings of the United Kingdom Systems Society Conference on Systems Science: Addressing Global Issues, held July 27–30, 1993, in Paisley, Scotland, United Kingdom

ISBN 0-306-44522-0

Printed in the United States of America

PREFACE

As we approach the end of the 20th century we can look back upon the achievements that have been made in a variety of human endeavours with pride. Enormous strides have been made to improve the quality of life of millions of people through the application of the scientific discoveries made during this and past centuries. The 20th century will be remembered as much for the mass exploitation of scientific discovery as for the discoveries themselves.

The technological age has meant that the human being is able to contemplate activities which "defy" nature. For example, some of the work involved in the preparation of these proceedings has been done whilst travelling at over 500 miles per hour seven miles above the surface of the earth. It is not difficult to conjecture about the effect that this relatively recent technology has had upon a number of "systems". Air transportation has provided a number of benefits including such disparate examples such as enabling holidays, famine relief and the cross fertilisation of cultural practices from other lands. Equally, there have been undesirable effects such as enabling the means of mass destruction, interference in other cultures and the speedy transportation of disease. Moreover, the physical presence of the aeroplane itself represents the consumption of fossil fuels, a source of pollution and a change in the way think about life.

The view expressed here is of course the view of an inhabitant of the "western world". The changes to the way in which the entire inhabitants of the planet live are impossible for one person to express. Many of our neighbours may have different views about the changes that the technological age has brought upon their lives. It is not difficult to imagine the fear and lack of control that some feel as technological progress alters their lives and, in some cases, as the output of technology pollutes their environment.

The need to be aware of the implications of our actions has never been greater than it is now. No time is more appropriate for us to review the way in which we introduce change and weigh carefully the implications of our actions upon the wider system. Technology has provided us with the means of assessing the effects of our actions before we take them and our awareness of the mistakes of the past should provide the motivation to ensure that we do.

Most of us are concerned with day-to-day problems which need to be solved in order to make some progress in our particular area of interest. It is "Global" in this sense that is the inspiration behind the theme for the 3rd UKSS international conference. The conference has provided the opportunity for Systems Scientists from many nations to debate the contribution that Systems Science is making to specific areas of human endeavours and to consider what further contributions might be made.

This text contains papers which are concerned with issues of a truly global nature but, equally important, it contains a larger number of papers in which Systems Thinking has been used to assist in problems which are more commonplace and within the competence of an individual to control. There are papers within this publication which deal with practical problems which serve as examples for those concerned with problems which need to be addressed here-and-now. It is such commonplace problems that makes the issues discussed within this text truly global - in the sense that many of us have or are experiencing similar difficulties. The book contains examples of the application of Systems Science to a variety of problem domains and the way in which the kind of thinking that this epistemology demands has shaped the outcome. The reader must judge to what extent those outcomes represent an improvement over alternative approaches and the value to them.

The division of the papers within this text reflects the way that the streams were organised for the conference as much as an attempt to classify papers into topic areas. Not surprisingly there is overlap to be found between topics. Systems Science is dedicated to the notion of holism and should not be considered a specialist branch of intellectual pursuit but a kind of thinking which may be appropriate to all domains of knowledge.

Some readers may find that they will classify the papers into "Hard" and "Soft" systems thinking. It is a classification which may provide intellectual satisfaction for some but is considered of limited value within the context of the purpose of this text. The conference was devoted to an exposition of the practice of Systems thinking and to raising the awareness of practitioners about diverse problems in which the thinking is used. Any debate about "Hard" or "Soft" systems thinking seems subordinate to that purpose.

On the whole the view taken in this preface is an optimistic one which expresses a naive belief that, not withstanding the problems which we are now experiencing, there will be a solution found for them. The support shown for the conference and the papers printed within this volume will, it is hoped, make a small contribution to the continued use of systems ideas in addressing problems of all kinds.

Frank Stowell
Information Systems
Department of Computing Science
University of Paisley

Programme Committee

Professor F A Stowell
Dr D West
Mr J G Howell

Acknowledgements

The Committee would like to express their thanks, and the thanks of the United Kingdom Systems Society, to the members of the Department of Computing Science and the University of Paisley for the help and the strong support given to the Conference throughout. In particular, the Committee wishes to acknowledge the support given by the Principal, Professor Richard Shaw, and the Secretary to the University, Mr James Fraser. Special thanks and a debt of gratitude is expressed to Irene McKeown for her unfailing support, clear thinking and total involvement in helping organise this Conference and in the preparation of these proceedings.

The Committee and the Society would also like to express their thanks to those who have sponsored this event:

Renfrew District Council
Renfrewshire Enterprise
Compaq Computer Manufacturing Ltd.
and
The University of Paisley

International Advisory Committee

Professor P.B. Checkland, University of Lancaster
Professor F.A. Stowell, University of Paisley
Professor R. Vallée, France
Professor R. Rodriguez-Delgado, Spain
Professor A. Ghosal, India
Dr P. Marshall, Australia
Dr Elohim, J.L., Austria
Professor G. de Zeeuw, Amsterdam

CONTENTS

PLENARY PAPERS

BUSINESS SYSTEMS

COMPLEX SYSTEMS

CRITICAL SYSTEMS

CYBERNETICS

EDUCATIONAL SYSTEMS

ENVIRONMENTAL SYSTEMS, SOCIAL SYSTEMS, AND HEALTH SYSTEMS

INFORMATION SYSTEMS

MANUFACTURING SYSTEMS

STRATEGIC IS AND SYSTEMS METHODOLOGIES

THEORY AND CONCEPTS

SYSTEMS SCIENCE - ADDRESSING GLOBAL ISSUES

David Smith

University of Edinburgh
Old College
South Bridge
Edinburgh EH8 9YL

INTRODUCTION

This keynote address is given from the standpoint of a reductionist scientist who is deeply concerned about global issues, but who accepts that the problems they present are far too complex to be solved by the methods and approach of reductionist science. My purpose is to illustrate the huge dimensions of these issues, and to underline their crucial importance to the future of human society. They are problems which must be solved, and their baffling complexity indicates that they are firmly in the realm of systems science.

Any discussion of global issues as they affect human society must be set in the context of the global environment itself as a large, complex system.

THE EARTH AS A SYSTEM - THE GAIA HYPOTHESIS

Many believe that the earth's environmental conditions are entirely determined by inanimate geological and climatological factors, and that life must adapt to them or perish. This is not correct. In reality, the biosphere exerts a major influence on key features of the environment: composition of the atmosphere, global temperature, availability of essential plant nutrients, and aspects of the reactive chemistry of the earth's surface such as oxidation state and acidity.

The essence of the 'Gaia' hypothesis is that 'life' - the totality of living organisms and their activities both present and over the past 4 billion years - has modified, maintained and in certain respects created environmental conditions favourable to its continuation on this planet. The hypothesis postulates that the earth behaves as if it were a single large living organism. It does not confer mystical or sentient properties on the earth, as some wrongly believe. It is merely a way of saying that like a living organism, the global environment behaves as a single integrated system with numerous subsystems and a very complex network of feedback mechanisms.

The Gaia hypothesis was developed by James Lovelock (1979, 1988, 1992; and see also

Allaby, 1989), initially from observations that the earth's surface temperature and the composition of its atmosphere were highly anomalous for its position in the solar system. Comparison with our neighbouring planets of Mars and Venus led to a straightforward prediction of what the composition of the earth's atmosphere *ought* to be like: about 95% carbon dioxide, 2 - 4% nitrogen and virtually no oxygen. Such a large amount of carbon dioxide would give a surface temperature of about 300°C, and hence no liquid water.

The biosphere plays a key role in determining the anomalous composition of the earth's atmosphere. The high oxygen content results from photosynthesis, and until about 2 billion years ago, little or no free oxygen was present. Oxygen continually turns over, and feedback processes maintain its level at 21% of the atmosphere. If it rose to 25%, damp vegetation could burn and widespread conflagration would result; at 30%, even wet plant material could burn. At levels significantly below 21%, biological processes requiring high energy (such as flying or prolonged rapid movement) would be difficult. Because oxygen is very reactive, it participates in interactions which help keep the atmospheric composition constant and the level of certain natural pollutants at a minimum. It also gives rise to ozone which filters out potentially lethal UV radiation from the sun.

Atmospheric nitrogen is mostly produced by denitrifying bacteria in the soil. As an inert gas, it 'dilutes' oxygen, and its abundance (79%) is important in maintaining atmospheric pressure, without which water would boil away and life could not exist. If atmopheric pressure fell by 10%, the earth's temperature would fall by 4°C.

The biosphere is important in maintaining carbon dioxide at its low concentration of 0.03% - the principal reason why, compared to neighbouring planets, the earth is much cooler than would be expected from its proximity to the sun. It is a key component of the relatively thin blanket of greenhouse gases (nearly all of biological origin) which maintains the present average surface temperature of 13°C; without any blanket, the temperature would be 33°C lower. If life was eliminated from the earth, the disappearance of green plants would result in CO_2 levels rising to over 95% and a runaway greenhouse effect would ensue (Margulis and Lovelock, 1974).

Even water vapour, the most abundant greenhouse gas, is also affected by the biosphere, through evapotranspiration from large tracts of vegetation on land and, according to Lovelock, through release over the oceans by plankton of critical amounts of dimethyl sulphide, which is converted to sulphates which stimulate the condensation of water vapour into clouds. Clouds have a cooling effect by reflecting solar radiation back into space, and thus act counter to the effects of greenhouse gases.

The components of the atmosphere turn over at very different rates, but a variety of feedback mechanisms maintain them in constant proportion to each other. The 'residence time' (half-life) of nitrogen is 500,000 years, oxygen 6,000 years, nitrous oxide 150 years, methane 7 - 10 years, CO_2 2 - 5 years and ammonia 1 week.

This complex system has a remarkable capacity to respond to external changes. For example, the amount of heat reaching the earth from the sun has increased by 25% - 30% over the last 4 billion years (part of the natural development of the sun as a main sequence star, growing slowly brighter until it changes into a red giant in about 5 billion years time). The sun was significantly cooler when life began about 3.8 billion years ago. The oceans should have been frozen solid if the atmosphere had contained the same low amount of CO_2 as today, but the geological evidence is clear that they were not. It is inferred that CO_2 levels were then much higher. As the sun slowly warmed, CO_2 levels progressively reduced, partly through processes of rock weathering, and partly through activities of living organisms - not only photosynthesis, but also through formation by myriads of microscopic organisms of shells and other calcium carbonate structures which then became deposits such as chalk and the deep ocean sediments.

It is important to be aware of the huge dimensions of the system that maintains the earth's environment. The vital process of photosynthesis occurs over both land and oceans, and each year it involves taking in about 8 million cubic kilometres of air so that 60 billion tons of carbon dioxide can be fixed (Piel, 1992). Terrestrial vegetation draws about 60,000 cubic kilometres of water through its tissues to help feed the needs of photosynthesis, which produces about 120 - 180 billion tons of new plant material each year (balanced by a comparable amount digested or decayed, consuming oxygen and releasing carbon dioxide in equivalently large volumes). The microbial production of gases such as nitrogen, methane and nitrous oxide occurs at rates of 1 - 2 billion tons per year.

Against the background of a huge global system which has maintained life for billions of years, it is nonsense to speak of a 'fragile' earth. What is 'fragile' is the continuance of the very narrow and specific conditions needed to sustain human civilisation as we know it. Never before in its long history has the earth seen such an explosive growth in the population of an animal of the size and resource demands of a human being. Never before has there been such rapid change in the land surface of the earth through urbanisation, agriculture, deforestation and desertification. Never before has just one component of the biosphere (humans) caused such rapid changes in the atmosphere as the present rise in CO_2.

It is important to stress that it is the *rate* of change, rather than its amount, which poses the real threat to human society. For example, let us put the matter into perspective over the burning of fossil fuels. If all the proven reserves of such fuels were burnt to exhaustion over the next few centuries, the CO_2 content of the atmosphere would increase about six to ten-fold, probably resulting in a ten degree rise in global temperature. This would return us to the climate of the mid-Cretaceous about 100 million years ago when dinosaurs roamed the earth, there were no polar ice-caps and sea levels were greatly different from the present (Hoffert, 1992). It is very unlikely that civilisation could withstand such a rapid and traumatic change. However, many other forms of life would continue, and the system would slowly return to its present stable state as it did after the Cretaceous.

The real problem confronting us is not just the fact that we are changing the environment, but that the rate of change may be beyond our capacity for management.

THREE MAJOR ISSUES CONFRONTING HUMAN SOCIETY

According to Doll (1992) three major hazards to human society may be identified, and Winkelstein (1992) conveniently labelled them as the '3 Ps':-

Pollution by greenhouse gases with consequent climate change
Population growth
Poverty

In evaluating how best society should manage them, each of these hazards poses a different kind of problem.

POLLUTION BY GREENHOUSE GASES

Human activities are increasing the level of greenhouse gases in the atmosphere (principally CO_2, but also others including methane and CFCs). All are agreed that this will result in *some* rise in the mean surface temperature of the earth and *some* change in climate, but there are a bewildering array of important uncertainties about what will actually happen, as the following list illustrates.

(i) Some of the major factors determining global climate are only poorly understood, especially the role of the oceans, ocean currents and the ocean-atmosphere interface.

(ii) Forecasting the rate and extent of global warming and climate change is very imprecise: apart from ignorance of key factors such as the role of the oceans, we still lack sufficiently powerful computers to model global climate accurately, and even if such computers were available, it is not known know how to model properly the effects of clouds, which are important determinants of climate. It is for such reasons that current forecasts of global warming over the next century vary between 1.5 and 5°C.

(iii) Nearly all present models of global climate change ignore the effects of feed-back loops, most of which are likely to exacerbate global warming (e.g., methane release from the large frozen reservoir of arctic methane hydrates; melting of polar ice making surface ocean waters less saline and hence less ready to sink and carry C02 to the depths).

(iv) Even if the rate and extent of global warming could be forecast precisely, the resulting climate changes in specific regions cannot be safely predicted. Hence, the impact on agricultural food production cannot be forecast, especially since this depends on a complex of factors such as seasonal distribution of rainfall, droughts and temperature, as well as soil nutrient cycles.

(v) The extent to which global warming will increase the severity and frequency of extreme climatic events such as droughts, storms and hurricanes is not known.

(vi) The extent to which global warming will cause sea levels to rise is not known, and it is impossible to predict consequences such as the increase in numbers of refugees and the directions in which they will migrate.

(vii) The global carbon cycle itself is not fully understood because the amount of excess carbon dioxide released to the atmosphere by human activity significantly exceeds the actual increase which is measured; there is clearly some hitherto undetected 'sink' in the system.

(viii) Finally, nearly all forecasts are limited to the next century, even though greenhouse gas emissions will have substantial effects beyond that. There are no widely accepted ethical principles to guide our obligations to our descendants for more than 2 or 3 generations beyond the present.

There is widespread agreement that the world ought to take some action to arrest the current increase in greenhouse gas output. However, because of the numerous uncertainties listed above, there is no general agreement on the size of the reduction targets, nor on the means and costs of achieving them. Even if the targets were agreed, it is far from clear how to get all the different countries of the world to accept and act on them. Some form of carbon tax is frequently advocated, but there are many uncertainties about how it should operate. Should it be a tax at the point of production or of consumption? Should it be additional to or a replacement for, other taxes? Should taxes be collected nationally or internationally? At what level will a tax start to have disadvantageous consequences? If global warming proves to be slow, is it better to use scarce resources to adapt to climate change rather than to preventing it?

What does systems science tell us about managing uncertainty?

POPULATION-GROWTH

By contrast, there are few factual uncertainties about the hazards of global population growth. Demographic projections set reliable upper and lower limits for growth into the next century for the different regions of the world. The growth will be inevitably associated with increasing levels of pollution; population growth accounted for almost two-thirds of the increase in CO_2 between 1950 and 1985, while future population growth in the Third World alone will almost double present CO_2 levels by 2025 (Newby, 1991).

Global food productivity should *in theory* be able to sustain a world population double its present size, but this is predicated on assumptions such as improved agricultural efficiency, radical changes in systems of world economics and trade, and greatly reducing the large area of land allocated to producing meat. However, we must recognise that of the 120 - 180 billion tons of new plant biomass produced each year on the planet, about 40% is already being appropriated for human use. The capacity of the earth's surface for photosynthesis ultimately sets an absolute limit on human population size.

There is no conceivable advantage to the human race of further substantial population growth, and all commentators agree that it constitutes the most serious threat of all.

The problem of preventing it lies both at the level of the individual human being, where powerful and basic instincts surrounding the reproductive process swamp any weak sense of global social responsibility, and at the level of society, where deep cultural and religious sensitivities inhibit direct political action to limit population growth.

However, past experience suggests two acceptable ways of indirectly limiting population growth. Firstly, the level of education of females should be increased, since female literacy and enrolment in education correlates well with 'contraceptive prevalence'. And secondly, the phenomenon of the 'demographic transition' underlines the importance of raising living standards in the poorer parts of the world.

Can systems science throw any new light on the problem?

POVERTY

The hazard of poverty refers primarily to the gross imbalance of wealth between advanced and developing countries. The poverty of the latter is closely correlated with their very substantial rates of population growth. Further, such opportunities as they may have for using their own natural resources to escape poverty will be largely based on escalating the use of forms of energy such as coal and oil which will further pollute the atmosphere with greenhouse gases.

The distribution of wealth between the so-called rich North and poor South of the globe must be made more equal; resources should be substantially diverted so that the population growth and pollution potential of the South is markedly reduced.

For this to happen, the complex and intricate systems of world trade and the economic policies which govern them must change. Political questions must be faced such as whether wealthy countries should continue subsidising their agricultural products so that poor countries cannot compete with them in world markets. An added complexity is that major environmental issues important to poor countries are inadequately treated in conventional economics. For example, while the depreciation of factories and other components of the manufacturing base is a conventional part of national accounting, the depletion of a natural resource which feeds that base - such as a tropical forest - is omitted from the calculations.

Areas of global poverty pose other serious threats: they can be reservoirs of pandemic diseases such as AIDS; politically, they are often dangerously unstable; and they generate substantial refugee migrations which are difficult to assimilate.

Poverty is probably the most intractable of all the global issues confronting us.

REFERENCES

Allaby, M., 1989, "Guide to Gaia", Macdonald Optima, London. ppix + 181.

Doll, R., 1992, Health and the Environment in the 1990s, *American Journal of Public Health*, 82, 933-941.

Hoffert, M.I., 1992, Climate Sensitivity, Climate Feedbacks and Policy Implications, *in:* "Confronting Climate Change", I.M. Mintzer, ed., Cambridge University Press, Cambridge, pp33-54.

Margulis, L. and Lovelock, J.E., 1974, Biological modulation of the earth's atmosphere, *Icarus*, 21, 471-489.

Newby, H., 1991, The Social Sciences and the Environment, "Robbins Lecture", University of Stirling.

Piel, G., 1992, "Only one world: our own to make and keep", W.H. Freeman, New York and Oxford. ppxii + 367.

Lovelock, J.E., 1979, "Gaia: a new look at life on earth", Oxford University Press, Oxford. pp xiii + 154.

Lovelock, J.E., 1988, "The Ages of Gaia: A Biography of our Living Earth", Oxford University Press, Oxford. ppxx + 252.

Lovelock, J.E., 1991, "Gaia: The practical science of planetary medicine", Gaia books. London and Stroud. pp192.

Winkelstein, W., 1992, Determinants of Worldwide Health, *American Journal of Public Health*, 82, 931-932.

SYSTEMS SCIENCE

Peter Checkland

Lancaster University
Lancaster, U.K.

I

This note outlines the skeleton of an argument concerning the nature of systems science and what its role might be in relation to the theme of this conference: global issues.

II

The systems field, presumably because of its assumed commitment to some kind of holism, has always attracted adherents from many disciplinary backgrounds. Not all of them are driven by a desire for the kind of intellectual rigour which any discipline needs if it is to be taken seriously, and systems conferences are rarely short of well-meaning papers ('here is a systems model of X') which are virtually content-free. Part of the difficulties for systems as a subject must stem from its nature as a meta discipline. The systems ideas can be applied <u>within</u> any other field, and the existence of systems thinking biologists, geographers, economists, psychologists, sociologists, engineers etc. itself suggests that systems as a discipline will not find it easy to define itself except in very abstract terms. Any meta discipline, in fact, ought to be rather circumspect in its claims, simply because of this abstraction; but the systems movement has rarely been scrupulous in its scholarship or circumspect in its claims (Checkland, 1992). No one was surprised that the Silver Anniversary Meeting of what was then the Society for General Systems Research had as its title no less than: "Improving the Human Condition"! But grandiosity does not build a discipline.

This argument examines systems science as a strand of thought emerging within the tradition of natural science, examines what its feasible current role might be, and examines briefly the question of global issues which is the theme of this meeting.

III

First, a caveat. For many people "systems science" would refer to a branch of mathematics, one illustrated by the English abstract of a paper from a current Ukrainian journal, <u>Kibernetics</u>:

Systems Science, Edited by F.A. Stowell *et al.*,
Plenum Press, New York, 1993

The estimates of stability characteristics for linear stochastic differential-difference systems of retarded type are considered. The optimization problem of stability characteristics in a class of positive definite matrices is solved.

This note does not concern such work; it focuses on the use of systems ideas in understanding and/or intervening in the real world.

Van Gigch and Kramer (1981), proposing a taxonomy of systems science, define two dimensions whose combination gives them a four-part matrix of the kind so common in the social science literature. Their dimensions cover firstly whether or not systems are taken as ontological or conceptual entities; their second covers the nature of the work: theoretical or applied. This yields:

	Ontological	Conceptual
Theoretical	1. Systems Philosophy	2. Axiomatic Systems Science
Applied	3. Living Systems Theory	4. Systems Methodology

I would wish to argue that in any subject area the ultimate concern lies in the interaction between our lived experience of the world and our conceptual representations of it. This makes cells 1, 2 and 3 in the matrix the possible location of work which ultimately serves cell 4, which is hierarchically prime precisely because of its complexity, which results from its focus on the interaction between theory and practice, which create each other in a never-ending learning cycle. (Checkland 1983, 1984, 1985, 1990).

IV

We can understand the nature of systems as a subject area by seeing it in relation to the method of natural science, to which it is a reaction and of which it is a subset. Natural science is concerned with the use of rational thought in making sense of our experience of the natural phenomena of the world. Much of it is devoted to contriving the special kind of experience which is 'the experiment'. There we first put limits on what is examined, and then seek evidence that what is observed can be repeated by others. If repeatable every time the experiment is carried out, then the observed happenings (but not necessarily our explanations of them) become 'scientific' evidence. Finally, in a sequence of experiments we can make a kind of negative progress by the refutation of hypotheses,(Popper, 1959) . Thus the three great principles of the method of natural science are:

Reductionism
Repeatability
Refutation.

(See Checkland, 1981, for this argument in more detail, where it is derived from the history of the development of the method of natural science.)

Of the three principles the first is of great interest because of the wide-ranging way in which it permeates the scientific approach. Scientific experiments are reductionist in the sense that any one experiment excludes most of the phenomena exhibited by the world; scientific explanation is reductionist in (very sensibly) operating according to the philosophical principle of Ockham's Razor - always working with the simplest explanation (of which there is only one); finally many scientists (for example the Nobel prize winner Crick in Of Molecules and Man (1966)) are reductionist in accepting the 'nothing but' principle: the belief that social phenomena are really 'nothing but' psychology, which is really 'nothing but' biology, which is really 'nothing but' chemistry, which is really 'nothing but' physics!

V

Now, we can make sense of systems as a subject if we see it as a reaction against the extreme reductionist position. It seemed to the 'organismic' biologists that science should focus on the organism as a whole (rather than on its chemistry and its physics). They developed general ideas concerning processes (of growth, decline and death, for example) characteristic of organisms as wholes. Later, in the 1940s, von Bertalanffy argued that these ideas about organisms could be applied to wholes of any kind - which, borrowing a term from everyday language, he called 'systems' (Bertalanffy, 1968). (We can now see how much confusion would have been avoided had he at that point made up the word Koestler (1967) later suggested for the abstract idea of a whole, namely 'holon' (Checkland, 1988).)

Systems, then, emerges as a subject area in which what is being constantly tested is the value of using the concept 'holon' to provide explanations of experience of the world and guidelines for intervening in the world.

The step from the world of natural science to the world of systems thinking is marked by the importance of a crucial idea: that of 'emergent properties'. An emergent property is a property of a whole which is meaningless in terms of its parts (eg the smell of ammonia, meaningless in terms of the nitrogen and hydrogen which form ammonia molecules). If you accept the relevance of this concept, then you are committed to considering wholes as wholes, to some kind of 'systems thinking'.

VI

The fundamental systems ideas have been developed in many fields, making systems itself a meta-discipline, from ecology to engineering, from anthropology to management. In the course of these explorations two fundamental complementary stances have emerged within systems work. (These parallel the two basic strands, positivistic and phenomenological, which have emerged in the attempt to use the method of natural science in the study of the social world (Checkland, 1981, Chapter 8).) In systems science the earliest position adopted was uncomplicated: to treat the world as a complex of systems which could be 'engineered' to achieve objectives. This obviously maps the positivist stance of experimentalists in natural science. The complementary position sees the world as problematic but assumes that the process of inquiry into it can be systemic. This shift of systemicity, from the world to the process of inquiring into it, marks the emergence of a strand of thinking which parallels the phenomenological strand in social science. This distinction is, in vulgar parlance, the difference between 'hard' (systemicity in the world) and 'soft' (systemicity in the process of inquiry) systems thinking (Checkland, 1983). Within the process of inquiry of soft systems thinking it is of course possible for the user of the ideas consciously to decide to take a part of the world to be a system. Hence the relation between the two strands is that the 'hard' stance is a special case of the 'soft' (Checkland, 1985).

Thus systems thinking now has an understandable history and structure. Its task remains critically to examine whether, how, and to what extent systemic inquiry enables us better to understand and intervene in the world. We should strive to see that this is done in the neutral spirit which the logic of scientific inquiry calls for, even though, as in natural science, what users of the ideas actually do will not infrequently fall short of that ideal.

VII

Finally, given the theme of this conference, it may be useful briefly to examine the concept of systems thinking being relevant to 'global issues'.

There are two senses in which that phrase might be used. It might be taken to refer to the big issues of global significance, the issues which affect all of us on this planet as 'the crew of spaceship earth'. It is not unknown for systems conferences to make grand claims here (see II above), but there is little evidence that policy making at the highest levels is ever affected by systems thinking. It would be better to focus on the second interpretation of 'global issues', namely issues which are ubiquitous, in different circumstances in different societies; I refer to such issues as how to get the most out of health care expenditure when limitations on resources available require overt or covert rationing, or how to make large cities more inhabitable. There, useful comparative studies might be made with some hope of contributing to human knowledge.

In general, the apparently all-embracing scope of systems thinking offers to many people a heady elixir. We should decline to drink it; we should stick to the more serious scholarly task of patiently trying to relate our epistemological devices to our experienced ontology (Checkland, 1992).

REFERENCES

Bertalanffy, L. von, 1968, "General System Theory", Brazillar, New York.

Checkland, P., 1981, "Systems Thinking Systems Practice", J. Wiley, Chichester.

Checkland, P., 1983, OR and the systems movement: mappings and conflicts, *J. Opl. Res. Soc.* 34, (8), 661.

Checkland, P., 1984, Systems Thinking in Management: the development of soft systems methodology and its implications for social science, in Ulrich and Probst, q.v.

Checkland, P., 1985, From optimizing to learning: a development of systems thinking for the 1990s, *J. Opl. Res. Soc.*, 36, (9), 757.

Checkland, P., 1988, The case for 'holon', *Systems Practice*, 1, (3), 235.

Checkland, P., 1992, Systems and scholarship: the need to do better, *J. Opl. Res. Soc.*, 42, 1023.

Crick, F., 1966, "Of Molecules and Men", University of Washington Press, Seattle.

Gigch, J. P., van and Kramer, N.J.T.A., 1981, A taxonomy of systems science, *Int. J. Man - Machine Studies*, 14, 179.

Koestler, A., 1967, "The Ghost in the Machine", Hutchinson, London.

Popper, K.R., 1959, "The Logic of Scientific Discovery", Hutchinson, London, (published in Germany 1934).

Ulrich, H. and Probst, G.J.B., eds., 1984, "Self-Organisation and the Management of Social Systems", Springer-Verlag, Berlin.

THE ACTOR AS A PERFECT CITIZEN

Gerard de Zeeuw

Center for Innovation and Co-operative Technology
University of Amsterdam
Grote Bickersstraat 72
1013 KS Amsterdam, Netherlands

INTRODUCTION

Reaching heaven, paradise, utopia, perfection, or as Fukuyama (1992) states, the end of history, is often seen as man's birthright. Sometimes it was assumed that mankind's future state of happiness had to be guaranteed by some external, hence eternal influence. Sometimes one was sure that improvement would come automatically, as part of man's destiny. Confidence in such matters seems to have increased, meanwhile. Many people now appear willing to accept that our existence is an 'achievement', though 'luck' and opportunism may play an important part in it - such as the piece of rock that, 65 million years ago, wiped out 70% of all species and allowed mammals their ascent.

A few centuries ago the banner for further such achievements - though on a more mundane level - was taken over by research (and by the technology based on it). It came to be trusted as an infallible activity, producing knowledge as required. Due to its successes, it often was taken for the end of history itself (Carnap, 1955; Popper, 1959). More recently, its role has become tainted, however. Its a-historical pride is being shaken: it turns out to need 'luck' and opportunity as well as any other human activity.

This trend is borne out by the many criticisms of the basic tenet of research, that is, that it is possible to know 'what will happen', and, later, to act on this knowledge. Many people now find themselves dissatisfied with such results. Too often, their use seems to abrade, rather than strengthen, what used to be normal activities such as the utilisation of resources, the disposal of waste, and growing food. People still dream, but often no longer in terms of research only. They appreciate Zen teaching, manual healing, management training and psychotherapy, and revel in the deeds of heroes representing the 'larger' view: Superman, Agent 007, Hans Brinkers, Maigret[1].

In fact, never before has there been seen such a need for additional information on the

[1] If one may speak of any new heroes, produced by research, they are an often lacklustre lot. They are not the common people, suddenly blazing into deeds of heroism. They are the professionals, claiming special status often only on the basis of their long periods of training. Research has not shown how to 'become an educated person', though it has made available amazing amounts of material to teach.

use of science. Research itself has largely failed at this point. Though it may have unravelled some of the secrets of nature, it falls short of unravelling the secrets of human meaning, which it even seems to have reduced in many ways. While looking for causes, many people have learned to blame. While looking for diagnoses, they have learned to label and stigmatise. While looking for levels of invariance, they seem to reach their levels of incompetence. While looking for objects, they have started to objectify, and to denounce people's right to be actors.

The paper considers possibilities to resolve this paradox. It identifies the major difficulty as the gap between the ideal user of results and the actual user of results. To bridge this gap the concept of a user language is introduced. It provides an instrument to change present theories of research, to have them pay heed to the increasing need to create leadership and actorship.

USERS

Most theories of research carefully distinguish between two activities, those are, on the one hand, identifying some objects, to determine their properties, and on the other, using the results of the first activity in some other set of activities. To be able to do so results of the first activity have to be 'invariant' under various kinds of use, and hence independent of such uses. The law of gravity, for example, is not assumed to change when used to improve shooting or kiting, or by kings, subjects or professionals.

This kind of independence implies testing the invariance against some 'perfect' user, someone who can understand results and their conditions, and who can combine them. The perfect user is the perfect observer, and also the perfect actor. The perfect user is a model of the user, as perfect results are a model of what is to be used. He or she is also a model citizen, knowing all societal constraints on usage.

Neither the perfect user nor the perfect observer is the same as actual users or actual observers. Both need some way to bridge the gap. The perfect observer is constructed by having actual observers' observations repeated, and averaged in some ways. The perfect actor cannot be constructed in a similar way. He or she cannot be averaged - but rather needs to be as varied as possible to get the most out of perfect observations.

Many methods and techniques have been devised to bridge the gap and construct perfect users out of actual users. One such method is by training and schooling, sometimes needing a long time, as in the case of the medical or the engineering professions. Another method has been to provide some kind of manual, guide or prescription for use, as is legally required in the case of medicaments, and is considered normal practice in the case of most technical instruments. A third method is to find some way to select those users that are nearly perfect, in terms of some particular piece of knowledge to be used[2].

The study of methods such as these is not part of most theories of research. Formally that is correct. There is a catch, however. First, one could say that such methods would and should be worthy of closer scrutiny, when the need would arise. Second, and interestingly, if such scrutiny would reveal some 'perfect' results, it would again create a gap - between the perfect user of those results and actual users. Knowing more about some part of the

---

[2] Most methods have been found quite effective. An example is the finding that Freudian therapy was more effective when practised on 'perfect' patients, that is patients with a strong training in Freudian theory (Merkies, 1988). Another example is provided by the need for manuals in the case of most technical instruments. A third example is the finding that behaviour therapy is most effective when patients can be pre-selected on the basis of their being well-educated, linguistically gifted, and aware of the nature of the therapy - in short, on the basis of their being 'educated' (Emmelkamp et al., 1989).

world, even if that part concerns the use of such knowing, apparently does not imply knowing more about the effective use of this type of knowing[3].

This paradox indicates that something is missing in theories of research as indicated. Training, teaching, providing manuals and instructions *cannot* be studied as any other type of 'object', nor can becoming an educated person or becoming an actor. An interesting failure is reported by Fischhoff (1992), on how to advise women to avoid rape. More than a thousand suggestions are provided, but it turns out impossible to model them in the usual way. For systematic scrutiny in such a case to be possible, special precautions and changes in present theories of research appear necessary.

USER LANGUAGES

What form these changes can take can be deduced as follows. Our formulation of research and its use implies that the variation generated by the perfect user is constrained by the results of research so as to precisely maintain their own invariance (and their quality, that is perfection). Actual users, however, may be able to provide more variation, and, hence, threaten this invariance. Results may become useless (even though perfect, as in the case of the untrained user), or their invariance may become broken. To prevent this from happening, and to continue reaching for perfection, additional constraints on actual users are necessary.

These constraints do not guarantee actual perfection. They can, however, help actual users approximate the perfect user. They can assist in determining 'what actual users need to do', *as if* they are perfect users and already know 'what will happen'. The set of such constraints, therefore, acts as a linguistic structure, to be called, appropriately, a *user language*. Such a language supports actual users to behave as perfect users, and hence, as actors and educated persons.

As argued above, the distinction between perfect results and perfect use is insufficient to support perfect usage by actual users. It must be supplemented by systematic development of additional constraints, which, however, do not represent knowledge. They make knowledge accessible, or conversely, indicate the conditions under which such knowledge can be used (to reach perfection). Systematic development of user languages has not been part of most theories of research.

A simple example of a user language would be the set of instructions or statements the coach would use, for example, to help an archer behave as if he or she is the perfect user of the laws of gravity and of the law-based influences of wind, temperature, etc. The language of such a coach obviously does not include the said law of gravity itself, though it will be constrained by it.

Constructing languages of this nature is not a simple task, as any sports coach will admit. It is, however, not an impossible task. As an example, one may think here of the 'language of decision making'. This language was purposefully designed to support actors to behave *as if* they are perfect users of what is (or can be made to be) known about a certain situation.

---

[3] In fact, even after decades of systematic research into education as one of the main methods to bridge the gap, there still is a dearth of usable theories for learning and teaching. The same applies to the results of many other studies, geared to be used in activities like psychotherapy, policy, planning, management, design, etc. These studies are not taken to be blamed directly themselves. It is contended in this paper that there is a fundamental defect in what they use as their basic theories of research. The latter are in need of change.

PROPERTIES

User languages imply paying attention to actual users, that is people *in situ,* people in the here and now, each with his or her own history and his or her own peculiarities. Such peculiarities have to be transformed into resources for action, to approximate perfect use. Differences between users, therefore, should not be masked in terms of some general model (as often is the case in constructing perfect observers). User languages will emphasise differences between actors, in terms of the intended use.

This implies that becoming an actor (and approximating the perfect user) is a social phenomenon. Though one may impose additional constraints in the sense of a user language on one's own, this does not seem to have much surviving value[4]. Generally speaking user languages will imply the possibility to create a number of actors at the same time, by interaction[5].

In fact, the latter property seems to provide one of the main advantages of user languages: they allow for the undisturbed and parallel use by a number of actors. User languages function as *co-ordination languages:* they not only help create individual actors, but also collectives, or collective actors, the activities of which result from the co-ordinated activities of a - possibly large - number of individual actors (or agents, Gasser, 1990)[6].

It may be noted here that standard theories of research do not allow for such variation. There is, indeed, only one perfect user, of whom there may be many identical copies. There are, therefore, not enough differences between such copies to make interaction potentially painful or even interesting. As actual users differ from perfect users, this expectation is refuted time and time again. Such users often resist the imposed reduction on interaction.

As indicated, user languages should help create actors and support people to behave as if they are perfect users, using perfect knowledge. As such perfection may not be attainable in practice, user languages can only help to approximate perfection. As there may not even be such a point of perfection (see the next section), it seems useful to define other criteria for quality, but the difference between the present state and some (future) point of perfection.

One such quality criterion is competence, defined on two levels. The first is a qualification procedure. It relates to the collective definition of contributing actions, and indicates the kind of events that can constitute an action. The second is a procedure for combining events, e.g. by linking them via some algorithm, or production scheme. Algorithmic compression can help to improve on the results of the second procedure, and will represent levels of skill (Barrow, 1991).

An interesting property of user languages is that each statement in such languages must contain a qualification part and a skill part, *though* their functions must remain undecidable in the collective, however, as each actor may have to use other actors as qualifiers as well as sources of skill enhancements.

In most theories of research this aspect of undecidability is negated. The skill part is reduced to high quality (perfect) observations, that is the observation of data. The

[4] In the text it is assumed that actual users are never close to being perfect users (unless some special measures are implemented). An exception might be what are called 'idiots savants', who, apparently by accident, can turn out to be perfect users of some piece of knowledge.

[5] Though not necessarily the same, rhetoric in the original sense, as developed in ancient Greece, fulfilled the same role, in a particular arena: it served citizens to access the powers of the state (Scholten, 1990).

[6] Reaching some form of agreement or consensus is a special case of this kind of co-ordination.

qualification part is reduced to the formulation of theories that have to cover the data, i.e. explain them (De Zeeuw, 1992), while their relation is taken as decidable.

RESEARCH IN PRACTICE

The property of undecidability often will blur the distinction between (perfect) knowledge and user languages, and, hence, will generate two tendencies, which can be clearly recognised in present day research.

The first tendency is still to reduce the role of user languages as much as possible. This is exemplified by attempts at reifying or objectifying user languages, and to interpret them as descriptions of what people do, that is as (possibly) 'perfect' knowledge (Swanborn, 1991). Such attempts are supported by a strong belief in the representational nature of knowledge (Simon, 1991), which has users replaced by their - preferably perfect - models. This tendency is also supported by findings that individuals make 'mistakes' by adopting user languages as their own general languages, e.g. in the context of decision making (e.g. Tversky, 1981; De Zeeuw en Wagenaar, 1974).

The second tendency is to start doubting the possibility of perfect knowledge itself, stimulated by the fact that many invariants keep being broken, notwithstanding the use of user languages. Such invariants must have been based, therefore, on earlier user languages that have become reified, or have become stable only in a particular group of actors. Compensating for wider usage in different groups of actors may therefore require *extensions* to these earlier languages, rather than the design of new user languages. Eventually one may come to doubt the existence of any kind of invariant.

Though both tendencies appear to have some advantages, there are drawbacks too. Reification may go too far, as in the case of SOAR (Newell, 1991), which objectifies even the 'language of problems'. Doubt about the existence of any kind of invariant may lead to radical forms of nominalism, constructivism and anthropomorphism[7]. A middle way seems useful, therefore. Most natural sciences seem to profit from some tendency to reification, while most social sciences seem to be dealing only with user languages, with no invariant remaining unbroken. One is left to base oneself on temporary invariants, and start from there.

An example of events that seem to warrant both types of approaches is provided by the AIDS epidemic. Most medical researchers may want to stick to trying to find 'perfect' knowledge, possibly to be completed later by the development of some user language. Most others may be interested mainly in extending presently involved user languages, to support people to remain actors as long as possible (Riper, 1990), to deal with negative social effects.

TO CONCLUDE

In the last three decades the standard practice of research has had to face fundamental challenges, such as increasing pollution and depletion of resources, all due to the increased interaction between many actual users, in stead of perfect ones. Many responses were tried, such as changing the organisation of research, e.g. in Universities, or improving training in research. Effects have been rather disappointing. There is still a strong tendency to avoid the interactive, and hence, historical nature of good research.

---

[7] One may be reminded here of the dictum that nothing is as practical as a good theory - a statement with seems to apply best when theoretical statements that help to improve the quality of activities coincide with similar statements in user languages.

In this paper an adaptation of such theories was argued, in the form of the construction of user languages. User languages make it possible for users to behave as if they are perfect users of research results, but without the properties of perfect users: time-independent understanding, no side-effects from multiple parallel usage, etc. By doing so, they help create both individual and collective actors.

A necessary complement to the definition of user languages is the development of methods for their construction. This has the undivided attention in our Center for Innovation. There is a general interest in the various structures that provide the additional constraints necessary when the constraints of 'perfect knowledge' appear insufficient (Vahl, 1992; Witte, 1991).

Specific interest is with the design of extended languages, as defined above. The work of Van Linder (1993) provides an example. He developed a number of games, aiming to extend the 'language of problems', using concepts such as initial and terminal states, problem spaces, selection rules to chose trajectories, etc.

Players of each game are induced to identify certain problems to be solved, in the context of the game - which provides some necessary and relevant knowledge. Most resources, however, are distributed over other players, and are therefore inaccessible at the start. During the game players must become aware of this uneven distribution, and start negotiations to acquire or even create their 'own' resources.

Playing the games can be demonstrated to achieve two things. First, the language of problems is shown to be extendable - most of the time - into a new language, which itself is a user language to the original language of problems. Second, players do use statements in this language, to approximate perfection. They lead to the development of new invariants, which help co-ordinate earlier, broken invariants.

In sum, to be able to create actors - and 'educated persons' - one needs to construct user languages. This is possible in a systematic way.

REFERENCES

Barrow, J.D., 1991, "Theories of everything", Vintage, London.

Carnap, R., 1955, Foundations of logic and mathematics, *in:* Int. Enc. Unif. Science, Vol. I. Univ. of Chicago Press, Chicago.

Emmelkamp, P., Bouman, Th., and Scholing, A., 1989, "Angst, fobieen en dwang: diagnostiek en behandeling", Van Loghum Slaterus, Deventer.

Fischhoff, B., 1992, Giving advice: decision theory perspectives on sexual assault, *American Psychologist* 47: 577-588.

Fukuyama, F., 1992, "The end of history and the last man", Free Press, New York.

Gasser, L., 1990, Co-ordinated problem solvers, *Annual Review of Computer Science* 4: 203-253.

Linder, B. van, 1993, "Spelen op variatie", Thesis Publishers, Amsterdam.

Merkies, Q.L., ed., 1988, "Psychotherapie in kontekst", SPCP, Amsterdam: 353-369.

Newell, A., 1991, "Unified theories of cognition", Harvard University Press, Cambridge.

Popper, K.R., 1959, "The logic of scientific discovery". Harper & Row, New York.

Riper, M.M., 1990, Individual and collective strategies in the fight against AIDS. *in:* Proceedings of the Seropositive Ball, Amsterdam: 18-23.

Scholten, J., 1990, "Retoren en demokratie", Ph.D. thesis, Groningen.

Simon, H.A., 1991, Methodological foundations of economics. *in:* "Praxiologies and the philosophy of economics", J.L. Auspitz, W.W. Gasparski, M.K. Mlicki, K. Szaniawski, eds., Transaction Publishers, New Brunswick.

Swanbom, P.G., 1991, Er waren eens drie theorieen ... *in:* "Aspecten van onderzoek. Theorie, variabelen en praktijk", P.G. Swanborn, J. De Jong Gierveld, T.G. van Tilburg, A.E. Bronner, G.W. Meijnen, eds., ISOR, Utrecht.

Tversky, A., and Kahneman, D., 1981, The framing of decisions and the psychology of choice, *Science* 211/ 1: 453-458.

Vahl, M., 1992, Stories and scientific enquiry, in: "Cybernetics and systems research '92", R. Trappl, ed., Vol.2. World Scientific, Singapore: 1063-1070.

Witte, M., 1991, Naar een andere wiskunde-opvatting. *in:* "Inspiratie uit de informatica. Beschouwingen over onderwijsvernieuwing", J.B. de Jong, and M. Witte, eds., Thesis Publishers, Amsterdam.

Zeeuw, G. de, and Wagenaar, W.A., 1974, Are subjective probabilities probabilities? *in:* "The concept of probability in psychological experiments", C.A. Stael von Holstein, ed., Reidel, Dordrecht.

Zeeuw, G. de, 1992, Is chaos te zien? *in:* "Gamma Chaos. Onzekerheid en orde in de menswetenschappen", C. van Dijkum, and D. de Tombe, eds., Aramith, Bloemendaal: 36-51.

INFORMATION SYSTEMS AND SYSTEMS SCIENCE

Frank Stowell

Information Systems
Department of Computing Science
University of Paisley
Scotland

INTRODUCTION

There is little doubt about the relationship between the success or failure of business in the market economy and the wealth of the population. The interrelationship between the two is equally relevant at every level of recursion whether that be at global, national or local level. The relationship of business effectiveness and national wealth has resulted in a competitive ethos in the market place as new businesses attempt to enter a market and old businesses defend their market share. This is not a new phenomenon of course since competition between businesses has been fierce throughout the 20th century. Increasingly, business enterprises throughout the world have had to improve their efficiency and quality of output in order to compete in the world market and one major contributor to the improvement of business efficiency has been the increased reliance upon Information Technology (IT) (e.g. Guy and Georghiou 1991; Esprit '90, 1990).

It is not just the business system that has been effected by IT, Information Technology itself has a significant impact upon many other areas of human decision-taking (Guy and Georghiou, 1991). Information Technology can be argued to have created the second industrial revolution (Friedrichs, G. and Schaff, A., 1982; Forrester, 1985) and hence a topic relevant to the theme of this conference. During the past decade or so, as was the case with new technologies introduced in the 19th century, there has been a decline in the need for some skills and the emergence of a new set of specialist domains have emerged to cater for the demands made upon this late 20th century technical phenomenon. One such domain of specialism is that of Information Systems which is claimed by some to be an intellectual discipline. (N.B. By Information Systems I mean technology supported human communication systems). If the effectiveness of communications is accepted as being important to global endeavours, such as business operations, then the way in which IT supported communication systems are developed seems also to be important. To this end this paper is concerned with the question of whether the domain of Information Systems can rightly be claimed to be a discipline and if so what influence Systems Science might have upon its development.

Systems Science, Edited by F.A. Stowell *et al.*,
Plenum Press, New York, 1993

INFORMATION TECHNOLOGY AND "INFORMATION"

Information Technology enables a large volume of data to be processed quickly, accessed easily and presented authoritatively. A mass of data is available to a variety of users who need no qualification to manipulate it outside that of being able to gain access. Easy-to-use software enables any user of IT to access and manipulate vast quantities of data and present them in such a way as to create the impression of the data being unquestionable fact. IT has enabled communication between human groups to take place faster than ever before and provides the opportunity to act swiftly. However, this facility may not always be advantageous. The data which is manipulated is often without context and may arrive too quickly. The speed and "authority" which the data is presented may have an urgency about it that may force hasty decisions and effectively make the decision-taking system behave as a positive feedback system.

It is not always possible for a decision taker to be able to deal with the intricacy of the data itself nor be fully aware of its source or integrity. The user is often left in a position of unconsciously influencing the behaviour of the wider dimensions of a problem situation in a way which may be counter-productive. For example, consider the effect that IT has had upon the celerity of stockmarket transactions and the knock-on effect that the sheer speed of these transactions have had upon national economies over the past decade. The increased use of IT in stockmarket transactions has made the market behave in unexpected ways. There have been instances where precipitate action has created undesirable effects on the global economic system which are of such magnitude so as not to have been contemplated either by the designer nor the user of the information technology.

A computer is a tool which facilitates improvements to information systems, intrinsically through the speed and accuracy in which data is processed. In a similar way a telephone system of communication provides a means of voice communication between remotely located individuals or groups. Neither computer nor telephone can be accurately defined as an information system but each is an important technical component supporting an information system.

INFORMATION SYSTEMS AND INFORMATION TECHNOLOGY

Long before a computer was used as a prime information system provider other forms of Information Technology were incorporated within and between business enterprises to facilitate management control. For example, lampson tubes, telephones and telex have been successful technological means of supporting inter and intra-departmental communication.

The enthusiasm amongst some managers about the potential data processing power of computers increased as machines become more powerful, smaller, cheaper and easier to use. This continuing enthusiasm has been fuelled by the development of business computers and the variety of software which provides specific business applications. Computing Science, which evolved out of academic departments of physics and mathematics, has itself spawned a number of other specialisms which have become necessary to deal with specific areas of application arising out of the novel uses of computers (e.g. Expert Systems). It is important to note that these specialisms have arisen not just to fulfil a technological need but also because of the requirements perceived by the client, usually a non-computer specialist, who has seen the usefulness of computer-technology as a management tool. However, the fulfilment of the client's computing needs has been the responsibility of the computer scientist who may have had little or no knowledge about management or business.

Computer scientists, or more correctly technologists (Bullock et al 1990), have developed ideas which, through the manipulation of voltage levels, provide the basic building block of indited human communication. By using advances in electronics such as the transistor and, latterly, the integrated chip the technologist has provided a generation of computers which are able to manipulate this building block into a variety of forms useful to business management. In a short space of time this variety of requirements has, not surprisingly, produced a variety of specialisms e.g. Information Systems Analysts, Software Engineers, Knowledge Engineers. The success of these experts is shown by the way in which they have made the computer easy to use. In many respects, the computer itself is now "transparent" to its user but this technological success carries with it a difficulty too. Whilst the technologist seeks to provide more effective technology, which is consequently more technically complex, the user of the technology is less concerned with the technology itself and more with the results that it might bring. In many respects the user of the technology places unrealistic demands upon the technologist by "asking" them to identify and then cure the problems of the enterprise through the use of computer data processing.

The advances in data manipulation provided by the computer scientist has provided mankind with an aid to communication more powerful than anything that has gone before and, by the sheer speed of this development, changed the loci of the problem from technology to the more difficult area of information system provision. Consequently, the role of IT is now subordinate to the definition and requirements of the information system.

THE WRONG CONCEPT?

The increased use of technology to support human decision making and the special requirements of technology mean that in each case where IT is used there are in fact two "information systems". The first is what may be described as an "intrinsic" information system, in which the analyst attempts to define the human communication and control system that is necessary to enable the particular decision-taking to take place, and, the second is the technological version of the "intrinsic" information system. Hitherto the responsibility for the whole design process involved in the development of technology supported information systems has been that of the technologists. Although as part of the design process the technologist may attempt to identify the needs of the user (e.g. Information Engineering. Finkelstein, 1989), ultimately most of these design methods are technology driven (Stowell, 1991).

It seems clear that the method by which we develop, and, the manner in which we use technology-based information systems will have a significant effect upon the way that it behaves and also upon the actions of those who access the data held. The lessons that have been learnt during recent times about the use of information technology and its effect upon major areas of decision-making suggest that the way in which we design information systems requires careful consideration. Examples exist which show that whilst a technology supported information system provides fast processing and response times it fails to provide the flexibility of a human communication system, sometimes with serious consequences (e.g. The London Ambulance Emergency Service failure in 1992). A dilemma for an information system designer is the realisation that whilst the technology can process data quickly and reliably an unquestioning acceptance of displayed results by a human operator may have unfortunate consequences.

Attempts by analysts to undertake a reappraisal of the IT design process have met with only partial success primarily because changes have been attempted at the practical level alone and ignored the philosophical underpinning of the whole domain. Why is this? the prime reason appears to be that the methods employed for the design and development of computer systems relate to what Winograd and Flores call the 'rationalistic tradition' (Winograd and Flores, 1988) and to what Hirschheim describes as the "posivitist" view (Hirschheim, 1985). Winograd and Flores refer to the "tradition of rationalism and logical empiricism" as the underlying philosophy from which computing has been developed. Such a development should not be seen as surprising; in fact it represents an historical development since as Winograd and Flores explain, today we are taught from an early age that the rationalistic tradition, with the prestige and success it has received through scientific method, has become synonymous with "what it means to think and be intelligent" (sic p16).

THE RATIONALISTIC TRADITION AND THE EVOLUTION OF COMPUTING SCIENCE WITHIN EDUCATION

It may be worthwhile to consider the educational evolution of computing science in order to gain a better understanding of the present difficulty. The conceptual basis of computing as a taught subject can be traced through engineering into reductionist science. The academic "home" of many computing science departments and computing science research until very recently has been in the faculty of science. Departments specialising in computing science often grew out of departments of mathematics and had close links with engineering, indeed many computing scientists today consider themselves to be engineers.

There is not the space in this brief paper to discuss in detail the philosophical premise upon which Engineering is based but a common view obtained from a variety of sources is that engineering is most often utilitarian and concerned with the utilisation of raw materials and with building things (e.g. Bullock et al, 1990; Ganic and Hicks, 1991).

In the 1980's there began to emerge IT courses that were less engineering orientated and more business orientated. This trend was further influenced by the proportion of computer science students opting for subjects concerned with systems analysis rather than the technological aspects of computing and also by the proportion of graduates taking posts related to the application of IT rather than working with the technology itself. Developments in technology reinforced this trend by reducing the need for an individual to posses skills in programming or microelectronics. The accent upon the application of IT is also reflected by the initiatives in the early 1980's and 1990's (NAB, 1982.; JUPITER, 1987). In both initiatives, which have been aimed at postgraduate IT education, the accent has been upon providing the students with a postgraduate programme which is aimed at the application of IT rather than a computing science specialism. Both initiatives have been designed for a target population of non-computing science graduates and, more recently, for the so called Hybrid Manager.

The evolution of "computing science" as a taught subject domain has affected the way in which the departments themselves have developed. Computing Science seems, over the past decade, to have reverted more to its electrical/ electronic engineering origins. During the 1980's a deliberate attempt has been made by departmental managers to recruit interdisciplinary teams in order to cope with the variety of topics which their portfolio of courses demanded. Many academic staff in departments involved in teaching computing science subjects possess both computing and non-computing qualifications. By the late 1980's the trend towards IT applications led to the formation of Business Information Systems Departments and the subsequent break-up of many computing science departments. In the latter case the computer scientists have moved into electrical and electronic engineering.

It is not unusual nowadays to find computing studies within a faculty of Business rather than Science or Engineering, particularly in the new Universities. Business studies is connected with those functions deemed necessary to operate a business with an emphasis upon Economics and Management. The intellectual underpinning of Business studies is varied but in general is dominated by ideas from Management Science. Many of the ideas within Management Science can be classified under the Functionalist Paradigm (see Burrell and Morgan, 1979). Generally, the methods of research and analysis in Business related subjects are inductive and although the actual method of research may be different from those used in engineering there are underlying assumptions which are similar (an interesting discussion can be found in Romm, 1991).

It may be worthwhile reflecting upon the apparent change of emphasis in educational direction from engineering to business applications of IT. Neither domain would seem to be appropriate in satisfying the needs of modern IT based information systems. Understanding about organisational behaviour and cultural aspects are substantive considerations in the information development process which require a more appropriate method of appreciation than those techniques suited to technical specification (Jayaratna, 1988). The information systems professional needs to have an interdisciplinary education (Stowell, 1990) if the specification and development of technology supported information systems are to avoid the difficulties presently experienced.

The problem that this requirement for multidisciplinary skills lies in the identification of an appropriate discipline to support the research and education of the information systems professional of the 21st century. A suggestion has been made above that reductionist science is an inappropriate paradigm for the investigation of problems related to human social systems and argument has been made elsewhere that the methods of management science are inadequate for intervention into social systems (Checkland, 1981-1991; Jackson, 1992; Flood 1988; Rosenhead, 1990; West, 1991). Perhaps Systems Science can provide the intellectual underpinning for Information Systems.

INFORMATION SYSTEMS AND SYSTEMS SCIENCE

Some commentators suggest IS has failed to become a discipline because those who are its practitioners recognise it only as the delivery of a technical system (Lyytinen and Klien, 1985). Many now recognise that IT changes the behaviour of any system in which it is incorporated. It seems logical that to understand the complexity of any system, paricularly those in which human actors are involved, the IS professional needs to posess both technological skills and relevant knowledge from appropriate areas of the social sciences.

An educational mix which will satisfy those elements defined as fundamental to a modern IS professional suggests a "Hybrid" degree. Such a degree may, by some, be considered unacceptable premise given the educational history of computing science and associated specialisms. A proposal for a degree which is part technology and part social science may be regarded as untenable even second rate compared to specialist degrees. Moreover, in a practical sense it might be hard to produce such a degree because of the difficulty in selecting appropriate module topics and then in finding a way to unify them. The solution to this difficulty might be to use the Systems epistemology as the intellectual basis for the development of Information Systems as a discipline. Many of the practical ideas that have developed within systems practice may provide useful contributions to information system definition, provision and implementation (e.g. Beer, 1972.; Checkland, 1981, 1990; Espejo and Harnden 1989; Wilson, 1984; SYSTEMIST, 1992), and, as a metadiscipline, Systems thinking may provide the means by which we can critically evaluate the lessons learnt and the methods and tools developed in Computing provision over the past decade.

In summary, this paper has attempted to provide a brief history of the development of the domain of Information Systems. This has been done because of the importance of IT based information systems to human affairs and because of a perceived deficiency in the concept upon which IT has hitherto been developed. Whilst recognising the ongoing work of others who are developing systems methods for specific areas of information system provision this paper goes one step futher by proposing that the discipline of Information Systems should be underpinned by the Systems epistemology.

REFERENCES

Boulding, K.E., 1956, General Systems Theory - The Skeleton of Science, *Management Science*, 2. (3). pp197-208.

Bullock, A., Stallybrass, O. and Tromby, S., eds., 1990, "Fontana Dictionary of Modern Thought", Fontana Press, London.

Checkland, P.B., 1981 "Systems Thinking, Systems Practice", Wiley, Chichester.

Checkland, P. and Scholes, J., 1990, "Soft Systems Methodology in Action", Wiley, Chichester.

Churchman, C.W., 1971, "The Design of Inquiring Systems", Basic Books Inc., New York.

Espejo, R. and Harnden, R., 1989 "The Viable System Model", Wiley, Chichester.

Esprit '90, 1990, "Conference Proceedings", Klewer Academic Publishers, Dordrecht The Netherlands,

Finkelstein, C., 1989, "An Introduction to Information Engineering", Addison Wesley, Wokingham.

Flood, R.L. and Carson,E.R., 1988, "Dealing with Complexity", Plenum, New York.

Forrester, T., ed, 1985, "The Information Technology Revolution", Blackwell, Oxford.

Guy, K. and Georghiou, L., 1991, "Evaluation of the Alvey programme for Advanced Information Technology", HMSO, London.

Jackson, M.C., 1991, "Systems Methodology for the Management Sciences", Plenum Publishers, New York.

Jayaratna, N., 1968, Guide to Methodology Understanding in Information Systems Practice, *International Journal of Information Management*, **Vol 8**, pp43-53.

JUPITER Consortium, 1987, "Management of Technology: The Invisible Advantage", Brunel Uxbridge, University Science Park.

Freidrichs, G. and Schaff, A., eds, 1982, "Microelectronics and Society", Pergamon Press, Oxford.

Lyytinen, K.J. and Klien, H.K., 1985, A Critical Theory of Jurgen Habermas as a Basis for a Theory of Information Systems, *Research Methods in Information Systems*, eds. Mumford, E. Hirschheim, R.A., Fitzgerald, G. and Wood-Harper, Elsevier Science Publisher BV, A.T. Amsterdam, pp219-231.

National Advisory Board, 1982, Postgraduate Conversion Course Initiative, *Pamphlet.*

Rosenhead, J., ed, 1990, "Rational Analysis for a Problematic World", Wiley, Chichester.

Romm, N.R.A., 1991, "The methodologies of Posivitism and Marxism", Macmillan, Hong Kong.

Stowell, F.A., 1990, Systems Analysis and the Design of Information Systems, Paper presented at the *8th International Congress of Cybernetics and Systems*, City University New York City. p116, p119.

Stowell, F.A., 1991, Towards client-led development of information systems, *Journal of Information Systems*, **1**, pp173-189.

Systemist, eds. Stowell, F.A. and West, D., 1992, "Information Systems Special Edition", Vol. 14, No. 3, pp99-106.

van Gigch, J.P., 1981, "Decision Making about Decision Making: Metamodels and Metasystems", Abacus Press, Tonbridge Wells.

Winograd, and Flores, 1988, "Understanding Computers and Cognition", Addison-Wesley, New York.

West, D., 1991, Knowledge Elicitation as an Inquiring System: Towards a Subjective Knowledge Elicitation Methodology, *Journal of Information Systems*, **2**, pp31-44.

Wilson, B., 1984, "Systems: Concepts, Methodologies and Applications", Wiley, Chichester.

ENGINEERING SYSTEMS

Malcolm K. Crowe

Department of Computing Science
University of Paisley
Paisley, PA1 2BE
Great Britain

INTRODUCTION

The purpose of this paper is to attempt to respond to the systems approach from an engineering standpoint, in particular that of a computing scientist. Most computing scientists, and even most engineers, spend regrettably little time reflecting on the nature of the systems they work with.

A typical Systems Analysis textbook definition (Senn, 1985) of *system* highlights the problem:

> "A system is simply a set of components that interact to accomplish some purpose."

That is, there is nothing beyond the set of components, except for the stated aspects of interrelationship (internal) and purpose (extrinsic). It is worth considering what analysis will then achieve: subdivision of purpose? greater precision in the nature of the interactions between components? or of the component set?

Very few books have a definition in agreement with Checkland's (Checkland, 1981, 1988) in highlighting emergent properties: an exception is (Waring, 1990). Oddly, though, Waring's rather complex definition requires that some of the emergent properties should be unpredictable, and has a separate requirement that

> "a system does something (there are outputs)"

Now is what the system does an emergent property or not?

From an engineer's viewpoint Checkland's remark that for man-made systems, emergent properties are the key to systemhood, is very reassuring. Such questions as ownership of the system can be left to the sales and marketing division while engineers get on with designing and implementing useful systems.

Systems Science, Edited by F.A. Stowell *et al.*,
Plenum Press, New York, 1993

THE ENGINEERING APPROACH TO SYSTEMS

First, it is common ground that systems are characterised by the emergence at various levels of properties not possessed by their component parts.

Engineering is *techne* (Finniston, 1980):

> "Engineering is the use of mathematical, scientific, and managerial techniques to produce cost-effective artefacts suited to their purpose."

The concept of "purpose" also appeared above. System requirements are a specification of the properties of the desired system. Design is seen by engineers as the fundamental process of their discipline (e.g. Duderstadt et al., 1982):

> "Engineering design is the development of a physical system or process to perform a required function.. a deductive procedure to develop a specific solution to a given problem from a general set of principles .. requires a synthesis of technical, economic and social factors."

Duderstadt et al distance the notion of engineering design from research, which is inductive, looking for general principles, and from invention which is about discovery, and therefore too dependent on happenstance.

In an important sense, it is this design process which sets apart engineering systems from other types of systems. In human systems, for example, it is more difficult to start with just the system requirements, from nothing; more common to identify and incrementally change an existing system.

Thus it can be asserted that design is that part of the process of engineering in which an abstract model of the system is constructed from the system requirements. The model specifies the components and describes the required properties in terms of properties of these components and or emergent properties of the system. Not all the components need to exist in advance: the design process may identify subsystems (with certain properties) that in turn need to be implemented. Not all the system properties need be implemented as emergent properties, some may appear directly or indirectly as properties of a component or subsystem.

The notion of subsystem is very important, especially in complex systems such as software products. Subsystems should be identified as part of the process of analysis and design. One approach, which agrees well with the systems view, could be called vertical partitioning, where a subsystem is identified as the part of the system needed to support a subset (e.g. one) of the properties of the system. For example, in a car, there is a subsystem to provide seating space for the occupants (supporting the emergent properties of passenger capacity, comfort, accessories, etc), a subsystem to provide the lighting system, a subsystem to provide the motive power, and so on. Existing subsystems, or subsystems for which a fully-worked-out design exists, can often be successfully included in a design.

In computing, a highly successful approach that has been developed over the years could, by contrast, be called horizontal partitioning. In this approach, layers of abstraction are identified which develop emergent properties stage by stage. This is often achieved by analysis of the concepts contained in these properties.

An example may be helpful. Consider the requirement to develop a relational database management system. The system consists of a set of software components, the hardware it runs on, and its state is determined by the data it has stored. The emergent property that refers to relational database management needs to be defined, in terms of the management functions of a database, and in terms of what constitutes a relational database. To define a database, we need a sequence of terms such as table, record, field, data type, that can be made to form a series of layers of abstraction. Each layer is implemented in terms of the lower layers, and a constraint imposed by the design method is that the lower layers should know nothing of the use being made of the functionality they provide.

As with vertical partitioning, existing layers, or layers whose design has been fully worked out, can often be included in a design. Most of the breakthroughs in computing science amount to the successful identification of a layer that can be included in many designs: for example the machine instruction layer (Turing, 1946), the operating system layer (Ross, 1956).

THE DESIGN PROCESS

The design process, from the above discussion, should steadily implement more and more of the system requirements by identifying a potential structure for the system in terms of components (or subsystems) and their interaction. The emergent properties of the system are implemented directly as properties of a subsystem (by renaming of parts, specialisation of function, or identification of data), or by means of a combination of properties of subsystems similarly adapted.

For example, the colour of the car is the same as the colour of the bodywork, the engine capacity of the car is the same as the engine capacity of the engine, the width of the car is the length of the front bumper. On the other hand, the maximum speed of a car will be implemented in terms of (a) the torque to revolutions per minute characteristics of the power unit, (b) the drive gear ratio, (c) the wheel size, (d) friction characteristics of all of these subsystems, (e) the weight of the car and its load. The designer will not in general be able to predict exactly the maximum speed achievable by the car, but will apply a series of estimates, rules from experience, formulae and professional judgement to fulfil the system requirements. The actual maximum speed will be measured using a prototype.

In designing a car, the engineer will have a repertoire of available components: engines, wheels, etc to use in the design. From simple cost/benefit analyses, it is generally only justifiable to develop new subsystems if the available materials do not allow fulfilment of the system requirements.

In software engineering, the designer has, apparently, a much freer hand. It is comparatively unusual to reuse existing software components. But software designs are frequently recycled.

In fact, there has been for years a sterile controversy about top down versus bottom up design. In reality, neither is sufficient on its own, although many engineers seem to think of bottom-up development as rather shameful. Top-down design generally takes a pride in not reusing existing designs.

Top down design focuses on the principal functions being provided (for which one might read emergent properties), and proceeds to analyse this into subfunctions, each of which will be provided by a subsystem. A problem with top down design is that the subdivisions may not be supported by any existing system, may involve wasteful duplication of effort, may be misdirected.

Bottom up design instead concentrates on reuse of existing designs. A problem with bottom-up design is that the initial focus is often too low level and may develop the wrong thing.

In practice, a good designer has a firm grasp of a repertoire of components and properties whose realisation is achievable: she will present analysis and design in a top-down way, but with a feasible way of realising that design always in view. Design using layers of abstraction is usually basically bottom-up, since it is most successful when new upper layers can be built on common lower layers. Thus, successively over the years, successful layers for hardware operation ("machine instructions"), peripheral and memory management ("operating system"), file system and network management, graphical user interfaces etc have become de facto standards across the industry.

PROBLEMS WITH ENGINEERING DESIGN

It may be helpful to review some of the main problems with system design in engineering:

The Linguistic Defect

The most tricky is the linguistic defect. This means that the means of expressing ideas imposes constraints on the ideas being expressed. This problem is not limited to engineering, of course, as many examples could be taken from philosophy, where distinct national movements are easy to identify.

Thus engineers working in different languages may arrive at different designs for no other reason than the language they use for their work. For example, the approach to layers of abstraction defined above may be affected by the availability of technical terms in the national language.

The linguistic trap is not confined to the natural language used by the designer, of course. The systems of signs and sets of functions used to express the existing layers of a design constrain the designer's ability to express the next layer. In computing, for example, different machine architectures provide different low-level functions, but to provide industry-standard interfaces at the next layer many of them must remain unexploited.

In engineering, however, there is a further trap. The language used to express the system requirements frequently has embedded within it the lineaments of a design. This can happen in a variety of ways. It could be that the system requirements were written with an existing system in mind, or are based on a feasibility study, and the design of these systems has leaked into the statement of requirements. This problem is particularly insidious and hard to detect, and even when detected it is hard to solve.

Fundamentally, and especially in computing, development of concepts is required in order to provide a language for description of a system, and the resulting design is then a prisoner of this collection of concepts. Which comes first, the concepts or the design?

Murphy's Law

A second major problem in engineering design has to do with the crucial requirement that a system must be suited to its purpose. Murphy's law is well known - systems (especially safety critical systems) become unreliable just when they are most needed, as the system suddenly encounters abnormal conditions, which by their nature may not have been fully explored during evaluation. A second problem is that such a system must be trusted. This is not a new problem (Turing, 1946):

> "It would clearly not be satisfactory if the checking system in fact prevented all errors, but nobody had any confidence in the results. The device would come to no better end than Cassandra. In order to inspire confidence the checking must leave some visible manifestations."

This principle has other applications. It also means, for example, that to be seen to be effective, quality assurance procedures should cause some inconvenience.

Speed of Change

A third major problem in computing science is the speed of change. Over the last two decades computing systems have changed, more than once, beyond all recognition. This means that in the multi-layered abstraction model that permits the development of complex systems, many of the older layers contain fossils of obsolete design paradigms. One of the most dramatic of these fossils is the concept of "the main-frame computer", but others, for example relating to time-sharing, the process model or networking, have more subtle effects. Dramatic changes can be expected where signs of strain are most evident.

COST-EFFECTIVENESS

As noted earlier, a crucial aspect of engineering is the focus on cost-effective solutions, or cost-effective artefacts suited to their purpose. One of the major tasks of engineering, and an area that attracts the greatest effort, is the attempt in each field of engineering to find objective ways of establishing that an artefact is in fact suited to its purpose, and to quantify both the costs and the effectiveness of any proposed solution.

Thus structural engineers have sets of tables establishing the appropriate thickness of walls for various purposes, and the necessary materials. Criteria have been established over the years which enable the extra costs, over the lifecycle, of "cheap" solutions to be identified, and allow the customer (or engineer?) to decide the best solution. As an example, buildings may be made cheaper to construct by using thin walls and single glazing; but the extra costs of heating need to be weighed up. Many customers have been happy to accept a smaller capital cost even if the life-cycle costs are higher.

INFORMATION SYSTEMS DESIGN

One of the problems with a young discipline like computing, or software engineering, is that such decisions are much harder to make. There are no engineering tables to tell the designer what speed of processor, or size of disk, is needed for a given system. The decision made by the designer is often based on little more than a gut feeling, or back-of-the envelope calculations.

An increasingly important issue is the decision whether to adapt or make do with an existing package, or develop a tailored product. The existence of many different packages and development paths makes this sort of decision very complex. Among the aspects to be considered are (a) any loss in facility or increased costs of use through use of a generic package - they are often much less efficient, have dollars instead of pounds, etc. (b) the usefulness of in-house expertise to modify the system - this point cuts both ways, of course; (c) the likelihood of teething troubles (bugs) with a newly-developed system; (d) the costs of training and maintenance.

Another vexed area is that of testing and verification. Different implementation strategies might call for widely differing amounts of validation/verification or testing. The right place to stop depends on the cost of failure of the operational system, but contractual aspects complicate the judgement. As an extreme case, the following reported exchange at the ICSE '87 conference is memorable:

> "D. Parnas: ... Star Wars software cannot be built to this specification.
> An acceptance test would be impossible to agree.
>
> General: We thought that with such a good team, especially headed by

Parnas, testing would not be necessary.

Parnas: If you are not going to test it, I can build it for you in five minutes!"

But it is clear that the design of an information system, whether carried out using a soft systems methodology or in any other way is an engineering process. At the end of the day, the client is going to pay a lot of money (in his eyes) for a new system, or to modify an existing one, or to change the working practices of the business. These are the costs of the product. How well-established are the techniques for determining the effectiveness or the suitability for its purpose of the proposed system?

The truth is that, compared with other branches of engineering, the mathematical techniques are not yet very useful. But they are being slowly developed, and it is worth remembering that the other branches of engineering are not flawless: bridges and aircraft of new design do fail from time to time.

ENGINEERING SYSTEMS PRACTICE

Engineering thinking, engineering practice... Since good engineering is so much a matter of professional judgement, it is not surprising that engineers are required to undertake a large practical component including several years industrial experience after graduating, before they are considered competent, or admitted as full members to the professional associations. The content of an engineering course then must consist of a thorough grounding in the scientific and mathematical disciplines required for this practice, together with a graded programme of Engineering Applications. These are traditionally called EA1, EA2, etc, and the first two stages must form part of any engineering degree.

Thus every computing course introduces the standard layers of abstraction, the standard components, some standard techniques and tools (EA1), before going on to tackle real-life problems (EA2) of engineering practice.

Abiding Principles

But it is worth considering what the "abiding principles" of computing are. What is the equivalent of Ohm's law, the periodic table, ionisation potentials, Maxwell's equations? No doubt the theoretical stock-in-trade of every computing scientist should include such important theoretical results as Turing's Halting Theorem, the notion of non-polynomial completeness, Shannon's Theorem. And yet these have never been seen as essential components in a course in computing science. Many whole subjects, such as software verification or artificial intelligence, can be studied optimistically only by people who do not know about these principles.

Worse still, if research really were the inductive process of discovering new general principles of systems, as Duderstadt et al claim, there would be very little research going on, or (arguably) possible. How can we characterise computing research?

Themes of Research

From the systems viewpoint, one strand of research in computing might be to re-engineer a system, identifying layers of abstraction or marketable subsystems in its design. This has been done very successfully with operating systems, graphical user interfaces, and communications subsystems, in addition to processors, memory and other hardware systems. Once subsystems have been isolated, they can be re-implemented more efficiently, more cheaply or more robustly. This kind of engineering research is into finding better designs and more cost-effective products for a given system requirement.

Another strand is to explore the area of new paradigms or layers that might build on top of existing well-accepted layers of abstraction. Much research into high-level languages, semantic databases, object-oriented stores, and the like comes into this category. To be useful engineering this sort of work must have a basis in tentative system requirements emerging from one or more sets of users or customers, who have a real-life problem that the proposed development may solve. Engineering research of this kind is about finding good design approaches to novel system requirements.

A third type of engineering research is into the development of new tools for information systems development (in the broadest sense). There is work of this kind in every university computing department, including Paisley. But who is researching into new theoretical results?

REFERENCES

Advanced Network System Architecture (ANSA), 1989, Architecture Projects Management Ltd, Cambridge, England.

Checkland, P.B., 1988, Information Systems and Systems Thinking: Time to Unite? *Int. J. Information Management*, Vol. 8, 1988, pp239-248; reprinted in Checkland, P.B., Scholes J., 1990, "Soft Systems Methodology in Action", Wiley, Chichester.

Checkland, P.B., 1981, "Systems Thinking, Systems Practice", Wiley, Chichester.

Duderstadt, J.J. et al., 1982, "Principles of Engineering", Wiley, New York.

Finniston, M., 1980, "Engineering our Future: report of the committee of inquiry into the engineering profession", HMSO, London.

Open Systems Interconnect (OSI), 1986, Basic Connection-mode Transport Service, ISO 8702, International Standards Organisation.

Ross, D.T., 1956, Servomechanisms Laboratory Requirements for Computing Facilities, MIT Servo Lab Memoir, *in*: Goldberg, A., ed., 1988, "A History of Personal Workstations", Addison-Wesley, Reading, Mass.

Senn J.A., 1985, "Analysis and Design of Information Systems", McGraw-Hill, New York.

Turing, A.M., 1946, Proposal for Development in the Mathematics Division of an Automatic Computing Engine (ACE), *in*: Carpenter B.E., Doran R.W. (eds.) 'A M Turing's ACE report of 1946 and other papers', MIT Press, Cambridge Mass, 1986.

Waring, A., 1990, "Systems Methods for Managers - a practical guide", Blackwell, Oxford.

GIVING REQUISITE VARIETY TO STRATEGY AND INFORMATION SYSTEMS

Raul Espejo

Aston Business School
Birmingham, B4 7ET

INTRODUCTION

In this paper I want to pursue a general problem of concern to all organisations, whether in the public or private sectors, large or small. This is the problem of how organisations relate themselves effectively to the environment in which they evolve and operate.

The prevailing view about this relationship is that of an organisation planning its way into the future, anticipating hurdles and difficulties, improving its information about external changes and in general taking advantage of environmental opportunities. All these are the concerns of strategic management, that is of senior managers and their support staff who share an overview of the organisation, its vision and mission. Implicit in this view is the widely accepted distinction between strategic and operational management, in which some people are responsible for the organisation's relations with the environment, while the rest are responsible for the internal operations.

The view developed in this paper is that organisations need far more than top level strategic management and the supporting management information systems to achieve effective interaction with their environment. They need to have the structures to involve all of their people, to the best of their abilities, in the creation of possibilities and responses to external disturbances.

It is argued that strategic planning and information systems are inherent in the structure of effective organisations. It is effective organisation that allows people to answer questions such questions as: "How do we know whether a new idea is likely to be appropriate to the organisation's longer term viability?" And "How can we stay 'on course' when change is the only constant feature in our surroundings?"

The argument of the paper is as follows: First, the organisation's interaction with the environment is related to the idea of task. Task is defined as the set of activities necessary for the purposeful implementation of a change in the environment. It is argued that the quality of this implementation depends, among other factors, on the management of complexity; in particular on the quality of the balance between the attenuators of environmental complexity and the amplifiers of managerial complexity. These are the complexity operators constituting the manager-task loop. The concept of organisational closure is then developed at a very general level. It is argued that effective information systems involve much more than the technically efficient production of data: they are constituted by the organisation structures which allow the effective flow of information. This is followed by a section in which two radically different views about an organisation in its environment are presented. These contrasting interpretations of the environment lead to very different views about organisational strategy. The implications of these views are illustrated by describing different organisational situations.

TASK CLOSURE

If we think of a craftsman and the relationship that he develops with his craft, we can perceive that closure is inherent in the activities he performs. The craftsman sees the product of his activities in real time and he is able to modify or alter the product in progress in such a way that his vision is finally realised in the finished product. This loop between the craftsman and the task of his concern - where task is defined as the set of activities necessary for the purposeful implementation of a change in the environment - is closed in real time; **task closure** is natural to his work. The craftsman may have the support of a whole range of tools, the purpose of which is to increase the scope of his action, allowing him to achieve outcomes that otherwise would be impossible. Some of these tools, like the chisel and hammer, may **amplify** his action capacity; others, such as focused lights, may **attenuate** the complexity of the total work, thus enhancing his sensibility to some of its aspects.

For managers the idea of task closure is as important as for craftsmen, yet for them tasks are inherently beyond their observational capacities. We may say managers, in general, are accountable for tasks possessing far more complexity than they themselves can handle. In other words managers do not have the benefit of direct closure: they depend upon other people, and additional resources, to close the task-manager loop (Espejo & Watt, 1988). This loop is constituted by managers in one form or another, whether they are aware of the fact or not. The problem is that if its constitution is left to chance then unnecessary costs may be incurred. Indeed, it is important to know how to design such a loop with minimum damage to people and to cost (Beer, 1979). Unfortunately, an inadequate design of this loop is common experience, and nowhere this is more evident than in the development of information systems and the planning of organisational activities.

Indeed, these developmental activities often increase the imbalance between attenuation and amplification. For corporate managers they tend to make environmental attenuation stronger - thus enabling them in principle to make more distinctions about trends, customers, competitors and the like - and their amplification capacity weaker - thus making it increasingly difficult for them to respond to the implications of more environmental distinctions. For the rest of the people in the organisation, these systems generally provide inadequate environmental attenuation, since they are usually skewed towards the information needs of the corporate level people. This increases the chances that the capacity of the rest of the organisation for action will be poorly used, thus making the motivation of people to provide amplification support weaker.

It should be apparent from this discussion that at an aggregated level the organisation should achieve the same quality of task closure as the craftsman achieves with his production. This implies that to be effective an organisation must have:

1 Vision, to provide clarity of task;

2 Cohesion, that is an ability to handle disturbances in an organic fashion, making a balanced use of the total system; and

3 The ability to operate in real time, that is be able to recognise and produce change as interactions and events unfold.

These characteristics are indeed difficult to find in most of today's organisations. However, as our understanding of human interactions and organisational phenomena grows and as technological developments increase the chances of better communications and information flows, we may expect significant improvements in the individual-task and the organisational-task loops. This is the concern of the rest of the paper.

ORGANISATIONAL CLOSURE

Autonomous organisations have closure in their operational domain, that is, are constituted by participants who define a closed network of interpersonal relationships. This network defines the organisation's space for action. Without closure there is no autonomy; organisational ultimate responses to disturbances would depend on 'external' participants, thus denying the assumption of autonomy.

The implication of 'closure' is that organisations are structure-determined. Organisational change is the outcome of a closed network of multiple adjustments taking place among the organisation's participants. These make continuous adjustments in their operational domain as they adapt to each other's position and to other external perturbations.

Change is thus determined by the internal coherence of the system and not by information about external events. Change may be triggered by information, but is not determined by it. Therefore it is misleading to say that organisations are 'responding' to external 'information' as this is construed in common language. That view may be useful in the informational domain of an observer, ie one who is observing the response and construing what is taking place between the organisation and its environment. However in the operational domain, the domain of the organisation's autonomy, participants are adjusting their stability vis-a-vis each other as a result of environmental perturbations. The complexity of the organisation's operational domain is much larger than that of the observer's informational domain. Whatever is her/his observation is a viewpoint about the organisation-environment interaction, but does not constitute the interaction itself.

This makes apparent that the concept of information is not enough to understand change; what organisations receive from their environment are perturbations that are absorbed in different ways by different structures. To a large degree it is not the content of a message that determines the response but the structure that absorbs the message. People in organisations are constantly receiving information (i.e. messages that are supposed to change their actions) about aspects for which they find no structure to be 'in-formed'. However valuable these messages might be for the organisation's viability they are not 'heard'. These messages simply are dissipated in the complexity of the organisation. The same message with a different structure would have produced a very different response.

A consequence of this appreciation of information is that **information systems** are structures that allow the effective flow of information. Their effectiveness is assessed with reference to a tacit purpose or legitimate declaration of purpose for the organisation. This is an important insight, with important design implications; different purposes imply different organisation structures and therefore different information processes (Espejo, 1989).

It is rare, to say the least, for information system and planning specialists to pay attention to the organisation's operational domain while designing their systems and defining their plans. While it is relatively easy to work out the information requirement of managers, it is far more difficult to work out the structural implications of these requirements. The lack of a language to recognise these structural implications makes it very difficult, if not impossible, for these specialists to take into account this aspect. Hence the relevance of a language such as that of the Viable System Model (Beer, 1979; 1985; Espejo et al., 1989). This is the language of complexity.

ORGANISATION AND ENVIRONMENT

It is part of our everyday language to talk about 'an organisation in its environment'. Such a statement may imply very different things depending on the assumptions that we make about the constitution of an organisation. The concept of organisational closure helps us to work out the implication of this statement. For most people the environment is the source of external information to the organisation, for a few the environment is the medium in which the organisation realises itself.

The first meaning is that of the external circumstances as well as the trends and issues perceived as relevant by an observer to the enterprise's success. On the other hand, the second meaning, that of a medium, has the connotation of the substrata, or surrounding, in which an organisation exists. The first view pertains to the informational domain of observers; it is a strategic concept and its focus is on planning response strategies, based on intelligence information gathered by the observers. Strategies are in the informational domain of these observers. The second view pertains to the autonomy of the organisation itself. It entails the instant-by-instant communications (i.e., structural coupling) between all the viewpoints constituting the organisation and its medium (Maturana, 1987).

Communications take place in the operational domain of the organisation. An emphasis on 'source of information' rather than on 'medium' leads to an emphasis on symbols and representation (i.e., information), rather than on structural coupling (i.e. communications and complexity).

The notion that a few individuals can make use of, and respond to, the environment, namely those scanning the environment and working out the strategy, is replaced by the idea that everyone is responsible for the organisation's balance with its medium. Every single person should develop an appreciation of how his/her actions, as members of the organisation, affect the environment. In other words, the idea of a boundary, that is, of a membrane separating the enterprise (i.e. its people) from the environment is useful only in the informational domain of particular observers (e.g. the corporate staff). In the operational domain it is misleading; everyone is receiving perturbations from the medium. How these perturbations are absorbed is structure-determined.

Dealing with the environment only from the perspective of the informational domain of its corporate level is likely to be inadequate for an organisation's effective stability with its environment. Everybody needs to learn how to communicate effectively with the environment - how to manage the 'environmental' impact of his/her actions. Indeed, the more the interaction with the environment is understood as only concerning the corporate level's relationship with the environment, the more this level will need information and planning. It will be responsible for 'planning' the responses of all the other organisational members, which not only increases the information needs of strategists, but most importantly, reduces the flexibility of the total organisation. On the other hand, if all individuals in the organisation are managing their interactions with the environment, as well as their internal interactions, then the relevance of the informational domain of corporate managers decreases. At the same time, the overall effectiveness of the organisation depends increasingly on achieving a close integration of its operational domain. In other words, the emphasis moves away from a centralised all embracing and detailed strategy towards an effective structure with distributed strategies.

We expect that the successful company will have well developed strategic planning methods and sophisticated systems and observe in disbelief if we encounter situations denying these expectations. It is perfectly possible to encounter successful, viable enterprises with apparently limited strategic planning and not much evidence of sophisticated corporate information systems. For as long as they are adjusted to their environments, with people in the organisation operating in effective interdependent task loops, enabling their organisational units to absorb disturbances, then centralised strategic plans and information systems are redundant. Such organisations have natural distributed information and planning systems built into their organisation structure. However, the effectiveness of this natural 'adjustment' will depend on whether the organisation has a clear sense of identity and a genuinely shared vision as to the enterprise's future.

This is a point that perhaps a well known trouble shooter was unable to see in his visit to the Morgan Motor Company, a successful old cars manufacturer in the UK, with orders extending for the next 10 years (Harvey-Jones, 1992). His interpretation of their behaviour was that they were a company lacking in ambition, afraid of change, unable to accept the challenge of dramatic growth. The diagnosis of the company was based on management's apparent lack of concern about possible competitors and their apparent lack of strategy, and not on the strength - or lack of it - of a company well adjusted to its medium, able to maintain an independent existence over time. Even with no direct knowledge of the company one can hypothesise that it was not growth or competition that really mattered to Morgan; it was the fact that they had selected a market segment in which they could perform well, in which the total human group was at ease and could operate in an effective balance with their total environment. This view is in contrast with the rational arguments, based on the information domain of an 'observer', who did not attempt to diagnose the operational domain of the company with regard to its autonomy, but limited his view to the strategic level.

But perhaps the most worrying, and common, experience is that of organisations with sophisticated but centralising strategic processes and information systems. These are cases in which the multiple natural loops between people and the environment are likely to be in a non-learning state. Not only may corporate people find themselves dealing with potentially relevant information for which they have no response capacity; they may also concern themselves with plans and information that are relevant to others within the organisation without recognising this fact, thus further overloading their limited information processing capacity.

As explained previously, at the core of the manager-task loop is the balance between attenuation and amplification. The overwhelming emphasis on information, and the information domain of some observers, reduces the attention paid to the operational domain, and in particular to the amplification needs of managers. This problem is illustrated by the following examples.

In a recent international seminar organised by the Colombian Presidential Office (Santa Marta, November 1992) on the 'Definition and follow through of investment projects', two of the most powerful contributions were "An indices system to follow through development programmes" and "An indices system for the focalization, follow through and assessment of social programmes for the youth, women and family". The first explained an executive information system to follow through the main programmes in the 'national plan' in order to alert managers about implementation problems. The second explained a methodology and related information system, based on geographic indices, to direct investment resources to social groups most in need. Both were good examples of high quality information systems, however they suggested no appreciation of the operational domain in which they where supposed to operate. The first project failed to recognise the basic social variables in the target population affected by the development programmes; the second failed to recognise the complexity of the organisational processes entailed by the focalization. Both aspects are discussed in more detail in what follows.

The system to follow through programmes puts the emphasis on following through investment programmes, regardless of their effect on the critical variables that triggered the investment needs in the first place and that will be affected by the implementation of these programmes. In other words, it decouples the actuality and capability of these variables (e.g. housing, education, health achievement indices) from their potentiality, as defined by the investment programmes (Espejo, 1992). This lack of connectivity between actuality and potentiality is likely to make the system irrelevant. The people with operational responsibility for measuring educational, health, housing achievements - those closing the related task-manager loops - are likely to be left out of the follow through system; at the same time, those receiving the investment information are unlikely to be directly related to the relevant manager-task loops. But even if the information were sent to the relevant people, its emphasis on the programmes themselves makes it difficult for them to effectively integrate the strategic and operational aspects of their management.

The focalization project also pays attention only to attenuation. The project is an attempt to design a better system that, at least in theory, should allow policy makers to recognise in great detail geographic areas for priority investment. However, it pays no attention to the operational implications of this higher level of resolution. Only for a few geographic areas, the 47 indices used to assess priorities provide uncontroversial investment foci; for the great majority of areas, this is unlikely to be the case. For those with borderline indices, the system will not produce clear cut decisions. Soft aspects, that is organisational processes, are not given consideration. The system's emphasis is on the hard, data processing aspects, not on the soft aspects of conversational processes in which people express their interests, values and beliefs. Further, the system does not address problems of time lags and delays; the indices are based on a census that is at best two years out of date, and at worst 10 years out of date. The concept of real time needs to be elaborated in any planning system. The focalization system satisfies neither the criteria of 'real time' nor the need for 'soft negotiations'.

The emphasis on information leads to delusions, or worse, to unqualified interference with the discretion of people within the organisation. The focalization project problem is no different from that of multiple Executive Information Systems (EIS), which provide managers with 'drilling down' facilities. It is a delusion to believe that because you can see more detail, you can manage more effectively; it is unqualified interference if you deny the complexity of organisational processes and proceed to act upon this information.

In other words, the lack of grounding in the complexity of the organisation leads to the design of systems that over-emphasise the attenuation side at the cost of the amplification side. The concrete expression of this imbalance is often a lack of alignment between the organisation's information systems and its management/operational processes. Indeed this is a problem that goes beyond EIS. It is common to find operational systems having the same effect, for example the **centralised** scheduling system which forces people in the organisation to work dysfunctionally. These systems often lie at the heart of the organisation's marketing-production interface, dictating operations right down to the detailed local level of machines on the shopfloor. In the case of a multinational business unit, the result was inbuilt inflexibility and inability to cope with the continuous changes in customers' specifications as well as with the changing operating circumstances on the shopfloor. The information system was not recognising the natural business processes, which required conversations at three structural levels, one of which was consistently being by-passed by the scheduling system. It was only after this alignment had taken place that several of the dysfunctionalities were dissolved.

CONCLUSIONS

Understanding the meaning of effective organisation-environment interaction requires understanding the complementarity between strategy and structure. Strategy is developed in the informational domain of some participants, structure is the expression of the complexity of the multiple interactions in progress in an organisation. We are talking about the complementarity between the informational and operational domains. These are the domains of observation and action, of the observer and the actor. Both are necessary for effective performance, yet it is apparent that our managerial culture is dominated by the informational domain, with a very limited understanding of the operational domain, and of the complementarity between the two domains. Nothing makes this point more evident than most people's claim that they understand this complementarity, while their actions deny this claim.

However, we are developing an appreciation of the operational domain. This seems to be the intuitive underpinning of movements like total quality management, ecological management and others. A 'responsive' organisation is one that is sensitive to the full complexity of its environment, rather than to information about that environment. The 'responsive organisation' is aware of far more complexity than the 'strategic organisation'. A move towards 'responsive organisations' seems to be an encouraging cultural change now at last taking place.

REFERENCES

Beer, S., 1979, "The Heart of Enterprise," Wiley, Chichester.

Beer, S., 1985, "Diagnosing the System for Organisations," Wiley, Chichester.

Espejo, R., 1989, Method to study organizations, *in:* "The Viable System Model," R. Espejo and R. Harnden eds., Wiley, Chichester.

Espejo, R., 1992, Strategy, structure and information management, *in:* "Proceedings of the Operational Research Society Seminar Series on Managing Complexity", sponsored by ESRC (forthcoming).

Espejo, R. & Harnden, R. eds., 1989, "The Viable System Model", Wiley Chichester.

Espejo, R. & Watt, J., 1988, Information management, organisation and managerial effectiveness, *Journal of the Operational Research Society*, Vol 39, No. 1, pp.7-14.

Harvey-Jones, J., 1992, "Trouble Shooter 2," BBC Books, London.

Maturana, H., 1987, Everything is said by an observer, *in:* "Gaia: A Way of Knowing," W.I. Thompson, ed., Lindisfarne Press, California.

UNITED KINGDOM HIGHER EDUCATION - A SYSTEMS FAILURE?

Geoff Peters and Joyce Fortune

Centre for Technology Strategy
Open University
Milton Keynes
UK

INTRODUCTION

The questions to be addressed in this session are whether UK higher education can usefully considered as a system, and if so, whether the system that is perceived is failing or likely to fail, whether weakness can be identified in current arrangements and what areas for further change can be identified.

BACKGROUND

In spring 1991 the UK Government issued a White Paper entitled "Higher Education: a new framework" which proposed changes to the administration of higher education throughout the United Kingdom. The prevailing structures had been established as a result of an earlier White Paper (1987) and legislation in 1988. In the autumn the Government introduced a Bill for England and Wales and a separate Bill for Scotland, which legislated for the changes. In the last few days of that Parliament modified versions of these Bills were passed. Subsequently many of the changes have been implemented.

In the context of this paper, the most relevant aspects of these changes were the Government's declared intention in the second White Paper to expand (with efficiency) the numbers of students in higher education, and to abolish the "binary" line between the old universities and all the other colleges, polytechnics and Scottish central institutions which provided higher education. The non-universities group had until the 1987 White Paper been known as the "public sector", a term which the Government found "unhelpful and inconsistent with its desire to see all higher education institutions do more to attract private funding". The 1992 legislation allowed, for the first time, the polytechnics, the Scottish central institutions and colleges which fulfilled certain criteria, to use the title University and for them and some other colleges to award their own degrees.

Prior to the legislation the funding of higher education had been complex. In England, polytechnics and colleges had their own funding council (PCFC), and most Universities were funded by a Universities Funding Council (UFC). The exceptions were, the Open University, Royal College of Art and Cranfield Institute of Technology, which were all DES funded, and

Systems Science, Edited by F.A. Stowell *et al.*,
Plenum Press, New York, 1993

the privately funded University College Buckingham. In Scotland and Wales the universities were funded by the same UFC, but the funding of public sector colleges in Wales was determined by the Welsh Office on the advice of the Welsh Advisory Body for Local Authority Higher Education (WAB) and in Scotland directly by the Scottish Office Education Department (SOED). For the future there was to be a funding council covering the whole of higher education in each of Scotland (SHEFC), England (HEFCE) and Wales (HEFCW). Northern Ireland was not affected by the legislation.

The legislation restricted the Government's ability to intervene in the running of individual Higher Education Institutions [HEIs], and the funding councils act as a buffer between the two. However as can be seen from Figure 1 these HE Funding Councils have a very close relationship with their relevant government departments. The Secretary of State selects their members, he can send a representative to any meetings, his department issues guidance and requirements, and with the exception of the chief executives who were all academics, many of the funding councils' senior staff were civil servants on secondment.

An initial comparison of the system perceived from a Government stand-point in Figure 1 with the Formal System model in Figure 2 indicates that they have fulfilled all the requirements of the wider system including specifying in more detail than expansion targets the performance information they require. Similarly the funding councils have, through funding agreements, both provided resources and made known their expectations in relation to student numbers. They have, also through their requirements for student data, and their individual quality assessment units, set in motion the process of monitoring performance.

The Government's insistence on efficient expansion has also been achieved by a simply mechanism. A formula in England and Wales gives funding for additional student places preferentially to HEIs with below average HEFC funding per student, thereby enabling less than the average to be paid for extra places, and simultaneously reducing the average for the future.

In would appear that in broad terms, if the wider system is perceived to be only the Government, the requirements of the Formal System model have been satisfied and there may

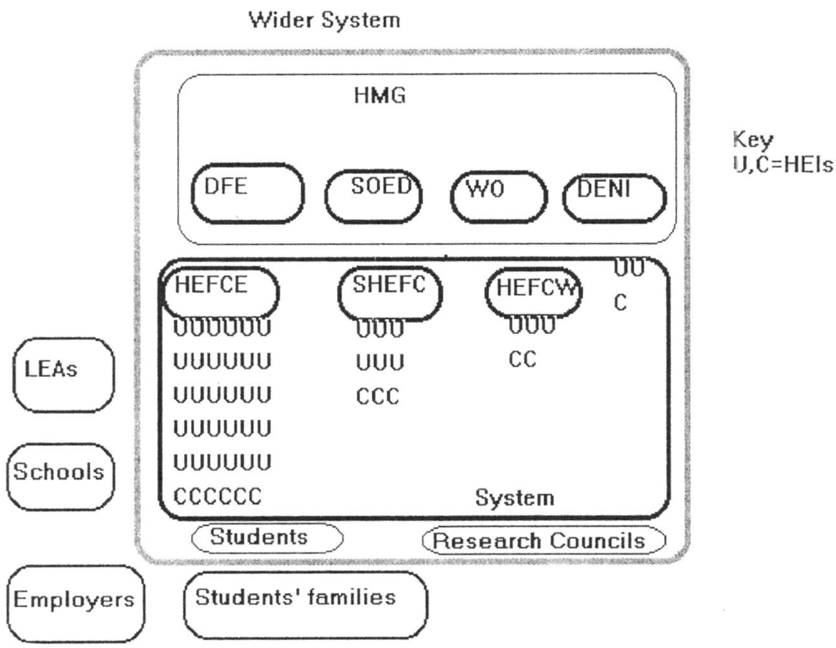

Figure 1. A systems map of UK higher education.

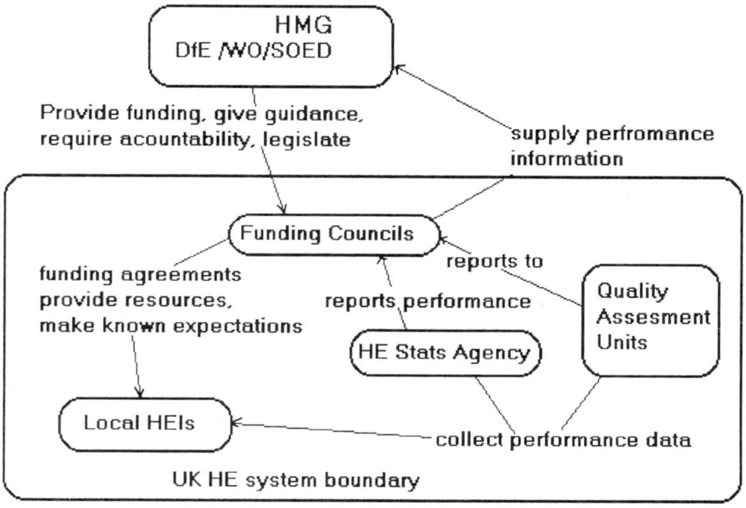

Figure 2. A Formal System model of UK Higher education.

be no prima facie case to expect failure. The one important area which has not been examined is whether the provision of resources will be adequate. It may be noteworthy that very similar conclusions was arrived at in relation to an earlier examination of the provision of primary science education and the National Curriculum (Fortune, Peters and Rawlinson-Winder 1992). The similarity in findings must pose questions about the extent to which an essentially mechanistic model such as the Formal System model is also replicated in standard policy planning practice or indeed common sense.

A CLOSER EXAMINATION

As always, the extent to which a situation can be considered a failure is a personal one, as is the judgment as to where and whether systems can be perceived which warrant comparison with the Formal Systems model. The Secretary of State for Education in England seemed well pleased when in Sept 1992 he addressed the Committee of Vice-Chancellors and Principals (CVCP) "In recent years universities have been able to increase student numbers and reduce unit costs while maintaining the quality of education they provide. This is an impressive achievement..."

However On 1 Nov 1992, a correspondent to the Observer (R.J. Ilic) wrote:

> "My son aged 30 has just started a combined honours course at Cardiff University. There are 200 students in his classes, packed in to lecture halls too small for them. He has twice returned home from lectures having been unable physically to get into the hall. On another occasion he got in but his arms were pinioned to his sides by the crush and he was unable to write. One of his subjects, philosophy is supposed to be studied in an atmosphere of debate, questioning and discussion but the lecturer told the students this would be impossible as there were so many of them. Apparently for financial reasons, the university has admitted far more students than it can cater for. My son receives a £2,265 grant and has taken out a £700 student loan. This is his total income and there are no jobs available to augment it. The books he needs are not available from the library (too few copies), there are no second hand copies available and the average price of a new textbook is about £16.
>
> It seems that the only people who are able to take advantage now of a university course are the rich and those young students who are able to live at home during the vacations. Why are others being excluded from higher education?"

Students in the wider system

The concerns of parent Ilic reflect a fear expressed by many teachers in higher education. Therefore it is necessary to consider the wider system to be not simply the Government, but also potential and current students, their parents (even for the majority of students who are mature entrants), and employers. On this basis there are several ways in which the current arrangements may prove to be deficient.

It is very hard to see many ways in which students and parents have had any real impact so far in any of the activities which link the system with the wider system. Largely as a result of activity in the 1960s by their predecessors, students do have some involvement in the governance of individual institutions. The Government might argue that the electoral process and the Citizens Charter are influencing mechanisms. It could also be argued that students make known their expectations when they select where they choose to study. However, although there is competition for students in some subjects, the expansion in higher education places has not kept pace with demand and the number of disappointed applicants for full time places is at record levels. An upturn in employment prospects and many more letters like the one above might change this position.

Full time students provide few financial resources to the HEIs directly. For almost all EC students who are studying full-time the Government sets maximum fees levels which most institutions adopt, and the student's fees are paid from public funds usually via the Local Education Authority. For students on part-time courses, it is exceptional for fees to be paid from public funds, but except in the biggest cities, part-time students have little choice but to select from the courses available from a few local HEIs or via distance education.

There are potential developments which may provide some new resource linkages. The Government might re-organise the mandatory grants and fees scheme to require parents or students to contribute more themselves directly to HEIs. Such a move might prove electorally problematic, so it would probably be attempted early in a Parliament. Politically easier would be the continued reduction in resources provided by the state for higher education to a point where HEIs decided that they need to charge a supplementary fee. Whether such changes would strengthen the relationship between the wider system and the system is arguable.

Of potentially greater significance might be the full implementation of a credit accumulation and transfer scheme (CATS or SCOTCAT in Scotland) which provided easy routes for students to acquire credit towards a qualification from more than one HEI. Although some schemes exist already in the UK there has been little transfer of credit. The major exception has been between the Open University and the rest of HE where, by 1992, 175,000 students had been granted credit towards an Open University qualification on the basis of work completed elsewhere. It is estimated that a large but smaller number of students transferred credit from the OU. A fully implemented CATS scheme coupled with a breakdown of the full-time part-time divide and a CATS based fees and funding method would, as well as adding much needed rationality to the current arrangements, allow students to vote with their feet far more, and hence to make their expectations more clearly known.

Employers

Employers are repeatedly quoted as being uneasy about the ex-students they recruit, but the government has gone to considerable lengths to ensure that the funding councils membership includes " a strong industrial and commercial element", and the HEFCs propose to use student employment patterns as one measure of performance when judging the quality of teaching in an HEI. HEIs include senior figures from business and local employers in their decision making and consultative processes. Nevertheless there have been increased signs of companies arranging there own higher education, but so far activity levels are tiny when compared with north American experience.

National and International boundaries

The most obvious query that arises from the decision to consider higher education throughout the whole UK, is the appropriateness of separate funding arrangements for each UK country. Undoubtedly primary and secondary education are organised on different lines in Scotland, but there is a considerable trade in students across the border, with Scotland being a net importer. There are therefore questions about the arrangements for ensuring that parents, employers and potential students from outside Scotland have opportunities to make their expectations known if the system which is discerned is a " Higher Education Institutions in Scotland system". On the other hand if the higher education of Scots is central then there are even more marked omissions associated with provision by other institutions, including those outside Scotland. In either case there are potential weakness associated with the handling of the Open University in Scotland which has over 13000 students, all with local home addresses, but whose places have been funded via the HEFC (England).

The position in Wales is more marked still. The school system is virtually identical to England, the majority of full time students go to England to study and the University of Wales has a majority of students from English homes.

An additional complication is the United Kingdom's place in Europe. The European Community has until recently, largely avoided explicit involvement in education (with the exception of vocational training and education designed to bring the community closer together), but the Treaty of Rome provided for the free movement and equal treatment of students. In the UK that means that other EC students are also funded both through mandatory fees and the funding councils.

A question of system

The earlier comparison with the Formal System model appeared to show that a collection of Higher Education Institutions and their local funding council could be considered as a system. But if it were possible to have degrees of systemsness then the judgment here might be that a system can only just be discerned. Taking a simpler view, to what extent can organised complexity be observed? If the funding councils were not present and students paid their own full cost fees, or simply turned up with a higher education voucher which the HEI could cash, there would be a "market" but the interrelation amongst the components would be weak. In such a case, the major links would be activities which the HEIs decided to undertake collectively and on a United Kingdom wide basis such as: full time undergraduate applications processing, statistical data collection, quality audit, and lobbying.

The addition of the local funding council provides more linkages through the use of academic staff from HEIs in policy development, quality assessment and research selectivity. Furthermore, at least in the early days, the very existence of a funding council drew its HEIs together to influence its development and thinking and to respond to its consultations. This was particularly marked in Scotland where even when the funding council was no more than a designated chair the Principals of Scottish HEIs were meeting regularly and had set up groups to consider policy issues the council might or should address. The funding council also tries to orchestrate the activities of the HEIs through its separate funding methods for teaching, research and equipment and by requiring strategic plans, and earmarking funds for special initiatives. It is not difficult to view an individual university as a system and to consider the funding council as a key part of its wider system. It is more difficult to discern the extent to which the funding council can or is dealing with the collective result of these individual interactions.

Research and scholarship

So far the deliberate emphasis has been upon the teaching functions of higher education. However, for the Universities, the advancement of knowledge through the undertaking of research and the pursuit of scholarship is an equally important part of their mission. Research is an international activity which is funded from a variety of sources and not just in HEIs. It is generally co-ordinated by topic rather than by type of institution funded. At the time of the White Papers and the subsequent associated legislation general government-funded research was co-ordinated by the Department for Education and Science (DES). Funding for specific projects was routed through broad based Research Councils which were co-ordinated by an Advisory Board for the Research Councils (ABRC). However in the United Kingdom universities were also funded for their academic staff to undertake research and for them to have basic infrastructure and facilities such as laboratories and equipment to undertake some basic and strategic research. These latter funds were dispensed by the UFC. Effectively, PCFC institutions received no such funding. The 1991 White paper maintained this dual support system but indicated that the Government wished to clearly separate the funding of research and teaching. For the immediate future, research funding was to be the responsibility of the Higher Education Funding Councils, but the Government did not rule out at some time in the future taking "the next logical step in research funding" which was to route all the research funding through a new research funding agency or the existing research councils.

After the General Election the DES became simply the Department for Education (DFE) and the Science and Research Council briefs of the DES were moved to a new department, the Office for Public Service and Science (OPPS). At the time of writing no proposals have been issued for the future handling of research, but some interesting issues will need to be addressed. It must be questionable whether the DFE will continue to have responsibility for funding any element of research. If it does not then the new OPPS has at least three options. First it could choose not to take "the logical step" and to dispense dual funding through the HEFCs. In terms of Figure 1 and the formal system comparison, little would have changed except for the introduction of another government department. Secondly OPPS could set up a new agency. However, it would be unlikely to set up separate ones for the four UK countries, so the UK systems depicted in Figure 1 would become markedly more complex and the rationale for separate country funding councils for teaching might need to be revisited. Thirdly OPPS could dispense with dual funding. Readers with experience of the United Kingdom may wish to decide for themselves what the implications of such a move would be for the success or failure of higher education.

REFERENCES

Fortune, J., and Peters, G., 1991, The Formal System Paradigm for studying failures, *Technological and Strategic Management* 3:1

Fortune, J., Peters, G., and Rawlinson-Winder, L., 1992, Science education in English and Welsh Primary Schools - a systems study, *J of Curriculum Studies*, (forthcoming).

Peters, G. and Fortune, J., 1992, Systemic methods for the analysis of failure, *Systems Practice*, 5,5.

White paper, April 1987, "Higher education meeting the challenge", HMSO, London.

White paper, May 1991, "Higher education a new Framework", HMSO, London.

SYSTEMS PRACTITIONERS:

FACILITATORS OF THE BRINGING - FORTH OF SOCIAL SYSTEMS

Michael C. Jackson

Centre for Systems Studies
School of Management
University of Hull
Hull, HU6 7RX
United Kingdom

INTRODUCTION

This paper restricts itself to some thoughts provoked by reading Heidegger's essay 'The Question Concerning Technology' (delivered as a lecture in 1953 and reproduced in Heidegger, 1978). The first set of thoughts centre on the adequacy of the systems tradition in producing technologies which can assist in the creation of social systems. The various strands of systems thinking are reviewed and their contributions assessed. The second set of thoughts centre on the notion of 'enframing'. Enframing defines modern technology, dominates existing systems approaches and, through the drive for complete and utter mastery which it encourages, threatens both man and nature.

HEIDEGGER ON TECHNOLOGY

In questioning concerning technology, Heidegger takes us back behind the current view of technology as a means to an end and arrives at a more fundamental meaning in which technology is the cause of something brought-forth, or revealed, or coming into being. According to ancient philosophy, Heidegger argues, there are four causes implicated in any such technological achievement. These he illustrates using the example of the coming into being of a silver chalice to be used in some sacrificial rite.

The first cause (the *causa materialis*) is the material, the silver, out of which the chalice is made. This is partly the cause of the chalice coming into existence. The second cause (the *causa formalis*) is the form into which the silver enters - the idea of chaliceness which leads the silver to take on the aspect of a chalice and not a sword or jewellery. This form or aspect is also co-responsible for the sacrificial vessel. The third cause (the *causa finalis*) is what gives bounds to the chalice by, in advance, defining its aim and purpose as part of some sacrificial ceremony. It is the end ".... for example, the sacrificial rite in relation to which the chalice required is determined as to its form and matter". The fourth cause is the silversmith who brings about the finished vessel by considering and co-ordinating the other three causes during the production of the chalice. The four causes occasion, or are jointly responsible for, the coming into being or the bringing-forth of the silver chalice.

Having discussed technology as a bringing-forth or 'revealing', Heidegger turns to consider the particular form of revealing that lies at the heart of modern technology. He finds that the revealing that rules modern technology "...... has the character of a setting-upon, in the sense of a challenging-forth". In this mode of thinking, everything becomes reduced to a 'standing-reserve' - it is ordered to stand by so that it can immediately be called upon for a further ordering. The earth is a potential source of mineral wealth, the soil a part of the mechanized food industry, the Rhine a resource for the provision of hydro-electric power. Even man, who through technology seeks to become ruler, risks being reduced to the 'standing-reserve' (as the recent fashion for the discipline of 'human resource management' confirms). As Heidegger remarks "The forester who measures the felled timber in the woods and who to all appearances walks the forest path in the same way his grandfather did is today ordered by the industry that produces commercial woods, whether he knows it or not".

The tendency of modern technology to do its revealing by bringing-forth a 'standing-reserve' results from its commitment to 'enframing'. Enframing is a particular 'destining' of revealing which privileges ordering and risks driving out any other form of revealing. This, for Heidegger, constitutes the supreme danger because "The rule of enframing threatens man with the possibility that it could be denied to him to enter into a more original revealing and hence to experience the call of a more primal truth".

SYSTEMS PRACTICE AS THE BRINGING - FORTH OF SOCIAL SYSTEMS

We cannot straightforwardly draw upon Heidegger's account of the four causes that enter into revealing the silver chalice in order to discuss the creation of social systems - not least because social systems are not objects. Nevertheless we can use his ideas to provoke our thinking. We do this by considering the various strands of work in the systems tradition and by asking how each of these can assist in the process of bringing-forth social systems.

A start might be made with 'hard systems thinking', which embraces such methodologies as classical operational research, systems analysis and systems engineering. These approaches assume that there is some pre-defined goal towards which the social system can be aimed and set themselves the task of discovering the most efficient arrangement of means in order to realise the goal. The system in which the problem is located is represented in a quantitative model. Experiments carried out on the model are used to identify an optimal solution which can be implemented in the real-world. The underlying rationale of this kind of thinking is that a system should be designed in the form of a machine if it is to be efficient and effective.

Another category of systems approach concentrates on the design of social systems to be self-regulating and self-organising. Work within this strand would include general system theory, Miller's (1978), 'living system theory' and Beer's (1979) 'organisational cybernetics'. Insights derived from studying organisms and brains contribute to an understanding of the shape social systems need to take if they are to be viable and effective over time. According to Beer, in order to become or remain viable, an organisation has to achieve 'requisite variety' with the complex environment it faces. It must be able to respond appropriately to the various threats and opportunities presented by the environment. Of course, the potential variety of the environment always threatens to overwhelm that of the system. Complexity, therefore, has to be managed. This is described by Beer as 'variety engineering'. Beer's 'viable system model' sets out the necessary relationships that must obtain between essential organisational elements (Systems 1-5) and information channels, so that the intricate variety equations can be balanced in a satisfactory way. One important recommendation is to make the operational elements of the system (the parts of System 1) autonomous in their own right, so that they can absorb some of the massive environmental variety that would otherwise flood higher management levels.

The strands of systems practice so far discussed are clearly concerned with the form into which the social system needs to enter in order that it can best serve its purpose. They are technologies which can be related to the *causa formalis*. As such their role in systems practice is significant and they cannot be ignored in the overall task of 'bringing-forth' social systems. It is only right, however, that they should be criticized if they set themselves up as being alone capable of creating social systems; for they have little to say about the other three causes implicated in revealing.

If, in considering systems practice oriented to social systems, we want to find a concern with the material out of which organisations are constructed (the *causa materialis*) or the aims and purposes towards which they are pointed (the *causa finalis*), we are forced to turn to other bodies of work. The socio-technical systems literature seeks to give attention to the people and social groups out of which organisations are made as well as to the arrangement of tasks in the work organisation. People, however, tend to get reduced to a set of psychological characteristics to which managers must pander, and there is little discussion of how the ends and purposes of social systems can be decided. Only with 'soft systems thinking' do we get proper attention being given to the role of people, as conscious actors, in constructing social systems and to the facilitation of debate about the aims and purposes of organizations.

Within soft systems thinking we can include the work of Churchman (1979), Ackoff (1981), Checkland (1981), Checkland and Scholes (1990) and the 'interpretive systemology' developed by Fuenmayor and his colleagues at Los Andes University, Merida, Venezuela (see *Systems Practice*, Vol. 4. 5). Soft systems thinkers seek to explore social systems by working with different perceptions of them. Systems are seen as the mental constructs of observers rather than as entities with a real objective existence in the world. Multiple views of reality are admitted and their implications are examined. Values are included rather than excluded (in theory) from the methodological process. The privileged role of experts is questioned and an attempt made to include problem - owners and other concerned individuals in carrying out the process of exploration and finding ways forward. The aim is to encourage debate involving a large number of those interested in the decision and its implementation.

The emphasis on people's perceptions and values, on corporate culture and on questioning objectives, which is the hallmark of soft systems thinking, is a big step forward for the systems tradition. Naturally, however, soft systems practice should be criticized if it represents itself as the sole element in the bringing-forth of social systems. Virtually no consideration is given to the *causa formalis*, the form into which effective and efficient social systems should enter. Moreover, no attempt is made to grapple with the constraints that, in social systems, often enter into and distort debate about ends and purposes. There is a tendency to accept at face value, and work with, existing perceptions of reality. No attempt is made to unmask ideological frames of reference or to uncover the effects of 'false-consciousness'. Further, there is a willingness to take as given 'compromises' and 'accommodations' achieved within the confines of prevailing power structures.

The silversmith did not really come onto the scene in systems practice until the birth of critical systems thinking in the 1980's (see Jackson, 1991a). The silversmith, it will be remembered, brings about the finished vessel by considering and co-ordinating the other three causes during the production of the chalice. Critical systems thinking is the conscious element in systems practice, committed to the complementary and informed use of systems methodologies in practice , and to the complementary and informed development of all varieties of systems approach. Complementary and informed use of systems methodologies requires an understanding of the strengths and weaknesses of available systems methods, technologies and methodologies, and a social awareness of the constraints on and the consequences of their use - two further commitments of critical systems thinking. Critical systems thinking is guided by a commitment to human emancipation or, what Heidegger might term, ".... the call of a more primal truth"; but that takes us towards the topic of the next section.

Provoked by Heidegger's work to consider the point of different strands of systems practice, we can take some pride in the growing maturity of our profession. At this stage of development it is possible for texts to appear that present systems practitioners as facilitators of the bringing-forth of social systems (Ulrich, 1983; Flood and Jackson, 1991a,b; Jackson, 1991). And this is a bringing-forth or revealing which respects all four of the causes. Yet, just as Heidegger was driven on in questioning technology to ask whether in essence it was, in fact, only a particular (and very dangerous) destining of revealing, so perhaps we need to view systems practice as a potentially serious servant of the thrust to 'setting-upon', 'challenging-forth' and 'ordering'.

ENFRAMING AND SYSTEMS PRACTICE

The destining of revealing that is enframing reduces everything to a 'standing-reserve'. Enframing 'sets upon' nature. As Heidegger says, in one of his examples, "The field that the peasant formerly cultivated and set in order appears different from how it did when to set in order still meant to take care of and maintain". According to this logic, the Rhine is set to work to produce power. Indeed, as Heidegger argues, even as a river in a landscape, it exists only as " .... an object on call for inspection by a tour group ordered there by the vacation industry". Man too, as we mentioned earlier, risks being reduced to the 'standing-reserve', in the service of order and system. And, of course, systems practice does play some small part in promoting the danger inherent in enframing. This is the case whether its main concern is with ordering, as in 'hard' and 'cybernetic' approaches, or whether, as with soft systems thinking, its commitment to encouraging genuine discussion of ends and purposes is so underdeveloped that it slips easily back into becoming an adjunct of enframing; readjusting the ideological status quo by engineering human hopes and aspirations is a manner which responds to the system's needs, so ensuring smoother functioning. Critical systems thinking cannot escape the charge, except that by being forced to consider and co-ordinate the other three causes, to effect closure to the bringing-forth of social systems, it cannot but consider the grounds on which it does this, and might yet still play heed to" ... the call of a more primal truth".

So how might it be possible to get in touch with another mode of revealing, an alternative way of bringing-forth social systems? Heidegger reminds us that the name 'technology' stems from the Greek *technikon* (that which belongs to *techne*) and that *techne* was used to refer not just to the instrumental activities of the craftsman but also to " ...... the arts of the mind and the fine arts". *Techne* referred to bringing-forth generally, to *poiesis*, and was something poetic. By contemplating the particular destining of revealing, enframing, that dominates modern technology, and by recognising the extreme danger it poses, we might be able to conceive of "..... a more primally granted revealing". Perhaps the fine arts, Heidegger speculates, which were once a part of *techne*, can lead us back to a more poetic revealing.

CONCLUSION

Technology, according to Heidegger, is an enframing which reduces nature and man to a 'standing-reserve'. Here surely is a global issue worthy of discussion at a conference.

Systems practitioners who, because of the development of their discipline, are on the verge of being able to present themselves as facilitators of the bringing-forth of social-systems, cannot avoid a particular responsibility. Only critical systems thinking can know of that responsibility and can make a choice. Will that be to reinforce enframing or to seek a more poetic revealing? In the words of Hölderlin, Heidegger's favourite poet,

But where danger is, grows
The saving power also.

REFERENCES

Ackoff, R.L., 1981, "Creating the Corporate Future", Wiley, New York.

Beer, S., 1979, "The Heart of Enterprise", Wiley, Chichester.

Checkland, P.B., 1981, "Systems Thinking, Systems Practice", Wiley, Chichester.

Checkland, P.B., and Scholes, J., 1990, "Soft Systems Methodology in Action", Wiley, Chichester.

Churchman, C.W., 1979, "The Systems Approach", Second Edition, Dell, New York.

Flood, R.L., and Jackson, M.C., 1991a, "Creative Problem Solving: Total Systems Intervention", Wiley, Chichester.

Flood, R.L., and Jackson, M.C., eds., 1991b, "Critical Systems Thinking: Directed Readings", Wiley, Chichester.

Heidegger, M., 1978, The question concerning technology, *in* :"Basic Writings", Routledge and Kegan Paul, London.

Jackson, M.C., 1991a, The origins and nature of critical systems thinking, *Systems Practice* 4:131.

Jackson, M.C., 1991b, "Systems Methodology for the Management Sciences", Plenum, New York.

Miller, J.G., "Living Systems", McGraw-Hill, New York.

Systems Practice, Vol. 4. 5., Special Issue on "Interpretive Systemology".

Ulrich, W., 1983, "Critical Heuristics of Social Planning", Haupt, Bern.

A THEORETICAL FRAMEWORK FOR AN INTERPRETIVE-SYSTEMIC STUDY OF PUBLIC HEALTH SERVICES IN VENEZUELA

Ramsés Fuenmayor

Departamento de Sistemología Interpretativa
Escuela de Ingeniería de Sistemas
Universidad de Los Andes
Mérida 5101
Venezuela

INTRODUCTION

Venezuelan law states that, "all people have the right to health care. Public officials are responsible for the maintenance of public health and will provide the means for prevention and assistance to those who do not posses those means." (Art. 76 of the Venezuelan National Constitution that embodies the rights of Venezuelan citizens --my translation).

Nevertheless, according to public opinion (news papers, interviews, etc.) and to some field research performed by a team of graduate and undergraduate students of our research programme (Sanchez, 1991 and 1992; Marin, 1992; Medina, 1992; Mendoza, 1992; Palomo, 1992; Rivero, 1992), the state does not provide the means for health assistance to many poor citizens. However, this apparent violation of a constitutional right does not bring about a social revolution, nor even legal actions against the state. Furthermore, interviews with poor patients that go to a public hospital reveal that, even though they would like to receive free treatment (many of them cannot receive it at all if it is not free), they do not really consider it a *right* (Fuenmayor and Cuinas, 1993). On the other hand, according to 95% of the adult inhabitants of a rich neighbourhood of Merida, there should be no such thing as free public health services (Mendoza, 1992). This opinion seems to match that of many government officials. Indeed, the official policy of the present government, adapted to International Monetary Fund (IMF) rules, has a clear radical liberal bias: privatisation of public services and no more free services (health, education, etc.).

The former raises the following judgments concerning public health services in Venezuela:

1) The legal right that poor Venezuelans have to unpaid public health services is being continuously violated.

2) Survey of public opinion seems to indicate that, although poor people have a legal right to free preventive and curative medicine, such a right is not accepted as **legitimate** by many people in Venezuela.

Systems Science, Edited by F.A. Stowell *et al.*,
Plenum Press, New York, 1993

The relationship between both judgments leads us to question the social sense (or mission) and the legitimacy of public health institutions in Venezuela. In legal terms, the state is bound to provide free preventive and curative medical services to poor people. But, laws can be changed. So the question comes back to the fore: "ought the Venezuelan state provide such public health services? Why?" On the other hand, although actual public health institutions do not seem to be accomplishing their legal mission, they do exist and, generally, provide a cheaper service than private health centres. What sort of actual role are they playing in Venezuela? How do they contribute to the power structure of this country?

The former questions concern both **possible** social senses of public health institutions and feasible interpretations of the **actual** sense of this sort of institutions in Venezuela. This is the sort of questions to which Interpretive Systemology's research programme on organisations is addressed; namely, the question about the holistic interpretive sense of organisations.

The aim of this paper is to briefly describe the present state of a theoretical framework of a research project about the sense of public health institutions in Venezuela.[1] This inquiry is being conducted along the methodological guidelines of Interpretive Systemology (Fuenmayor, 1991c). Although I do not have enough space to provide either a detail account of such methodological guidelines or, even less, of their theoretical foundation,[2] in the following I shall briefly comment on the main methodological idea behind interpretive systemology's organisations studies.

THE PROBLEM OF THE SOCIAL SENSE OF AN ORGANISATION

As stated above, interpretive systemology's research programme on organisations is addressed to "comprehending" the holistic interpretive sense of organisations. In methodological terms, such an inquiring purpose can be translated into the construction of an "interpretive systemic framework" for the discussion of the social meaning of an organisation. An interpretive systemic framework is constituted by a debate between a plurality of "thematic interpretive contexts" (or, "thematic contextual systems") and their different interpretations about the social sense of the organisation under study. A thematic interpretive context is a theory (or model) which makes meaningful and justifies a particular social mission for the organisation under study. In the present study about public health institutions, the interpretive contexts are theories about the role of the state in health services. Each thematic interpretive context produces a "thematic interpretation" concerning the organisation under study. This discourse both puts forward a desired (according to the normative content of that particular context) state of affairs for the organisation in question and judges the present state of the organisation (again, according to each interpretive context's normative content). Obviously, the judgment about the present state of the organisation has an empirical content. A debate is established between the different thematic interpretations. Each one defends its point of view in terms of normative and empirical claims based on its supporting thematic interpretive context.

The purpose of the systemic-interpretive framework and its debate is to open a space of possibilities and their justifications concerning the organisation's social sense. Due to the critical character of interpretive systemology, special attention is given to "oppressed

[1] This theoretical framework is also relevant for discussing the social sense of other public institutions.

[2] See Fuenmayor (1991a, 1991b, 1991c).

interpretations"[3] and their supporting thematic interpretive contexts (Fuenmayor and Lopez-Garay, 1991).

LOGICAL INTERPRETIVE CONTEXTS FOR UNDERSTANDING THE SOCIAL MEANING OF PUBLIC HEALTH SERVICES IN VENEZUELA

Ought the state to pay for health services? If so, to what extent? How? These are questions concerning the role that a (modern) state should play in health services. Logically speaking, such a role can be thought within a range that goes from a state that does not intervene at all in health services to a state which is totally responsible for and in absolute control of health services. Those logical possibilities (possible interpretations) concerning the role that a (modern) state plays in health services obtain their meaning in terms of the general role of the state in a society. In the late modernity, that general role is mainly understood in economic-politic terms.

Following this line of argument, the "thematic interpretive contexts" concerning the role of the state in health services must borrow their leading ideas from the theories developed under the science of political economy. Four "thematic interpretive contexts" or interpretive models were thus designed: *"socialism," "socialised liberalism," "radical liberalism"* and *"social Darwinism."* They correspond to four different theories about the **economic**-politic mission of the state in a modern society *(social Darwinism* being a limiting case).

Each interpretive model presents two parts: 1) a conception (theory) about the general (economic-politic) role of the state; and, 2) the role of the state concerning health services. Obviously, as stated before, the second part is derived from --and is meaningful in reference to-- the first part. In the following, I will first, briefly summarise the main ideas of each model;[4] and, second, present some general considerations concerning the boundaries between the models.

Social Darwinism

General Role of the State. The state does not exist (that is why this is a limiting case). Society is considered as an ecological niche ruled by the survival of the fittest. A systemic order that strengthens the stability of the human ecological niche is generated by competition among individuals in a restricted environment. That order must not be disturbed by any sort of centrally imposed order (as that represented by the state).

Role of the State Concerning Health Services. There is no role because the state does not exist. Access to private health services is restricted to those individuals who can pay for it.

Radical Liberalism

General Role of the Radical Liberal State. The state watches over the competitive process so as to reject any form of illegal competition (e.g. violence) which is not given in terms of low prices and high quality and to preserve the legal right of any citizen to have access to the market.

Role of the Radical Liberal State Concerning Health Services. Health is considered

[3] "Oppressed interpretations" are those which, from the point of view of the researcher, are rather concealed from public debate. The relationship between interpretivism and the issue of power is briefly discussed in the Concluding Remarks of this article.

[4] For a more complete version of these models see Fuenmayor R.L. and Cuinas D. (1993).

as any other service or good subject to market laws (under "healthy" competition) which must not be provided by the State. Official intervention would hinder the "natural" process of improving health services. Consequently, there should be no such thing as public health services. Access to private health services is restricted to those individuals who can pay for it. Obviously, as with any other service, the state watches over competition in health services and provides the legal means for citizens to defend themselves against malpractice.

Socialised Liberalism

General Role of the Socialised Liberal State. In addition to the general role ascribed to the radical liberal state, the socialite liberal state must provide health, education and other basic social services to all citizens that cannot pay for the corresponding private service. This is necessary in order to avoid monopoly and other forms of exaggerated concentration of power in society.

Role of the Socialised Liberal State Concerning Health Services. Concerning **private** health services, the socialised liberal state would play the same role as the radical liberal state. However, it would also ensure a free service for those citizens who cannot pay for it. This second function can be performed in two different ways: either by directly providing a free public service, or, when there are only private services, by providing insurance to those who cannot pay for them.

Socialism

General Role of the Socialist State. The state owns all means of production and manages all production and service organisations of the society.

Role of the Socialist State Concerning Health Services. The state owns and manages all types of health services (there are no private health services).

Comments on the Boundaries of the Interpretive Contexts.

The Common Region Between Liberalism and Socialism from the Perspective of Social Darwinism. *Liberalism and socialism (the political theories of Modernity) are theories or doctrines concerning the economic-politic role of the state whereby maximum social welfare is reached through a process of economic growth and industrialisation.* There are two important points that must be highlighted in the former statement:
1) The final end guiding both the liberal and socialist discourses is "social welfare."
2) Both theories are based on the assumption that social welfare can only be reached by means of an accelerated process of economic growth and industrialisation. Hence, the discourse generally looses sight of the final end (social welfare) and refers solely to the role of the state so that the highest economic growth rate can be achieved.

On the other hand, social Darwinism represents the negative limiting case concerning the role of the state: there is no state. Society is viewed as an ecological niche ruled by the law of the survival of the fittest.

The Boundaries Between Liberalism and Socialism. The basic principles of **liberalism** can be summarised thus:

* Modern urban societies can only reach a high status of social welfare by means of an accelerated process of economic growth and industrialisation (as stated before, this basic principle is also shared by socialism)

* * The best way to reach a high degree of economic growth and industrialisation is by means of a "healthy" competition among private producers (owners of the means of production) of goods and services. "Healthy competition" means that competition is given in terms of low prices and high quality. Buyers will buy those products and services which offer

the highest quality and the lowest prices. Such "healthy" competition brings about an economic process of increased productivity and, hence, economic growth. The main economic role of the liberal state is thus to watch over this free and healthy competition. It must, on the one hand, provide the infrastructural conditions for a healthy competition; and, on the other, reject any form of competition which is not given in terms of low prices and high quality (e.g. violence). It is obvious that the higher the number and variety of competitors ("competitive variety"), the healthier the competitive process. The state must, therefore, avoid monopoly and oligopoly, which are the negation of a healthy competition.

According to the **socialist interpretive context,** private property of the means of industrial production must not exist. There must be common ownership (represented by the state) of the means of production and distribution. In this way there would not be a surplus value that is alienated to the worker by the owners of the means of production. "The alienated character of the work for the worker appears in the fact that it is not his work but work for someone else, that in work he does not belong to himself but to another person." (Marx, 1844, pp85-86). On the contrary, under common ownership of the means of production and distribution, the worker "would feel at home" (ibid) and would thus be more productive. This increase in productivity would render more economic growth which, under a fair distribution system, would bring about more social welfare.

The Boundaries Between Radical and Socialised Liberalisms. The main principles of liberalism stated before are shared by these two forms of liberalism. The difference between these two interpretive contexts rest on two different interpretations about the role of the state in the preservation of a "competitive variety." According to radical liberalism, state intervention must be restricted to the rejection of any form of competition which is not given in terms of low prices and high quality (e.g. violence) and to preserve the legal right of any citizen to have access to the market.

In addition to the previously described role of the radical liberal state, the socialised liberal state must also watch over the necessary conditions that citizens must posses in order to be skilful competitors. Accordingly, the state must provide education, **health** and housing services to those citizens that cannot pay for them. In this way, with minimum basic needs being satisfied, individuals can become productive citizens. It is the claim of socialised liberalism that if the state does not watch over this minimum living conditions of all individuals, the number and variety of competitors will diminish and the threat of monopoly will endanger economic growth. On the other hand, radical liberalism would deny this assumption and hold that maintenance of a high competitive variety will be a "natural" outcome of a "healthy" competitive system, for it will --"naturally" and in the long term-- provide welfare for the whole society.

HISTORICAL INTERPRETIVE CONTEXTS FOR THE VENEZUELAN SOCIO-ECONOMIC-POLITICAL SITUATION

The former interpretive contexts provide a general interpretive framework for discussing the meaning of certain public services like health and education in any modern (or would-be modern) society. However, in this study we are interested in placing such a discussion on a more specific ground; namely, on the social meaning and legitimacy of public health services in Venezuela at present. We need, therefore, an interpretive framework that furnishes different arguments supporting and rejecting the idea that "the state must provide a free health service to those citizens that cannot pay for it," not just in any modern society, but in a particular society living a particular historical situation. In other words, we must bridge the general arguments of the *logic-type interpretive contexts* outlined above with a particular socio-historical situation. However, in terms of interpretive systemology's theoretical foundation (Fuenmayor 1991a, 1991b, 1991c) there is no such thing as the particular historical situation of, say, Venezuela. A historical account is always a possible interpretation. Hence, the idea expressed above of "bridging the general arguments of the logic-type interpretive contexts with a particular socio-historical situation" means to relate the *logic-type interpretive contexts* with *historical-type interpretive contexts*. Let us see how such a relationship between both types of interpretive contexts looks in the interpretive-systemic study of health public services in Venezuela. Thereafter we can comeback to discuss its general methodological meaning.

Put in its most concrete terms, the point we are discussing in the study about health services can be stated in the form of two questions:
1) Should there be public health services for poor people in Venezuela now?
2) What is the actual meaning of public health services in Venezuela?

Concerning the first question we have heard the general "opinion" of the four interpretive contexts outlined above. In particular, we have witnessed an abridge version of a general discussion held between social liberalism and radical liberalism. Now, if we allow judgments of fact in the discussion, new arguments come forward. For example, suppose that both positions (socialised and radical liberalism) accept as a "fact" that 44% of Venezuelan families have a monthly income lower than the amount needed to buy the minimum quantity of food to be properly fed (in the following, this indicator of poverty will be called "critical poverty"). Suppose that they also agree that 47.8% of the school age population do not go to school (OCEI, Anuario estadistico de Venezuela). Under this assumed agreement, the argument of socialised liberalism becomes stronger: Under such figures, competitive variety is in real jeopardy. Forty-four percent of the population does not have access to the market, not even as consumers of many basic goods. If competitive variety is in jeopardy the whole liberal system is in jeopardy. Suppose now that, again, an imaginary defender of social liberalism finds some "facts," accepted as such by defenders of radical liberalism, that are supportive of the following hypothesis: "The formation of private capital in Venezuela does not behave according to the accepted laws of liberalism. Private capital originates rather from the gifts that corrupted governments have given to a few Venezuelans. Such "gifts" have been paid with oil incomes which, by law, belong to all Venezuelans." Such an hypothesis would show the illegitimacy of both the "democratic" state (represented by would-be democratic governments) and the whole capitalist structure of the country. This "illegitimacy" would be given both in terms of present legislation and in reference to the basic terms of liberalism. Hypotheses like these would weaken, not only radical liberal policies, but, in general, any liberal policy to be applied in a country under those hypothetical conditions.

What these two examples are showing is that we need a new "plane" for our systemic interpretive framework in order to enrich and particularise the discussion concerning a particular public institution. This new "plane" would be constituted by interpretive contexts which account for the formation of capital, its distribution and the degree of legitimacy of the whole Venezuelan capitalist system. Two possible interpretive contexts fulfilling this need are outlined below.

After the failed coup d'etat attempted by a faction of the Venezuelan army on February 4, 1992, there have been strong discussions about the legitimacy of democratic governments in Venezuela. The opponents in this debate seem to believe that their political positions are radically different from each other. Nevertheless, a possible hermeneutic examination of main public discourses (from politicians entrepreneurs, economists and other scholars, priests, leaders of community organisation, etc.) that participate in this debate shows a surprising result: Putting aside some circumstantial differences, all seem to agree in a basic interpretive context concerning the historical development of the country's economic-political situation. Obviously, in order to see and draw the boundaries of such an alleged widespread interpretive context, it is necessary to draw another context that can serve as a contrasting background for the first one. In the following, an outline of both interpretive contexts is presented. The first (widespread) interpretive context will be called "Legitimating the Private

Sector and illegitimating the Public Sector." The second (contrasting) context will be called "Illegitimating the Whole Economic-political Structure." Let us start by presenting the common "region" or common discourse of both contexts; i.e. a normative and empirical discourse which is shared by both interpretive contexts. Thereafter we will outline the differences. However, before introducing the "discourses" of the interpretive contexts, a clarifying note concerning the presentation of these discourses is due.

The presentation of the discourses of the interpretive contexts and their thematic interpretations can be done under two different discursive styles. On the one hand, the discourses might be presented as if they were seen from outside by a sort of commentator or arbitrator. The logic interpretive contexts and their thematic interpretations were presented under this type of **external discourse.** On the other hand, the discourses of the interpretive contexts might be presented from inside, as if they were delivered by an imaginary advocate or defender of that particular interpretive context. Call this second style, **internal discourse.** Concerning the following historical interpretive contexts, I will use both styles.

The Common Region Between the Historical Interpretive Contexts

Venezuela has lived a period of uninterrupted would-be democracy (government appointed by means of public polls) since 1958. During this period, approximately 95% of Venezuelan National Income has come from the sale of petroleum, which, according to Venezuelan law is public domain. During this "democratic" period, the official discourse has pivoted around the intention of using petroleum revenues to finance an accelerated industrialisation process within a capitalist or liberal system (Baptista and Mommer, 1987). Such an intention is grounded on the belief that, if such a process is accomplished, the country's economy would pass from a weak monoproducer rentist economy to a strong, diversified, capitalist, industrial economy.

Since the advent of democracy in 1958, government has been in the hands of two political parties with similar ideologies. They have developed a strong "clientelist" mechanism of securing votes. By means of a very efficient hierarchical organisation, these two political parties involve an important portion of the population in a complex system of individual compromises with the political party. Each individual feels that, due to his/her relationship with the party, he/she has received, is receiving or will receive a special "gift" (a job, a legal permission, a pile of bricks, a house, etc.) which he/she would not receive otherwise. Hence, the vote of each individual is conditioned by the possibility that he/she obtains a special advantage through his/her personal relationship with another person who is attached, directly or indirectly, to the party. Obviously the gifts and favours paid by the party have their origin in public domain. The nature and cost of gifts varies enormously according to the "importance" of the person.

Apart from the corruption involved in the clientelist system, many government officials have embezzled some state funds for their private benefit.

Below I will present the distinguishing features of each interpretive context.

First Historical Interpretive Context: Legitimating the Private Sector and Illegitimating the Public Sector

Democratic governments have embezzled and wasted a good portion of oil revenues. This corruption process has taken place in three main ways: 1) Giving money away to the population for demagogic purposes; 2) accepting bribes from individual citizens and private companies; 3) using state resources for private purposes.

Although the first way of corruption is not legally punishable because it takes the official form of "welfare programmes," it has been very harmful to the country, both in an economic sense and in a social sense. In the economic sense, it has been a way of squandering the oil revenues that, otherwise, could have been invested in developing an appropriate industrial infrastructure. In the social sense, Venezuelans have been spoiled so that they have become lazy and have not learned to be hard and competitive workers. This bad quality of human resources has had terrible consequences for economic development.

The other two forms of corruption, although legally punishable, have not been punished. Many politicians and officials have got away with their pockets full of money. Apart from those forms of corruptions, public enterprise has been terribly inefficient. Among other things, it employs many more people than what is needed. Employment is a way of paying for votes.

Under these circumstances, the main recommended course of action is **privatisation of public enterprises.** If public institutions and other state enterprises (health, education, water supply, post office, oil exploitation and other mining activities, etc.) were owned and managed by private enterprises, corruption would be considerably diminished and, in general, those organisations would become much more efficient. With more efficient organisations there would be higher economic growth. Higher economic growth would bring about higher labour demand which, according to the "market laws," would bring about more employment and higher salaries. Hence, in the long term, the privatisation process would bring about greater social welfare. Otherwise, the country's economic growth and industrialisation process would never take off. Without economic growth and industrialisation there will always be poverty. Obviously, if public services are privatised, people would have to pay for those services (in this way, they would learn their value). Unfortunately, at present, some people will not be able to receive these services for they cannot pay for them. This sacrifice is, however, a necessary one for the economic development process to take off and bring about, in the long run, social welfare. Public charity programmes, disguised under a mask of "social welfare programmes", do nothing for the welfare of the population in the long term. They are nothing but a way of perpetuating the state of poverty with clear demagogic purposes (getting votes from the majority of the population in order to maintain the power of political parties).

In consequence, public health services should be another service subject to the market. Due to the "healthy" competence among the different private health centres, the quality of this service will improve. At the beginning some people will not receive those services because they are not able to pay for them. In the long term, however, most of the population will be able to pay for them with the product of their own effort.

Observe that the policy recommended by this historical interpretive context is founded, on the one hand, on a radical liberalist approach, and on the other, on the historical idea of the illegitimacy of the public sector versus the legitimacy of the private sector.

Second Historical Interpretive Context: Illegitimating the Whole Economic-Politic Structure

This second historical interpretive context is presented as a reaction against the first one. As its name indicates, the first historical interpretive context claims the illegitimacy of public management. However, according to the second context, such a claim is grounded on a dogmatic assumption; namely, the legitimacy of the private sector, i.e. the legitimacy of the formation of private capitals in Venezuela. According to the second context, such a dogmatic assumption made by the first context can clearly be seen in its recommendations for action, where the illegitimacy of the public sector is attacked from the platform of the legitimacy of the private sector. On the contrary, the case of the second interpretive context is that **the illegitimacy of the public sector both is founded upon and has led to the illegitimacy of the private sector.** How is this recursive relationship of illegitimacy possible? Witness a summary of the argument of this second historical interpretive context.

Venezuelan law and all its official discourse clearly state that the development of Venezuelan capitalism should take place according to the principles of liberalism. [This declaration of principles is supported by the two historical interpretive contexts]. However, the formation of private capitals in Venezuela is illegitimate in terms of the liberal principles upon which it is pretended to be justified. Such formation of private capital has led to a concentration of power which allows a relationship of domination from the private sector to the public sector and to the rest of society. The pole of domination constituted by those private capitals has determined the political course of the country so that the illegitimacy of the whole system has been perpetuated.

Venezuelan private industry has never been competitive. Its wealth has not been the result of a healthy competitive system. Private capital has been accumulated under two mechanisms quite opposed to the principles of liberalism:

1) Over the last Venezuelan democratic period, specially in the seventies when oil prices suddenly increased, the state lent huge amounts of money to a few private entrepreneurs in order to make overnight purchases of big and highly sophisticated industrial structures. These loans were given under very low interest rates and, in many cases, they were never paid back. Due to the protectionist policies of would-be liberal governments, the owners of these "key in hand" installed industrial structures did not have to worry about any foreign competition. On the other hand, the main sectors of the market were distributed among a few monopolistic enterprises; hence, they did not have to worry about internal competition either.

2) The main client of many of these gift-industries was the government itself. Juicy contracts were negotiated through bribery so that the "client" paid much higher prices than it would if the selection of the company had been performed through legal bidding.

This was the way in which the oil revenues were used "for industrialising the country." The state's wealth was just given away to a few Venezuelans. Some figures are enough to show the result of this process. At present, a quarter of national income is used to pay the service (interests) of Venezuelan public external debt *(El Diario de Caracas,* 28 October,1992). Such a debt is approximately $34.000 millions *(Country Report of The Economist Intelligence Unit, No.* 1,1992). On the other hand, **24 Venezuelan entrepreneurs have accumulated $58.000 millions in foreign banks during the last 15 years** --almost twice the whole external debt *(El Nacional, 19* October 1992). In the mean while, 44% of family households live in conditions of critical poverty *(Ultimas Noticias,* 2 June, 1991). The point is that most of private capital formed under the official policy of using oil revenues to industrialise the country has emigrated out of Venezuela. As a matter of fact, according to Baptista (1991), whereas private benefits have increased throughout the history of Venezuelan "democracy," private investment has decreased in the last 14 years and the real salary of labour has decreased to half of its value during that period.

Now we can, from the perspective of this reactive historical interpretive context, discuss the course of action recommended by the first historical interpretive context.

The source of legitimacy in the normative discourse of the first context is the private sector. If that source of legitimacy is shown to be mere illusion, the whole normative content of that discourse looses its legitimacy. Obviously, from this perspective, even accepting the principles of liberalism, the idea of a present sacrifice of the poor for the sake of the future is totally meaningless. The argument supporting such an idea is based on the assumption of the existence of a competitive industry (which is the country's source of wealth) and of a modern rational state which protects the rights of its citizens. In Venezuela, however, there is neither such thing as a competitive industry that provides wealth nor a modern rational state based on the notion of "right". On the contrary, the state has only been an instrument of very few Venezuelans to expropriate most of the oil income (which, by law, belongs to all Venezuelans). Under these circumstances each dollar that the state does not spend on a public services will be used, not for reinforcing a productive industry, but to fill the pockets of the few economic trusts' owners and of some corrupt politicians.

The only way out of this general situation of illegitimacy seems to be a drastic social revolution of the poor masses in order to reinstate a modern state with equal rights for its citizens. However, a social revolution of the poor would not be allowed by the international power structure. If a social revolution takes place, the country would be immediately invaded by United States army, supported, passively or actively, by the governments of other developed countries. Something similar has already happened in Panama and Iraq.

Observe that the argument of the second historical interpretive approach could be based on either socialism or liberalism. If it is based on liberalism, a **socialist revolution might be seen as a first step towards reaching the social conditions of competitive variety necessary for a liberal system.**

CONCLUDING REMARKS

There is no space in this paper for further enriching the so called stage of "thematic comprehension" of interpretive systemology methodology (Fuenmayor, 1991c). After presenting the interpretive framework constituted by logical-type and historical-type interpretive contexts, a more "interactive" discussion among those contexts is due.[5] An interactive discussion provides a more precise demarcation of the boundaries between the interpretive contexts. The arguments that constitute such a discussion relate the normative contents of the interpretive contexts with "facts" (which are claimed to be such by each interpretive context). Obviously, the "facts" chosen as such by each interpretive context will provide more power to the arguments in which they are used if they can also be regarded as valid "facts" by other interpretive contexts.[6]

A final point can be raised in these concluding remarks concerning the methodological relationship between the two types of interpretive contexts.

As far as I understand, one of the important outcomes of the famous Gadamer-Habermas debate[7] was Habermas recognition of his work as an attempt to reinforce a "depth hermeneutics." In a few words, Habermas claims that hermeneutics is in need of a notion of power to be critical. This critical hermeneutics is called "depth hermeneutics." The relationship between the two types of interpretive contexts (logical and historical), illustrated in this paper, constitutes a methodological form by which interpretive systemology incorporates the notion of power within a hermeneutic horizon. This can easily be seen if we pay heed to the general structure of the relationship between the two types of interpretive contexts.

Logical interpretive contexts are designed in order to discuss possible general meanings of an organisation or institution. In the present study, the general sense of a public health **service in any would-be modern society** can be discussed in the light of those general theoretical models presented in the first part of this paper *("socialism," "socialised liberalism," "radical liberalism"* and *"social Darwinism").* However, it is only by crossing these types of models with historical interpretive contexts that we can discuss the specific social sense of a "now and here" organisation. If we are to discuss the social sense of institutions like public health services **in a particular society at a particular time,** we must draw interpretive contexts that reveal those particular circumstances of a historical nature which question the

---

[5] This discussion can be seen in Fuenmayor R.L and Cuinas D. (1993).

[6] Concerning the inquiry into public health services, field work leading to the construction of these arguments with empirical content has been developed by Indriago (1991), Marin (1992), Medina (1992), Mendoza (1992), Palomo (1992), Rivero (1992), Sanchez (1991 and 1992).

[7] See Holub (1991).

contexts that reveal those particular circumstances of a historical nature which question the "would-be" modern character of a specific society. Historical interpretive contexts must project the fundamental ideas of the logical contexts on the "particularity" of a "now and here" society where the organisation has its "time" and "place" (as viewed by each context). **That "particularity" is constituted mainly by a social power structure** (in each case). The logical type models provide, in turn, the basis over which power structures are legitimised or illegitimised.

REFERENCES

Baptista, A., and Mommer, B., 1987, "El Petroleo en el Pensamiento Econ6mico Venezolano," Ediciones IESA, Caracas.

Baptista, A., 1991, La cuesti6n de los salarios en Venezuela, *SIC*, May 1991, 187-189.

Fuenmayor, R., and L6pez-Garay, H., 1991, The scene for interpretive systemology, *Syst. Pract.* 4, 401-418.

Fuenmayor, R., 1991a, The roots of reductionism: A counter-ontoepistemology for a systems approach, *Syst. Pract.* 4, 419-447.

Fuenmayor, R., 1991b, The self-referential structure of an everyday-living situation: A phenomenologica ontology for interpretive systemology, *Syst. Pract.* 4, 449-472.

Fuenmayor, R, 1991c, Truth and openness: An epistemology for interpretive systemology, *Syst. Pract.* 4,

Fuenmayor R. and Cuinas D., 1993, "El papel del estado venezolano en los servicios de salud," Talleres Graficos de la Universidad de Los Andes, Merida (forthcoming).

Holub, R.C., 1991, "Jurgen Habermas: Critique in the Public Sphere," Rutledge, London.

Indriago, F., 1991, Estudio del problema de la demanda insatisfecha en la unidad de emergencia del HULA, B.Sc. dissertation, School of Systems Engineering, Universidad de Los Andes, Merida.

Marin, P., 1992, Estudio sistemico interpretativo de la unidad de cardiologia del HULA, Research Report, Postgraduate Programme in Interpretive Systemology, Universidad de Los Andes, Merida.

Marx, K., 1844, Economic and philosophical manuscripts, *in*: Marx, K., 1963, "Selected Writings in Sociology and Social Philosophy," T.B. Bottoms and M. Rube, eds., Penguin, London.

Medina, 0., 1992, Historia cllnica familiar en Mesa Seca, Research Report, Postgraduate Programme in Interpretive Systemology, Universidad de Los Andes, Merida.

Mendoza, M., 1992, Historia clinica familiar en la urbanizacion Santa Maria, Research Report, Postgraduate Programme in Interpretive Systemology, Universidad de Los Andes, Merida.

Palomo, T., 1992, Una contribucion a un estudio sistemico interpretativo de la unidad pediatrica del HULA, B.Sc. dissertation, School of Systems Engineering, Universidad de Los Andes, Merida.

Rivero, L., 1992, Bases para un estudio sistemico interpretativo de la unidad de pediatria del HULA, B.Sc. dissertation, School of Systems Engineering, Universidad de Los Andes, Merida.

Sanchez, L., 1992, Estudio sistemico interpretativo de la unidad de neurocirugia del HULA, Research Report, Postgraduate Programme in Interpretive Systemology, University of Los Andes, Merida.

Sanchez, L., 1991, Historia clinica familiar en tres sectores urbanos de Merida, Research Report, Postgraduate Programme in Interpretive Systemology, Universidad de Los Andes, Merida.

INTRODUCTION TO BUSINESS SYSTEMS STREAM

The Systems movement seems to be gaining both adherents and credibility at a steadily increasing pace. So does the Systems view of the world.

The number of Systems Societies in existence world-wide bear witness to the breadth of interest in the subject. And the rate of publication of articles rooted in Systems thinking provides the evidence for the depth of interest.

The fact that this conference itself has attracted well over 100 papers is a tribute not only to the organisers and to those who have submitted the papers but also to the strength and virility of the Systems community today.

Inevitably the bulk of Systems thinkers to date have owed their prime allegiance to academia. Many of the active members of societies are University based and the majority of papers published to date have been authored or co-authored by academics.

I believe that this trend is likely to continue and that we need to be alert to the potential consequences. I believe that the systems movement needs to harness the vigour and the resources of the public and private sectors.

Somehow the Systems community has to build, and sustain, powerful links into the business community. Academic analysis on its own is not enough, as Peter Checkland has said there needs to be a continual thread woven from theory and practice.

The business arena is where many of our systems approaches can be used and used beneficially. We have to help businessmen and women to understand what we can offer and the potential benefits to them. The benefits to the world at large come when academic learning makes the crossover into the wider community.

Only if we can migrate our ideas and views into the commercial arena can we trigger a spiral of benefits. The next phase of our development will depend on transplanting our seedlings into the fast growing business culture. We must find businesses which can benefit directly from our approaches and our interventions.

Then they will become advocates for us and will support investment into our fields of research and interest. The funding of projects, of research, and in many cases of senior academic appointments today is increasingly dependent on finance from commerce and industry. In institutions competing for limited funds we need to ensure that we can attract funds that will otherwise be spent on alternative options.

The challenge for all of us is to find ways of winning practical support from the business world for our beliefs. And to do this we have to deliver tangible results back to businesses when they give us the opportunity to help them.

I am delighted that at this Conference we have the opportunity to have a business stream. I believe that this is a key step.

It gives us the chance to focus on the issues of the interrelationships between the business world and the systems world. It provides the opportunity to exchange experiences of working in the business world and to explore our philosophies about the business world. It creates a forum for debate of different ideas and beliefs.

In sum it gives us the opportunity to develop a platform from which we can address

Systems Science, Edited by F.A. Stowell *et al.*,
Plenum Press, New York, 1993

the wider task of building and sustaining valuable, two way, relationships with the business community.

The truly global span of the contributors to this stream and the diverse range of our papers is excellent news. I believe we have an exciting and stimulating event in prospect and I look forward to sharing views and experiences at the seminar with colleagues both old and new, from academia and from the business community.

Colin van Orton
Digital Equipment Co Ltd

METAGAME ANALYSIS AS A TOOL FOR LONGITUDINAL ANALYSIS OF MULTI-ACTORIAL DECISION PROCESSES

Andrea Debernardi

Politecnico de Milano
Dipartimento di Scienze del Territorio
via Bonardi 3
20133 Milano
Italy

INTRODUCTION

Generally, the carrying out of decisions taken by public administrations involves a variety of operators, with different intervention opportunities. The actions of these subjects can transform the objects of decisions, often producing unexpected outcomes. An intervention in such complex situations can easily require a method to describe the decision process emphasizing each actor's part.

Many tools for the description of processes of change are available in the field of longitudinal analysis (for a review, see Miller and Friesen, 1982; Van de Ven and Huber, 1990). The majority of them, however, has been developed particularly for the analysis of single organizations, and has rarely been applied to multi-actorial situations.

On the other hand, methods such as the *metagame analysis*, proposed by Howard (1971, 1987, 1989), and the *conflict analysis,* proposed by Hipel and Fraser (1984, 1991), supply several elements suitable for representing multi-actorial interactions. If integrated in a longitudinal framework, these methods can be used as a tool for modelling multi-actorial decision processes, on the basis of structural characteristics of reciprocal interactions, *i.e.* of actors' intervention power. .

I used the *metagame analysis* to model a complex multi-actorial process, concerning the location of a waste incinerator, promoted by the Regional Government of Piemonte - Italy (see Debernardi, 1991). During the period analyzed (March 1988 - May 1991), this process involved (in different times) more than 25 individual or collective actors, having different interests, dissimilar institutional tasks and unequal intervention powers. The study was established mainly in a retrospective way, by means of interviews with the actors and document analysis (direct observations of the multi-actorial context were possible only in the last months of the process).

Systems Science, Edited by F.A. Stowell *et al.*,
Plenum Press, New York, 1993

A LONGITUDINAL FRAMEWORK

At a first modelling step, the process was represented as a sequence of events or actions, concerning the decision on the incinerator site (Hirsch et al., 1978; Abbott, 1990). Each action was accomplished by one or more specific actors that, in this way, could influence the decision. This influence could be either direct or indirect, according to the structure of the multi-actorial system at that moment (Howard, 1989).

Some groups of events or actions in the sequence were particularly important, because established an irreversible transition of state in the multi-actorial system (for example, introducing a new actor) and/or in the object of the decision (for example, modifying the incinerator project). According to Hirsch et al. (1978), the comparatively stable interval between two transitions of state was called a phase of the process.

Seventeen phases were picked out, whose length ranged from a few days to nine months. This was emblematic of a process that alternated long intervals of latency with short, highly conflictual periods in which the question come to the surface. These phases regarded five consecutive or simultaneous proposals for a specific site in two municipalities near Turin, and at least two important changes in the the incinerator project.

Each phase was interpreted as a problem situation, in which the actors didn't carry out any action, but managing acts only (*actes de gestion*; Hirsch et al., 1978). Each set of acts could be seen, on the whole, as a bargaining among the subjects involved in the situation. For each problematic situation, this bargaining was studied with the methods of *metagame* and *conflict analysis*.

MODELLING PROBLEM SITUATIONS

The first step to describe a problem situation was identifying the set of the options of each actor. An option is an action that an actor can decide to undertake (unilaterally or cooperatively) for influencing the decision (see Howard, 1989, Hipel and Fraser, 1991). The options showed the differences among the conditions of different actors involved in the bargaining. Some actors could unilaterally decide to undertake a variety of actions, which be used as resources in the negotiation; others were in worse conditions, and could not take on any autonomous initiative.

As an example, a particular situation can be seen, concerning the phase in which the Regional Government had just communicated to a specific Municipal Government the decision to locate the incinerator in its territory (see table 1). The Municipal Government had to decide if accepting or not the location, and this decision could be taken in a reserved or in a public way, either involving the Town Council (and then the opposition parties) or summoning a press conference. It could be seen how the unpopularity of the Regional Government's decision gave the minority groups an opportunity to intervene against the Municipal Government, expecially if it would have tried to hide the problem.

For all situations, the options specification was a crucial step, since it related to *possibilities* of actions only, and not to actual events, and therefore always implied a non negligible degree of arbitrariness. Only a rather deep and circumstantial knowledge of the problematic context permitted the specification of the actually relevant options.

The options were expressed in a binary form as "to undertake or not to undertake the action A". A specifical choice between the two alternatives of an option was modelled by ascribing it the value 0 or the value 1 (Howard, 1989). Each combination of values ascribed to the different options formed a scenario (Howard) or outcome (Hipel and Fraser), *i.e.* a feasible future state of the system (see table 1).

Since the options generally were not mutually exclusive, a few of them could generate a large number of scenarios. The purpose, however, was not generating all the scenarios with a combinatorial technique, but recognizing the pertinent prospects for the analyzed situation (Simos 1990). Although in some situation it would have been possible to generate thousands of scenarios, the number of recognized relevant ones was always included between 5 and 19.

Table 1. The problem situation 1.1. was represented as a set of three actors with six options in all. Ten relevant scenarios were generated

Actors and options	Scenarios										
	s110	**s111**	**s112**	**s113**	**s114**	**s115**	**s116**	**s117**	**s118**	**s119**	**s1110**
MG - Municipal Government											
- to disagree with the Regional Government's decision	0	1	1	1	1	0	1	1	1	1	1
- to submit the question to the Town Council	0	0	1	1	1	0	0	0	1	1	1
- to summon a press conference	0	0	0	0	1	0	0	0	1	1	1
TC - Town Council (minority groups)											
- to disagree with the Regional Government's decision	0	0	0	1	1	1	1	0	0	0	1
- to summon a press conference	0	0	0	0	1	1	1	0	0	0	1
RG - Regional Government											
- to withdraw the decision	0	0	0	0	0	0	0	1	0	1	1

For each situation, the relevant scenarios were represented as nodes of a strategic map (Howard 1987, 1989; see figure 1). The nodes were connected by arcs representing the options. Each arc was controlled by one or more actors. The strategic map showed how the actors controlled the decision: which scenarios they could unilaterally drive the system into, the areas that could not be reached without cooperating with other actors, etc... For example, it can be seen in figure 1 how scenarios 1.1.5 and 1.1.6 could not be reached against the will of the minority groups in the Town Council.

The strategic maps supplied a basis for the study of the bargainings that occurred among the different actors during every phase of the process.

DESCRIPTIONS OF BARGAININGS

In studying a problematic situation on the strategic map, each scenario was considered as a dynamic guideline (Howard 1989:246) orienting the interactions between actors. The scenarios were firstly analyzed one by one, avoiding to consider them in specific sequences. When a single scenario was considered, the focus was not on the probability for it to become actual, but on the part that it played in orienting the process (Howard 1989: 247).

One of the principal aspects in classifying the scenarios was the comparison among each actor's preferences. The scenario that was considered as the most satisfactory by an actor was called the position of that actor (Howard 1989: 244). In the situation showed in figure 2, the position of the Regional Government was scenario 1.1.0, that of the Municipal Government scenario 1.1.7, and that of the minority parties scenario 1.1.5. The preferences of each actor was showed putting a direction on the arcs controlled by him. This way was not sufficient, of course, to show *all* the actor preferences, but supplied a clear image on the strategic map.

The "distance" among the different actors' positions was used as a measure of the conflict level. This "distance" has not to be confused with the geometric distance on the strategic map: actually, scenarios 1.1.0, 1.1.5 and 1.1.7 were rather close referring to the map, nevertheless they very differently judged by different actors, and so the situation was classified as a conflictual one. This fact can be easily seen, if one examines scenario 1.1.7: it was the Municipal Government's position, but in no way the Regional Government would have agreed to reach it.

The bargaining was thought as a sequence of different declarations about scenarios, starting from the *status quo* (Howard). The scenario that was talked about at a specific time was called the present scenario (Howard). A specific negotiation was described as a path on the strategic map. In figure 1, one of these paths is showed. Since scenarios 1.1.0, 1.1.5 and 1.1.7 were all highly conflictual, the bargaining avoided them all, reaching scenario 1.1.4, which was something like a compromise between the Municipal Government and the minority groups, against the Regional Government's decision. It can be seen that the Regional Government had no means to prevent this agreement, since it didn't control any arc of the path from scenario 1.1.0 to scenario 1.1.4.

According to the actors' preferences, the bargaining could be either a cooperative one, with a search of an agreement, or a conflictual one, with a tendency to trials of strength (Simos 1990, Hipel and Fraser 1991). The cooperativeness degree of the negotiation was showed by the congruence degree between the path followed and the direction of the arcs crossed.

TRANSITIONS AND STABILITY

In every problematic situation, the set of the actors' declarations was studied as an exploration of different parts of the strategic map. This approach helped me understand the

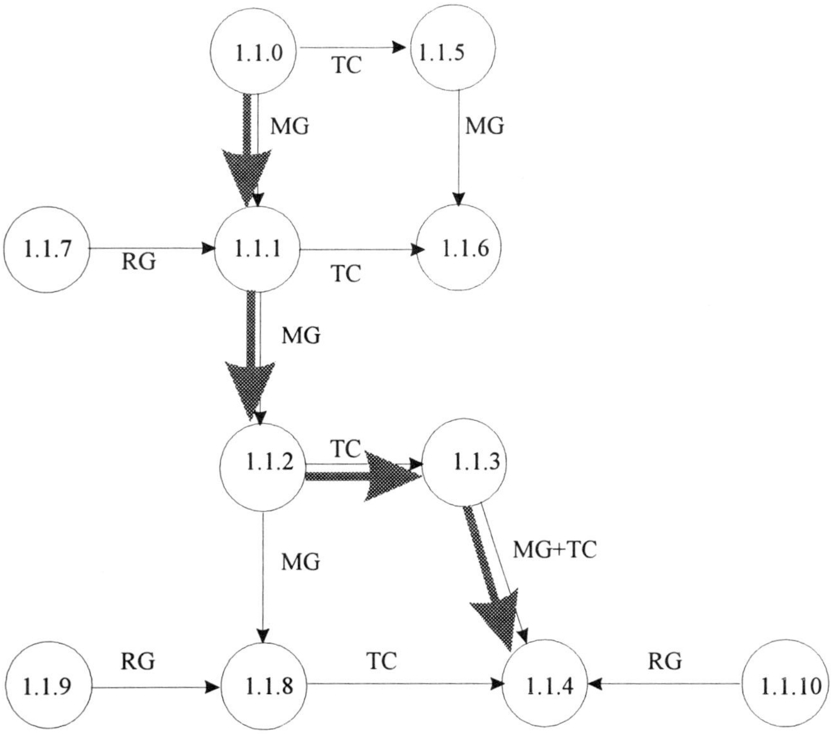

Figure 1. The strategic map of situation 1.1. Each node is a scenario, each arc an option, controlled by an actor. The direction of an arc show the preferences of the actor controlling that arc. A sequence of declarations going from scenario 1.1.0 to scenario 1.1.4 is showed.

interaction between cognitive and structural aspects of the process. For example, the strategic map was a good tool to analyze the credibility of different declarations (Howard, 1989:255-260, Hipel and Fraser 1991:14-16).

Another important task was the stability analysis (Howard 1989:248-251; Hipel and Fraser 1991:13). I considered this part of *metagame* and *conflict analysis* in a non-deterministic way, assuming that there was no particular reason for an unstable scenario not to happen and using the idea of the scenarios' stability only as a dynamic indication about the process (see Howard 1987:11-13). Actually, in the process I analyzed, unstable scenarios occurred more than once.

In general, unstable scenarios occurred more easily in presence of actors tending to act without bargaining their choices. In figure 1, scenario 1.1.2 was unilaterally reached by the Municipal Government to avoid scenarios 1.1.0, 1.1.5 and 1.1.6. Only after the system had been moved to the lower part of the strategic map, an actual negotiation with minority groups occurred.

Each time an effective group of events took place, the multi-actorial system passed into a new problematic situation, with new options and new scenarios. This transition could imply a change in the actor's positions and therefore in the conflict level. The scenario corresponding to the occurrence of these events became the new *status quo*. If it was unstable, the scenarios that could still be reached were preserved in the new problematic situation.

The possibility of sequences of unstable scenarios make the idea of a process dynamics completely included in its starting conditions untenable. On the contrary, it was seen how great the influence of unforeseen events and actorial freedom of change is.

On the whole, I obtained a succession of states, traversed by the system to regulate its internal and/or external perturbations. According to the variations in the conflict level at each transition, the process could either continue or stop (relating to the analyzed context).

The process didn't necessarily converge to a collective decision, but it could diverge increasing the conflict among actors, or it could reach a stalemate or fluctuate around some problematic point, or break into a variety of unilaterally incompatible decisions, etc... Actually, a conclusive decision was never taken. The end of the study corresponded to a long stalemate, characterized by the co-existence of a site formally approved by a theoretically sufficient number of actors and of an official (but feeble) decision not to build the incinerator at all.

CONCLUSIONS

The study of a real case allowed me to highlight some potentialities and some limits of this way of using *metagame* and *conflict analysis*. From a formal and computational point of view, *metagame analysis* was an effective method, suitable for modelling complex situations, involving a rather large number of actors. The dynamics of the multi-actorial system could be described in a non-deterministic way, as a continuous succession of choices among many possible futures.

In particular, the effectiveness of the method consisted in showing the structural differences between actors in specific situations. This fact allowed me to outline the pertinent prospects in the different phases.

Moreover, some critical actions or events towards the final decision could be identified. For many of these events, their function in the process was not clearly noticeable before the analysis took place. On this basis, it was possible to outline specific multi-actorial process typologies (see Abbot 1990) and consistent approaches to conflictual situations.

Among the difficulties, I could firstly point out the narrowness of *metagame's* stability analysis, which didn't allow the analyst to consider declarations different from threats and promises. This made the method unable to understand a number of evolutionary cognitive factors that were present in every situation.

A consequence of this problem was in some way the stationariness of situation modellings. It was possible to reach a satisfactory coherency between the descriptions of different situations and the longitudinal framework only by splitting the most difficult situations into two or more parts. Therefore, there was a numerical increase of very short phases, and a growth of computational complexity. Modelling short phases can be greatly exacting, since very detailed (often not immediately available) information and efficient criteria are needed to put in order a variety or close and indipendent managing acts.

Finally, the preference modelling was not supported by a satisfactory methodology. *Metagame* and *conflict analysis* are lacking on this point, and an integration is necessary to make the method mentioned complete, especially with regards to the stability analysis.

These difficulties, however, don't seem big enough to make the method inapplicable. With some integrations and cares, it can be used to describe multi-actorial decision processes, especially if they are highly conflictual and oriented towards trials of strength.

REFERENCES

Abbott A., 1990, A primer on sequence methods, *Organization Science* 1:375-392.

Debernardi A., 1991, Pianificazione territoriale in contesti multi-attoriali: la localizzazione dell'inceneritore Torino-sud (Regional planning in multi-actorial context: the location of the "Torino-south" incinerator), Graduation thesis, Politecnico di Torino.

Hipel K.W., Fraser N.M., 1984, "Conflict Analysis", North-Holland, New York.

Hipel K.W., Fraser N.M., 1991, Cooperation in conflict analysis, Cahier du Lamsade n.103 - Univ. Paris, Dauphine.

Hirsch G., Jacquet-Lagrèze E., Moscarola J., Roy B., 1978, Description d'un processus de decision. Quelques concepts, Cahier du Lamsade n.13 - Univ. Paris Dauphine.

Howard N., 1971, "Paradoxes of rationality", MIT Press, Cambridge (Mass.)

Howard N., 1987, The present and future of metagame analysis, *European Journal of Operational Research*, 32:1-25.

Howard N., 1989, The manager as politician and general: the metagame approach to analysing cooperation and conflict, *in*: "Rational Analysis for a problematic world: problem structuring methods for complexity, uncertainty and conflict", Rosenhead J., ed., John Wiley & Sons, New York.

Miller D., Friesen P.H., 1982, The longitudinal analysis of organizations: a methodological perspective, *Management Science*, 28: 1013-1034.

Simos J., 1990, "Evaluer l'impact sur l'environnement", Presses Polythecniques et Universitaires Romandes, Lausanne.

Van de Ven A.H., Huber G.P., 1990, Longitudinal field research methods for studying processes of organizational change, *Organization Science*, 1:213-219

CULTURAL CONTRADICTIONS IMPEDING GLOBAL COMPETITIVENESS OF AMERICAN HIGH-TECHNOLOGY FIRMS

David B. Keever

Science Applications International Corporation
Advanced Technology Group
McLean, Virginia 22102 U.S.A
and
George Mason University
Department of Communication
Fairfax, Virginia 22032 U.S.A.

INTRODUCTION

Several authors have examined the factors that have led to one of the U.S.'s most critical issues of the past two decades -- the steady erosion of high-tech competitiveness as measured by the loss of global market share and diminishing productivity growth. High technology firms are those that rely on state-of-the-art research and development (R&D) to maintain or capture market share in traditional or emerging fields. The loss of high-tech prowess has been a major factor in reducing the comparative standard of living with high-value jobs being increasingly captured by foreign-based international competitors. Productivity growth is considered the method with which to reverse these trends.

Two commonly espoused, primary stimuli to productivity growth are the application of technology and the development of human resources with higher knowledge and greater skills. By many measures, comparative studies have concluded the U.S. has fallen behind its competitors in these dimensions (Iacocca Institute, 1991). In addition to the push created by scientific and technological advances, there are a number of other factors placing increasing pressure on U.S. high-tech industries, including: i) Customer-Driven demands; ii) perceptions and realities of an uneven "Playing Field" (trade, regulations, etc.); iii) a propensity for short-term versus long-term focus; and iv) organizational characteristics. The customers in high-tech markets require agility, speed, flexibility and quality in meeting their needs.

Identifying the organizational symptoms of the U.S.'s loss of leadership position (the "what"), is easier than identifying the underlying causes (the "why"). The answer may be more subjective in nature, as it is intimately tied to individuals' attitudes, behaviors, beliefs, and roles -- the culture. Cultural contradictions appear to impede significantly U.S. firms in their efforts to regain leadership position in the global high-tech community. The combination of a lack of teamwork, communication, and flexibility, in addition to a short-term

Systems Science, Edited by F.A. Stowell *et al.*,
Plenum Press, New York, 1993

focus, and individualistic, military-metaphor-centered attitudes and beliefs, is hampering the ability of U.S. firms to compete in today's agile, and quality- and customer-oriented markets.

CORPORATE CULTURE AND ORGANIZATIONAL EFFECTIVENESS

Organizational culture is a metaphorical term that has come from the sociology and anthropology fields. It represents the set of behavioral variables that refer to the underlying values, beliefs, and principles that serve as a foundation for an organization's management system as well as the management practices and behaviors that exemplify and reinforce those basic principles. These beliefs and principles endure because they have meaning to members of an organization (Burrell and Morgan, 1985).

Studies of organizational culture have focused on beliefs, values, theories, and propositions that govern the organizational members' behaviors by which resources are translated into actions and outcomes. Numerous studies have shown that organizational culture significantly influences organizational effectiveness (Deal and Kennedy, 1982; Schein, 1988; Kanter, 1992). Organizational effectiveness is a direct function of the practices and policies used by managers and held (or not held) by members of the organization over time. Effectiveness is also a function of translating core values and beliefs into policies and practices in a consistent manner that is sensitive to the organizational environment. Particular environments may create or require a particular type of culture for an organization to survive.

The value of using culture as a basis for analyzing and predicting organizational competitiveness has been difficult. As Fiol points out, the reasons for this are that research efforts have bifurcated into disparate directions and conclusions (Fiol, 1991). Strategic management theorists, seeking to employ culture as a key variable in competitiveness, have traditionally been concerned with observable organizational phenomena. Their studies have tended to be prescriptive, suggesting ways to improve those observable aspects of organizations for strategic competitive advantage. Cultural theorists, in contrast, have been concerned with unobservable (interpretive-symbolic or psychological) phenomena. Their studies have tended to be descriptive, primarily suggesting ways to better understand those unobservable forces in organizations that influence actions. Each group views culture in fundamentally different ways: the first group views culture more pragmatically, namely a variable or key that can unlock organizational commitment, productivity, and profitability (Smircich, 1983); the second views culture from the anthropologically purist perspective, that culture is an emergent process that has as its source a set of deep, underlying values (Lucas, 1987).

In this paper, the first view is adopted. It is posited that three fundamental cultural characteristics of U.S., high-tech organizations impede global competitiveness. These impediments are: Research versus Performance-Based Attitudes; Continued Use of Myopic Design and Manufacturing Practices; and "Lone Wolf" Versus Participative Behavior.

RESEARCH VERSUS PERFORMANCE-BASED ATTITUDES

Those individuals or organizations that conduct research purely for the sake of research without regard for its potential utility for the user community can inadvertently impede global competitiveness. It is "widely recognized that a purely market-led approach can be every bit as limiting as a purely technology-driven one. The art, as most commentators have agreed, is to combine the two; or, to use a phrase adopted in two of the key studies in the field, to ensure the 'proper coupling of R&D with marketing'." (Hendry, 1989, p.20).

One of the main purposes of employing participative practices with senior managers at the beginning of a technical project is to establish with the team members the organization-wide project parameters, e.g., managers' and users' goals, time and budget constraints, and expected outcomes. Many case studies have shown this be an effective first step in successful innovations, especially in overcoming the organizational politics of innovation (Hendry, 1989). The translation and interpretation of these high-level parameters and mandates show one of the first, observable potential dissonances between an organization's formal planning system (goals, objectives, strategies, etc.) and the cultural norms (the cultural maps, behaviors, attitudes, and values). Depending on the degree of dissonance, technical experts in the organization can independently develop two distinct perspectives about the overall nature of a project. One perspective is a research-orientation perception -- the project is an extended laboratory -- yielding state-of-the-art products or processes. The other observable perspective is more pragmatic -- an opportunity to assemble tried-and-true components into a customized configuration addressing deficiencies of past designs and satisfying current customers' needs. These two perspectives create numerous opportunities for group cohesion, i.e., consensual views among the technical staff about the overall project design parameters. However, at the technical level, the value placed on the degree of technical design innovation may not guarantee shared priorities.

These perspectives and underlying value differences are usually expressed differently during the technical design phase. Gregory used an interpretive framework and extensive industry interviews to characterize these two orientations for the computer science field. He observed computer scientists (and software engineers) are more interested in the research or academic side of projects and will want to write a paper about what they did and don't really care deep down in their hearts whether customers have any use for anything they're doing. In development work, the engineer's/ designer's point of view is to use well-know techniques that are sure to work. Only if there is a problem that no one knows how to solve do you have to research or invent a new thing (Gregory, 1983, p23-25). In short, companies' efforts are misdirected away from activities that make them internationally competitive because the managers and technical staff do not include or adequately address the cultural orientations of research versus performance-based activities. Without revised incentives and rewards to realign this dual orientation, competitiveness suffers.

CONTINUED USE OF MYOPIC DESIGN AND MANUFACTURING PROCESSES

A group can be said to have a strong culture to the extent that its members have rule-based expectations, based on the experience of the members of the culture, that are congruent with their shared value or ethical systems (Schall, 1983, p283). As these rules become more formalized, they gain additional strength through institutionalization, i.e., a code of conduct or accepted design processes. New technologies often require new organization designs, new channels of information transfer, and different distributions of authority if they are to be used effectively. For most companies this means a flattening of the traditional hierarchy to five or four levels or less, and the elimination of departmental competition. Simplifying the organization will lead to the establishment of an infrastructure of integrated business functions participating as a team and supporting the strategic vision of the company. The successful operation of the new technology may require roles and knowledge not found in the old organization, i.e., renewal versus transformation. Management must be prepared to define broader and more flexible job categories and commit to the implementation of on-going educational opportunities aimed at reskilling employees at all levels (Ross, 1991). Jaikumar found in his comparative studies of Flexible Manufacturing Systems (FMS) between the United States and Japan, that American systems are characterized by "an astonishing lack of flexibility compared with similar systems in Japan." (Jaikumar, 1986, pp69-76). He attributed the difference to technical training (in which Japanese workers received 75 percent more than American workers) and investment by Japanese companies in upgrading worker skills (by as much as 67 percent). Manufacturing technology by itself cannot realize its full potential without significant attention to the people organization.

Too often ethnocentric attitudes or the NIH (Not Invented Here) syndrome drive employees to put blinders on to the external environment. Internally developed strategies will not only backfire in the high-tech industry, but ethnocentric attitudes will exacerbate this condition. As a result, competitors will far surpass these companies technologically by providing products and services that meet market needs. Sometimes in prestigious companies, hiring practices focus strictly on credentials and past work experience. However, more subjective measures (e.g., team, cooperative, self-motivated orientations and other shared values) would provide employers with a workforce that could truly operate in a best-in-class organization. Healthcare organizations whose culture -- and financial performance -- keeps them well at the top of their industry, hire selectively. Employees are hired for 'shared values,' not just their credentials or even the extent of their past work experience (O'Donnell, 1989).

"LONE WOLF" VERSUS PARTICIPATIVE BEHAVIOR

Underlying a lack of collaborative behavior is a lack of understanding among members of an organization. Organization behavior and development researchers have long recognized the marketing and R&D functions are typically characterized by strongly contrasting organizational subcultures, with different values, motivations and goals, differing status structures and reward systems, and different concept of procedure and control and political interests (Hendry, 1989, p23). U.S. firms are learning from Japan the value of improved vertical and horizontal relationships. In 1985, Ford conceded that it could not match Excel Industries, Inc.'s (Elkhart, Indiana) team-based methods to make car windows. Excel bought Ford's window factory, Ford acquired 40% of Excel and agreed to buy 70% of its windows through 1993. By 1991, Excel's sales had nearly quadrupled since 1985 and Ford was saving millions. Additionally, Excel has reinvested its profits to better its manufacturing systems. Today Excel "can equip new Ford models a year faster than before, helping Ford shorten its new-product cycle." (Kelly, 1992, p52).

While organizations are flattening the traditional pyramid structures, employees should assume a wider variety of functions and master additional skill sets. With the growing use of cross-function project teams, a member might be a leader of one team and a participant in another, calling for a flexible employee. Appropriate compensation to reinforce behaviors and continuous skill development and enhancement are necessary. Unless participative behaviors are rewarded, they will not be adopted or sustained. Traditional pay reward systems find that most employees progress in pay by moving to a job requiring a higher skill grade, usually in a different work group. As a result, if individual employees wanted a promotion, the shop system often meant moving away from jobs they were good at. And since team members received no additional compensation for their expanded responsibilities, the old compensation system failed to meet the needs of the new team organization (Womack, et. al., 1990).

SUMMARY AND CONCLUSIONS

Table 1 summaries the highlights of the cultural contradictions discussed above. Organizational goals to achieve American high-tech competitiveness in the global market cannot be matched and integrated into better business and management practices without real changes to cultural values and attitudes. True world-class organizations require an operating culture concerned with such key concepts as teamwork, value-added products and services, ongoing education and training, flexibility, a longer-term focus, and an environment characterized by open communication and information sharing between people and functions. Driving these changes is the need for a radical change in how organizations measure performance. Managers must be prepared to refocus their efforts away from splintering their companies into narrowly defined departments designated by convenient accounting codes and measured by tracking departmental variances and detailed costs. Instead, a balanced micro-macro vision of total productivity and total costs should form the standards of success for global competition (Ross, 1991). In short, innovative, competitive organizations require different structures, different resources and different evaluation systems than do businesses operating in a stable environment. Organizations and individuals will be required to experiment, accept more risk, and exhibit flexibility.

Table 1. Cultural contradictions that dispute professed attitudes and manifest themselves through behaviour

PROFESSED ATTITUDES	CULTURAL CONTRADICTIONS	ACTUAL BEHAVIOR
• Self-Managed, Empowered Valued, Multi-Functional Individuals • Team-Focused, Customer-Driven • Champions • Facilitation of Technology Transfer and Adaptation • Continuous Best Practices	• Research Versus Performance-Based Attitudes - Coupling of R&D with Marketing - Lack of Technical Knowledge by Managers	• Incentives/Rewards that Encourage or Reinforce Research-Based Behavior; Withholding of Key Information to Support Individual's Interest Rather Than Collective Good; NIH • Managers Do Not Evaluate on Technical Productivity; Manage by Spurious Goals or Performance Measures (insufficient knowledge base)
• Enterprise Integration and Flexibility	• Continue Use of Myopic Design & Manufacturing Processes - Lack of or Sensitivity to Concurrency	• Business Practices (accounting, structure, task decomposition, information channels, training) that reinforces splintering, non-systemic short-term focused performance • Empire Building
• Team-Oriented Performance Measures • Learning Organizations - Continuous Training - Systemic Sensitivity - Customer Quality Focus - Multiple Roles • Development and Hiring of a Diverse, Talented Workforce • Customer-Focused, "Global Localization"	• "Lone Wolf" versus Participative Behavior	• Individual Bonuses Based on Inflexible Job Classifications and Accounting Constraints • Organizational Structures and Information Sharing Practices that Reinforce Individualistic, Protective Behavior • Team Training Not Reinforced Outside the Classroom (matrix management confusion) • Disconnect Among Market Managers Who Have Little Knowledge of Design/Production Process and Technical Managers Who Have Little Knowledge of Market Needs • Hiring Processes Focus Too Narrowly on Past Experience and Credentials; Little Sensitivity to Desired Attitudes • Few Formal Programs For Cultivating and Developing Champions • Insensitivity to Environmental Factors; Incentive Systems (Bonuses) Reinforcing Individualistic Behavior

REFERENCES

Burrell, G. and Morgan, G., 1985, "Sociological Paradigms and Organizational Analysis," London: Heinemann.

Deal, T. and Kennedy, A., 1982, "Corporate Cultures", Addison-Wesley Publishing, Massachusetts.

Fiol, C. M., 1991, Managing Culture as a Competitive Resource: An Identify-Based View of Sustainable Competitive Advantage, *Journal of Management*, 17(1) March, 191-211.

Gregory, K., 1983, Native-Views Paradigms: Multiple Cultures and Culture Conflicts in Organizations, *Administrative Science Quarterly*, 28(3) September, 359-376.

Gupta, A. K., Wilemon, D. L., & Singhal, A., 1991, Bringing New Products to Market Speedily, *Product & Process Innovation*, 1(4) July/August, 5-10.

Hendry, J., 1989, Barriers to Excellence and the Politics of Innovation, *Journal of General Management*, 15(2) Winter, 20 - 31.

Iacocca Institute at Lehigh University, 1991, *21st Century Manufacturing Enterprise Strategy* Draft paper funded by the Office of the Secretary of Defense Mantech Program.

Jaikumar, R., 1986, Post-Industrial Manufacturing, *Harvard Business Review*, 64(6), November-December, 69-76.

Kanter, R.M., Stein, B., and Jick, T., 1992, "The Challenge of Change", Simon and Schuster, New York.

Kelly, K., Port, O., Treece, J., DeGeorge, G., & Schiller, Z., 1992, January 27, Learning from Japan, *Business Week*, pp. 52-60.

Lucas, B., 1987, "Political-Cultural Analysis of Organizations", *Academy of Management Review*, 12(1) January, 144-156.

O'Donnell, K. P., 1989, November 3, "Shared Values, Corporate Culture Foster Good Hiring", *Modern Healthcare*, p44.

Ross, D. F., 1991, "Aligning the Organization for World-Class Manufacturing", *Production and Inventory Management Journal*, 32(2) Second Quarter, 22-26.

Schall, M. S., 1983, "Communication Rules Approach To Organizational Culture", *Administrative Science Quarterly*, 28(4) December, 557-581.

Schein, E., 1988, "Process Consultation, Volume I", Addison-Wesley Publishing, Massachusetts.

Smircich, L., 1983, "Concepts of Culture and Organizational Analysis", *Administrative Science Quarterly*, 28(3) September, 339-358.

Womack, J. P., Jones, D. T., and Roos, D., 1990, "The Machine That Changed The World", Harper Perennial, New York.

SYSTEMS METHODOLOGIES AND ENABLING TECHNOLOGIES: A ROLE FOR GROUP SUPPORT SYSTEMS?

Judy McKay and Peter Marshall

School of Information Systems
Curtin University of Technology
Perth, Western Australia

INTRODUCTION

Managers in contemporary organisations have little option other than to do business in increasingly turbulent environments. As a result, they are more frequently confronted with large, highly interconnected and complex problem situations for which no obvious or appropriate course of action exists. Problem situations should not be interpreted as referring solely to negative, potentially damaging situations, but should also be viewed as including opportunities for improvement, requiring reflective and considered problem solving and decision making in order to take advantage of the opportunity. Crucial decisions taken in these environments are often based on highly uncertain information, and are often subject to time constraints, yet these decisions may have a big impact on the future stability and viability of the organisation.

Modern management science has evolved to the point where it now embraces a number of systems methodologies which can enable groups of managers, or composite teams of managers and consultants, to understand and manage the complexity and interconnectedness involved in the problems and opportunities mentioned above. The methodologies suited to dealing with complexity and interconnectedness would normally provide assistance in bringing about an intersubjective understanding of the different perspectives of the complex problem situation.

An important development in management science came with the recognition and acknowledgment that the systems methodologies referred to above tended not to be equally suited to a number of different types of problem contexts: a methodology that was reported to work effectively in one particular context, for example, was sometimes deemed less appropriate in a different context (Flood and Jackson, 1991). This resulted in the development of a matrix classification of systems methodologies: the system of systems methodologies (see Jackson and Keys, 1984; Jackson 1990a; Flood and Jackson, 1991).

As systems methodologies grow in their ability to handle increasing complexity and/or increasing disagreement amongst participants, it seems that they become more general, more

Systems Science, Edited by F.A. Stowell *et al.*,
Plenum Press, New York, 1993

flexible, and more powerful in the sense of being able to tackle large-scale, pervasive problems. However they thus become more difficult for the methodology user to implement as they tend to lack specific step-by-step guidelines for application within a particular context. It is in the support of the implementation of the various stages of different systems methodologies that technologies such as Group Support Systems could be argued to have a significant role to play.

GROUP SUPPORT SYSTEMS

Group Support Systems (GSS) is a broad category of information systems which includes a number of tools and techniques, typically computer-based, that have proved useful in supporting groups of decision makers or problem solvers in the clarification and resolution of certain tasks and problem types. Some examples of GSS would include:

* group decision support systems
* computer conferencing
* cognitive mapping
* decision conferencing
* negotiation support systems
* visual interactive modelling and simulation

This list is representative, but by no means comprehensive, of available GSS. Given the confines of this paper, it is planned to focus on group decision support systems and computer conferencing, although it is our contention that other GSS could equally usefully be employed in supporting the implementation of systems methodologies.

GROUP DECISION SUPPORT SYSTEMS

A group decision support system (GDSS) has been defined as "an interactive computer-based system which facilitates solution of unstructured problems by a set of decision makers working together as a group" (DeSanctis and Gallupe 1985). The general aim of a GDSS session is to remove communication barriers and to provide structure to the decision making process, thus promoting the generation, storage, retrieval, exchange and analysis of information amongst group members (Huber 1984; DeSanctis and Gallupe 1987). The purpose of GDSS is to enable groups to make higher quality decisions, whilst decreasing the time taken and other resources required to reach a decision (Kraemer and King 1988).

A simple GDSS session, for example, might consist of electronic brainstorming, facilitator-guided structured discussion, followed by voting, rating or ranking of ideas. Participants would each have access to a computer terminal at which they could simultaneously and anonymously input ideas (Lewis 1992). They would typically be situated in a room replete with good audio-visual technology and one or two large screens for the display of the whole group's ideas, resulting from a brainstorming session for example.

In addition to its major aim of more effective decision making, there are significant and important benefits of GDSS that apparently result from the structure that is imposed upon group processes, the facilitated group discussion, and the benefits derived from the employment of information technology. Briefly, these are:

i) increased depth of analysis of the issues comprising the decision or problem
ii) increased task oriented communication
iii) increased efforts to clarify the problem situation and the group processes

iv) increased participation by group members
v) decreased domination of the group by one or a few group members
vi) increased task focus and consensus reaching

(Pinsonneault and Kraemer 1988)

In addition, large GDSS facilities have been shown to enable the bringing together of representatives of many or all hierarchical levels of an organisation for decision making and problem solving activities. This has been demonstrated to have a number of advantages, including a reduction in time to gain organisational approval for decisions, and an improvement in organisational communication (Dennis et al. 1988).

It would appear that simple GDSS sessions like the one outlined previously could be used to support the implementation of a number of systems methodologies. Take, for example, the planning committees which are a feature of all stages of Ackoff's Interactive Planning methodology. Empirical studies suggest that GDSS help to equalise and maximise participation, promote the focus on ideas rather than the individual putting forward the idea, and reduce the dominance of powerful and articulate group members (Nunamaker, Applegate and Konsynski 1987). The electronic recording and display of ideas should also help prevent important ideas from being 'lost' in the general group discussion (Bostrom and Anson 1988), and can also provide participants with an appropriately detailed record of the group's deliberations and progress towards amelioration of the problem situation.

The electronic brainstorming and topic commenter capabilities of GDSS are particularly helpful in aiding generation of ideas (Dennis et al. 1990), and also allow for multiple brainstorming sessions. These two features should be particularly helpful to support the implementation of a methodology such as Ulrich's Critical Systems Heuristics (CSH). In addition, the effect of GDSS on group process, in particular anonymity and lack of individual domination, seems totally appropriate and helpful in achieving the emancipatory intent of CSH. The GDSS is thus more than aiding efficiency only, and may also contribute to the effectiveness of the implementation of the methodology.

COMPUTER CONFERENCING

Computer conferencing functions in a similar manner to GDSS. A significant difference is that rather than being in the same room, participants can take part in group deliberations from geographically dispersed locations. This could range from scattered points throughout a single location (eg an office block) to multiple, dispersed locations throughout the world (eg national offices of a large multinational company). Currently available computer software facilitates the exchange of text, voice, graphics, and images or photographs (Rapaport 1991). Hence, while enjoying nearly all the benefits of the face-to-face GDSS sessions, computer conferencing can aid in the alleviation of many of the problems associated with distance and asynchronous communication (Belitsos 1988). Computer conferencing should not be viewed solely as an alternative to GDSS: the two systems can be used in a complementary fashion.

Whenever systems methodologies are being employed in a company that operates from more than one site (perhaps internationally), computer conferencing clearly has a role in the emulation of face to face planning meetings. The communication possible facilitates wide participation, rapid responses, offers a rich variety of communication media, and should help increase intersubjective understanding and commitment to plans generated (Piturro 1989). The 'anytime' aspect of computer conferencing could also help individuals capture good ideas that occurred to them 'out of office hours' or when they were engaged primarily in another task. Once captured, computer conferencing enables these ideas to be immediately available for others to view and perhaps enhance.

Computer conferencing may also facilitate the emancipatory intent of methodologies such as CSH, when applied to dispersed groups. It would allow expert planners to input and exchange ideas over a wide time frame, and collect public opinion and feedback on social issues over large geographic areas with little inconvenience. Computer conferencing would permit individuals in the community to have direct input into the social systems planning that affects their lives, sometimes in contexts where perhaps without computer support, the large numbers of people involved would have made direct public input a very difficult or cumbersome task.

CRITICAL SOCIAL THEORY AND THE USE OF GSS

In this paper, a number of reasons have been offered to support the use of GSS as an aid in the implementation of systems methodologies. However, it would appear that a more compelling justification for the use of GSS can be offered if one considers the philosophical underpinnings of some of the systems methodologies, especially those with a critical or emancipatory intent.

Critical systems thinking first emerged during the 1980s. However, it is only in recent years that it has been developed and clarified to the point of providing theoretical and philosophical robustness to the foundations of the new, critical management science.

Jackson (1991a) identifies five commitments of critical systems thinking: critical awareness, social awareness, methodological complementarism, theoretical complementarism, and human emancipation. Many of the ideas underlying these commitments are drawn from the social philosophy of Habermas (Jackson 1990b; Jackson 1991b). It is from Habermas's critical social theory, more specifically from his theory of language and communicative competence, that the justification for and the inspiration of the use of GSS to support the implementation of systems methodologies can be drawn.

For Habermas, an important aspect in the quest for human emancipation is to achieve intersubjective understanding and genuine consensus on appropriate actions through discourse (Held 1980). This is rarely achievable in reality, Habermas suggests, because of barriers and distortions which serve to obstruct the communicative exchange between participants (Thompson 1981). These impediments can exist at an individual level, caused by personal neuroses or psychological repressions. They may also exist at group or societal level, caused by conflict, competition, power inequalities and repressive social organisations (Mingers 1980; Thompson 1981). However, Habermas describes the "ideal speech situation" as an environment in which barriers to true and free communication are absent. He argues that:

> "the structure of communication itself produces no constraints if and only if, for all possible participants, there is a symmetrical distribution of chances to choose and to apply speech-acts." (in Thompson 1981).

Thus, the ideal speech situation requires social arrangements that not only enable equal opportunity to participate in a discourse, but also remove power differentials that may inhibit an individual's ability to question, challenge, and make counter-claims to ideas currently under consideration. Although acknowledging the idealism inherent in the notion of an ideal speech situation, Habermas apparently believed that the concept was not impossible to actualise, and that it could be transformed into reality (Thompson 1981).

CONCLUSION

The major features, impacts and benefits of some examples of GSS have been detailed in this paper. These included the equalisation of power within the group, decreased dominance, anonymity, and equal opportunity to input ideas. It is our contention that the techniques and tools offered by GDSS and computer conferencing, for example, each represent steps towards the kind of institutional arrangements that Habermas conceived of in his ideal speech situation. This is not to suggest that the ideal speech situation has been realised through the use of GSS. However, used appropriately, GSS can help to provide in part, an environment conducive to undistorted communication.

REFERENCES

Belitsos, B., 1988, A foundation for groupware, *Computer Decisions*, November 1988: 31-33.

Bostrom, R. and Anson, R., 1988. A case for collaborative work support systems in a meeting environment, to appear, *Information Executive*, September issue.

Dennis, A.R., George, J.F., Jessup, L.M., Nunamaker, J.F., and Vogel, D.R., 1988, Information technology to support electronic meetings, *MIS Quarterly*, 12:591-624.

Dennis, A.R., Heminger, A.R., Nunamaker, J.F., and Vogel, D.R., 1990, Bringing automated support to large groups: the Burr-Brown experience, *Information and Management*, 19: 111-121.

DeSanctis, G. and Gallupe, B., 1985, Group decision support systems, *DataBase*, 16: 3-10.

DeSanctis, G. and Gallupe, R.B., 1987 A foundation for the study of group decision support systems, *Management Science*, 33: 589-609.

Flood, R.L. and Jackson, M.C., 1991, "Creative problem solving: total systems intervention", Wiley Chichester.

Held, D., 1980, "An introduction to critical theory: Horkheimer to Habermas," Hutchinson, London.

Huber, G.P., 1984, Issues in the design of group decision support systems, *MIS Quarterly*, 8: 195-204.

Jackson, M.C., 1987, Present positions and future prospects in management science, OMEGA International *Journal of Management Science*, 15: 455-466.

Jackson, M.C., 1990a, Beyond a systems of systems methodologies, *Journal of the Operational Research Society*, 41: 657-668.

Jackson, M.C., 1990b, The critical kernel in modern systems thinking, *Systems Practice*, 3: 357-364.

Jackson, M.C., 1991a, Five commitments to critical systems thinking, *in:* "Systems thinking in Europe," Plenum, New York.

Jackson, M.C., 1991b, The origin and nature of critical systems thinking, *Systems Practice*, 4: 131-149.

Jackson, M.C. and Keys, P., 1984, Towards a system of systems methodologies, *Journal of the Operational Research Society*, 35: 473-486.

Kraemer, K.L. and King, J.L., 1988, Computer-based systems for cooperative work and group decision making, *ACM Computing Surveys*, 20: 115-146.

Lewis, L.F., 1992, Computer software to support group decision making, in: "Computer-aided decision analysis: theory and applications," S. Nagel, ed., forthcoming.

Lewis, L.F. and Keleman, K.S., 1988, Issues in group decision support system (GDSS) design, *Journal of Information Science*, 14: 347-354.

Marshall, P., Atkinson, D., Ee, L., and McKay, J., 1992, Computer-aided group problem solving, to appear, *The Environmental Professional.*

Mingers, J.C., 1980, Towards an appropriate social theory for applied systems thinking: critical theory and soft systems methodology, *Journal of Applied Systems Analysis*, 7: 41-49.

Mingers, J.C., 1992, Recent developments in critical management science, *Journal of the Operational Research Society*, 43: 1-10.

Nunamaker, J.F., Applegate, L.M. and Konsynski, B.R., 1987, Facilitating group creativity: experience with a group decision support system, *Journal of Management Information Systems*, 3: 5-19.

Pinsonneault, A. and Kraemer, K.L., 1989, The effects of electronic meetings on group processes and outcomes: an assessment of the empirical research, *European Journal of Operational Research*, 46: 143-161.

Piturro, M.C., 1989, Computer conferencing: brainstorming across time and space, *Management Review*, August 1989: 43-50.

Rapaport, M.J., 1991, Computer conferencing, bulletin boards, and information professionals, *Online*, May 1991: 33-37.

Thompson, J.B., 1981, "Critical Hermeneutics," Cambridge University Press, Cambridge.

THE EVALUATION OF IT INVESTMENT

Annet Nottingham

Information Systems
Department of Computing Science
University of Paisley
Paisley, PA1 2BE

INTRODUCTION

Even in the present recessionary climate investment in information technology (IT) continues to increase.[1] Organisations may decide to make such investments because they believe that real opportunities and benefits are to be gained from information systems, but increasingly the prevailing feeling seems to be that they have to make these investments in order not to fall too far behind their competitors.

Inevitably senior managers find themselves faced with the complexities and uncertainties involved in estimating the benefits of projects in which large investments have to be justified in terms of future benefits.[2] Some of the benefits of IT can be quite clearly defined in monetary terms and hence present little difficulty. For example the gains to be made from early computer systems could be expressed in terms of cost reduction, increased efficiency and effectiveness: quantifiable principles, which could be projected over future years, and so be used to justify their initial investment.

Other benefits may be less tangible and much more difficult to quantify. Where the introduction of information systems is expected to change the way in which business is conducted, or improve the organisation's market position, it is not easy to make even rough forecasts. This problem becomes acute when proposals for investment in IT projects are competing with other project proposals and a ranking of these proposals is required. To make matters even worse, IT projects have a reputation for exceeding their initial cost prognosis. Over the years, therefore, senior managers have been forced to look for methods and techniques that will allow them to ascertain costs and evaluate benefits.

Traditional investment evaluation techniques are only helpful in some situations, and when the proposed information systems are based on a new technology the estimation of their potential becomes even more complicated. A different approach to investment in new

---

[1] See for instance Willcocks (1992b)

[2] A case study of 106 companies by Bjornsson and Lundegard (1992) shows 46% of the decisions to invest in IT are taken by Company management.

Systems Science, Edited by F.A. Stowell *et al.*,
Plenum Press, New York, 1993

technology is expressed by the concept of competitive advantage, where the innovative exploitation of a technology could lead to an improved market position, leaving competitors behind.

This paper will review some of the techniques available to senior managers to evaluate investment in IT projects, and discuss how the concept of competitive advantage can be applied to the introduction of a new technology.

TECHNIQUES FOR EVALUATING IT INVESTMENT

It is clear then, that as computerised information systems have spread into areas such as decision support and executive information, the benefits they yield have become more difficult to express in monetary terms. These systems incorporated analytical models which improved the understanding and control of the organisation. They could provide a wealth of information to the manager, which was previously not readily available, and which increased the quality of the manager's decision making. These benefits, although perceived as significant by the users of these systems, were harder to anticipate and define at the planning stage, and more difficult to justify against competing investment needs within the organisation.

The search for evaluation techniques to assist senior managers in situations like these, has pointed in the direction of established cash flow analysis techniques. A number of adaptations of these techniques have been developed in attempts to quantify more or less intangible benefits. However, none of these attempts have resulted in the formulation of a technique which can be applied to a wide variety of situations. Angell and Smithson (1991) go so far as to compare this search process to the quest for a philosopher's stone.

Farbey et.al. (1992) have done extensive research in methods adopted by companies to appraise their investment in IT. They review a number of approaches and methods that have been developed over the years, many of them attempting to deal with the problem of expressing the more intangible costs and benefits in some measurable unit.

Their case study of sixteen companies describes how investments in major IT projects were evaluated. In contrast to their extensive listing of evaluation techniques, their findings are that of these sixteen companies, seven had no formal justification for their investment, mainly because of the absence of evaluation techniques, and the remaining nine companies (which represents 56%) did make an attempt to justify the investment, and relied on established techniques such as Return on Investment and cost-benefit analysis. The latter group was aware that these cash flow analysis methods concentrate on improvements in efficiency, while the actual benefits in other areas were more significant.

Farbey et.al. conclude that since a great number of evaluation techniques are available to the decision maker, it is crucial to find the right technique for a particular set of circumstances. To this end they have designed a matrix, which considers specific dimensions of the IT project, and matches these to available evaluation techniques.

Willcocks (1992a,b) comes to similar conclusions. A survey conducted of fifty organisations suggested that 62% used cost/benefit analysis as their predominant criterion in the evaluation process. Only 16% used more than four criteria on which to base their evaluation. Willcocks observes that, by adopting these narrow evaluation approaches that do not clarify and assess less tangible inputs and benefits, organisations may not only be missing opportunities, but are also taking large risks. Indeed, David Flint, Head of Research at Butler Cox, has warned that investment in IT may only exaggerate the impact of other factors: "If you're in a very strong strategic position, spending on IT may improve it. If the converse is true, it will get you into a hole faster." (Independent, 1992)

Couger (1987) describes techniques that his research into the estimation of intangible benefits of information systems has showed to be successful.

The main method he describes is the Bayesian analysis, (after Thomas Bayes), which relies on the assumption that an experienced and skilful manager is able to make an informed guess of potential benefits. A system analyst will help the manager to identify improvements which the information system may bring and in this way the expected return on the investment can be calculated.

The drawback of this approach is that senior management has once more to rely on the impartiality/objectivity and integrity of this experienced and skilful manager. Although experience and expertise is valuable, it may not be the right approach in a situation where competing projects have to be ranked. As Angell and Smithson (1991) point out, it is necessary to bear in mind *who* is evaluating, *why*, and *for whom*? The intricacies and significance of office politics will need no further elaboration.

COMPETITIVE ADVANTAGE

Moving away from the mostly unsuccessful attempts to express information systems benefits, the ideas and terminology developed by Professor Michael Porter have been adopted as a useful alternative approach. Porter (1985) developed the 'value chain' principle, which divides a company's activities into the technologically and economically distinct activities it performs to do business, and describes the interrelationship between these 'value activities'.

Porter identifies nine typical activities of the value chain: five primary activities involved in the production: inbound logistics, operations, outbound logistics, marketing and sales, and services; and four support activities: procurement, technology development, human resource management, and firm infrastructure. Porter also described three generic corporate strategies for achieving competitive advantage: cost leadership, product or service differentiation, and niche strategy in which a specific narrow market is targeted.

Porter and Millar (1985) have applied the 'value chain' concept to IT. They illustrate how IT can affect not only all the separate 'value activities' and thus offer a considerable scope for competitive advantage through product differentiation and cost leadership, but also how IT can affect the way in which these activities are linked. Examples of the successful application of IT in these linkages can be found in the adoption of Electronic Data Interchange technology and Just In Time deliveries. By focusing on the competitive advantage IT can help to provide, it offers a strategic perspective.

The success stories of how IT has been exploited in a new and imaginative way are manifold. American Airline's Sabre computerised reservation system, and Citibank's Automated Teller Machines are frequently quoted in this context. In some cases it has proved possible to lock-in customers through the use of technology. An equally well-known example is American Hospital Supplies, who provided on-line order entry terminals to hospitals, which made it inconvenient for their customers to change over to competitors. However, as increasingly computer-literate users favour freedom of choice and demand international standards, it is becoming more difficult to achieve advantage in this way.

However, competitive advantage can be a short-lived phenomenon. Often it is just a matter of getting there sooner than your competitors, and, more often than not, their arrival is imminent. For competitive advantage to be sustainable Wiseman (1988) identified two key criteria: the advantage must be sustainable if the environment in which it operates changes radically; and it must be sustainable when challenged directly or indirectly by competitors.

Clemons (1991) points out that sustainable competitive advantage is quite rare and rather difficult to achieve. Competitive advantage is likely to be sustained only if it leverages key non-technological assets not readily available to competitors. In these cases the advantage is the result of the information content, stored or generated by the information system, rather than the combined hardware and software. It is important to bear this in mind when considering investment in a new technology such as Multimedia. For example, a Danish estate agent increased its market share by putting its portfolio on a multimedia system. First of all it attracted new customers through its novelty value, and secondly the success to be rate of visits paid by prospective buyers improved dramatically. However, this may prove to be a Multimedia application which can easily be duplicated by competitors. The competitive advantage is in this case based on technology and not on the information content.

Porter and Millar (1985) have outlined five steps for senior managers to follow in their pursuit of competitive advantage through the use of IT. The first of these steps, and in the light of Clemons' findings the most important step, is to asses the information intensity, both in the organisation's value chain as well as in its products.[3] Finding new ways to exploit existing or newly created information may lead to an advantage that can be sustained over a longer period of time.[4]

IT investments that affect for instance the linkages between the various activities within the organisation, or between its suppliers and customers, generally result in a temporary advantage. Competitors can, and often will, also exploit these infrastructural technologies, such as Electronic Data Interchange, and the investments they make in order to catch up will then be of a kind they feel they cannot afford not to make.

CONCLUSION

The often disappointing results of traditional evaluation techniques, and the difficulties in applying more specific techniques, has led to a growing awareness that the answer lies not in techniques, but in a wider, organisational approach towards the formulation of an IT strategy.

Moreover, the complications in the achievement of sustainable competitive advantage necessitate a company-wide approach, which strengthens the need for an IT strategy. A discussion of IT strategy formulation must go beyond the scope of this article, but can for instance be found in Earl's multiple methodology (1989).

Paul Strassman (1990) also moves the emphasis towards the strategic management approach in concluding that there is no correlation between IT investment and improved business performance other than good management. He is of the opinion that the exploitation of computers for profit is a responsibility that should be shifted from the technology specialists to managers. He argues that information systems management should be an inseparable element of every manager's daily work.

The number of recent publications in this area testifies to the importance and pertinence of these issues, and indicates further that it is unlikely that a single best technique will be found. Each case will continue to require distinct analysis and the selection of an appropriate technique will demand expertise and understanding. Most instances will involve the use of both quantitative and qualitative techniques. Both perspectives have an important if variable contribution to make.

---

[3] The remaining four are to determine the role of IT in the industry structure; ways in which IT may create competitive advantage; ways in which IT may spawn new businesses; and finally to develop a plan to take advantage of IT.

[4] For techniques to assess the potential of information value to an organisation see for instance Porter's information intensity matrix (1985) and Ward et.al.(1990).

REFERENCES

Angell, I.O., and Smithson, S., 1991, "Information Systems Management - Opportunities and Risks", Macmillan, Basingstoke.

Bjornsson, H., and Lundegard, R., 1992, Corporate competitiveness and Information Technology, *European Management Journal*, Vol.10, No.3, pp341-7.

Clemons, E.K., 1991, Making the investment decision: evaluating strategic opportunities in information technology, *Communications of the ACM*, Vol.34, pp22-36.

Couger, D.J., 1987, Techniques for Estimating System Benefits, *in:* "Information Analysis, Selected Readings", Addison-Wesley, Workingham.

Earl, M.J., 1989, "Management Strategies for Information Technology", Prentice Hall, Hemel Hempstead.

Farbey, B., Land, F., Targett, D., 1992, Evaluating investments in IT, *Journal of Information Technology*, 7, pp109-122.

Porter, M.E., 1985, "Competitive Advantage: Creating and Sustaining Superior Performance", The Free Press, New York.

Porter, M.E., and Millar, V.E., 1985, How information gives you competitive advantage, *Harvard Business Review*, Vol.63, No.4, pp149-60.

Strassman, P., 1990, "The Business Value of Computers - An Executive's Guide", The Information Economics Press, New Canaan.

Ward, J., Griffiths, P., Whitmore, P., 1990, "Strategic Planning for Information Systems", John Wiley, New York.

Willcocks, L., 1992a, IT Evaluation: managing the catch-22, *European Management Journal*, Vol.10, No.2, pp220-30.

Willcocks, L., 1992b, Evaluating information technology investments: research findings and reappraisal, *Journal of Information Systems*, 2, pp243-268.

Wiseman, C., 1988, "Strategic Information Systems", Irwin, Homewood Ill.

HEURISTIC METHODS AS AN INSTRUMENT TO DEVELOP A MANAGER

Kazimierz Perechuda[1] and Wojciech Cieslinski[2]

[1] Academy of Economics
Institute of Organization and Management
Wroclaw 53-345
Komandorska 118/120

[2] Academy of Physical Education
Department of Sport
Wroclaw 51-617
Witelona 25a

INTRODUCTION

The paper presents a concept of management training which uses heuristic methods in a manager's development process.

A GENERAL MODEL OF IMPROVING MANAGEMENT TRAINING USING CREATIVITY TRAINING AND CREATIVE PROBLEM SOLVING

Looking for possibilities of improving management training should begin with development and improvement of creative abilities potentiality and creative problem solving skills in every subject of training system (e.g. trainers, supporting staff), that is, in the personal subsystem (Figure 1). The way to achieve this goal is to use creative problem solving and creativity training. The essence of those activities is to change attitudes of people who participate in creativity training and creative problem solving, that is, of all the subjects of management training (Figure 2).

In creativity training, as well as in creative problem solving, the most important element of acting is to make changes in the participants' emotional sphere. The character of attitude may be positive or negative. The point is to develop such an attitude in managers and trainers that would make them positive about changes in general.

Creativity training and creative problem solving are permanent transgression of established patterns and stereotypes, which makes the change possible.

In cognitive sphere, the aim of creativity training and creative problem solving is to develop the potentiality of creative abilities in managers and trainers, and to develop knowledge of creative processes, which mostly concerns trainers and supporting staff.

Systems Science, Edited by F.A. Stowell *et al.*,
Plenum Press, New York, 1993

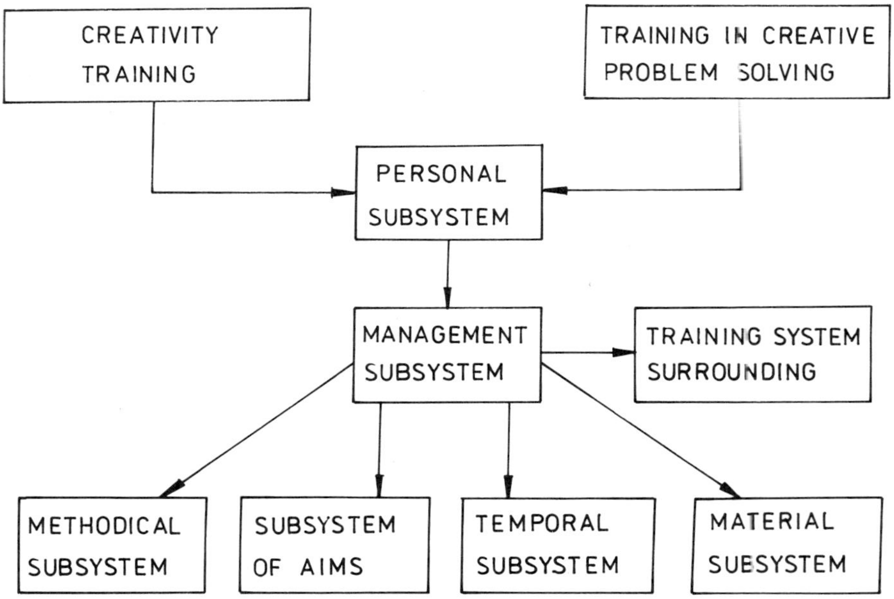

Figure 1. Developing management training presented as a participation of all the personal subsystem elements in creativity training and creative problem solving, and their influence on other elements of the training system.

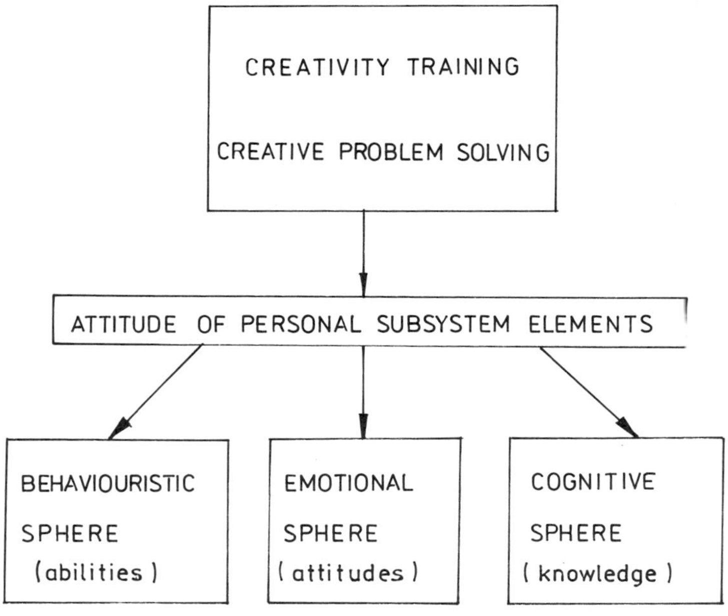

Figure 2. A graphical model of influence of the creativity training and creative problem solving on attitude of the personal system elements in management training.

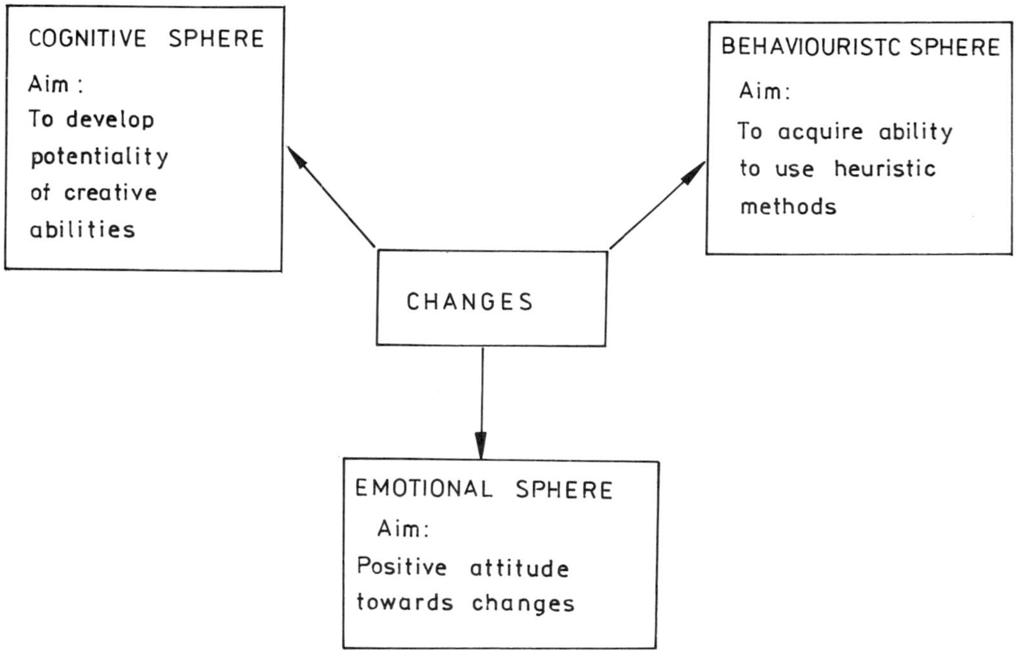

Figure 3. Graphical model of aims which have to be reached through changes in attitudes of personal subsystem elements participating in creativity training and creative problem solving.

Management training as a process of making changes originates from training system subjects. Thus the following practical indications may be formulated:

1. All subjects of management training should participate in creativity training and creative problem solving. Such a participation influences training system surrounding and other training subsystems in an active way (see Figure 1).

2. Participation of all the subjects of management training in creativity training and creative problem solving arouses their positive attitude to changes in training process (emotional sphere), as well as competent application of heuristic methods in certain situations (behaviouristic sphere) or developing knowledge about creative processes (cognitive sphere) - see Figure 2 and 3.

3. Participation of managers, trainers and supporting staff in creativity training and creative problem solving makes it possible for them to break the established patterns of perception in problematic situations and improve aptitude for using not only intellect but also intuition (understood as a result of many years' experience in certain area of activity).

A PROGRAMME-AND-ORGANIZATION MODEL OF CREATIVITY TRAINING

The programme of creativity training concerns:

1. Presenting topics of creativity training (basic intellectual operations, elements of interpersonal training including motivation)

2. Detailed definition of the necessary notions

3. Examples of exercises performed during particular stages of the training.

The organization of creativity training concerns:

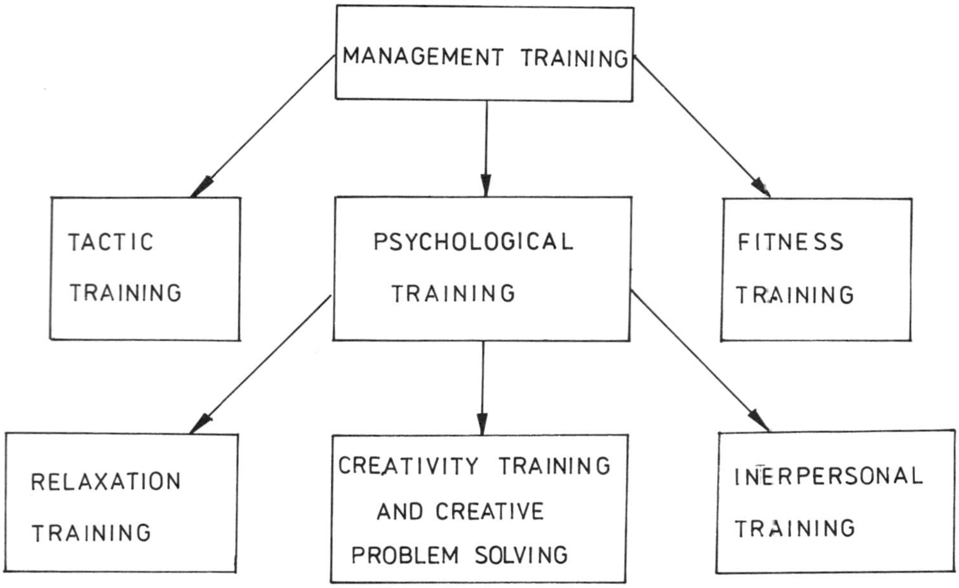

Figure 4. Place of creativity training and creative problem solving in various kinds of management training.

1. Composition of the group :
- 5 to 15 participants
- criteria of selection to the group (different sexes, experience, relatively the same level of age, intelligence and motivation)
- every group should have a person with friendly character, a person busy with speculations and a person relieving the tension
- every group should have a so called client - a person having a problem - and an expert - a person knowing how to solve the problem
- the trainers should have very high qualifications but the most important thing is experience in conducting a group and having passed an interpersonal training.

2. Arrangement of exercises (criteria of selecting exercises):
- interpersonal exercises (including motivation exercises)
- beginning with simpler exercises involving little emotions
- exercises should be conducted in the following order: interpersonal exercises, motivation exercises, developing skills, theoretical exercises and exercises in overcoming difficulties
- every exercise should finish with a comment concerning the aim, point and meaning of the exercise
- the participants should apply particular techniques in practice.

3. Duration of the training : depending on the aim, the training may be carried out in 12 sessions during 6 days or in 12 weeks with 1,5 hour per week.

The presented programme-and-organization model of creativity training in the process of improving management training makes it possible to formulate the following indications:

1. The group should consist of as many as 15 people.
2. Creativity training may be conducted by an experienced trainer, pedagogue or psychologist, depending on the needs of people participating.

Table 1. Model of applying heuristic methods in organized activity cycle in management training.

HEURISTIC METHODS \ ORGANIZED CYCLE STAGES	SYNECTICS	LAMBDA	CIRCEPT	FUNCTIONAL ANALYSIS	MORPHOLOGICAL ANYLYSIS	DELPHIC METHOD	PROS & CONS	VALUATION METHOD	CRITICAL ANALYSIS & EVALUATION	BRAINSTORMING	BRAINWRITING
FORMULATING THE AIMS	0	0								0	
PREPARING MEANS AND CONDITIONS TO ACHIEVE THE AIMS	0		0	0	0					0	0
PLANNING OF SYSTEM	0	0		0		0			0	0	0
ACHIEVING SYSTEM AIMS					0				0		
SYSTEM CONTROL				0	0	0	0	0	0	0	0

3. Creativity training of trainers and supporting staff may be conducted in two outgoing sessions of 5 to 6 days in a year.

A MODEL OF ORGANIZED ACTIVITY CYCLE WITH APPLICATION OF HEURISTIC METHODS

The basis for activities improving management training is organized activity cycle, which consists of five stages :

1. Formulation of aims
2. Examination of means and conditions
3. Preparing means and conditions (activities planning)

4. Reaching the aim
5. Controlling

The detailed analysis of the stages shows the similarity between organized activity cycle and problem solving process. Table 1 presents usefulness of heuristic methods on particular stages of organized cycle. The practical indications which can be formulated as a result of the above presented information are following :

1. Each stage of organized activity cycle used in preparing training activities is an individual problem to be solved.

2. Rationalization of the cycle may be increased by using heuristic methods on each stage.

EXPECTED PRACTICAL CONSEQUENCES OF USING THE MODEL OF CREATIVE ACTIVITIES IN MANAGEMENT TRAINING DEVELOPMENT

Basing on the concepts described above one can formulate the following practical indications concerning improving of management training process using heuristic methods :

1. Management training may be improved through using creativity training and creative problem solving by the participants.

2. Participating of management training subjects in similar forms of self development makes it possible for them to change their attitude to a desirable direction. The desirable effects of changing subjects' attitude are:

- positive reaction on changes (which is a purpose of development - emotional sphere of attitude)
- ability to use the heuristic methods in particular situations of training process and its betterment - behaviouristic sphere
- development and improvement of knowledge about creative thinking processes - cognitive sphere.

3. Showing the application of heuristic methods to particular stages of organized activity cycle, one indicates the ways of solving problems which appear on stages of training organization.

REFERENCES

1. E. de Bono, "Lateral Thinking for Management", Penguin Books, (1990).
2. W.L. French, "Human Resource Management", Houghton Mifflin Company, Princeton. (1990).
3. H. Levinson, "Organizational diagnosis", Harvard University Press, Cambridge, MA., (1975).
4. J.A.F. Stoner and Ch. Wankel, "Management", PWN, Warszawa, (1992).

SUPRAHUMAN SYSTEMS AND MANAGEMENT:

STEERING IN JEOPARDY?

Fenton F. Robb

The University of Edinburgh
Edinburgh EH8 9JY, U.K.

INTRODUCTION

Suprahuman systems (here to be called 'institutions') emerge from human "conversations" and "communications"[1]. Understanding how this emergence comes about, and what its consequences are, may help to explain some of the problems of coordination of activities within organisations.

SUPRAHUMAN SYSTEMS

The conversations which give rise to institutions are about recursively defining, describing, explaining and evaluating things and events. By generating interpretive schemes, conversations "make realities" and define what is "unreal"; they create "goods", and define what is not good; they conjure up moral agendas, and describe "causal" relations; they define what is true, false and uncertain; and they generate expectations about what the future may bring if certain actions are taken or avoided.

The communications, which contribute to the emergence of institutions, transmit definitions and evaluations as instructions, advice and exhortations, and they explore the world of action by discovering what is and what is not the case, and what does and does not happen. They feedforward intentions, expectations and prescriptions for action, and they feedback reports of the experiences and the perceptions of those observing the apparent outcomes of actions. These experiences are framed by the prior conversations. Particularly important is the experience which comes from the comparison between intentions and the expected effects of particular courses of action, and the actual outcomes as they are perceived to have occurred. Communications record the differences between these and thus generate further conversations to reduce tensions, to explain and describe the differences between expectations and outcomes and to describe and evaluate once again what is real, good, true, possible and desirable.

The cycle, conversation - communication - conversation, is thus closed recursively,

Systems Science, Edited by F.A. Stowell *et al.*,
Plenum Press, New York, 1993

and, as it continues, an interpretive scheme relating understandings with perceptions of the real world emerges. This is an "institution". The objectives of conversation and communication, expressed in the terms of the participants, are those of explaining and relating theories about, and experience of, the world, but with the emergence of an institution, a new objective emerges at a (logically) higher level, that of the institution itself. This is the objective of preserving the continuity of the cycle of recursive conversations and communications and thus of preserving the boundary between the institution's interpretive scheme and all else in the domain of human interactions. The organisation of the recursive cycles of self-production is preserved by constraining human interactions within the institution's interpretive scheme. Thus institutions define and sustain themselves.

How exactly any one institution achieves organisational closure and self-definition is not accessible directly to human perception because no interpretive scheme in the domain of human conversations can transcend that domain. This is a consequence of institutions being realised at a higher logical level than that domain. Any communicable description of an institution has to be made in terms of another institutionalised interpretive scheme. The world, as seen through the interpretive scheme of the physical sciences, is, to some (rather limited) extent, predictable. Few, if any, other interpretive schemes in common currency in our society have such pretensions. Hence, by construing institutions to be "systems of a natural kind"[2,3,4,5] we may be able to come to some understanding of how the processes of self-production of institutions are constrained and thus make some useful predictions about how institutions, as a class, are likely to behave in certain conditions.

In this view, human interactions are to be seen as giving rise to real world entities which are operationally open and which entrain and dissipate energy in the form of human lifetime through their organisationally closed processes of self-production. The emergence of institutions is no casual accident but a determinate, and possibly determinable, causal process. Institutions entail processes which are directed solely towards self-production and that it is their ability to entrain and dissipate lifetime, not that of explaining, or controlling, the world to meet human needs, which is the measure of their evolutionary success.

If institutions 'learn', then what they learn is about making themselves more viable, not about enabling cooperative action among humans to make human societies, and organisations, work "better", however that may be defined. If institutions "converse" with each other, their conversations are <u>not</u> about the negotiation of a shared meaning, nor about the formation of compromise over the contents of their schemes. The contents of their schemes are irrelevant. They are about nothing other than the mutual stimulation of auto-catalytic processes which enhance the self-productive processes of each of the institutions involved[6] by reconstructing, rather than translating, communications from the language of one institution into that of another. Rather than forming new interpretive schemes of use to participants, or enabling the convergence of interpretive schemes of different origins, institutions simply accelerate their own trajectories by forming and adopting their own interpretations of each other's utterances. In such ways do they amplify themselves and 'defend the boundaries' of their processes of self-production. Their modes of behaviour, their agendas, are not ours.

MANAGERIAL ORGANISATIONS

Managerial organisations are observer-defined patterns of intensive conversations performed by people interacting to make sense of complex and often conflicting evidence of problems which demand solutions, and of communications resulting in actions which, all too often, produce unintended outcomes. However any observer cares to attribute purposes to organisations and however any observer defines and bounds the elements and relations of organisations, organisations lie in the meet of the interpretive systems of many institutions.

In organisations, institutions are manifest (i) in the beliefs and actions of people, performing the various activities of the organisation, (ii) in their spontaneous emergence within the organisation itself and (iii) in the environment in which any organisation operates.

The Beliefs and Actions of People

Over time, different institutions have emerged from various conversations and communications addressing particular aspects of the activities in managerial organisations. This has resulted in the partitioning of managerial activities and the specialisation of these into discrete functions dividing the organisation "laterally" so that mutually dependent activities have become separated from each other by institutionalised distinctions[7]. Concurrently institutionalised interpretive schemes have divided the organisation 'vertically' so that hierarchies of operatives, supervisors, managers and executive directors have developed. Although they are often taken for granted, these classical functional and hierarchical divisions, and the institutionalised specifications of the interpretive schemes and communicative actions appropriate to each, are sanctified only by the history of the emergence of institutions, and do not necessarily reflect the operational needs of any organisation. There are more ways than one of managing complexity[8], but institutions bury critical thought. Management theories tend to look away from the assumptions of the interpretive schemes which support them and, narrowly, towards the phenomena of interest defined by their particular institutionalised specialisms. This may be, in part at least, why change in organisations is so difficult to execute.

The growth and increasing complexity of each of these interpretive schemes and the emergence of new specialisms add further to the difficulties of keeping the assembly of schemes together and of steering it towards the organisation's objectives.

Emergent Internal Institutions

Before information became a cheap commodity, functional and hierarchical management culled its information from a limited range of numerical measurements and from "walking about". The emergence of the institutionalised notions of what is "normal" and of the all-pervading belief in the explanatory powers of statistical patterns, noted by Hacking[9] to have gained a hold on scientific reasoning, are manifest in management also. Now it is assumed that almost all that it is important to know about the organisation and its situation, in each function and at each level in its hierarchy, can be represented in "statistical" and "accounting" terms. Both of these, previously almost extraneous, systems of understanding have now pervaded the organisation both vertically and horizontally. Neither has really been assimilated into the interpretive schemes of management. For example in regarding statistical operations, "average" is often mistaken for "normal", "correlation" is confused with "causation", "clusters" are interpreted as real-world classes and so on. There are similarly confused reconstructions of accounting terms.

The resultant potential for the progressive divergence of views of the organisation is becoming very great as functional and hierarchical institutions assimilate and (mis)interpret statistical and accounting concepts in the processes of enhancing their own self-production; some of the consequences have been examined elsewhere[10,11]. They include the misreading of complex situations and the formation of self-fulfilling prophesies and the monocular self-serving vision of institutionally-entrained "experts".

The satisfaction of the resource demands of the growing number and complexity of such conversations in the organisation becomes a necessary condition for the viability of the whole. If these demands are not satisfied, and if the activities of the functions and levels are not adequately co-ordinated, the specialised functions and levels in the organisation drift apart, become secluded and even effectively closed against inputs which do not conform with their specifications of reality. This tendency, too, greatly adds to the already substantial management tasks of integrating and co-ordinating its operations.

Whereas it might have been adequate to carve up the organisation as suggested above, into lateral and vertical divisions, into functions and levels of hierarchy, now such is the complexity of organisations that even the most sparse representation of them requires many dimensions of distinction.

Emergent External Institutions

In the world outside the organisation there is an increasing tendency for institutions to emerge to support the conversations of various specialised interest groups and of the multiplicity of differing beliefs about politics, economics, the environment, ethnic minorities and other "people categories", nation states, international collectivities, agricultural practices, food habits, the environment, the climate, certain species of animals, plants, consumer interests and so on.

Often originating in the conversations of the "chattering classes" and groups of "concerned" people, many of these emergent institutions can now exert very real power over management practices through the state, law, professions, communities, and in their own right[12]. Sometimes this power exerts pressure on managers to behave in ways which need not be in the interests of the organisation as a whole. Whatever management does, it cannot afford to ignore the emergence of these ever more diverse views of the world, the new realities which they invent and the purported causal relations and values which they espouse. This continually increasing complexity makes the task of management no easier.

POSSIBLE SOLUTIONS TO GROWING INSTITUTIONALISATION

There seem to be three common responses to the added complexities discussed above, (i) the formation of special groups to deal with aspects of the situation, (ii) the formation of a counter-culture, an organisationally founded institution and (iii) self-dissipation. Space precludes no more than a brief discussion of these.

Task Forces, Think Tanks and the Like

In the past, it was the responsibility of individual managers to define and solve problems for themselves. The pluralism of organisational objectives is now coming to be recognised. Much stress has to be laid now on coordination as a (potentially institutionalised) function, and on the formation of cooperating groups of interested people representative of adjacent and related functions and so on.

Task forces, working parties, consultative groupings, think-tanks and the rest are portrayed as recruiting the brain power of the organisation in constructive and cooperative ways which are intended to reduce inter-functional, inter-regional, inter-departmental and inter-personal rivalries and frictions. This they may well do, but that is not what is of concern here.

Such groups are often seen to nucleate so as to become "power bases". Often formed to deal with some transient error or accident, they can become "black holes" sucking in information and people to be dissipated in debate and in the generation of new prescriptions for action, and firmly integrated into the bureaucracy of the organisation and assimilated into its managerial structure.

There are doubtless great benefits to be had from activities of this kind; problem-solving groups can solve problems, sometimes even inventing and solving problems that no one had dreamt existed, but the warning here is against letting them run free lest they take on a "life of their own". Instead of reducing complexity, such "solutions" to the co-ordination problem may amplify it.

Counter-Cultures

Management may try to synthesise a corporate "image" to induce interpretations of itself which will shape the views of others more favourably towards its interests. But the "mission" statements, corporate codes of conduct, commitments to customers, to investors and to employees, protestations about care for the community, the environment and so on, often signal little more than the recognition of "good practices" promulgated by external institutions. More seriously, an organisation may attempt to reduce the variety which has been discussed above by addressing the divisive effects of institutionalised thought on its operations, and it may try to reduce, if not eliminate, these by developing its own interpretive scheme, its own definitions, distinctions and evaluations, and its own means of communicating these to all the participants in its activities.

In effect, an organisation may so attempt to explain itself to itself that it sets up the recursive cycle of conversations and communications needed to make itself into an institution "with a life of its own". Recruiting newcomers may be confined to the young and inexperienced, training may be given in only those aspects of specialisms that are thought relevant to its operations, preferment may be given only to those who express undivided loyalty and understanding of its interpretations of the world and who regard expertise from external sources with suspicion. Some large American and Japanese firms and military organisations have pursued this course of action. Some of these may have become thoroughly developed institutions but they may serve no purposes other than their own self-production.

Self-Dissipation

This consists in breaking down the whole organisation into much smaller, "more manageable" parts, some of which reflect narrowed aspects of institutionalised distinctions and values. For example headquarters units with all statistical, accounting, administration and other specialisms may be concentrated in one location, while others, say, the "operating arms" of the organisation, less affected by institutional complexity, are farmed out to sub-contractors, franchisees; or parts for the organisation with potentially incompatible interpretive schemes may be divided from each other[13]. This may not divorce the organisation entirely from institutional influence, but it does appear to contain it. The effect of this arrangement, from the institutions' viewpoint, is to segregate their conversations and to limit them to comparatively small groups of people, some being multi-disciplinary and others very highly specialised and unperturbed by what they come to perceive as "noise" from other institutional viewpoints.

There are many examples of this move, but only time will tell whether or not it will be successful in freeing the organisations adopting it from the coordination problems which are so patently besetting more conventionally structured organisations.

CONCLUSIONS

The conclusions reached from this brief discussion are not exactly original, but they are rarely taken very seriously. They are to the effect that institutionalised thinking, the uncritical acceptance of "good advice" from, or the unthinking compliance with, institutionalised prescriptions is potentially dangerous. As we have noticed above, institutions, being what they are, are not noted for the clarity with which they expound any theories which they may have. Each one thrives on ambiguity, incoherence and incompleteness, and each converses with the others, not to solve organisational problems but to further their own processes of self-production.

Because a genuine sharing of interpretive schemes between institutions would threaten their boundaries and inhibit their self-production, there are good reasons to doubt that there can ever be a real convergence of the different institutionalised interpretive schemes employed by differently specialised managers being achieved in the same organisation. This is to say that the adoption by the individual managers in an organisation of different institutionalised practices, together with the unqualified acceptance of the rationality (or otherwise) of their supporting theories, may be respectable, but näive, and potentially destructive of organisational coherence. Even the division of functional activities and the pervasive arrangement of hierarchies of management are themselves a production of institutions and should not be taken for granted.

It has been suggested that "solutions" to the problems of steering organisations in the face of the complexity thus generated must entail the reduction in some way or another of the invasion of interpretive schemes and the adoption of organisational arrangements which reduce the possibilities of institutional emergence and growth both within and outside organisations. Two organisational developments may be moving in this direction, and one common response may not. Failure to respond appropriately to the possibilities of institutional emergence may put organisational steering in jeopardy.

ACKNOWLEDGEMENTS

The author is greatly indebted to Dr. R J B Munro and Simon Lilley for the very generous help they gave in the preparation of this paper.

REFERENCES

1. G. Pask, "Conversation Theory", Elsevier, Amsterdam and New York (1976).
2. F.F. Robb, Are institutions entities of a natural kind? *in*: "Handbook of Systems and Cybernetics", C. Negoita, ed., Dekker, New York, 149-16 (1992).
3. R. Swenson, Emergent attractors and the law of maximum entropy production, *Systems Research*, 6:3 187-97 (1989).
4. R. Swenson, Order, evolution and natural law, *in*: "Handbook of Systems and Cybernetics", C. Negoita, ed., Dekker, New York 125-148 (1992).
5. R. Swenson, Order, evolution and natural law, *in*: "Handbook of Systems and Cybernetics", C. Negoita, ed., Dekker, New York 125-148 (1992).
6. G.C.M. Teubner, Autopoiesis and steering, *in*: "Autopoiesis and Configuration Theory", R.J. in T'Veld, L. Shaap, C.J.A.M. Termeer and M.J.W. van Twist, eds., Kluwer, Dortrecht (1992).
7. F.F. Robb, The information revolution in managerial organisations, *Systems and Cybernetics in Management*, 18:12 5-13 (1989).
8. R.L. Flood and E.R. Carson, "Dealing with Complexity: An Introduction to the Theory and Application of Systems Science", Plenum, New York & London (1988).
9. I. Hacking, "The Taming of Chance", Cambridge University Press, Cambridge (1990).
10. F.F. Robb, In defence of conversation, *Systems and Cybernetics in Management*, 9:1-2 15-25 (1990).
11. F.F. Robb, Accounting: a virtual autopoietic system?, *Systems Practice*, 4:3, 251-35, (1991).
12. S.R. Clegg, "Modern Organizations", Sage, London (1990).
13. R. Evans, A global gas company: vision and action, *RSA Journal*, 5434 (1992).

USING AN ORGANISATION WIDE CONSULTATION APPROACH FOR THE CREATION OF INFORMATION AND INFORMATION TECHNOLOGY STRATEGIES

Alan Warr

Henley Management College
Greenlands
Henley on Thames
Oxon RG9 3AU

INTRODUCTION

Increasingly, organisations both in the public and private sectors[1,2] are creating strategies for the development of their information and information technology (IT) resources. These strategies "aim to align Information (Systems) with the aims and objectives of the business Strategic planning for Information (Systems) is high level and deals with resources at the aggregate level. Typically it looks five to ten years ahead"[3]. The purpose of this paper is to outline an approach to the development of these strategies which uses extensive consultation across the organisation. This is set within the general developments and approaches within this field and the strengths and weaknesses of the approach are discussed. Finally the need for fit between the strategy development processes selected and the organisational context in which they take place is discussed.

THE EVOLUTION OF STRATEGIC PLANNING FOR INFORMATION AND INFORMATION TECHNOLOGY

From the earliest deployment of computer technology in business organisations the need to plan was recognised[4]. As was to be expected, the early planning was simple and used unsophisticated techniques. As organisations learned more about using the technology and as the technology itself became more sophisticated, more formal approaches emerged from management academics[5] and computer suppliers[6]. Over time these have become increasingly sophisticated[7]. Today there is a wealth of different techniques available[8,9,10,11,12,13]. It is worth noting that the increasing sophistication in the information/IT strategy planning techniques has followed the increasing sophistication of business strategy planning generally[14].

INVOLVEMENT AND CONSULTATION WITHIN INFORMATION AND IT STRATEGY PLANNING

The focus of the increased sophistication in strategic planning for information and IT has not simply been on the techniques. Over time the types of personnel involved in the processes have also evolved. Early techniques were technically orientated and involved mainly the data processing professionals with vendor consultants. Later internal and external business consultants were added and top management involvement became recognised as a key to success. Today it is also common for senior business managers from those business areas using computerised information to be involved. To some extent this increase in business involvement has been mirrored in strategy development generally with strategy moving onto the agenda of staff at lower and lower levels in the organisation.

However today it is still rare for ordinary information users to be consulted directly during information and IT strategy formulation projects. Those involved are most usually from the management elite of the organisation.

This leads to the question as to whether involvement from the lower levels of an organisation, in the information strategy processes in particular, would be beneficial. An information strategy is most usually a plan which impacts the whole organisation and therefore brings about significant organisational change. This is not always the case with other functional strategies where the pace of change is generally slower. Involvement through consultation can be expected to lead to easier management of the changes that will result from such a strategy[15,16]. In particular it can be expected to lead to lower resistance to change from those consulted since they have contributed to the strategy process and consequently have a stake in the resultant strategy as well as a better understanding of its origins. Additionally the involvement of lower levels of the organisation would contribute additional knowledge and experience to the information strategy development process[17].

THE CONTEXT FOR THE CASE STUDY

The exploration of whether the involvement of lower organisational levels leads to effective information and IT strategies is in its early stages and the ground is being broken by practitioners rather than researchers. This paper describes one such case.

The example is the information and IT strategy development of a UK county police force. It is a large sized organisation providing a wide variety of police services to an urban and rural population approaching 1 million people. The strategy was created from April to December 1991, a period of significant challenge for the UK police service generally. At this time UK crime was increasing at 24% per annum whilst public expenditure generally, including that spent on the police service, was under pressure.

As is often the case the information and IT strategy project was part of a wider strategic initiative. This initiative resulted from the arrival of a new Chief Constable and comprised a review of all aspects of the county police force's activities. Although initiated by the Chief Constable it was to be undertaken by 12 project teams known as working groups who were each assigned major areas to review. The underlying philosophy for this "Force Review" was widespread involvement and consultation to create, through the review, a police force designed by the force as a whole.

One of the working groups was assigned to review the force's information and communication systems and create a formal strategy for their development. Since they were part of the overall Force Review they had to adopt an organisation-wide consultative approach. The context therefore had a particularly significant effect on the participants in the information and IT strategy study and upon the processes they could use.

THE PROCESSES USED TO CREATE THE INFORMATION STRATEGY

Some of the processes were imposed upon the information and IT strategy study by the Force Review. For example, the composition of the working group was dictated by the Review. This had to reflect the composition of the Force as a whole, so was made up of different ranks from police constable to superintendent, included both civilian and police staff and also representatives from the three main staff associations. The group was lead by a traffic police constable and included no IT professionals, though they could be seconded into the group if the group decided to do so, as they very quickly did. The project management process was also inherited from the Force Review. In this, all working groups reported through a Review Steering Group to the top management team of the Force.

The higher level processes used were fairly standard. They were required to conform to standards laid down for the UK government sector by a Treasury body called the Central Communications and Telecommunications Agency (CCTA). The processes used were therefore broken down into four stages: (1) a Scoping Study to plan and scope the project, (2) the formulation of an information strategy, (3) the formulation of an information technology (IT) strategy to support the information strategy, and (4) implementation planning for the combined information and IT strategy.

What was far from standard was the personnel involved in carrying out these stages and the techniques used to execute these processes. The working group was described above. This is a very unusual body to carry out such a project. It was lead by a low level business user, not a senior IT manager or senior business manager which is most commonly the case. The group had little experience of IT let alone information/IT strategy formulation. This they overcame by obtaining training from the author on the options available to them. However the 10 staff in the group were representative of the organisation as a whole, to the extent that a group of 10 could be. They understood the organisation, particularly the concerns of those at the "sharp end", they worked on a day to day basis with the force's systems, actively engaging in service delivery. Therefore they knew onlv too well how the force's information and communication systems were performing. They quickly added the 2 computer managers and the communications manager to the team, but in an advisory capacity only.

Another unusual aspect was the way that the developments and policies for the force's information resources were determined. To conform with the ethos of the Force Review it was necessary to consult with the force at all levels. This contrasts markedly with normal practice where only senior managers in an organisation are consulted. An approach exists for this whereby a questionnaire is used to survey user needs but it is rarely used in the UK. However many of the working groups were using questionnaires so sending out yet another to the force was likely to be met with a poor response. The working group therefore decided to adapt another technique commonly used, namely the management workshop[18]. Except that in this case they ran workshops at every level throughout the organisation from police constables to the top management team. In all, 26 workshops were run with 6 to 12 attendees lasting 3 to 4 hours each. Nearly 10% of the organisation was consulted over a six week period to determine their information needs. All functions were included and all levels of the organisation. From this, all the information needs were recorded and then pulled together by the whole working group during two days of intense effort to form the information strategy. Other processes were also undertaken. An audit of existing information and communication systems was completed, some external liaison was possible with partner organisations and other forces and a review was undertaken of the information and IT strategies of other county forces. The result of all this effort was an information strategy comprising: (1) applications requiring development, (2) necessary information policies, (3) management changes and (4) training needs.

The next stage, the IT strategy, addressed technology issues which were, in the main, beyond the expertise of the force as a whole. This stage was effected using a small group comprising the three technology managers and three of the working group members. The IT strategy was formulated by the technology managers. The three working group members assisted where they could, however their main role was to ensure that the IT strategy was formulated to support the information strategy and that the technologists did not re-interpret the information strategy in terms that suited the technology. An external technology consultant was used to add a wider view of the technology currently available and likely to become available during the life of the strategy. The IT strategy stage comprised four sets of activities: (1) making decisions on how the applications and training identified by the information strategy stage should be supplied, (2) deriving any necessary infrastructure developments, (3) checking the overall architecture of the proposed information and IT resources, and (4) implementation planning. This stage therefore did not include wide consultation since the collective wisdom and experience of the mainly technical issues involved was held within the IT function.

Once the information and IT strategies were completed they were made available throughout the force for comment and feedback and passed to outsiders for quality assurance.

OUTCOMES, STRENGTHS AND WEAKNESSES OF AN ORGANISATION-WIDE CONSULTATION APPROACH

Obviously the main outcome of this process was an information and IT strategy for top management consideration and approval. However the consultative nature of the processes used lead to benefits which would probably have not resulted from other approaches. Firstly the level of involvement lead to a "head of steam" building up behind the strategy. This means that now that the strategy is well into implementation there is considerable organisational motivation for the changes required. Secondly the strategy was realistic, since those who were suggesting changes and developments were, in the main, highly aware of the capabilities and limitations of the organisation. Thirdly the process lead to a raised awareness of the importance of information as a resource which has general operational benefits outside the strategy planning processes. Finally the information and IT strategy has considerably more credibility than would have been the case had it been created by the IT function alone or by a small management team. It was an information/IT strategy for the force created by the force itself.

Some weaknesses were also encountered. The process was inward looking, with most of the effort focused on internal staff, with limited consultation with customers or partner organisations. This is typical of information/IT strategy generally, but also mirrors the slightly embattled culture of the organisation at this time. The approach was risky for both the top management team and the information technologists. Once started the process meant that the outcomes were out of their control. In the case of the top management there is little doubt that the Chief Constable understood this, however the information technologists were not given a choice and at least initially they were very apprehensive about the outcome of consulting users on such a scale. The workshops were more effective at identifying operational information and communication's systems than management information needs. This is a common problem and points to the use of a specialised technique such as Rockart's CSF analysis9 which is tailored for the identification of management information needs, perhaps as preparation to inform the management workshops. Finally workshops found it difficult to prioritise the applications identified. This also mirrored to some extent the culture of the organisation which was experiencing general difficulties in coping with the tough decisions resulting from increasing demands on fixed resources. In the end the working group was left to prioritise the portfolio of applications using a Delphi technique.

CONCLUSIONS

The case therefore identifies the fact that wider consultation is possible, particularly in the processes undertaken to formulate information strategies. Indeed some benefits arose which would have been difficult to achieve with more traditional approaches. However what is also clear is the situational nature of what was undertaken. A major influence on the approach used was the Force Review, indeed it is unlikely that such widespread consultation would have been undertaken other than for this review. This points generally towards the need to fit the information and IT strategy processes into the organisational environment. Systems theory would predict that a poor fit between the overall processes used and the environment in which they take place will inevitably lead to poor outputs if not failure of the strategy development process. Many important dimensions of the environment need to be considered. Some of these include both internal elements such as culture, management style and the other strategy planning systems in operation; and external elements such as the importance of information to the organisation, the extend of any need to integrate with other organisations and any requirements placed upon the organisation by super-ordinate organisations. The opposite is also true that a good fit between the processes used to formulate an information and IT strategy and the organisational context will contribute to the likelihood of success. The case is certainly an example of this, and by using an approach tailored to its organisational context it has, perhaps inevitably, extended the boundaries of practice in this field.

REFERENCES

1. T.D. Wilson, The implementation of information systems strategies in UK companies: aims and barriers to success, *International Journal of Information Management*, 9, pp245-258, (1989).
2. N. Bacon, Information systems strategies in government: recent survey evidence, *Journal of Information Technology*, 6, pp94-107, (1991).
3. K. Consodine, "An Introduction to Information Systems in Government: Setting the Scene," CCTA/John Wiley & Sons, Chichester, (1989).
4. R.F. Osborn, GE and UNIVAC: Harnessing the high speed computer, *Harvard Business Review*, Vol. 32, No. 4, pp99-107, July-August (1954).
5. W.M. Zani, Blueprint for MIS, *Harvard Business Review*, 48, 6, pp95-100, (1970).
6. IBM Corporation. "Business Systems Planning: Information Systems Planning Guide", IBM Document GE20-0527-3, White Plains, New York (1981).
7. C. Sullivan, The changing approach to systems planning, *Journal of Information Systems Management*, Vol. 5, No. 3, pp8-13, (1988).
8. R. Nolan, Managing the crisis in data processing, *Harvard Business Review*, Vol. 57, No. 2, pp115-126, March-April (1979).
9. J. Rockart, Chief executives define their own data needs, *Harvard Business Review*, 57, 2, pp81-93, (1979).
10. M.E. Porter and V. Millar, How information gives you competitive advantage, *Harvard Business Review*, Vol. 63, No. 4, pp149-160, July-August (1985).
11. C.K. Carlson, E.P. Gardner and S.R. Ruth, Technology driven long range planning, *Journal of Information Systems Management*, Vol. 6, No. 3, pp24-29, (1989).
12. J. Martin. "Information Engineering: Book 1 Introduction", Prentice Hall, Englewood Cliffs, New Jersey (1989).
13. R.J. Benson and M.M. Parker, "Enterprise wide information management: an introduction to the concepts", Report No G320-2768, IBM Research Center, Los Angeles (1985).
14. M.E. Porter, Corporate strategy, the state of strategic thinking, *The Economist*, May 23rd (1987).
15. E. Brooks. "Organisational Change: The Managerial Dilemma", Harper and Row, New York, (1980).
16. E.A. Johns, "The Sociology of Organisational Change", Pergamon, Oxford, (1973).
17. J. Child, "Organisation: A Guide to Problems and Practice", Harper Row, New York, (1984).
18. A.C. Boyton and R.W. Zmud, Critical success factors: a case based assessment, *Sloan Management Review*, 25, 4, pp17-27, (1979).

INTRODUCTION TO COMPLEX SYSTEMS STREAM

Human activity systems are perhaps the most intractable of all complex systems to model. Practitioners in the social, management and information sciences concerned with such systems continue to operate without the kind of methodological and theoretical framework that has contributed to the perceived success of the physical sciences and of the engineering disciplines. Practitioners have had to reflect on whether useful formal models for such systems can be constructed, or even whether there are aspects of such systems that are susceptible to formal treatment. Each paper in this stream describes a different approach to extending the formalization of models of human activity systems.

Dunderdale's "Systems Synthesis" is an attempt to formalize the procedures that one should follow in examining organizations. The approach is based on a premise that all systems have a common structure, and aims to produce an abstract description of the organization in terms of this structure.

Korn, Huss, and Cumbers seek a predictive theory that can be applied to human activity systems, and describe an approach using natural language to represent knowledge about the world and fuzzy logic to deal quantitatively with possibility. Dynamical models generally admit only a single possible trajectory from a given state; the use of a fuzzy semantics seems to offer some promise in allowing situations with uncertain outcomes to be modelled.

Gregory raises some interesting issues concerning formal models. Specifically, his paper continues an earlier argument that the conceptual models of soft systems methodology are insufficient to represent causal relations in the real world, but that a sufficient "logico-linguistic model" can be produced, through iterative debate, from the conceptual model. In passing, the paper draws a distinction between the sense of a model and its reference, and applies Popper's criterion on the scientific status of hypotheses to justify the objectivity of the result. These latter issues are themselves matters of considerable debate and a consensus awaits the development of a coherent philosophical framework for models in the human activity domain.

Certainly, the goals, limits, and range of formal approaches to complex systems remains a central and increasingly vital topic for discussion. The papers in the current stream give much food for thought on what, why, and how of formalization.

Dr Richard B. Beeby
Univeristy of Paisley

Systems Science, Edited by F.A. Stowell *et al.*,
Plenum Press, New York, 1993

SYSTEMS SYNTHESIS

Peter Dunderdale

Federation of Information System Centres (Bristol)
Sheffield Science Park
Sheffield S1 2LX
UK

INTRODUCTION

Complexity - the Problem

The requirement to understand is paramount to survival; to understand complexity requires that it can be quantified. This paper seeks to explain one of the effects contributing to the complex nature of organisations. The brain efficiently handles and processes information to retain a continuous picture of the relationships between events. This provides the holistic view of organisations. Complexity occurs when the structure of an organisation is too large for personal retention. The problem is then one of making information available to assist in retaining a coherent and uniform perception of the organisation.

To understand any organised activity you require at least two views, one examines the detail of the organisation (analysis), the other examines the conceptual system which provides a framework for the detail (synthesis). Analysis seeks to model an event or an operation, with a precision that is useful to Managers or Technologists.[1] This paper makes a clear distinction between analysis, the study of physical detail, and synthesis, the study of a conceptual system behind that detail. **The requirement is for a conceptual model which accurately describes a system study area, and has a precision which allows a detailed analysis to be mapped into it.**

Systems Synthesis: Background and Origin

Systems Synthesis is introduced as a modelling technique that utilises the holistic view to produce a systems model. It is a Craftsman's technique. Its first effective use was in 1965 and it has been used and developed concurrently ever since. The tool carried is a perception of the nature of systems; this can be used to move between reality and concept using a single set of references. There is a requirement for a more formal approach and a linking to the established field of systems thinking.[2] The technique is based on the hypothesis that objects, defined as anything that on examination might have some purpose, may be described using

Systems Science, Edited by F.A. Stowell *et al.*,
Plenum Press, New York, 1993

a limited set of characteristics which are common to all objects, and are the attributes of a system. Each characteristic may be used to identify a structure that corresponds to that particular attribute of the system.

The original development work posed the basic question 'what promotes variations in equipment designed to perform the same function?'

The initial search was for sets of attributes that were common to a wide range of objects. It was felt that a definitive way of classifying objects using a limited set of common attributes might lead to some understanding of the design problem.[3] A sub-set, present in all objects, was identified; it does not refer to an object's physical properties and is directly related to its existence or behaviour. The statement, **objects are the regulated and maintained interaction of objects**, explicitly identifies four of the characteristics in the sub-set. Others are implied. These characteristics are felt to be those that provide a systemic nature. This statement is referred to as the **Root Statement**.

The characteristics, Function, Regulation, Procedure, Demands and Licence have been named for convenience. While these terms have been selected for the closeness of their English meaning, more detailed descriptions are provided later.

Outline Practice

The **Root Statement** is used as a basic framework for examining the structure of a system. When used in practice it gave rise to the idea of a whole systems structure (system's domain) that can be associated with a particular example. This domain illustrates that a system has a conceptual structure which demands resources from the physical environment. Design by whatever means is the process of assigning resources to satisfy the system's demands; this lies between the conceptual system and the physical example formed by design.

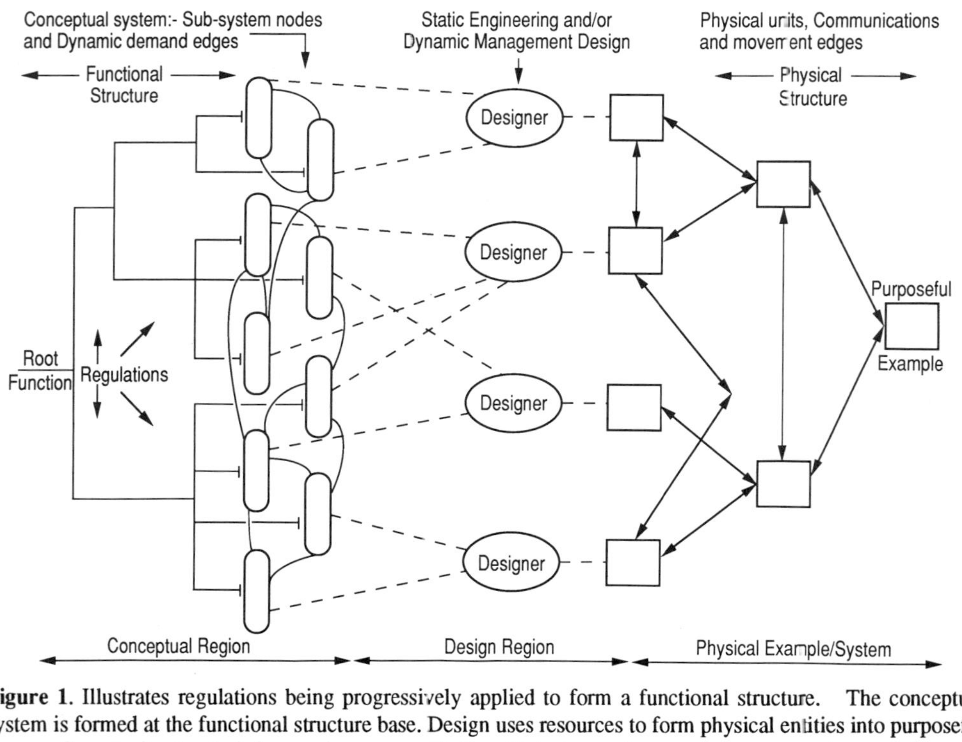

Figure 1. Illustrates regulations being progressively applied to form a functional structure. The conceptual system is formed at the functional structure base. Design uses resources to form physical entities into purposeful groups. Complexity due to influence increases with the efficient use of resource.

While Figure 1 is helpful in explaining the structure of a system, the art lies in visualising this structure when examining the parts of a system's physical example. The basis for this structure can be found in the Root Statement about objects. Systems Synthesis was developed from an understanding of how this statement applied to the real world. With practice it is possible to view the world through a mind filter that allows it to be seen freed from its physical form. With such a view the observer may see the common basis for organisations, that a single conceptual engineering design matches all the machines carrying out that process, and that natural systems have well developed structures that can guide our designs.[4,5] In the following I have tried to retain the original order of some significant experiences so that the technique's development can be followed.

DEVELOPMENT

Functional Structures and Licence

The initial application of the **Root Statement** was aimed at understanding the design of a Carpet Printing Range[6], so that a processing technique could be established. The first step was to define the Range as a single object. The Print Range became an 'object that relocated dye stuff', I assumed that the natural physical laws were its basic regulations. To this basic function I then added designer regulations that imposed the minimum constraints possible. These imposed regulations were seen as different from the natural laws.

As the regulations were added they suggested the need for other objects to support the main function, thus the Print Range function with regulations became, 'to control the relocation of various dye stuffs from a dry powder store to defined relative locations on a carpet material'. This suggested the objects/functions 'to move dye from one location to another', and 'to control relocation'. Relevant regulations were added to these, and a new level in the object/function structure appeared. One basic rule developed was; functions must be singular and with their regulations guide the identification of the supporting functions required. These support functions must also be singular so that the process can be made recursive. Each function had to be satisfied by the regulated and maintained interaction of other functions, and the need for these prompted the identification of other regulations and required supporting functions.

I proceeded to design a processing Range without defining any physical parts and the result was a hierarchy of imaginary objects/functions that would satisfy the overall function. I stopped the process wherever I could see a simple way of substituting a real object for one that was defined as a function. There were more functional objects than real ones on the machine and part way though the process I spotted where one real object on the equipment was being used to fulfil two functions. Interference between their uses was making the process unstable. This provided a first view of why systems are complex.

It was also clear that the designer had assumed a process method and then designed the Range as a response to this method. In doing so, much of the design was not related to the process but to resolving conflict caused during design. Complexity lay in the designer's apparent arbitrary but sensible use of single physical parts to satisfy several functions, and the need to accommodate any conflicts caused by these assignments. This made it impossible to logically work backwards through the design process. Later when I needed to examine 'the competition's equipment' it became clear that the functional design was a general case and allowed the design of a particular case to be understood.

In producing the functional structure it was clear that regulations did not always absolutely constrain the scope of activity. When physical systems were examined it was still evident that this scope allowed other systems to influence behaviour. Since it was impractical to design without allowing for possible variation, I tended to regard all systems as inherently licensed and saw good engineering design[7] as overcoming the effects of licence. Licence became far more evident when I applied the technique to management[8,9] rather than engineering processes.

Licence and influence account for much of the complex behaviour in a system. It is essential to distinguish between a system satisfying its regulated requirement and the apparent disorder in the way this is achieved. It may be impossible to predict the internal behaviour of a system, but possible to understand the nature of its disordered behaviour. Lately I have begun to see licensed behaviour and influence that becomes regulation as a means of modelling the structure of intelligent systems. The problem being to model the difference between machine, animal, organisation and human intelligence.[10]

The Conceptual System

The next major step occurred when I designed the model of a company.[11] Producing a functional hierarchy showed how a Company started with the function 'to control the balance of relationships'[12] and this provided a compound control structure with a response 'to benefit from the provision of goods or services'. The base of this structure held the set of functions required to service corporate activity. It did not indicate how this would be converted into reality, nor did it appear that continuing to go farther down the hierarchy would produce an answer.

The **Root Statement** suggested that the answer must lie in how the system interacted and was maintained. The base of the functional structure had two sets of demands. One was produced when the structure was divided to show supporting sub-functions, and could be seen as dynamic exchanges between functions when these were sub-systems. The other was a demand to make each sub-system real. The conceptual system was the collection of sub-systems that formed at the base of the functional hierarchy and the dynamic demands that bound them into a total system. Each sub-system had a set of static demands which when satisfied converted it into a physical object.

The Design System

Design presented a particular problem because it was generally associated with an intellectual process; however experience had taught that 'chance events' were evident in the formation of some organisations. If design is only an intellectual process, it argues that natural organisations (ant hill) and human organisations (Company) have a basic difference in overall structure when viewed as systems domains. This suggests that the **Root Statement** is either incomplete or not generally valid. Flattering as it may be to set humanity apart I feel this is not the case.

Design is a system that seeks 'to control the assignment of available physical resources'. Human intervention makes resources and/or creates their availability. Control can be natural;[2] the flow of water out of a tank is regulated by physical laws and responds in relation to any change that occurs. Control when used in this natural sense allows for natural design control which will apply to human designers when either there are no natural resources or new resources cannot be fabricated to meet a demand. Thus design as an intellectual process can continue until it cannot make or find a resource, then it shares the same type of control as a natural system.

A tree's design system (natural evolution) requires it to produce a licensed range of design variations (seeds) and allows the environment (resources and demands) to determine good responses. The time scale is different and its use of licence is effective. Generalising the design system may appear unnecessary for every day problems, but to argue a general structure for different types of organisations it is necessary to show that the conceptual structure of the organisation is present prior to its formation, and that formation is a design process. Since this appears to be correct it argues that management can be modelled as a continuous design process. Understanding the design process shows that conceptual systems are far more stable than the design response. Long term organisational strategy must be based on the conceptual structure and seek to design an example of that system based on the resources available.

THE DEVELOPING RULE BASE

Figure 1 represents a general systems structure and the following are used to associate a physical example with that structure or to design total systems.

The **Root Statement** (objects are the regulated and maintained interaction of objects) holds good in both the physical, design and conceptual regions of the diagram.

Function, System, and Object interchange to suit the region being examined. Object without purpose is function; with structure but no physical form it is system; with physical attributes it is object. As an object it is both systemic and functional. For example; 'to relocate' has function but suggests no purpose; 'to relocate (regulated by) dye stuff to locations on a carpet' has both function and purpose but no form; relocation will be achieved by an 'object' designed to fulfil this purpose. The word object is assigned to anything that can be experienced, including that which can be imagined. An 'idea' is an object in the mind. The first reference to object in the **Root Statement** implies both existence and function. It is only the function attribute that is recognised as common.

Objects, the second reference, implies that they are subordinates which group to satisfy the first reference to object. Depending on their use these may be seen as component functions, systems or objects.

A systems study needs a practical end point. At some level down the hierarchy it is convenient to cease sub-division and substitute a 'physical example' as the component object. Continuous descent finishes with the assumption that an object may be single and yet exist in several states. This object is maintained by a demand 'to change state', thus interacting with itself. The most accurate structures are formed when a single function is related to a system or object. A functional structure is the hierarchy formed by applying regulations to a function and then satisfying those regulations with sub-functions. A systems structure is the conceptual system that lies at the base of the functional hierarchy. An object's structure is the arrangement of design assignments.

Regulation lies outside the system's domain, it is the Natural, or Designer created laws, which determine (to a certain extent) the selection of, and interaction between the components of a system. Regulations reduce the scope of a function and promote sub-functions to accommodate the reduction in scope.

If a system's regulations alter then it is replaced by a new system that includes the alteration, this new system may use the resource released by the old system. It is necessary to distinguish between regulation and influence, the latter adjusts behaviour within the scope of the former. A regulation affects all below its point of application, thus a sub-system's regulations are those in its hierarchical path (not all regulations at a higher level). Where regulations conflict the sub-function must be 'to control ...' .

Procedure, from 'interaction' in the **Root Statement**, is the correlated activities of the system's components. The procedure appears as dynamic demands in the conceptual system and as activity in the system's example.

Demands, from 'maintained' in the **Root Statement**, are that which is required to give support to a function, structure to a system and form to an object. The attributes, static and dynamic are a convenience that allows a conceptual system to be illustrated, static refers to demands within a system's domain, and dynamic to demands required to sustain that domain.

Licence is implied by the inexplicit nature of the **Root Statement**. The function of an object is to some extent a designer assignment. Regulation may be complied with and yet not exactly determine how a function is satisfied. Maintenance does not imply that the demands will be efficiently or correctly satisfied. Thus licence recognises the degree of flexibility that exists within systems. Partially regulated behaviour leaves the system open to the influence of other systems. Influence is that which might disturb the system without altering its regulations. This allows low level disordered behaviour. High level disordered behaviour occurs when influence becomes regulation.

CONCLUSION

It would be unrealistic to offer a general modelling technique without indicating how people are accommodated. People are a resource that can be assigned to organisational systems. They (all categories of personnel including the Directorate) are component units assignable to conceptual sub-systems, and their behaviour is complex because they are assigned many systems, typically work, social, domestic etc.

Very simplified and excluding all the physiological parts, people have a sub-function 'to control' (resolve) the regulation sets that determine behaviour in the systems in which they participate. Unrealistic, conflicting, incorrect, or absent instruction sets cause the component to become stressed. Regulation sets provide expertise, dexterity and other qualities, these are improved by education, training and experience. People are complex because their personal level of control over and/or influence between these regulation sets is difficult to predict.

It is the efficiency with which responses are designed that makes the world complex, yet our survival requires a far more efficient use of resources. While some functional structures can be complex, I believe that the basic nature of a system is simple, even when its behaviour cannot be predicted. Systemic thinking should confer the understanding required for control, but it will not confer control any more than control confers understanding.

ACKNOWLEDGEMENTS

I wish to thank Professor R.D. Milne of the Department of Engineering Mathematics, University of Bristol, U.K., for his guidance in the presentation of this concept.

REFERENCES

1. B. West, ed., "The Systems Training Library", NCC Blackwell, Oxford. (1984)
2. D. Hitchins, "Unified Systems Hypothesis", UKSS Workshop, City University, London. (Dec.1990).
3. F.N. Mangill, ed., "Masterpieces of World Philosophy in Summary Form", George Allen & Unwin Ltd, London, (1963).
4. F.H.C. Crick, "The Genetic Code: III", *Sci.Amer.*, 215 (4):55-62, (1966).
5. K. Pearson, "The Control of Walking", *Sci.Amer.*, 235 (6):72-86, (1976).
6. Associated Weavers. "Carpet Printing - a New Era", Supplement to Carpet Review, (1970).
7. P. Dunderdale, Brit.Pats: 1256893 (D1) Oct 1971-May 1972, 1300754 (D1,B8), Dec. 1972-July 1973.
8. P. Dunderdale, "The Design of a Technical Service for a Group of Companies using a Systems Theory Design Approach", Internal Report for Carrington Viyella, (Nov. 1976).
9. H.D. Koontz and C.J. O'Donnell, "Principles of Management", 3rd ed., McGraw-Hill, (1964).
10. H.A. Fatmi and R.W. Young, "A Definition of Intelligence", *Nature*, 288:97, (1970).
11. P. Dunderdale, "Basic Model Structure of Bristol Information Technology Centre Limited", Internal Report for Bristol City Council Steering Committee, (1983).
12. Companies Act, H.M.S.O., (1948).

MAPPING CONCEPTUAL MODELS ON TO THE REAL WORLD

Frank H. Gregory

Warwick Business School, University of Warwick
Coventry, CV4 7AL, United Kingdom

ABSTRACT

SSM conceptual models must be capable of mapping on to the real world if they are to be used as a basis for information system design . An abstract model is used to illustrate how a mapping attempt can be evaluated. The logical methods used might be capable of replacing traditional IS modelling methods.

THE MAPPING PROBLEM

The conceptual models (CMs) of human activity systems used in Soft Systems Methodology are notional. They are not intended to represent a real world state of affairs. They are models of what the stake-holders in the organization concerned consider to be desirable. We can distinguish two uses of these models: the first associated with the name of Peter Checkland the second with that of Brian Wilson.

With Checkland it is the building of the model as much as the model itself that can resolve organizational problems. Checkland is concerned with changing the way the stake-holders think about their problem rather than with producing a detailed plan for a problem solution. In practice the problem sometimes simply disappears during the model building process. Other outcomes might be the identification of an organizational *raison d'etre*, or the definition of a new role for a department. In the context of information systems Checkland's models tend to be models of the organizational context or models of the development context for an information system, they are not models of the information system itself.

Wilson takes the models further and uses them as the basis for information system (IS) design. With Wilson the CM is developed until it becomes the information system design. While Checkland's models are fairly well accepted as being a useful front end to IS design, Wilson's work is more contentious. There are those that think it is not possible to produce an IS design from a CM (see Mingers, 1992). It is not difficult to see why they are concerned. Wilson starts with a notional model and ends up with a system that handles information about real world entities such as stock in a warehouse (p 195 - 208, Wilson 1984). It is common sense that our notions do not always correspond to reality and that what

Systems Science, Edited by F.A. Stowell *et al.*,
Plenum Press, New York, 1993

is desired cannot always be achieved. This is the problem of real world mapping. Another problem is that Wilson's information systems need to take account of causal sequences. They must contain, explicitly or implicitly, an account of what causes changes in real world events such as stock levels. However, the logical connectives in traditional SSM models are insufficient to represent causation (Gregory 1991). Thus, Wilson's projects, but not Checkland's, have to contend with two major theoretical problems.

The problem with causation can be overcome by increasing the number of logical connectives in the stake-holder constructed models (Gregory 1992a, 1992b), the result is a Logico-linguistic CM. The power of these models is such that they can also be used to address the real world mapping problem. Traditional SSM models are taken as giving definitional attributes of a desirable state of affairs. When expanded into a Logico-linguistic model they will become full definitions which can be expressed formally in the predicate calculus. At this stage the terms in the model will have a fully established sense (connotation, intension) but not an established reference (denotation, extension). The problem of real world mapping then becomes one of establishing a real world reference for the terms of the model.

THE GOR TONKING MODEL

When the idea of Logico-linguistic models was presented at the UKSS SSM and IS seminar (Proceedings in *Systemist* Vol. 14, No 3, Aug. 1992) Prof. Checkland suggested that one be constructed on the basis of the Gor Tonking model. This model is constructed from an imginary root definition: "A dag-owned gor tonking system which, within legal constraints, tonks those gors which meet criteria gog". The activities in the model include *2. Know/decide*

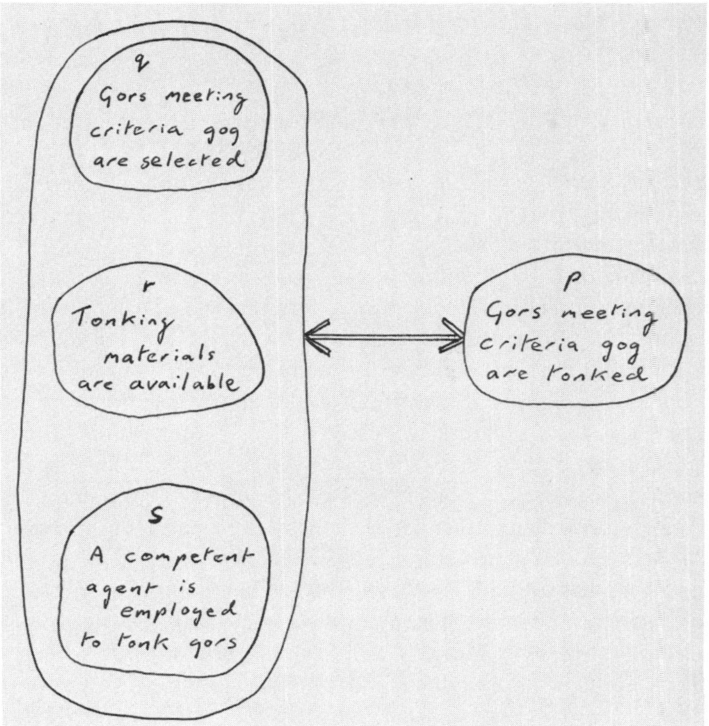

Figure 1

how to tonk and 5. *Ascertain which gors meet criteria gog* these have arrows leading to *6. Tonk gors meeting the criteria.*

This is particularly pertinent because "The model...cannot possibly include unjustified real-world knowledge since it is, deliberately, an RD [root definition] without meaning. It is included to show how a defensible logical structure for a model can be created from an RD, even though the RD does not refer to the everyday world." (Checkland 1989). Another way of putting this is to say that the model has sense but no reference. All CMs are like the Gor Tonking models in this respect; at the time of their completion (stage 4 in the learning cycle) they will have sense but reference will have to be established later (stage 5).

The real world mapping problem is concerned with whether the model can have reference. In the following the Gor Tonking model will be used to illustrate a procedure to determine whether a model can have reference. We will not assume that the model does not have reference but imagine that we are facilitators building a model for stake-holders who use a slang that we do not as yet understand. There are many circumstances which might suggest a mapping idea. A verbal description, ostensive definition, or observation of behaviour could suggest the reference of a term. In principle the facilitator could come up with the mapping idea on his own initiative, in practice it would be better to have the stake-holders construct the entire model as they will know how their ideas are intended to map on to the real world. In the example we will try to map the Gor Tonking model on to a system to re-spray scratched cars. The model that will be produced will enable us to determine if this mapping works.

To convert the CM to a Logico-linguistic model the first step is to convert the expressions from commands into statements. Thus the command "Tonk gors meeting the criterion" is replaced with "Gors meeting the criterion are tonked". When expressed as statements the models become truth functional and standard logic can operate. The next stage is to introduce two new logical connectives. The connectives (arrows) in traditional CMs indicate logical dependency. In the Gor Tonking example 6. is logically dependent on 2. and logically dependent on 5. This corresponds to the relation of implication such that if 6. is true 2. and 5. must be true. In the Logico-linguistic model a broken arrow indicates the opposite relation; a broken arrow from x to y indicates that x implies y which means that if x is true y must be true. A double headed arrow stands for mutual implication i.e. x implies y and y implies x which means that if x is true y must be true and if y is true x must be true.

A partial Logico-linguistic development of the Gor Tonking model is shown in figure 1. This can be read as if q and r and s are true p is true, and if p is true q and r and s are true. It expresses an identity and in this case is a comprehensive definition. A comprehensive definition is a logical identity such as "all bachelors are unmarried men", this can never be false. But other identity statements are contingent, that is true but not true by definition. For example "John Major is the Prime Minister" is a contingent identity statement, true at the time of going to press but likely to become false in the near future. While CMs are definitional and logically valid (Gregory 1992b), statements about the real-world are contingent. They will be true or false depending upon how the real world actually is, not upon how we define our terms. This distinction is made in modal logics. In the figures a simplified system of modal flagging is used: "N" indicates logically necessary relations i.e. definitions and what follows from them, and "C" indicates contingent relations.

A "SCIENTIFIC" MODEL

Figure 1 shows what must be true, as a matter of logic, if the desired state of affairs is to come about. Figure 2 is an attempt to describe a real world state of affairs that will bring it about. This state of affairs may happen to exist. If it does not, the next stage, stage 6 in the learning cycle (see Checkland 1989 for the learning cycle), is to make changes in the real

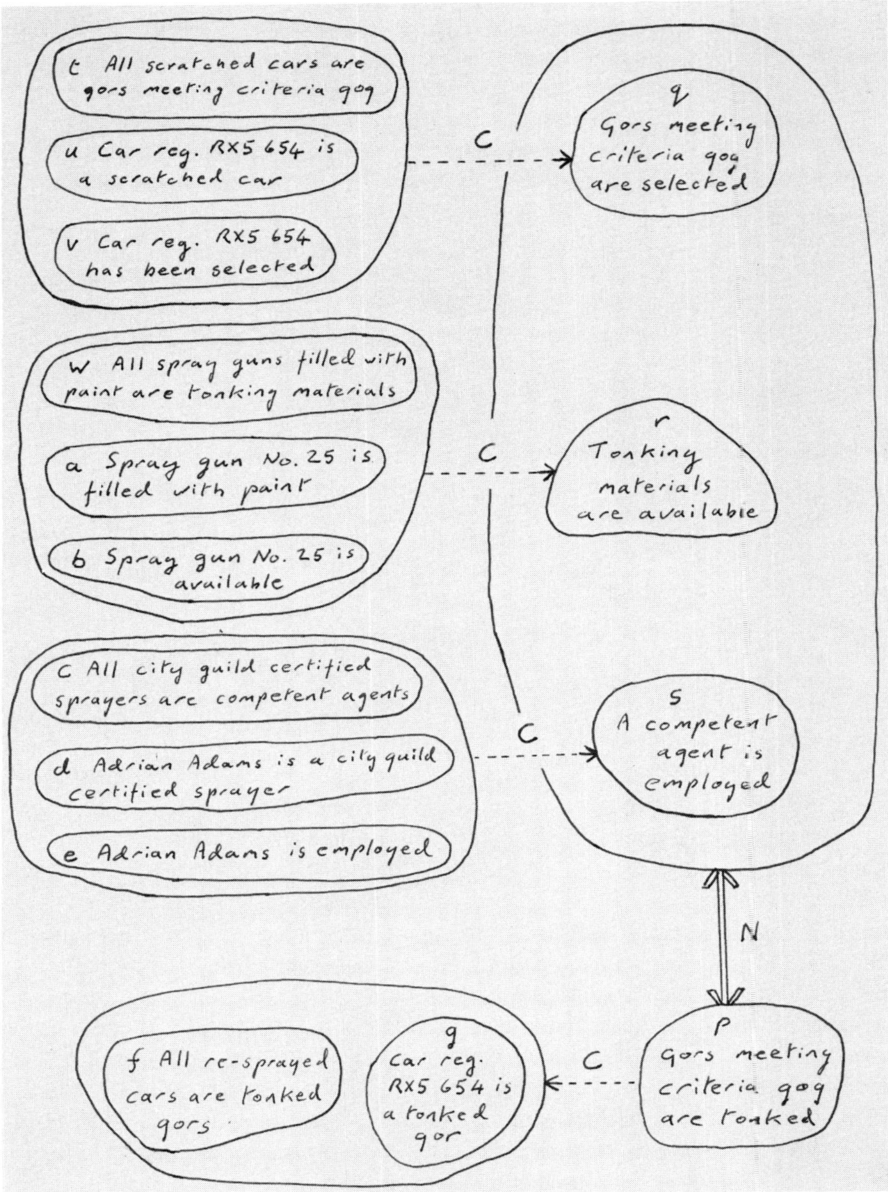
t All scratched cars are gors meeting criteria gog
u Car reg. RX5 654 is a scratched car
v Car reg. RX5 654 has been selected
C
q Gors meeting criteria gog are selected
W All spray guns filled with paint are tonking materials
a Spray gun No. 25 is filled with paint
b Spray gun No. 25 is available
C
r Tonking materials are available
c All city guild certified sprayers are competent agents
d Adrian Adams is a city guild certified sprayer
e Adrian Adams is employed
C
s A competent agent is employed
N
f All re-sprayed cars are tonked gors
g Car reg. RX5 654 is a tonked gor
C
p Gors meeting criteria gog are tonked

Figure 2

world that will actually bring it about. The figure 2 model will tell us when and if the desire state of affairs exists.

The figure 2 model consists of particular statements *u*, *v*, *a*, *b*, *d*, *e* and *g* which are putatively true and have real world reference. In addition there are contingent universal statements *t*, *w*, *c* and *f*, these are inductive hypotheses. In contrast to definitions, inductive hypotheses are always open to falsification by particular events. For example, if *u*, *v*, *a*, *b*, *d*, *e* are true and *g* is false then we know that at least one of the inductive hypotheses, *t*, *w*, *c* or *f*, must be false. We know this because the fact that "*q* and *r* and *s* imply *p*" is true by definition and so cannot be false.

The inductive hypotheses allow contingency into the system and this gives the model the logical flexibility to map on to a contingent world. It is also "scientific" in Popper's sense of the word because the figure 2 model is always capable of being falsified by particular events. This solves the problem of how a CM that is true by definition can map on to a contingent world - the answer is that the mapping itself is contingent.

This has profound consequences for IS design. IS design methodologies do not distinguish between inductive hypotheses and definitions. What actually happens is that they treat inductive hypotheses as though they were definitions - they make them fixed in the system. When this is done statements of the type *t*, *w*, *c* and *f* become necessarily true and the broken arrows in figure 2 change from having the "C" modal flag to having a "N" modal flag. When this happens the particular statements *u*, *v*, *a*, *b*, *d* and *e* cannot be true when *g* is false. Given that a computer systems configured in this way accepts that *g* is false it will refuse to accept *u*, *v*, *a*, *b*, *d* and *e* as all being true, even when they are in fact all true. In other words it will only map on to the real world when it wants to; but in this case it does not really map on to the real world at all. The only reason that such computer systems can ever work in a changing world is because the human operators apply informal mapping techniques.

It might be thought that it is the inductive hypotheses that are the important thing and that the definitions can be dispensed with, but this is not so. There is an argument (Gregory 1992b), too lengthy to repeat here, that definitions are logically required in the formulation of any inductive hypothesis.

CONCLUSIONS

This paper has attempted an argument to show how CMs can be mapped on to the real world. It is not claimed that the method used is simple but it is claimed that any method of IS design must address considerable complexity. The building of Logico-linguistic model will involve a lot of extra work at the stake-holder driven front end of an IS design project, but it will save work at later stages. A Logico-linguistic model can be expressed in the predicate calculus and this can be readily converted into Prolog (Gregory 1992b). The predicate calculus also represents the logical form of a relational data base. The construction of a Logico-linguistic model could, therefore, stand to render many of the devices used in traditional IS design, such as data flow diagrams, structure English and entity relationship modelling, unnecessary.

REFERENCES

Checkland, P. B., 1989, Soft systems methodology, *in*: "Rational Analysis for a Problematic World" J. Rosenhead, ed., John Wiley & Sons, Chichester.

Gregory, F. H., 1991, Causation and soft systems models, *Systemist*, 13, (3), pp105-112.

Gregory, F. H., 1992a, Cause, effect, efficiency & soft systems models. *Warwick Business School Research Paper No. 42, Journal of the Operational Research Society* (forthcoming).
Gregory, F. H., 1992b, Logic & meaning in conceptual models: implications for information system design. *Warwick Business School Research Paper No. 62.*
Mingers, J. 1992, SSM and information systems: an overview. *Systemist*, Vol. 14, (3). pp82-88.
Wilson, B., 1984, "Systems: Concepts, Methodologies and Applications," John Wiley & Sons, Chichester.

SYSTEMS MODELLING WITH NATURAL LANGUAGE AND FUZZY LOGIC

J. Korn, F. Huss and J. Cumbers

Middlesex University
Bounds Green Road N11 2NQ London UK

INTRODUCTION

Providing explanations and predictions of states of affairs has been of great interest to many people from times immemorial. The interest has been driven by curiosity, a desire for guidance to succeed and to survive, and has proved to be a reliable means of generating knowledge about the world[1].

Non-scientific explanation and prediction had been practised by people like the ancient Egyptians and Greeks and, in various forms of divination, such practice is still going on at the present time[2]. Such explanation and prediction is much concerned about issues directly related to human interests. However, its methods appear to be only very weakly related to aspects of the world in the form of observables and they have no mechanism for confirmation and refutation of their explanatory and predictive hypotheses. Scientific explanation and prediction in its pursuit of "certain" knowledge, strives for the establishment of universal statements arrived at by inductive generalisations. Such statements then form one of the premises of a deductive and valid argument, the source of "certainty"[3]. Scientific explanation and prediction also makes use of the notion of controlled experiment, or observation, as a means of confirmation and refutation of its hypotheses. In such a pursuit, there is a need for an inferential structure expressed in precise terms which can be provided by the appropriate branches of mathematics.

The operation of scientific explanation and prediction is thus, by its very nature, restricted to those part of the world which are amenable to its methods, ie the natural world and aspects of man-made objects, man-machine and human activity situations. Science also tends to be compartmentalised and prefers to deal with static situations, and as such appears to be ill at ease when attempts are made to extend its application to multi- and interdisciplinary situations and problems of dynamics[4].

Accordingly, there is a large field of experience of human activity situations of intense interest which is not accessible to theoretical treatment by conventional scientific explanations and predictions. Also the application of nonscientific explanation and prediction would not be acceptable at the present time. The object of this paper is to outline a qualitative approach leading to quantifiable possibilities through the use of fuzzy logic. The approach is based on the use of natural language suitably processed by linguistic analysis, as a means of

Systems Science, Edited by F.A. Stowell *et al.*,
Plenum Press, New York, 1993

representation of empirical knowledge which is used to connect observations with theoretical constructs. The analysed linguistic text is then expressed in terms of basic constituents which are seen as the building blocks of a semantic diagram. From such a diagram, an inferential, or reasoning structure can be deduced in the form of predicate logic. The appropriate terms can then be expressed as fuzzy sets and, using computing means such as Prolog, observable conclusions can be drawn regarding the possibility of the occurrence of outcomes[5].

THE NATURE OF CAUSE

Basically, but not exclusively, the subject matter of linguistic modelling is the dynamics of situations involving people[5]. In accordance with the generalised form of Newton's first law, no object can change its state unless acted on by another object in the form of interaction in terms of either physical power or influence[6]. Thus, a cause must be generated.

A cause is thought to consist of

1. Two sets of conditions, one referring to the object involved in generating the change, and the other to the object which is seen to be affected.
2. A process which is the realisation of interaction.

In linguistic terms, states can be expressed by means of adjectives or adjectival phrases, and processes are described by verbs qualified by adverbs or adverbial phrases. In the case of statements with one-place verbs, the action of the single object is subject to at least one condition[5].

The method of scientific explanation and prediction and the present approach are thought to be linked through the notion of cause as it occurs in a hypothetico-deductive argument. This method is essentially interested in a cause which appears to be necessary and sufficient for the occurrence of a change of state ie the appearance of a new property. The present approach cannot guarantee either necessity or sufficiency. The linkage is also strengthened by the recognition that an assembly, or a system, of physical objects is involved in the generation of a cause itself[6].

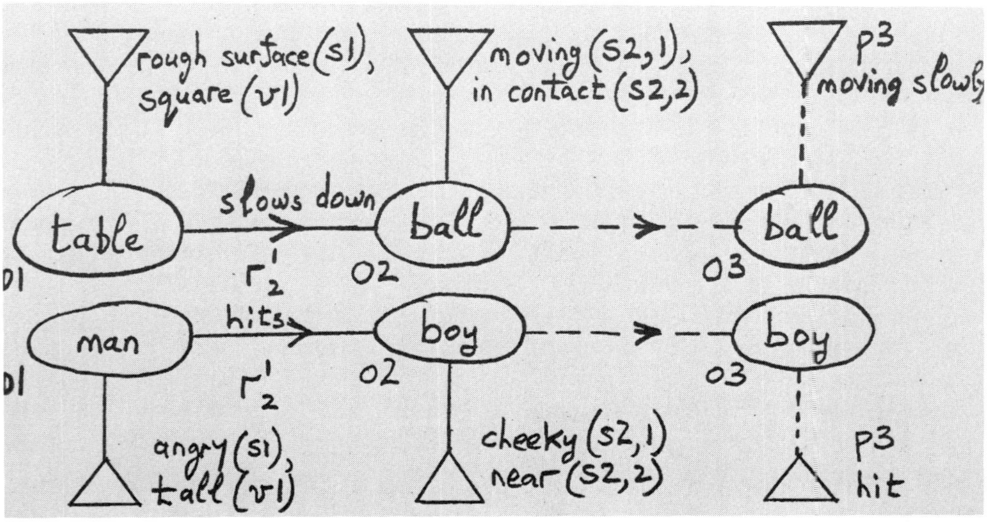

Figure 1. Diagram of a pair of two-place statements

The aforesaid is illustrated in Fig. 1 as a diagram of statements involving two-place verbs. Two statements are combined in one diagram, to show the difference between causes, one of which is necessary and sufficient for the occurrence of outcome, ie the property p3, and the other which is not. Symbolically,

$$\begin{array}{r} o^2_3 \rightarrow p3 \\ r^1_2 \rightarrow o^2_3 \\ s(2,2).s(2,1).s1 \rightarrow r^1_2 \\ v1 \end{array} \qquad 1.$$

All this can then be directly expressed in Prolog terms[5].

AN EXAMPLE

An example is used to illustrate some of the points in the application of linguistic modelling to the formation of an inferential structure involving fuzzy logic[7]. The following is a description of a dynamic, human activity situation:

"Due to an expansion in the number of small businesses, employment opportunities in a city has increased, leading to migrants of various types being attracted into the urban area. Population growth on some scale caused by the influx of migrants, tends to encourage construction of a range of accommodation, which in turn makes less land available, which results in a decline in employment opportunities. Once migrants perceive the lack of such opportunities, they move elsewhere and the population declines in the urban area".

This text is then subjected to linguistic analysis leading to the production of a series of one- and two-place statements which can be represented as a semantic diagram as shown in Fig. 2. From this diagram the following causal chains can be derived by considering the paths from the objects representing the final outcome (o15) and those which prompt further changes (o12, o14):

o15,8,7,6,5,4,3,2,1
o12,4,11,10,16,9
o14,13

Using the causal chains, a series of logical implications, similar to eq.1, are written which can then be expressed in Prolog terms. The Prolog program can be used to investigate the existence or otherwise, of outcomes as a consequence of specific disturbances.

Alternatively, in order to introduce quantification into linguistic modelling adjectives, adverbs and verbs are expressed in numerical terms in accordance with the rules of fuzzy set theory[7]. This is exemplified in Fig. 3 showing the quantisation which gives a subjective view, of the adjective "small".

With reference to Fig. 2 we use the first implication "If the businesses are small then their number expands" as an example. Quantising the verb "expands" and employing the appropriate relations for fuzzy implication we have,

$$s1 \rightarrow q^1_1 = \{(x,y,\mu(x,y,)) \mid x \in X, y \in Y\} \qquad 2.$$
$$s1 \rightarrow q^1_1$$

and

$$(x,\mu_{s1}(x)) \text{ and } (y,\mu q^1_1(y)) \qquad 3.$$

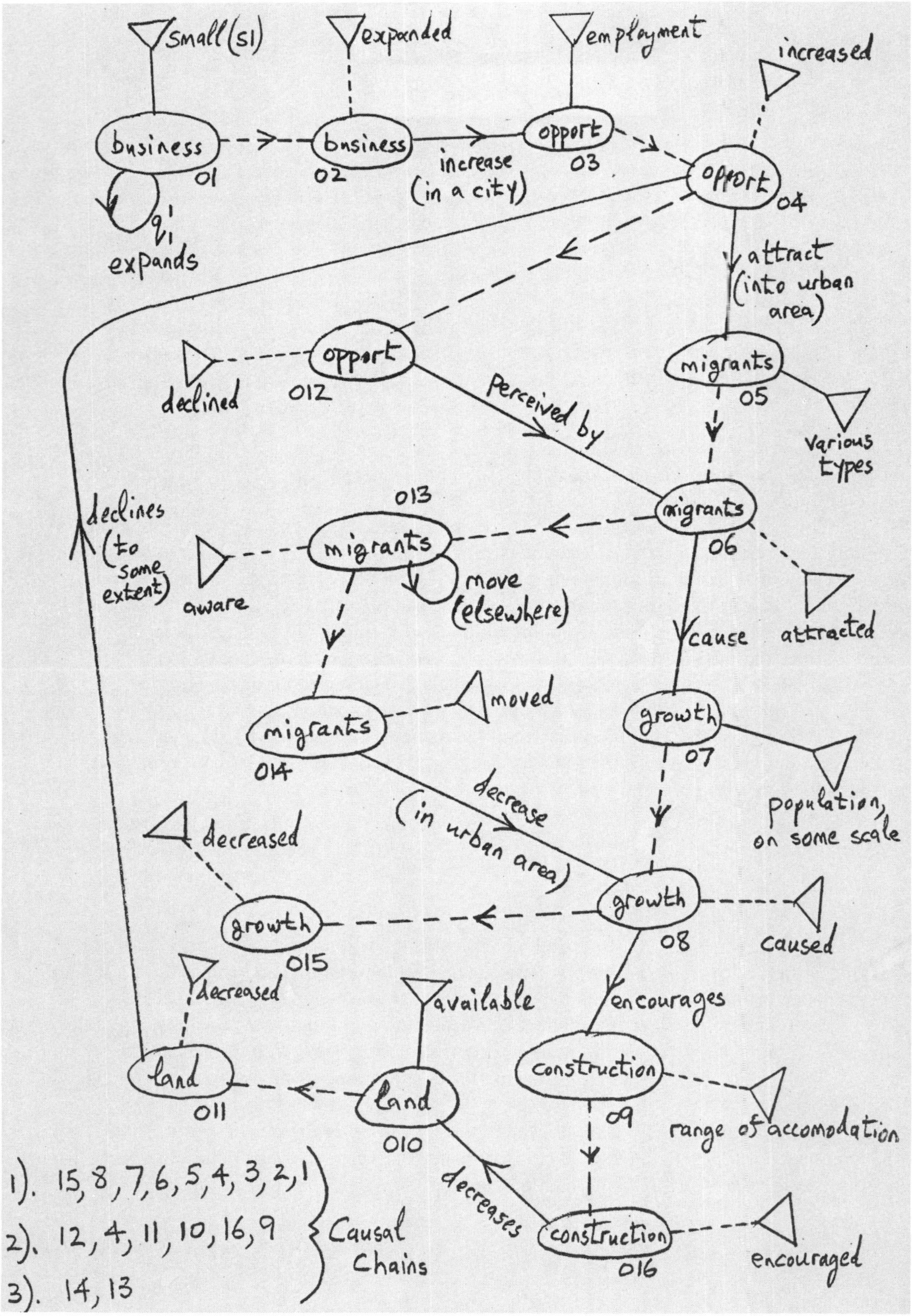

Figure 2. Semantic diagram

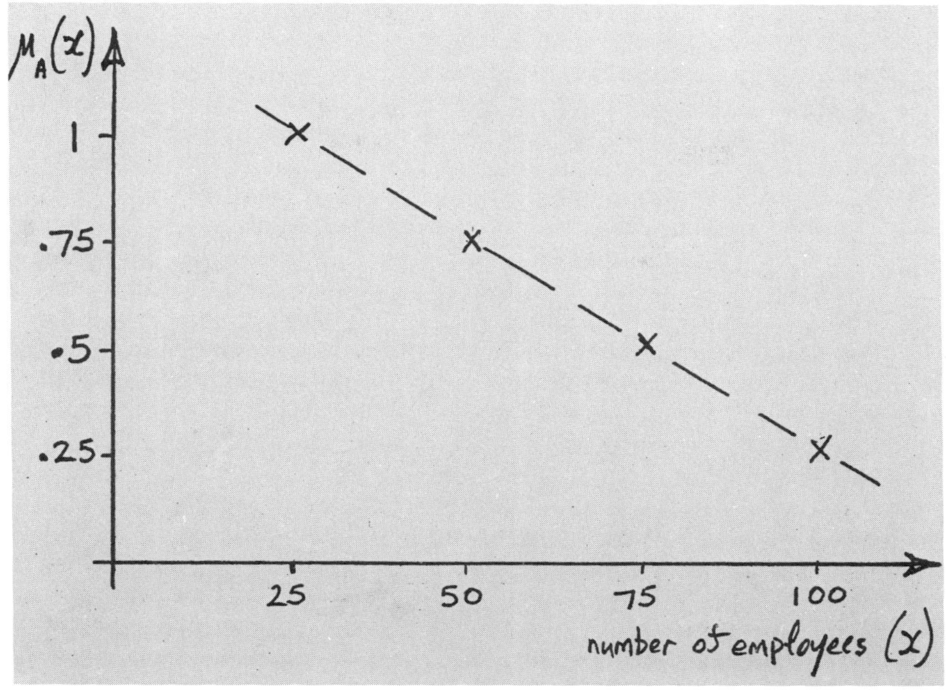

Figure 3. Quantisation of "small"

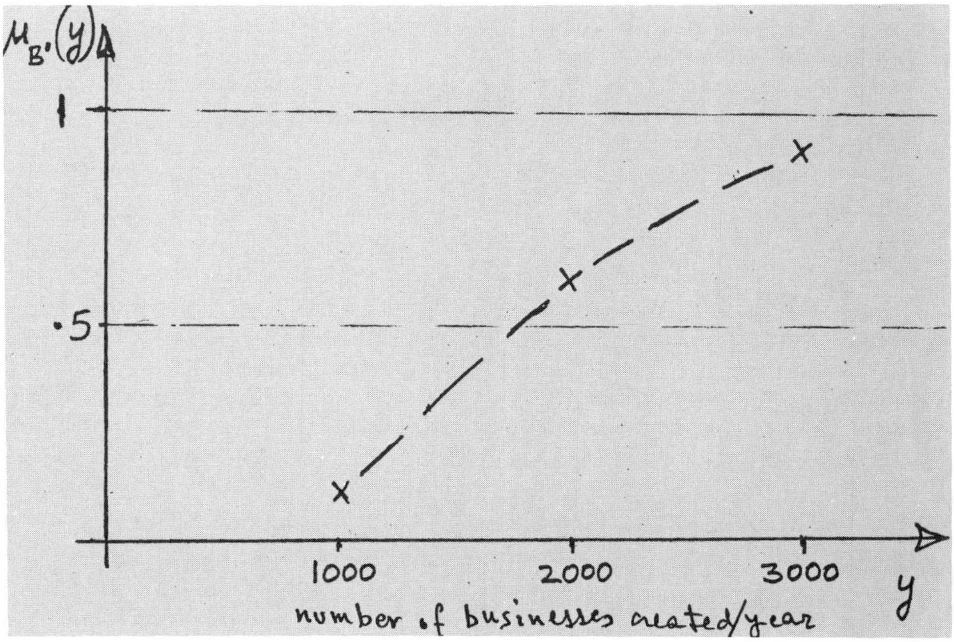

Figure 4. Possibilities for business expansion

which are fuzzy sets in X and Y. We can now work out the numerical form of the implications in eq.2

$$s1 \rightarrow q^1_1 = \begin{matrix} .2 & .6 & .9 \\ .45 & .85 & 1 \\ .7 & 1 & 1 \\ .95 & 1 & 1 \end{matrix} \qquad 4.$$

For the question: "For a given fuzzy relation as in eq.4, given there are small businesses with about 50 employees, what is the corresponding business expansion ?", the answer can be calculated as a possibility distribution depicted in Fig.4. Fig. 4 says that "If the businesses are small, with about 50 employees, then the number of businesses expands in accordance with the graph" in Fig. 4.

REFERENCES

1. Anon. "Scientific Method", Encyclopaedia Britannica, W. Benton, London, (1968).
2. Anon. "Divination", Encyclopaedia Britannica, W. Benton, London, (1968).
3. C.G. Hempel, "Aspects of Scientific Explanation", The Free Press, New York (1965).
4. J. Korn, Integrated theory of dynamics and control, *Syst. Sci.* 12, 1 (1986).
5. J. Korn, F. Huss, J. Cumbers, Linguistic modelling of situations, UKSS Conference, University of Hull, 12-15 July, pp324-333, (1988).
6. J. Korn, Systems and design as the basis of engineering knowledge, *IEE Proc.* 136, Pt.A, 2, (1989).
7. L.A. Zadeh, Outline of a new approach to the analysis of complex systems and decision processes, *IEEE Trans.* SMC-3, 1, (1973).

INTRODUCTION TO CRITICAL SYSTEMS STREAM

Since the first United Kingdom Systems Society International Conference in 1988 Critical Systems Thinking has spawned a significant number of papers and texts. This stream adds testimony to the progress that has been made in this area of systems thinking.

Critical Systems provides the problem solver with a framework through which to evaluate the impact of the chosen systems methodology upon the problem at hand. Simply stated, Critical Systems enables the practitioner to gain an awareness of the problem situation itself and of the analytical method chosen to aid in the evaluation of the problem. This is achieved through five commitments which, summarised, are concerned with the evaluation of the assumptions and values embedded in existing system design and the ramification of subsequent proposals, the emancipation of individuals through the process of their control over their environment and, finally, by making explicit the concept and practice of the systems approach that is used.

Fundamental to Critical Systems thinking and practice is the notion of complementarism which underlines the importance attatched to the tacit acceptance of other strands of systems thinking. The acceptance of other ideas, however, is not to the extent of ignoring the conceptual differences that may exist between them neither does the approach allow these differences to hinder their use. Critical Systems can be argued to provide a basis for a variety of applications. This section includes papers concerned with the political dimension and of freedom, ecological concerns and post modernism. Elsewhere Critical Systems thinking has been applied to the problems of information system development and to problem structuring. The potential for the useful application of this new strand of systems thinking is an exciting prospect for the future.

Frank Stowell
University of Paisley

Systems Science, Edited by F.A. Stowell *et al.*,
Plenum Press, New York, 1993

PRACTISING FREEDOM

Robert L. Flood

Department Management Systems and Sciences
University of Hull
Cottingham Road
HULL HU6 7RX
UK

INTRODUCTION

Arguments about who does and does not practice freedom persist in systems thinking. There are many sides. Each side is partially correct. Freedom can be *designed*, achieved through *debate*, and caused by *disemprisoning*. Each one is discussed below.

DESIGNING FREEDOM

In 1973 Beer pre-empted and smashed the arguments to be levelled by the critics of his organisational cybernetics who still fear autocratic dimensions to his work. Unfortunately Beer's 1973 masterpiece, explaining that autocracy does not have to be the case, has been left in the wilderness. Beer has made further responses but has got 'too tired' of mindless social and systems theorists to force his practical argument. But I want to bring home his unique understanding of the contribution that cybernetics can make to management and human freedom.

The message of Beer's cybernetic genius was captured for all time in the following passage from *Designing Freedom*:

> "Civilisation is being dragged down by its own inefficiency. We cannot feed the starving; we cannot stop war; we are in a terrible muddle with education, transportation, the care of the sick and old; institutions are failing, and often we feel unsafe in the streets of our own cities. All this is inefficient. *Then it cannot be the case that the only way to preserve liberty is to be so damned inefficient that freedom is not even threatened.* We have to become efficient in order to solve our problems; and we have to accept the threat to freedom that this entails - and handle it."

Beer's plea is for us to understand that cybernetics will enable human activity to become

Systems Science, Edited by F.A. Stowell *et al.*,
Plenum Press, New York, 1993

more efficient, and will help to resolve many modern day problems in society. In this it will help to win freedom. He acknowledges "...the threat to freedom that this entails." But the point is, why criticise and want to be rid of cybernetics, when by so doing we dispose of the design dimension to freedom? Succinctly, Beer is asking the critics of cybernetics, how can we be inefficient in the modern day complex world and successfully promote freedom at the same time? The structure of Beer's argument was engraved in 1973 in *Designing Freedom.* We will now review argument.

The last 150 years, or so, has seen development from relatively simple lives in small communities with a low level of communication between them; to highly integrated national and international societies run by powerful pieces of social machinery. But the question remains, does the apparatus left over from the old civilisation, that has become established in industrial, post-industrial and information-based society, actually work in these modern times?

Old variety reducing methods involve tight structure, with organisations broken down into small chunks, all controlled from a single central source. They are governed by rules. They are rigid. As the variety that they have to handle increases, they respond by increased breaking down of the organisation into smaller and smaller chunks, with more rules to keep each chunk in place. In this machine-like cybernetic structure people become cogs operating without freedom. There is no room for freedom, not even to make local relevant decisions. Mindful people must become mindless parts. Creativity is suppressed by restricting commands and extrinsic control. This type of pyramidal shape can only work if, as a manager(ess) rises up the pyramid, their heads and brains get bigger to cope with escalating amounts of variety. As this does not happen, senior people remain out of touch with events on the ground, and their decisions become increasingly bizarre. People, already acting a cog-like role, have to turn, not only against their individuality, but now against their rationality. They become exponentially less free. Further, organisations of this sort are static and thus non-adaptive, moving only toward catastrophe, and in this sense too they threaten freedom of people belonging to them because they are not viable.

The old apparatus is catastrophic for human beings today. They are destroying the institutions that enshrine everything that we hold most dear to us. They threaten the family based on love and mutual support, the cohesion of neighbourhoods, the community, churches, businesses and growth of prosperity; all exemplified and protected by the state and international communities. They need to survive, adapt and change. What actually happens is that they clam-up and rigidify. They need to know the nature of *dynamic surviving systems*. They need to know about the *science of effective organisation*. Organisations fail because they disobey the laws of effective organisation, or variety attenuation, which in general their administrators know nothing about. The loss of freedom is a consequence of a loss of control over variety attenuation.

Only variety can reduce variety. Various components of an organisation, according to Beer's organisational cybernetics, should cope with variety relevant to them. There are five main management functions set out in recursive form to do this (e.g. see *Designing Freedom*).

Now think of a human being in a viable system organisation. To have a viable organisation does mean that people receive instructions to do things, because they cannot know and understand everything that is going on everywhere else. People have to loose some potential freedom so that the whole organisation can be viable, can achieve quality and can be run efficiently. This is where freedom becomes subordinated to efficiency. The alternative that we are left with is anarchy.

We must choose between organisation or anarchy. If we wish to achieve freedom, then no choice exists. Although anarchy is the fight in principle for total individual freedom, in modern day affairs it turns out to be self-defeating. Anarchy leads to disorder, chaos, catastrophe, mass inefficiency and a huge loss of freedom. Organisation is needed, not anarchy.

A science of effective organisation is essential. This does not, Beer says, bring with it mind-blowing techniques that are already beyond most people's understanding. It is not a 'big brother' that will alienate us further from monstrous electronic machinery that governs and rules our lives.

We depend today on electronic machinery. Computers and telecommunications are vital in any organisation to help to absorb variety. Such equipment is misunderstood to be monstrous only because they contribute to the wrong side of the variety equation. If they do, they become variety amplifiers rather than absorbers, creating more variety to be managed. Such machinery can bring with it new demands that were unforeseen and hence cause rather than absorb variety. Only when electronic machines contribute to the wrong side of the variety equation are they monstrous.

With Beer we have a model of any viable system. In order to maintain viability and quality the total system must have a central regulatory model. The model ought to be created by democratic consultation but cannot dodge the truth that it will constrain variety in the parts. Some freedom has to be given up. Recursion dictates that the precise form of variety attenuation is a matter for local decision makers. We have a say in what we are prepared to give up. In short, people only need to give up as much freedom as is necessary to maintain viability of the whole. The whole returns a greater amount of freedom to people because of its efficiency, cohesion and guarantee of continuity - in short, I mean viability.

Beer's main contribution has been to explain how to design freedom. He is aware that other issues exist. Beer expressed a concern for a continuous process of liberating our minds from the programmes implanted in them. He makes reference to Vicker's *Freedom In A Rocking Boat*, where a mind-trap is described as "... only a trap for creatures which cannot solve the problems that it sets." In this, Beer sets us up nicely for the next section. As we shall see, he links designing freedom to freedom through debate.

FREEDOM THROUGH DEBATE

Vickers' book just mentioned has stamped on its front cover, "If we can decide upon our priorities we can use our new machines rather than be used by them". Vickers might as well have said "any technical system" in place of "machines". Beer and Vickers then make a good pair. Beer wants to design technical systems, whilst Vickers wants to work out priorities for them.

Priorities, of course, are not worked out in isolation with single minds processing away. Far from it. Priorities are established through communication. The necessary interaction ideally comes in the form of dialogue and debate, Vickers suggests.

Debate can lead to a wholly different kind of freedom, a freedom of the mind. This involves making escapes from trap-like preconceptions. Vickers' lobster pot metaphor is a brilliant way of illustrating mind freedom. He pictures mind-traps as if they were man-traps working like lobster pots, and says in his book,

> "Man-traps are dangerous only in relation to the limitations on what men can see and value and do. The nature of the trap is a function of the nature of the trapped. To describe either is to imply the other."

Vickers' mind-trap can be extended to illustrate 3 things. First, that whichever mind-trap we are in, in principle there is a means of escape assuming that there are insufficient forces wanting to concretise them. Second, that it does not matter how Houdini-like we are, escaping from one mind-trap leads us straight into another. We never escape from our minds. And third, we are freer in the sense that we have a memory of all the mind-traps that once trapped us, and our means of escape from them, which have enriched our minds and our ability to make decisions with greater scope. We can pan through our history of great escapes to our advantage. We are freer. But we must never forget that we are still trapped creatures, trapped in our current thoughts. Each bit of freedom we secure, in the process described above, releases the grip of the traps, but our minds are still trapped.

The most adequate means by which we are able to work out how to escape from mind-traps, Vickers argued, is through debate. Each participant holds part of the combination for each other participant's current mind-trap. Participation in debate enables people to see more clearly the horizons of those involved. Ideally, these horizons will be brought together in one rich map that brings out new possibilities for all. In general, debate offers real possibilities for participants to learn and understand.

A great range of methods urging the use of debate have been formulated. Each one aims to release our minds from traps that otherwise restrict possibilities for innovative change proposals. Works of the following authors can be consulted, Ackoff, Checkland and Mason and Mitroff.

We may ride with Beer and his design for freedom, but as he says, we may be ambushed if "the threats to freedom this entails" overrun our efforts. The viable system idea designed to win freedom can easily be used to fillet freedom from an organisation.

With Vickers we may become freer people by debating, learning and understanding, for as long as no other person holds the secret combination to the traps we are in. Unfortunately, the ability to create mind-traps and to lock us in them with virtually no way of escape, may be within the power of others. Our traps may be concretised. Debate can only be effective if it is open and conciliatory. When information is purposely distorted or other political activities enter into play, the process of freedom by debate breaks down. This method of practising freedom then turns in on itself, to have the reverse effect. It either locks people inside the traps within which they already exist, or, leads them on to some other trap to the advantage of the leader (i.e., the powerful).

Designing and debating may achieve the exact opposite to the intent built into models and methods. Intentions do not guarantee desired actions. It is possible that nothing more or less than coercion is achieved. *Designs and mind-traps can become prisons.* We therefore need another dimension to freedom, that is to disemprison.

FREEDOM BY DISEMPRISONING

When designing freedom and/or when freedom by debate fail and imprison people, it is because they are being used to serve particular people's interests. The lack of freedom is structural and conceptual respectively. People may wittingly or unwittingly use humans as instruments for their own purposes and are able to secure this because they have sufficient 'resources' to bring to bear on and to control situations. The imprisoned need to be released if we are to practice freedom.

There is a need to have practical ways of addressing whose interests are being served, to question this and if appropriate to change what is found. The key to opening the prison is to empower people at least with the knowledge that they are subject to the interests of others. This is not always self-evident. Such knowledge may help to break concretised conceptual traps that support coercive structures in society. A mode of critical reflection is required, to reflect upon the goals attained and means used through cybernetic models and upon the nature of consensus achieved through debating methods. Unlike designing and debating, however, disemprisoning is not flush with models or methods. Very little practical work has been done on disemprisoning.

Complicating matters further, other less well understood coercive forces exist. Whole civilisations operate on basic premises derived from currently accepted knowledge. For example, at the moment Western civilisation is strongly influenced by traditional science, coming to the fore in society in technology and mechanistic cybernetic management practices. This dominant knowledge base is so deeply ingrained in everyday conceptual thought that convincing people their thinking and acting is shaped by some greater social consciousness is difficult indeed. But there exists a forceful argument proposed by Michel Foucault which explains that this is the case.

Much of what people think and do and the meaningfulness of it to them, rests on a set of unquestioned assumptions about the nature of and our knowledge about the world. These are invisible except to the most critically trained or minded people. They are continually reinforced by procedures established in civilisations. Institutions are central to this. Science and modern Western civilisation are our example. Schools, colleges and universities use material that rest on scientific assumptions and are prime in installing a scientific way of thought. Libraries are full of books that accept and promote the scientific way. Journals, magazines, television and radio are all caught up in this scheme of things. In fact, all sources of information in society are dominated by the underlying wisdom offered by science. Society also has its heroes, 'the great men of science', who we are taught to strive to emulate. If we challenge the scientific way and the people of wisdom then micropolitical procedures switch to a more active method of operating. The challenger is subjugated. Their new ideas are prevented from being publicised. Articles are not published. Keynote talks are reserved. Ways of getting the ideas out are closed down. But if somehow the ideas break out then the dominant group have to respond. Tactics become a matter of exploitation of privileged and respected positions. The defence is simple but effective. The new ideas are 'shown' to have no sense in terms of the established rationality of science and are therefore 'shown' to be nonsense.

In terms of Vicker's mind-traps, we find a further degree of complexity to conceptual disemprisoning. As we have understood things so far, we can in principle escape from mind-traps using appropriate intervention methods. This leads us inevitably into another trap from which we may then escape. Freedom is achieved in the sense that we develop ever richer appreciation by continually escaping, although we remain trapped. But now we have to admit that this process is set only within some much larger and greater trap that captures the consciousness of civilisation. Individuals and organisations can escape from one trap into another and then another trap, but make no progress in escaping from the greater trap of their social consciousness. This trap is laid by micropolitical forces capturing the minds of whole civilisations. Our argument must then be that conceptual disemprisoning has to happen at all hierarchical levels in civilisations; from single minds, to organisational minds, to the mind of civilisations. It must tackle all types of political force.

Unfortunately, very little work has been done in this area of 'problem solving'. Ulrich has developed the only available method - but it remains on the breeding grounds and has matured very little over the last decade.

Even less developed as a practical tool is the element of Liberating Systems Theory (LST) that includes the idea of freeing knowledge that is suppressed by dominant 'institutions' in society. LST attempts to get to the hierarchy at the level of the consciousness of civilisation. It shows the importance of releasing suppressed knowledge to create a greater diversity of available knowledge from which we can develop and practice systems thinking. It asks, How can we practice freedom adequately when so much human knowledge with all its unleashed potential is hidden?

Much more work needs to be done on CSH and LST before we can claim to have adequate methods to disemprison. Without further substantial efforts on these two most pressing developments our hopes of practising freedom as such will remain just good ideas.

CONCLUSION

Each dimension of freedom is linked according to 3 human interests. We need to have efficient cybernetic designs so that society can operate effectively. The amount of efficiency realised from the cybernetic designs depends upon the following - developing learning and understanding about how to operate the designs, roles to be played by people according to the designs, how each role contributes and fits in to the whole design, appreciating the benefits of the whole, etc. Achieving learning and understanding requires open and free debate. When debate is subject to coercive forces a means of revealing, challenging and overcoming the forces is essential. Disemprisoning promotes genuine open debate for learning and understanding which leads to realisation of maximum efficiency from designs, which all adds up to maximum freedom. This is the complementarist ideal that we must strive for if we are serious about practising freedom.

REFERENCES (LEADING TEXTS ONLY)

Ackoff, R.L., 1974, "Redesigning the Future", Wiley, New York.
Beer, S., 1973, "Designing Freedom", Canadian Broadcasting Company, Toronto.
Checkland, P.B., 1981, "Systems Thinking, Systems Practice", Wiley, Chichester.
Churchman, C.W., 1979, "The Systems Approach and its Enemies", Basic Books, New York.
Flood, R.L., 1990, "Liberating Systems Theory", Plenum, New York.
Foucault, M., 1980, "Power/Knowledge: Selected Interviews and Other Writings, 1972-1977", C. Gordon, ed., Harvester Press, Brighton.
Mason, R.O. and Mitroff, I.I., 1981, "Challenging Strategic Planning Assumptions", Wiley, New York.
Ulrich, W., 1983, "Critical Heuristics of Social Planning", Haupt, Berne.
Vickers, G., 1970, "Freedom in a Rocking Boat", Penguin, Harmondsworth.

ECOLOGY AND EVALUATION:
THE MACRO-QUALITY PERSPECTIVE

Amanda Gregory

Centre for Systems Studies
Department of Management Systems and Sciences
University of Hull
Hull, HU6 7TQ

INTRODUCTION

This paper seeks to examine what the adoption of an ecological perspective implies for the practice of organisational evaluation. It will be argued that an ecological approach to evaluation involves judging the quality of an organisation on the basis of its contribution to society and the natural environment. Such an approach, it will be claimed, begs an holistic, rather than an analytic, mode of enquiry. This type of approach to evaluation has been termed macro-evaluation and has arisen as a consequence of the negative effects of evaluating organisations solely in terms of a narrow range of variables known as micro-quality variables. Based upon a review of the negative effects of the dominance of micro-quality based evaluations, it will be argued that initiatives to develop and promote macro-evaluation should be cultivated. This paper will seek to show that macro-quality initiatives should reject the 'scientific facade' of micro-quality based evaluation, acknowledge the essentially political nature of evaluation theory and practice, and critically appraise the issues surrounding decisions about boundaries and levels of analysis.

APPROACHES TO AN ECOLOGICAL APPROACH

The term ecology is derived from the Greek words 'oikos' meaning house and 'logos' meaning understanding. Environmentalists have been keen to promote a positivist definition of ecology as "the means by which we understand our planet" (Porritt, 1984, p3) but this is a rather narrow interpretation of ecology as it fails to recognise that the 'house' is shared by interacting participants whom effect the nature of the 'house'. A preferred definition of ecology, the one which will be adopted throughout this paper, is "the study of relationships among communities of organisms and between communities and their inanimate environment" (Allaby, 1989, pp89-90).

Systems Science, Edited by F.A. Stowell *et al.*,
Plenum Press, New York, 1993

It has been claimed that an ecological approach is not achievable in a capitalist system, nor in a socialist system if operating under the same technology as capitalism (Gorz, 1987). This has been termed the ultra-ecological approach, the deadliest enemy of the ultra-industrial approach. Whilst such extreme approaches may be theoretically pure, the major changes they imply would only result from major societal and environmental upheaval, hence the approach adopted in this paper promotes "an emerging world view, occupying the developing middle ground between the ultra-industrial and ultra-ecological perspectives" (Elkington, 1992, p13).

CONCEPT DEFINITION: MICRO-QUALITY V. MACRO-QUALITY

In a paper entitled "Proposed Changes in the Organisation of Health-Care Delivery: An Overview and Critique", Reinhardt distinguished two distinct concepts of quality: micro-quality and macro-quality. In defining the two concepts, Reinhardt drew on the subject of his study as illustration (Reinhardt, 1973, p177):

"Given the present and prospective resources of the American health-care sector, any meaningful definition of "quality care" involves an implicit trade-off between two quite distinct concepts of quality, i.e.,

1. The quality of the services delivered to those to whom such services are actually available (hereafter referred to as micro-quality);

2. The effectiveness of the health-care sector as a whole in maintaining or improving the health status of the United States population as a whole (hereafter referred to as macro-quality)."

It can be seen from the above, that in defining macro-quality Reinhardt rejected the organisation as the unit of analysis, in favour of the 'the sector', and the organisation's clients as the boundary for the evaluation, in favour of 'the population as a whole'. Whilst it will be claimed later in this paper that this relating of macro-quality variables to the sector rather than the organisation is counterproductive, the way in which Reinhardt has defined the two concepts does force explicit critical reflection upon boundary decisions and levels of analysis.

BOUNDARY DECISIONS AND LEVELS OF ANALYSIS

Ulrich (1991) reflects critically upon the process of defining boundaries, or the making of justification break-offs as he terms the process, when stating "Frequently, models of "systems" are presented as if the boundaries were objectively given, and the model itself does not tell us whether the boundaries in question have been adequately chosen .... the implicit criterion (in adopting this type of approach) is that everything that cannot be controlled or is not known falls outside the boundaries of the model, so that the model itself looks neat and scientific" (Ulrich, 1991, p106). This would seem to be the rationale behind the narrow, highly defined boundary implied by micro-quality based evaluation. Conventional micro-based evaluations may commonly be located within the 'positivist scientific paradigm' which refutes the implicit nature of decisions about boundaries and the essentially value-laden nature of evaluation theory and practice. With a macro-quality based evaluation, however, boundary decisions become problematic, as they should be, and the issue of to whom the organisation should be of value, and hence who should be involved in the evaluation, is forced. The explicit treatment of such issues in a macro-based evaluation contrasts markedly with the way such issues are dealt with in a micro-based evaluation. If macro-evaluation is to remain true to its ecological foundations it is essential that the making of boundary judgements remains problematic; once the making of boundary judgements is unquestioned then the essentially critical nature of the evaluation process is sacrificed.

Whilst Reinhardt's distinction between the two different concepts of quality is useful in that it highlights the problematic nature of boundary decisions, his relating the two concepts to different levels of analysis, the organisation and the sector, is more questionable.

Following Reinhardt, Nord justified the relation of micro-quality variables and macro-quality variables to different units of analysis on the grounds that:

"It encourages us to go beyond the criteria of micro-quality analysis and hence, to be more sensitive to outcomes that are not of concern to any one organization but which are meaningful to members of the community as a whole" (Nord, 1983, p101). Whilst Nord may relish the theoretical challenge of inventing/discovering quality variables which relate to the community as a whole, it is argued here that it is perhaps more purposeful to get organisations to realise that whilst the achievement of macro-quality may be a sector wide responsibility it is only through organisations' individual actions that it will be realised (collective responsibility is no responsibility at all). It is the failure of evaluation theorists to relate macro-variables to the organisational level which has repressed organisations recognising their responsibilities to society at large. Hence, whilst it is desirable that micro- and macro-based evaluations should relate to different boundaries, the organisation should be maintained as the unit of analysis for both types of evaluation.

THE DOMINANCE OF MICRO-QUALITY

There can be little dispute of the fact that for most organisations during the 1970s and 1980s micro-quality or, to be blunt, profit was sovereign. In the late 1980s, though, the effects of long-term irresponsible behaviour on a mass-scale became evident. The phrase 'a disaster waiting to happen' echoed across the world as one environmental disaster after another occurred. Furthermore, the societal consequences of post-industrial society, for example mass unemployment and the ubiquitous nature of marginalised groups of individuals such as 'the homeless', led many to ask whether we had 'gone too far' and the call for a return to Victorian values was often made through the mass media.

It is in response to these negative consequences of the privileging of micro-quality variables that macro-quality has developed. The 1980s saw the rise of the conscientious consumer and the growth of the ecological pressure group whose influences, it may be said, have been a major motivating factor in organisations' pursuit of macro-quality. The discerning customer will actively seek out those organisations which act responsibly, in other words those organisations which pursue macro-quality. Hence, the pursuance of macro-quality can have a knock-on effect in terms of micro-quality or profit for the organisation. Many organisations have been quick to realise that "the green consumer is simply the visible part of a green iceberg" (Elkington, 1992, p18) hence the ecological initiative, especially at a time of economic depression, could be seen as more of an opportunity than a cost.

MACRO-QUALITY: FACT OR FICTION?

Whilst a rigorous investigation has not been undertaken, preliminary evidence, for example from newspapers, advertising, etc., would seem to suggest that many organisations have grasped the ecological mettle. Whilst some organisations' motivation for the achievement of macro-quality may be suspect, more attributable to advantageous knock-on effects to profit levels than pure concern for ecological issues, it is heartening to see that ecological issues are at least on the organisational agenda.

However, it would seem that consideration of macro-quality variables is one instance in which organisational theory has lagged behind organisational practice. Scott recognised this state of affairs in stating "we analysts have emphasized structural or process measures of effectiveness - the types of measures preferred by organizational directors and participants - to the neglect of outcome measures - preferred by clients and the public. Both ideological and economic factors help to produce this bias" (Scott, 1977, pp88-89). In acknowledging the lack of attention which has been paid to the development of macro-quality based evaluation Scott's comments also highlight a commonly neglected fact, that evaluation theory, equally with evaluation practice, is the product of a political process (Wandesforde-Smith and Kerbavaz, 1988, address this issue in their appraisal of the co-evolution of politics and environmental impact assessment policy in the USA). Indeed, theorists who have concerned themselves with the development of macro-based evaluation, commonly based on critique of the value-free facade which grounds micro-evaluation, have failed to appreciate the insidious nature of the political process and its effect upon their own work. For example, in formulating a political-economic approach to evaluation Nord adopts a factor analytic approach to macro-quality and proposes that Hage's 28 quality of life variables are relevant dimensions in a macro-based evaluation (Hage, 1978). In adopting a factor analytic approach, proponents of macro-evaluation fail to make explicit the political-economic processes which have influenced the selection process of the variables and hence such an approach is not an appropriate means for the realisation of macro-quality: an approach to macro-evaluation is needed which forces debate about variables, boundary decisions and so on. The essentially emergent and holistic nature of macro-quality implies a form of evaluation which shall be referred to throughout this paper as multi-actor based evaluation. This form of evaluation parades under many guises, such as the participant satisfaction model (Keeley, 1978) and fourth generation evaluation (Guba and Lincoln, 1989).

Following Zammuto (1982) multi-actor based evaluation revolves around the notion that quality or effectiveness cannot be determined at a single point in time but emerges from the organisation's ability to satisfy interested parties' wants over time. The multi-actor form of evaluation has very much grown up in the shadow of micro-quality evaluation, hence it has been usual for a conservative approach to be taken to the definition of interested parties. With the demise in influence of micro-quality though, it might now be feasible for multi-actor evaluation to embrace the opinions of many more groups of interested party, consistent with the expanded boundary for macro-quality based evaluations, than would previously be considered appropriate. The process of defining such groups and the means for surfacing their opinions are emerging issues of concern to organisations which are worthy of theorists' attention.

It may be argued, however, that the failure of theorists to grasp the notion of macro-quality has been constructive in motivating organisations to question what the achievement of macro-quality implies for them. Indeed, it is through this process of constant questioning that organisations have come to appreciate that macro-quality variables are time and context specific and quality itself is an emergent concept. Organisations have learnt to appreciate that what macro-quality implies for them cannot be definitively specified but is revealed through a process of investigation. This pragmatic age in which practitioners have stumbled in the dark in pursuit of macro-quality without the lead of theorists will surely serve to benefit evaluation theory in the long run. It is now time for theorists to catch up with practitioners and reunite theory and practice so that one may enrich the other.

THE REALISATION OF MACRO-QUALITY

As has been said, what is meant by the achievement of macro-quality is context specific, so no attempt will be made at the definition of variables, instead examples will be given of initiatives aimed at the achievement of macro-quality. The examples will be divided into two categories, consistent with the definition of ecology adopted in this paper, those which are concerned with the organisation's responsibilities to society and those which are concerned with the organisation's responsibilities to the natural environment.

Firstly, if we turn attention to the organisation's responsibilities to society, the most obvious group of variables in this category are those which relate to the organisation's responsibilities to its employees. In recent years, the call has been to recognise that man's motivation to work is more than merely financial and to appreciate that man has responsibilities and ties outside of the work-place. It may be claimed that the organisation has a responsibility to society to enable man to fulfil responsibilities to his family and friends. Hence, in order to achieve macro-quality in terms of its employees, many organisations have introduced such initiatives as the introduction of flexi-hours and home-working systems, the provision of creche facilities, the provision of paternity leave, the facilitation of staff secondment schemes and so on.

Whilst the organisation's responsibilities to its employees and their direct dependants are quite readily accepted by most organisations, a more controversial matter is the organisation's responsibilities to those groups of individuals which most see as not falling within their sphere of direct responsibility (this may be seen as an example of a problematic boundary decision). Such marginalised groups of individuals might include school-children, the unemployed, the homeless, the handicapped and so on. Given restricted resources how do organisations decide the priority status of these groups and, hence, how to devote their efforts? In the absence of theoretical guidance on how to define the groups of individuals who might be involved in the evaluation process and on how direct access might be gained to these groups, some organisations have sought to draw on the opinions of their staff, as representatives of society as a whole, as guidance. One of the more inventive organisations has implemented an initiative in which the company matches £ for £ money raised by its staff for charity. Thus employees are being used to help define the organisation's macro-quality variables.

If we now turn attention to those macro-quality factors which relate to the organisation's responsibilities toward the natural environment, it is pleasing to report that examples of environmentally responsible action on behalf of many organisations are too numerous to mention and many such organisations are taking a leading hand in encouraging consumers to act responsibly. Some examples include the use of alternative energy sources such as solar power, recycling of packaging products, development of environmentally friendly delivery vehicles to combat noise and air pollution, etc.

CONCLUSION

In this paper the implications of an ecological approach to evaluation have been addressed. It has been argued that such an approach involves consideration of macro-quality, as opposed to micro-quality, issues and an holistic, rather than an analytic, approach. It has also been claimed that the essentially political nature of the process through which all evaluation theory is generated be explicitly acknowledged with the dropping of the scientific facade. The surrendering of scientism would mean that decisions about variables, boundaries, levels of analysis and so on would, rightly, be seen as problematic issues for discussion and debate.

Commonly held assumptions about the nature of organisations would seem to suggest that they would find such a loose and messy definition of quality as that suggested by macro-evaluation hard to accept, but the opposite would seem to be true. Superficially, at least, many organisations appear to be in hot pursuit of macro-quality. Whether or not such organisations value macro-quality for its own sake or for its beneficial knock-on effects on micro-quality is an emerging issue but does this matter? Is it not enough that organisations are doing the right thing as regards ecological issues without demanding that they be doing them for the right reasons as well?

Despite such reservations about the motives underlying organisations' efforts, suffice to say that for whatever reasons, organisations are leading the way when it comes to macro-quality based evaluation. However, the time has come for an end to this period of pragmatism. The practitioner may know what works when it comes to macro-based evaluation but this may be the result of an unnecessarily long and painful process of trial and error. Whilst the definition of macro-quality may be unique for each organisation, the means by which this state may be defined and achieved are not unique. Hence, the results of this period of pragmatic experimentation by practitioners needs to be collected and critically reflected upon so that theories can be developed and implemented on how macro-quality can best be realised. It is time for theoretically grounded methods of macro-evaluation.......then the future may look brighter and not just because of the hole in the ozone layer.

REFERENCES

Allaby, M., 1989, "Guide to Gaia," Optima, London.

Elkington, J., and Knight, P., 1992, "The Green Business Guide. How To Take Up - and Profit From - The Environmental Challenge," Victor Gollancz Ltd, London.

Gorz, A., 1987, "Ecology as Politics," Pluto Press, London.

Guba, E.G., and Lincoln, Y.S., 1989, "Fourth Generation Evaluation," Sage, London.

Hage, J., 1978, Toward a synthesis of the dialectic between historical-specific and sociological-general models of the environment, *in:* "Organization and Environment," L. Lappik, ed., Sage, London.

Keeley, M., 1978, A social justice approach to organizational evaluation, *Admin. Sci. Q.*, 23: 273-292.

Nord, W.R., 1983, A political-economic perspective on organizational effectiveness, *in* "Organizational Effectiveness. A Comparison of Multiple Models," K.S. Cameron and D.A. Whetten eds., Academic Press Inc., London.

Porritt, J., 1984, "Seeing Green. The Politics of Ecology Explained," Basil Blackwell Ltd, Oxford.

Reinhardt, U.E., 1973, Proposed changes in the organization of health-care delivery : an overview and critique, *Milbank Memorial Fund Quarterly*, 51: 169-122.

Scott, W.R., 1977, Effectiveness of organizational effectiveness studies, *in*: "New Perspectives on organizational Effectiveness," P.S. Goodman and J.M. Pennings, eds., Jossey-Bass, London.

Ulrich, W., 1990, Critical heuristics of social systems design, *in*: "Critical Systems Thinking. Directed Readings," R.L. Flood and M.C. Jackson, eds., Wiley Ltd, Chichester.

Wandesforde-Smith, G., and Kerbavaz, J., 1988, The co-evolution of politics and policy: elections, entrepreneurship and EIA in the United States, *in*: "Environmental Impact Assessment. Theory and Practice," P. Wathern, ed., Unwin Hyman, London.

Zammuto, R.F., 1982, "Assessing Organizational Effectiveness," State University of New York Press, Albany.

'NON-TRADITIONAL' LOGISTICS - A STEP FORWARD ?

Bridget Mears-Young

Department of Management Systems and Sciences
University of Hull
Hull. HU6 7RX

INTRODUCTION

This paper proposes that logistics research will not make any significant progress if logisticians continue in their limited use of research methodologies.

I intend to outline how research into logistics has evolved from a 'traditional' perspective to a 'non-traditional' perspective. Having defined what is meant by the 'traditional' view I will identify how this outlook is limited and how a new approach to logistics is necessary. This argument is not new, many authors have reached this conclusion categorising it as 'non-traditional' logistics (although such a view is not widely accepted by the practitioner community).

However, this paper then proposes that this 'non-traditional' approach appears to do little but expand the boundaries of the logistics concept. Through a wide literature search it highlights that logisticians are single-minded in their use of hard systems methodologies. Finally it proposes that the way forward for logistics research is to detract from a hard systems approach and move towards soft systems.

TRADITIONAL LOGISTICS

Since the '50s and '60s a growing number of academics have become involved in the sphere of logistics research. Their work has traditionally focused upon the implications involved in matching the performance of the supply chain to customers' demands. This is in accordance with the definition given by the Council of Logistics Management (1986) "The process of planning, implementing and controlling the efficient, cost-effective flow and storage of raw materials, in-process inventory, finished goods, and related information from point-of-origin to point-of consumption for the purpose of conforming to customer requirements.

Logistics as this operational concept stemmed from the discipline of physical distribution management. However, where physical distribution was concerned with the planning, co-ordinating and control of information and products for the outbound market,

Systems Science, Edited by F.A. Stowell *et al.*,
Plenum Press, New York, 1993

logistics has gone beyond this to include the total material flow through the entire supply chain, from raw materials to end product, from purchasing through production to warehousing to distribution to the end customer.

Thus the priority of logistics is a systems orientation focusing on information and its control in order to satisfy customer needs on a longer term basis. As such, with this traditional view the level of customer service to be attained is constrained by the company's overall strategy, and, as logistics is purely considered an operational function it is not involved at this level of decision making. Therefore, logistics managers are continually striving for efficiency within goal posts they have not been involved in setting i.e. efficiency although not necessarily effectiveness. Hence, if emphasis could be redirected away from this operational concept perhaps the full benefits of logistics could emerge; it could be seen to figure at the corporate strategy level, and help to create a competitive advantage for a firm in its market place.

NON-TRADITIONAL LOGISTICS

Hence in 1990(a), James Stock, editor of the International Journal of Logistics and Materials Management stated that although there is still a great deal to be learnt from this research paradigm, the limits of logistics growth and influence are finite if the traditional norms are followed. He had determined through his research that logistics was generally viewed by outsiders as a separate and distinct area of activity, which was inconsistent with the fact that since the sub-components of logistics are functions within themselves, logistics most certainly has implications for, and interfaces with, other disciplines.

Thus Stock (1990a), along with many others (e.g. Bowersox and Daugherty, 1987; Christopher, 1986; Lalonde, et al., 1989; Lambert and Cook, 1990; Schary, 1985; Stevens, 1989; Stock, 1990b; Witt, 1986) proposed that the way forward was an expanded view, that of 'non-traditional' logistics. Stevens (1989) advocates, that "to turn the supply chain into a weapon for gaining competitive advantage requires the development of an integrated supply chain driven by the needs of the business." Porter (1985) illustrates that logistics can have value in strategic management and strategic planning. Such an expanded perspective also incorporates the notion that logistics is a strategic ally that is closely interrelated with the other functional areas of the firm. Witt (1986) supports this view stating "everyone in a company, in marketing, development engineering, distribution and in all functional areas of manufacturing, in one way or another establishes, controls or influences the logistics process.... All of the organisation must work in an integrated team approach for total product success."

In consequence, non-traditionalists view logistics as an 'integrating concept' where by logistics not only creates costs savings and efficiencies, but also adopts a managerial philosophy that considers and incorporates the elements of corporate strategy and planning, competitive advantage and other more global concepts.

PROPOSED INTEGRATION

To date there is limited evidence of non-traditional research, and of that apparent, a great deal purely stresses the benefits of this 'new' logistics. Shapiro (1984) contends that competing on logistics performance could provide a differential advantage and hence a means to protect market share and profitability. Schary (1985) ventures that target market control could be achieved through a synchronised marketing and logistics orientation. Both Yanacek (1989), and Gattorna and Day (1986) state that including logistics at the strategic level would engender a clearer corporate mission statement; and Kenderdine and Larson (1988) are bold in their statement that logistics and quality are synonymous.

While visioning of this kind may be evident, the immediate question that springs to mind is 'How can this integration actually be achieved ?' A few authors have actually tackled this issue and they have identified four ways of achieving integration, either through changes in organisational structure, advances in information technology, reward systems or adopting customer service as a corporate philosophy; others appear to believe that an integrated strategy will just happen (e.g. Remmel, 1991)!

Organisational Structure

The major barrier to total integration of logistics activities, according to Brewer and Rosenweig (as discussed in McGinnis and Kohn, 1991) was the lack of organisational co-ordination among logistics functions. Thus, management should now be engaged in organisational analysis and design so that a smooth transition to a broader based logistics concept is ensured (Vorhees et al, 1988).

Indeed, there has been a trend recently towards centralising logistics activities (Germain and Droge, 1989) with the view to increasing the likelihood of formalisation of both a logistics mission statement and a logistics strategic plan, within the company's overall strategic plan. That is by increasing the likelihood of participatory strategic decision making, then an increase in the impact of logistical input can be expected.

Another approach is a matrix organisation, with logistics personnel brought together from functional areas in a project atmosphere, pursuing the message of efficient and effective logistics. The use of this joint task force should remove the interdepartmental communication obstacles, aiming to develop a better understanding of the logistical capabilities of the company and create mutual task dependence (Lambert and Cook, 1990).

Advances in Information Technology

Kerr (1989) believes that many strategic opportunities are being provided to logistics managers through advances in computer and telecommunication technology. Further, as the need evolves for information systems and logistics to interact in order to use information more effectively, so it is imperative that information is gleaned from other corporate areas and from external sources. Indeed, it is proposed that IT can be used to maximise the integration among logistics, marketing, manufacturing and other functional areas of the firm (Stock, 1990b).

Reward Systems

Reward systems that compensate management for combined performance can be instrumental in creating a corporate effort (Lambert and Cook, 1990). This may inhibit functional area managers from trying to maximise their function's output at the expense of total systems output. In addition it may also reduce the drive for a larger piece of the budget and more power by each functional area, which would in turn reduce conflict.

Customer Service as a Corporate Philosophy

According to Williamson et al (1990) fully integrated logistics requires management to accept customer service as a corporate philosophy.

LIMITATIONS OF NON-TRADITIONAL LOGISTICS

A statement by Muller (1989) epitomises the state of logistics' research: "Everyone shares the company's goals, but they're not sure how to proceed.. . Logistics, since it cuts across nearly all of these areas, is in a perfect position to direct traffic.'' Nobody appears to have stopped and asked 'Does everyone see logistics the way we do ?' There is still evidence to suggest that logistics is not a world wide initiative (Yanacek, 1987) and more so that its potential is definitely not appreciated (Christopher, 1982), so how can logisticians continue to take the same line of approach ? The preceding four proposed ways of integration are similarly staid.

Organisational Structure

The idea of centralising logistics activities does nothing to increase logistics awareness rather it purely enforces the concept of integration on the rest of the firm. While such autocracy may have some positive impact, the full benefits of this proposed new logistics will not be realised.

Similarly, the suggestion of a joint task force, while removing interdepartmental obstacles does not immediately mean that all parties involved are necessarily speaking the same language; logistics is still being enforced.

Advances in Information Technology

The concept of information flow is imperative for integrated logistics, and if systems are in place to support this then integration will be easier. However, before these systems can be implemented each function needs to determine what information needs to be exchanged, and the barriers to communication that exist within companies need to be broken down. Therefore, while information systems might accelerate integration they cannot initiate it.

Reward Systems

Reward systems are no different, all they do is bribe managers to accept the concept of integrated logistics. Perhaps this method increases logistics visibility but without a shared objective this form of logistics is equivalent to turning the tables, purely constraining other functions to be efficient although not necessarily effective.

Customer Service as a Corporate Philosophy

The tendency of all these integration policies seems to be to focus too much on the outcome of logistics (i.e. customer satisfaction), and not enough on the stakeholders involved. How can a company expect all functions to accept customer service as the corporate philosophy without identifying what exactly is involved for each function, how it will affect their actions, and if they actually believe it to be a positive step forward? This method is tantamount to believing that an integrated strategy will just happen.

NON-TRADITIONAL LOGISTICS DOES NOT GO FAR ENOUGH

Williamson et al (1990) stated that people concerned with logistics management are more interested in the level of integration than they are in their perceptions of the essence of the process. Indeed, while Stock (and others) may have identified the weaknesses in 'traditional' logistics, this proposed 'non-traditional' logistics appears to do little but expand the boundaries of the logistics concept. While they recognise that one of the main aims of logistics management must be to help remove the intra-functional drive towards protectionism by looking at the supply chain as a single entity they continue to tackle logistics with what following Checkland (1981) might be regarded as a 'hard systems approach', attempting to solve the 'problem' of logistics as they see it; a system of interacting parts. However, since logistics evidently interfaces with so many other functions each player may not necessarily perceive the 'problem' to be the same. Different participants may well have different views as to what logistics has to offer them and in turn what they have to offer logistics.

Hence, the approach taken so far ignores all these issues of subjectivity; issues that cannot necessarily be modelled. Moreover, this is substantiated by the fact that although logisticians may view logistics as a " 'boundary-spanning' activity" (Middleton, 1992), it is still viewed by outsiders (defined as those not directly involved in logistics) as a separate distinct area of activity. Consequently, worries about the field being in danger of becoming very insular (Stock, 1990a) may well hold true if researchers continue with this hard systems approach.

NEW DIRECTIONS IN LOGISTICS RESEARCH

Logistics is like a machine that has been developed but remains incompatible. If the relationship between logistics and other functions were investigated, problems encountered could be understood and appreciated. Only once this happens, and given the correct attitude, will a total integrated logistics be feasible.

For successful logistics it appears imperative to understand the points of view and intentions of all those involved. Participants in logistics need to be encouraged to look beyond the boundaries of their own function towards establishing mutual ground on which inter-functional initiatives may be based. Hence, cross-functional barriers need to be broken down and a common logistics' culture instilled. In short the creation of a new attitude of mind is required. However, hard systems methodologies require concrete problem definition and are not capable of taking into account elements of subjectivity . Since we are faced with a situation where by it is impossible to define desirable ends or objectives, then to develop new directions in logistics the methodology needs to be adjusted (Checkland, 1978).

Soft systems are process driven rather than expert driven as hard systems are, thus they work on values and perceptions aiming for all parties involved to pull in the same direction. Through soft systems methodology, multiple perceptions of logistics may be taken into account therefore incorporating the views of all members of the supply chain. Although individuals' perceptions, "worldviews" or "weltanschauung" (according to soft systems terminology, (Checkland, 1981)) may be decidedly restricted, they are usually highly resistant to change (Jackson, 1991). Hence they will not be seriously challenged just by stating the benefits of integrated logistics, changing the organisational structure, or implementing new information or reward systems, as each in turn will simply be interpreted according to their fixed preconceived ideas. However, these different perceptions could be used as a platform for debate in an attempt to bring about a more objective appreciation of logistics, creating the mutual ground necessary to encourage integrated logistics. That is, to develop logistics on the basis of an attitude of mind created by systems thinking rather than a system of interacting parts.

CONCLUSION

There is a need to radically reorientate thinking within logistics research ! Complex logistics activities in complex settings with an increasing array of interfaces suggest that univariate/unidiscipline approaches to logistics research will continually fail to grasp the relevant issues and insights needed to make progress in logistics research. A soft systems approach to logistics could provide the next step forward.

REFERENCES

Bowersox, D.J., and Daugherty, P.J., 1987, Emerging patterns of logistical organisation, *Jrnl BusLog*, 1 :46-60.

Checkland, P.B., 1978, The origins and nature of 'hard' systems thinking, *Jrnl App.Sys.Think*, 2:99-110

Checkland, P.B., 1981, "Systems Thinking, Systems Practice", John Wiley & Sons Ltd, Chichester.

Christopher, M., 1982, Remorseless logic in logistics, *Hand & Ship Mgmnt*, Nov:3-6.

Christopher, M., 1986, Implementing logistics strategy, *Mgmnt Today*, 1:52-62.

Council of Logistics Management, 1986, "Council of Logistics Management Yearbook", Council of Logistics Management, Oak Brook.

Gattorna, J., and Day, A., 1986, Strategic issues in quality, *IJPD&LM*, 2:3-42.

Germain, R., and Droge, C., 1989, The impact of the centralised structuring of logistics activities on span of control, formalisation and performance, *Jrnl Acad Mktg Sci*, 1:83-89.

Jackson, M.C., 1991, "Systems Methodology for the Management Sciences", Plenum Press, New York.

Kenderdine, J.M., and Larson, P.D., 1988, Quality and logistics: a framework for strategic integration, *IJPD&LM*, 6:5-10.

Kerr, A., 1989, Information technology - creating strategic opportunities for logistics, *IJPD&LM*, 5:15-17.

Lalonde, B.J., Ellram, L.M., and Weber, M.M., 1989, Retail logistics, *IJPD&LM*, 10:29-39

Lambert, D.M., and Cook, R.L., 1990, Integrating marketing and logistics for incresed profit, *Bus*, 3:22-29

McGinnis, M.A., and Kohn, J.W., 1990, A factor analytic study of logistics strategy, *Jrnl BusLog*, 2:41-64.

Middleton, A., 1992, Logistics: yesterday's horizons are today's reality, *Focus*, 5:15-16.

Muller, E.J., 1989, Linking logistics to corporate goals: the corporate crossroads, *Chiltons's Distn*, 9:2834.

Porter, M.E., 1985, "Competitive Advantage: Creating and Sustaining Superior Performance," The Free Press, New York.

Remmel, U.M., 1991, Integration of marketing and logistics: a way to competitive advantage in South Africa, *IJPD&LM*, 5:27-31 .

Schary, P.B., 1985, A strategic problem in logistics control, *IJPD&LM*, 5:36-50.

Shapiro, R.D., 1984, Get leverage from logistics, *HBR*, May-June: 119-126.

Stevens, G.C., 1989. Integrating the supply chain, *IJPD&LM*, 8:3-8.

Stock, J.R., 1990a, Logistics thought and practice: a perspective, *IJPD&LM*, 2:3-6.

Stock, J.R., 1990b, Managing computer communication and information technology strategically: opportunities and challenges for warehousing, *Log & Trans Rev*, 2:133-148.

Vorhees, R.D., Teas, R.K., Allen, B.J., and Dinkler, E.T., 1988, Changes in the marketing-logistics relationship, *Jrnl BusLog*, 1:34-50.

Williamson, K.C., Spitzer, D.M., and Bloomberg, D.J., 1990, Modern logistics systems: theory and practice, *Jrnl. Bus.Log*, 2:65-85.

Witt, P.R., 1986, "Cost Competitive Products," Reston Publishing Company, Reston.

Yanacek, F., 1987, The logic behind logistics, *Hand&Ship Mgmnt*, 8:30-34.

CRITICAL SYSTEMS THINKING, POST-MODERNISM AND THE PHILOSOPHY OF RICHARD RORTY

M.W.J. Spaul

Department of Technology
Anglia Polytechnic University

INTRODUCTION

Jackson (1991a, 1991b) has identified and elaborated a view of the post-modern challenge to Critical Systems Thinking (hereafter CST). He concludes that the post-modern world-view offers little support to any major strand of systems thought. This paper examines one particular 'post-modern' stance - that of the philosopher Richard Rorty - and considers the possibility of a compromise with CST. The linkages between the five commitments of CST (Jackson 1991c) are considered in the light of this examination. A tentative framework for the investigation of post-modern systems thinking is also considered.

THE POST-MODERN CHALLENGE

Jackson considers, broadly, three facets of post-modernism which constitute a direct challenge to CST and its basis in critical modernism. The first facet, expressed by Lyotard (1984), is the characteristic post-modern incredulity towards meta-narratives; in particular, the enlightenment meta-narrative which underwrites logic, rationality and order as a permanent basis for theory. The hope of making all discourses commensurable within some embracing scheme is abandoned, with knowledge fractured into incommensurable, localised 'language games'. The second facet, which may be regarded as the first projected into a social context, is the loss of belief in progress and the ability of human agency to secure social improvement. This is expressed vividly by the despairing vision of Baudrillard (1984), that in the post-modern world we are sitting amongst the ruins of past cultures and projects, playing with the pieces. The third facet is that of the retreat from the seriousness of high modernism in the arts and media (Harvey 1989, Part I), and its replacement by a playful eclecticism and the primacy of the superficial image.

These three facets, taken together, are clearly inimical to the aspirations underlying systems thinking, and CST in particular, with its spirit of progressive optimism. Seen in this light, post-modernism appears as a trend to be resisted by the systems movement - but such resistance may take different forms. Habermas has adopted a position of aggressive opposition to post-modernism (Habermas 1980, 1989) upholding the values of the 'unfinished project of the enlightenment'; a Habermasian reading of post-modernism regards it as a means of legitimating neo-conservatism. An alternative strategy is to decouple the three facets above, and attempt to forge a recognisably post-modern position which preserves the key values and aspirations of critical modernism and CST. Such a strategy is made possible by the absence of a strong connection between the different strands of post-modernism, since it is simultaneously a theoretical movement and complex cultural phenomenon (Featherstone 1988). As will be shown below, the loss of faith in the enlightenment meta-narrative need not lead to a loss of faith in progress, or the devaluation of a serious, scholarly approach to human problems and needs.

One approach to the decoupling and accommodation of aspects of post-modern thought is exemplified by Flood (1990, Ch.3). Flood absorbs the genealogical method of Foucault (see, e.g., Foucault 1970) within the framework of CST, using it to liberate suppressed knowledges. An alternative form of this strategy, to be elaborated here, is to give an interpretation of the commitments of CST in terms compatible with a post-modern world-view in such a way that their practical consequences (as motivations and guides for improved systems practice) are preserved. The pragmatic philosophy of Rorty offers a promising candidate for such a world view, a philosophy which combines a post-modern conception of knowledge and enquiry with a strong commitment to liberalism and social progress. It should be noted that Rorty's combination of post-modernism and progressive optimism are not unique: Taylor (1985) has constructed a post-modern view of the social sciences with a view to improving practice; Saul (1992) argues strongly that a rejection of enlightenment rationalism is a pre-requisite of some forms of social progress. However, Rorty is a useful figure to consider in the train of argument initiated by Jackson, since his avowed intention is to 'split the difference between Lyotard and Habermas' (Rorty 1991b, Part III), who are the central figures in Jackson's discussion. An extended examination of other aspects of the debate between modernism and post-modernism, as it impinges on modern systems thinking, is clearly required; but this will not be attempted here.

RORTY AND POST-MODERNISM

Rorty works within a philosophical tradition rooted in the pragmatism of Dewey and James, adding elements of European traditions, notably that of the hermeneutics of Heidegger (1962) and Gadamer (1975), the genealogy of Foucault and the deconstructionism of Derrida (see Norris 1987). A characteristic of his thought - its thoroughgoing pragmatism - is that he parts company with any tradition which adopts a 'metaphysical' turn, be it philosophy in the Descartes-Kant mould, or the more obscure aspects of Heidegger or Derrida. A synoptic introduction to (and critique of) his thought may be found in Malachowski (1990).

From this pragmatic viewpoint Rorty has launched a post-modern assault on analytical philosophy and the ontological and epistemological problematics which motivate it. Most notably, Rorty (1980) carries out a large-scale deconstruction of the epistemological problematic derived from Descartes and Kant, and suggests its replacement by a distinctive version of hermeneutics which he labels 'edifying discourse'. More recent works (e.g. Rorty 1989) elaborate this central theme, re-casting traditional philosophical problems of language, ethics and the self as 'edifying discourse' with other areas of culture. Rorty's undermining of the claims of philosophy to be the court of appeal in which the claims of other disciplines are adjudicated is of especial relevance to the ongoing quest in the systems community for an adequate epistemology and ontology for systems practice (see, e.g., Flood and Ulrich 1990, Fuenmayor 1991).

Rorty has an ambivalent status as a post-modernist; he is labelled as such by several commentators (Cooper and Burrell 1988, Best and Kellner 1991, Jameson 1991), but resists the epithet (see Rorty 1991b, Introduction), finding the periodisation of culture implicit in the term unhelpful and implausible. His anti-foundationalism is closely allied to Lyotard's 'incredulity towards meta-narratives'; but it is cautiously expressed using a strategy designed to make the vocabulary of enlightenment rationalism look unproductive, rather than constructing a full-scale theoretical alternative (Rorty 1980, Part II). He thus escapes the charge levelled at Lyotard (Kellner 1988) that a theory of post-modernity is just one more meta-narrative. Rorty is post-modern to the extent that he believes that first-order narratives (and dialogues between them) are all one needs, or can have, in the pursuit of knowledge.

Despite the post-modern alliances sketched above, Rorty is strongly antipathetic towards aspects of French post-modernism. In particular, its tendency to theorise about lives and cultures from an antiseptic distance collides with his Deweyan social concern and New England liberalism. For Rorty, a sympathetic identification with the oppressed and under-privileged is a pre-requisite of any concerned social action; and social action (as for Heidegger) is, on his account, prior to social theory (Rorty 1991a Part III, 1991b Part I). In this respect, his position approaches that of the Frankfurt School, in practice if not theory; a point conceded by Habermas (Dews 1986, Ch.6).

THE CRITICAL COMMITMENTS EXAMINED

The five commitments which underlie CST have been set out in various sources; the account in Jackson (1991c) is used as the basis for this discussion. Principal consideration will be given to the commitments to theoretical and practical complementarism, and to emancipation; the commitments to critical and social awareness will be discussed briefly.

Various linkages exist amongst the commitments of CST; for the purposes of a post-modern critique the most important is the foundational linkage between the commitment to complementarism at the level of theory - construed as an adherence to Habermas' (1972) interest constitution theory - and the commitments to practical complementarism and emancipation. As Jackson (1991a) points out, the post-modern assault on the tenets of critical modernism bears upon these commitments via their Habermasian basis.

One form of this post-modern assault is set out in Rorty (1980, Ch.8). Habermas presents his interest constitution theory as the result of a reflection upon the logic of enquiry in the natural and cultural sciences, with the results having a 'quasi-transcendental' status, a universal validity. For Rorty, this is an illegitimate move: an attempt to portray the creation of our discourses in the same way as we portray issues within our discourses, to describe ourselves and our activities in some way which is prior to that of science or cultural anthropology. He urges that there is no ahistorical or neutral standpoint to which we can retreat to describe ourselves, and hence no way of 'grounding' interest constitution theory. All attempts to provide 'foundations' for knowledge (a perennial theme of Western philosophy) will suffer the same fate, since they will be cast in some discourse - a contingent instrument which history has delivered into our hands (Rorty 1989, Ch.1).

Within Rorty's framework, however, there is another sense which we can give to 'theoretical complementarism' - and one which is capable of motivating and informing practical complementarism. For Lyotard, a language game - a theoretical discourse, say - is isolated and incommensurable with other discourses; and there is no universal scheme under which discourses may be gathered. Rorty concedes the second point, but not the first: cultural history abounds with examples of the clash of supposedly incommensurable discourses which resulted in peaceful and fruitful accommodation; not by repairing to a universal framework, but by the contingent messiness of free and open encounters (Rorty 1980, Ch.7, Rorty 1991a, Part III). Rorty's pragmatism embraces not only the simple formula of 'what works in practice' (see the pragmatic alternative to complementarism in Jackson 1991b, Ch.10) but also 'what works in theoretical dialogue cleansed of metaphysical pretensions'. Rorty's vision of such 'post-philosophical' dialogue is related to that of current cultural anthropology (Rorty 1982, Introduction). Recent writings in interpretivist (Geertz 1983) and post-modernist (Clifford and Marcus 1986) cultural anthropology are characterised by 'theoretical' discussion being conducted in terms of a densely-connected web of experience, examples, historically-located theories and literary allusion. Under such an interpretation of 'theory' there can be no clear distinction between practical and theoretical complementarism, since the concept of a realm of pure theory is abandoned. A possible instance of such an encounter in the systems sphere is the current absorption of soft systems thinking by the information systems community, with its traditionally hard view of systems development. This encounter has not generally been characterised by a philosophical retreat to a theoretical framework in which both discourses may be accommodated, but rather by a dialogue - informed by experience - in which both discourses have undergone subtle change (see Holwell 1992, Prior 1992, and other papers in the same volume).

An important feature of Habermas' interest constitution theory is its attempt to establish emancipation - freedom from domination - as a universal human interest, based on the idea that it is a universal presupposition of human communication (Habermas 1979 Ch.1). The post-modern arguments against interest constitution theory sketched above cast doubt on the legitimacy of some universal sense of 'emancipation'. Rorty has written extensively on this issue (Rorty 1989, Ch. 3, Rorty 1991a, Part III), being concerned to establish that we do not need a transcendental (or quasi-transcendental) foundation for emancipation in order to be fully committed to it. His argument, patterned on his general adherence to first-order discourse, is that to have been brought up in a liberal democracy provides all the commitment to emancipation which one needs - the contingent commitment to a culture group of which one is a part. He urges the rejection of Habermas' rationalist means of ensuring that we are not deluded, happy slaves, and that we restrict ourselves to the first-order political discourse of our culture. Rorty bases his arguments partly on those of Berlin (1969) that to realise the relative validity of one's commitments and yet stand for them unflinchingly is a mark of political and moral maturity. A further point made by Berlin, to the effect that there is no basis for the common conviction that all the positive values in which men have believed must be compatible with each other, casts doubt upon even the possibility of a universal sense of emancipation.

The commitment to emancipation characteristic of CST need not be weakened by exchanging a critical modernist view of freedom and human nature for a post-modern one - commitment to an emancipatory culture is all that is required to motivate practice. Theoretical ideals, such as Habermas' ideal of free encounter between individuals exhibiting communicative competence (Habermas 1984, 1987), provide little in the way of guidelines for exposing domination in practice. Emancipatory practice is based rather on an open-ended set of contrasts and examples familiar within our culture: Socratic dialogue or parliamentary debate versus hypnotic suggestion, brainwashing, media hype, economic blackmail in trade disputes and 'false consciousness' (Rorty 1989, Ch.3). Examples of activities which, within our liberal culture, constitute coercion; but not activities which can be characterised as transgressing against the ahistorical 'unfettered human nature' of the enlightenment.

CST's commitment to critical and social awareness hangs naturally with Rorty's post-modern view of knowledge. All theories, all discourses - under this conception - are simply instruments constructed for human purposes, historically and culturally located (Rorty 1989, Ch.l). The idea that any such theory or discourse can be legitimated or secured for all time evaporates with the realisation that there can be no retreat to a neutral standpoint from which theories can be judged. With that realisation comes the requirement to continually reassess theoretical instruments against the needs and purposes of our present historical and cultural location - which is exactly the attitude enshrined in the commitments of CST. These remarks on the general aspect of critical and social awareness (the questioning of theoretical instruments) carries over directly to the specific aspect: that of questioning particular applications of our theoretical instruments. Post-modernism emphasises the complex and diverse micro-structures of cultures, and rejects 'totalisation' in the social sphere. Such an attitude motivates the careful justification of every application of methodologies, theories and value-schemes.

POST-MODERN SYSTEMS THINKING?

There is a clear sense in which post-modernism and CST are incompatible: a post-modern standpoint must clash with the tenets of critical modernism. However, the values which underlie CST need not be lost with the adoption of a post-modern view of knowledge. In particular, practical values and motives - a desire for progress and emancipation, a reflective and informed attitude to practice - may be retained, and even strengthened. Cynicism and despair need not be viewed as a necessary characteristic of post-modernism; such an attitude may be seen as one of the cultural themes of the 1980's but not as a concomitant of the retreat from enlightenment rationalism. Many post-modern works have a euphoric and celebratory quality (Harvey 1989, Part I).

The post-modern loss of Habermas' interest constitution theory does, however, mount an important challenge to CST as a systematic framework in which the diverse strands of systems thinking may be organised. Habermas' framework underwrites the systematic separation of methodologies appropriate to each of the three human interests; but post-modernism is eclectic and unprincipled in its adoption of theoretical instruments, an eclecticism being felt throughout the human sciences (Marcus and Fisher 1986). For example, Rorty considers the social novel and the ethnography to be key instruments of social analysis and progress, citing Kundera's (1988) manifesto that the novel is uniquely qualified to portray the essential relativity of human problems. The pressure to absorb such diverse instruments is being felt throughout the systems movement; in addition to Flood's absorption of Foucault's 'genealogy' there is the 'stream of cultural analysis' (with the complex web of anthropological issues which that exposes) which appears in later versions of SSM (Checkland and Scholes 1990).

The principal challenge of post-modernism may perhaps be the problem of characterising 'the systems approach' in the face of the explosive growth in complexity of the theoretical framework of the human sciences. Post-modernism need not compromise the values which underlie CST; but it does compromise its ordered theoretical framework, and the set of assumptions which motivate the construction of such a framework.

REFERENCES

Baudrillard, J., 1984, On nihilism, *On the Beach* 6:38-39.
Berlin, I., 1969, "Four Essays on Liberty," Oxford University Press, Oxford.
Best, S. and Kellner, D., 1991, "Postmodern Theory: Critical Interrogations", Macmillan, London.
Checkland, P. and Scholes, J., 1990, "Soft Systems Methodology in Action," John Wiley, Chichester.
Clifford, J. and Marcus, G., 1986, "Writing Culture: The Poetics and Politics of Ethnography," University of California Press, Berkeley.
Cooper, R. and Burrell, G., 1988, Modernism, postmodernism and organizational analysis: an introduction, *Org. Stud.* 9:1, pp91-112.
Dews, P., 1986, "Autonomy and Solidarity: Interviews with Jurgen Habermas," Verso, London.
Featherstone, M., 1988, In pursuit of the postmodern: an introduction, *Th. Cult. and Soc.*, 5:2, pp195-215.
Flood, R.L., 1990, "Liberating Systems Theory,", Plenum, New York.
Flood, R.L. and Ulrich, W., 1990, Testament to conversations on critical systems thinking between two systems practitioners, *Sys. Prac.* 3:1, pp7-29.

Fuenmayor, R., 1991, Truth and openness: an epistemology for interpretive systemology, *Sys. Prac.* 4:5, pp473-490.
Foucault, M., 1970, "The Order of Things: An Archaeology of the Human Sciences," Tavistock, London.
Gadamer, H-G, 1975, "Truth and Method," Sheed and Ward, London.
Geertz, C., 1983, "Local Knowledge: Further Essays in Interpretive Anthropology,", Basic Books, New York.
Habermas, J., 1972, "Knowledge and Human Interests," Heinemann, London.
Habermas, J., 1976, "Communication and the Evolution of Society," Heinemann, London.
Habermas, J., 1980, Modernity versus postmodernity, *New German Critique 22*, reprinted in "Interpretive Social Science: A Second Look", 1987, P. Rabinow and W.M. Sullivan, ed., University of California Press, Berkeley.
Habermas, J., 1984, "The Theory of Communicative Action, Vol. 1," Heinemann, London.
Habermas, J., 1987, "The Theory of Communicative Action, Vol. 2", Polity Press, Cambridge.
Harvey, D., 1989, "The Condition of Postmodernity," Blackwell, Oxford.
Heidegger, M., 1962, "Being and Time," Blackwell, Oxford.
Holwell, S., 1992, SSM information approach: a critique, *Systemist* 14:3, pp93-98.
Jackson, M.C., 1991a, Modernism, Post-Modernism and Contemporary Systems Thinking, in: "Critical Systems Thinking: Directed Readings," R.L. Flood and M.C. Jackson, eds., John Wiley, London.
Jackson, M.C., 1991b, "Systems Methodology for the Management Sciences," Plenum, New York.
Jackson, M.C., 1991c, Five Commitments of Critical Systems Thinking, in "Systems Thinking in Europe," M.C. Jackson et al, eds., Plenum, New York.
Jameson, F., 1991, "Postmodernism: The Cultural Logic of Late Capitalism," Verso, London.
Kellner, D., 1988, Postmodernism as social theory: some challenges and problems, *Th. Cult. and Soc.*, 5:2, pp239-269.
Kundera, M., 1988, "The Art of the Novel," Faber and Faber, London.
Lyotard, J-F, 1984, "The Post-Modern Condition: A Report on Knowledge," Manchester University Press, Manchester.
Malachowski, A., 1990, "Reading Rorty", Blackwell, Oxford.
Marcus, G. and Fisher, M., 1986, "Anthropology as Cultural Critique: An Experimental Moment in the Human Sciences," University of Chicago Press, Chicago.
Norris, C., 1987, "Derrida," Fontana, London.
Prior, R., 1992, Linking SSM and IS development, *Systemist* 14:3, pp128-132.
Rorty, R., 1980, "Philosophy and the Mirror of Nature," Blackwell, Oxford.
Rorty, R., 1982, "The Consequences of Pragmatism," Harvester, Hemel Hempstead.
Rorty, R., 1989, "Contingency, Irony and Solidarity," Cambridge University Press, Cambridge.
Rorty, R., 1991a, "Objectivity, Relativism and Truth: Philosophical Papers Vol. 1," Cambridge University Press, Cambridge.
Rorty, R., 1991b, "Essays on Heidegger and Others: Philosophical Papers Vol. 2," Cambridge University Press, Cambridge.
Saul, J.R., 1992, "Voltaire's Bastards: The Dictatorship of Reason in the West," Sinclair Stevenson, London.
Taylor, C., 1985, "Philosophy and the Human Sciences: Philosophical Papers Vol. 2," University of Cambridge Press, Cambridge.

INTRODUCTION TO CYBERNETICS STREAM

According to the subtitle of Wiener's seminal book, *Cybernetics*, the term refers to "control and communication in the animal and the machine." Although, nearly half a century later, the word is not so widely used in these disciplines, the importance of the discipline of cybernetics for computing scientists and biologists cannot be questioned. As almost all the papers in the present stream illustrate, however, the range of systems to which control and communication theory can be applied are not limited to organisms and machines, and the focus here is on human organizations.

Perhaps the single largest influence on such application has been the Viable Systems Model of Stafford Beer. Here we will see proposals to use VSM in the management of companies, quality , information systems, and the performance of universities. The idea of combining VSM with soft systems approaches is explored in more than one paper, and will doubtless contribute to the debate (particularly felt in information systems) of how best to move from the present situation to the desired one that the use of SSM has identified.

Cybernetics is concerned with general formal models of communication and control. Although quantitative models are important in the control theory used by engineers and in biochemical systems, much of cybernetics is qualitative. Certainly this is the case for applications to human activity systems! In contrast, studies of dynamical system have generally involved quantitative models. The possibility of a qualitative system dynamics that can be used in management situations is an exciting one, and it seems appropriate to include Professor Wolstenholme's discussion of this possibility in the present stream. An example of a qualitative non-linear investigation of the risk of flashover in fire assessment concludes the collection.

Richard B. Beeby
University of Paisley

Systems Science, Edited by F.A. Stowell *et al.*,
Plenum Press, New York, 1993

NONLINEAR DYNAMICS AND FIRE RISK ASSESSMENT

Alan Beard[1], Dougal Drysdale[1], Paul Holborn[2] and Steven Bishop[2]

[1]Unit of Fire Safety Engineering
University of Edinburgh
Edinburgh EH9 3JL

[2]Centre for Nonlinear Dynamics and its Applications
University College
London WClE 6BT

ABSTRACT

As part of decision-making with regard to fire risk, mathematical models of fire growth are starting to be developed and used. In order to further this effort, work is under way to try to understand the nonlinear behaviour of fire in a compartment and the instabilities which may exist. In particular, the conditions for the existence of a mathematical jump may be interpreted as a 'flashover'; a potentially disastrous event in which a small, localized, fire comes to involve the whole of a compartment.

INTRODUCTION

Decision making with regard to fire risk is a complex matter which, in general, involves a consideration of many different interacting systems. This basic complexity has been discussed already to some extent, together with the role which mathematical models might play in understanding features of the entire system [1]. In particular, the part played by theoretical models in understanding fire growth in a compartment has been raised. Of special concern is the ability to predict the phenomenon of flashover, a potentially disastrous event. The ideas of nonlinear dynamics have been applied in an attempt to understand better the conditions which may result in flashover [1,2,3].

FIRE GROWTH AND FLASHOVER

Fire growth in a compartment may be divided into several stages : 1 Ignition, 2 Early Growth, 3 Flashover, 4 Fully Developed Fire, 5 Decay; see Figure 1. Not all stages

Systems Science, Edited by F.A. Stowell *et al.*,
Plenum Press, New York, 1993

necessarily occur. In a 'flashover jump' a relatively small, localized, fire suddenly increases in intensity to involve the entire compartment. This is associated with a dramatic increase in both the temperature of the hot gases and the burning rate of the fire. The nature of the flashover jump suggests that a nonlinear dynamical process is at work.
The problem addressed in this work has been to construct one or more theoretical models which would enable flashover to be predicted, given geometrical and thermo-physical properties. Thereby designs which are more likely to result in flashover may be avoided; at least in principle.

THEORETICAL MODELS OF FLASHOVER

Theoretical models intended to be used to predict flashover have been devised [2,3] which use the concepts of nonlinear systems theory [4]. In such models differential equations which are expressive of conservation principles may be used to describe the system. In particular, the models constructed have been 'zone models' according to which a compartment is divided into a number of zones within which conditions are assumed to be uniform. This simplification has been necessary in order to allow current concepts of nonlinear dynamical systems theory to be applied.

Especially, two main zones have been assumed : a hot upper smoke layer at a uniform temperature and a cooler lower layer; see Figure 2. A single ventilation opening has been assumed. Energy passes into the upper layer from the fire and is removed by hot gases flowing out of the opening and by conduction through the walls and ceiling. Also, air flows into the room through the lower part of the vent.

A key equation, corresponding to conservation of energy for the hot layer, may be expressed in a form identified by Thomas et al [5]

$$dE/dt = G(T,t) - L(T,t) \qquad (1)$$

Here, G(T,t) and L(T,t) are respectively the rate of gain and loss of energy of the upper layer and E includes both internal energy and work done terms [6]. The energy of the smoke layer, and hence its temperature, are governed by the forms of the gain and loss, which are themselves functions of the upper layer temperature, T, and the time, t. The energy balance is shown schematically in Figure 2. The term $Loss_1$ represents energy lost through the boundary surfaces whilst $Loss_2$ represents energy lost by convective flow of hot gases through the ventilation opening. (In general a ventilation opening may be either a door or a window.)

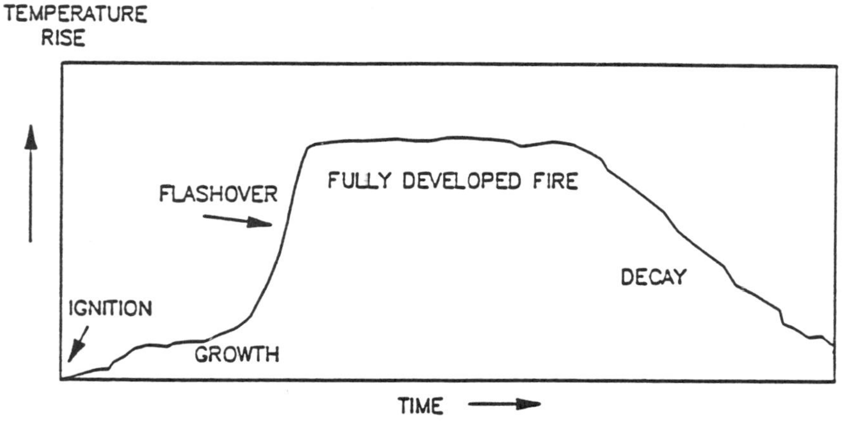

Figure 1 . Stages in a Compartment Fire.

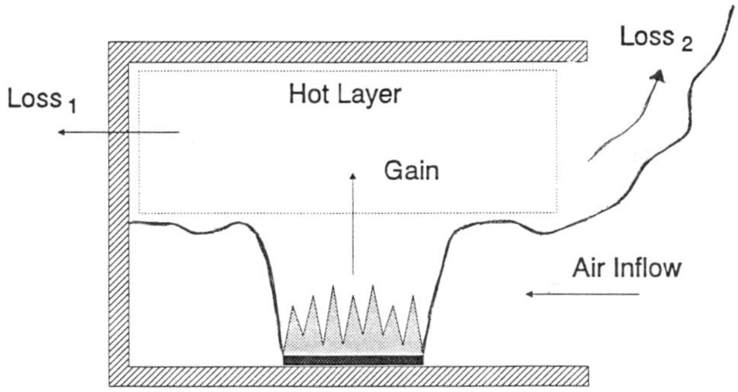

Figure 2 . The Zonal Arrangement Assumed.

The rate of gain of energy of the hot layer is determined primarily by the burning rate of the fire, which itself is governed by two factors :
(a) Whether or not the fire is fuel-controlled or ventilation-controlled. In the former case there is sufficient air to fully burn all fuel volatiles produced, in the latter case there is not.
(b) Radiation feedback from the hot layer to the fuel surface. An increase in the temperature of the upper layer results in more radiation to the fire base, increasing burning rate and so the energy going into the hot layer. That is, there may be a positive feedback loop.

A MECHANISM FOR A JUMP

Given equation (1) as a key element of a model it then becomes necessary to ask what are the forms of the G and L curves as functions of temperature and time.

Considering G first; as temperature increases the thermal radiation feedback mechanism exerts its influence resulting in an upward curvature of the gain function as a function of temperature, at a given time. This continues until a temperature is reached at which the fire becomes ventilation-controlled. In the simplest case, where a constant flow rate of air into the compartment is assumed, the G function becomes constant.

The loss function, L, resulting largely from out-flowing hot gases and conduction through boundaries, may be assumed to be essentially linear as a function of temperature. More realistically, it would tend to curve upward. Figure 3 shows three different loss curves, L1, L2, L3, which could, for example, correspond to different ventilation conditions, room sizes or wall materials. It is seen that for the curves given there are three types of intersections indicated by A_1, A_2, B and C_1, C_2. An intersection at point A_1 or A_2 corresponds to a large ventilation-controlled fire whilst point C_1 or C_2 corresponds to a small, localized, fire. Both states are stable with respect to temperature fluctuations. The intersection at point B, however, is unstable with respect to temperature fluctuations since a small change will result in a fall to C_2 or a rise to A_1. The gain and loss curves shown in Figure 3 correspond to a particular time; that is they are essentially 'snap-shots'.

In general the G and L curves shift relative to each other with time. For example, the loss curve may drift down relative to the gain curve as time passes. This is illustrated in Figure 4. If such a relative movement is assumed to be slow by comparison with changes in the absolute magnitudes of gain and loss terms then the fire may be regarded as being in a 'quasi-steady' state at any given time. The left hand side of equation (1) is then approximately zero and the fire state may be assumed to correspond to an intersection of the G and L curves. For example, a fire represented by the intersection at C in Figure 4 might tend to move up the gain curve as time passes. Simultaneously the intersection at B, corresponding to an unstable state, would tend to move down the G curve. As C and B meet, the state would become unstable and the fire would be expected to jump to the ventilation-controlled state at

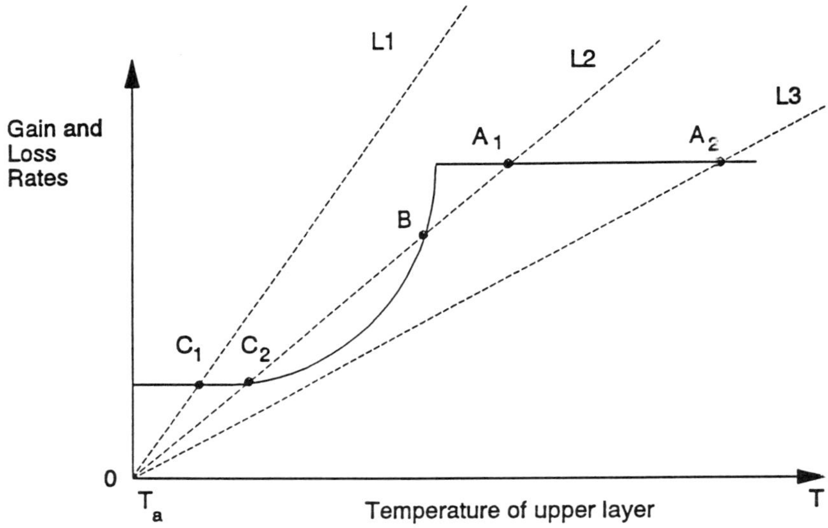

Figure 3 . Gain rate curve, G, and three loss rate curves, L1, L2 and L3; showing three possible types of intersection.

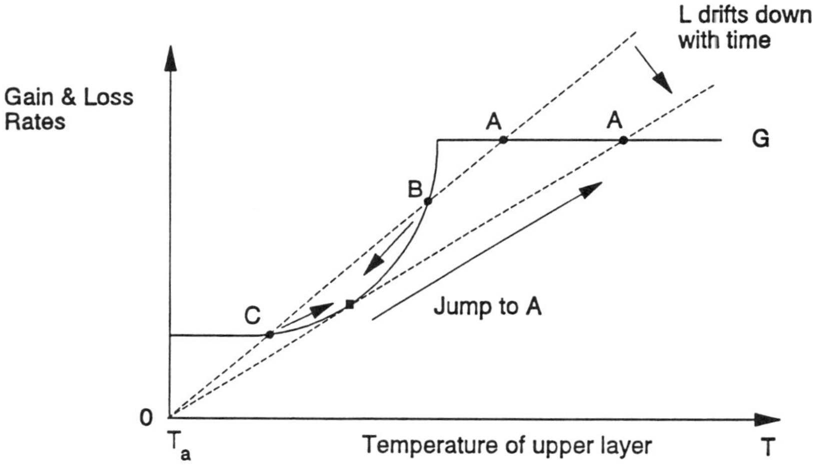

Figure 4 . A mechanism for a jump.

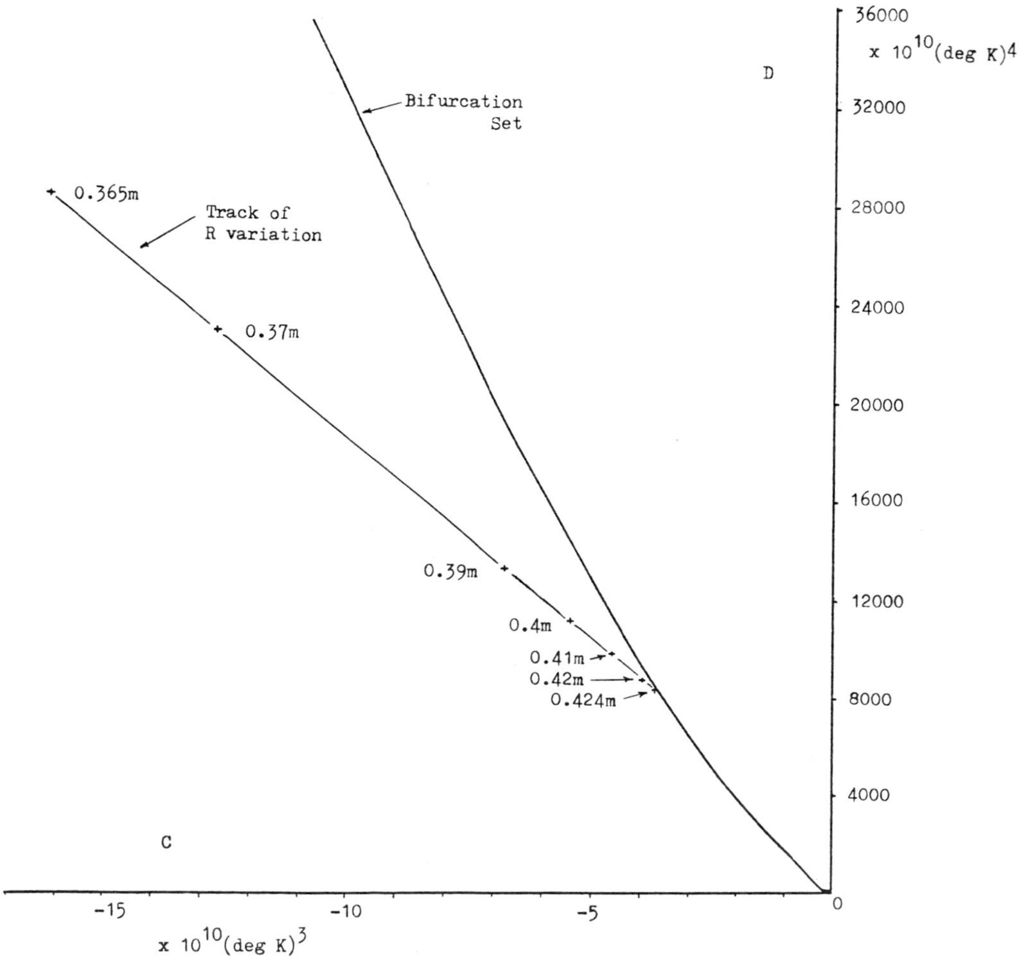

Figure 5 . Juncture of bifurcation set and track of fire radius variation.

A. This may be regarded as a model for a transition from a small, localized, fire to a generalized fire in a compartment. That is, it would correspond to a flashover jump.

This discussion, albeit simplified in nature, has served to show qualitatively how a sudden transition in such a system may be conceptualized in terms of the relative changes of gain and loss curves. Whilst the forms of the G and L curves in the fore-going discussion have been idealized, the essential qualitative features would be expected to be present even with more realistic forms. For example, a loss function might be expected to curve upwards rather than change linearly with temperature because of thermal radiation losses. If such an upward curvature were sufficiently great, relative to the gain curve, then the G and L curves might well continue to intersect at a relatively low level (corresponding to point C in Figure 4) rather than the intersection sliding up the G curve to coalesce with point B.

In the present project, theoretical models have been developed which employ the ideas of nonlinear systems theory to predict sudden transitions in state and these transitions have been associated with flashover. One of the models produced has resulted in a cusp catastrophe and one has resulted in a swallowtail catastrophe [2,3]. The emergence of two different types of catastrophe, however, is not contradictory but complementary. Each of the models makes different assumptions and tends to illustrate different features of the system. In the next section, as a specific example of how a jump may emerge in a model, a particular case is given using the 'swallowtail model' of reference [3]. This model is also briefly described in reference [1].

A SPECIFIC EXAMPLE

In the model of reference [3] a three dimensional space is defined, each axis of which corresponds to a 'control parameter' being B, C or D. (B and C here must not be confused with the intersections B and C described above.) Each control parameter is defined as a combination of geometrical and thermo-physical parameters. A given case corresponds to a point in this space. A surface within the space (which tends to resemble a swallow, hence the name) defines the 'bifurcation set' [3] for the system. If a point corresponding to the state of the system hits the bifurcation set then this would imply the existence of a transition. The plane defined by B = O corresponds to the case in which there is no heat loss through the upper surface (ie the ceiling and upper part of the walls). The bifurcation set in this plane is shown in Figure 6.

Simulations have been conducted for the case of a room 4 metres by 4 metres by 3 metres high with a ventilation opening from floor to ceiling of width 0.4 metres. Results showing the effect of increasing radius, R, of the fire area are shown and it is seen that the track of variation in radius hits the bifurcation set just after R = 0.42 metres. According to this model, therefore, at this point the state would become unstable and a jump to a stable (ventilation-controlled) state would be predicted.

CLOSING COMMENT

An indication of how a model of a flashover jump in a compartment fire may be constructed has been given. Such a model may be used to help to identify the conditions which would be expected to lead to a jump. Thereby such a model may be used to help to assess a design.

REFERENCES

[1] A.N. Beard, D.D. Drysdale, P.G. Holborn, & S.R. Bishop, 'Instabilities in Systems associated with Fire Risk Assessment', submitted to: *Systemist.*

[2] S.R. Bishop, P.G. Holborn, A.N. Beard, & D.D. Drysdale, 'Nonlinear Dynamics of Flashover in Compartment Fires', To appear in *Fire Safety Journal*, Vol. 20 or 21 (1993).

[3] A.N. Beard, D.D. Drysdale, P.G. Holborn, & S.R. Bishop, 'A Non-Linear Model of Flashover'; submitted to: *Fire Science and Technology.*

[4] J.M.T. Thompson, 'Instabilities and Catastrophes in Science & Engineering', Wiley, Chichester, (1982)

[5] P.H. Thomas, M.L. Bullen, J.G. Quintiere, & B.J. McCaffrey, 'Flashover and Instabilities in Fire Behavior', *Combustion & Flame*, 38, pp 159-171, (1980)

[6] M.W. Zemansky, 'Heat and Thermodynamics', McGraw Hill Inc., New York, (1957).

A CYBERNETIC VIEW OF QUALITY MANAGEMENT

Diane Bowling

Syncho Limited
Aston Science Park
Aston Triangle
Love Lane
Birmingham B7 4BJ

INTRODUCTION

Where problems are found in an organisational setting, the structure of the organisation will have a considerable bearing on the way that the problem is tackled and potential solutions that may emerge. All too frequently we address the problem without consideration of the organisational context.

This paper looks at quality management and the effect of organisational structure on it. The Cybernetic Methodology[1,2] of Raul Espejo provides a framework to integrate structural influences on problem solving with learning about the problem itself. Stafford Beer's VSM[3,4] is used to provide a model of the organisation which highlights the relationships necessary for the viability of the organisation. The methodology and model are not described here.

QUALITY IN THE ORGANISATION

The quality of goods and services produced by an organisation is not an intrinsic attribute, but is an outcome of the interactions of that organisation. These interactions include both those within the organisation and those with other outside organisations. Through these interactions relationships develop that define the structure of the organisation. The general structures of our commercial and industrial organisations have undergone some significant changes over the last century[5]. The progression from Craft Industry, through Mass Production to Lean Production summarises these changes.

Table 1 illustrates these three structural types through which organisations are moving, with examples of characteristics for each type. The Mass Production form has spread and is now the norm within most organisations. Knowledge based occupations such as accountancy, marketing, production management and inspection are now found in most organisations, often in functional departments. Many of these occupations came into being with the development and refinement of the Mass Production structure. They are found in the service and public, as well as the manufacturing sectors.

Systems Science, Edited by F.A. Stowell *et al.*,
Plenum Press, New York, 1993

The perception of product and service quality and how this may be achieved differs amongst these forms. In the craft form, quality was built into the product through the close relationship between the craftsman and customer. This direct supplier-customer interaction largely disappeared in the Mass Production era and arm's length relationships developed. A new profession of inspection was needed to ensure that the items produced were in accordance with a written specification that passed from the customer's purchasing department to the supplier's sales team, via the production department to the operator. Non-conforming items were 'inspected out' and reworked or scrapped.

Table 1. Three general styles of organisation

Craft	Mass Production	Lean Production
• Group of individual producers	• Functional departments	• Multifunctional teams
• Multiskilled individuals	• High division of labour	• Multiskilled teams
• Low inventory	• High inventory	• Low inventory
• Individually made parts	• Standard parts	• Standard parts
• Products tailored individually to customer requirements	• Limited range of standard products	• Large range of targeted products
• One-off	• Long production runs; long change-over times	• Short production runs; short change-over times
• Quality maintained by craftsman's standards	• Quality maintained by inspection	• Total quality management
• Individual customer as user	• Mass marketing	• Differentiated customers
• Problems solved by craftsman	• Problems referred upwards	• Problems solved by local teams

Current moves in quality management attempt to bring quality back to the relationship between customer and supplier. This relationship now takes place through a large number of individuals and teams, each representing a different aspect of the customer and supplier organisations, rather than the single relationship of the craft organisation. Through their relationships, large organisations must reach agreement on what quality is.

The rest of this paper looks at a small component producing company. The focus is on the way in which the company was organised and how this affected the quality of the company's products.

QUALITY MANAGEMENT IN GENERAL COMPONENTS LTD

General Components Ltd made a number of pressed items for other subcontractors in industries including: white goods, automotive, DIY and the traffic furniture industry. It had been structured along the lines developed for Mass Production, with a high degree of division of labour and a functional structure including a Quality Department populated by inspectors. The family tree, in figure 1, shows some of the relationships within the organisation, but these are predominantly who reports to whom. It does not throw light on the relationships necessary to define the job, get it done and satisfy the customer. Relationships or daily interactions between, for example, point X and point Y on the tree may be necessary, without going via all the higher levels until the two branches meet. If these 'informal' relationships are not permitted to develop the company will be the poorer for it. Referring problems upwards leads to delays in tackling the problem as each higher level is made aware of and investigates the problem. The higher levels are not in daily contact with the details involved; their daily concerns are of a different nature. Decision making is most effective at the level at which the

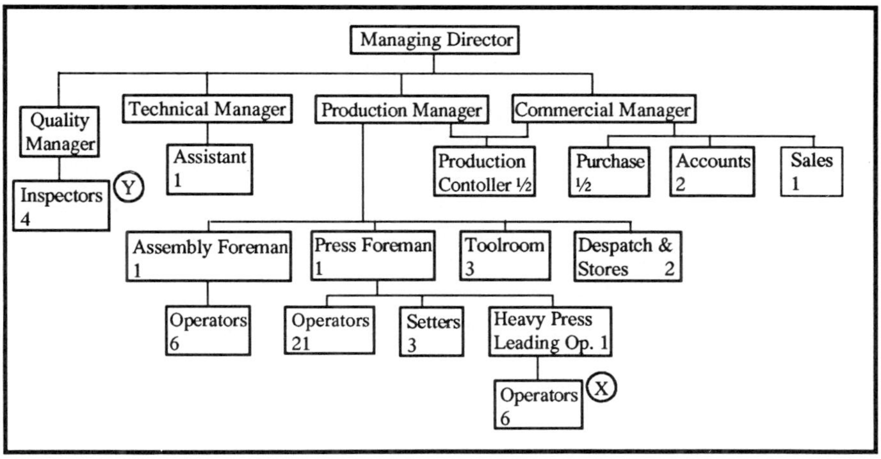

Figure 1. Family Tree for General Components Ltd.

problems occur. It is the role of the higher levels to deal with problems which affect the whole of the organisation and to provide the overall context for the lower levels to solve local problems as they arise.

Relationships within an organisation are much more complex than those shown in a family tree. The need to manage the complexity of the relationships that exist within a large organisation leads to the grouping of individuals into units and departments. The development of functionally based hierarchies has served us well in the past, allowing us to manage more complex organisations than small groups of craftsmen. It enabled the spread of goods and services to a much wider customer base. However, alongside this gain, we lost the closeness of contact with the customer, we reduced the variety of goods and services on offer and many abilities of the work force were stifled. The distancing of the customer from the supplier has led to an internal focus for the definition of quality. It became the conformance to written specifications, policed by inspectors.

This functional approach to organisation is now failing. We need a new way of looking at the relationships required to carry out complex tasks, if we are not to rely solely on trial and error or the catch-up tactics of copying of current best practice. We need to be creative.

The Quality Problem

The company produced a wide range of pressed metal components in small numbers for other sub-contractors. The pattern of customers' orders at General Components did not lend itself to dedicated machinery set up to carry out long production runs.

The company, in common with many others, held large amounts of inventory. As the orders it received tended to be small, but regular, it frequently made batches larger than the quantities ordered, putting the excess into stock. This stock was particularly useful when customers found deliveries faulty and returned them. It was advantageous when either customers required parts at short notice, or deliveries were late and customers were screaming for them. The stocking policy also avoided short runs punctuated by long set-up times. This had worked well for many decades, even though much stock sat on shelves for long periods, sometimes until it was obsolete.

The environment was changing. Increasingly customers expected only good parts, delivered on time. The Japanese example made clear that this was possible. In the future, business would be awarded increasingly to those who were able to deliver what the customer wanted on time, at an agreed price. Thus high inventories, late deliveries and rework would all have to become things of the past.

Tackling The Problem

The company tackled the problem on a number of fronts. BS 5750 was becoming increasingly important; the Quality Department led the initiative to gain the standard. Improvement teams from management and the work force examined the high inventory and made efforts to reduce it. The causes of late delivery were investigated. Rejected parts were examined more closely and the origins of the rejects were traced rather than just reworked.

These initiatives traced the origins of almost all quality problems to one area within the Pressing Unit, the Light Pressing Section. This section took up almost all the Pressing Foreman's time. It also had the benefit of three setters. Heavy Pressing, whilst nominally under control of the Foreman, was run entirely by the Leading Operator, without the assistance of a Setter. Also, the inspectors spent a disproportionate amount of time in examining Light Pressing and still were not picking up the rejects before they left the factory.

Deming[6] and others suggest that over 90% of defects are systemic rather than controllable by the operator. It appeared that one part of the organisation was producing many more defects than the other two. Examining the structure of the organisation provided some useful insights into the underlying systemic forces that were producing this outcome.

The Structure of General Components

Figure 2, a Viable System Model of General Components, gives an alternative view of the company's relationships. This puts the emphasis on the relationships for viability that provide the structural context for the problem areas highlighted above.

Whilst the family tree showed Heavy Pressing as part of the pressing unit, in practice the Pressing Foreman concerned himself only with Light Pressing: he claimed no control over Heavy Pressing and the Leading Operator confirmed this. Thus, in practice, Components consisted of: Heavy Pressing, Light Pressing and Assembly. These are the units shown in the VSM.

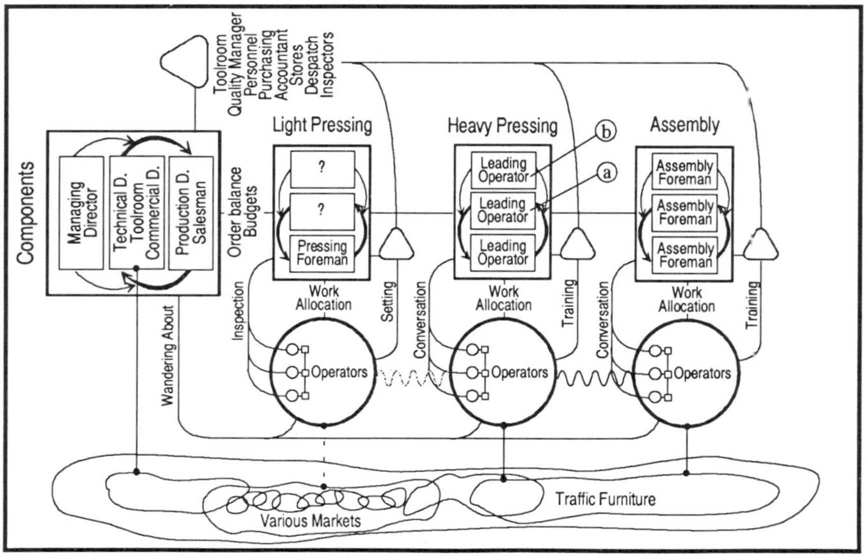

Figure 2. VSM for General Components Ltd

Development provide 'intelligence' for the company as a whole.

Heavy Pressing produced all of the main parts for Assembly. Operators within Heavy Pressing understood the key customer requirements for their products, set their presses themselves, knew the schedule and inspected their own work. This left the Leading Operator time to improve his relationships with customers and to develop improved ways of running his section. In figure 2, this is illustrated by the presence of the Leading Operator in the 'intelligence' and 'policy' boxes, labelled 'a' and 'b'.

In contrast, Light Press operators only operated machines. Setters set the machines so that operators could move directly from one job to the next, as allocated by the foreman. They knew little of the customers' needs, therefore they did not know which parameters were critical. Inspectors checked their work. With a much larger group of people to control, all of whom only used physical skills, the Foreman's time was fully committed ensuring that everyone was working at all times. Therefore 'intelligence' and 'policy' were poor, hence the question marks in figure 2. Sales, Finance, Purchasing and Quality Management were all centralised. They are shown in figure 2 as co-ordinating functions supporting the overall management of the company. Marketing and Technical Development provide 'intelligence' for the company as a whole.

Management of Complexity

There was no difference in calibre between the people employed in Heavy and Light Pressing. However, the complexity of the two units and the ways in which this was managed was entirely different. Heavy Pressing used the abilities of all the work force to manage the complexity of the unit. Light Pressing, used only the Foreman to manage much greater complexity, reducing most of the rest to lever pullers.

Exhorting the Light Pressing Operators to improve would not bring about a change. They needed to know why the parts were failing. Some quality problems were not due to operator errors, but to poor setting, purchasing or production control decisions. Exhorting the Foreman to do better would have had an equally limited effect.

Use of the VSM highlighted the absence within Light Pressing of relationships enabling the people in the unit to understand their customers' needs. Equally they had no relationships through which to understand the technologies they used and how to improve them. Some individuals knew something of these customers and technologies, but the structure neither permitted these to be shared and used by the members of the unit, nor encouraged the enhancement this knowledge.

Once the structural difficulties of the two pressing units were recognised, it became possible to consider the necessary changes in relationships to assist the improvement in quality. Assembly and Heavy Pressing were closely connected, so these were put into one unit focused on Traffic Furniture. Light Pressing was split, focusing on broad customer categories. These changes reduced the complexity of the foreman's role in Light Pressing by using the abilities of the people already working in the company; the foremen, leading operator and setters were given new roles leading these new units.

The structure of these newly defined units could not be allowed to develop along the lines of the old Light Pressing section. Following the example of Heavy Pressing, the operators were all to be trained to set their own presses, understand better their customers' needs and move towards solving more of their own problems. This took place in teams. The foremen and setters became increasingly involved in finding improvements in the running of the whole of their new units, working closely with both the operators and the service units of the whole company. Light Pressings' previous aim of 'keeping the presses running' slowly gave way to 'providing the customer with what he wants'.

The Move Towards Total Quality Management

These changes were long term. They required training and commitment. However, without these changes in structure the solutions developed by the improvement teams could not take place. The structure originally found in the company was dependent on high inventory, operators working physically regardless of whether or not what they were producing was currently needed, and talking was viewed as idle time. Yet investigating the causes of rejects and developing and implementing ways of reducing them all required conversations to take place. It is through conversations that relationships develop.

Changes were not confined to the shop floor, but also involved the relationships between the shop floor and the services of the company including Despatch, Production Control, Purchasing and Quality. Total Quality Management involves structuring a company such that the relationships necessary to produce the goods or services required by the customer are re-inforced, not inhibited.

CONCLUSIONS

Improving the quality of a company's goods is not simply a matter of trying harder. Studying the structure illustrated clearly that part of the underlying cause of poor quality lay in the way the company was organised. Simply providing better instructions is not enough - people must have the information to understand the problem and this can only be gained through the relationships they develop.

Light Pressing had the typical functional structure of a Mass Producer form. It could not deal adequately with the complexity of many different products serving a varied market. This had to change to be successful in the new climate.

The VSM is a useful model in both the analysis of the relationships for viability and the design of improved structures. In this example, changes were made from the typical structure of a Mass Producer, towards those found empirically in Lean companies. Management Cybernetics provided a means to understand the effect of structure in a practical situation and to generate creative change in the organisation. This approach enables organisations to move away from copying current best practice to design company specific structures, that offer a real competitive advantage.

REFERENCES

1. R. Espejo, Management of Complexity in Problem Solving, *Trans Inst MC*, Vol. 14, No. 1, pp8-16, (1992)
2. D. Bowling and R. Espejo, An intervention with the Cybernetic Methodology in Regent Engineering, *Trans Inst MC*, Vol. 14, No. 1, pp17-28, (1992).
3. S. Beer, "Heart of the Enterprise", Wiley, Chichester, (1979).
4. R. Espejo, The VSM Revisited, *in:* "The Viable System Model", R. Espejo and R. Harnden, ed., Wiley, Chichester, (1989).
5. J.P. Womack, D.T. Jones, and D. Roos, "The Machine that Changed the World", Rawson Associates, New York, (1990).
6. W.E. Deming, "Out of the Crisis", Cambridge University Press, Cambridge, Mass., (1982).

INNOFUSION: FEEDBACK IN THE INNOVATION PROCESS

James Fleck

Department of Business Studies
The University of Edinburgh
50 George Square
Edinburgh EH8 9JY

INTRODUCTION

Innovation - "the process of bringing inventions into commercial use" (Freeman, 1982) - is an important subject, crucial to the health of modern economies. There is currently much discussion in policy circles about how best to encourage innovation. For instance, the first UK Innovation Lecture, sponsored by the Royal Society, The Fellowship of Engineering and the DTI featured Mr. Akio Morita, the Chairman of the highly innovative Sony Corporation (DTI, 1992). There is also an emerging focus on the role of the user in technology development, with a long-running debate centering on the role of the market versus various forms of planning or organizationally mediated transactions (Williamson, 1975). These issues essentially concern the function of different forms of feedback in the innovation process.

At Edinburgh, we have been examining innovation processes, especially the social influences on the development of technology, under the rubric of "the social shaping of technology" (Edge, 1986). This approach complements and contrasts with the conventional focus on the impact of technology on society. Two of us are developing a systematic framework for understanding the range of different technologies and how they variously come into being (Molina and Fleck, 1992). We hope thereby to devise practical analytical tools showing how policy makers and managements can improve their own practices. For example, by focussing on the necessity for feedback of various sorts under certain conditions, we have outlined how policy makers can harness and complement existing forces to enhance the innovation process.[1]

Innovation has been analyzed from many different perspectives: the history of technology; economics; sociology; management; policy studies; and technological practice itself. As such it perhaps offers an ideal target for attack by the unifying approach of systems analysis (von Bertalanffy, 1973). However, although there have been various cybernetic and

[1] See for instance Fleck (1990) and Molina (1990), analyses which have been used by the DTI and the CEC respectively.

Systems Science, Edited by F.A. Stowell *et al.*,
Plenum Press, New York, 1993

systems analysis treatments of innovation, these are usually either verbal discussions making analogical rather than technical use of systems concepts, or over-formalized and inaccessible analyses, inadequate in their empirical grounding.

In this paper I outline various broad types of innovation process, paying particular attention to the part played by feedback under different conditions, and indicating how these issues have been previously treated. I believe that useful insights are to be had from a more thorough going cybernetic analysis which may help to provide objective means for distinguishing different types of innovation and for measuring the extent of different types of feedback. I hope, therefore, that this account of the rich empirical domain of innovation may alert interested parties to a fruitful opportunity for further cybernetic modeling effort.

MODEL 1: THE LINEAR MODEL

The linear model (Figure 1), perhaps reflects the intelligent lay person's view of how things are. The processes of creation, production and diffusion are seen as quite separate and distinct, and indeed are perceived as the prerogatives of different functions or even organizations: creation or invention being the exclusive concern of R&D departments; production, the business of manufacturing divisions, with some organizations (such as Sony for example) specializing in innovation, i.e., the first commercialization of a new invention; and diffusion being the concern of sales and sometimes constituting the main business of market-oriented organizations which focus on delivering high-volume low-price offerings (as, for example, with Alan Sugar's Amstrad).

In this simple model, there exists only the ultimate feedback mediated by success or failure in the market place. Such mobilised market forces can provide adequate information for the evolution of innovations in an essentially Darwinian process, provided an appropriate industry structure exists to channel and harness those variety matching forces. This model is a good approximation of the situation for consumer product innovation, where there exist mature, highly segmented and specialised industry structures, and products which are also highly discrete and generally highly standardised. Only a minimal role is played by consumers (the users in this case), primarily via their choices in the market place. However, even here linearity is mitigated by the rise of market research which seeks to enrich information feedback beyond pure price/demand signals. This model is relatively readily quantified, and patterns of development over time (i.e., diffusion) can be captured in terms of product life cycles described by a variety of logistic or epidemiological models.

The linear model is widely thought to be more generally applicable. It is implicit, for example, in the recent influential analysis by Fukuyama (1992). He sees modern science (which he does not distinguish from technology) precisely in its capacity as a generator of

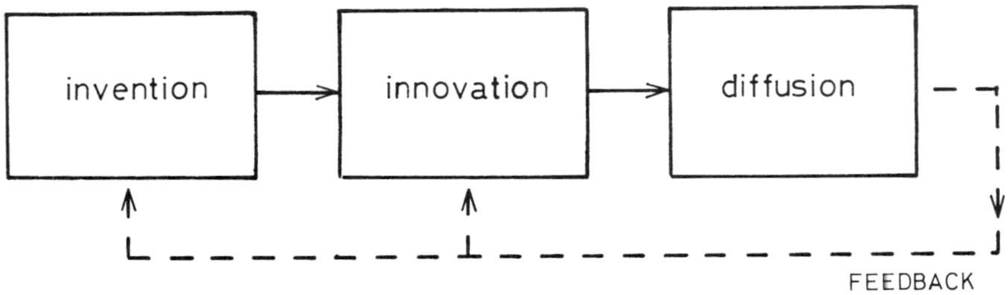

Figure 1. Schematic depiction of the linear model of innovation.

consumer goods as the major driver of the world-wide trend towards the universal adoption of liberal democracy. It is also assumed in much governmental policy towards research and development. In particular, it is often argued that "spin offs" into wider civilian application will naturally flow from military research expenditure, although this is coming into question with the demonstrated superior performance of the Japanese and German economies, both of which focus effort more directly on civilian applications.

However, after considerable empirical research on innovation (Langrish et al., 1972; SPRU 1972; Gold et al., 1980; Sahal, 1981; Georghiou et al., 1986) it is now abundantly clear to innovation studies specialists that the linear model is strictly limited in its scope, and functions more as a "straw man" in critical discussions of technology development (Freeman, 1992).

MODEL 2: EXPERT USER CONTRIBUTIONS

In certain situations, the inadequacies of the simple linear model are very apparent. Von Hippel (1976, 1978), showed that users are often the originators of innovations such as scientific instruments. In the creation of specialist tools for scientific, surgical or medical, or industrial use, the users play a crucial role - in many cases actually making the instruments or at least prototypes themselves. In such situations the market information feedback loop of the linear model is replaced by direct user action.

Once such novel products have been developed, new industrial structures may form over time to specialize in the development and production of variants. Contributions by specialist "expert users" continue to be an important input, transformed by the service capability of the sector into practicable devices. Such innovations are usually discrete and relatively small-scale in nature.

Von Hippel has sought (1988) to elaborate an economic theory explaining the conditions under which users contribute. In one of the most theoretically sophisticated analyses of innovation yet devised, he explains the propensity of users to contribute to innovation in terms of the basic economic rents they are able to derive from their inputs; i.e., he explains their involvement in terms of the economic system in which they are agents.

MODEL 3: TECHNOLOGICAL SYSTEMS AND TRAJECTORIES

Modern society, however, is not comprised of discrete technologies alone. Large technological systems, such as rail transport, electricity generation and distribution, and most recently telecommunications are typical. More compact, though no less complex systems such as modern aircraft, computers, and military weaponry are also common. Technological systems are too complex for any one institution to handle by itself. Extensive, mature and segmented industry structures have evolved - as the technologies themselves have evolved - and comprise many separate companies working in close complementary relationships. Complex systems of agreed standards ensure that the component innovations produced by different suppliers fit together to make up the overall systems.

Development in technological systems is constrained by the need for the whole complex to work effectively as a coherent whole, leading to "natural trajectories" of development (Nelson and Winter, 1982; Dosi, 1982). High-level system exigencies set problems which component innovators have to solve: as Hughes (1983) observes, "reverse salients" (i.e., critical areas in which lack of progress holds back overall development) attract the attention of would be innovators.

In such situations, very complex patterns of feedback exist at a variety of levels. These are articulated via evolved structures of industrial relationships (especially between user and supplier *institutions*), and via the recursive structure of engineering knowledge itself. Only at a very restricted local level does anything like the simple linear model apply. Under these conditions, events in the early history of an industry or technology may be crucial, and thus path dependencies become important for an understanding of development (Arthur, 1987 and 1990).

MODEL 4: CONFIGURATIONS AND INNOFUSION

Certain large-scale systems, however, are not characterized by any high-level dynamics of development, and therefore follow no clear trajectories. Rather, each installation is a more or less unique adaptation to the local contingencies of application. Extensive implementation effort (Leonard-Barton and Kraus, 1985; Voss, 1988) is consequently required to achieve successful operation. In such situations, often explicitly referred to as *configurational* (Fleck, 1993), extensive user inputs at all levels are required, to such a degree that in-house development within the user organization is the rule rather than the exception. Organizationally-convoluted, company-wide information technology installations such as branch networks in banks, computer integrated manufacturing systems and computer aided production management systems, are prime examples, as are production lines which frequently have to be specifically reconfigured to meet the requirements of a particular production run.

Configurations are made up out of a wide variety of both technical components such as computer hardware or software, and non-technical components such as particular patterns of work organization or models of motivation (McGregor's theories X and Y and Ouchi's theory Z). Associated with the lack of a system-level long-term dynamic of development, there are emergent industrial structures in which the processes of invention, innovation and diffusion are not easily separable: suppliers are often at the same time important users, as was the case with industrial robots for instance (Fleck, 1984). Sometimes configurations turn out to have wider applicability beyond the specific contingencies which gave them birth. In these cases, generic technological systems as already discussed above emerge, accompanied by an evolving industrial structure characterized by increasing separation between suppliers and users, a series of incremental improvements to the overall systems, and eventually the development of industry-wide standards.

But in many cases each configuration remains essentially unique, and the implementation effort required to achieve operation constitutes significant innovation in its own right (Leonard-Barton, 1988). In such cases the processes of invention, innovation and diffusion are collapsed together in a process of *innofusion* (Fleck, 1987), far removed from the simple linear model of innovation. Feedback is primarily through the internal learning processes involved in each exercise of implementation, rather than via transactions in the external market place. With the advent of home automation requiring considerable configurational effort by users (Cawson et al., 1990), these processes can involve domestic consumers, though in general they are more typical of various forms of process innovation.

SYNTHESIS AND POLICY IMPLICATIONS

If an inappropriate model of the underlying situation is applied, unsuitable and ineffective if not harmful policies may be followed. Indeed, the attempt by many organizations, usually relatively inexperienced with technology adoption, to straightforwardly *install* complex configurations, rather than recognizing the innovative implementation effort required, underlies the very high rates of failure typical of high technology adoption.

The above four models of innovation are related in several ways, and together indicate ideal-type boundaries of a graduated space of possibilities.

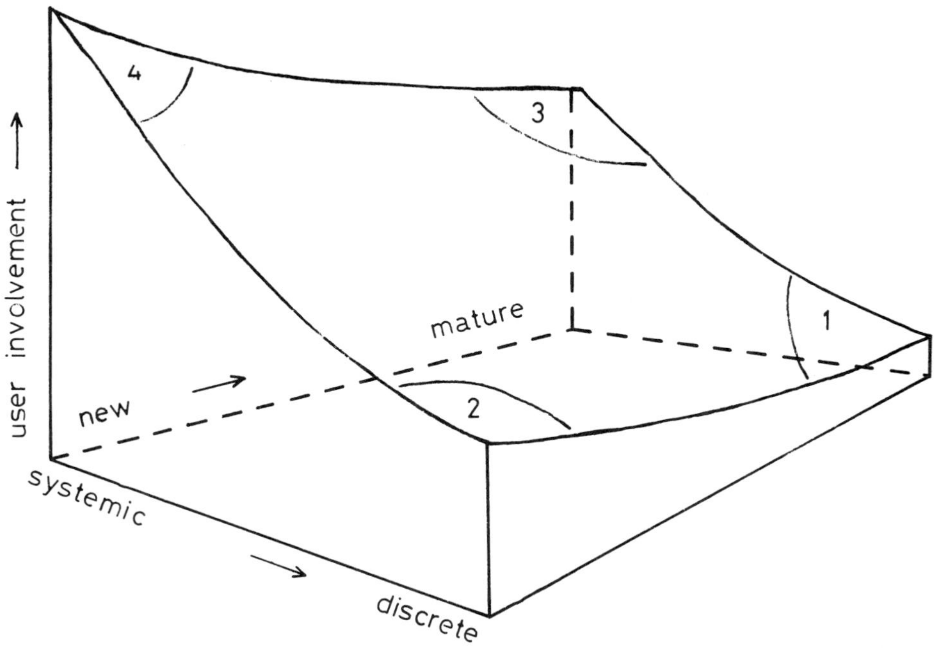

Figure 2. Schematic depiction of the space of broad types of innovation process.

Figure 2 indicates how they can be portrayed in terms of three parameters: (a) the *maturity* of the technology and its associated industrial sector; (b) the *complexity* of the technology - i.e., the extent to which it is systemic or discrete and stand-alone; and (c) the *degree of user involvement* (ratio of local process knowledge to generic specialist expertise). As discussed above, the *type* of user involvement (whether end-user, specialist-user or institutional-user) also varies across the delineated space. This schematic depiction is of course, capable of much further refinement and development, and is currently the focus of our research efforts.

CONCLUSION: THE CONTRIBUTION OF SYSTEMS ANALYSIS

In the above I have outlined various models for the innovation process, ranging from a simple linear model to an extreme non-linear process, *innofusion.* In looking at the situation from a systems perspective:

(a) would it be possible to devise a unifying cybernetic model or series of models for the different forms of innovation?

(b) would it then be possible to devise simple objective measures for distinguishing the various types of innovation described above, or even identifying other key types?

(c) would it be possible to simulate the results of policy initiatives in terms of such models, thus providing an alternative to conventional economic modeling which tends to be couched in terms of rather general economic variables and which lacks discrimination in terms of the specific innovation processes occurring?

(d) given that by definition innovations are implicated in rapidly changing systems, would it be possible to derive (and substantiate) any general results about the dynamic stability (homeorheosis) of the different forms of innovation process? In particular, would it be possible to derive any general conclusions about the fragility of innovation diffusion when it proceeds too quickly, as suggested by Robert May's analysis of the chaotic region of the simple logistics equation (Gleick, 1987)?

REFERENCES

Arthur, W.B., 1987, Path-dependent processes and the emergence of macro-structure, *European Journal of Operational Research*, 30:294-303.

Arthur, W.B., 1990, Positive feedbacks in the economy, *Scientific American*, February:80-85.

Bertalanffy, L. von, 1973, "General System Theory," Penguin Books, Harmondsworth.

Cawson, A., Haddon, L. and Miles I., 1990, "New Consumer Technologies and the Home," paper presented at ESRC New Technologies and the Firm Dissemination Conference, The Science Museum, London (mimeo).

Dosi, G., 1982, Technological paradigms and technological trajectories: a suggested interpretation of the determinants and directions of technical change, *Research Policy*. 11:147-162.

DTI, 1992, "The UK Innovation Lecture," (Video and transcript), DTI, London.

Edge, D., 1986, "The Social Shaping of Technology," Edinburgh University PICT Working Paper No. 1, Edinburgh.

Fleck, J., 1984, The introduction of the industrial robot in Britain, *Robotica*, 2:169-175.

Fleck, J., 1987, "Innofusion or Diffusation? The Nature of Technological Development in Robotics," Edinburgh University PICT Working Paper No. 4, Edinburgh.

Fleck, J. (with Maguire), 1990, "Information-Integration and Industry (a digest of Fleck (1988) The Development of Information-Integration: Beyond CIM?)," commissioned report prepared on behalf of the DTI, London. Available as University of Edinburgh Department of Business Studies Working Paper Series 90/13.

Fleck, J., 1993, Configurations: crystallizing contingency, *Int. J. of Human Factors in Manufacturing*. (forthcoming)

Freeman, C., 1977, Economics of research and development, *in* "Science, Technology and Society. A Cross-Disciplinary Perspective," I. Spiegel-Rösing and D. de Solla Price ed., Sage Publications, London, 223-275.

Freeman, C., 1982, "The Economics of Industrial Innovation," Frances Pinter, London.

Freeman, C., 1992, The Concorde syndrome in medicine? *EASST Newsletter*, 11(3):3-7.

Fukuyama, F., 1992, "The End of History And The Last Man," Hamish Hamilton, London.

Georghiou, L., Metcalfe, J.S., Gibbons, M., Ray, T. and Evans, J., 1986, "Post Innovation Performance," Macmillan, Basingstoke.

Gleick, J., 1987, "Chaos: Making a New Science," Penguin Books, Harmondsworth.

Gold, B., Rosegger, G. and Boylan M.G. Jr., 1980, "Evaluating Technological Innovations," Lexington Books, Lexington, MA.

Hughes, T.P., 1983, "Networks of Power: Electrification in Western Society, 1880-1930," Johns Hopkins University Press, Baltimore.

Langrish, J., Gibbons, M., Evans, W.J. and Jevons, F.R., 1972, "Wealth From Knowledge," Macmillan, Basingstoke.

Leonard-Barton, D. and Kraus, W.A., 1985, Implementing new technology, *Harvard Business Review*, 63 (November-December):102-110.

Leonard-Barton, D., 1988, Implementation as mutual adaptation of technology and organization, *Research Policy*, 17:251-267.

Molina, A.H. and Fleck, J., 1992, "TechMaPP: Proposal for a Technology Management and Policy Programme," Edinburgh University Department of Business Studies, Edinburgh.

Molina, A.H., 1990, Transputers and transputer-based parallel computers: sociotechnical constituencies and the build-up of British-European capabilities in information technologies, *Research Policy*. 19:309-333.

Nelson, R.R. and Winter, S.G., 1982, "An Evolutionary Theory of Economic Change," Harvard University Press, Cambridge MA.

Sahal, D., 1981, "Patterns of Technological Innovation," Addison-Wesley, Reading MA.

SPRU (Science Policy Research Unit), 1972, "Success and Failure in Industrial Innovation," Centre for the Study of Innovation, London.

von Hippel, E., 1976, The dominant role of users in the scientific instrument innovation process, *Research Policy* 5:212-239.

von Hippel, E., 1988, "The Sources of Innovation," Oxford University Press, Oxford.

Voss, C.A., 1988, Implementation: a key issue in manufacturing technology: the need for a field study, *Research Policy*, 17:55-63.

Williamson, O.E., 1975, "Markets and Hierarchies," The Free Press, New York.

ORGANIZATIONAL CYBERNETICS

Arnulf Hauan and Jon-Arild Johannessen

Bodø Graduate School of Business
8002 Bodø, Norway

ABSTRACT

Heterarchical logic transcends the traditional understanding of rationality in organizations. In this paper, we present a conceptual model of heterarchical organization. Implications of the model are derived based on observations in a Norwegian shipyard that has attempted to design and implement a "joint management" concept of information exchange. Addressing Jackson's (1986) criticism of the use of cybernetical models in organizations, productivity gains obtained are attributed to cybernetical rather than cultural factors. The holistic, integrative qualities of heterarchical organization,however, hinges on the organization being able to (somehow) mold together crucial alliances that are willing and capable of interacting in ambidextrous ways (logico-rational and emotional-intuitive). Special attention is given to the process through which competing values, i.e. instrumental rationality and playfulness, may be integrated in ways that generate problem-structuring that circumvent some of the anomalies of traditional hierarchies and protects the organization against its own shortsightedness (Johannessen, 1991, Brown et Duguid, 1991).

BACKGROUND

The company "Sterkoder" Ltd. in Kristiansund, Norway, is a medium-sized shipyard employing about 650 people that specializes in constructing advanced trawlers, special tank-chemical carriers and offshore-supply-vessels. In 1989, the company signed a contract with a Soviet agency for a total of 15, later to become 20, trawlers to be completed by the end of 1992. This contract triggered a major technical reorganization of the yard and the launching of an ambitious productivity campaign aimed at a 50% overall increase in the productivity of work by the end of the campaign period.

The present discussion is prompted by the interpretation of various outcomes of this process in terms of hierarchy/ heterarchy-transitioning reported in Johannessen & Hauan (1992). The investigations were undertaken in the spirit of "naturalistic inquiry" (Lincoln & Guba, 1985) through a highly interactive, collaborative approach (Glaser & Strauss, 1967) involving people on all levels in the company through extensive and continuous dialogues.

Systems Science, Edited by F.A. Stowell *et al.*,
Plenum Press, New York, 1993

The paper aims at capturing some of the texture of complex organizational realities in order to develop more sensitizing concepts and models for investigating and theorizing about the "logics of organizing". The model shows how an information system based on communicating structurate "creative fields" that facilitate utilization of "smart-tech" ("locally optimal") solutions to the organization's endeavors to enhance its performance.

HETERARCHICAL LOGIC

Fig. 1 shows central features of a heterarchical organization based on heterarchical logics of organizing. Work-groups are organized in accordance with the principle of "requisite variety" based on multi-disciplinary skills within a context of lateral power-structures. Groups function and are interconnected through the creation of "consensual domains" (Maturana & Varela, 1980, von Krogh & Solem, 1991) and relational norms within a conceptual framework of circular causality. The relational norm of reciprocity is based on the principle of complementarity (Bateson, 1972, 1979). Heterarcichal integration is seen as a process of ambiguity-reduction (information-creation) through which the particular causal impact of actions and events are interpreted in terms of holistic, generative effects on the system, especially in terms of self-renewal. As Nonaka (1988) rightly observes," "anti-rationalistic" models suffer from a lack of focus on the dissolution and creation of order such as the self-renewal of organizations". Developing and elaborating on models of heterarchical organization may contribute towards remedying this.

THE POLITICAL LEVEL

To create a change momentum in organizations, it is necessary to weld together networks of people pulling in the same direction. Networks may be loosely or tightly coupled. Since changes in policy are associated with changes in interests, people form alliances to enhance or impede the progress of change processes. These alliances may be tightly coupled both within and across hierarchical levels depending on the motivational strength associated with the risks perceived to be at stake. Thus, the configuration of alliances that develops throughout a transitional process becomes crucial to the outcomes of the change efforts. In the company, power as an organizational variable is synonymous with two different kinds of crucial alliances (Weick, 1979:16): One that connects, and one that separates. The former emerged through then interplay between the "Change Agent" cum Technical Director and local trade-union-leaders who exerted a considerable influence on shop-floor- as well as strategy-level (as members of the Board of Directors). Top-management created the sort of "hierarchical" legitimacy needed to make the crucial alliance become the "dominant coalition" in the company. The alliance that separates (basically at the staff- and foreman stratum), often in subtle ways, would act as "contrapreneurs" to the entrepreneurial ambitions of the connecting forces.

Positive economic and social outcomes were rapidly generated that were attributed to the "Sterkoder-Model": downward-up, action-oriented, minimum staff, operational-level-focusing and "co-knowledge-worker"-orientation, and attracted the attention of potential investors. Negotiations with large scale "institutional" investors; large, diversified companies strongly committed to a philosophy of engineering-oriented managerial hierarchies, failed. When the connection was made, however, between an entrepreneurial investor with considerable political and strategic-marketing clout, long-term-ambitions with regard to industrial renewal and the dominant coalition in the company, the project could be provided with the type of resource-flow in terms of psychological, political and financial support that it needed to succeed.

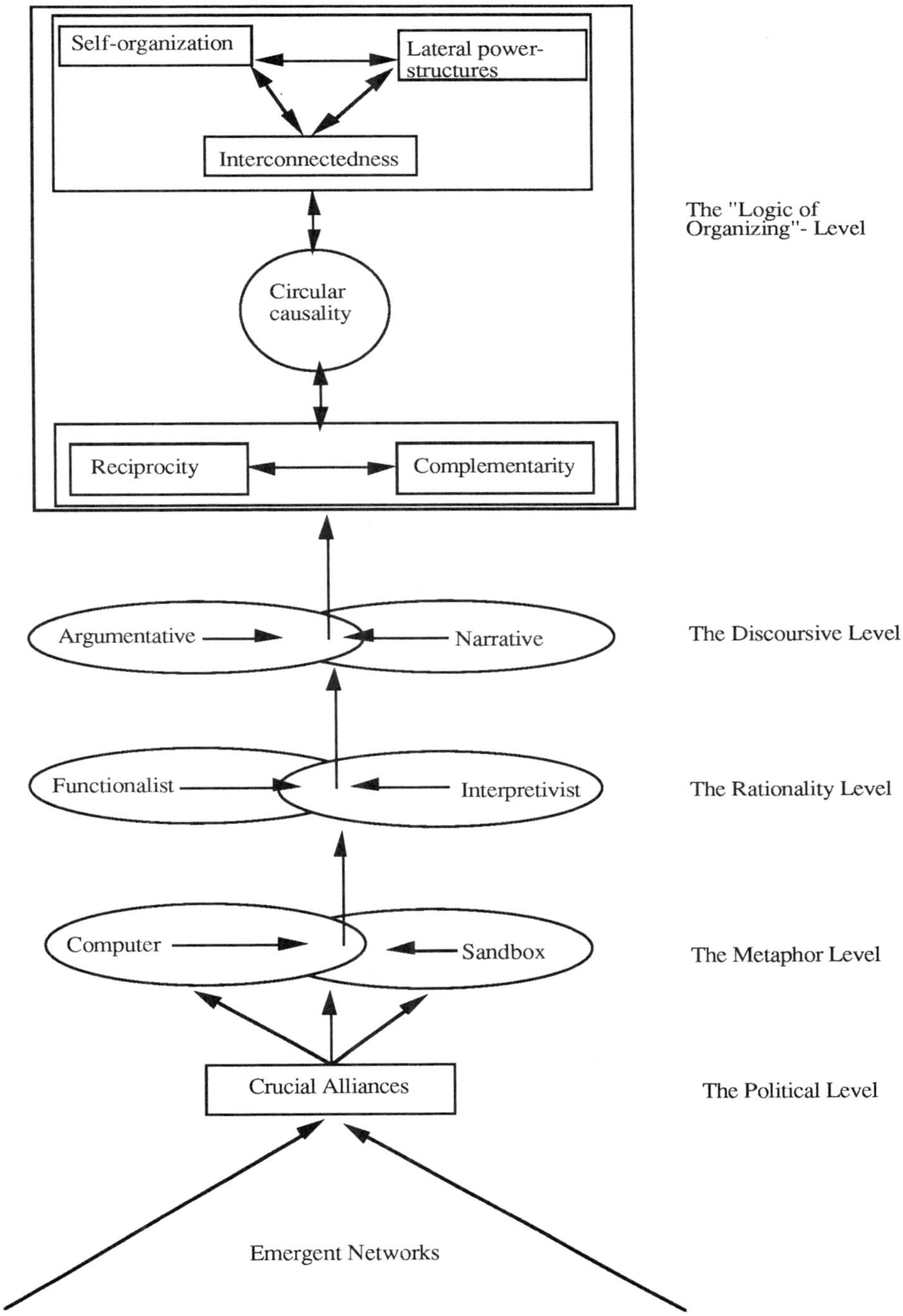

Figure 1. The Dynamics of creating heterarchical organizing.

THE METAPHOR LEVEL

"By studying metaphors of the field, we can learn much about what we always wanted to know and never new how to ask: about the values and beliefs held by organizational members" (Czarniawska-Joerges,1988:27). Metaphors are important because they say something about how things are, they relate, imagize, give life and model peoples interpretations of what is going on in the company, thus shaping their actions. The Managing Director (MD) and the Technical Director (TD), while in agreement about the change-processes, used different metaphors to convey their messages. The MD, anchored in a tradition of management science, operation research, first-order cybernetics, MBO, planning, programming and budgeting, were inclined to use metaphors related to the computer. He would refer to technological changes as changes in hardware, and changes involving employees as changes in software. Largely responsible for functions related to here-and-now-management (Beer, 1981) and stability, his metaphors tended to be frozen (Czarniawska-Joerges, 1988: 30), more like labels depicting integration of known procedures. His metaphors reflected a strategic-management-concept rather than a creative-management-concept (Olaisen & Johannessen, 1990). The dominant metaphors of the TD, on the other hand, were richer in content symbolized by the "sandbox" indicating improvizing, technology -of-foolishness-experimental behaviour, serendipity and the spontaneous rationality of the child at play or the bricoleur combining whatever tools might be at hand towards whatever ends may develop. Play can be said to contain a "certain element of foolishness that is difficult to justify locally, but are important to the broader system" (March, 1981: 572). The duality of the logics of the machine and the logics of flux (Morgan, 1980) were reflected in the complementarity of the metaphors of the MD and the TD. The metaphors of the countervailing forces were of a different nature. Towards the new background (metonym) of interpretive reality (smart-tech-industrial-plant), the family-circus metaphors of the old-guard had an impact more like nostalgic platitudes, as reminiscences of a bygone era of "amicitia" (Zaleznick, 1989: 230) in which rival groups spent much of their energy fighting each other. High in dramaturgical value, their metaphors nevertheless were politically dead and widened the gap between the competing alliances.

THE RATIONALITY LEVEL

The turnaround efforts in the company had a dual focus: it involved a major change towards more assembly-line-like production (increasing technological interdependence) and trying to establish a mutual understanding and consensus for the implications of the changes on the social organization of work (increasing emphatic bonding). Habitually, the engine were expected and inclined to view their role in terms of functionalist experts and technical puzzle-solvers providing blue-prints for work groups to guide construction. Engineer/ Construction-interaction was largely limited to crisis-intervention in cases of severe misfit between technical specifications and operational capabilities. Misfits that could be corrected locally were largely not fed back systematically as a correctional device. The relationship between the different groups was characterized by a system of "acting-on" rather than "acting-with" each other (Zuboff, 1989). Thus, through mistrust and lack of organizational interconnectedness, learning took on a single-loop rather than double-loop or deutero-character. To overcome this lack of coherence, the change efforts were explicitly founded on the requirements of the "operators": All procedures (technological or otherwise work-related), were to go through a crucial test based on their compatibility with what the system was doing (regardless of its "objective" merits). That meant: transforming engineers into "co-workers" and "workers" into "knowledge-workers" by generating a framework of joint management.

More specifically: By demanding (not offering) participation in identical management courses of all members of the company, a meta-context was established within which subjective experience and interpretive-schemes could be grinded against each other producing a "creative-field" of intersubjectively legitimated actions. Within this framework, specific routines were implemented to secure a continuous flow of communicative interaction, notably a change report routine through which corrections, suggested alterations and new ideas could be channelled to the "appropriate" person(s) responsible. Action (or non-action) on the basis of reports were not left to the discretion of the receiver(s) but subject to intersubjective validation between all parties involved regardless of function or position. Thus, dialogues evolved between people who would not normally otherwise be connected. Consequently, new knowledge were created "that would not otherwise have existed in the world" (Jantsch, 1980). It is in the tension-field between ideal functional requirements and intersubjectively constructed "local" meaning that the organizational dynamics emerge. Furthermore, when "local" and "global" knowledge is linked, higher-order-learning (organizational effectiveness) is enhanced and implementation (organizational efficiency) facilitated as the context sends more messages.

THE DISCOURSIVE LEVEL

Argumentative rationality focuses on the technical context (Farrell, 1976). It is associated with basic assumptions of linear causality, zero-sum-games and objectivism. Narrative rationality is coupled to "reason,value and action". (Fisher, 1985: 30) It is associated with circular causality, plus-sum-games and subjectivism. "Narration, much like metaphor has power precisely because it captures complex experience that combines sense, emotion and imagination" (Weick & Browing, 1986 :250). Stories are coupled to narrative rationality with ontological status, i.e. stories are not only a reflection of something underlying, it is the underlying (Smircich & Morgan, 1982). Expanding communicative consciousness to encompass non-technical arenas of understanding, more common schemes of interpretations were generated. Much of organizational reality is hidden because of the predominance of technically related communication in organizations of the sort we are dealing with. Focusing on understanding, circular causality, interconnectedness and personalism elicit "hidden knowledge" and facilitates diffusion of multi-contextual learning-experiences. Competing values create paradoxes that need to be interpreted and "resolved" to guide action. Creative applications result from the interaction of competing narratives to the extent that there exist communicative routines ("market places"/"creative fields") that legitimize and connect idiosyncratic, organizationally embedded knowledge. Also, tacit knowledge may be elucidated and epiphenomena (e.g."trained incapacities" and "aspect-blindness"), endemic to highly differentiated hierarchies, may be alleviated.

CONCLUSION

Organizational performance may be increased by structuring activities within systems of feedback/feedforward based on heterarchical logic. A crucial alliance is needed to establish a sustainable framework within which different epistemologies may be grinded against each other to produce higher-order learning and local innovation. Focusing on meso-level communicative routines, arenas of ambiguity-reduction may be created that tap deeper into the inherent potential of organizational capabilities and stretches the limits of organizational performance.

REFERENCES

Bateson, G., 1972, "Steps to an Ecology of Mind", Intext Book, London.

Bateson, G., 1979, "Mind and Nature", Banton Books, Toronto.

Beer, S., 1981, "Brain of the Firm", John Wiley & Son, New York.

Brown, J.S. & Duguid, P. 1991, Organizational Learning and Communities-of-Practice: Toward a Unified View of Working, Learning and Innovation, *Organization Science*, Vol.2, No.1.

Czarniawska-Joerges, B., 1988, To Coin a Phrase, "The Study of Power and Democracy in Sweden Stockholm".

Farrell, T.B., 1976, Knowledge; Consensus, and Rhetorical theory, *Quarterly Journal of Speech*, 62, 1-14.

Fisher, W.R., 1985, The Narrative Paradigm: In the Beginning, *Journal of Communication*, 35, 74-89.

Glaser, B.G. & Strauss, A.L., 1967, "The Discovery of Grounded Theory", Aldine, Chicago.

Jackson, M.C., 1986, The Cybernetic Model of the Organization : An Assessment, *In*: R. Trappl, ed., "Cybernetics and Systems", Reidel, Dordrecht.

Jantsch, E., 1980, "The Self-Organizing Universe", Pergamon Press, Oxford.

Johannessen, J-A. 1991, Levels of Problem Structuring and Problem Definitions, *In*: M.C. Jackson, et.al. "Systems Thinking in Europe", Plenum Press, New York.

Johannessen, J-A.& Hauan, A., 1992, Reaching out for Heterarchy, *Cybernetics and Systems : An International Journal*, 23: 169-189.

Lincoln, Y.S. & Guba, E.G., 1985, "Naturalistic Inquiry", Sage, Beverly Hills, C.A..

March, J.G., 1981, Footnotes to Organizational Change, *Administrative Science Quarterly*, 26: 563-577.

Maturana, H. & Varela, F., 1987, "Kunnskapens tre", Ask, Aarhus.

Morgan, G., 1986, "Images of Organizations", Sage, London.

Nonaka, I., 1988, Creating Organizational Order out of Chaos : Self-Renewal in Japanese Firms, *California Management Review*, Vol XXX no. 3 : 57-74.

Olaisen, J. & Johannessen, J-A., 1991, From Strategic Management to Creative Management : Dynamic Learning and Information Processing in six Norwegian Banks, *Information Resources Management: An International Journal*.

Smiricich, L. & Morgan, G., 1982, Leadership: The Management of Meaning, *The Journal of Applied Behavioral Science*, 18, 257-273.

von Krogh, G. & Solem, O., 991, Theories of Autonomous Systems: A Comparative Analysis, *In:* M.C. Jackson et al., ed., "Systems Thinking in Europe", Plenum Press, New York.

Weick, K.E., 1979, "The Social Psychology of Organizing", Addison-Wesley, London.

Weick, K.E. & Browing, L.D., 1986, Argument and Narration in Organizational Communication, *Journal of Management*, 12,2, 243-259.

Zaleznick, A., 1989, "The Managerial Mystique : Restoring Leadership in Business", Harper & Row, New York.

Zuboff, S., 1988, "In the Age of the Smart Machine: The Future of Work and Power", Heinemann, Oxford.

MOVING MANAGEMENT AROUND THE ORGANISATION OR HOW TO MAKE JOBS FOR THE BOYS

Simon Lilley

Department of Accounting and Business Method
University of Edinburgh
William Robertson Building
50 George Square
Edinburgh EH8 9JY

INTRODUCTION

Robb[1] provides us with a conception of organisational life as a complex of institutionalised interpretive schemes competing for survival. The "fittest" institutions are, according to Robb, those with the greatest capacity for the dissipation of energy, in the form of human lifetime. Thus, the most successful interpretive schemes, such as those provided by Accountancy and Human Resource Management, are not necessarily the "best" at explaining the world. Moreover, institutions "entail processes which are directed solely towards self-production"[2]. However organisational participants may choose to rationalise and calculate their activities, a large proportion of their energies will be devoted to the perpetuation of institutionalised interpretive schemes.

Robb draws our attention to the difficulties such a situation presents for those who seek to steer organisations. Much of their attention must necessarily be devoted to reconciliation of conflicting views of organisational reality. The intention here is to extend the argument to partially invert its conclusions. By characterising the interpretive scheme that constitutes management as another lifetime dissipating institution, one starts to see the situation as an opportunity for management, rather than a threat. Mutual survival of management and other "job creating" institutions may be guaranteed by higher level institutional co-operation. The organisations in which these institutions are **currently** embedded may not prove so fortunate.

The paper draws on Miller's[3] insights on the subject of government generally, and management specifically, to consider recent changes to the refining business of one of the "big seven" oil companies. Management of activities and particularly the management of changes to the way in which activities are carried out, are seen to be "congenitally failing"[4], ensuring a constantly renewable source of jobs for the boys and the perpetuation of the institutional within the organisational. The paper concludes with a brief examination of the oil company's attempts to curtail the expansion of management. This move is conceptualised

Systems Science, Edited by F.A. Stowell *et al.*,
Plenum Press, New York, 1993

as an emergence within the organisation of "the politics of the product"[5] and an attempt is made to reconcile this view with Robb's[6] "Possible Solutions to Growing Institutionalisation."

GOVERNING THE ORGANISATION

Government, and management - its specific instantiation in enterprises, is characterised by Rose and Miller[7] as a "problematising activity". The obligations of managers derive from the specification of problems to be addressed. The institutions and practices of management are "bound to the constant identification of the difficulties and failures of [management]". "Programmes of government" of the enterprise are elaborated around articulated problems but these programmes are not determined by the managerial rationalities that they purport to represent. Rather, they are "translations" between the "morals, epistemologies and idioms" of those rationalities and the practicalities of the management or government of a particular problem space. Translations establish "..a mutuality between what is desirable and what can be made possible through the calculated activities of [managerial] forces"[8].

Programmes do more than articulate plans. They entail a legitimation of the right to intervene predicated upon claims to a certain knowledge of the problem domain. "Governing a sphere requires that it can be represented, depicted in a way which both grasps its truth and re-presents it in a form in which it can enter the sphere of conscious political calculation"[9]. Conceptualisation in these terms brings an entity into being in a form that both enables its characterisation as problematic and identifies those "best" able to provide a solution. Esoteric expertise is deployed both to explicate the indispensability of management and to provide managers with a vehicle for self instantiation[10].

Programmes themselves are not, however, directly implementable. They are deployed through an assemblage of heterogeneous techniques and procedures that constitute the "technologies of government"[11]. Technologies of government are necessarily realised within a system of complex contingencies and interdependencies. They are nothing more than a temporary stabilisation of elements in a network and are thus amenable to both systematic and "random" distortions. By the time managerial expertise is instantiated in the technologies of government it may be virtually unrecognisable.

THE CONGENITALLY FAILING ACTIVITY OF MANAGEMENT

Managers colonise the right to manage through the perpetuation of a particular "moral fiction". That is, they claim to "possess systematic effectiveness in controlling certain aspects of social reality"[12]. Through such processes problems of management are removed from the province of the moral and deposited within the realm of the technical. This apparent absence of a "rational" justification for managerial expertise has, however, had little effect on the proliferation, deployment and transformation of managerialism. Realisation of the inadequacies of the practices of management does not seem to lead to critical reflection on the philosophical underpinnings of the expertise deployed therein. Rather, the ineffectiveness of management serves to produce more management purporting to open up "those other issues" that brought about failure, to the "solutions" of managerial intervention. Thus, the identification of the failure of one form of management provides a justification for the implementation of another. "Managerial expertise, and the government of the enterprise more generally, is a congenitally 'failing' activity to the extent that a succession of programmes is the norm rather than the exception"[13].

We now go on to consider recent changes to the management of oil refining in one of the world's largest enterprises through the lens of institutionalised failure.

CHANGING MODALITIES OF MANAGEMENT AT THE WORLD FUEL ORGANISATION[1]

An oil refinery transforms crude oil into "products". For a refinery to function, arrangements must be made to ensure the availability of crude as an input and the possibility of the (profitable) distribution of the output products. Examination of the activities of World Fuel's (WF) management reveals how problematisation of these arrangements led to the deployment of a series of programmes at various points in the supply-refine-distribute system. These programmes tended to generate increased complexity, both planned and unintended. As the serial nature of the programmes suggests, further opportunities for management were "revealed" by earlier interventions, often at other sites within the system. The interrelatedness of supply, refining and distribution meant that the work of management today guaranteed the "necessity" of management, elsewhere in the system, tomorrow.

Up until relatively recently, oil companies, including WF, relied upon a highly vertically integrated organisation of their activities. They tended to bring their own crude, through their own transportation means, to their own refineries before products were dispatched through their own distribution means to their customers. This situation was finally disrupted by a coupling of management expertise, in the form of routinised forecasting, and recalcitrant political reality in the middle east.

> *...there was a vast expansion in crude oil processing capacity in the seventies in anticipation of a [continued] rapid growth in oil demand. Now what then happened was we had Yom Kippur in 1973, we then had the Iranian crisis in 1978/79 and because the crude price jumped in two very large steps from effectively $3 a barrel to $30 a barrel what you then had was a dramatic slowdown in growth of demand. The capacity had already been built and hence people found themselves with surplus capacity. What they then had to do if they wanted to move crude through it was get rid of the products.* (Market Trader, WF)

Thus, during the early nineteen-eighties intermediate markets started to become much more important to the oil refining industry, particularly with regard to product trading. The desire of oil companies, such as WF, to use assets in an overcapacity area, resulted in refineries pumping out surpluses and the generation of third party markets. As a consequence, "there was a whole new commercial scene to manage" (Market Trader, WF).

The emergence of this commercial scene was particularly marked at one of WF's major European refineries. Although the refinery had a relatively small local market, it's geographical location was such that it seemed well placed to exploit North European trading on the newly developed spot market. However, the benefits of playing in an intermediate market did not come without costs, the main one of these being the consequent increase in the complexity of oil management at refineries. This situation was exacerbated by the geographical separation of supply and distribution management, carried out at WF's corporate centre, and manufacturing management, carried out at the refinery. Whilst WF was acting as a vertically integrated company this arrangement did not seem to be problematic. However, with the emergence of third party markets, the activities of those individuals concerned with commercial dealings in refinery products, and later supplies, became more dynamic and consequently had a much more considerable effect on the refinery's activities. The proximity of the refinery to an important trading area established a highly unstable mutuality between manufacturing management and the decisions of buyers and sellers on the spot market.

An information systems solution to the separation problem was proposed and

[1] The "World Fuel Organisation" is a pseudonym.

implemented. WF traders working within the corporate centre supply and distribution function needed information on the refinery's stocks in order to allow them to play the market. The desire to maintain centralised commercial control led to the introduction of a Supply Information System (SIS) at the corporate HQ, that provided remote refinery information to satisfy the traders' requirements. SIS sought to increase the potential for "action at a distance"[14].

The SIS was seen to provide a partial solution to the problem of controlling production in relation to the demands of an intermediate market, but the resulting enhanced trading potential had some unpleasant side effects on the complexity of the refinery production management task. Tensions developed as the SIS facilitated the traders' ability to trade and these tensions were compounded by the increasing importance of immediate electronic trading. The refinery's proximity to the spot market's main distribution route, coupled with the traders increased ability to trade resulted in the refinery using up to fifty different crude supplies a year. Each cargo of crude was throughput in approximately two to three days. Thus, every two to three days the refinery had to respond to changes in its supply side, whilst at the other end output was extremely variable as it was market driven. As a consequence, WF was frequently unable to honour agreed deals from its own refinery manufacturing. Instead they were forced to buy stocks on the market to meet customer needs to which they were committed. The organisation was being "forced" to act as a "distressed" purchaser, leaving itself in a vulnerable position and providing opportunities for the deployment of new management programmes. These problems were apparent at many of WF's refineries but they were undoubtedly most pronounced at the site of the SIS implementation. Attempts to solve the problems at this refinery were to serve as exemplars of appropriate intervention at the group's other refineries.

The "distressed purchasing" was seen as symptomatic of the poor quality and lack of timely availability of information with which to manage, particularly for the refinery production planners and schedulers. A need was created and perceived for a new system that would allow "managers at [the refinery] to manage their plant more effectively" (Senior Manager, Business Development Unit). As refineries were closed during the eighties to remove overcapacity and "realign supply and demand", this need became all the more pressing as oil companies started to enter into collaborative relationships where product exchanges were made in an attempt to maintain market visibility in geographical areas that were no longer equipped with an "in-house" supply base. Exchanges, as well as the new commercial scene required management.

These examples of the results of management are not unrepresentative, but they are, perhaps, presented in a rather "cynical" fashion. The intention is not to claim that the accounts provided by managers to rationalise their own activities are disingenuous, but merely to demonstrate that management begets management.

INTRODUCING THE "POLITICS OF THE PRODUCT"

More recently, life seems have become a little less rosy for the institution of management within WF. The organisation has embarked upon a new programme which, whilst allowing increased deployment of managerial expertise in the short term, holds out the promise of a future of curtailed managerialism. Layers of management are being removed and a change programme is being deployed that seeks to inculcate a new culture in the organisation. The recent provision of a **refinery based** information system is reflective of the trajectory of the change process. The changes envisaged, and currently underway, seem largely consistent with the "Possible Solutions to Growing Institutionalisation" described by Robb in this volume. They may also be seen to reflect the wider emergence of a "new commercial agenda"[15] predicated upon "the politics of the product"[16].

Management and the other institutionalised interpretive schemes that inhabit the organisation have been so successful in establishing bodies of esoteric expertise that are insulated from the moral minefield of the political realm that the resulting combined hierarchies of professionalised specialisms purporting to direct organisations have been revealed as lacking in contact with the real world. The politics of the product, supported by a wider political mentality of neo-liberalism, requires the freeing of talent and the introduction of entrepreneurship to the fabric of the enterprise. With customer as king and the market as his messenger "expertise, activities, and outcomes are to be subject to the incessant demand that they contribute to the product"[17]. This framework has enabled the indictment of managerial expertise and the instigation of a programme of reform "to be undertaken within a broader transformation of modes of government of the enterprise...The authority and legitimacy of managerial expertise is to be reconceived and reconstructed. Excellent management seeks to allow and reinforce only those forms of expertise and activity that shape the required characteristics of the product"[18]. The reformation of management is seen as essential for its survival.

CONCLUSIONS

There seems little doubt that managerial understandings and the programmes that seek to instantiate them are changing, but only, perhaps, in order to maintain an adequate supply of the organisational resources on which they have come to depend. Changes may indeed be profound but they may equally be superficial and a range of outcomes to the latest collection of programmes are possible. Moves towards self management in organisations may just as easily represent an increased efficiency in the workings of the relays of power as a move towards empowerment and freedom, particularly with regard to their relative invisibility[19]. Moreover, work is seen to be required of managers in new modalities to provide systems to support the new ways of working, enabling them to build **their** understandings into the artefacts of the organisation. With fewer specialised managers but more managerial activity overall, the institution of the interpretive scheme of management is not dead, it is simply more effectively deployed and more deeply embedded.

REFERENCES

1. F.F. Robb, "Suprahuman Systems and Management: Steering in Jeopardy", this volume, (1993)
2. Robb, op.cit.
3. see, for example, N. Rose and P. Miller "Political power beyond the State: problematics of government", *British Journal of Sociology*, 43, Iss. 2, pp173-205; and P. Miller and T. O'Leary (forthcoming, 1993) "Accounting Expertise and the Politics of the Product: Economic Citizenship and Modes of Corporate Governance", *Accounting, Organizations and Society*, 18, Iss. 2&3
4. P. Miller and T. O'Leary, op. cit.
5. P. Miller and T. O'Leary, op. cit.
6. F.F. Robb, op. cit.
7. N. Rose and P. Miller (1992) op. cit.
8. N. Rose and P. Miller (1992) op. cit.
9. N. Rose and P. Miller (1992) op. cit.
10. see M. Foucault (1978) "The History of Sexuality, Vol. 1: An Introduction, London, Allen Lane, for an expression of the process of self instantiation.
11. N. Rose and P. Miller (1992) op. cit.
12. A. MacIntyre "After Virtue", 2nd ed., London, Duckworth, (1985), cited in P. Miller and T. O'Leary, op. cit.
13. P. Miller and T. O'Leary, op. cit.
14. see, for example, M. Callon, Some elements of a sociology of translation, *in*: "Power, Action and Belief", J. Law, ed., Routledge and Kegan Paul, London, (1986).

15. see, for example, R. Munro and D.J. Hatherly (forthcoming) Accountability and the New Commercial Agenda, *Critical Perspectives on Accounting*.
16. P. Miller and T. O'Leary, op. cit.
17. P. Miller and T. O'Leary, op. cit.
18. P. Miller and T. O'Leary, op. cit.
19. M. Foucault, "Discipline and Punish: The Birth of the Prison", Allen Lane, London, (1977).

CYBERNETICS AND THE SOCIAL: CONVERSATIONS WITH UNSPEAKABLE MACHINES

Rolland Munro

Department of Accounting and Business Method
University of Edinburgh
50 George Square
Edinburgh EH8 9JY

INTRODUCTION

The terms cybernetics and the social stand uneasily together. Their respective laws reflect an antinomy. The one appears to exclude the other, the technical against the human, the rational against the emotional, the future against tradition. The antipathy is one of deletion. The growth of cybernetics heralds the 'end of the social'[1]. The Luddites on the one side of the millennium and on the other side the 'cyborgs'[2]. And yet the rhetoric of one attack invariably draws from the ground of the other. This might stir second thoughts.

The dichotomy of cybernetics and the social is not complete, nor final. There are overlaps enough to call for some reconsideration. For example, both terms draw on holistic thinking. This is not a matter of noting similarities in vocabularies, say governor and government. Rather it is to recognise how deeply theories of cybernetics and thinking about society share a common heritage in a systems thinking that goes back to Plato's Republic and a division between parts and wholes which is embedded in Aristotelian logic.

It is, for example, as easy to talk quickly of Cybernetics as if it were an entity, just as it is too easy to talk of Society as if it were a whole, forgetting those who are displaced and marginalised. As terms, cybernetics and the social are both subject to *reification*. The problem of reification is one *of false attribution*. Certain properties which are thought only to belong to the parts become attributed to the whole. To keep asking to see the University even when one has seen the library, the lecture rooms and the college halls, is according to Ryle[3] to commit a category mistake.

In organization theory it is common enough to talk of the organization 'thinking this' or 'doing that'. And it was exactly to attack this type of reification that Silverman[4] launched his attack against systems theory. Before abandoning all systems work, however, some words of wisdom from Mrs Thatcher might be borne in mind:

> There is no such thing as society. There are individual men and women and there are families[5].

An odd remark from someone in Government. But a useful reminder how avoiding one absurdity can quickly lead one to embrace another.

Systems Science, Edited by F.A. Stowell *et al.*,
Plenum Press, New York, 1993

Humans as machines. Machines as humans. Little is to be gained in a game of inversions. In what follows I want to explore what is entailed by a care to avoid reification *and* heroification. The problem which Silver-man and the Iron Lady share (apart from the metallic quality of their names) is that they make heroes of individuals *as* individuals. In order to avoid one form of reification, they commit a further category mistake in attributing all properties of organizations to individuals. Flattering but fallacious.

In re-considering the social and its relations to cybernetics I examine first the strong view of the social from anthropology. Essentially this position argues against any displacement of the social by the cybernetic view, of which economics is perceived to be the paradigm case. Next I consider the socio-technical view of scholars working in Science, Technology and Society. This sets out a terrain which substantially helps to diminish tensions between cybernetic theory and social theorists. In the following section I raise the problem of *membership* by way of giving some reservations about extending agency to non-human entities. Membership, I argue, retains a strong view of the social but develops the anthropological view to accommodate *trading* between members and machines. What I propose prevents the social of being denuded of all animation and radically changes its geography.

CONSUMPTION AND CULTURE

Following Levi-Strauss, Douglas[6] and Douglas and Isherwood[7], among others, argue for an inversion of the traditional displacement of the social by the economic. Against Levi-Strauss's division of social life into three communication systems of goods, women and words, Douglas & Isherwood[8] claims to synthesise these into a theory of consumption.

> The meanings conveyed along the goods channel are part and parcel of the meanings in the kinship and mythology channels.

By dissolving false boundaries, Douglas challenges the reification of economics. Implicit in Douglas's theorising of systems as instantiated through *consumption* is the notion of boundaries being constituted by a process of structuration. This raises possibilities for understanding 'the social' without recourse to a cultural determinism similar to Durkheim's.

In drawing on Levi-Strauss, Douglas does not eschew systems notions. Instead she *grounds* them by theorising consumption in terms of meanings. Douglas and Isherwood[9] insist that 'any choice between goods is the result of, and contributes to, culture'. By arguing that goods are consumed because they are good to think, Douglas rescues 'the social' from its displacement by economic thinking in western societies.

Against its marginal status in the production frame of economics, where society has no place in analysis beyond its status as another good, Douglas reasserts the centrality of the social. The social, in this view, is much more than the economist's sum of revealed preferences. Reimporting the social *as* social turns around so-called irrational consumption. Again Douglas & Isherwood:[10]

> Theories of consumption which assume a puppet consumer, prey to the advertiser's wiles, or consumers jealously competing for no sane motive, lemming consumers rushing to disaster, are frivolous, even dangerous.

Who sleeps with whom, who feeds with whom and who *is* whom, these are not matters only for economics. These matters are recursively organized through the social:

> Studies of the ancient Israelites..contemporary Thai villagers..and contemporary Lele tribesmen in the Congo show how the world is organized in a recursive system of metaphors dealing with admission to bed, board and cult.

Trading reveals the social which constitutes them and is instantiated by such tradings. Douglas & Isherwood theorise 'the social' as being *active* in the sharing of goods:

> No one likes to recognise that the capacity to share all three [bed, board and cult] is socially endowed, a result of current decisions, and not an ineluctable fact of nature.

What economics simply deletes becomes for the social theorist of utmost importance.

In drawing attention to the recursive effects of meanings Douglas's theorising of consumption achieves more than a distancing from Durkheim's cultural determinism. She shows how notions of society need not rely on 'shared values' or 'shared meanings'. Instead Douglas points to the *effects* which are sustained as a consequence of people's beliefs about sharing values. Of principal importance here is the effect of *exclusion.*

It is important to note how intimately for Douglas exclusion is tied to the consumption thesis:

> Sharing goods and being made welcome to the hospitable table and to the marriage bed are the first, closest fields of inclusion, where exclusion operates spontaneously long before political boundaries are at stake.

But more is entailed. The 'naturalness' of exclusion has also to be asserted:

> ..whenever exclusion is operated to define a category of outsiders, the segregated category tends to be accredited with a different nature..

Exclusion is always 'natural' and the consumption thesis helps to explain why this appears so. 'Communication gaps' and the unspeakability of the Other are intimately related.

The relation between exclusion and assertions of the different nature of the Other will be discussed further below when the question of membership is examined. First, however, the unspeakability of machines is considered.

SCIENCE, TECHNOLOGY AND SOCIETY

For those concerned with Science, Technology and Society, a persistent drawing on technology interpenetrates social conduct. According to Callon[11], and Law[12], the *nature* of the social is changed and extended by machines.

What makes the Science, Technology and Society view of interest and different from 'evolutionist' views is that no voluntaristic or deterministic perspective is involved. Society does not wait upon human discovery, but neither is it mechanically altered by technical innovation. The social is not eliminated by the technical. Instead there is an interpenetration and a reweaving around each other. Interpreting the position crudely, change is an effect of *agents*. But the important rider is that agents may be *either* humans *or* machines. A more correct statement might go further and drop the distinction, recognising that professions such as sociology and accounting create 'experts' whose thinking cannot be delimited to the simply human. According to Law[13], to be an expert is to be a 'social-technical' creature.

The nub of the Science, Technology and Society argument is that we are all experts, all 'social-technical creatures'. Or at least those who use cash machines, kitchen mixers, desktop terminals. All betray expertise. Being part of the *distribution* of knowledge is inescapable as is a reliance on that distribution. The point is more general, however, than say Poster[14] is making and does not require a new epoch to be theorised. Being networked with databases is not a new phenomena, although drawing cash from a wall may seem more impressive than following a cooking recipe.

What has to be given up are boundaries which fix the self as a single entity or system. The body is extended multifariously through different networks. This is what is entailed by the 'actor-network' view. What also is to be abandoned is the idea of self having an essence; there is no constant defining core. If this is puzzling think of Wittgenstein's[15] point about games. No one definition will suit all games. Instead, Wittgenstein argues, there are family resemblance's. It is not that there is nothing to be called self. Rather there are too many socio-technical arenas, too many actor-networkings to sustain a singular definition of self.

THE SOCIAL AND MEMBERSHIP

Although Douglas's consumption thesis has received scant attention in the systems literature, elsewhere the theme has become a familiar one. For example Friedman[16] develops the thesis by drawing on the relations between the construction of identity and consumption:

> Following a line of argument that began with the recognition that goods are building blocks of life- worlds, we have suggested, as have others, that they can be understood as constituents of selfhood, of social identity.

In terms also reminiscent of Bourdieu[17], Friedman goes on to argue that, by further assuming a global historical frame, it is possible to:

> ..detect and even to account for differences among broad classes of strategies of identity and therefore of consumption and production.

These themes of identity being constructed through consumption have some overlap with identity considerations in actor-network theory, especially where they provide arguments for a fragmentation of selves. However, whereas Friedman tacitly relies on an animation through the social, no such reliance has been made by Law for actor-network theory. Treating machines as inseparable from humans certainly allows for a greater heterogeneity in the social, but leaves the social strangely attenuated. That networks work well for some and effect pain on others can be accounted for only by *positions* in the network. Acknowledgement of the importance of distribution seems necessary but not sufficient. What is lacking is Douglas's thrust of the social as *organizing* inclusion/exclusion.

Drawing on Harold Garfinkel, 'membership' offers one possible way of overcoming this lack of animation. Membership in Garfinkel's sense depends on recognition in the eyes of others. Membership is centrally concerned with affiliation, a sense of belonging with each other. Membership therefore should not be confused with formal acceptance to a 'fictional' body, such as UKSS, since affiliation is never fixed but depends, as Goffman[18] recognised, on performance. For example, my turning up as a newcomer at a UKSS conference raises different expectations to that of say Peter Checkland. My electing myself for membership does not entail others affiliating to me.

Membership is something the would be member has to keep happening. We have to keep 'making it'. To keep qualifying for membership even to very broad categories such as 'human' requires more than expertise in speaking a 'natural' language, like English. As Goffman's examples of 'face work' reveal, conduct also has to fit with what is *expected.* Membership is a moveable feast.

Garfinkel[19] relates this continuous performance to making the everyday accountable:

> ..the activities whereby members produce and manage settings of organized everyday affairs are identical with members' procedures for making those settings "accountable".

Producing and managing settings is a matter of expertise. Expertise is required to 'do jury', to 'do nurse', to 'do manager'. But the question of performance here is not a mechanical one. Knowledgability extends to knowing *what* performance to give, *when* to give it and for *whom* the performance is given. What members of an organization know for example, in addition to much else, is how to *accomplish* hierarchy. By 'doing manager' and 'doing employee' they sustain superior-subordinate relationships. As someone 'doing newcomer' I may meet at the UKSS someone who will 'do old hand'. The performances are not easily reversible.

On these arguments the nature of the social may change less from the *fact* of expertise and more from its *gradient.*[20] This is certainly a view that those concerned with Science, Technology and Society might agree. It is the gradient of expertise which may determine inclusion and exclusion. But this gradient is animated through the social-actor networks. Importation of cybernetic forms of accountability entails more than the social *accommodating* changes in accounting techniques as, say, an extension of economic practices. Rather the animation of the social changes as more and more 'normative labour' is turned away from face work and embedded in artefacts. For example, as the prevalence of accounting numbers extends into the day to day, what is 'new' about accounting is the extent to which the *conduct* of society is being rewoven around its techniques. As accounting numbers and accountability become more intertwined in ways which make, and define, membership of society, it is the *moral order* which is being changed and carried by artefacts.

Consider, for example, the following person who has started rising swiftly through his changing organization. He wasn't making his performance indicators and so was sent on a time management course. Now as you enter his room, there is no comfortable chair to sit down for a chat. A single upright chair is hard against the wall and farthest from the door. As he walks to the door, your body follows. Soon you are back in your own office. Blocked from the chance to chat with your colleague, perhaps you too will now network with your daily priority list, ticking off the major items as 'you' go.

Cybernetic forms of accountability are central to current change agendas sweeping organizations[21]. But the technologies alone do not make people jump before they are pushed. It is not the closing of 'communication gaps' which are at stake in the drive to change social conduct. Certainly any technology like accounting offers an opening up of some routes for communication and closing down of others. But this is not Robb's[22] point. By generating interpretive frameworks, conversations:

> ..'make realities' and define what is unreal..they create 'goods' and define what is not good.

Conversation changes members' *readings* over the desirability of making certain moves.

This is the target. Conversations do not just take place in the workplace, they take place, for example, *with* the workplace[23]. To understand the interpenetration of the social by the cybernetic, the notion of conversations has to be widened from those between members to include those between members and agents. The secret exchange of the social goes beyond that of a nod and a wink into feeling pressure from five items being left on a priority list instead of four.

CONCLUSIONS

Ours is indeed a well populated planet. This seems particularly so when machines are admitted to the population of agents, a re-counting which may balance fears of a population explosion in an ever-breeding third world. While moral issues remain, they have to be rethought against the density of agents in the West and the Pacific Rim.

The purpose of this paper, however, has not been to suggest that machines simply add to the terrain. Nor is the point one of simply recognising that actor-networks *change* the terrain. While both these positions are true they miss the point. Coming to terms with a cybernetics of the social involves reworking many of our deepest epistemological structures.

To heroize the individual does not to make society human, rather it simply ignores the social as it exists today. And to talk of cybernetics in cybernetic terms is to bracket the social as background, to treat the social as an ever-receding context. The social cannot be recessed into 'context'. The social is not a container-like material into which cybernetic techniques such as accounting are occasionally, if somewhat persistently, drawn.

Within change agendas sweeping across all organizations, an increase in the gradient of expertise affects the social *and* is effected through the social. Cybernetic technologies, such as accounting or auditing may get the credit (or more likely the blame) but the effects are more subtle than this. The expectation is that people alter their everyday conduct in ways which radically affect membership through evolving new groupings and affiliations.

Most puzzling is the question of whether much of our sense of belonging, if not actually conducted by machines, is conducted through these wider forms of agency. Perhaps drawing on these wider forms of agency helps sustain multiple notions of self? In this sense there may appear to be a 'liquidation' of the social. Nevertheless this paper suggests some care in celebrating a postmodern future too soon. A 'universe of signs' as predicted by Baudrillard may usher in more than freedom in the care of the self. Where the gradient of expertise is raised, displacement, exclusion and pain are likely to become more and more self-enacted through reading, almost literally, the writing on the wall.

Those interested in the formation of the social ask themselves questions about *how* central aspects of it, such as community or expertise, are constituted, particularly in the everyday practice of its members. By considering the anthropological view alongside that of Science, Technology and Society, what has been suggested is that notions of the social have traditionally overlooked how actors network with machines. A new geography of the social lies waiting to be explored.

The social has to be rethought. The *conduit* of conduct have to be re-imagined. Including machines as agents is a first step. As a second step, widening the notion of conversation to include talk *within* the actor-network makes visible the vast amounts of consumption which are overlooked from the trading accounts of the economic and excluded from view by the humanists. A third step might be to weigh up the cost of all this normative labour in terms of displacement, exclusion and pain.

REFERENCES

1. J. Baurillard, "Simulations", P. Foss et al, trans., MacMillan, New York (1983).
2. D. Haraway, A Manifesto for Cyborgs: Science, Technology, and Social Ferninism the 1980s, in "Feminism/Postmodernism," L.J. Nicholson, ed., Routledge, New York (1990).
3. G. Ryle, "The Concept of Mind," Hutchinson, London (1949).
4. D. Silverman, "Theory of Organizations," Heineman, London (1970).
5. Quoted in M. Strathern, "After Nature: English Kinship in the Late Twentieth Century," Cambridge University Press, Cambridge (1992).
6. M. Douglas, "Implicit Meanings: Essays in Anthropology," Routledge, London (1975).
7. M. Douglas and Baron Isherwood, "The World of Goods: towards an Anthropology of Consumption," Penguin, Harmondsworth (1980).
8. Douglas and Isherwood, op cit, p.88
9. Douglas and Isherwood, op cit,p.76
10. Douglas and Isherwood, op cit,p.89
11. M. Callon, Struggles and Negociations to Define What is Problematic and What is Not: the Sociologic of Translation, in: "The Social Processes of Scientific Innovation," K.D. Knorr, R. Krohn, & R.D. Whitley, ed., pp.197-219, Reidel, Dordrecht and Boston (1980).
12. J Law, Introduction: monsters, machines and sociotechnical relations, pp.l-23 in: "A Sociology of Monsters: Essays on Power, Technology and Domination," J. Law, ed., *Sociology Review Monograph 38*, Routledge, London (1991).
13. Law, op cit.
14. M. Poster, "The Mode of Information: Poststructuralism and Social Context," Polity. Cambridge (1990).
15. L. Wittgenstein, "Philosophical Investigations," Basil Blackwell, Oxford, (1958).
16. J. Friedman, Being in the World: Globalization and Localization, *Theory, Culture & Society* 8 (4) p.327 (1990).
17. P. Bourdieu, "Distinction: A Social Critique of the Judgement of Taste," R. Nice, trans., Routledge, London (1984).
18. E. Goffman, "The Presentation of Self in Everyday Life," Social Sciences Research Centre, University of Edinburgh, Edinburgh (1958).
19. H. Garfinkel, "Studies in Ethnomethodology," Prentice Hall, Englewood Cliffs, N.J. (1967).
20. J. Law, privately circulated paper.
21. R. Munro and D. Hatherly, Accountability and the New Commercial Agenda, *Critical Perspectives on Accounting* (forthcoming).
22. F.F. Robb, Suprahuman Systems and Management, this volume.
23. R. Munro, Enabling Participative Change: the impact of a strategic value, *International Studies in Management and Organization* 21(4) pp.52-65 (1991).

LINKING SSM WITH VSM FOR INFORMATION SYSTEM MANAGEMENT

Graham Paton

Army CIS Agency
Blandford Camp
Blandford Forum DT11 8RE

INTRODUCTION

Wilson (1990) describes a number of years spent grappling with the problem of defining and designing what we commonly refer to now as "information systems". Strictly speaking of course since both "information" and "system" are intellectual constructs we can't really build a physical "information system". This linguistic tantalogical nicety doesn't stop us trying however and therein lies the focus of our difficulties in managing the process. But more of that later. Wilson has described the systems movement as rather like an extensive flat country with the odd isolated hill. Now it seems to this observer that within this 'country' there are two particularly high points of especial significance. These are Stafford Beer's Viable System Model (VSM) and Peter Checkland's Soft Systems Methodology (SSM). This paper describes one way in which these too high points can be linked to facilitate the management of information systems. The very idea of linking a methodology with a model may seem perverse but it is hoped that what follows will justify such a link and provide a rationale for it.

However the setting which this paper addresses (information systems management) is itself a complex area and because of the very wide range of usage associated with all three terms ('information', 'systems' and 'management') it is important to define the approach followed here.

DEFINING THE TERMS

Shannon & Weaver (1949) defined information as the quantification of the relationship between a particular communicated message and the associated set of possible messages. Shannon specifically excluded any consideration of the meaning of a message: "These sematic aspects of communication are irrelevant to the engineering problem". Although he does not return to this issue, his co-author Warren Weaver does. In the introduction to a detailed discussion of information, he is careful to point out that "The word information, in this theory, is used in a special sense that must not be confused with its ordinary usage. In

Systems Science, Edited by F.A. Stowell *et al.*,
Plenum Press, New York, 1993

particular, information must not be confused with meaning. In fact, two messages, one of which is heavily loaded with meaning and the other of which is pure nonsense, can be exactly equivalent, from the present viewpoint, as regards information....... this word information in communications theory relates not so much to what you do say, as to what you could say".

Shannon's definition remains a fundamental underpinning of our present understanding of communications theory. It is, in the communications context, a powerful and useful intellectual construct. However the meaning of 'information' in 'information systems' is significantly different. Here it is more akin to the general language usage of 'information'. This usage of 'information' is close to the concept of knowledge. We can think of information as knowledge which has been passed to us. Something we have been, or become, <u>informed</u> about. We should remember that such an intellectual commodity is highly valuable. Seneca said 'knowledge is power'; information too then can be seen as a crucially important political commodity.

Let us now turn to consider 'system'. This is a complex word which has been widely used and abused in the twentieth century. This is not the place for a detailed consideration of its status. Many, including the author here regard it too as an intellectual construct; a way of thinking about the world rather than a description of real world objects. This is the way in which Checkland uses the word and hence it is the basis of Soft Systems Methodology (SSM). Nevertheless a common usage of 'system' as referring to any organised, interconnected set is now pervasive and must be considered legitimate as a facto linguistic development. So we should consider the implications of both these uses here.

We are not here talking of information alone as a commodity however, but of 'information systems' (IS). What are these? Well in one very clear sense as we have already observed, an 'information system' is an intellectual construct rather than a physical entity. In spite of this many people are engaged today in the specification design, implementation and use of what they call computer-based 'information systems'. So IS follows the same course as 'System', having concurrent meaning in two quite distinct domains. We have already proposed to consider both. However, because the general position of this paper is to assert that the domain of system (and IS) as intellectual construct is of far greater significance, we shall reserve that domain for the moment, and look first at the "IS are real world entities", domain. One final definitional task is necessary before we do however; that is to state what is meant here by 'management'. In comparison with our two earlier essays in definition this is an easy task. The term is used here in a very broad sense, to encompass all the activities involved in the specification, design, implementation and use of IS. Within this broad definition however it is intended to focus on the issue of what the IS is, and what it should be, rather than on operational details of day to day administration and use.

DEVELOPING THE BASIC FUNCTIONAL MODEL

Let us first consider what SSM can contribute to this management task. There is not an extensive literature of case studies of the application of SSM to IS. It has been suggested previously, Paton (1992), that this may be because of reticence on the part of organisations to reveal details of internal analysis. Those case studies which are available all show SSM used in the now classic way, after Wilson (1990), of creating a Primary Task Model (PTM) of the IS to be developed and then (somehow) making the transition into design. Now traditional structured analysis and design methodologies have been strongly criticised for moving too readily into a specific design without real discussion of whether an IS is really needed. Using SSM to create a PTM does allow, indeed promotes, a debate with key actors in the situation on the issue of whether the activities in the PTM are correct. This is very useful, (better than the traditional "quick march" from stated user requirement into design), but does not adequately address the question of whether an IS is truly needed. To do that one has to step back from the immediate "problem" of defining the IS, to understand the functional organisational activities which the IS is to support.

So the sensible first step would seem to be to create a primary task model of the enterprise for which an IS is proposed. This business PTM can then be the vehicle for discussion of the sufficiency and adequacy of present business activities and the closely linked debate about which can be assisted by computers and how. Such modelling can be mapped onto Wilson's ISM and this has been done, see Paton (1989).

However classic PTMs (that is those developed using the classic SSM) of Checkland (1981) do not alone provide an adequate description of an enterprise. This is because they focus on the functional activities of the enterprise. They provide no understanding of management, or control mechanisms. In particular they are ill-equipped to show how any particular system fits into the hierarchical management structure common in large organisations.

INCORPORATING HIERARCHY

It is of course possible to show a degree of management activity explicitly within a PTM. This often leads to the rather cumbersome inclusion of the 'management spider' in the PTM; a 'monitor and control' subsystem with links to, and from, all other subsystems. This is not a particularly useful convention as it tends to obfuscate the PTM.

Checkland and Scholes (1990), in the latest and most developed version of SSM, use several onion-like layers around the conceptual model to show successive levels of management activity. This formulation does allow some representation of management and control, but does not fully describe the complex linkages involved. However, a very good mechanism already exists which is specifically designed to display hierarchical management. This is Beer's Viable System Model (VSM). Beer (1985) provides a good overview of how to apply the VSM. Let us now look at the contribution this can make to IS Management.

THE RELATIVE STATUS OF SSM AND THE VSM

As a first step we have to consider the relationship between SSM and the VSM. The VSM was specifically created to portray the hierarchically-linked management which must exist for a system to be viable. Beer accepts that systems "exist" in the real world and can be engineered. This ontological position is, at first sight, in total opposition to that of SSM. In SSM Checkland is careful to point out that conceptual models are independent of the real world; to be used as intellectual tools to understand the real world and inform a debate about changes (in the real world). However the position is not quite so stark as it appears. SSM itself makes few formal proposals about how the actors in a so-called "problem situation" should go about implementing any changes which the process of debate has caused the client to decide on. It is therefore reasonable to suggest that one possible outcome is to take the conceptual model, and change its status from analytical tool to that of blueprint for a real-world organisation. Such a move is, of course, only possible when the conceptual model concerned is a task-based model. It is especially easy to adopt this course of action when the conceptual model has been developed as a PTM. So, in these particular circumstances, it is possible to make the ontological shift from the rigorous position which SSM presents, (of the conceptual model as intellectual construct), into the "model as blueprint" in the real world. Once this shift has been achieved we are working with a PTM which is of the same ontological status as the VSM.

DEVELOPING THE MODEL SET

At this point we have a specifically context-related PTM. We need next to develop a complementary context-specific hierarchical model following the form of the VSM.

In doing this we can use the component sub-systems from our PTM as base operational systems. There is no advantage to be gained from attempting the tricky diagramming task of developing a pictorial model in which PTM and an associated context-related VSM (C-VSM) are combined. The two models, as a set, however, provide a comprehensive schematic representation of both operational activities of, and managerial control on, our system under study. This is a more powerful and useful representation of the system under study than either a primary task conceptual model or context-related VSM alone.

Up to this point we have focused on the development of models, both PTM and C-VSM, which are closely linked to the real world. The underlying assumption we have worked with so far is that something called an IS can be engineered, probably largely computer supported, and does actually exist.

A MORE RADICAL DEFINITION OF IS

Let us consider the more radical view that IS do not actually exist, but are constructs of our minds. Well perhaps the most important thing to say, straight away, is that this does not alter the practicality of creating the PTM and C-VSM model set. It does alter the way in which we view the model set however. We can both develop and utilise the set with a clear understanding that they are abstract models and have no correspondence, actual or proposed, in the real world. For this we are accepting that the model is not necessarily the best (or indeed the only) set which we could create; merely a useful one. From this perspective the value of the model set is that it allows us to hold an informed discussion about how computing support could be introduced to or expanded within the situation. This debate is most useful at the point of initial introduction of computing support, however it is also relevant at points of expansion or upgrade and in assisting with whole-life management.

Only by relating actual computer support to the "business" activities and management constraints (especially resource constraints) can really cost-beneficial computer support be provided and maintained. Without some sophisticated model of the business on which to map this support, achieving such an understanding is very difficult. Thus the PTM and C-VSM model set can be a powerful aid to effective management of computer support.

GENERAL WAYS OF USING THE MODEL SET

Whichever of these two rationales of IS is being adopted, it is difficult to demonstrate how such a PTM + C-VSM set can be used without providing the contextual background material of a specific case. This is not feasible in this short paper. However some general ways of using the model set have emerged from the author's attempts at a small number of applications. Checkland in describing how a CM can be compared with the real world situation, suggests assessing whether and how far activities in the model are actually present in the real world. This process can be applied equally to a PTM, not withstanding the fact that by its method of creation as a consensus model by the key actors in the situation, it will necessarily be more closely linked with the real world than a conceptual model created in a more objective manner by some (independent) analyst. The PTM activities alone are not the only elements of the model which can usefully be analyzed in this way however. It can be very instructive to undertake the same exercise with each of the logical link arrows of the

PTM. For each of them the analyst can ask how (by what procedures and methods) is this link logic instantiated in the real world. Often the apparently transparent link logic implies a whole series of information activities and data transfer operations which have form and substance in the real world. Systems paradigms (especially the communication and control models) can be used to examine such real world activities and check their effectiveness and efficiency. A similar exercise can be conducted using the C-VSM logic links.

The recursive nature of systems is superbly demonstrated within the VSM. However the VSM, in providing a cybernetic perspective of the control of a system, can in complex organisational situations obscure more easily comprehensible forms of presentation. Comprehensible, that is, to the actors in a situation; especially those who are concerned with making decisions based on the analysis. SSM can contribute to the effective presentation of human activity related to control and management by providing an alternative view of elements within the C-VSM. For example a conceptual model can be constructed of one of the management levels in the C-VSM, to show how this activity could be logically conducted in this particular situation. Such modelling is especially useful when the subject of the analysis is such that there are questions about how extensive an IS is to be developed; that is where the division of activities between (computerised) IS and human actors is particularly problematic.

CONCLUSION

The combination of SSM and the VSM described in this paper provides a rich model set which can be utilised to assist in the specification, design or management in operation of an IS. The model set provides both analyst and IS designers or managers with a more comprehensive understanding of the context within which the IS must operate than either use of a primary task model or application of the VSM can provide on its own.

REFERENCES

Beer, S., 1985, "Diagnosing the System for Organisations", Wiley, Chichester, England.

Checkland, P.B., and Scholes, J., 1990, "Soft Systems Methodology in Action", Wiley, Chichester, England.

Checkland, P.B., 1981, "Systems Thinking, Systems Practice", Wiley, Chichester, England.

Paton, G., 1989, Applying Wilson's ISM to a Planning Agency, in "Proceedings of the 33rd Annual meeting of the international Society for Systems & Sciences" (P.W.J.Leddington, Ed), Edinburgh.

Paton, G., 1992, "Information Systems - not artifacts but models", *Systemist*, UK Systems Society, Vol 14, No 3., pp120-122.

Shannon, C.E., and Weaver W., 1948, "A mathematical theory of communication". University of Illinois Press, Urbana, 1949.

Wilson, B., 1990, "Systems: Concepts Methodologies and Applications" John Wiley, Chichester, UK.

CYBERNETICS IN MANAGEMENT

Angus J. Quin

Information Systems
Department of Computing Science
University of Paisley
Paisley
Renfrewshire PA1 2BE

INTRODUCTION

This paper argues that the viable systems model can be used by managers to understand and control their organizations.

The success or failure of any organization largely depends on its ability to grasp the opportunities available to it. This in turn depends on its internal management structure, systems, culture and the resources of the organization. In the U.K. many of the traditional industries are in a state of decline. New industries are emerging but they are being adversely affected by the recession and competition from overseas. Competent managers are critical to the success of these industries. In order to be more effective managers should have a solid grounding in organizational structures and system theory.

Developments in systems theory and cybernetics over the last 40 years have provided new opportunities for managers to improve the way they work. Managers no longer need to rely on intuition, instead they can build and refine models to understand and control their organizations. This paper is concerned with the support that control theory or cybernetics can give to managers in order to ensure the effectiveness of their organizations. In particular it is argued that Stafford Beer's Viable Systems Model (VSM) is a suitable tool for applying cybernetic methods and that VSM can be applied to any size and type of organization.

THE NEED FOR SYSTEMS THINKING

There has been a dramatic increase in the use of information technology (IT) in many organizations but results have been disappointing. IT has also taken the brunt of the blame for disasters such as the stockmarket crash on Black Monday in 1987, (The Economist, 1987). London's ambulance crisis was blamed on the computer system though it is claimed that the system was working exactly the way it was designed, (Watts, 1992). The lesson learned is

Systems Science, Edited by F.A. Stowell *et al.*,
Plenum Press, New York, 1993

that new technologies must only be introduced after careful thought about the context in which they are expected to operate.

Commercial and political organizations have existed in every period in recorded history. Their life span varies enormously. Some organizations have survived for centuries whilst others die quickly. There has been a great deal of speculation about the reasons for this phenomenon, however what is becoming increasingly clear is that a holist or systemic approach should be adopted when developing new organizational structures. Systems theory has been suggested as one possible conceptual basis for this (Stowell and West, 1992).

Early systems theory has its roots in biology though systems concepts were applied in both military and economic systems in the 1940's (Tustin, 1953). The systems approach is based on the notion of holism for example an organization cannot be properly understood outside its environment. Cybernetic systems are systems that use feedback and are designed. This definition includes organizations. Waring (1989) claims that systems are designed solely to produce the useful subset of their emergent properties. After all if it were possible to obtain the desired results from individual components of the system there would be no need for the system.

The problem for the manager is to design organizations that achieve these desirable emergent properties and avoid unpredictable side effects. In addition should the manager try to eliminate unwanted emergent properties? Frequently complex systems do not behave in a predictable way (Waring, 1989). This is the basis of Goodharts law (Goodhart, 1984). Beer addressed these problems and came to the conclusion that there are fundamental principles of control that apply to all large systems (Beer, 1981). Beer argues that these principles apply to the human brain, computer systems, social systems and economic systems. Beer (1981) is concerned with the contribution cybernetics can make to management.

STAFFORD BEER'S VIABLE SYSTEMS MODEL (VSM)

According to Beer, (1991), the cybernetic model of any viable system consists of five necessary and sufficient subsystems which interact in order to ensure the survival of the organization in a given environment (see Figure 1). The VSM is fully described elsewhere (Beer, 1981; 1991).

Beer does not claim that the VSM optimises the operation of an organization, however he proposed that it can be used to analyze any type of organization structure and help in any redesign found necessary.

WHY BEER'S VSM MIGHT HELP IN IS DESIGN

The IS practitioner using VSM would probably break the existing information system down into viable subsystems which both the client and systems analyst can agree upon before discussing the clients needs with respect to these subsystems. It has been suggested that the resulting specification can be rigorously tested for validity against the system context and a formal specification drawn up (Swatman and Swatman, 1992). The VSM can support many different views including those of sociologists, systems analysts, managers, network engineers and operatives (Espejo and Watt, 1988; Espejo and Harnden, 1989).

VSM IS NOT WITHOUT CRITICISM

It has been argued that the VSM does not model issues such as shared culture, trust and tradition (Checkland, 1986; Espejo and Harnden, 1989).

Espejo and Watt (1988) discusses how to apply the VSM to social organizations.

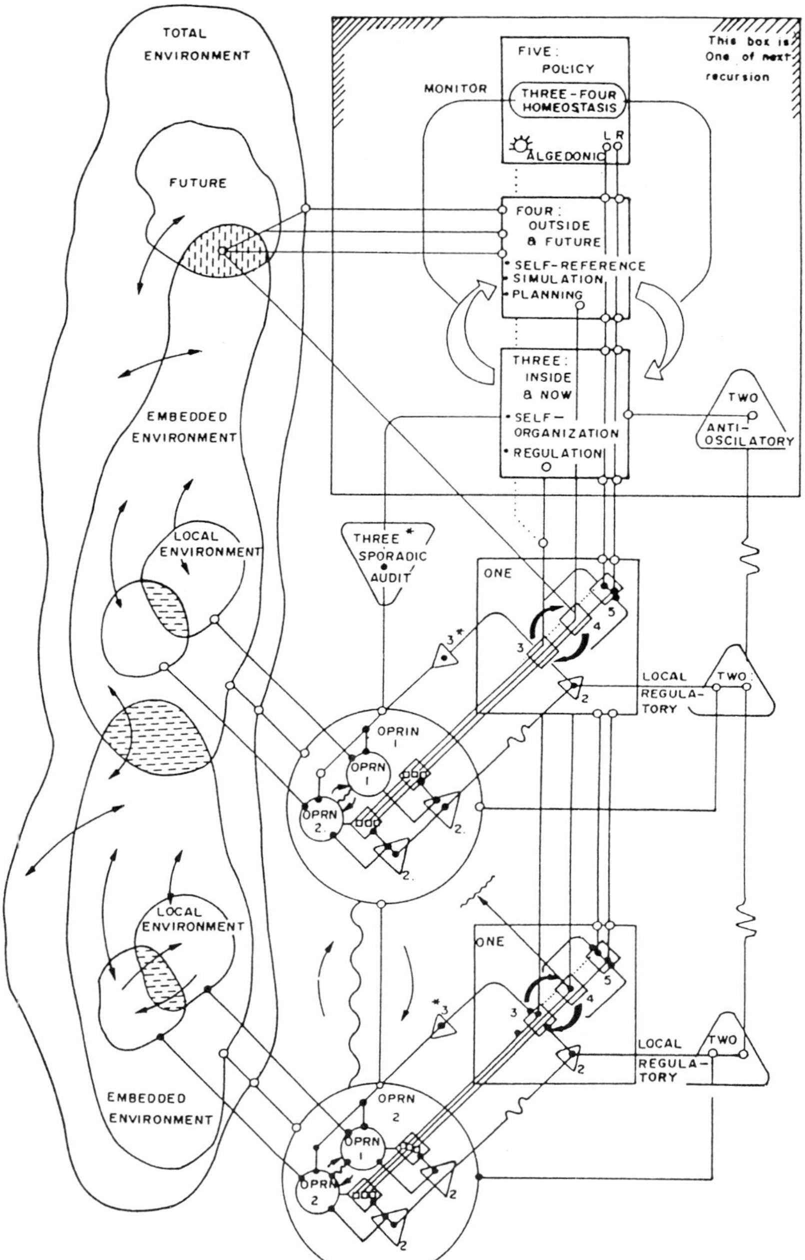

Figure 1. The Viable System Model (V.S.M.) showing recursive embedments (Reproduced from the *Journal of Operational Research Society*, 35, 1984)

Espejo argues that there is not one organization 'out there' but a set of viewpoints and describes a methodological approach using VSM for both as a template for analysis and design.

The VSM supports many different views of the organization from many different perspectives. For example, managers might identify their positions in the management hierarchy together with their span of control, similarly a production engineer introducing simultaneous engineering or a just in time (JIT) system identifies the operational circles involved and which ones can operate in parallel along with the regulatory centres needed.

VSM AND TRADITIONAL MANAGEMENT STRUCTURES

The structure of any organization is determined by its strategy (Galbraith and Kazanjian, 1986; Wheelen and Hunger, 1992). This should change with major changes in the environment. If V.S.M, is a universal approach then it must be able to model any realistic organizational structure. Morgan (1989) lists 6 models which can be used to classify types of organization. These models form a continuum starting with a rigid functional bureaucracy (Weber, 1947), to a product model passing through the critical balanced matrix organization. One of the problems with the balanced matrix is that valuable resources are often wasted because of the continuous battle for power between product and functional managers. This can be seen from the VSM model which would require two senior managements for any one strategic business unit (S.B.U.) operating in a balanced matrix. Larson and Gobeli (1987) suggest a solution to this is to operate either a product or functional matrix. In a product matrix the senior management in system 5 in the S.B.U will be product managers and the functional managers could operate in system 4. The reverse would apply to a functional matrix.

The last of Morgan's models is of the loosely coupled organic network in which the organization has made the decision to subcontract rather than employ large numbers of people. This model includes professional partnerships such as medical and legal practices. Senior management set a strategic direction and provide the operational support necessary to sustain the network. Even key operational activities are contracted out to individuals and organizations. The system of organizations constituting the organic network is directed by senior management in the firm issuing the contracts. This contracting firm may have a name and image, however it contracts out the manufacturing, market surveys, distribution, product design etc. VSM supports this model (Quin, 1992).

It can be argued that the VSM could be used to analyze any organizational structure which may be under consideration. In this respect it can be used to verify the effectiveness of any proposed changes to the organizational structure before implementation.

VSM AND STRUCTURED SYSTEMS ANALYSIS AND DESIGN

Early systems analysis and design methodologies assumed the organizational structure was correct and that the information system should closely parallel the organizational structure (Shave and Bhaskar, 1982). This was a reasonable assumption as the existing system appeared to be working correctly and all that was needed was some way of encouraging the flow of consistent, accurate and relevant information to the top of the hierarchy, Shave and Bhaskar (1982). This proved to be a misconception (Waring, 1989). If the organization structure is dysfunctional no amount of managerial or financial skills will save it (Beer, 1991).

SSADM recognises that the analysis phase could reveal that the system needs to be redesigned and therefore includes the business options step to allow for change. Indeed the primary purpose of analysing the current system is to identify problems in it. This allows for recommending changes to the organization but it is of limited use.

On the other hand VSM requires managers to think through their ideas about their organization, diagnosing faulty structures in depth. Once the organizational structure has been created or redesigned then SSADM can be used to define the information system.

Later versions of SSADM support the rapid redesign of existing information systems since the boundaries of an organization often change as it adapts to an ever changing environment. In VSM these changes will be the loss or gain of other VSM's or the organization itself may be acquired by a larger organization.

VSM AND OBJECT ORIENTED ANALYSIS

Stability is essential for any organization prepared to incorporate a changing environment (Coad and Yourdon, 1991). The object-oriented paradigm keeps the problem domain framework intact (Coad and Yourdon, 1990). Objects such as structures, other systems, devices, roles, operational procedures, sites and organizational units do not change rapidly over time. This stability is seen as the key to reusable analysis. Details may change but the organizational framework remains stable.

As with SSADM, OOA methodology does not include any mechanism for checking the viability of the organization, it only attempts to model the existing structure.

It can be argued that VSM has all the advantages of an object-oriented methodology. In addition VSM provides a cybernetic framework which ensures the organisational structures adopted are viable. A viable system can be constructed from reusable objects such as senior management, chief executive officer, auditors, planners and operatives. Each of these objects can be constructed from the same rules as objects in object-oriented systems.

CONCLUSIONS

Any of the organizational models suggested by Morgan (1989) can be analyzed by VSM. VSM can help identify structural problems such as in the balanced matrix and team approach. This is obviously an advantage if major structural problems can be identified early in the design.

It is believed that cybernetics and VSM in particular also fit the requirements for a paradigm shift suggested by Kuhn (1970) firstly by having a growing group of adherents and secondly by providing a framework for further research. In addition it is found to appeal to users of many different backgrounds.

VSM as a method of analysing structures, recording experience, investigating theories and ideas for managers, seems to provide a suitable framework. Much research work has still to be done. On balance VSM is seen as a versatile management tool.

REFERENCES

Beer, S.A., 1981, "Brain of the Firm", 2nd edition, Wiley, Chichester.

Beer, S.A., 1991, "Diagnosing the System for Organizations", 2nd edition, Wiley, Chichester.

Coad, P. and Yourdon, E., 1990, "Object-Oriented Analysis", Yourdon Press, New Jersey.

Coad, P. and Yourdon, E., 1991, "Object-Oriented Design", Yourdon Press, New Jersey.

Checkland, P.B., 1981, "Systems Thinking, Systems Practice", Wiley, Chichester.

Checkland, P.B., 1986, Are organizations machines? *Futures*, Vol. 12, pp421-424.

The Economist, 19th December 1987, What caused the meltdown, Vol 305.

Espejo, R. and Harden, R., 1989, "The Viable Systems Model: interpretations and applications of Stafford Beer's VSM", Wiley, Chichester.

Espejo, R. and Watt, J., 1988, Information Management, Organization and Managerial Effectiveness, *Journal of Operational Research*, 39, pp7-14.

Galbraith, J.R. and Kazanjian, R.K., 1986, "Strategy Implementation: Structures, System and Processes", 2nd edition, West publishing Company, St. Paul, Minn.

Goodhart, C.A.E., 1984, "Monetary Theory and Practice: The UK experience", MacMillan, London.

Kuhn, T., 1970, "The Structure of Scientific Revolutions", University of Chicago Press, Chicago.

Larson, E.W. and Gobeli, D.H., 1987, Matrix Management: Contradictions and Insights, *Californian Management Review*, Vol. XXIX, No. 4, pp126-138.

Morgan, G., 1989, "Creative Organization Theory: A resourcebook", Sage Publications, London.

Quin, A.J., 1992, Designing Organizations, *in:* "Proceedings of the third International Conference on Information Systems Developers Workbench", Sopot, Poland.

Shave, M.J.R. and Bhaskar, K.N., 1982, "Computer Science applied to Business Systems", Addison-Wesley, London.

Stowell, F. and West, D., 1992, SSM as a vehicle for client-lead design of information systems: utilizing "ideal type" mode 2, *Systemist*, 14, 3, pp99-106.
Swatman, P.A. and Swatman, P.M.C., 1992, Formal specifications - an analytic tool for (management) information systems, *Journal of Information Systems*, 2, pp121-160.
Tustin, A., 1953, "The Mechanism of Economic Systems", W. Heinemann, London.
Waring, A., 1989, "Systems methods for managers: a practical guide", Blackwell Scientific Publications, Oxford.
Weber, M., 1947, "The Theory of Social and Economic Organization", Free Press, New York.
Wheelen, T.L. and Hunger, J.D., 1992, "Strategic management and business policy", Addison-Wesley Publishing Company, Reading, MA.

A WORKSHOP METHODOLOGY BASED ON THE "VIABLE SYSTEM MODEL" OF STAFFORD BEER

M. Haynes and N.D. Stewart

Brian Wilson and Associates
Hampson, Lancaster, LA2 0HY

INTRODUCTION

This paper covers one specific project and supporting case studies and research which has the objective of creating an accessible workshop methodology based on the "Viable System Model" ("VSM") of Stafford Beer. "Soft Systems Methodology" ("SSM") has been used to access the model. Both VSM and SSM are referred to in the text by their initials.

The paper describes the intellectual foundations of the work from the viewpoints of both management science and systems thinking, moving on to describe the generation of the workshop and some of its outcomes.

ORGANISATIONS

Much has been written on organisation and yet, as Goold and Campbell point out from a survey of senior corporate managers, there is "little that has proved of practical value to managers". (Goold and Campbell, 1987). There has been a clear shift in academic literature away from the machine analogy most associated with Taylor's scientific management, towards an understanding that organisations of human beings need to confront such issues as identity, motivation, freedom etc. While this shift has been reflected in management practice, managers retain their strong allegiance to the "organigram" and clearly defined goals. No methodology has succeeded in practice to provide an alternative structure for practical organisation.

Management literature does indicate major trends in organisational design. In the sixties and seventies authors such as Chandler (1962), Lorenz and Lorsh (1969) and Galbraith (1973) and many others charted the movement towards more devolved structures of management. The increasing complexity of business, called "task-uncertainty" by Galbraith, obliged managers to push power down to the operational units and create cross-functional teams to cope with issues which require a coordinated response. The ultimate result of this is the matrix, an unpopular organisational design, but one seen as necessary to cope with complexity.

Systems Science, Edited by F.A. Stowell *et al.*,
Plenum Press, New York, 1993

The preoccupation of seeing environmental complexity as a driver for organisational form is clear in more modern writers such as Morgan (1986) and especially in his description of "autopoiesis", Pfeffer (1982), Quinn Mills (1991) and especially in his description of "clusters", and, addressing the complexities brought about by global operations, by Bartlett and Ghosal (1989). The current preoccupation with delayering organisations charted by Drucker (1988) is surely just a logical extension of this.

The categorisation advanced by Scott (1987) will be used in this paper. He discusses organisation theory in terms of three dominant perspectives; these are the "rational system", the "natural system" and the "open system" viewpoints. While the dominant paradigm of managers is apparently the rational system ethic based on the metaphor of organisation structures, objective measurements and goals, the real-life decision making process in organisations is frequently very intuitive. What managers lack is a coherent framework for action based on logic combined with intuition.

Space does not allow a repeat of Scott's definitions or diagnosis. To satisfy the purpose of this paper it is sufficient to note that an adequate methodology for organisational design needs to be able to draw from the three schools of thought; that organisations are simultaneously goal-seeking (rational system), composed of human beings with specific individual needs and viewpoints (natural system) and yet need to be in harmony with their environments (open system).

Waelchi (1989) states that "we particularly need a translation of our reasonably good theoretical understanding of....management cybernetics, into active prescriptions that a manager can use". By combining the cybernetics of the VSM with the learning power of SSM, the objective of this work is to produce just such prescriptions.

SOFT SYSTEMS METHODOLOGY AND THE VIABLE SYSTEM MODEL

SSM and VSM are very fully described by respectively Checkland and Beer and their various disciples. Descriptions of SSM are given by Checkland (1981, 1989), Checkland and Scholes (1990), and Wilson (1990). Descriptions of VSM are given by Beer (1979, 1981, 1985), a selection of papers published by Espejo and Harnden (1989), and it is included in a review of systems techniques by Flood and Jackson (1991). These are the main references.

The philosophy behind SSM in (a) the construction of models of purposeful human activity, (b) the use of differing perspectives ("Weltanschauungen") to inform model building and (c) the comparison of models with the real world to facilitate learning, offer a generic methodology which covers most if not all systems applications. It offers the basis for the coherent expression of systems ideas seen as necessary by Checkland (1991). As Checkland points out in a review of one of Beer's books, VSM is a model and SSM can "cheerfully accommodate the VSM model". (Checkland 1986). It can be used to inform model building in SSM. Thus SSM can be used to interpret the VSM and to explore its considerable complexity. An example of this approach being used as a tool to restructure organisation has been published separately by the authors. (Haynes and Stewart, 1992).

The VSM used with SSM provides a model which addresses organisation structure; as Anderton (1989) points out the VSM has the potential of offering "operational usability with a clear theoretical base" and stands alone in doing so. The VSM is grounded in rational system thinking and is "machine-like" (Checkland 1980) and "instrumental" (Checkland 1985), which gives it its power as a model. Organisations must deliver; they must be goal-seeking, but must be many other things as well. However use of SSM to interpret and explore the VSM removes the need for very detailed analysis which often characterises the VSM and helps understand the variety of human responses so much needed in real organisational practice.

Additionally, as Jackson (1989) also points out, one beneficial feature of the VSM is that it pictures organisations in close interrelationship with their environments. One of the important steps in using it lies in understanding the structure of the interface between the organisation and its surroundings. Thus in combining SSM and VSM a methodology emerges in which rational, natural and open systems perspectives can effectively be addressed.

WORKSHOP DEVELOPMENT AND PRACTICE

The main task confronted in this endeavour is to make the VSM accessible. The approach used was to produce materials which would allow managers to understand the important messages of the VSM in workshops, helped by the authors acting as guides or "facilitators".

The VSM itself, as it is often displayed by Beer and others, would, in our view, be perceived by many potential users as too complex for use in workshops; its apparently layered structure would unnecessarily confuse managers used to modern-day flatter organisations. Thus, as described above, the VSM itself and its sub-systems were used in this work reproduced according to the modelling processes used by SSM. This is a vital step in making the VSM more accessible.

The VSM is described by Beer in terms of five systems but many of the channels and transducers of the model are also very important and constitute systems in themselves. Beer thus attaches considerable importance to the important "audit" channel (called by him system 3*) and the homeostat between systems 3 and 4, and, from practical application of his ideas, our work has followed this guide. Thus seven sub-systems from the VSM have been chosen for attention in the workshop, these being

the five systems of the VSM.
the audit channel, called by Beer system 3*.
the homeostat between systems 3 and 4, (called by us system 4*).

Precise choice of words to describe the systems concisely, has proved difficult and descriptions have changed over time, but the seven subsystems have been named:-

Interacting with external environment (1).
Stabilising internal operations (2).
Auditing operations (3*).
Providing necessary resources (3).
Creating relevant strategy (4*).
Creating appropriate visions (4).
Providing legitimacy (5).

As preparation for the workshop, the VSM itself and each of the seven subsystems were expressed as human activity systems in logical form using SSM modelling techniques. Space does not allow a detailed description of this, but as example, the conceptual model for the system "auditing operations" is shown as Figure 1.

The workshop therefore involved exploring each of the seven subsystems of the VSM so created. This was done by representing each of the subsystems as generic activities in picture form ("visual metaphors") and preparing a checklist of open questions ("prompts") from the respective models of human activity. An example of the visual metaphor used to explore the "system to audit operations" is shown as Figure 2.

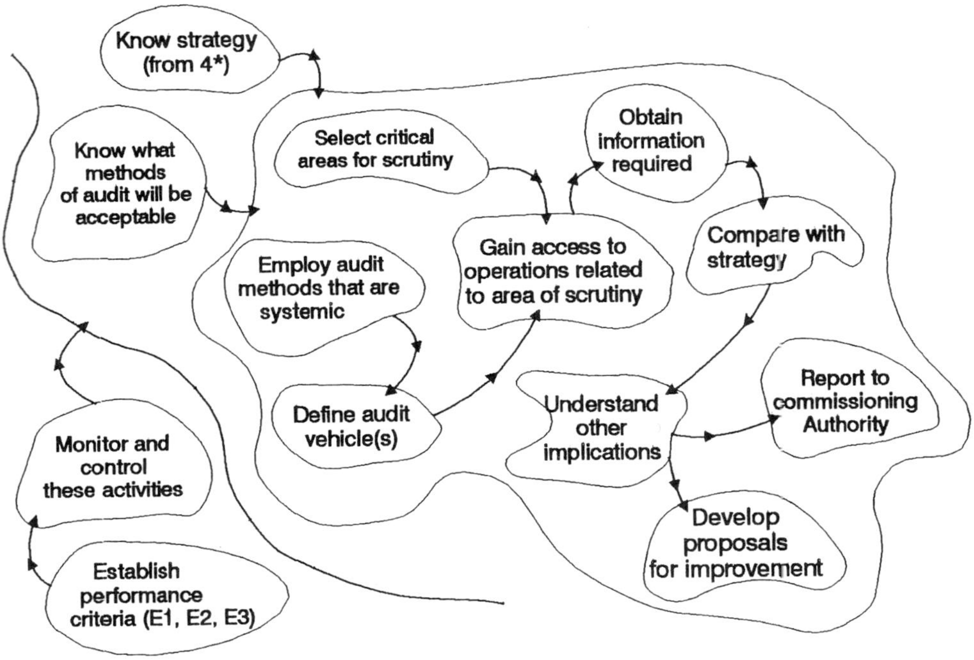

Figure 1. A system to audit operations.

Figure 2. Auditing operations.

After some description of the subsystem using the visual metaphor and relating it to the other subsystems, the managers who participated in the workshops were invited to choose questions from the checklist which they felt most challenged their organisation. The issues so raised were discussed, possible actions were debated and, as appropriate, turned into proposals for organisational change.

Creation of the VSM, a model grounded in the rational system tradition, in the conventions of SSM, recasts the issue of recursion; recursion becomes hierarchy. The VSM models produced in this way cover what would be called "one layer of management" and the focus is on, say, the manager and his/her team and the conditions for their "viability" as part of the environment in which they work. Beer (1985) calls this the "system-in-focus". In workshop mode, this works quite naturally. Exploring other systems hierarchically related to the "system-in-focus", as is often essential, requires no special explanation. Managers are used to it.

Visual metaphors have very high variety and the facilitators, having familiarity with the cybernetic principles on which they are built, can enhance this in helping to explore specific issues which they stimulate. An example of this lies in the concept of control in the VSM described by Espejo (1989). The facilitators must understand this, and reveal the essence of the thinking at appropriate times during the exploration of the model.

OUTCOMES

Using VSM in this way to inform management thinking has some general consequences. A key element in this is "Beer's Law"; the critical observation that viability implies "minimum vertical intervention". The operational units need to develop autonomy to respond to an ever faster-moving environment with minimum command from "above".

This was especially powerful in the main project on which this paper is built, as the manager involved was especially anxious to persuade his managers to take more responsibility. This was termed "realising an achievement culture".

The model puts sharp focus on precisely this by:-

Focusing attention on the nature of the operational interface with the outside world, which drives the model. ("It is System 1 and not senior management which produces the viable system", (Beer 1985)).

Focusing attention on how the various components of systems 2 and 3 create cohesion in the activities of the operational units, without invoking an authoritarian mode.

The model stimulates the creation of organisational forms responsive to the environment yet cohesive in this response. The model also puts focus on the need to sharpen attention on the critical activities of vision creation, strategy creation and adaptation. This it does by:-

Giving a very broadly-based view of the process of putting the organisation into the context of the world around it in systems 4 and 5, and thus generating real future options. (Vickers (1970) describes the essence of this process in the chapter on "The Human Context").

Showing how these visions can be turned into effective strategies in system 4*

Clarifying the essential activities of management control (system 3) to produce a creative response to external events.

In short, the messages derived from the workshop methodology described in this paper are not only very relevant to today's management world; they depict on a coherent theoretical platform, what today's managers are trying to achieve.

REFERENCES

Anderton, R., 1989, The need for formal development in the VSM, in: "The Viable System Model", R. Espejo and R. Harnden, ed., Wiley, Chichester, UK.

Bartlett, C. A. and Ghosal, S., 1989, "Managing Across Borders, The Transnational Solution", Harvard Business School Press, Boston, Mass.

Beer, S., 1979, "The Heart of the Enterprise". Wiley, Chichester.

Beer, S., 1981, "Brain of the Firm", Wiley, Chichester.

Beer, S., 1985, "Diagnosing the System for Organizations", Wiley, Chichester.

Chandler, A.D., 1962, "Strategy and Structure", M.I.T. Press, Cambridge USA.

Checkland, P.B., 1980, Are Organisations Machines?, *Futures*, October 1980, 421.

Checkland, P.B., 1981, "Systems Thinking, Systems Practice", Wiley, Chichester.

Checkland, P.B., 1986, Review of 'Diagnosing the Systems for Organisations' by S. Beer. *Journal of Operational Research*, 23:269.

Checkland, P.B., 1989, Soft Systems Methodology, in: "Rational Analysis for a Problematic World", J. Rosenhead, ed., Wiley, Chichester.

Checkland, P.B. and Scholes, J., 1990, "Soft Systems Methodology in Action", Wiley, Chichester.

Checkland, P.B., 1991, Towards the coherent expression of systems ideas, *Journal of Applied Systems Analysis*, Vol 18, 25, pp25-28.

Drucker, P.F., 1988, The Coming of the new Organization, *Harvard Business Review*, Jan-Feb 1988, p45.

Espejo, R., 1989, The VSM revisited, in: "The Viable System Model", R. Espejo and R. Harnden, ed., Wiley, Chichester.

Espejo, R. and Harnden, R., 1989, "The Viable System Model", Wiley, Chichester.

Flood, R.L. and Jackson, M.C., 1991, "Creative Problem Solving", Wiley, Chichester.

Galbraith, J., 1973, "Designing Complex Organisations", Addison-Wesley Publishing Company, Reading, Mass. USA.

Goold, M. and Campbell, A., 1987, "Strategies and Styles", Blackwell, Oxford UK.

Haynes, M. and Stewart, N.D., 1992, Improving the Capability of Organisations. *Creativity and Innovation Management*, 1(4):194.

Jackson, M., 1989, Evaluating the managerial significance of the VSM, in: "The Viable System Model", R. Espejo and R. Harnden, ed., Wiley, Chichester.

Lawrence, P.R. and Lorsh, J.W., 1969, "Developing Organisations: Diagnosis and Action", Addison-Wesley Publishing Company, Reading, Mass. USA.

Morgan, G., 1986, "Images of Organization", Sage Publications, London, UK.

Pfeffer, J., 1982, "Organizations and Organization Theory", Pitman Publishing Inc., Marshfield, Mass. USA.

Quinn Mills., D., 1991, "Rebirth of the Corporation", Wiley, New York.

Scott, W.R., 1987, "Organizations: Rational, Natural and Open Systems", Prentice Hall International, Stanford, California.

Waelchi, F., 1989, The VSM and Ashby's Law as illuminants of historical management thought, in: "The Viable System Model", R. Espejo and R. Harnden, ed., Wiley, Chichester.

Vickers, G., 1970, "Freedom in a Rocking Boat", Allen Lane, London.

Wilson, B., 1990, "Systems: Concepts, Methodologies and Applications", 2nd ed., Wiley, Chichester.

INTRODUCTION TO EDUCATIONAL SYSTEMS STREAM

The poet, painter and visionary William Blake maintained: "Without contraries is no progression." In his own work and life he focussed on intuited, energetic, imaginative elements, existing in opposition to a world of rationality, deduction, measurement - a state (for him) of "single vision and Newton's sleep."

Within the sphere of education we tend to have a rather limited notion of "progression" although its significance, especially as political and economic pressures are currently brought to bear, needs little emphasis. However, progression, if we take it in the sense of developing to a new level of achievement or understanding, is not the term which gives us pause here. Rather it is the deliberate provocation of "contraries," and the context of things apparently at odds with each other, that offers the main stimulus in Blake's unsettling motto.

Approaching the selection of papers within the education stream of the conference, it is interesting to note how often "contraries," explicit or implicit, figure in the arguments presented: whether touching on debates between alternative proposals, or dealing with the marriage of heterogeneous, even supposedly "incompatible," elements. We find, for example:

- the coming together of "hard" and "soft" systems.
- advocates in the colloquy between "direction" and "emergence."
- the translation of "interpretive" approaches to knowledge into "technology-based models.
- objections to "output criteria" defined simply as competences.
- a paradigm which aims to embrace two "complemental" and "differentiated" factors such as selectivity and creativity.
- designs to educate a variety of hybrid specialist, equally capable and responsive in the fields of technics and ethics, intellect and practicality, insight and management.

Within the framework of planned change or redesign in education, and going beyond that to a broader spectrum, we see an effort to establish clear links between a science of systems and the intelligibility of complex phenomena. Whether we can "progress" by these means in the sense that Blake meant is open to question, though surely the ground is prepared for fruitful discussion. In that regard it's worth a reminder that he also said: "Opposition is true friendship" - in itself another good hedge against complacency or facile accommodation.

Dr A.N. HUTCHISON
Educational Development Unit
University of Paisley

Systems Science, Edited by F.A. Stowell *et al.*,
Plenum Press, New York, 1993

READING AS A COMPLEX PHENOMENON

E. Andreewsky, F. Cochu, C. Parisse and D. Orsoni-Maillard

INSERM - TLNP
La Salpêtrière
47, Bd. de l'Hôpital
75013 PARIS - FRANCE

INTRODUCTION

Reading is both a simple and a highly complex phenomenon. Simple, since most of us are able to read without too many problems. Nonetheless complex, in spite of scientific attempts to *reduce* reading mechanisms to terms of independent *modules* (or "components"). We will present data reflecting the complexity of a specific component devoted to word-meaning, the *mental lexicon*, which far from appearing as an encapsulate module, could very well embed the whole cognitive system.

The "mental lexicon" is traditionally viewed as an inner "Webster", supposed to contain the psychological "meaning" of words. These "meanings" are internal representations reflecting the outside world, and cutting it up into elements. They are hence stable and independent of the reader's or listener's history or experience. How does one "transmit" information? Linguistic stimuli such as written words are not supposed to "transport" their meanings (despite the frequent use of metaphors such as *the idea "conveyed" in this sentence* ...). The role of these stimuli in the process of comprehension is, classically, perfectly clear : they define the "address" of the part of memory where the representation of word meaning can be found. Within this theoretical framework, the understanding of a lexical stimulus therefore consists in retrieving its meaning (from the mental lexicon) from a place defined by morphological data, resulting from purely perceptual processing for identifying such an "address".

In reading, this identification is taken to be either a straightforward morpho-visual step, or a step enabling the reader to revert to speech identification, through grapheme-to-phoneme conversions. The latter involves some difficulties, for instance, coping with the utterence of the homograph *lead* in sentences such as : "the lead was taken by a Peugeot" vs. "the lead is to sale for scrap".

More generally, examples such as the following do not favour such bottom up modular views of reading mechanisms, nor any "encapsulate" perceptual processing, since to perceive either "went" vs. "event" or "clear" vs. "dear" (cf Figure 1) when confronted with *identical*

Systems Science, Edited by F.A. Stowell *et al.*,
Plenum Press, New York, 1993

handwritten drawings obviously means that many "high-level" cognitive mechanisms are at work, even at this "earliest" stage of letter perception.

Such examples give an indication to the tremendous difficulties of automatic handwritten recognition research (Parisse et al., 1987) which assumes that the identification of a given item is no more than a pattern recognition problem. They demonstrate that, for both Artificial Intelligence and Cognitive Psychology, pattern processing and cognitive domains should be conceived as *interdependent* for modeling reading.

We will present, from several view points, behavioural data which substantiate this claim; this data notably leads us to revisit the postulated "interface" between perceiving and understanding - that is, the mental lexicon. It will be shown that the results of the first two experiments contradict the structural characteristic of this lexicon, specifically the "all-or-none" property of word meaning "retrieval". Further data will contradict the traditional view of the lexicon in terms of a (rather static) word-meaning "storehouse"; indeed, this data requires cognitive resources for the dynamic definition of these meanings.

IS WRITTEN WORD UNDERSTANDING AN ALL OR NONE PHENOMENON?

In the framework of the traditional mental lexicon, a stimulus is understood if the representation of its meaning is accessed ; otherwise, it is not understood. A number of phenomena are not in keeping with such a view of lexical comprehension. Specifically, the *all-or-none* property of word comprehension is incompatible with the following examples :

"Subliminal" phenomena and the status of lexical meaning:

If word meanings stored in the mental lexicon are considered as rather stable and constant, the time needed to "mobilize" these meanings (the lexical access) depends on the conditions under which written words occur. This has been the subject of a lot of research. Experiments on "lexical decision", in which subjects are asked whether an item (such as "table" or "mirpe") belong to the language or not, have shown that the more common (hence familiar) a word is, the faster subjects answer. A further interesting result of these experiments is the following : the time a subject takes to decide whether a given item a is a word or a non-word (his "reaction time") depends on the items displayed before a ; it is shorter when these items are semantically "close" to a (e. g., if for a = "pen", ink or pencil, for instance, was presented first). This phenomenon occurs even when the presentation of these "inducing" items is *subliminal* (i. e., so short that the reader cannot identify the displayed item, or even notice it). This means that a lexical item presented in such a way, though *not identified*, can determine a behavioural *effect* linked to its meaning ...

One can imagine the conceptual difficulties of dealing with such data using classical approaches in terms of an "all-or-none" retrieval of lexical meanings. Indeed, it is difficult

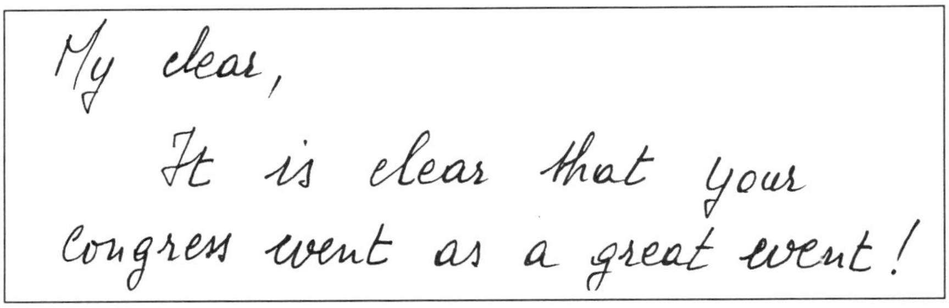

Figure 1. Handwritten examples.

on the one hand, to explain why the "reaction time" is shortened without claiming that the retrieval of the subliminal item meaning has occured (in other words, that it has been understood), and, on the other hand, why subjects have not identified an item if it has been understood ... This type of contradiction calls for new theoretical formulations, throwing into question the "all-or-none" retrieval from mental lexicon, the idea of access to this lexicon and, finally, the very concept of the existence of such a lexicon (Henderson, 1987).

"Alexia" and the status of lexical meaning:

The following example is similar to the above lexical decision task; it deals with the "mental lexicon" properties suggested by the behaviour of some "alexic" patients, that is aphasic patients[1] whose main problem is reading. "Global" alexia concerns patients who seem totally incapable of processing any written material. This is assessed by ensuring that patients are no more able to recognize any isolated letter or to identify any written word (Cf. Deloche et al, 1981). Written lists of words were presented to such patients, with one odd-word such as "hat" in : cat, dog, pig, hat, cow,. The patients, who first claimed that the task was impossible for them (since they have "lost" all reading abilities), were nevertheless perfectly able to point to the odd word (hat) without, however, being able to explain their choice. As in subliminal experiments, this phenomenon, when analyzed in the classical framework of the mental lexicon and the "all-or-none" retrieval of lexical meanings, obviously implies that alexic patients *have* and *have not* understood the meanings of the words in the list ...

Such data challenges the traditional approaches of reading comprehension, since it implies that subjects may rely on the "meaning" of written words they have been *unable to identify* ...

MENTAL LEXICON AND THE "STOREHOUSE" METAPHOR

Classical approaches deal with sentence understanding in terms of "construction" out of "bricks" - that is, the meaning of words belonging to these sentences - hence the theoretical requirement of a "storehouse" (such as the mental lexicon is granted to be) in order to provide access to these bricks.

The trouble with these approaches is that such bricks may look somehow metastable, for instance, when a given sentence (and therefore the words in it) could present a number of different meanings, according to external circumstances. How to "build", for example, the meaning of a sentence such as *It is better to give than to receive*, uttered first during a boxing match, and then, during a sermon in a church, without being first rooted in one or the other circumstance? A similar example is the meaning of : *to integrate is our main problem.* Uttered by the mayor of an overcrowded, highly cosmopolitan suburb, this sentence has very little in common with ... the same sentence, uttered now by students anxiously embarking on their studies at University! The meaning of "to integrate" is obviously closely linked to what seems likely to be relevant (Sperber and Wilson, 1989), given each circumstance. Both for this example and for the previous one, the evaluation of relevance requires a cognitive model

---

[1] "Aphasia" is the name of neuropsychological disturbances of language. It is not an impairment of all patients' language performances. On the contrary, their performances may present specific "breakdowns", suggesting certain properties of normal psycholinguistic mechanisms.

of the speaker's socio-cultural status[2]. Such a dynamic model could not be considered encapsulated in the modular mental lexicon framework ...

Similarly, several if not all cognitive models seem to be involved in word understanding. Let us, for instance, stress the importance of a given title on the meaning of what follows this title. It is the case, even for an everyday word such as water, in *There is no more water* ! Under a title such as *Supermarket*, this sentence obviously means that the mineral water shelves are empty - and accordingly, "water" means : "bottles of mineral water". Under a title such as *Ten days in Sahara*, "water" becomes "a vital element for survival", and the sentence switches to a scream of despair. Under the new title : boat race, the sentence is metamorphosed in "alleluia", since it now means that the team has no longer to bail out the boat; "water" here is "garbage to be thrown away". Again, it will mean "heavy-water" under a title such as *drama in an atomic energy center*; or "vegetable soup" if it is a *household drama*, in which the soup becomes carbon, etc. All these meanings of "water" are obviously closely linked with what seems likely to be relevant, given the circumstances "induced" by each title. Such circumstances are induced through cognitive processing of these titles; in other words, cognitive processing - that is *cognition* - is dynamically embedded in written word comprehension ... It is as if the mental lexicon is "transmuted" from a traditional static "storehouse" into a *complex dynamic cognitive system.*

CONCLUSION

The whole cognitive system is called on to resolve even a seemingly "simple" task such as written word understanding, a fact which had not appeared evident, given the isolated hypotheses put forward in experimental psychology. Indeed, psychology does not recognize other methods of analyzing a complex unity except by reducing it to its most simple elements. The result of this attitude conforms, today more than ever before, to what was for Vygotsky (Vygotsky, 1982) the Psychology of the 30's : *a grandiose atomistic panorama of the disarticulate human mind.*

The complexity of reading and cognition calls for the development of hypotheses on their interrelations, as inferred by our examples, and obviously requires a systems science approach, insofar as such approaches aim at the intelligibility of complex phenomena.

REFERENCES

Andreewsky, E. ed., 1991, "Systémique et Cognition", Dunod, Paris.

Deloche, G., Andreewsky, E. and Desi, M., 1981, Lexical meaning : a case report, some striking phenomenon, theoretical implications. *Cortex*, Vol. XVII, 147-152.

Henderson, L., 1987, "Word Recognition: A Tutorial Review". *in:* "The Psychology of reading", M. Coltheart, ed., Lawrence Erlbaum, Hove and London, 171-200.

Parisse, C., Rosenthal, V. and Andreewsky, E., 1987, An approach to machine recognition of handwritten texts. *in:* "Cybernetics and systems: present and future", J. Rose, ed., Vol.1., 185-188.

Sperber, D. and Wilson, D., 1989, "La pertinence. Communication et Cognition", Edition de Minuit, Paris.

Vygotsky, L. S., 1982, "Oeuvres complètes. Problèmes de théorie et d'histoire en psychologie", Pedagogika, Moscou.

---

[2] Both emergence and properties of the cognitive design of such a model are obviously related to our culture, our history and our experience.

DESIGN OF VOCATIONAL EDUCATION SYSTEMS: A SYSTEMS SCIENCE CONTRIBUTION TO THE "COMPETENCES" DEBATES

Sylvia Brown

The Open University
School of Management
Walton Hall
Milton Keynes MK7 6AA

INTRODUCTION AND BACKGROUND

Vocational education sits uncomfortably between viewpoints between which there are deep philosophical schisms. On one side have been liberal educators (e.g. Leavis, 1943), on the other instrumental trainers to whom training is the set of "formal procedures which a company utilizes to facilitate learning so that the resultant behaviour contributes to the attainment of the company's goals and objectives" (McGehee & Thayer, 1961). Within this debate, training is of groups or, less often, individuals, for the employer; education is of the individual for the individual.

"Changing relationships between higher education and working life" was identified for the 1980s as one of four major issues (OECD, 1983). By 1991 a "manpower approach" to higher education for economic competitive advantage had been elided with an incompatible "social demand" approach - as long as access to a route to a lifetime of helping their country be economically competitive is improved, social justice will have been distributed (OECD, 1991).

References to "training" abounded by 1991, thus: "In several national systems, an additional reason for the broad acceptance of TRAINING programmes from the Non University Sector (NUS) is the shorter period of time which it takes to get a degree or other qualification"; thus again that NUS are designed to meet "needs of the economy *not met adequately by universities* (my emphases); "vocationalizing" of education became an international trend (Dale, 1985; Lauglo & Lillis, 1988). However, education and training are not synonymous and in certain respects are incompatible. Goals and outcomes differ, ergo process must also differ.

Unprecedented UK governmental intervention in education at all levels from budget cuts to direction of policy and removal of Non University Sector establishments from LEAs increased centralised control (White paper, 1981; Jarratt report, 1985 etc). Training saw similar levels of intervention, e.g. ITBs disappearing; policy has been for greater employer

Systems Science, Edited by F.A. Stowell *et al.*,
Plenum Press, New York, 1993

involvement in both education and training (Ryan, 1991). A shift from student demand-led to vocational systems and from humanities to business and applied science was demanded by the CBI in 1980 and endorsed by government. Then, as now, employers were confused on goals for vocational education, recording complaints that technical subjects fail to produce candidates as "good" as those from the arts and wanting to see "personal qualities" given more weight, i.e. non-discipline-specific attributes.

This lack of clarity is at the nub of this paper; is the demand for a system to produce certain kinds of process-facilitational skills, vocationally relevant knowledge or possessors of relevant job skills? The confusion is rooted in differences between education and training: employers, educators and policy-makers all seem unclear what they want.

"SYSTEM1" HIGHER EDUCATION

This is "a system to produce holders of a qualification". It is the output-oriented model most commonly found in all education systems and a fortiori to be expected in vocational education. It couples selection for employment to education in advanced 1st world countries; it aims to prepare aspirants to enter employment markets rather than to train them for a trade or vocation (Dore & Oxenham, 1984).

Principle stakeholders in the wider education system have wished it thus, namely parents who wish for "the best chance in life" for their student offspring, employers (who have become locked into a frame that regards qualifications, especially degrees, as "harder currency" than other forms of evidence of ability) the institution that provides the educational processes and its staff.

Systeml leads inevitably to "Diploma Disease", included amongst the symptoms of which are the following:

- a constraining influence of examinations on curriculum specification and delivery (Lewin, 1984)
- a focus on assessment, to the detriment of education
- usual forms of assessment militating against advances in curriculum design and pedagogic process, the status quo being supported by students and anxious parents alike (Brooke & Oxenham, 1984).

Accepting for the time being that such a system is what is desired by all major stakeholders, there are certain logical consequences for its design if good systems engineering practice is followed. The first of these is control; Systeml control operates by adjusting inputs at the first feedback loop, by adjusting the goals of the system at the second and at the third by redesigning the transformation process. It aims to maximise outputs (qualifications) by optimising existing contributory factors, creating pressures to reduce student variability and to make criteria for award of the qualification easy to measure and attainable by a majority.

Politically imposed efficiency (input-output) relationships (e.g. staff-student ratios, shorter courses) are alleged to produce few adverse effects on outputs of qualifications (Bosworth, 1985).

Outputs: "The Dependent Variable".

At the first loop, the dependent variable "success" is normally defined as "assessment results". For any defined system, the relation between inputs and outputs is parsimonious and logical; there can be no "left-over" inputs that never emerge as outputs and no "extra" outputs that cannot be accounted for by looking at inputs. Traditional reliance on assessment results gives evidence only of the extent to which students met some pre-set criteria, i.e. it is a measure of shortfall - what they did not do. From a systems perspective, this tells me only that I have mis-identified the system; if it were "a system to produce A", it would produce

A. Assuming a pass\fail boundary at 50%, the fact that it produces, averagely, A-50% shows that, it is "a system to produce Ax50% + not-AxS0%" but doesn't tell me what the 50% that is not-A is.

Systeml for professional education presents further paradoxes; a system to regulate the rate of flow into professional bodies must set comparative criteria as normative ones will not achieve this function; maximisation of the wider system is precluded as a goal hence as a criterion for success. In such cases, therefore, candidates cannot be prepared for a vocation but for entry to the competition for employment within a particular field.

Business Studies and Management candidates are a stage even further removed, being prepared to compete for entry into professional entrance competitions and/or entry into direct competition for non-professional employment: maximisation of outputs is more easily defended, with the attendant risks of "Diploma disease"

The second loop has been less researched. Debate centres on "defence of standards", erstwhile maintained by CNAA, as a block to qualification-stock increases. "Variety", on US lines, is weighed against NVQ aims for comparability (OECD, 1987b; 1991; Eustace, 1991).

It is only very recently that attention has been paid to what criteria for system outputs might be appropriate (appropriate for whom has not been debated), hence the establishment of the Management Charter Initiative (MCI), Industry Lead Bodies (ILBs), Technical Education Councils (TECs), all under the control of non-educators and all pursuing the Holy Grail of "competences", currently a mish-mash of logical categories and defined in anti-Saussurean or Chomskyan terms of performance. The third loop was and is "live" in other ways, as the scrabble for "performance criteria" accelerates. Locations, instructional activity, staffing and "ownership" of course content and methods, collaborations, course design, student expectations and demographic trends have received research attention, (e.g. Jaques & Richardson, 1985). This operational level activity has, in the main, preceded strategic considerations; root definitions of the desired "process to" have been ill or unconsidered.

IF NOT SYSTEM1, WHAT?

SYSTEM2 is still an output model but a less "hard" version than Systeml, with a somewhat stronger emphasis on process; outputs include relevant inputs (abilities) to other, employment systems as well as or instead of qualifications that stand as "second order evidence" foroff these abilities; it contains more "soft complexity". It can be entitled "A system to produce holders of concepts and principles that are transferable to a wide range of appropriate situations". An alternative version is "A system to shift cognitive frames from initial state I to target state T" (and sustain them there). Either is compatible with Bloom's "Mastery System" (Bloom, 1974; Block & Burns, 1976).

Other options than Systemsl and 2 are education and process-oriented, for example a system to facilitate individual intellectual development in whatever directions this growth tends or a system to channel individual intellectual growth in preferred directions (raising the major question of whose preferences should be paramount). In the present political climate System3 opportunities are very restricted.

TRAINING SYSTEMS

Training psychologists advocate closed systems, incorporating both feedback and knowledge of results and beginning with goals that "must be made ... specific before systematic instruction can be designed to attain them" (Gagne et al., 1988), i.e. a process to move candidates from Initial state I to Target state T. The design process is iterative. Gagne et al compare education and training systems and propose as many as fourteen more factors for education. Educational goals are seen as "those human activities that contribute to the functioning of a society". Gagne and others assert that potential learning outcomes from training systems fall into at least five categories, of which only one maps directly onto "competences" (i.e. ends), the others being inputs or means (process). They are:

Intellectual skills
Cognitive strategies
Verbal information
Motor skills
Attitudes

The irony is that training design includes more traditional educational objectives (knowledge and understanding domain) than some educators currently do.

Training and vocational "competences".

Trainers' standard learning curves track progress towards Experienced Worker Standard (ESW), i.e. "competence". Curves illustrate early, substantial gain attributable to cognition, whether or not accompanied by practise. Practise and feedback sustain and improve the gain (Annett & Kay, 1957). Valid evidence of gain is pre- and repeated measurement against ESW; at the "validation" level this can be "off-the-job. At any of the "evaluation" levels workplace measurement is essential. "Competence" systems must be similarly output-driven. Goals must be unambiguous, concrete statements of habitual performance at or above a known standard, measurable against comparable criteria for NVQ (etc) endorsement. Trainers did not always achieve the required clarity and concreteness and MCI et al are experiencing similar difficulties. Opportunities for demonstration of ESW depend on job-cycle; trainers often had to resort to off-the-job simulations, whereafter some trainees failed to transfer skills. Vocational "competence" development is more difficult, since the added dimension of ability to extrapolate to new situations is implied. Progress up the training learning curve depends heavily on instructor-trainee interactions (ratio typically in the range 1:8 to 1:15) precluded by efficiency requirements imposed on vocational education. "Competence" assessment is most vexed of all: by definition it has to be work-based, hence the O.U separation of course and portfolio assessment.

From the above points can be constructed two arguments, i.e. that "competence" development is closer to training than education and that establishments of higher education cannot and should not provide either - their proper role is "mastery", within the knowledge domain.

THE EXPERIMENTAL PROGRAMME

It was known in advance that the delivery system in operation at Oxford Polytechnic was highly successful in producing holders of qualifications; the number of 2:1 degree classifications had attracted close scrutiny from CNAA on several occasions. Graduates were highly regarded by employers and had excellent prospects. At entry the majority was WASP, in the 18-22 years range, had high UCCA points, was Conservative and instrumental in approach (Bines & Watson, 1992).

Between 1984 and 1989 an experimental programme aimed to describe more precisely the then current business studies delivery system as a stage towards change. The main focus of interest was system-outputs, especially "not-A". The suspicion was that, whilst highly successful in producing holders of qualifications, the system was less able to produce in them sustained "concept mastery".

Longitudinal and latitudinal studies used a battery of measures on various cohorts, including BA honours and BTEC. "Mastery" of applied psychological concepts was investigated both by examination of outputs in non-traditional ways and by non-interference-producing experimental interventions in the process. The aspect presented here separated ability to apply concepts taught on the BA course to "real" situations from ability to pass conventional assessments of the same concepts via cued (e.g. exam questions) versus uncued responses (e.g. videotaped simulations of work situations) as repeated tests, to which subjects reacted by producing written accounts. Stimuli were presented before and after the relevant module (i.e. beginning and end of academic yearl) and again after the "sandwich" year3. Results from T0, T1 and T2 were compared with each other, with traditional output measures in the form of assessments and with a set of background measures taken at the point of entry. Validity was established by triangulations. Amongst the measures used to assess recognition of instantiations of the concepts were admission into the professional language register of psychology, of business and management, "parroting" and correct/incorrect use of "technical" terms used in the videos, stereotypical account classifications, subjective understanding of the concepts modelled and others. Since virtually all controllable aspects of the delivery process were common, Systeml would predict very high correspondence between inputs and outputs.

Discussion of Relevant Findings

Firm relationships between input characteristics and various traditional outputs were sought and compared across cohorts. Three Personal Style profiles proved significant, all other measures failing to reach required confidence limits. One explanation is that stringent selection (typically up to 4000 applicants for 60-70 places) had succeeded in reducing input variability; many means and standard deviations were very similar but course selectors clearly favoured the Extroversion and Feeling ends of those two dimensions and employers the highly Judgmental. An interesting finding was that input measures of "a scientific approach" that would normally predict success in studying applied psychology did not, reinforcing suspicions of pre-entrance "cramming".

Movements were seen on a range of measures of video response. Typically T0 and T2 did not differ significantly whereas T1 showed positive shift of varying degrees. Admission into the targeted professional language register (psychology) was rare and seldom sustained; use of "management" language and stereotypes increased over time.

A high degree of relationship was not found between ability to do well on conventional assessments and concept mastery.

CONCLUSIONS AND RECOMMENDATIONS

Since "scientific" attributes scores did not predict success in Psychology, Personal Style effects probably emerge from interaction with uncontrollable variables like tutor and employer approval to create "self-fulfilling prophesy" effects, rather than directly. Few initial and even fewer sustained System2 "mastery" outputs or Training system skills were seen. The system was thereby a modified version of Systeml - " a process to produce "the right type" of holders of qualifications".

The study sheds some light on vocational education systems. Government is reluctant to declare its agenda too explicitly; employers and education designers are just confused, failing to appreciate differences between possible systems as well as to define goals or use the term "competences" consistently.

Ulrich's twelve boundary questions are apposite, in the interests of clarity: "who ought to be the beneficiary?", "who ought to be involved in design?" and "who ought to represent stakeholders" are all the stuff of manifesto; "what ought to be the purpose?" is the stuff of root definition; "what ought to be the system's measures of success?" is the stuff of this paper.

REFERENCES

Annett, J. & Kay, H., 1957, Knowledge of results of skilled performance, *Occupational Psychology*, 31, 69--79.

Berdahl, R.O., Moodie, G.C. & Sitzberg, I.J. Jr., eds., 1991, "Quality and Access in Higher Education", SRHE & Open University Press, Buckingham

Bines, H. & Watson, D., 1992, "Developing Professional Education", The Society for Research into Higher Education & Open University Press, Buckingham.

Block, J.H. & Burns, R.B., 1976, Mastery Learning, *in:* "Review of Research in Education, 4", L.S. Shulman, ed., Peacock, Itasca, IL.

Bloom, B.S., 1974, An introduction to mastery learning theory, *in:* "Schools, Society and Mastery Learning", J.H. Block, ed., Holt, Rinehart & Winston, New York.

Bright, B., ed., 1989, "Theory and practice in the study of adult education", Routledge, London & New York.

Brooke, N. & Oxenham, J., 1984, The Influence of Certification and Selection on Teaching and Learning, *in:* "Education versus Qualifications?", J. Oxenham, ed., George Allen & Unwin, London .

Eustace, R.B., 1991, Gold, Silver and Copper: standards of first degrees, *in:* Berdahl et al., q.v.

Dore, R. & Oxenham, J., 1984, Educational Reform and Selection for Employment, *in:* "Education versus Qualifications?", J. Oxenham, ed., George Allen & Unwin, London.

Gagne, R.M., 1985, "The conditions of learning", (4th edn.), Holt, Rinehart, Winston, New York.

Gagne, R.M., Briggs, L.J. & Wager, W.W., 1988, "Principles of Instructional Design", (3rd edn.), Holt, Rinehart and Winston, New York.

Jaques, D. & Richardson, J., eds., 1985, "The Future for Higher Education", SRHE & NFER Nelson, Guildford.

Jarratt, Sir A. (Chairman), 1985, "Report of the Steering Committee for Efficiency Studies in Universities", Committee of Vice-Chancellors and Principals of Universities, London.

McGehee, W. & Thayer, P.W., 1961, "Training in Business & Industry", John Wiley, New York.

OECD, 1983, "Policies for Higher Education in the 1980s", Paris.

OECD, 1991, "Alternatives to Universities", Paris.

Public Finance Foundation Paper No. 13, 1986, "Management and Efficiency in British Universities: The implications of the Jarratt Report", PFF, London.

Schon, D.A., 1987, "Educating the Reflective Practitioner", Jossey Bass, London.

Scott, P., 1985, The Leverhulme Recipe, *in:* Jaques & Richardson, q.v.

Ulrich, W., 1987, Critical Heuristics of Social Systems Design, *European Jnl. of Operational Research,* 31:276-283.

Unger, J., 1984, The Sobering Experience of China's Urban Schools, *in:* J. Oxenham, q.v.

Youngman, F., 1986, "Adult education and socialist pedagogy", Croom Helm, London.

Zeichner, K., 1990, Changing directions in the practicum: looking ahead to the 1990s, *Journal of Education for Teaching,* 16, (2), pp105-132.

DESIGNING AN MBA CURRICULUM FOR INFORMATION SYSTEMS MANAGERS: AN ANALYSIS OF CONTEXTUAL FACTORS

Prof Sadri Gilani[1] and Paul Northcott[2]

[1] Department of Management Systems
[2] Centre for Teaching and Learning Services
Bournemouth University
Poole House
Talbot Campus
Fern Barrow

INTRODUCTION

This paper is based on the detailed and extensive research that was undertaken to support the curriculum development of an MBA programme in Information Systems Management. The contextual, or "environmental" issues, are discussed under two main headings: organizational/professional and educational/ethical. Organizational and professional issues are seen to be so interrelated that we consider it appropriate to address them under a single heading. Likewise, we address the educational and ethical issues jointly, in both the process and content aspects, of the curriculum. We pay close attention to the needs to be considered in an MBA programme, especially one that is being studied on a part-time, work-related basis, that will suit the needs of information systems professionals who aspire to senior managerial positions. We refer to this as the education of **hybrid** managers. The purpose of this paper is to provide a focus for the discussion of guidelines in curriculum development for senior, professional information systems managers.

KEY WORDS: Information systems hybrid education; MBA curriculum; educational contexts and process; professionalism; ethics.

Organizational/professional contextual issues

From the mid 1980s, the business community has made widespread use of personal computers. The availability of accessible technology and software, for increasingly information intensive companies, has created and filled, business demands. Increased pressure has occurred to provide the right information at the right time. Rockart, as early as 1982, confirmed from his interviews with a sample of executives, that the information systems executive was no longer merely one of a group of service individuals solely concerned with a limited accounting-centered data set which was located at a central location. The information systems executive was now set in an environment in which computing power was distributed throughout the company. The computer systems included a diversity of applications and involved a full range of company employees at all levels. The need to manage the growing computing capability so that it met corporate requirements and so that interdepartmental conflicts did not dissipate the advantages of the growing ability to use information across the whole company, led Wyman (1985), amongst others, to identify the emergence of a new position: Chief Information Officer, a counterpart to the well-established

Chief Financial Officer. Whilst such a post was often seen as desirable, filling the post with a suitably skilled and able executive was not easy. Just as a board of directors require of their Chief Financial Officer a broader array of managerial skills than those confined to financial operations, so they would expect of their Chief Information Officer more than experience and skills in data processing. Wyman stressed that Chief Executive Officers should do more than make appointments in information systems management; they needed to convince their companies that their futures depended on the successful application of technology and to this end they should set-up organizational support structures that would ensure that information technology applications were both well-implemented and apposite.

By the mid 1980s, therefore, the need for a strategic view of the use of information systems had become a dominant theme in the development of information systems thinking. Madnick (1987), clarified the changes taking place in the role and status of information systems personnel within the company. He recognised that the role of the top information systems executive was changing fundamentally. In the 1960s and 1970s, the role was that of a technical specialist who routinely implemented a given system. In the 1980s and 1990s, he recognised the need for a new breed of top information systems executives. These executives would not only have technical skills of a high order but they would also need to have business insight and the managerial skills to lead and motivate. Such an executive would need to develop effective relationships with other managers and top executives to be fully effective.

From the above it can be seen that there is now a need for information systems professionals who have the knowledge and skills necessary to lead in the effective use of appropriate information technologies to cope with rapidly changing business requirements. It is in this context that we analyse the postgraduate programmes that will serve the needs of the education of such managers. The term currently used for this new professional is **information systems hybrid.** Doubtless, some hybrids will occur naturally within certain companies, perhaps more by accident than design. The research commissioned by the British Computer Society (1990), however, suggested that the large number of information systems hybrids required by companies in the 1990s, and beyond, cannot be met in this way. Some formal educational support is needed. There are, it is recognized, many highly skilled, highly intelligent, technically-based leaders who are employed in the information systems function who have the potential to manage but currently they do not have the skills or the experience to perform well at high management levels due to insufficient, or inadequate, integrated management education. Given the appropriate support from a suitable postgraduate information systems programme, preferably at MBA level, these information systems personnel could develop into the hybrids now needed by business.

We turn to the professional issue and professional development. Abraham Flexner (1915), has long been quoted almost as an "Aristotelian authority," concerning the definition of profession and professionalism. We do not wish to cover this ground fully here except to remind ourselves of his criteria (1):

> Professions are therefore intellectual and learned; they are in the next place definitely practical. No professional can be merely academic and theoretic; the professional man must have an absolutely definite and practical object...(p.579)

...professions involve essentially intellectual operations with large individual responsibility; they derive their raw material from science and learning; this material they work up to a practical and definite end; they possess an educationally communicable technique; they tend to self-organization, they are becoming increasingly altruistic in motivation...(p.581)

Given these criteria, the form of professional development needed to create the information systems hybrid, as described above, will require managerial competencies that encourage intellectual rigour in the formulation of solutions to problems. This will allow **applied intellect and creativity** in decision-making and strategic analysis to predominate in the critical development of information systems strategies at the highest levels in the company. To achieve this level of skill, we argue, requires immersion in an appropriate intellectual environment. The evolution of the information systems hybrid professional, from an information systems technician, requires continuous, rigorous study and this is unlikely to be achieved by a short course.

Having analysed the organizational and professional context, we can now move to explore the educational and ethical considerations in designing an appropriate MBA curriculum for the professional development of information systems hybrids.

Educational and ethical contextual issues

In developing the curriculum we need to consider the psychological, sociological and professional contexts in which the learning takes place. In our particular case we considered that we needed to consider especially carefully the mode of study and the work environment of the students. With this context in mind and in discussion with sponsoring companies, we designed a course that is run over two years on a part-time, student-centered, work-related basis. An extra year may be taken to complete the dissertation component of the programme. During the two years, students undertake eight block release study sessions on campus of four days each. The course units are studied on an integrated theme basis. The themes studied in the first year are those identified as being important for the senior information systems executive. In the second year, the themes and skills related to operating as a director in information systems management are addressed.

From the extensive literature on learning theory, we considered aspects of the following: management learning; adult learning (with particular reference to learning contracts); professional learning and socialisation; and the use of the workplace as a place of learning. From this corpus of knowledge we consider of crucial importance is the notion of **critical thinking and action skills**.

Critical reflection on the role and nature of management, we believe, has to be studied in association with the practice of management. This is what Schon (1983; 1987), has described as "reflection in action." By this he means the identification and solving of problems which are often surrounded by "...uncertainty, complexity, instability, uniqueness and value conflict." The essential context, therefore, of the MBA programme under study is one of vocational and professional education of an applied and critically reflective nature.

Brookfield (1987,141-142), has commented on the unsuitability of the behaviourist approach to teaching and learning in management studies:

> In the study of high-learning managers by Lowy, Kelleher, and Finestone (1986), the lack of goal clarity, absence of regular feedback of performance and ambiguity surrounding high-learning managers' functions and responsibilities contrasts markedly with behaviourist notions of successful learning. According to behaviourally oriented writers...some of the most important characteristics of successful learning are that learning goals are specified previously in precise

behavioural terms, regular feed-back on learners' progress is provided, and ambiguity is avoided as much as possible. Instructional processes are sequenced according to their complexity, so that learners are taken through a series of progressively difficult tasks.

Jarvis (1987,16), in exploring the social context of adult learning, defines learning as: "the process of transforming experience into knowledge, skills and attitudes." He also makes the important point that a person may learn but be loath to practice what he or she has learnt when confronted by a socially threatening environment. Thus, although the academic environment may be supportive, it is not the academic environment and culture that is pervasive for part-time students who spend only a fraction of their time on campus. Furthermore, students run the danger of being frustrated if they are thwarted in being able to apply the new knowledge and skills they have developed in their studies to their work. We have also found that they suffer stress if they are unable to balance time demands for their work, study, family and social commitments. Thus, whereas part-time MBA programmes provide the powerful opportunity to study in the context of work and work in the context of study, there are potential psychological and social barriers to success with this mode of study. In designing the curriculum, which was a collaborative venture between academic staff and a selection of senior company staff with a background in information systems management, we have attempted to address these problems by using: block release; the use of integrated themes and learner support services, viz:

1. **The block release** attempts to overcome the time pressure constraints. This, however, has proved only partially successful. Students still find difficulty in finding the amount of time that they would like to have to complete their learning contracts to their satisfaction.

2. The use of integrated themes, rather than separate "strands of subjects," makes the study programme more manageable, efficient and effective. (2) Furthermore, it promotes the opportunity for visioning and strategic thinking. We strongly discourage viewing the organization as discrete functional units with clear boundaries. Business processes rather than functions are stressed. Effective management in the context of corporate leadership and creative problem solving is emphasized. Personal and business skills required of flat organizational structures are also stressed.

3. The learner support services provide a major contribution to the effectiveness and success of the programme. We have constructed what we term "networks for discourse" between the students, faculty and the students' companies. Faculty work through a company coordinator and undertake industrial visits. In turn, company representatives, as adjunct faculty, play a major part in the study sessions conducted at the block release residential schools. The networks aim to make the

It is important to recognise, we believe, that there is an important distinction to be made between the practical behaviour of individuals and their belief in the theoretical model they are using. Argyris and Schon (1974), and Schon (1983), distinguish between what they term **theories in use** and **espoused theories.** Theories in use can be inferred from action and are known as " cognitive maps by which people design action." Such theories are privately developed, are idiosyncratic, and are often not mentioned in textbooks of professional practice. In contrast, espoused theories are those that are publicly agreed upon norms. They are the theories that people often claim to follow even if in practice they do not. This being the case, there is a real danger of the discourse breaking down between faculty and students if faculty are unable to understand why their students are apparently incapable of appreciating the value of academic theories to their practices. We believe that a constructive way out of this problem is for faculty to appreciate that their students do theorise in use about their practice and should be encouraged to do so in a critically reflective manner. Faculty also need to be tolerant concerning the process of assimilation that students have to undergo in reconciling "external theory" with their own personal theorising. This is not only a question of time but of empathy, respect and negotiation. To this end, we have used learning contracts (see Knowles, 1986), and integrative case studies which encourage wide-ranging debate, as components of our MBA programme. In essence, our view is that the ethos of our programme is about encouraging students to engage in critical inquiry about their own professional role and practices. We also believe that management practice itself is a basis for conceptualizing and forming acceptable theories about the nature of managerial work if it is undertaken critically and in an informed manner by challenging assumptions and by exploring alternatives in a creative manner. It is especially evidenced by an ability to advance, support and justify a reasoned position in the face of sustained opposition.

1. A broadening of the scope of the managerial activities the students undertake both within the students' current responsibilities in information systems and through a voluntary expansion of interests into other departments or functions. Many information systems staff fail to realise the potential scope of their jobs and the impact their creativity can have on the company. Thus, the MBA programme addresses the middle management agenda of strategic management in anticipation of running an organisation and the senior management agenda of the strategic and policy skills to run an organisation more effectively as an information systems hybrid.

2. An increase in confidence in the information systems professional to exercise managerial skills and effective leadership which come with a greater understanding of management issues.

3. Improved communication between the information systems professional and the functional specialists by an appreciation of their specialist fields of operation.

4. An holistic view of the organisation and its goals thereby allowing a fuller formulation of problems and improvement of decision making.

5. A greater understanding of management style, both in the cognitive and affective domains, so as to improve the ability of the information systems professional to manage.

6. A conscious promotion of learning, unlearning, and re-learning both in the personal and organisational domain.

7. View short-term objectives critically and their implications on long-term corporate goals and vision so as **to foster strategic thinking at all times and levels.**

8. An holistic view of the role of information - information management - information technology, in supporting the achievement of corporate goals at all levels.

CONCLUSION

In this paper we have reported on the research that went into formulating and implementing an MBA in information systems management. The contextual issues were discussed under two broad dimensions: organizational/professional and educational/ethical. This analysis is important, we believe, as it provides a more profound approach than that of just following slogans such as marketing-led, vocational needs, demand-led (as distinct from supply-led) or student-centered format. We saw that the curriculum development needed to pay close attention to both the rapid changes that are occurring in information systems management in organizations and the professional demands on and the status of information systems professionals. The educational issues were related to both the mode of study and the organizational/professional issues. We discuss the ethical issues, not in a moralistic sense, but so as to provide a theoretical framework from which we could work in devising and implementing the programme. We hope that the issues discussed in this paper are of interest to faculty members who are engaged in teaching MBA programmes-especially those who are responsible for teaching information systems management in such programmes.

REFERENCES

Argyris, C. and Schon, D. A., 1974, "Theory in Practice: Increasing Professional Practice", Jossey Bass, San Francisco.

British Computer Society, 1990, "The Hybrid Manager", BCS, London.

Brookfield, S. D., 1987, "Developing Critical Thinkers", Open University Press, Milton Keynes.

DeGeorge, R. T.,1990, "Business Ethics", Macmillan, New York, third edition.

Dunfee, T and Robertson, D., 1988, Integrating Ethics into the Business School Curriculum, *Journal of Business Ethics*, November, pp847-859. (See also Dunfee, T. W., 1986, "Integrating Ethics into the MBA curriculum", The Wharton School of the University of Pennsylvania, Philadelphia.)

Flexner, A., 1915, Is Social Work a Profession?, *in*: "Proceedings of the National Conference of Charities and Connections", Chicago, pp576-590.

Isenberg, D. J., 1983, The Structure and Process of Understanding: Implications for Managerial Action, *in*: "The Executive Mind: New Insights on Managerial Thought and Action", S. Srivastva and Associates, eds., Jossey Bass, San Francisco.

Jarvis, P., 1987, "Adult Learning in the Social Context", Croom Helm, Beckenham.

Knowles, M. S., 1986, "Using Learning Contracts", Jossey Bass, San Francisco and London.

Kolb, D.D., Rubin, I.M., and McIntyre, J.M., 1984, "Organizational psychology, an experiential approach to organizational behaviour", Prentice-Hall, Englewood Cliffs, New Jersey, fourth edition.

Locke, D., 1986, A Psychologist among the Philosophers: Philosophical Aspects of Kohlberg's Theories, *in*: "Lawrence Kohlberg. Consensus and Controversy", S. Modgil and C. Modgil, eds., The Falmer Press, Philadelphia and London, pp21-42.

Lowy, A., Kelleher, D., and Finestone, P., 1986, Management Learning: Beyond Program Design, *Training and Development Journal*, 40, 6, pp34-37.

Madnick, S.E., ed., 1987, "The Strategic Use of Information Technology", Oxford University Press, Oxford.

Rockart, J.F., 1982, The Changing Role of the Information Systems Executive: A Critical Success Factors Perspective, *Sloan Management Review*, Vol 24, No 1.

Schon, D.A., 1983, "The Reflective Practitioner: How Professionals Think in Action", Basic Books, New York.

Schon, D.A., 1987, "Educating the Reflective Practitioner: Toward a New Design for Teaching and Learning in the Professions", Jossey Bass, San Francisco.

Rawls, J., 1971, "A Theory of Justice", Harvard University Press, Cambridge, MA.

THESEUS - A MODEL FOR GLOBAL CONNECTIVITY

Roger J. Harnden[1] and Roy Stringer[2]

[1]117 High Street
Blaenau Ffestiniog
Gwynedd LL41 3AX
North Wales

[2]Interactive Designs Ltd
2A Princes Boulevard
Bebington, Wirral L63 5LN
England

INTRODUCTION

According to Greek myth, *Theseus* had an eventful life, performing many heroic deeds to the benefit of humankind. Born of the union between Aegeus (the sea, derived from *aisso* "to move quickly like the waves"), and Aethra (the pure air), one of his feats was to free the citizens of Athens from the tribute imposed on them by the Cretans. King Minos of Crete, consequent on his victory over Athens, demanded that each year seven youths and seven maidens be delivered as a sacrifice to feed the Minotaur. This terrible monster was lodged in an intricate labyrinth which, even were it possible to slay the monster itself, none would be able to escape. Even its architect, Daedalus, was unable to find the way out on its completion. He was forced to resort to constructing the wings which were to bear his son, Icarus, to his fate.

Well, on hearing the heartrending cries of youngsters being transported to their doom, Theseus inquired, and was told this recent history. Hearing the story, he resolved to do something about the situation, and offered himself up to join the condemned youngsters. With the help of Ariadne, daughter of the King of Crete, Theseus was able to slay the Minotaur. Ariadne provided him with a sword to kill the monster, and a ball of thread which he used to find his way out of the maze.

Theseus provides a model for orienting around an information space. Like any true hypermedia system, the Theseus software concerns itself with the 'labyrinth' and the 'ball of thread' - how to keep track of where one is, and how to get back to the start point. But, more than that, it is inspired by the broader-brush implications of the original myth, leaning towards a particular understanding of creativity and conceptual processes (the waves and the air - freedom and light over retribution and darkness). Our discussion of 'global' is influenced by this 'creative learning' dimension, and how it relates to general, biological fundaments of human cognition and language.

Systems Science, Edited by F.A. Stowell *et al.*,
Plenum Press, New York, 1993

Structure of paper

Having dealt with the mythical past, it's time to come up to date. We will consider Theseus along three dimensions:

a: The technological infrastructure for a global, in the sense of world-wide, system;
b: The Theseus methodology;
c: The *space* of expression, or *semiotics*, provided by electronic delivery methods and presentation formats, approached from the perspective of global, in the sense of *biologically human*.

THE TECHNOLOGICAL BASE

The immediate deliverables of the Theseus System, to be discussed under 'methodology' and 'space of expression', are not a function of a technological variable, but already exist. At this historical juncture, the crucial thing about *conversational* or *enabling* tools (Harnden & Stringer, in press), of which Theseus is one, is the coherence of the underlying logic in the light of technological capability.

This logic has to do with the direction of the design task. Such tools approach problems to do with increased channel capacity and processing power by focusing on the *conceptual space* of their users, rather than on technological mechanisms. Notably, through encouraging *interactivity* in a variety of ways, they enrich and expand the cognitive space of the individual or the shared understanding of the group. Instead of focusing on provision of information per se, such systems aid in the construction of meaning.

Bearing in mind these general remarks, which we will return to, it is useful to highlight the technological limitations of networked multimedia, and to indicate how the Theseus model deals with such constraints.

Computer technologies already enable us to efficiently store pictures, high definition video, sound and other media in random access digital formats, along with the computer-based text with which we are all becoming familiar. However, transmission of these data types from a storage device to a remote access point makes huge demands on the carrying medium. Copper wire networks currently carry two-way data at 800 bytes per second over standard telephone lines, and one million bytes (1 megabyte) per second is common over local area computer networks. By comparison, current television signals require transmission of analogue data at an equivalent of around 30 megabytes per second.

Allowing for data compression techniques, networks will probably have to deliver point-to-point transmission rates of about 5 megabytes per second in order to meet the needs of real-time video transmission and video conferencing. At present, this sort of transmission rate can only be achieved with optic fibre cabling over small local area networks. To provide such services to the home would require an order of magnitude greater, a nationwide optical data network with a bandwidth of perhaps 200 to 1000 megabytes per second. In certain nations this might be possible in the early part of the next century but even allowing for these bandwidths, real-time, point to point multimedia will become increasingly sluggish as the number of simultaneous 'subscribers' increases.

With this in mind, Theseus is designed to separate the transmission of data across the network from the user's interaction with that data. In effect, a 'chunk' of data, (see *entity* later), is requested and loaded into the user's workstation in its entirety where it can be manipulated in real time. This approach enables us to create meaningful real-time experiences for users even with very low network bandwidths. The cost is high end user terminals with perhaps 5 to 10 MIPS (million instructions per second) of CPU power and 20 megabytes of RAM, but such systems will become commonplace over the next few years while high bandwidth public networks are at least a decade away. Already IBM is developing their Power PC architecture, which will offer desktop processing power at about 20 times current standards. DEC has announced microprocessor developments which exceed the IBM effort by a factor of four, aiming for a massive 400 MIPS of processing power, approaching the most powerful Cray supercomputers of just a few years ago. David Nagel, Director of Apple Computer's Advanced Technology Group and Director of both Taligent and Kaleida (the new joint ventures set up by Apple and IBM), predicts the disposable pocket Cray before the end of this decade.

Given the rapid speed of change, the Theseus design allows for its development and implementation on a modular basis, in a way that fits financial and technological constraints of the moment. For instance, the first full-scale implementation of Theseus was as a TEED funded project (U.K. training agency initiative), to provide a stand-alone multimedia training package for cytology screeners. For reasons of immediate practicality, this package is lodged in a Macintosh environment and utilises laser-disk storage. Neither of these platforms reflect the nature of the Theseus model, which is being developed as cross-platform software at the operating system level, designed to be able to take advantage of novel developmental tools as they appear. Thus, the next version of the cytology package will be CD-ROM based using QuickTime digital video technology, while moves are under way to develop file-serving capacities on a local network, as a pilot for macro systems such as SUPERJANET (a new network for education which will provide 20 Megabytes of bandwidth by the second half of this decade).

THE THESEUS METHODOLOGY

As frequently noted, storing huge volumes of information material offers enormous opportunities for education and personal development, but does not in itself guarantee provision of useful knowledge. Finding our way quickly and easily through such a bulk, to just that information relevant to our needs and interests, and then finding the way back to our immediate area of study when we have wandered too far, poses major problems.

The pressing need is to devise a viable method for shaping and organising these novel forms of information in a controlled and coherent way. Like Theseus in the labyrinth, we need both to be able to 'slay' the Minotaur and be able to retrace our steps. This idea lies at the heart of the Theseus Project, which proposes a set of data structures and a teaching/learning methodology for the long-term development of a distributed public database of audio-visual information, criss-crossed with subject-related pathways, that enable readers to discover new areas of knowledge and rich associations, while pursuing a structured programme of learning.

Theseus outlines a hypermedia model (see below), and its long term success depends on its availability and usefulness to institutions and individual users who, at one and the same time, utilise and contribute to its evolving resource base. Its effectiveness and attractiveness to both teachers and learners, will be a function of the richness and diversity of an underlying *Hyperbase*, or collection of interlinked multimedia objects. This resource is not some fixed historical residue like the fossil record (or the conventional library of today), but takes account of the 'active reader', whose take-up and referencing of stored materials, impacts upon the future shape and quality of the whole system (Bolter 1991 - see below). A key criteria for such a resource, is that its 'atoms' should both appeal to curiosity and respond interactively to a user's interests.

The general requirement was for a readily understood and easily exploited data structure. Thus, although opening up a truly electronic hypermedium, with benefits of cross-referencing and linking across different works and disciplines, the model builds upon the well understood *linear* or *sequential* structure that underlies traditional linguistic or textually lodged methods of presentation and teaching (such as 'the lecture', or 'the essay').Its success, and the attractiveness to users, is that, despite the familiar sequential presentation of ideas and theses, these intersect with other theses, both sympathetically and antagonistically, in a multidimensional space offered by the *entities* which make up the mediabase.

At an operational level, the model contains two discrete layers, each with its own type of structured information, the *subject path* and the *mediabase*. Together, these two layers give rise to the *hypermedium*.

a: The Mediabase Layer

A mediabase is a flat-file database containing records stored as text, sound or images. Each discrete item is referred to as an *entity*. An entity may be interactive. The terms of its interactivity will be determined within it, rather than be defined by a general set of rules for the mediabase as a whole. A single entity might be as simple as a photograph called "London Bridge"; or as complex as a virtual environment or the representation (i.e. the simulation) of a complex piece of machinery that can be dismantled and rebuilt.

The point about a mediabase entity is that it should be as *categorical* or *objective* as possible. Any written or spoken descriptors related to such a record should refer to generally accepted facts or agreements. The functional significance of an entity, is that it can be 'picked up' as a totality - in other words, it is something that can be referred to in its own terms. This says nothing about an *intrinsic* or transcendental objectivity. Its objective status is a matter of recognisability in terms of the understanding and behaviour of a particular community of observers, or culture. In other words, its objectivity is a relative quality, something defined and reaffirmed through conjoint interaction.

b: The Subject Path

The significance of an entity in a given instance is not something pre-given, but is shaped or coloured by a particular context - an argument which unfolds significance for some purpose or end. In respect of such an argument or theme, an entity is referred to, perhaps as an illustration or perhaps simply as an adornment. This is what braids subject paths into broader consensual horizons. A reader may interrogate an entity about which subject paths have referenced it to illuminate their differing arguments. In this way, although following one particular subject path, the reader can move *orthogonally* into any other subject paths.

c: The Hypermedium

It is the combination of these two distinct layers which constitutes the dynamic and evolving *hypermedium*. This is a collective name for the set of mediabases (the hyperbase), together with any subject paths that cross through them, accessible to the user from any terminal at a given moment. Thus, entities may be said to allow the *orthogonal intersection* of distinct viewpoints or understandings (Maturana, 1991), constituting a nexus between otherwise distinct pathways, or subjective perspectives. In this way, the system encourages emergence of what, in the domain of human consensual interaction, Humberto Maturana calls the *multiversa*.

To pick up the theme of creativity: the engagement of the reader takes place along two distinct paths and due to quite different mechanisms. Firstly, the user as an author can lay down subjective pathways that depict a personal thesis or point of view, or can follow and reflect upon the subjective viewpoint of other people. Secondly, the 'entity' is a self-sufficient object in whose terms the user can interact. By 'self-sufficient', we mean that each entity has within it rules for its own 'plasticity' - for how to respond to particular usage. An entity is not a passive object, lying 'open' before a reader's gaze. An entity is, and will increasingly become, a product of art, competing for a user's interest through its intrinsic seductiveness. This will be secured as much by aesthetic and dramatic criteria, as by some claimed for accuracy or truth. Claims for accuracy or truth reside in the argument explicated by the subject path. This is something demonstrated by the *situatedness* of an entity, as this indicates or gives evidence of the perspective of an author.

Despite this simplicity, the model provides a conceptual framework that enables people to work with a very large volume of information whose overall structure is not only indeterminate but is constantly changing. It succeeds by distinguishing between commonality in cultural expression (represented in the mediabase), and the uniqueness of individual human experience and expression (the subject path). One of the reasons for the effectiveness of this model, we believe, is its close correspondence to contemporary insights into language, cognition, and the way the nervous system functions (Dupuy & Varela 1992; Varela 1992).

SEMIOTICS

As indicated at the start, given an outline of the necessary technological infrastructure and underlying methodology, the concern of this paper is to pinpoint the implications of Theseus as a primitive (in the sense of early) realisation of 'enabling' technology. This will be described in terms of the 'space for expression' provided by an electronic medium, and its significance highlighted with reference to contemporary developments in other fields and disciplines.

As far as a user is concerned, Theseus is not a computer system, but rather an 'electronic writing space' (Bolter 1991). Bolter's attention is not upon the written form, but upon the *space for expression* provided by the electronic medium. In noting that the electronic medium more nearly resembles the oral tradition than that of the printed book, Bolter points out that the distinction between the page and the electronic medium concerns not just two different neutral surfaces for the written word, but different conceptual spaces. Notably, the electronic space puts the 'reader' in the driving seat, where the book maintains the relationship of the reader as that of a passive observer of the writer's activity.

Bolter's thesis is that the canon of western literature (consisting of a hierarchical arrangement of 'great works') entails an implicit philosophy of worshipful reading in which the writer *as writer* is granted authority over the reader. In the electronic medium, the reader is able to literally 'touch the text' in a process of active reading very different from the written tradition we have come to take for granted (Bolter 1991: 144ff). In a sense - and here Bolter references Derrida and others - the electronic medium threatens the whole edifice of a static, fixed tradition. Instead, texts become part of a process of 'intertextuality', and gain their significance and meaning in the network of texts related through the experience of the individual reader of them (ibid: 163). For Bolter, this holds out the promise (or threat), of embodying the semiotic views of language and communication posited by people such as Peirce, Saussure or Eco.

From a very different background - that of physics - the late David Bohm and F. David Peat consider scientific creativity, and the creative spark in general (Bohm and Peat 1989). This follows up Bohm's earlier insight into scientific creativity, when he first postulated the idea of 'natural intelligence'. Crucial to the authors' thesis are several themes compatible with our own model: the co-dependence of the observer and that which is observed; the social nature of knowledge; a *variety* in the natural order of things as well as embodied in human individuals, which is not to be treated as oppositional or fragmented, to be smoothed out or banished, but which, through the encouragement of creative intelligence comes to be realised in intellectual, emotional and social practice, as “unity in diversity” through a process of dialogue.

In studies as to the biological bases of language and cognition there are increasingly explanations of *neurological* processes which serve a function of grounding the organism in its effective domain of action by enabling an orthogonal intersection of phenomenal domains which might well be intrinsically discrete and distinct from one another (Maturana, 1991). Cognitive processes, including language, then become described in terms of recursions of *coordinations of actions* in a community, in the course of which different perspectives and subjective understandings become braided together while maintaining their integrity and differences.

In computer studies derived from insights into cognition itself, there is a shift away from the Artificial Intelligence programme which attempted to consider higher functions when abstracted and isolated from their biological and common sense grounding. Instead, there is a significant return to early cybernetic investigations as to just what constitutes *autonomy* in an organism (Bourgine and Varela, 1992). This focus on what is technically referred to as 'Artificial Life' is closely allied to insights into cognitive processes in terms of *embodied knowledge* (Varela 1992; Bourgine and Varela 1992). In this view, the focus of research has shifted from the problem of 'representation', to the issue of 'effective action' in an historically grounded process (i.e. ontogeny). Relevance or not, signified by effective action in a particular behavioural or experiencial context, is not pre-given, but *enacted* or *brought forth*, in an ongoing process of selection and punctuation, from an historical background (Varela 1992).

CONCLUSION

The above strands of contemporary thinking, and indeed many others, are consonant with the underlying Theseus model. None of them form the foundations for Theseus itself, and in the space permitted we have not been able to develop the links in any depth, but perhaps their mention provides the reader with an echo or resonance that serves to confirm the pertinence of the Theseus model, in the light of contemporary findings.

Theseus is an environment for structured yet creative learning, an 'infrastructure for connectivity'. Its ongoing development is on a modest, immediately practical scale, while the design allows potential for implementation on any scale, however large. This is at least in part because the underlying data model is being advanced and logically refined in full awareness of contemporary research in compatible yet disparate fields, which itself will contribute to the rigour and flexibility of the final computer-mediated environment, whatsoever that might eventually turn out to be.

REFERENCES

Bohm, D., and Peat, F.D., 1989, "Science, Order, and Creativity", Routledge, London.

Bolter, J.D., 1991, "Writing Space: the Computer, Hypertext, and the History of Writing", Lawrence Erlbaum Associates, London.

Bourgine, P., and Varela, F., 1992, Towards a Practice of Autonomous Systems, *in*: "Towards a Practice of Autonomous Systems", P. Bourgine and F.J. Varela, eds., MIT Press, Cambridge, MA.

Dupuy, J.P., and Varela, F.J., eds., 1992, "Understanding Origins: Contemporary Views on the Origin of Life, Mind and Society", Kluwer Academic, Dordrecht.

Dupuy, J.P., and Varela, F.J., 1992, Understanding Origins: an Introduction, *in:* J.P. Dupuy and F.J. Varela eds., op. cit.

Harnden, R.J., and Stringer, R., Theseus - the evolution of a hypermedium, *Cyb. and Syst.*, in press.

Maturana, H.R., 1991, Response to Jim Birch, *Jour. Fam. Ther.* 13: 375-393.

Varela, F.J., 1992, Whence perceptual meaning? A cartography of current ideas, *in*: J.P. Dupuy and F.J. Varela eds., op. cit.

PROBLEM-SOLVING SKILLS WITHIN THE CURRICULUM: A CASE FOR A "SOFTER" APPROACH

Brian Hopkins

Anglia Polytechnic University
Victoria Road South
Chelmsford
Essex CM1 1LL

INTRODUCTION

As this paper is being prepared a survey of the "problems" confronting the world affords us an insight into both the scale of the obstacles to be overcome and the paucity of our collective response to them.

The paper argues that, endemic to the cultural outlook of the dominant nations is a belief in the feasibility and viability of technologically-based " solutions" to problems. It further contends that we can trace this philosophical thread through our problem-solving methods at all levels, including the development of Information Systems (IS).

After an exploration of the background to this phenomenon and its manifestations in different environments, an experimental approach (which is currently being employed within a pedagogical research project) is described and the results so far are discussed. The approach has as its foundation a belief in the need to broaden our view of the scope and impact of problem situations if we are to achieve greater success in ameliorating those situations. It relies essentially on a judicious blending of both "soft" and "hard" approaches in ways which are appropriate to the specific problem situation and in which the combined elements complement each other to optimum effect.

PHILOSOPHICAL ENVIRONMENT

The emergence of Europe from the Dark Ages heralded a philosophical transformation which was to exert an over-riding influence on patterns of thought and practical application for a further five centuries. We are still imbued with the fundamental concepts and outlook which were formed at that time. However, increasingly we are seeing that received wisdom being questioned and an approach growing which argues for a broadening of our conceptual horizons with regard to effective ways of dealing with problematic situations.

THE ENLIGHTENMENT LEGACY

The Age of Enlightenment witnessed the triumph of rationality over what was perceived as myth and superstition as the guiding intellectual principle within "civilised" societies. From this base was built the commitment to the "scientific" method which was to be the cornerstone of research and development programmes up to the present day.

Kuhn (1970) traces the developmental route for these ideas, identifies the inherent features of such developments and, most significantly for our purposes, raises to prominence the need for researchers "to ask new sorts of questions and to trace different, and often less than cumulative, developmental lines for the sciences". He has posed a series of fundamental questions about the very nature and validity of "normal science" the answers to which are still many and conflicting.

THE INDUSTRIAL REVOLUTION

The technological (and related social) transformations which occurred in the late eighteenth and the nineteenth centuries in Britain and elsewhere represented the pinnacle of achievement for rationalism and the scientific approach. The period witnessed dramatic changes in the technologies of production based around a supreme confidence that "problems" were capable of "solution" i.e. once solved they would remain solved. This arrogance permeated the economic hey-day for Britain - the Victorian era - and, it can be argued, contributed to the decline of that economy because the willingness to innovate was stifled by the prevailing cultural attitudes. It is possible to draw parallels between the dominant attitudes of a century ago and those prevalent today within, for example, certain (influential) sectors of the U.K. education system. Pronouncements in August, 1992, from the Secretary of State for Education concerning validity of examination results and their causal connection with teaching and learning methods are indicative of a nostalgia for those days of certainty.

THE INFORMATION AGE

Some commentators see the developments in information technology since 1945 in a similar light to those blissful entrepreneurs of a century ago. They see convincing evidence of equivalent radical technological and social change allied to an unshakeable confidence in the human ability to overcome all obstacles - once we have defined the causes of the "problem". These attitudes have often been graphically revealed in the information systems domain through the writings of some members of the artificial intelligence research community who have focused on "the computer as brain" (Evans, 1979, Jastrow, 1982) and have occupied an optimistic position, based upon certitude, reminiscent of some of their nineteenth century predecessors. This stance is countered by Winograd and Flores (1987) in a way which, while being intellectually powerful and convincing, also - more crucially and appositely for our purposes - points to a need for a breadth of vision in our approach to problem situations which is both welcome and alarming; welcome for its pointers to social and cultural realities, alarming because of the revealed complexity of those realities. They conclude that "...knowledge and understanding do not result from formal operations on mental representations of an objectively existing world. Rather, they arise from the individual's committed participation in mutually oriented patterns of behaviour that are embedded in a socially shared background of concerns, actions and beliefs".

This radical perspective informs the Anglia experiment, is central to our approach to problem solving and, we believe, holds out the eventual prospect of a heightened methodological awareness amongst ourselves and our students which could have wider, beneficial repercussions.

METHODOLOGICAL ENVIRONMENT

"HARD" APPROACHES

The traditional approaches to problem solving in both science and technology have relied heavily upon methodologies which are based on the rationalist philosophy outlined above.

We can trace these developments within management science and computer systems development through a series of methods including systems analysis (RAND), operational research, cybernetics and various proprietary versions of structured methods for systems and/or program development (JSP, SSADM, Yourdon, DeMarco and even Object Oriented approaches).

Common to all the above is a belief that there is a consistency of goal amongst all the partners, a unitary approach which renders the "problem" soluble by ensuring a disciplined, structured, efficiently engineered set of procedures. In this environment effective answers to problems are those which are perceived to be, primarily, technically efficient.

These techniques have served us well (apparently) in those areas of endeavour where we more readily accept a level of uniformity of objective and where either we have a neatly bounded problem area or we reduce a larger problem area to manageable sub-sets. These conditions are seen as typical of investigation and progress toward solution in science and engineering. Similar methods, when applied to problems of a human and social category proved to be ineffective in the main and, indeed counter-productive in many cases.

"SOFT" APPROACHES

Out of the experiments with "hard" methods in what we may now see as alien environments emerged a set of methods which were based on the premise that there were no clear objectives within many problem situations and that it was this very lack of clarity which constituted the root cause of the problem. Most people with experience of working with problems of this type find an easy identification with this assertion (a point which is of relevance in the Anglia experiment where many of the students do not find such identification at all easy).

The practitioners of the "hard" methods were inclined to place this awkward fact beyond their boundary of consideration. Indeed, early researchers and consultants in this field used the only methods then available to them and only slowly and painfully came to recognise their inherent weaknesses and to develop more appropriate approaches attuned to the subtleties of the problem domain (for further illustrative detail see Checkland, 1981, Checkland and Scholes, 1991).

This acknowledgement of the "messy" nature of many of the human systems in which we intervene (and of which we are, significantly, a part) was a major breakthrough in our approach to problem-solving and was the start of a systemic process of investigation and learning which has provided fresh insights into the characteristics of such systems and tools to help in intervening in them.

COMPLEMENTARISM

Building upon the developments in the "soft" area certain researchers (Flood, 1990, Jackson, 1991, Jackson and Keys, 1984) have extended the initial concepts in imaginative ways bringing to their work perceptions from several disciplines and a range of authors (Habermas, 1972, 1974, 1984; Morgan, 1986, Ulrich, 1991).

Driving the work is a perceived need to address emancipatory and pluralist issues in the systems practice field. The movement represents a project to optimise the propensity of owners of problems to become full participants in interventions and to encourage the use of methodological approaches (possibly in combination) which are appropriate to the problem domain and its wider environment. One product of these efforts is Total Systems Intervention (TSI) which "... represents a new approach to planning, designing, "problem solving" and evaluation. The process employs a range of systems metaphors to encourage creative thinking about organisations and the difficult issues that managers have to confront" (Flood and Jackson, 1991).

The thinking behind TSI represents a further extension of systems practice away from the classical, scientific approach to problem-solving towards a more comprehensive, sensitive, truly responsive and, thus, hopefully, effective means of tackling society's problems.

While welcoming creative initiatives of this sort it is important to recognise (as do the initiators) that such developments are further steps in a journey and are not, in any way, the final destination. To argue otherwise would be a negation of the philosophical foundation upon which the initiative rests.

PEDAGOGICAL ENVIRONMENT

The progression of ideas outlined above has been embedded within our culture and has surfaced prominently in the education systems of the European countries in which the ideas originated, and in those other countries over which they exercised imperial, and hence, cultural, hegemony.

We are currently witnesses to global problems of a truly frightening nature and of an awesome scale. Within this we are also engaged in processes which can, with some justification, claim to be educating more people to a "higher" level of attainment and competence. Yet the problems remain; the pessimists would argue that they are worsening.

It is imperative that we derive some fresh insights from this paradox and then act upon them urgently. The principles of "soft/critical" systems thinking described earlier are submitted as a blueprint for this process of renovation.

TRADITIONAL

For several generations of students the educational framework in the European countries has been typified by a uniformity of philosophy and attitude even where the actual practice has varied. Central to this philosophy has been the concept of a received body of knowledge (Kuhn,1970) and, more significantly, a methodological conformity, allied to a tutor/student relationship which restricted rather than liberated both partners.

More specifically, in higher education in the U.K. there has been the added ingredient injected by the market for our graduates - a demand for trained people. This has exacerbated an already present negative feature (the over-structured, prescriptive, linear attitude) and further distorted (from a "liberating" standpoint) an already unbalanced approach.

These ideas were incorporated in a persuasive series of BBC Television programmes called "The Transformers" in 1990, in which it was asserted that "...there are many reasons for this crisis and why the educational system cannot help us out of it. One fundamental reason is that our education system still operates largely according to a particular way of understanding .... development and its relationship with teaching. The present view, the existing "paradigm" ..... is that minds grow, like bodies, in a kind of biological unfolding. All education can therefore offer is a means to fine-tune a process which will happen anyway" (Minnis, Overell and Sutton, 1990).

EXPERIMENTAL - THE ANGLIA EXPERIENCE

At Anglia a small group has been experimenting over a period of three years with a radically different approach to problem-solving. Our main focus has been within the information systems (IS) domain but the essence of our approach is to inculcate and nurture a broad vision of the nature of social and organisational problems (which subsume the IS area) and multiple perspectives of their causes and possible solutions.

We have worked, over this period, with groups of widely varying academic background and life experiences pursuing differing levels of outcome; the range extends from HND in Computer Studies to postgraduate programmes at Diploma and Masters level in general management and business and systems management.

We have documented elsewhere the background and processes of this experiment (Hopkins, 1991, Hopkins and Spaul, 1992). Over this period we have moved from grafting the systems thinking as elements within existing modules - with very mixed results - to a realisation that we needed to provide a basic foundation in this type of conceptual approach. We concluded that such an approach was essential in order to make a fresh start, to "clear an area" intellectually and culturally for new growth. The foundation module embodies an introduction to basic systems ideas, considers (and analyses causes of) some system failures, looks at some allied concepts and intellectual tools - game theory, gestalt, lateral thinking, ethnography, anthropology - and draws together the various strands in an assignment centred on the Channel 4 Equinox programme "New York, New York". Parallel to this we are also (for expedient reasons connected with increased student numbers) injecting sub-sets of this foundation programme into other modules.

We are now rigorously, on a weekly basis, monitoring response to this initiative from both students and tutors, both of whom, in most cases, are being newly exposed to these ideas. A picture is building of illuminating generalised lessons regarding experiential start points, reactions to the merit of the ideas and practical evaluation of their validity and utility.

At this point (November, 1992), we can report progress on each of these fronts in respect of both of the "client" groups referred to above. In particular, we have found that, although the grafted approach continues to cause problems to all parties, the foundation module is being well received, its themes and rationale are sympathetically engaged with, its practical potential is being realised (in both senses) and its central tenets are being seen as having relevance and possible application across a range of problem areas.

It would be fair to state that the outcomes of this, so far, limited experiment have been extremely encouraging. We have been able to establish constructive dialogues with both students and colleagues which are enriching our mutual views of problem-solving and, we believe, are a sound base for constructing and using a system of methodologies with which to begin to address the global problems referred to earlier. Many conceptual obstacles (and accumulated intellectual stances) are being, initially recognised, explored and then tackled by all of us. It is a fruitful learning experience which is exciting, and, potentially, extensively beneficial.

CONCLUSIONS

This paper has taken a philosophical journey of observation and discovery. The journey is a continuing one and the destination is by no means certain.
We would welcome fellow-travellers, convinced as we are that improved techniques in this field can emerge only from the widest possible collaboration and from exposure to a rich variety of experiences.

REFERENCES

Checkland, P.B., 1981, "Systems Thinking, Systems Practice," John Wiley, Chichester.

Checkland, P.B. and Scholes, J., 1990, "Soft Systems Methodology in Action," John Wiley, Chichester.

Evans, C., 1979, "The Micro Millenium," Viking, New York.

Flood, R.L., 1990, "Liberating Systems Theory," Plenum, New York.

Flood, R.L. and Jackson, M.C., 1991, "Critical Systems Thinking: Directed Readings," John Wiley, Chichester.

Habermas, J., 1972, "Knowledge and Human Interests," Heinemann, London.

Habermas, J., 1974, "Theory and Practice," Heinemann, London.

Habermas, J., 1984, "Reason and the Rationalisation of Society," Beacon Press, Boston.

Hopkins, J.B., 1991, Information systems training: the "soft" option, in: *Proceedings of the fourth UK conference on Information Systems Teaching: improving the practice*, Sunningdale, Berkshire, 1991.

Hopkins, J.B. and Spaul, M.W.J., 1992, Teaching soft methodologies: broadening the information systems curriculum, in: *Proceedings of the fifth UK conference on Information Systems Teaching: improving the practice*, Sunningdale, Berkshire, 1992.

Jackson, M.C. and Keys, P., 1984, Towards a system of systems methodologies, *J. of the O.R. Soc.*, 35, 473-86.

Jastrow, R., 1982, The thinking computer, *Sci. Digest*, 90:6.

Kuhn, T.S., 1970, "The Structure of Scientific Revolutions," University of Chicago Press, London.

Minnis, F., Overell,G. and Sutton, A., 1990, "The Transformers: the Art of Inspired Teaching," Broadcasting Support Services, London.

Morgan, G., 1986, "Images of Organisations," Sage, Beverley Hills.

Ulrich, W., 1991, Toward emancipatory systems practice, in: "Critical Systems Thinking: Directed Readings," R.L. Flood and M.C. Jackson, ed., John Wiley, Chichester.

Winograd, T. and Flores, F., 1987, "Understanding Computers and Cognition," Ablex Corporation, Norwood, NJ.

CREATIVITY AND SCIENCE ECOLOGY OF MIND: THE FUTURE SIGHT OF SCIENCE HAS BEGUN

Paolo Manzelli

Director of Educational Research Laboratory
University of Florence
Italy

Creativity is the basic principle of the improvement of personal intelligence and of progress of society Is one of the fundamental strategies of natural evolution.

Every system, no matter how complex, develops itself by means of transformation or achieves natural evolution by means of mutations.

There is no scientific definition of creativity because science is not developing a complete interpretation of systems transformation for understanding the evolution in nature.

Therefore, at present, science needs to conceptualise creativity as a natural strategy for transforming evolutive development. From our point of view, this means that we should be able to perform a new mental of science paradigm for addressing such global issue, to understand creativity as an expressions of system evolution in nature.

We can symbolise this new mental paradigm as, <E/I/M> , and include the notion of "Information" among the general fundamental variables that science needs for the general description of energy matter systems interactions.

To answer the question, "How many conceptual variables we need for a global description of nature ? Einstein considers "hidden" variables.

Attempting to understand such problem, we think of people's brain as part of nature.

For a general description of system theory, we need to accept that an explanation of the natural phenomena , including brain functioning, can be defined in relation to the two traditional parameters,"Energy" (E) and "Matter" (M) , and a third conceptual parameter "Information" (I).

<E/I/M> paradigm, in our concept, which inserts the thinking brain into the physical reality, is to be considered a "global systems paradigm", i.e. a conceptual context in which the meaning of physical objectivity is extended to include the subject . that is observing and interpreting the reality, into a global objective interpretation of system science. Using this meaning, we can develop the <E/I/M> paradigm of science understanding, that includes the person and his/her creative thought, as an objective part of natural evolution.

The <E/I/M/> paradigm, allows us to integrate into global theory Natural and Educational systems evolution. Including the brain of the observer, into the generalized concept of objective reality, enables us to refer not only to the external world, with its

Systems Science, Edited by F.A. Stowell *et al.*,
Plenum Press, New York, 1993

traditional paradigm of "mechanical" science, but to a new scheme of the cognitive reality of the world construction of the brain.

Using the <E/I/N> paradigm, we can look to the knowledge of the entire universe as a generalised system of learning, that evolve itself by means of creativity.

Creativity in science needs to invoke the "Occam's Razor": theories with the smallest number of arbitrary assumptions are to be preferred. Using the <E/I/M> paradigm the whole process of Universe can be discussed in an approach that does not require a fundamental distinction between subjective observer and objective reality. For this reason it is not more necessary to become involved in different arbitrary assumptions about the nature of observers and around the constitution of external world.

Henceforth referring to only one arbitrary postulate means that diversity of sciences can be reduced to an substantial unity overcoming the limited context in which sciences are normally formulated.

We apply this procedure to the new <E/I/M> paradigm, selecting only one fundamental postulate; Energy cannot be created or destroyed. A logical consequent of this postulate , the total Energy must be a constant, equal to say One at any time. If we considered the information as a parameter of the general description of the energy-matter transformation (inclusive of brain interactive function), the global variation (d) of the different aspect of energy , obtained from the sum of the free-Energy (Ef), and the codified-Energy (i.e. Matter - Em), and also (Ei), then Energy dissipated into the information process of the brain understanding, must be equal to zero. (A variation of a constant is zero for definition).

$$E\ \text{total} = (E_f) + (E_m) + (E_m) = 1$$

{i. e.} - at any time- $d\,[\,(E_f) + (E_m) + (E_i)\,] = 0$

Hence: $$d\,(E_i) = -d(E_f) - d(E_m) \qquad \{1\}$$

Hence the evolution can be seen as a program in which nature progressively transforms Energy and Matter interaction, to develop an increase of Energy linked to the process of information (E_i).

This line of reasoning permits us to argue The Heisemberg "uncertainty principle" hence is not inherent in the very nature of reality itself but is an expression of ignorance due to a lack of quantum mechanics to complete the sum of energy, forgetting the energy linked to the information process in nature.

The loss of "chaos" is the most important consequence of the change of the paradigm. The future of science has begun, within the context of the <E/I/M> paradigm, that will develop an creative impulse to the future science understanding.

The process of interactions described, in our mind, in terms of < E/I/M > paradigm, in which the time is related to the { E/I } relationships and the space is described by means the { I/M } correlations, can be seen as a evolutive process of the degree of the global system's quality.

We observe an equivalent procedure when the software becomes incorporated into the hardware of computers thus obtaining the evolution of computer's generations.

Likewise in nature, the progressive increase of embodied "information" into energy-matter transformation processes, provide a good intuitive expression to interpret the evolution that creates, during the natural evolution, the human brain.

Using formula {1}, we can explain the main direction of evolutive process and hence this line of reasoning permit us to argue that the Heisemberg "uncertainty principle", is a consequence of the ignorance of the consideration of the parameter "information in the explanation of physical phenomena".

In fact, we think that the "uncertainty principle" is not inherent in the very nature of reality itself, but is an expression of the lack of quantum mechanics to complete the sum of energy, forgetting the energy linked to the information Process in nature.

In truth, if we do not consider the objective existence of information as a physical aspect of nature, it will be impossible to refer the scientific meaning to a general and complete objective interpretation of the global events that produces the evolutive processes in nature.

Reasoning in this way, we can develop the new mental paradigm <E/I/M> of general system science education, in which evolution of natural and educational systems follows the same low of increasing the information's qualified energy (Ei). In our approach of the general system theory, the tendency to transform energy and matter to improve the degree of information exchange in nature is defined as creativity of the system evolution.

CONCLUSION

In conclusion, we can guess that, the selective processes, considered by Darwin, and the creativity processes described within the new <E/I/M> paradigm, are working as two complementary differentiated factors for the evolution of nature.

REFERENCES

Bohm, D., and Peat, F.D., 1987, "Science, Order & Creativity", Routledge, Great Britain.

Manzelli, P., 1991, Chemical Communication in the Brain, *in:* "European System Union- Handbook of System Science", M. Decleris, ed., Hellenic System Soc., Athens.

SPORT TRAINING AS AN OPEN SYSTEM

Kazimierz Perechuda[1] and Wojciech Cieslinski[2]

[1]Academy of Economics
Institute of Organization and Mgt.
Wroclaw 53-345
Komandorska 118/120

[2]Academy of Physical Education
Department of Sport
Wroclaw 51-617
Witelona 25a

SPORT TRAINING AS AN OPEN SYSTEM - PRELIMINARY ASSUMPTIONS

Sport training is a complex phenomenon characterized by a great number of elements which act on one another. Beside developing psychomotoric abilities, increasing sport knowledge and skills, it consists of elements of education and upbringing. Therefore this is not only an instructional process, but also a didactical one. The didactical and instructional reasons should mutually overlap and create a coherent system. Very often, however, only instructional objectives (among which competition results) are taken into account. The educational goals of such training are often neglected and sport training is regarded only as work.

Combining all its aspects during the process of sport training is a very difficult and responsible task. Finding the right ways of acting which join the above mentioned aspects of sport training, is presently the main problem of applied sciences.

In a closed system a process runs in a constant, or planned in advance way. If the initial conditions submit to any change and no corrections are made, then the final state (to which the process aims) will also change. However, in an open system the same final conditions can be reached in different ways and due to different initial conditions.

Elements of sport training

The preliminary characterization of sport training as an open system shows that all elements of the system form a certain structure and are exposed to the influence of the surrounding.

The main elements of the sport system are:

A. The surrounding of the system
B. The management subsystem
C. The personal subsystem
D. The subsystem of aims
E. The methodical subsystem
F. The temporal subsystem
G. The material subsystem.

Systems Science, Edited by F.A. Stowell *et al.*,
Plenum Press, New York, 1993

	A	B	C	D	E	F	G
A	*	1	1	1	1	1	1
B	1	*	1	1	1	1	1
C	1	1	*	1	1	1	1
D	1	1	1	*	1	1	1
E	1	1	1	1	*	1	1
F	1	1	1	1	1	*	1
G	1	1	1	1	1	1	*

Figure 1. A matrix of the training system structure.

The variety and complexity of relations presented in the sport training system (negative feedback) confirms the fact that sport training is a large, complex and open system. Next paragraphs will be devoted to a description of particular elements of sport training and their relation to a direct or indirect surrounding.

THE SURROUNDING OF THE SPORT TRAINING SYSTEM

The basic elements of the surrounding are the factors which have direct or indirect influence on training system elements. Those factors can be classified as:

A. Physical factors (temperature, humidity, noise, light, dustiness, colour, geographical location)
B. Economic factors (financial resources of the training group, club and association, material situation of sportsmen and coaches, sports equipment)
C. Social factors (cultural level of the society, regulations, directives, sport administration, qualifications of the staff, system of control, system of competition, school environment of sportsmen, the peer and training group, sportsmen's parents).
D. System of competition and sportsmen selection.

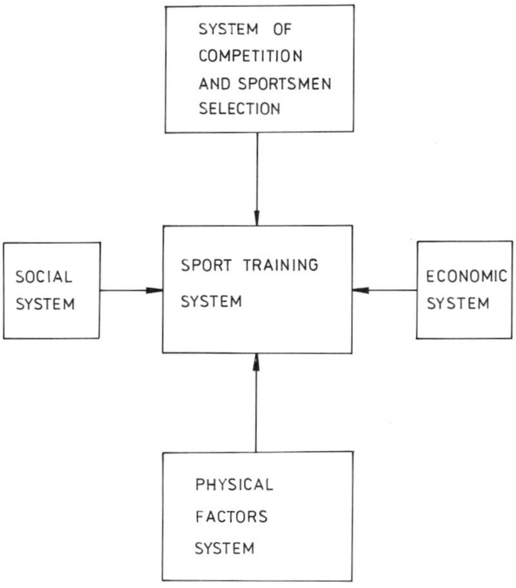

Figure 2. Relations between the training system and elements of its surrounding.

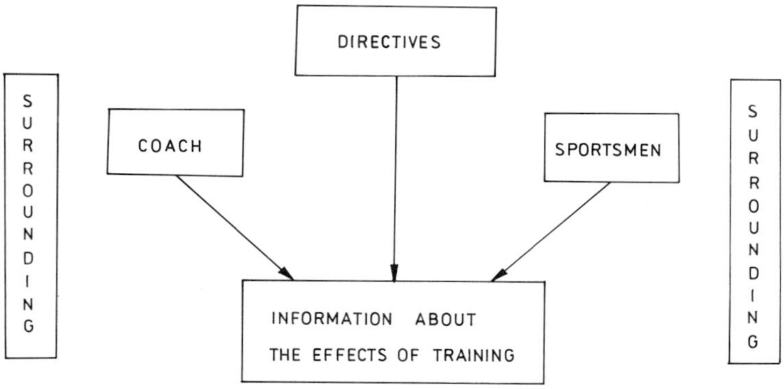

Figure 3. A simple model of coach-sportsman monitoring.

THE MANAGENENT SUBSYSTEN AND THE SURROUNDING

In an organized system, such as sport training, the coach, who controls the behaviour of the sportsmen, has to direct their behaviour at a certain aim. The coach tries to achieve the goal with help of the directives and feedback information about the effects of training from sportsmen (Figure 3).

THE PERSONAL SUBSYSTEM AND THE SURROUNDING

The personal subsystem, or so called human factor, is the basis for development of the sport training system. An open character of the system makes this factor (namely sportsmen) voluntarily integrate with the system after the preliminary selection. Therefore on the one hand the system receives an inflow of "energy", on the other hand it satisfies needs of the close and remote surrounding (the society).

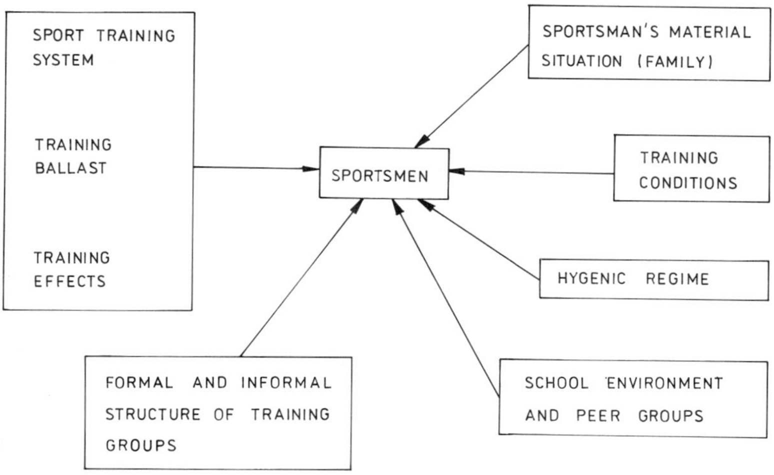

Figure 4. The factors influencing the personal subsystem of sport training.

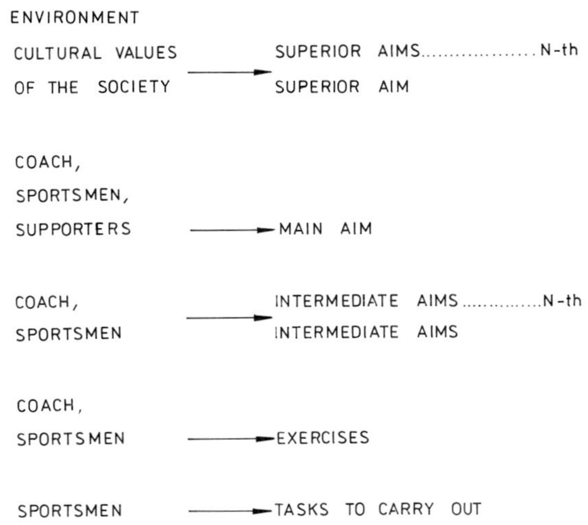

Figure 5. Influence of the environment on sport training aims.

THE SUBSYSTEM OF AIMS AND THE SURROUNDING

Sport training is a system which acts purposefully (teleological system). Thus the subsystem of aims is its very important element. According to the organized activity cycle, every action should start with setting goals. In sport training, there are superior, main and intermediate aims (Figure 5).

THE METHODICAL SUBSYSTEMS AND THE SURROUNDING

The methodical subsystem of sport training consists of training methods, forms and means which determine the structure of training ballast. This subsystem is dependent on the structure of aims determining what methods, forms and means should be applied for achieving the expected goals (Figure 6).

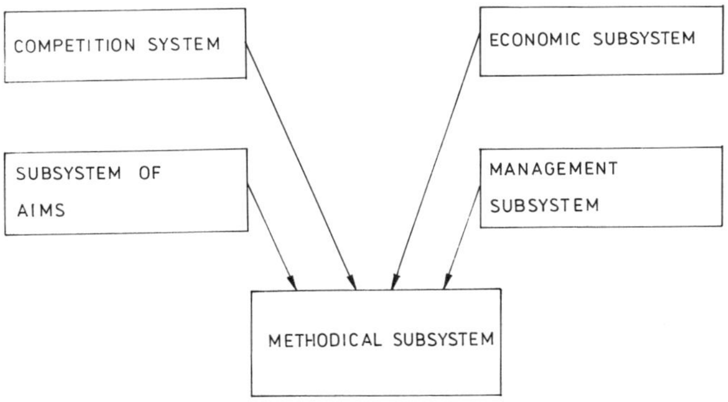

Figure 6. Direct and indirect influence of the surrounding on the methodical subsystem.

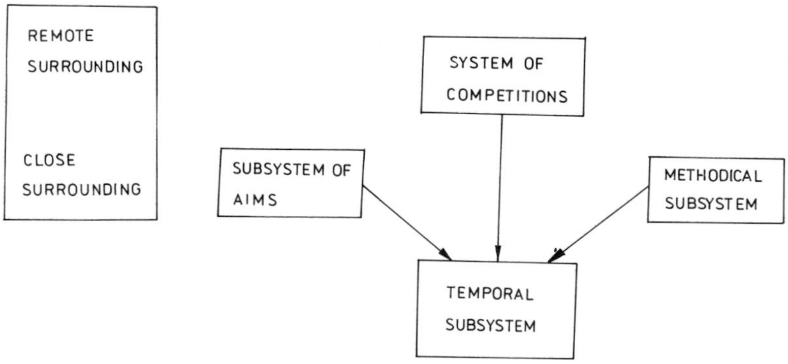

Figure 7. Influence of the remote and close surrounding on temporal subsystem.

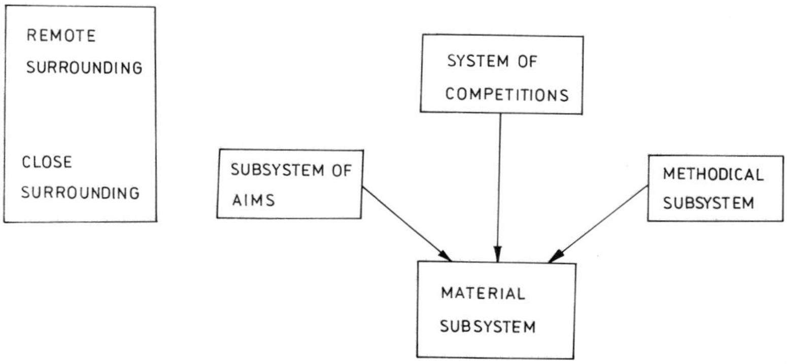

Figure 8. Influence of the remote and close surrounding on material subsystem.

THE TEMPORAL SUBSYSTEM AND THE SURROUNDING

Phasic character of sport form development, that is, building up form (preparation phase), stabilization (starting phase) and temporary losing form (transitional phase), is the essence nf functioninq of temporal subsystem (Figure 7).

THE MATERIAL SUBSYSTEM AND THE SURROUNDING

The effective usage of the material subsystem of sport training depends on financial resources of the club, training section or sportsmen, that is, of the econcmic system (Figure 8).

CONCLUSION

The description of sport training is an epistemological assumption made in order to find complementary research methods.

The interactions which are present in sport training system are based on negative feedback mechanism. Those interactions make sport training a dynamic process of cyclic changes, directed to bringing up a sport champion. However, there are more than one ways of reaching this championship. This state can be accomplished in an equifinal way. This means that a sportsman may reach the goal by various methods and training means and in various conditions.

The classical presentation of sport training as a system is insufficient. Sport training analysed as a system is characterized by significant contactability (feedback) with the outer environment. The latter conditions in a sense the work of the system and its effective functioning.

Basing on the characterization and description of sport training as an open system, a model of improving sport training by using heuristic methods may be presented. A practical illustration of such an approach is prevalence of methods of systems thinking, researching, designing and managing. In this context, systems approach to all kinds of problems must have acomplex, interdisciplinary and hierarchic character. This means that the problems have their practical, as well as theoretical dimension, which involves looking for the tools to solve the problems.

REFERENCES

1. L. von Bertalanffy, "General theory of systems", PWN, Warszawa, (1989).
2. W. Gasparski, ed., "Technologies of planning", WNT, Warszawa, (1988).
3. D. Katz and R.L. Kahn R.L., "Socialpsychology of organization", PWN, Warszawa, (1979).
4. J. Kisielnicki, "Systems methods", PWE, Warszawa, (1979).
5. A.K. Kozminski, "Modern concepts of management", PWN, Warszawa, (1985).
6. G. Lasinski, "The praxeological and systemology of the sporttraining research and efficiency", SiM, Wroclaw, (1988).
7. P. Sienkiewicz, "Systems of directing", WP, Warszawa, (1989).

INTRODUCING SYSTEMS THINKING INTO MATHEMATICS LEARNING

Neil Pitcher

Department of Mathematics & Statistics
University of Paisley
High Street
Paisley PA1 2BE

INTRODUCTION

Systems thinking is at the heart of mathematics and characterises the general approach of any experienced researcher in the subject. Yet for the student it can be an uphill struggle to attain to such a mature view of mathematical activity. In traditional mathematics courses, a large part of the problem has been the tendency to immerse students in techniques and procedures so that the essential nature of the discipline, namely posing problems and solving them, is not grasped.

This is not to say that techniques are unimportant. After all, algorithms and techniques are prerequisites to any degree of expertise and are necessary to obtain anything which might be called a solution. Initially, technology in the form of mainframe computers did not fundamentally alter things. True, one could now obtain a solution from the computer, but a program containing algorithms still had to be written and normally one had to wait before the results were available.

What is now changing things, however, is the personal computer, with the opportunity it affords of instant access to results and the widespread use of packages, whether highly sophisticated or less so. It is increasingly the case that algorithms and techniques are implemented quite easily on the machine. By using the computer appropriately within mathematics courses, it is possible quite readily to introduce in more meaningful ways higher level thinking, such as analysis and problem solving, in other words, systems thinking. In what follows a Soft Systems view of mathematics learning is proposed, and this is illustrated by a case study problem given to mathematics students which exemplifies the general approach.

MATHEMATICS LEARNING AS A SYSTEM

The process by which students learn mathematics can be viewed as a system. As far as society is concerned, this is an open system, as in Figure 1.

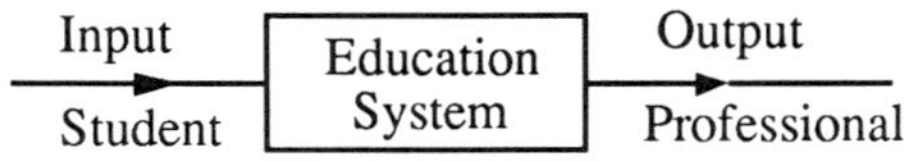

Figure 1. An open system.

In this the requirements implicit in the word "professional" set the agenda determining the make-up of the education system in question. As an attempt to specify these requirements, we suggest that, for mathematics graduates, good problem handling skills and analytical skills are of the essence. It is important, therefore, for mathematics degree courses to contain a good mix of both "problem solving" and "techniques". These are related and a well designed course will deal adequately with both. It has perhaps been the case however that traditional mathematics courses have tended to emphasise analytical skills to the detriment of problem handling skills. The key element which now enables us to rectify that imbalance is the computer, because it enables us to "short circuit" the calculation process and thus make problems more accessible to the students of mathematics. This makes it practically possible to introduce problem scenarios along the lines of the case study described below. In practice however there is a considerable task of management to provide a learning environment within which students can acquire enough of the skills necessary to be able realistically to tackle such a problem. In the author's department a considerable effort has been invested in recent years to produce such a learning environment. This takes the form of a mathematical sciences laboratory[1,2] which is run throughout the first two years of the degree course. Briefly these classes take the form of small group sessions using the computer as a tool to solve problems using mathematics. Beginning with fairly simple problem scenarios related to conventional algebra and calculus material covered in lectures, the students are given an experience of a succession of problems in which the standard of difficulty is carefully increased with time. It has been found that, with careful management, it is possible to take in students who initially have little confidence in their mathematical ability, and to bring them to a stage of real capability as graduates. It has been particularly heartening to find that, by using the approach of phasing in progressively harder case studies and positively encouraging group work, many of the problems associated with non-homogeneity of student intake effectively disappear.

Viewed as a soft system, this approach to mathematics course management is illustrated in Figure 2. A simplified version of this is shown in Figure 3. The elements of the system shown in Figure 2 can be grouped as Activities (laboratories, lectures, tutorials),

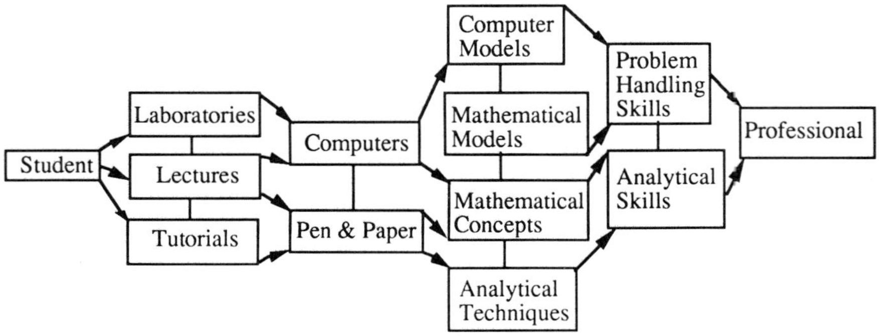

Figure 2. Mathematics learning as a system.

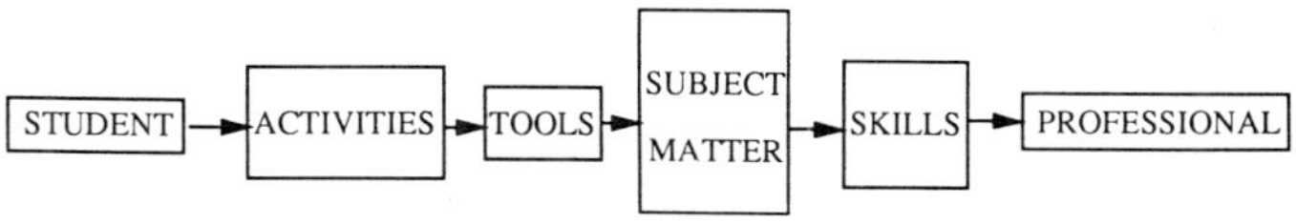

Figure 3. Simplified view of figure 2.

Tools (computers, pen and paper), Subject Matter (concepts, models, techniques, problems) and Skills (problem handling and analytical). In this a proper balance of skills is viewed as paramount. The subject matter dealt with in the course is matched with the skills required, whilst the activities are chosen so as to use available tools, including the computer, as means of delivering the subject matter. The activities which are visible within the courses are a blend of classes of traditional type (lectures and tutorials) with the more recent innovation of the mathematics laboratory. The result is an overall learning experience which goes a long way towards enabling students to acquire a healthy balance of the range of skills required to make a positive contribution as a mathematician to today's society. Traditional courses, not using the computer, have tended to promote the lower stream in Figure 2, emphasising analytical techniques and skills. The approach suggested retains this, whilst at the same time using the computer to impart problem handling skills also.

A CASE STUDY

An example of a problem given to students in a laboratory situation is now described. The problem concerns deer and lion populations within an area near the Grand Canyon, called the Kaibab Plateau. It is in fact a good example of a complex system, in which an "obvious" solution has unforeseen and undesirable consequences. As such it provides a valuable introduction for mathematics students to the whole notion of systems thinking.

In the years following 1907 deer hunting was banned and at the same time a hunting bounty was imposed on mountain lions, so as to reduce their population, the object being to make the area more habitable for humans. Unfortunately the planners at that time failed to foresee the catastrophe which would follow. The reduction in the mountain lion population led to dramatic increases in the deer population, which caused the deer food supply to be exhausted, and the final result was mass deer starvation, which occurred in 1924 and 1925. Specifically the deer herd grew from about 4000 in 1906 to about 100,000 in 1924, when around 60% of these starved. Ultimately the deer population settled out at 10,000. This scenario, which has been quoted previously as an example[3], was given to a group of 52 degree year students in mathematical sciences, as a case study exercise, to be tackled over a period of five days. The students were allowed to work in groups of 3 or 4. The groups spent days one to three building a mathematical model which could accurately represent the deer population dynamics. On day four the students examined the likely effects of different proposals for a deer control policy, finally concluding with specific recommendations which, if adopted in 1907, would have averted the disaster of 1924/25. The groups submitted work for assessment on each of the first four days, and finally on day five an unseen examination was set, in order to differentiate between individual students' performances.

On day one an initial model was formulated, given that the natural birth and death rates for deer are 50% and 20% per annum respectively. In addition the rate *dk*, at which deer are killed by lions, was specified as a function of *dd*, the deer density, in deer per square kilometre. Given that $dk(1.25) = 3$ and $dk(13) = 61$ and that the curve has the S shape shown, students identified the equation:

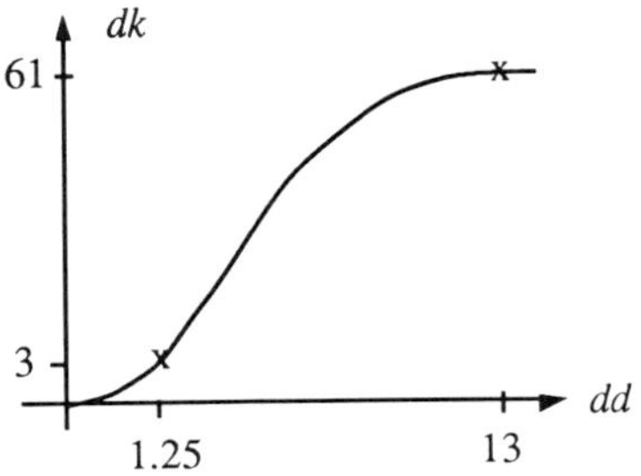

Figure 4. Kill rate as a function of density.

$$dk = 490257\ dd * dd/[6587\ dd * dd + 245050]$$

following the approach of Murray[4] (See Figure 4). The natural birth and death rates for lions were given as 20% and 10% respectively. Before the imposition of the lion bounty in 1906, the fraction of lions killed by hunting per year was 0.1. Subsequently to that, this increased to 0.3. Taking 1900 as year zero then, and given that the area of the plateau is 3200 square km, the following equations were obtained, for the functions d and l representing deer and lion populations. Here it is assumed that the bounty was imposed in 1906 and ceased in 1950.

$$d' = 0.5d - 0.2d - l.dk$$

$$dd = d/3200$$

$$l' = 0.2l - 0.1l - l\ [0.1 + 0.2\ (H(t - 6) - H(t - 50))]$$

$$H(t) = 1 \text{ if } t > 0;\ H(t) = 0 \text{ if } t \leq 0$$

Implementing this model using a simulation package gives results which indicate that the deer population grows beyond bounds. The catastrophic decline in deer population which occurred in 1924 however is not detected.

On day two ways of improving the model are investigated, by representing the deer food supply.

The more food is available, the greater the life expectancy of deer, and data are available as in Figure 5, to which students fit the relation

$$a = 40\ v * v/[5\ v * v + 3]$$

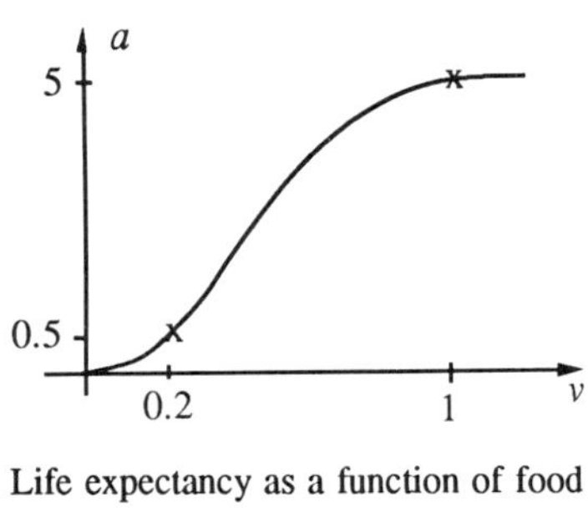

Figure 5. Life expectancy as a function of food available.

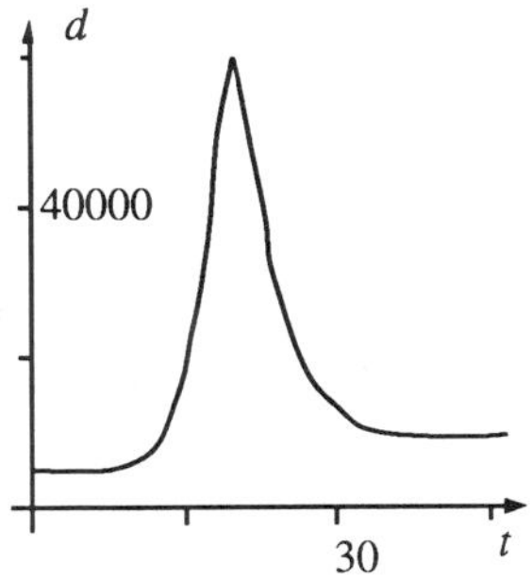

Figure 6. Deer population.

where a is the average lifetime of a deer and v is the number of units of vegetation available per deer for food. Given also that there are 200,000 units of vegetation on the whole plateau, it follows that $v = 200000/d$, and the differential equation for the deer is altered to

$$d' = 0.5d - d/a - l.dk.$$

Implementing this on the computer gives results which indicate an initial sharp rise in deer population, culminating in a levelling off to a high steady state value. This is an improvement on the previous model, but still fails to predict the disastrous decline in deer population of 1924.

The model is further refined in day three, in which consideration is given to the fact that the initial 200000 units of vegetation are in fact likely to be diminished as the deer population increases, whereas in the previous model the vegetation was assumed constant.

When the resulting model is run on the computer, it is found that the results do follow the observed events of 1924/25 quite well (see Figure 6).

Additionally graphs for l, a and vu enable one to see the general trends for these quantities also.

On day four amendments to hunting policy are considered. The most obvious thing to try is to reduce the time slot in which the lion bounty is allowed. The equation for l' can be adjusted easily for this. Another possibility is to introduce a deer bounty, to keep the deer population within bounds, so that decimation of food supply is avoided. By carrying out this kind of activity, students are able to formulate their own ideas and hence make specific recommendations for policies which would have avoided the catastrophe.

As a result students see that the ecological scenario in view is a complex system, in which an "obvious" solution was implemented, leading to disastrous consequences. The whole exercise is beneficial in the sense of students seeing that mathematics can be of value in comprehending a complex system and hence in enabling decision makers to apply appropriate policies. Moreover the students' general view of mathematics is elevated, so that they begin to comprehend the basis of systems thinking, and see the place of model building in dealing with complexity.

CONCLUSION

The advent of the personal computer is having a profound impact on mathematics instruction, mainly because it automates the techniques which characterise the solution process. To some this might seem like a threat to mathematicians - with fears that we might suffer the same fate as the redundant production line worker on the arrival of mechanisation. The truth is however that it is not a threat, but an opportunity to lift students from the slough of technique after technique, and into an awareness of mathematical thinking in its authentic sense.

The opportunity now presents itself within mathematics courses of introducing students to complex problems of the kind outlined. This kind of educational experience will be of great value in ensuring that the system of mathematics learning produces graduates who are in a position to apply their knowledge in mature and positive ways within society.

ACKNOWLEDGEMENT

The Kaibab Plateau problem was brought to our attention by Professor J J Gonzalez, of Agder College of Engineering, Norway, as an exemplar of computer based teaching using the training system SimTek, which is a product of ModellData, Inc. Further information can be obtained from the following address: ModellData, Inc., PO Box 42, N-4871 Fevik, Norway.

REFERENCES

1. R.R. Burnside, A.R.G. Macdivitt, N. Pitcher, and E.T. West, Mathematics Teaching for the Next Millennium, Proc., "The Fifth International Conference on Technology and Education", Vol. 1, pp 444-447, (1988).
2. N. Pitcher, A computer based laboratory course in mathematical sciences, *Computers Educ.* Vol. 18, No. 1-3, pp 135-141, (1992).
3. N. Roberts, D. Anderson, R. Deal, M. Garet, and W. Shaffer, "Introduction to Computer Simulation", Addison-Wesley, Reading Massachusetts, (1983).
4. J.D. Murray, "Mathematical Biology", Springer-Verlag, Berlin Heidelberg, (1980).

SYSTEMS EDUCATION: IS THERE A MASS DELIVERY APPROACH?

John Robson[1] and Graham Paton[2]

[1] School of Accounting and Financial Management
Luton College of Higher Education
Park Square
Luton, LU1 3JU

[2] Tutorial and Counselling Staff
Open University
South Region
Foxcombe Hall
Boars Hill
Oxford, OX1 5HR

INTRODUCTION

Even into the 1990's systems remains a complex and relatively inaccessible discipline with few texts or other media available which can usefully lead the inexperienced user to acquire a working knowledge of the skills embraced within it. Within this paper the authors draw upon their experience in teaching the Open University course T301 "Complexity, Management and Change: Applying a Systems Approach"[1] to discuss some issues concerning the value of open learning styles as vehicles for teaching systems methods.

A BRIEF SURVEY OF AVAILABLE TEXTS

At the hard end of the systems spectrum there are still a preponderance of O.R. texts which only pay lip service to methodology, considering it in an initial chapter before getting down to the bundle of mathematical techniques which are currently in vogue. The reader is usually left unaware of the nature and reasons for the adoption of a particular analytic frame. This is of course not dissimilar from the teaching of most science disciplines, where the nature of scientific enquiry is taken for granted. Most of the texts which do discuss the principles, concepts and theoretical frameworks are in the softer area but again it is difficult to find an introductory text. Indeed texts remain so rare that it is not unusual for an academic bookshop to stock none at all and, even where some are available, they are difficult to find because they appear to defy all known classification systems. With the systems discipline being established (even if not precisely defined) for at least 25 years this is a rather strange

Systems Science, Edited by F.A. Stowell *et al.*,
Plenum Press, New York, 1993

situation, particularly when contrasted with the enthusiasm of the practitioners for their areas of interest.

Some texts do exist but do not provide the best type of material for the independent learner. Peter Checkland's "Systems Thinking, Systems Practice"[2] is a masterly exposition of both philosophy and practice but is more appropriate to the practitioner who already has some knowledge of the methods discussed or of similar approaches. The later text "Soft Systems Methodology in Action" from Checkland and Scholes[3], although based on practical applications, is still rather scholarly and is more suited to existing practitioners with an academic bent than the newcomer who is trying to learn systems ideas. "Systems: Concepts, Methodologies and Applications" by Brian Wilson[4] is aimed at "the student, or would-be practitioner". The author presents his text as being complementary to that of Checkland's and it does contain much different material although it can also be regarded as a translation of some of Checkland's work into a more accessible language code. Even with Wilson's efforts to make it amenable to a wider audience, it is still a rather difficult text for a beginner. It is more useful for the highly able student who has access to lecture and tutorial support.

Moving to a simpler level problems again emerge. "Practical Soft Systems Analysis" by David Patching[5] is accessible (although not always entirely accurate in presenting the ideas it purports to describe) but somehow still feels rather unsatisfactory. Patching's attempts to simplify the presentation have at times prevented the reader from appreciating the value and power of some of the concepts covered. This text can certainly communicate many of the essential skills but many practitioners feel somewhat unhappy with the approach whilst at the same time as recognising this text as a very valuable introductory resource. (Maybe this is simply an observation that the NVQ/Competence approach will not be embraced by those in the system's movement).

The problem of providing easy explanations has long been recognised. For example John Naughton produced a "Reader's Guide"[6] to the Checkland Methodology[7] in order to help Open University students and tutors understand the essential principles of the Checkland approach. This document itself led to substantial discussions at course tutorials but again did not provide a document well suited to an independent learner working in isolation. Although this was a useful contribution (which interestingly was used by Lancaster MA students as well as those of the O.U.) it was again a text more suited as one component of a learning experience rather than a tool which could be used on its own.

THE CONVENTIONAL TAUGHT COURSE APPROACH

The lecture program plus dissertation approach to the teaching of systems would appear to be extremely successful. Certainly the Master's Program in Systems in Management at the University of Lancaster produced graduates who were both well informed and enthusiastic. However in today's terms this approach can hardly be regarded as a mass delivery method as it is labour intensive in its need for lecturer support and hence expensive. Even where the costs are felt to be justified it is difficult to see how it could be extended to a mass market. At Lancaster a typical student cohort numbered something like 30 to 40.

This type of approach is probably well suited to the training of specialists but would appear to be highly extravagant as a means of introducing the discipline to a wider audience.

PROBLEMS ENCOUNTERED BY THE SYSTEMS PRACTITIONER

The lack of a level of non-expert knowledge in the populations with which systems practitioners work causes many problems. In most professional relationships there exists sufficient knowledge of the skill areas of the various participants to legitimate the areas of intervention of each person without the need for complex negotiation of responsibilities. Although this can be limiting as well as empowering it does provide a useful initial definition of roles in a work group. Within the systems area the practitioner is still confronted with the question, "What is systems?" usually in a context of, "What do systems practitioners do?" Meaningful answers to these questions are of course extremely complex. It is much safer to be an accountant or a physicist. Our colleagues do not get asked, "What is accountancy?" or, "What do accountants do?" since there is at least an assumption (probably largely justified) that everyone has some idea of the answers. This problem is compounded by the meta-disciplinary nature of systems. Without an understanding of what the practitioner is trying to do or the need for it, the work of the systems practitioner will be perceived as an unjustifiable piracy of roles which have been attributed to other professionals. It is only too easy for the systems practitioner to be marginalised by the groups with which he or she works, as an understanding of the discipline is fundamental to establishing its purpose and application. This problem continues to exist for systems academics as well as practitioners in the field.

THE OPEN UNIVERSITY SYSTEMS COURSES

The Open University Systems Group has probably more experience in the teaching of systems than any other university or higher education institution in the U.K. Although there are substantial difficulties in the teaching of systems, and the limitations of text have been discussed above, the Open University has had considerable success with systems courses in its undergraduate program.

The O.U. has three courses on systems at CATS level 2, each worth 30 points:

T244 "Managing in Organisations"[8]
T247 "Working with Systems"[9]
T273 "Food Production Systems"[10].

T247 is an introduction to systems ideas which aims to develop the students' skills in using systems as descriptive models. T244 is essentially a management course which introduces systems ideas in providing an overall framework to support a range of practically oriented management ideas. T273 uses a systems based frame to support the analysis of food production.

The jewel in the crown is widely regarded (by both tutors and students) as the course at CATS level 3, worth 60 points. This is T301 "Complexity, Management and Change: Applying a Systems Approach". This is an essentially practical course which introduces the student to three systems methods:

1. A failures method[11] developed by a team led by Victor Bignell.

2. A hard method based on ideas similar to those of de Neuffville and Stafford (very similar to the Rand Corporation method) but with some important differences, notably a very much softened front end.

3. A soft method essentially based on the ideas of Peter Checkland.

T301 includes a project, worth approximately half of the continuous assessment marks, for which the student can use any of the three methods taught on the course. The course is complete in itself and has no prerequisites although students are often advised to study a systems course at level 2 as an introduction before studying T301.

For students who want to go further the O.U. offers a technology project course (T401) for which a more substantial systems project may be submitted. This is again worth 60 CATS points at level 3.

The systems courses as a whole are very successful on a wide range of metrics. Student numbers are high; the level 2 courses each have student populations consistently over 1000 and the level 3 course has populations of over 500. Clearly the Open University is producing more undergraduates whose degrees include systems than the rest of the UK higher education altogether. Students on T301 are usually highly enthusiastic about the ideas contained within the course and generally rate it as one of the best courses that they have included in their degree. Continuous assessment assignments produced by students are on the whole of a very high standard with some projects demonstrating well developed action research skills. The course is examined in a conventional way with a three hour unseen paper. This is in some ways not the ideal form of assessment for the type of material but some commendable work is produced, although it is noticeable (and maybe predictable under examination conditions) that students perform significantly better on the applied case study type questions than on the more philosophical ones.

THE POTENTIAL FOR WIDER APPLICATION OF OPEN UNIVERSITY SYSTEMS TEACHING

At first sight it looks as if T301, or a development of it, could be applied to much larger populations with little difficulty. This may well be the case but it needs to be carefully assessed why this course can successfully teach in an area where texts are often felt to be unsatisfactory. Like most texts, the course is fundamentally practical in intent but soon finds a need to discuss more philosophical issues. These issues are not avoided and the course is recognised as being of considerable difficulty for an undergraduate level. (It should be remembered that when this course was developed the O.U. did not have a very well developed postgraduate program so the question as to whether it might have been better regarded as part of a Masters program could not and apparently did not arise).

Without a lot of detailed research it is only possible to speculate on the success of T301 as a teaching vehicle. However the student population of any O.U. course is very diverse. Students, by necessity or choice, study in a variety of different ways, so some judgements can be made as to what are the particular strengths of different parts of the T301 learning package. In the written material itself, it has been felt by many tutors that the decision to use a selection of three methods was a good one. One of the previous courses which T301 superseded introduced nine different methods in the first block as an introduction to what systems methods were like. This caused much confusion as untrained students did not possess the skills necessary to distinguish where these methods were essentially similar and where they differed in important ways. On the other hand it is useful, when one is essentially teaching thinking, to be able to contrast one style with at least one other. If this is not done students tend to assume that the method being taught is common sense, essentially confusing plausibility and internal consistency of a particular method with an idea that they already knew it. It should be noted that virtually all systems courses and texts do use this comparative approach in which one method is compared with another systems or perhaps non-systems method. T301 is rather unusual in not identifying a preferred method. All three methods are presented as useful vehicles for practical analysis and development and are given broadly equivalent status within the course. The relevant course units are written by authors who are working with their own preferred approach.

THE NEED FOR INTERACTION WITH PEER GROUP AND TUTOR

It is easy to assume that an O.U. course by its nature depends primarily on the printed material as its most effective teaching mode. However John Martin identified strong correlations between some assignment grades and the use of discussion with others of course material. The influence of tutor led groups was strongest, followed by student only self help groups, but discussions with people outside of the course (for example work colleagues) were also found to be helpful. Tutors have also reported that some previous exposure to systems ideas makes the course more accessible. Students particularly report that they find the summer school associated with the second level courses to be very helpful preparation as it is based upon a rather fast but useful introduction to two of the methods included in T301. For the student who is completely new to systems the early blocks are indeed difficult. Although there is certainly a higher than reasonable workload for many of these students, most do come to understand the material, often drawing on the support of their tutors or other students. Although there are students who work in isolation with no contact with other tutors or students, their performance is often lower than their peers. The course can be successfully studied in this way but the course team encourage students not to use this approach. There is also evidence that students value a substantial amount of face to face support. Within the East Anglian region tutorial support is arranged so that a student will have access to 25 hours of face to face sessions with inputs from a team of four tutors. Attendance at a typical Saturday school will be at a level of 65-70% of active students. Given that O.U. students tend to have a wide range of other commitments this is a surprisingly high attendance. It should also be noted that this is in the face of long joumeys for many, as a Cambridge teaching location serves students from as far away as Watford, Kings Lynn, Yarmouth and Southend. Some years ago the East Anglian region used residential weekend conferences to provide an introduction to all students new to systems (whether at level 2 or level 3). Regrettably these were a victim of their own success, as an appropriately priced venue could not be found to cater for the high levels of demand as systems courses became more popular.

In response to student demand, the authors have provided revision courses sponsored by the Open University Travel and Study Society (an independent student group affiliated to the O.U. Student Association). This year these attracted about 15% of the national course population. This is remarkably high when it is considered that the courses were located in Stirling and Brighton, which are not ideal when national access is considered. The course team and regional office staff are aware of this student demand for higher than usual levels of face to face contact. Most staff tutors allocate higher tutorial provision to T301 than to other courses of nominally similar level and content, and the course team is planning to add a national Easter residential school to the course to help students. particularly those with no previous systems knowledge, to get off to a good start.

CONCLUSIONS

In T301 the Open University Systems Group have produced a remarkably ambitious and successful systems course. The quality of exposition in the texts make them very useful teaching material on their own. The course however is much more extensive than its textual component and students place great value upon face to face support provided by tutors and other students. This does mean that as presented at the moment tutorial support costs are high and. with planned initiatives, are likely to increase. The course is also dependent on finding sufficient tutors capable of supporting it and regional staff have identified T301 as one of the courses for which recruitment is most difficult, although recently former students have taken on tutorial staff positions with great success. Some of the demand for face to face contact is undoubtedly from the more able students who want to develop their skills and understanding beyond the boundaries of the written material and the demand for tutorial support might reasonably be interpreted as a measure of student enthusiasm rather than a more basic need.

Overall T301 must be seen as a considerable success in a discipline where questions about pedagogical problems remain. As a cheap mass teaching system a course on the lines of T301 might work, but its real success in current terms is as a high quality course with good cost effectiveness. T301 can be used to train many more students than conventional courses but the need for an introduction to systems which is both cheap and effective remains as elusive as ever.

REFERENCES

1. T301 Course Team. "T301 Complexity, Management and Change: Applying a Systems Approach," Open University Press, Milton Keynes (1984).
2. P. B. Checkland. "Systems Thinking, Systems Practice," John Wiley, Chichester (1981).
3. P B. Checkland, and J. Scholes. "Soft Systems Methodology in Action," John Wiley, Chichester (1990).
4. B. Wilson. "Systems: Concepts, Methodologies and Applications," John Wiley, Chichester (1984/1990).
5. D. Patching. "Practical Soft Systems Analysis," Pitman, London (1990).
6. J. Naughton. "The Checkland Methodology: A Readers Guide," Open University, Milton Keynes (1976)
7. P.B. Checkland. Towards a systems-based methodology for real world problem solving, *Journal of Systems Engineering*, 3:2 (1972).
8. T244 Course Team. "Managing in Organisations," Open University Press, Milton Keynes (1986).
9. T247 Course Team. "Working with Systems," Open University Press, Milton Keynes (1990).
10. T273 Course Team. "Food Production Systems," Open University Press, Milton Keynes (1979).
11. V. Bignell, J. Fortune, J. Robson, and L.Watson. "T301 Block II: A Systems Approach to Failures," Open University Press, Milton Keynes (1984).

COMPUTER-BASED METHODS OF KNOWLEDGE ACQUISITION AND ELICITATION IN TERMS OF THE SUBJECTIVE REPRESENTATION AND TEACHING OF COMPLEX DOMAINS

Mark H. Stansfield

Information Systems
Department of Computing Science
University of Paisley
Paisley, PA1 2BE
Scotland

INTRODUCTION

A complex domain of expertise can arguably be identified as one comprising a mixture of well-structured, clearly defined knowledge and heuristic, subjective knowledge. This paper discusses various computer-based approaches to knowledge acquisition (KA), and briefly highlights an investigation into the use of computer-based technology to support what can be termed an 'interpretive' approach to knowledge elicitation (KE), namely the Appreciative Inquiry Method (AIM) (Stowell et. al. 1991; West, 1992), which seeks to elicit complex knowledge directly from an expert at a subjective level of experience. KA usually refers to the acquisition of knowledge from both human domain experts and from other sources such as texts and documents. KE, on the other hand refers to the elicitation of knowledge directly from human domain experts. The rationale for this paper is based on the inadequacies of current computer-based approaches to KA in eliciting the more heuristical, experiential and 'tacit' features of expertise and in providing a subjective representation of those complex domains of expertise.

COMPUTER-BASED KA AND SOME OF ITS ASSOCIATED PROBLEMS

There are currently a wide range of computer-based support systems and approaches to the acquisition of knowledge; many focusing upon different aspects in the acquisition, elicitation and representation of domain knowledge. This section will concentrate on very briefly critically assessing some of the more widely known and developed computer-based approaches to KA in terms of: KA tools, KA workbenches and hypermedia-based tools.

Systems Science, Edited by F.A. Stowell *et al.*,
Plenum Press, New York, 1993

Computer-based KA tools which are based on individual tools can be divided into two groups; those which are based upon the repertory grid technique and those which provide knowledge editors which are tied to domain tasks for specialised knowledge bases. The KA tools which use repertory grids as the basis for the elicitation and analysis of knowledge include PLANET (Shaw and Gaines, 1987), KSS0 (Gaines, 1990) and KITTEN (Shaw and Gaines, 1988). These tools are used to interact directly with the expert, using a rating grid as the basis for eliciting knowledge. However complex phenomena within certain subject domains may be beyond measurement in terms of a bipolar scale and grid framework. Whilst such tools may be well suited to structured and less complex domains involving analytical tasks (Neale, 1988), when applied to highly subjective domains containing qualitative elements, attempts at quantification and measurement may be undesirable; in that, what are the respective sizes of the bipolar ranges, what are the limits of the ranges, who decides the scope of the subject domain being measured and are the measurements and limits for the domain universally acceptable?

Another category of computer-based KA tools are those which provide knowledge for specialised knowledge-bases. Such examples include SALT (Marcus, 1988) and MOLE (Eshelman, 1988). SALT and MOLE provide automated mechanisms for interviewing domain experts in the development of specialised knowledge bases. MOLE has been used to build systems that do differential diagnosis in automobiles and SALT has been used to build a knowledge based system which configures elevator systems. Thus, such computer-based KA tools are relatively domain dependent in that they provide knowledge about well structured, clearly defined and usually technical subject domains. Due to the fact that they have been designed to elicit specialised domain knowledge for specific knowledge bases, these tools are not generally applicable to many subject areas other than those within the clearly defined domain.

KA workbenches attempt to overcome some of the problems associated with KA tools in that they offer an integrated set of tools rather than a single technique and seek to provide ways in which the captured knowledge can be translated and represented within a functioning knowledge base. For example, AQUINAS (Boose and Bradshaw, 1990) combines a number of integrated tools sets and incorporates repertory grids as an interface with the expert for a variety of analysis tasks. Furthermore, AQUINAS presents hierarchical representations by decomposing problems into subsets. Another KA workbench KRITON (Diederich et. al. 1988) supports a number of tools such as automated interviewing, repertory grids, and laddering techniques. However, such KA workbenches present the problem of attempting to quantify a complex domain in terms of bipolar scale as well as breaking it down into smaller parts without artificially breaking the domain up or distorting its overall appearance.

Another widely documented KA workbench is that of the KADS methodology (Wetter, 1990) and KADS-II (Balder and Akkermans, 1992). Central to the KADS methodology are models of expertise which are based on interpretation models, and the decomposition of expert knowledge into various layers. A criticism levelled against the KADS methodology is that there is insufficient support in the construction and evolution of interpretation models (Neale, 1988). Furthermore, the use of formal methods for the modelling of knowledge, in terms of upholding the consistency and correctness of a functioning knowledge-based system (Balder and Akkermans, 1992) may be applicable for relatively structured, rule oriented domains; however, in terms of complex domains this may incapable of representing highly subjective and heuristical knowledge.

There are KA support tools which use hypermedia, amongst other means as a way of acquiring and subsequently structuring the domain knowledge. Examples of such support tools are provided by Gaines and Linster (1990) who have developed a highly integrated knowledge support system based on HyperCard, the KA tool KSS0 and an expert system shell BABYLON. Gaines et. al (1992) have developed another hybrid integrated KA support system based on HyperCard, a KA tool NEXTRA, a database Oracle and an expert system shell NEXPERT. However a problem which could be cited against such KA support tools is that the knowledge representation of a subject domain, particularly if subjective in nature, may be constrained by the conventions of a particular expert system shell or other software package.

NOTIONS OF COMPLEXITY

There are certain determinants and features of complexity which have to be addressed and recognised in order that complex knowledge can be adequately elicited and represented in some form of computer-based decision support system. Various commentators have placed different emphasis on the determinants which are argued comprise complexity. Some authors (eg. Simon, 1965; Pippenger, 1978) suggest two determinants of complexity; those being the number of parts and the intricacy of their interactions. However, whilst recognising those determinants which are quantitative and hence measurable, these commentators fail to take into account the more qualitative and hence subjective components which comprise complexity. This point is argued by Schoderbek, et. al. (1990) who suggest that the determinants comprising complexity should also include the attributes of the specified elements and the degree of organisation within the system, which they argue are qualitative. Thus it is important to recognise that complexity has an immeasurable aspect which may provide a number of problems for those seeking to elicit and represent complex knowledge.

The role of the observer in complexity is a crucial one. Ashby's analogy of the brain as being complex to the neurophysiologist and being simple to the butcher (Ashby, 1974) highlights the notion of complexity as being in the eye of the observer. However, there are a number of important points which arise from this analogy. Firstly, from the perspective of the neurophysiologist the brain is not always a complex phenomena in that if he were to purchase a sheep's brain from the butcher for the purpose of feeding his pet dog, then in this context the brain would be a relatively simple phenomena. Similarly, if one were to ask a butcher to explain how the neurons of the brain interact, then it is highly likely that he would say that this would be too complex a subject for him to explain and well beyond the bounds of his existing knowledge. Therefore, it is important that complexity must be viewed in terms of its context and perspective.

What is required is an approach to eliciting and representing complex knowledge in which the observer becomes part of the perceived complexity and addresses the problem from a systemic and holistic perspective which seeks not to reduce problems of complexity to merely rates and levels with artificial boundaries, but seeks to accommodate a variety of perspectives from within the complex phenomena, each viewed as an interacting whole.

INTERPRETIVE APPROACHES TO KE

A number of researchers and authors have criticised the adoption of what can be termed 'functionalist' problem oriented approaches to the design and development of computer-based information systems and knowledge-based systems (eg. Winograd and Flores, 1986; Dreyfus and Dreyfus, 1986; Stowell, et. al. 1991). It is argued that such approaches seek to provide rational and practical solutions to relatively structured problems and hence are inappropriate when applied to highly complex domains where knowledge and expertise cannot be readily translated into a rationalistic and rule-oriented functioning knowledge-base. Certain areas of knowledge have been recognised by a number of authors as being heuristical, experiential and tacit in nature (eg. Dreyfus and Dreyfus, 1986; Bell and Hardiman, 1989; West, 1992). This type of knowledge has been identified as being very difficult to articulate and measure using quantitative analysis and not accessible to the knowledge elicitor in the form of facts, rules and figures.

Examples of what can be termed 'interpretive' approaches to KE include teachback interviewing (Johnson and Johnson, 1987) and naturalistic knowledge engineering (Bell and Hardiman, 1989). However criticisms have been cited against these approaches in terms of difficulties operationalising these ideas and the functional nature of the techniques (West, 1992). However, the Appreciative Inquiry Method (AIM) (Stowell, et. al. 1991; West, 1992) is an interpretive approach to KE which offers a practical framework by which the knowledge elicitor and the expert can appreciate and learn about a complex domain in its widest sense rather than placing an emphasis on quickly developing prototype knowledge-bases which inadequately capture the essence of a domain. AIM comprises of four main iterative activities of identify, specify, model and extract. Detailed accounts of AIM are provided elsewhere (eg. West, 1991; 1992; Stowell et. al. 1991). To summarise the main activities, AIM uses a map technique which as well as determining important issues and elements which comprise the domain, also serves to aid the expert to learn about his experiences and views as well as helping to define boundaries around the domain. The second phase of AIM uses ideas from Soft Systems Methodology (Checkland, 1990) in using CATWOE and root definitions which are used to specify the elements and issues identified in the map. A conceptual model is used in the third phase to operationalise the activities as defined by the expert. Finally, an agenda is developed from the conceptual model which is used to specify structured elements suitable for a knowledge-base as well as subjective elements in which the complexity is viewed in terms of context and perspective, and may require further iterations of AIM to gain further clarification.

A COMPUTER-BASED METHOD OF KE BASED ON AIM

There is presently an approach to KE being developed by the author which seeks to incorporate the techniques as used in AIM into a computer-based method of KE. One of the problems in using a 'manual' non computer-based approach to KE is in translating the captured knowledge and expertise which may be of a subjective nature into a functioning knowledge-base without losing the essence or quality of the expertise and knowledge. The rationale behind developing a computer-based method of KE based on AIM, is that if the elicited expertise and knowledge is already captured within a computer, then the gap between developing this into a knowledge-based system will be a step nearer to being bridged.

The expert interacts directly with the computer using the techniques in AIM as a means of expressing his/her views, with the computer-based approach using hypertext as a means of linking the elicited knowledge into a framework which seeks to view the knowledge in terms of its context and perspective rather than in an amalgamation of views whereby the diversity and richness of a domain may be lost. The approach also has the advantage of having the capability of using sound, animated graphics and video if deemed appropriate, as a means of representing the knowledge.

This computer-based method of KE should be viewed as support tool for the knowledge elicitor in that it is not seeking to dehumanise the KE process by actually replacing the knowledge elicitor. This computer-based method is also domain independent in that it can be used for different subject areas, although only one is cited in this paper.

IMPLICATIONS FOR THE TEACHING OF COMPLEX DOMAINS

The computer-based method of KE incorporating hypertext features and based on 'interpretive' notions may provide knowledge to enable a critical examination of the ways in which information and advice can be presented so as to meet the demands of those attempting to manage, make decisions and learn about complex domains. One of the chosen complex domains is that of marketing in small manufacturing business enterprises, a domain where managers tend to rely on their experience and intuition, hence the implicit elements of expertise in making critical decisions rather than a rational 'text book' approach. Marketing is cited as one of the single most important problems areas which owner managers of small businesses think that training would be suitable (Stanworth and Gray, 1992). This may in turn contribute to the process of developing computer-based decision support systems for complex domains; in particular advising and teaching managers of small manufacturing business enterprises on the critical issues concerning marketing options and strategies in terms of its specific context and perspective from the experiences of managers and marketing experts, rather than an idealised view of the domain. Thus it is hoped that a more holistic appreciation of a domain can be achieved rather than a disjointed and fragmented view, for it is invariably the manager who has to make a decision and suffer the consequences of decisions based on his learning and knowledge of the domain.

CONCLUSIONS

This paper has sought to critically assess some of the major computer-based approaches to KA highlighting certain deficiencies in the acquisition and representation of complex phenomena associated with a subject domain. It has also attempted to highlight the need for adopting a computer-based approach to KE which adopts holistic and 'interpretive' principles as a more appropriate means of attempting to elicit and represent complex phenomena, without artificially breaking a domain up, but viewing it as an interacting whole.

REFERENCES

Ashby, W.R., 1974, Some Peculiarities of Complex Systems, *Cybernetic Medicine*, Vol. 9. pp1-8

Balder, J. and Akkermans, H., 1992, TheME: an Environment for Building Formal KADS-II Models of Expertise, *AI Communications*, Vol.5, No.3. pp136-147.

Bell, J. and Hardiman, R.J., 1989, The Third Role - The Naturalistic Knowledge Engineer, *in*: "Knowledge Elicitation - Principles, Techniques and Applications", D. Diaper, ed., Ellis Horwood Ltd., Chichester. pp49-85.

Bradshaw J.M and Boose, J.H., 1990, Decision Analysis Techniques for Knowledge Acquisition: Combining Information and Preferences Using Acquinas and Axotl, *Int. J. Man-Machine Studies*, 29. pp121-186.

Checkland, P.B., 1990, "Systems Thinking, Systems Practice", Wiley, Chichester.

Diederich, J., Ruhmann, I. and May, M., 1988, KRITON: a Knowledge-Acquisition Tool for Expert Systems, *in*: "Knowledge Acquisition Tools for Expert Systems: Knowledge-Based Systems Volume 2," J.H. Boose and B.R. Gaines, ed., Academic Press, London. pp83-94.

Dreyfus, H.L. and Dreyfus, S.E., 1986, "Mind over Machine: The Power of Human Intuition and Expertise in the era of the computer", The Free Press, New York.

Eshelman, L., 1988, MOLE: a Knowledge Acquisition Tool that Buries Certainty Factors, *Int. J. Man-Machine Studies*, 29. pp563-577.

Gaines, B., 1990, Knowledge Acquisition Systems, *in*: "Knowledge Engineering, Volume 1. Fundamentals," H. Adeli, ed., McGraw-Hill Publishing Company, New York. pp52-102.

Gaines, B.R. and Linster, M., 1990, Development of Second Generation Knowledge Acquisition Systems, *in*: "Current Trends in Knowledge Acquisition," B. Wielinga, J. Boose, B. Gaines, G. Schreiber and M. Van Someren, ed., IOS Press, Inc., Amsterdam, pp143-160.

Gaines, B.R., Rappaport, A.T. and Shaw, M.L.G., 1992, Combining Paradigms in Knowledge Engineering, *Data and Knowledge Engineering*, 9 (1992/93). pp1-18.

Johnson, L. and Johnson, N.E., 1987, A Knowledge Elicitation Method involving Teachback Interviewing, *in*: "Knowledge Acquisition for Expert Systems: A Practical Handbook", A.L. Kidd, ed., Plenum Press, New York, pp91-108.

Marcus, S., 1988, Taking Backtracking with a Grain of SALT, *in*: "Knowledge Acquisition Tools for Expert Systems: Knowledge-Based Systems Volume 2", J.H. Boose and B.R. Gaines, ed., Academic Press, London, pp211-226.

Neale, I.M., 1988, First Generation Expert Systems: A Review of Knowledge Acquisition Methodologies, *The Knowledge Engineering Review*, 3 (2). pp105-145.

Pippenger, N., 1978, Complexity Theory, *Scientific American*, Vol.19, No.4. pp90-100.

Schoderbek, C.G., Schoderbek, P.P. and Kefalas, A.G., 1990, "Management Systems - Conceptual Considerations," (4th edition). Business Publications, Inc., Dallas.

Shaw, M.L.G. and Gaines, B.R., 1987a, An Interactive Knowledge Elicitation Technique Using Personal Construct Technology, *in*: "Knowledge Acquisition for Expert Systems: A Practical Handbook," A.L. Kidd, ed., Plenum Press, New York. pp109-136.

Shaw, M.L.G. and Gaines, B.R., 1987b, KITTEN: Knowledge Initiation and Transfer Tools for Experts and Novices, *Int. J. Man-Machine Studies*, 27. pp251-280.

Simon, H.A., 1965, The Architecture of Complexity, in: "General Systems Yearbook, Vol. 10", pp63-76.

Stanworth, J. and Gray, C., 1992, Entrepreneurship and Education: Action-based Research with Training Policy Implications in Britain, *International Small Business Journal*, 10, 2. pp11-23.

Stowell, F.A., West, D. and Fluck, M., 1991, The Appreciative Inquiry Method: An Approach to Knowledge Elicitation Based Upon the Notion Knowledge Elicitation as an Inquiring System, *Systemist*, Vol.13 (4), pp154-165.

West, D., 1991, "Towards a 'Subjective' Approach to Knowledge Elicitation Methodology for the Development of Expert Systems," PhD Dissertation, Portsmouth Polytechnic (unpublished).

West, D., 1992, Knowledge Elicitation as an Inquiring System: Towards a 'Subjective' Knowledge Elicitation Methodology, *Journal of Information Systems*, 2. pp31-44.

Wetter, T., 1990, First Order Logic Foundation of the KADS Conceptual Model, *in*: "Current Trends in Knowledge Acquisition", B. Wielinga, J. Boose, B. Gaines, G. Schreiber and M. Van Someren, ed., IOS Press, Inc., Amsterdam. pp356-375

Winograd, T. and Flores, F., 1986, "Understanding computers and cognition: A New Foundation for Design," Ablex Publishing Corporation, Norwood, New Jersey.

DEVELOPMENTS IN THE AUTOMATION OF SSM TUTORIALS USING MULTIMEDIA TECHNOLOGY

Frank Stowell, Daune West and Valerie Javelaud

Information Systems
Department of Computing Science
University of Paisley

INTRODUCTION

Increasing interest has been shown over the last 2-3 years in the possibility of developing some information technology-based teaching aid for Soft Systems Methodology (SSM). The impetus for such research has been varied, as has the success of the different ventures, but a reason for pursuing this line of enquiry has been the recognised difficulties surrounding the teaching of SSM. In this paper the authors describe developments in the production of an automated tutorial package for SSM using current multimedia technology.

THE USE OF TECHNOLOGY TO SUPPORT THE TEACHING OF SSM

SSM has not avoided the seemingly frequent inclination of researchers to try to produce computer-aided versions of the methods, techniques and methodologies used in various disciplines, whether for the purpose of learning about the approaches themselves, for demonstrating the flexibility of information technology, or with practical or commercial purposes in mind. With regard to SSM, attempts to apply information technology have included packages to provide support tools for the production of Rich Pictures using a dictionary of icons held on a computer (Avison and Golder, 1991a; 1991b), and the provision of computer based "training-wheels" to help new practitioners apply the methodology (Dolling, 1992). To date, little literature has been forthcoming from such studies that critically evaluate whether SSM is a feasible subject for translation into a computer based format. Indeed, the relative lack of success in terms of commercial availability of such a product may well be evidence of the unsuitability of SSM to such treatment. With these difficulties in mind Stansfield (1990) explored the potential use of information technology, not to automate SSM, but to provide a technology-based tutorial package on the methodology using an expert system shell. Rather than to produce a commercial product the rational behind this early study was to discover how far a particular type of information technology (i.e. expert systems) could provide support to the teaching of the methodology. Recognition of the essence of SSM and belief in the importance of learning from those who have personal experience of

Systems Science, Edited by F.A. Stowell *et al.*,
Plenum Press, New York, 1993

practising the methodology meant that no attempt was made to develop a stand-alone teaching unit. Instead, the aim was to provide a readily available back-up programme of guidelines regarding the application of the methodology, together with examples and exercises, complete with "expert" answers and explanations (Stowell et al., 1991).

Apart from providing a limited tutoring aid for SSM, the project also proved to be a valuable learning experience about the teaching of the methodology. Whilst the technology adopted imposed serious constraints upon the scope of the teaching package, it also provoked further questions about the way in which SSM is taught and learnt and highlighted potential future areas of research. One such area was the investigation of the use of multimedia technology to support a more sophisticated tutoring package than had been possible using an expert system shell.

USING MULTIMEDIA TECHNOLOGY AS A BASIS FOR AN SSM TEACHING PACKAGE

The production of the expert system 'SSM-Aid' (Stansfield, 1990), highlighted potential avenues of research that might be followed in order to produce a technology-based teaching aid for SSM which did not compromise the essence of the methodology. However, the expert system shell presented certain limitations which centred around the fact that it was seen to be too rigid to represent satisfactorily a domain as complex and intellectually demanding as SSM.

In spite of the recognised philosophical and practical difficulties of using information technology to provide a teaching aid for SSM (Stowell et al., 1991), the idea of such a package for SSM has some interesting advantages that were felt worthwhile pursuing, such as providing new ways of learning, stimulating interest through the novelty of the approach, and providing a "tireless" tutor. However, the experiments with the expert system shell indicated that a more flexible media was necessary to tackle such a 'difficult' domain. The use of multimedia technology, therefore, appeared to offer a more realistic approach. The rationale of this postulate was that multimedia technology, being the seamless integration of data, text, images and sound within a single information environment, would enable users to explore and navigate through the domain in a non-linear fashion and hence give them 'greater control over the exploration of the domain and in a way which might be more closely aligned to human thought and learning processes'(Stowell et al., 1991).

Multimedia technology, as well as offering a high level of interactivity, brings a range of different media to bear on the issues of clarifying, communicating and informing, which, it was felt, might help enhance the student's understanding of SSM. Repetition has proved to be essential in the process of learning, especially when handling a 'difficult' domain and so multimedia technology, through the use of different types of media to teach the same concept, has been argued to increase user receptivity and motivation (Feldman, 1991).

Outline of the project

The multimedia package was designed to help the user to *understand* SSM by means of textual, audio and graphical explanations. Whilst at present the package is developed around the well-known seven-stage model of SSM as a way of explaining the process and models of the methodology, it also attempts to provide the more experienced user with a general understanding of the essence of SSM as articulated through SSM "ideal type" mode 2. The characteristics of multimedia technology were seen as an asset not to be neglected in the learning process. To this end, different materials were collected or produced in a number of different formats (e.g. text, diagrams, taped recordings of Checkland discussing the methodology, photographs, music, and recorded oral advice).

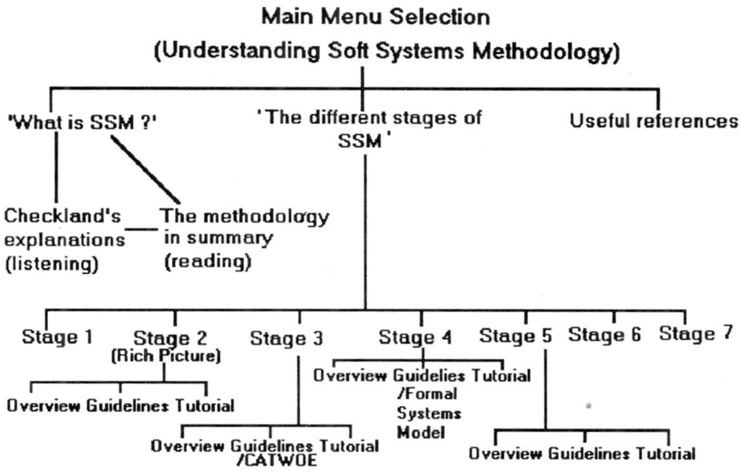

Figure 1. Overall structure of the multimedia package.

The user interface

The style of interface used in the package is that of a menu with buttons that can be activated by a mouse. This particular style was selected so as to ensure that the package was easy to use and so as to reduce the amount of input (and hence the likelihood of error) by the user. Sound was also fully integrated so as to provide the user with some general advice as well as with directions on how to carry out the tutorial.

AN OVERVIEW OF THE MULTIMEDIA PACKAGE TO HELP TEACH SSM

Figure 1 shows the overall structure of the package. The levels of hierarchy within the figure indicate the different menu screens presented in the package. The problem of accounting for different levels of user-ability was solved by either providing the user with the freedom to work directly on the stages of the methodology or advising him to work sequentially through the package in order to gain confidence in fully understanding the core of SSM.

"What is SSM?"

The main direction of the first item on the main menu ('What is SSM ?') is to discuss the methodology, not in terms of its familiar seven-stage model, but in terms of its general features. This section is designed for the most inexperienced user of SSM who needs to be aware of the main characteristics and general orientations of SSM before actually starting to learn how to use the methodology. First animation is used to place SSM within the context of systems methods and then the user is presented with two possible routes: either to read a piece of text describing ideas central to the methodology or to listen to Checkland's explanations on some major issues of SSM. This choice represents an attempt to (a) avoid placing a constraint upon the user as to what kind of media should be used to understand a particular point of the methodology, and (b) to try and enrich the learning experience by providing information in different formats.

"The different stages of SSM"

Activating the second "button" in the main menu takes the user into an explanation of the different "stages" of SSM. Each stage of the methodology is treated at three levels: level 1 - an overview of the stage, level 2 - guidelines on how to carry out the stage, and level 3 - tutorial exercises whereby users can apply their knowledge to specific examples and then compare their answers to that of an "expert". In order to emphasise that the different stages of the methodology do not have to be carried out sequentially the menu offered the user in this section (headed "The different stages of SSM") is a graphic showing the organisation of the methodology and the relationships between the different stages. The information given about each stage is described below.

Stage 1 of the methodology is the "fact-finding" stage and as little can be found in SSM literature about this aspect of the methodology, few general guidelines are given in the teaching package. The main direction of this stage, then, is to give the user textual advice on how to obtain information on the problem situation.

Stage 2, the "analysis" stage, is introduced by a "voice-over" explaining how Rich Pictures can be a powerful way of capturing the essence of a problem situation. The user is then presented with the choice of accessing any of the three parts of this section, namely, an overview of the stage, guidelines about its practical application, and a tutorial exercise. The "overview" is aimed at providing the user with a basic understanding of the role of Rich Pictures as a way of describing ones perception of a problem situation whilst the "guidelines" are aimed at providing the user with advice in drawing a Rich Picture. The user is provided with a selection of visual symbols for use within Rich Pictures as well as an example of a hand-drawn rich picture (scanned) about a simple problem situation. The user is encouraged to consider the significance of using each of the two methods (hand-drawn and computer graphics) for Rich Picture construction. The "overview" and "guidelines" are designed to provide the novice with sufficient information in order to carry out the tutorial exercises.

In the first tutorial exercise, the user is presented with three Rich Pictures (produced using Harvard Graphics and then scanned into the teaching aid), and asked to nominate which of the three examples best describes the problem situation scenario which is displayed as a piece of text to one side of the screen. Movement between the different Rich Pictures is achieved through activating "turn-the-page" icons. A facility to enlarge the Rich Pictures to full-screen size is also provided. On selecting one of the Rich Pictures the user is presented with an expert's opinion of their chosen example which is then compared and contrasted with the other two examples. In the second exercise of the tutorial the user is presented with a different problem situation (in text format) and asked to develop a rich picture to represent the situation. (This problem situation is then used as an example throughout the rest of the tutorial exercises at each stage as a way of providing a complete worked example.) The Rich Picture produced can then be compared with one produced by an "expert" (hand-drawn and scanned into the package).

Stage 3 : On moving to stage 3 the user listens to a description of the process of identifying problem themes from which relevant systems, deemed to be likely to yield some insight into the perceived problem situation, can be developed. Attention is given to the fact that the user is now engaged in "below the line" activities: that the modelling of Stages 3 and 4 take place strictly in the world of "systems thinking". Once more this section of the package is divided into an "overview", "guidelines" and "tutorial exercises". The "overview" section emphasises the notion of relevant system and describes the two major types of system which may be usefully identified, namely, "issue-based" and "primary task". At this point the user can choose to listen to Checkland making distinctions between these two "types" of system. The "guidelines" section centres around a description of how to develop Root Definitions and to "test" them for completeness using CATWOE. The user is then given the opportunity to listen to Checkland's views on how to carry out the CATWOE test.

The "tutorial" section is based upon the use of CATWOE to check given Root Definitions. Two Root Definitions are presented and the user required to identify the CATWOE elements. Having done this the user is presented with an "expert's answer". The user is then asked to write a Root Definition and carry out a CATWOE test for an identified relevant system taken from the problem situation used as an example throughout the tutorial exercises. Care is given to explain (using a voice-over) that the "expert answers" given in the exercises are meant merely as a "guide" to the type of answer expected: they are not meant as **the** correct and only answer.

Stage 4 is concerned with explaining and giving examples of the modelling activities that are undertaken in order to provide a description of the transformation process that is central to any Root Definition. This section of the package follows the previous division into an "overview", "guidelines" and "tutorial" exercises. The "overview" section aims to provide the user with some general characteristics of conceptual models. The "guidelines" section centres around offering a set of general "rules" that may be useful to bear in mind when building conceptual models. Although nowadays the Formal Systems Model seems to be little used as a way of checking that a conceptual model is well-formed, its use is discussed as a yardstick for considering the necessary activities are present to describe purposeful activity. A recording of Checkland's own explanations of how to "test" a conceptual model can be accessed at this point.

The "tutorial" aspect of this section is divided into two parts. In the first part, the user is required to produce conceptual models from given Root Definitions. The expert's graphically drawn "solutions" are presented as a guide to the sort of activities that should be included. In the second part of the tutorial the user is asked to apply the Formal Systems Model to a conceptual model of the example developed throughout the package and, once again, is given an "expert's answer" on this exercise.

Stage 5 : In the "overview" section of Stage 5 the information presented focuses upon the fact that the user is now moving back into the "real world", leaving the abstract world of "systems thinking" and offering the developed model as a means of comparing a logically-derived "ideal" with "reality". In the "guidelines" of this section four different ways of carrying out the comparison stage are described and the user can choose to listen to Checkland discussing the comparison stage. An example is given of an agenda, derived from the Conceptual Model, that can provide the basis for ordered questioning about the "real" situation. The "tutorial" exercise is based upon this use of an agenda (using a tabular display) in that the user is asked to draw up an agenda for the example developed throughout the tutorial exercises. The user can then compare their agenda with that of an "expert".

Stages 6 and 7 : Guidelines for carrying out stages 6 and 7 do not appear to be easily implemented within a software package. This seems to be due to the fact that relatively little advice can be found in the literature as to appropriate methods or techniques in carrying out these stages.In stage 6, the advice given is that the agenda should be used to structure the debate and that any proposed changes should be both systemically desirable and culturally feasible. In stage 7, it is pointed out that any changes to be made fall within three main categories of change, namely, structural, procedural and attitudinal.

"Useful references" : The third "button" that can be activated under the main menu ("Useful references") gives the user details and a precis of some useful references for finding out more about SSM. The importance of reading around the subject and becoming acquainted with different views on the methodology are all useful ways of increasing one's understanding of the subject. It is hoped to develop this section to include reviews and discussions of the named texts and articles.

SUMMARY OF RESULTS

In comparison to the expert system shell "SSM-Aid", the multimedia package for teaching SSM incorporates many additional features which may be seen as potential advantages for those wishing to learn about SSM. By exploiting the characteristics of multimedia a number of improvements were made on the original "SSM-Aid". First, user interfaces seemed to be more effective because of their diversity, colour, and interactivity. Second, the intensive use of oral advice and taped "conversations" proved to be an additional source of information that could be "over-laid" onto textual and graphical displays and provided a way of reducing the amount of text necessary.

It is hoped that future research in this area will add video sequences as a further means of providing information, particularly as a means of representing a "problem situation" for analysis and expression through Rich Pictures in Stage 2. As far as rich pictures are concerned, the use of both scanned hand-drawn pictures as well as graphically-produced pictures provided the opportunity for the user to critically evaluate the effect of each method. Although there are numerous area of the package to be further explored and experimented upon it is felt that multimedia offers a platform that can more successfully support the essence of Soft Systems Methodology and provide a valuable teaching aid.

In terms of lessons learnt from this experiment, it was discovered that excessive and uncoordinated mixing of media for the presentation of information and repetitive oral advice could distract from the purpose of the package.

The "technical" problems encountered in using the multimedia development tool include the fact that it cannot offer the user the opportunity to key in responses to the tutorial exercises as could be done using the expert system shell. The result of this deficiency is that the user is likely to feel distanced from the package and in this sense it is difficult to describe the application as being truly "interactive".

Whilst the use of Multimedia technology seems to offer the chance to enhance the learning of SSM and the opportunity to provide a sympathetic treatment of the methodology itself, to be of real use the package must be applied in the right context: it can only be of real value when used in conjunction with lectures, tutorials, and practical sessions.

CONCLUSION

Elsewhere the authors have discussed what they consider to be the theoretical underpinnings of both SSM and information technology, namely their "interpretive" and "functionalist", or "rationalist", origins. Such apparent diametrically opposed paradigms were argued to be likely to prevent current technology providing a computer-based form of SSM that is consistent with the essence of the methodology but that, planned with care, technology might provide a useful platform for the teaching of the methodology. However, the practical difficulties of trying to make SSM "fit" an expert system shell indicated that increased success might be had if information could be stored and retrieved using less formal and more naturally expressive media, such as sound, pictures, and animation.

REFERENCES

Avison, D.E. and Golder, P., 1991a, The need for tool support for soft systems, *in*: "Systems Thinking in Europe", M.C. Jackson, G.J. Mansell, R.L. Flood, R.B. Blackham and S.V.E. Probert, eds., Plenum, New York.

Avison, D.E. and Golder, P., 1991b, Tools supporting soft systems, *in*: "Systems Thinking in Europe", M.C.Jackson, G.J. Mansell, R.L. Flood, R.B. Blackham and S.V.E. Probert, eds., Plenum, New York.

Dolling, A.C., 1992, Electronic-mail communication.

Feldman,T., 1991, Multimedia in the 1990s, BNBRF, The British library.

Javelaud, V., 1992, "The development of a multimedia application to help the user of Soft Systems Methodology", MSc dissertation, University of Paisley.

Stansfield, M.H., 1990, "The development of an expert system to aid the user of soft systems methodology", MSc dissertation, University of Portsmouth.

Stowell, F.A., West, D. and Stansfield, M.H., 1991, The application of an expert system shell to an unstructured domain of expertise: using expert system technology to teach SSM, *European Journal of Information Systems*, Vol. 1, No. 4, pp281-290.

INTRODUCTION TO ENVIRONMENTAL SYSTEMS, SOCIAL SYSTEMS AND HEALTH SYSTEMS STREAMS

In his keynote address Sir David Smith reminded us of the complex environmental system upon which our survival depends. It is ironic that the so called progress which we have made in controlling our environment has itself created so many other problems. For example, the improvements in health care and in food production have themselves created health problems e.g. Pesticides, "Super" viruses. It would seem that despite the many warnings that we receive about the harm that is being done to our environment they seem to go largely unheeded. The balance of argument put forward about the environment and health matters by the richer nations often seem to be less convincing to those nations whose first priority is to feed their people. Arguments related to birth control seem to take on a different dimension when considered in countries with few social services and a high infant mortality rate.

The issues surrounding the topics within this stream could well fill a conference themselves, indeed have on many occasions. Not surprising this stream has attracted papers which address some of the important issues surrounding agriculture, social problems and health care. It may be that the way in which the problems contained within these papers will provide a source of ideas and make a small contribution to the thinking about some the environmental, health and social systems. Moreover, the underpinning philosophy and practice of the conference itself may help in reminding us of the wider ramifications of our actions and provide some practical assistance to addressing some of these important problems.

Frank Stowell
Information Systems
Department of Computing Science
University of Paisley

SYSTEMS SCIENCE AND THE ALTERNATIVES TO THE CHANGES OF SOIL IN AREAS OF TRADITIONAL AGRICULTURE

A.J. Hernández,[1] C. Jiménez,[1] and J. Pastor[2]

[1] Ecología, University of Alcala Henares, Spain
[2] C.C.Medioambientales, CSIC., Madrid, Spain

INTRODUCTION - AIM OF THE TASK

The ecological theory has been developing for many years in the same conceptual and methodological framework as the general theory of Systems (Hernández, 1991). Ecosystems, as complex systems, are in many cases difficult to study, but much more when they are subjected to human interventions during many centuries. Such is the case of Agrosystems. The agricultural landscape is the expression in space and time of human effort in search for his daily nourishment. But since the landscape is a complex system of relationships, it is necessary to analyse its components so as to understand the past and their processes of development.

The present trends about landscape knowledge, taking into consideration the ecological and systemic approach are expressed in Hernández and Pastor (1991). We have already talked about the high degree of resemblance that exists between "ecological thought" and "sistemic thought" (Hernández and Pastor, 1989). Now the question is if we can apply the most theoretical question to the realization of a project about the changes which some characteristic systems can suffer.

It is a question of giving importance to ways of relationship man-nature and of social interaction which can satisfy basic human needs, but looking for a new balance with the environment at the same time. Therefore we'll talk about agroecosystems. In short, to maintain the ability of production and respect for the environment.

To apply the knowledge of systems science to our study subject means that agroecosystems will be explained by means of some variables which is important to determine.

But our final aim is to be able to know the dynamics of these systems so as to intervene in them in a more reasonable and ecological way. Because of that, to describe them will not be enough. We have to find the variables of these dynamics and then to built a model which allows us to act taking into account the exploitation, the resources conservation and the decreasing of the erosion which these systems are subjected to. Then, next the main results

Systems Science, Edited by F.A. Stowell *et al.*,
Plenum Press, New York, 1993

that we have been obtained in the first stage of this project are exposed: characterization of the systems components and the main variables.

THEORETICAL BASE OF THE PROJECT

The conception of the system as an interrelationship between these elements has privileged the character of its organization. The central concern is not the "relationship" but the organization which shows the parts of the whole. We agree with Morin (1983) affirming that the organization is the category which explains the productive and dynamic strength that the concept of system has in itself. The properties for the assessment of Agroecosystem: productivity, stability, sustainability, autonomy, equitability are directly related to the organization of the system.

The productivity parameters will not only be of an economic nature, but they will also be based on the potential of the edaphic and biotic resources. The main questions which demonstrate what we understand by stability are expressed in Hernández (1991). The autonomy will take into account the self-sufficiency (autochthonous species against imported cultivation, farm work techniques suitable for the territory where the agroecosystems are located, etc.) as well as not promoting environmental impact (for example by agricultural practices which contribute to further erosion or pollution of the soil. As for the property of equitability, it will be necessary to take into consideration the evenness of distribution of benefits (Hence the importance that the agricultural collectives have in these systems).

One of the main aspects to consider in the development of sustainable projects is the correct management according to the environmental conditions of each region of the three natural and inter-connected resources the soil, the vegetation and the livestock. However, these resources are often considered independently over a short time, and without an integral view of them. In our country, big extensions of the mediterranean area with low rainfall and exclusive use for cereals and/or sheep (González and Hernández, 1992), are much damaged as a result of ignoring the complex system of soil-plant-herbivore.

Therefore, it is necessary to get to know the organization of the system so as to manage it reasonably. Firstly the components of the system will be identified and secondly the principal variables of each of them to be analyzed. The flow or the interrelation between the variables will be seen later.

STUDY AREA

Mediterranean agrosystems of siliceous soils are of great interest due to the heterogeneity of their structure and the high biodiversity in forage value grasses and legumes which are in many cases needed to resist erosion processes. At the same time these semiarid environments have a scarce capacity of regeneration which is sharpened by the negative influence of deforestation, misguided repoblation, fire, heavy grazing and mechanized agriculture. Recently, new problems like the abandonment of cereal growing lands and the loss of traditional agrarian activities have added to them. These disturbances facilitate the disappearance of plant cover and generate an almost bare soil with low organic matter content, nutrient loss problems and altered biogeochemical and biotic cycles. Irregularity and scarcity of rains, and the mild climate, which favours the quick disappearance of mediterranean agrosystem, as well as, the bad state of the plant cover which provides low soil protection.

According to the E.C. criteria, the mediterranean landscapes show the strongest proportion of impoverished areas. The productive base of land and work in this area not only

Table 1. Estimated parameters for the study of the system components

Components	Parameters
Physical	Soil particles, soil density, soil structure, soil humidity parameters.
Chemical	Ca, Mg, P, N, K, S; organic matter, soil pH.
Biological	Plant cover, soil seed bank, soil microfauna.
Productivity	Sheep production, cereal production, pastures production.
Human	Searchers-farmers workshops.

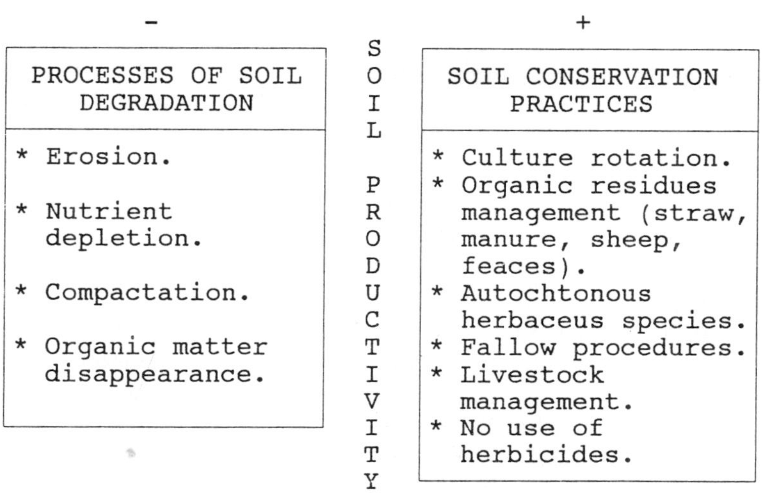

Figure 1. Main processes of degradation and conservation practices in the studied agrosystems.

reflects a relative economic underdevelopment, but also difficult agricultural conditions of productions.

The work has been carried out in two experimental farms located in the central area of Spain, with similar climate but different soils, frequently used for dry farming, grazing or fallow-cereal rotations with the sheep participation. The experimental design takes into consideration the different uses of these soils and the repercussions on soil structure modification, potential pioneer pasture plants and sheep stocking rate (Hernández et al, 1992).

Degradation processes of the soil due to the use of it just for cereal cultivation are described in Figure 1. Conservation practices which are being studied are also reflected in the same figure.

CHARACTERIZATION OF THE COMPONENTS AND SYSTEMS VARIABLES

The soil degradation is a process which lowers the present and potential ability it has to produce goods and services. This process does not necessarily in a continuous way, but may happen in a relatively short period between stages of ecological balance. This degradation is not only function of material factors, but that the type and use of the soil determines the degree and extension of degradation. In different territory sectors that have been studied, the several alternatives of use, we are analysing the same physical, chemical, biological, of productivity and human components.

The difficulty to look for the variables with a greater information in the dynamic of these systems is that they are always variables linked to the inner part of the subsystems, which -at the same time- interfere among them in the bosom of the global system. For example, the study of the productive component implies the simultaneous consideration between relationships, soil, climate, water, slope and the plant, as the interferences between that aforementioned set and other combinations of variables in the determination of different forms, order and election of the cultivated species.

The evaluation of the components of the soil-plant-herbivore system has been elaborated from the integrated analysis of the information. (See Table 1).

CONCLUSIONS

The recent reforms in the directives of the Community Agrarian Policy are affecting cereal agrosystems and also the systems of traditional production in our country. We are studying the compatibility between exploitation, conservation of resources and the reduction of erosion in these systems. The experimental design which seems to allow us to get satisfactory results so as to channel a sustainable development of the resources, soil, vegetation and livestock is based on the science of systems.

The first results obtained about this fact help us to assess the next variables: compactation, erosion and the nutrients in the soil; biodiversity generation through the soil seed bank; the plant biomass and the efficiency of crops and the ways of farmer intervention of the area in the research project.

Acknowledgement: This study is included in a project supported by the Community of Castilla-La Mancha.

REFERENCES

Hernandez, A.J., 1991, Ecosystems, global problems, in: "Handbook of Systems Science". European Systems Union, ed., Athenes, 1991: 253-259.

Hernandez, A.J. and Pastor, J., 1989, Relations between ecological thought and Systems thought, in: "European Congress on System Science", Lausanne, tome II: 643-652.

Hernandez, A.J. and Pastor, J., 1991, "El aprendizaje perceptual del entorno", Servicio de Publicaciones Universidad de Alcala de Henares.

Hernandez, A.J., Pastor, J., Gonzalez, T., Estalrich, E. and Urcelay, A., 1992, Soil-plant-herbivore relationships in mediterranean semiarid environments: experimental design, "14th General Meeting European Grassland Federation", Finland: 710-712.

Malassis, L., 1992, Pasado y devenir de los paisajes agricolas. in: "Paisaje Mediterraneo", Electa, Milan: 266-273.

Morin, E., 1983, "El método. La vida de la vida", Catedra.

A SYSTEMS APPROACH TO ENVIRONMENTAL MANAGEMENT

Margi Levy[1] and Michael Synnott[2]

[1] Operations & Systems Group
Warwick Business School
University of Warwick
Coventry CV4 7AL

[2] MBA Programme, University of Warwick
Formerly at
School of Architecture & Planning
Curtin University of Technology
Perth, Western Australia

INTRODUCTION

The spiralling third party costs of pollution and environmental degradation have provoked a search for effective management systems for the environment. Inevitably, the main stumbling block has been the role of human agreement and human action in achieving policies for a sustainable environment. On the one hand, individuals are motivated by different circumstances and imperatives and thus, agreement on problem definition is difficult. On the other hand, individuals may fail to understand the potential for concerted action implicit in their relationships with others.

To illustrate these difficulties this paper begins by focussing on the problem of water conservation in Australia. A brief case study highlights the diverse interests in one locality. An approach based on soft systems methodology is proposed as a means of reconciling the conflicts involved.

WHAT IS INTEGRATED CATCHMENT MANAGEMENT

Since the mid-1970s, Australian water managers have become increasingly concerned at the deteriorating state of the continent' s water resources. The unprecedented agricultural boom enjoyed by Australia in the post war years is now seen to have extracted a high toll on land and, subsequently, water resources. Such things as excessive land clearance and high fertiliser use have given rise to salinisation and pollution of water resources especially in the south-east and south-west of Australia. In other words, precisely where most Australians now

Systems Science, Edited by F.A. Stowell *et al.*,
Plenum Press, New York, 1993

live and work and where water resources are becoming scarce, pollution of the water systems is most prevalent.

Also, since the mid 1970s, Australian voters have become increasingly vocal in their concerns for conservation of the environment. Any politician seeking election to public office must expect heated debates about the environment at the hustings.

A complicating factor in seeking ways to stem pollution and meet-public concerns about the environment has been the diverse range of actors and motivations now involved. On the one hand, agriculturalists are under great pressure to strive for the highest level of productivity. It is not always clear to farmers how such an objective can be achieved while contributing to the social objective of a more sustainable environment. On the other hand, the catchment areas of key water supply river systems are not exclusively devoted to agriculture. Improvements in communications and transport coupled with more income and leisure time, mean that there is a continuing and expanding demand for rural retreats, rural and water-based recreation and small holdings. Coupled with industrial demand for land and resource use, these non-traditional rural land uses also exert a significant threat to the river systems. For example, the presence of more and more septic tanks as a result of increased private housing can result in excessive volumes of nutrients reaching the river.

In a bid to meet the challenge of reconciling these diverse elements and conservation of Australia's water resources, state and federal governments have promoted the concept of integrated catchment management (ICM). ICM is directed to an understanding of the inter-relationships between natural, human and economic systems. It seeks to achieve agreement of some common goals on the environment and natural resource use by co-ordinating the decisions of individuals and organisations. Specifically, ICM has three stages and as we shall see, each of these highlights a potential role for the information management professional.

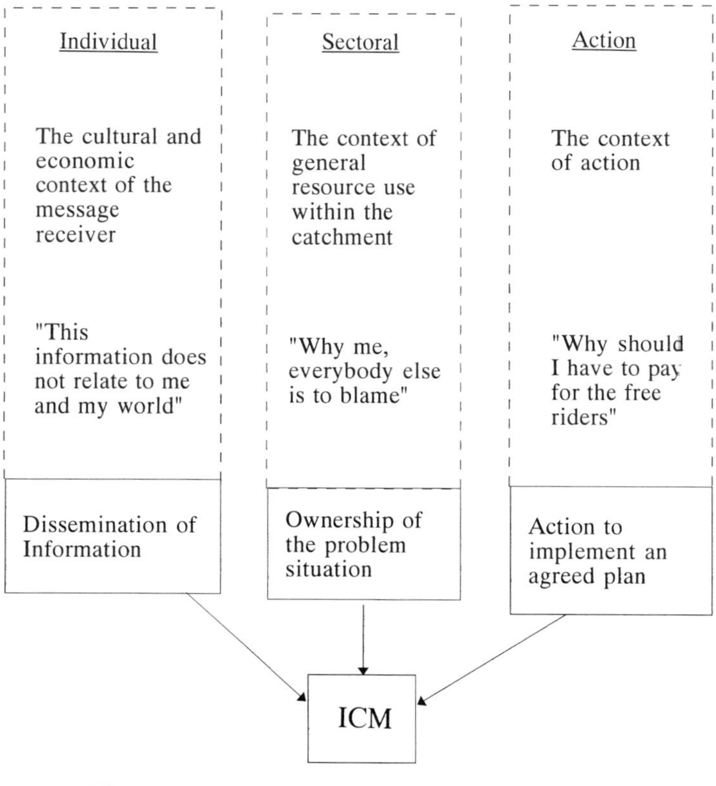

Figure 1. The Stages of Integrated Catchment Management.

First ICM seeks to raise general awareness about the total impact of different resource uses on the river system. Thus the context, form and language of disseminated information is a key issue - it must be relevant to those receiving the "message". Second, ICM seeks to gain landholder "ownership" of the issues within the catchment. Individuals are not easily disposed to see themselves as the cause of environmental pollution and will tend to see others as the real cause or to view any pollution they cause to be more than offset by their contribution to community welfare in the form of food and job creation. Third, ICM seeks to provide a framework for concerted action on catchment wide problems - unless landholders are confident that others are willing to act they will not risk time and money on conservation projects. As Figure 1 shows, these three stages give rise to issues of individual perspectives on the problem, their relationships with others and the concept of legitimate action.

AN EXAMPLE OF INTEGRATED CATCHMENT MANAGEMENT

A good example of the difficulties facing ICM is presented by the algae pollution of the Peel-Harvey inlet in Western Australia. Against all advice at the time, in the mid 19th Century, the surrounding land was cleared of trees for agricultural production. In order to overcome serious water-logging the pioneer farmers constructed a series of channels to carry surplus water to the Peel-Harvey inlet and subsequently, the Indian Ocean via a very narrow opening to the sea. However, in more recent times these same channels have acted as very effective conduits for transporting nutrient enriched run off from the land to the same inlet. The nutrient pollution arises from the powerful fertilisers applied by farmers to raise productivity. The ecological consequence has been algal pollution.

The socio-economic consequences have also been serious. Firstly, the Peel-Harvey inlet has served as a recreation area for the population of the city of Perth. Pollution of the inlet has led to a complete ban on swimming in the inlet. Second, Western Australia derives great competitive advantage in the tourist market by highlighting its clean environment, polluted inland waterways endanger this image. Third, the rotting algal bloom must be moved annually at a significant cost to the local community.

It is the case that action to stem the growth of pollution on the Peel-Harvey inlet has been underway for twelve years. Furthermore, ICM type activity has been a central part of the strategies deployed by the authorities during this time. However, real change has been slow and minimal. The stumbling blocks have been two-fold. Firstly, farmers have argued that deteriorating trade conditions would make implementation of conservation very costly and difficult to sustain. Secondly government departments and other interested groups have sought to portray the problem in ways favourable to their interests. For example, those anxious to deflect the argument away from local causes have insisted that the pollution has occurred because of the regional water management policies which have reduced the flow of fresh water to the inlet.

Failure to agree on any single explanation has led to the decision to adopt a technical solution. A new channel is being cut through the coast to enable the Peel inlet to be flushed by the sea and thus increase dilution of the pollution. Whether this will be a comprehensive solution is in doubt, but the price tag of AUS $1.6 million is certain.

THE FAILURE OF CURRENT APPROACHES TO ICM

What can be seen from the example of the Peel-Harvey inlet is, that despite the intention of ICM to lead to an outcome which is subscribed to by all interested parties, individual priorities and perspectives may lead to sub-optimal solutions being chosen. Indeed it could be argued that the measure chosen does not address the real problems of algal pollution but avoids conflict while appearing to limit the damage caused.

A concern about the ICM process is the boundary of interest in its outcome. Hence, the hypothesis is that it is not merely the physical boundary of the catchment which determines the outcome of any policy, but that the boundary is more intangible. There is a need to address regional, national and indeed global forces. Therefore we need to consider the information flows between ICM players when considering the 'real' location of the boundary.

WHY THE SOFT SYSTEMS METHODOLOGY IS A USEFUL APPROACH FOR ICM

The soft systems approach would appear to offer a way forward to enable a greater understanding of the issues and information needs of each group within the catchment. This increased understanding may then lead to a more sustainable outcome for land and water resource planning in Western Australia as all players recognise the needs of the others. The strength of SSM is its ability to model complex human situations from a number of different perspectives. The recognition that interpretation is important in determining Weltanschauung and that there is a need to make assumptions explicit (Checkland in Rosenhead, 1990) is clearly vital to the bigger picture that is ICM. The process of SSM described by Checkland & Scholes (1990, p29) which clearly shows the multiple criteria which must be considered during the analysis, while still focussing on the holistic nature of the problem, would seem to be a useful way to focus on addressing ICM. Wilson (1988, pp46-50) has also shown how SSM can be used to address the overlap between natural resource issues and social systems.

We need to be able to show participants in ICM the relationships between them and the impacts and effects of those relationships. There is also a need to be able to recognise and understand the information which different players need in order to make them better able to see the impacts which they can have on the resource system. In addition the different interpretations and uses to which that information is put needs to be modelled to increase understanding among all players. The objective of the models being to increase the understanding of each of the groups involved in ICM, such that they recognise the need to overcome the "prisoner's dilemma", by which it is in no-one' s interest to be the first one to act.

Figure 2 attempts to describe the elements which comprise an ICM scenario. It is interesting to see that the other catchment users (e.g. mining companies, forestry, recreation, urban development) do not have any immediate impact on the land-holders, which may limit the learning about the overall impact each has on the environment.

A CONCEPTUAL MODEL FOR ICM

In attempting to develop the conceptual model for ICM it became clear that there were a number of major relating systems. These systems need to be expanded further to show the detail of the relationships and information links. The model has been developed to give an understanding of the general principles underlying ICM. The objective is to show that only by an understanding of the relationships can progress be made to the development of strategies to implement ICM.

The following root definition has been developed which it is hoped gives a clearer picture of ICM and what it is trying to achieve. It was useful to use the CATWOE mnemonic to derive the definition:

Customers:	state government agencies; local government; land-holders; community; future generations
Actors:	state government agencies; local government; land-holders; community

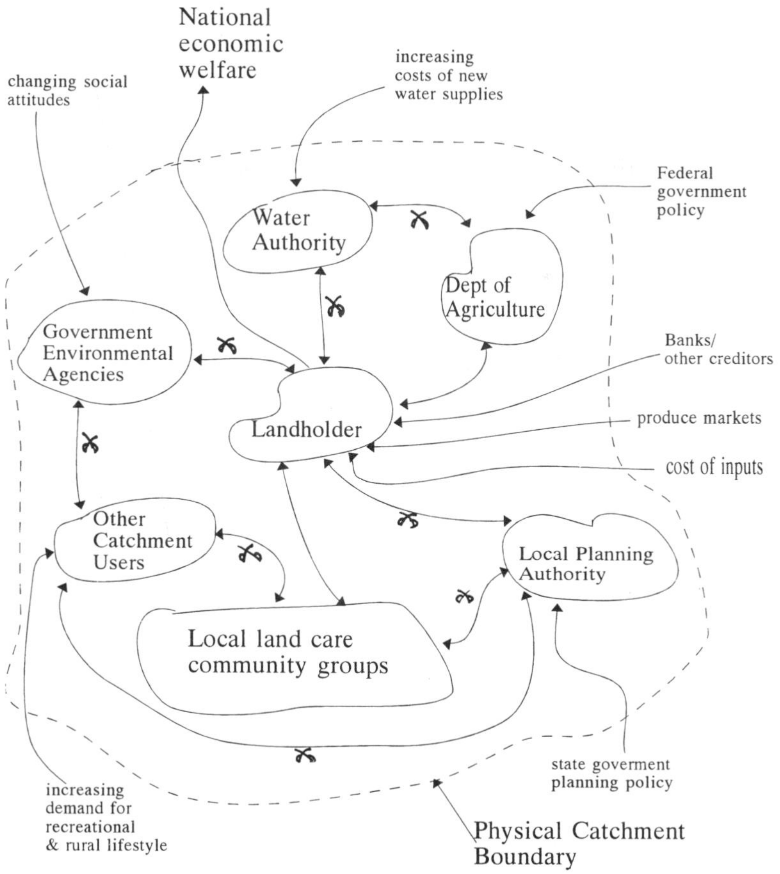

Figure 2. A Rich Picture for ICM from the land-holder's perspective.

Transformation:	the minimisation of all adverse impacts on land and water systems resulting from individual water resource use
Weltanschauung:	the recognition of the need for sustainable land and water resources within economic and political constraints
Owners:	state government agencies; local government; land-holders; community
Environment:	government policies; river pollution; global warming; sustainable development

Root Definition for ICM

A system subscribed to and owned by state and local government agencies, land-holders, and the community to implement ICM policies, recognising the need for sustainable development within economic and political constraints for current and future generations.

CONCLUSION

It is readily acknowledged that the models presented here are, in reality, the first stage of an iterative process involving all key stakeholders. However, the strength of SSM is

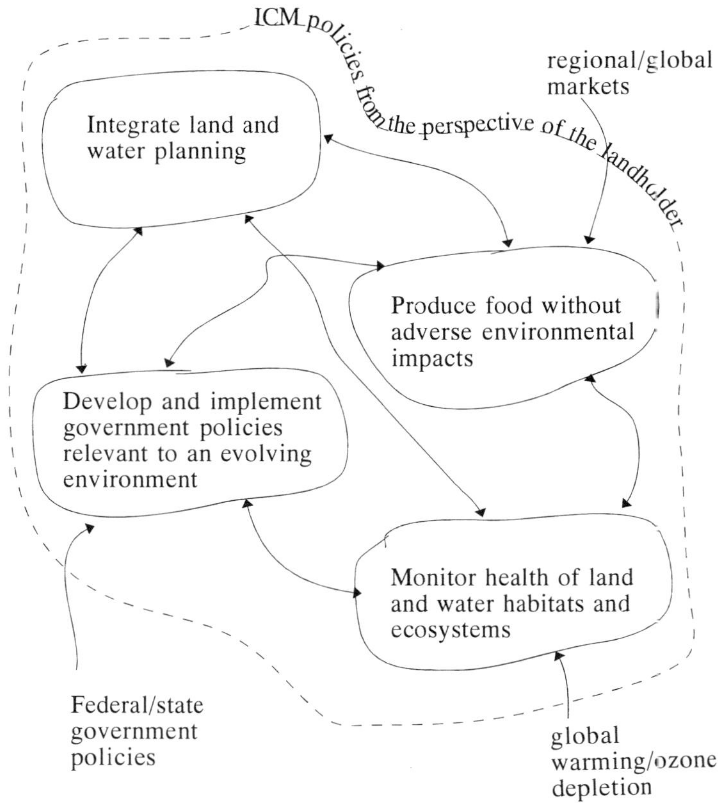

Figure 3. A Conceptual Model for ICM

precisely to provide a coherent yet flexible structure for achieving discussion about change. As may be gleaned from the example above the lack of such a structure lies at the heart of many contemporary environmental problems.

In the case of the Peel-Harvey inlet, an SSM approach may not have been able to remove all the barriers to change. However, the SSM method would have provided - and still can provide - a structured basis for considering alternatives and a means of making explicit the point of least unhappiness for all involved.

REFERENCES

Checkland, P., 1990, Soft Systems Methodology *in:* "Rational Analysis for a Problematic World", J. Rosenhead, ed., Wiley, Chichester.

Checkland, P.B. and Scholes J., 1990, "Soft Systems Methodology in Action", Wiley, Chichester.

Wilson, J., 1988, "Changing Agriculture: An Introduction to Systems Thinking", Kangaroo Press, Kenthurst.

A SYSTEMS APPROACH TO SOCIAL INNOVATION

Gerard Fairtlough

5 Belmont Grove
London SE13 5DW, UK

INTRODUCTION

Famine, ethnic conflict, ecological threats; these and many other global issues seem unlikely to respond to political or technological fixes. A better hope may be grass-roots social innovation. If so, a good understanding of the processes of social innovation is needed. The aim of this paper is to contribute to such understanding. Technological innovation is, of course, one kind of social innovation. I start by looking at the ways in which human institutions create, maintain and develop a capacity for technological innovation, whether at the level of firms, of industries or of whole economies. I then explore the system features which are needed for social innovation. I conclude by suggesting that community-based innovation may be the future basis for social problem-solving.

TECHNOLOGICAL INNOVATION

Freeman and Perez analyse technological innovation using four categories.[1] *Incremental* innovation is the kind which goes on continuously in any industry, much of it through learning-by-doing. This can be expressed in the learning-curve, which shows that productive efficiency improves steadily with experience. Incremental innovation is often firm-specific and is certainly industry-specific. Learning is often tacit, being hard to reduce to a codified form. After a decade of incremental innovation a firm's products and processes may be strikingly different, but the change is gradual and evolutionary. *Radical innovation* results from individual inventions and usually requires new production techniques and organisational changes, as well as the technical breakthrough, if it is to be industrially useful. An example might be the discovery of nylon, the first true synthetic fibre. Radical innovation can take place in a firm, or in a university, or elsewhere. In a firm, it can be an embarrassment, because the new ways of doing things which it requires may be disruptive to the firm's existing production methods.

Systemic innovation is the third category. This is far-reaching change, affecting several industrial sectors and combining radical and incremental innovations with a range of organisational and managerial changes. It is a 'clustering' phenomenon, where innovation in one firm affects what others do, through movement of people from one firm to another and through inter-firm emulation. The growth of the semi-conductor industry, mainly in the USA,

Systems Science, Edited by F.A. Stowell *et al.*,
Plenum Press, New York, 1993

is an example. The final category is *paradigmatic innovation,* in which changes in technology system are so far-reaching that they transform the whole economy and result in a new techno-economic paradigm, in a technological revolution. The way of thinking of technologists and industrial managers is transformed and the new paradigm becomes the 'natural' way of doing things for several decades. The macro-economic aspects of this kind of innovation are at least as important as the technological ones, as whole branches of industry may die, and others arise, as a result. The combination, dominant in the 1950s and 60s, of Henry Ford's assembly line production backed by supporting labour laws, with the corresponding creation of consumer demand by advertising and by stimulatory economic policies, is the most obvious example of a techno-economic paradigm.[2]

Freeman and Perez view industrially-useful knowledge as subject to decay, unless it remains in continuous use. This is because much of it is tacit (that is, it resides with the human actors involved and will be lost if groups of these actors do not continue their active association) and because it needs to adapt to changing economic conditions (due to competition, market shifts etc.) Hence, incremental innovation is a *homeostatic* process. Unless there is continuous innovation of this kind, the industrial activity will wither. This innovation depends on continuing activity, on on-going learning-by-doing, on the persistence of an innovating community.

The other three types of innovation are *morphogenic*.[3] They depend on the creation of new socioeconomic structures, which, once they have come into being, then persist with a degree of stability, sometimes for fifty years or more. These leaps of re-structuration are difficult to achieve, as many things have to change simultaneously if a stable new structure is to emerge. This is particularly so if the whole economy is being re-structured. (The present chaotic state of the world economy may be evidence for the difficulty of attaining a new structure).

It may seem paradoxical to describe any process of innovation, even incremental innovation, as homeostatic. But if we accept the idea that industrial knowledge decays unless it is constantly used and developed, then the paradox is resolved - the industrial unit must run to stay in (roughly) the same spot. There is a paradox of a different kind connected to the three morphogenic categories of innovation - radical, systemic and paradigmatic innovation. When a new structure has been established, through one or more of these kinds of innovation, then morphogenesis should stop and homeostasis should take over. 'Sticking to the knitting' becomes the right strategy for firms, industries and economies. Any further morphogenic innovation, with its departure from the familiar, then has its dangers, at least until the point is reached when the new structures no longer respond to maintenance through incremental innovation. At this point homeostasis fails and the structures (now no longer new) are swept away by the 'winds of creative destruction' and replaced (if we are lucky) by another round of morphogenesis. Thus the alternation, over time, between phases of stability and phases of change may be a necessary part of industrial life.[4]

FACTORS ON WHICH INNOVATION DEPENDS

One conclusion from the first section of this paper is that innovation, of different kinds, is necessary in both the stable, homeostatic and the unstable, morphogenic phases of industrial development. If it is as vital as this, we ought to make sure we understand as much as possible about the factors which enhance innovation. I suggest that the first factor is the innate capability of the people involved, their native wit. I suspect that native wit is spread around pretty randomly, and even if it is not there is not much we can do to increase its supply. The second factor is education. The evidence today is strikingly clear: good education and training for the whole of a country's population is essential for an innovative industry, Japan being the best example.

However, it is on two further factors that I wish to concentrate in this paper, because they relate most closely to a systems approach to social innovation. They are *democracy and openness*. All four factors interact, but again it is the interaction of the last two which I will deal with here. By democracy I mean the widespread dispersion of power in an industrial organisation. In the past there may have been various other reasons (including the fact that education was available to only limited numbers of people) why those whose job was task definition, the managers, were a different group of people from whose whose job was task execution, the workers. But one universal reason, then and now, is power. Those who define tasks gain more knowledge than those who execute them. This knowledge gives them more power than the task-executors, which leads them to cling onto the task-defining role. But as innovation becomes more and more competitive, the split between task definition and task execution becomes more and more dysfunctional, introducing unnecessary delays and losing the problem-solving capabilities of those who know the tasks best, namely the task-executors. By spreading task definition more widely, more people get involved in innovation and problem-solving. As well as gaining the innovative capabilities of a much wider group of people, spreading task definition to the people who are responsible for task execution is strongly motivational.

The first reason why openness is essential is that it supports democracy. When people know all the facts and there is no secret knowledge held by a limited group of people, then power-sharing becomes more certain. Even in a democracy, some people will have power through their expertise or charisma, but if there is full openness about what is going on, then it becomes harder to misuse this power. Openness encourages communicative competence and allow people to build their communicative skills, providing a further foundation for democracy. The second reason for openness is that it coordinates action in a non-hierarchical way and supplies a fertile field for innovation. If task definition is dispersed among all those concerned with task execution, then there is a coordination problem, except in the rare cases where the tasks are quite independent of each other. The answer to this problem is self-organisation, through communication between the people who are both defining and executing the tasks. However, self-organisation will only be effective if everyone is well-informed and that, in turn, depends on complete openness about all the tasks.

In the incremental, homeostatic phase of industrial innovation, democracy is perhaps the most important factor. Within a firm, a particular pattern of products and of production processes has become established, certain routines have been set, machines are in place. This established pattern in itself provides a good deal of coordination, especially if the people concerned are well trained and therefore have a solid understanding of the products and processes of the firm. The key factor leading to continuous, incremental innovation will be the empowerment and responsibility of task-executors throughout the firm. And empowerment and responsibility stem from democracy.

In the morphogenic phase of industrial innovation, there is no preexisting structure to provide coordination, because this phase necessarily requires the creation of new structures. Openness then becomes the key factor, both in the generation of new ideas and in the interconnection of these ideas to produce new systems and paradigms with far-reaching effects. It is only through the exchange of a diverse range of information and knowledge that people can come to imagine possible new structures. It is only through excellent communication that the implications of new ideas can be explored and their practicability tested. As I have said, democracy and openness reinforce each other, so both are needed in both the homeostatic and morphogenic phases, but democracy is particularly important in the former phase and openness in the latter phase.

Let me give an example of openness and democracy in action.[5] When a small group of people, led by myself, started the biotechnology company Celltech, we aimed for completely open and free communication within the firm. We tried to make sure that everyone knew all the time what was going on and that everyone was involved in key decisions, such as the choice of priorities among the projects on which Celltech was working. Openness and trust reinforced each other and by the end of our first year, when Celltech had around fifty people, a tradition had been started which was strong enough to involve newcomers quite quickly, stifling any temptation to withhold information as a power-play or to generate a petty scandal, and once established the tradition continued.

Here is a story which shows how strong the tradition became: in 1989 Celltech's largest shareholder got into financial difficulties and was forced to offer its shares for sale. This could have triggered a take-over bid for the whole company which would obviously have had a great effect on everyone working in it. It was necessary to keep this development secret from the outside world for about three weeks while various preparations were made. Some members of Celltech's board of directors felt that it should also be kept secret from people working in the firm because, as one of them put it, 'if we tell the staff it's bound to be in the Financial Times tomorrow'. I managed to persuade the board that this was not likely. I was willing to stake my own reputation as a chief executive who knew his company on this and I was right, as everybody in the company was given the information in confidence and kept it secret until a public announcement was made. At that time there were about four hundred people in the company.

How real was the democracy within Celltech given that there was a conventional management hierarchy, with greater rewards for senior people? One answer is that in formal terms there was no democracy; Celltech's legal status was that of a Limited Company controlled by its shareholders, who elected, and could dismiss its board, who appointed its management, who hired the staff. Formally it was the outside shareholders who had the power, not those working in the firm. But in actual terms the situation was different. The value of the company lay almost wholly in the skills of the people working there and in the combination of those skills in carrying out research, development and production activities. So shareholders and employees had to cooperate if anything was to be achieved. They had to have faith in each other, power really was shared and Celltech became a highly innovative company.

What I have tried to show in this discussion of industrial innovation is, first, that innovation has both homeostatic and morphogenic phases, second, that without continuous innovation industrial enterprises will perish and, third, that two critical factors for continuous innovation are democracy and openness. In the final section of this paper I will apply these conclusions from industry to social innovation in general.

SOCIAL INNOVATION

Industrial innovation, although technical, is also very much a social process and is therefore an example of social innovation. But is it a *representative* example? In my view it is, and for two reasons. Firstly, technological change is a huge influence on the life of everyone in the modern world, whether at work, as consumers or as victims of the side-effects of technological change, such as traffic or pollution. Secondly, industrial organisation is the prototype for capitalist organisation as a whole, and capitalist organisation has become now become the predominant model for social organisation throughout the world. Thus, industrial innovation is a central feature of the modern world and models of it may be expected to be useful in describing social innovation in general, providing the specific origin of the models is kept in mind.

So let us assume that social innovation has homeostatic and morphogenic phases and that democracy and openness are vital for its success. The kind of democracy which I have described is direct, day-to-day, power-sharing democracy, far removed from the kind which allows citizens to vote, every four years or so, for a small set of people who thereby get enormous power. Direct, rather than representative, democracy means that the scale must be small. So too does complete openness. It is simply impossible to aim for the complete sharing of information among anything other than a small group. The model of social innovation developed in this paper therefore implies that an open, democratic community will be a face-to-face one, on the scale of no more than a few hundred members, all of whom can interact directly with one another. Such a community will be able to coordinate its activities without a hierarchy and to develop the communicative competence and common language necessary for open communication between all of its members. The model implies that small is beautiful.

The model suggests that social innovation will be enhanced by a type of social organisation which I call the *creative compartment,* which is a group of people linked by geography, by contract, by tradition, by common purpose or by a combination of any or all of these.[6] Members of a creative compartment know each other, meet regularly with each other, often work closely with each other. Communication between a compartment's participants is effective because they develop a common understanding of their situation, a common set of concepts for thinking about that situation and a common language for talking with each other about it. There is freedom, openness and equality in communication within the compartment. There is a well-defined boundary to a creative compartment; free and general communication stops at this boundary although more specific communication takes place between the compartment and the outside world.

Let me give two examples of social arrangements which approximate to this description.[7]

Delancey Street is a centre in San Francisco, with several hundred people, dedicated to helping people from all over the world, who have serious problems with either drugs or alcohol or who have just come out of prison. Residence is for a minimum of two years. The first few months are spent in working very hard for the centre. After that, people join one of several training schools, which earn money as well as giving training. They include a restaurant, a graphic arts studio, a printing shop, a removal company and a construction business. There are no paid staff. The whole thing is run by the residents, who certainly learn. Graduates of Delancey Street have become teachers, truck drivers, sales people, doctors and policemen.

In a graduate's words:

"We start with basic things - how you dress, how you walk, how you speak...We push you to help the next guy - you're looking after someone else. And then pretty soon you feel good about yourself when you're helping someone else. And that kind of self-empowerment and self-government from running the business, is really a motivating factor. Because it's ours, and we want to make it work. It doesn't belong to someone else. It's mine. This is my home, this is my theatre."

About a hundred people belong to the in-shore fishing cooperative at Alanya, in Turkey, described in Elinor Ostrom's book *Governing the Commons* . In the 1970s, fishing was unrestrained and this led to over-fishing, conflict between the fishers and increasingly costly fishing practices. The cooperative decided to do something about this and after ten years of learning devised a set of rules to govern the fishery, applicable to all fishers in or out of the cooperative. Each year a list of eligible fishers and a list of fishing locations is prepared and, by drawing lots, fishers are each assigned a location. Because some locations are much better than others, the location assigned by lot is for one day only; every day each fisher moves to the east to the next location. Although the lists of fishers and locations are deposited with the mayor and the police, monitoring is by the fishers themselves, who mostly go out everyday on the sea, especially when it is their turn for a productive location, and who would therefore spot anyone trying to cheat. Apparently cheating is rare and any disputes are settled by the fishers themselves.

At the start of this paper I suggested that grass-roots innovation might be a promising theme. I have tried to show why this is so and have briefly described how creative compartments could be the way to organise such innovation. I believe that if there were to be a paradigmatic social innovation, on a global scale, which resulted in compartmentation becoming a major form of social organisation, then the prospects for socially-desirable innovation would be greatly enhanced. Such a paradigmatic change could only take place if the idea of the creative, democratic, openly-communicating compartment were to capture the world's imagination. But if it did, this would lead to the realisation of Rene Dubos' vision: 'Think globally, act locally.'

REFERENCES

1. C. Freeman and C. Perez, Structural crises of adjustment: business cycles and investment behaviour, *in:* "Technical Change and Economic Theory," G. Dosi *et al.*, eds.,Pinter, London: pp45-47, (1988).
2. On the connection between technological and socioeconomic change, and on how this connection may be the cause of economic 'long waves', see C. Perez, Structural change and assimilation of new technologies in the economic and social systems, *Futures,* October 1983, 357-375; G. H.Fairtlough, Can we plan for new technology? *Long Range Planning,* **17,** 3, 14-23 (1984); M. J. Piore and C. F. Sabel, 'The Second Industrial Divide: Possibilities for Prosperity,' Basic Books, New York (1984); R. Nelson, Institutions supporting technical change in the United States *in*: Dosi *et al, op cit:* 312-329 and P. Pelican, Can the imperfect innovation systems of capitalism be outperformed? *in*: Dosi et al, *op cit.*: 370-398.
3. A. Wilden. "System and Structure: Essays in Communication and Exchange," Tavistock, London XII, (1972).
4. On the relevance of Schumpeter's ideas (including 'creative destruction') to the picture of industrial innovation discussed here, see Freeman's introduction to Dosi *et al. op cit* : 1-8. On the need for alternation between stability and change see M. Maidique and R.Hayes, The art of high technology management, *Sloan Management Review,* 17-31, Winter (1984).
5. For more about experience at Celltech see G. H. Fairtlough, Systems practice from the start: Some experiences in a biotechnology company, *Systems Practice,* **2,** 4, 397-412. (1989) and Mark Dodgson, 'The Management of Technological Learning: Lessons from a Biotechnology Company,' de Gruyter, Berlin (1991).
6. G.H.Fairtlough, Creative compartments, *London Business School J.* **10,** 2, 2 -14 (1986) and G.H.Fairtlough, A model of capitalism derived from communication theory, *Futures,* **22,** 1, 69-77 (1990).
7. *Netview: Global Business Network News.* **3,** July 1991 and Elinor Ostrom, 'Governing the Commons: The Evolution of Institutions for Collective Action,' Cambridge U.P., Cambridge, UK, 19 & 180, (1990).

AUTOMATION TECHNOLOGY AS A HUMAN TOOL FOR ACCOMPLISHING HUMAN PURPOSES

J. L. Elohim[1] and Eduardo Oliva López[2]

[1] Cultural Aspects of Automation
Abteilung für Systemtechnik und Automatisierungstechnik
Wissenschaftliche Landesakademie für Niederösterreich
Postfach 15, A 3504, Krems/Stein, Austria

[1] Institut für Systemwissenschaften
Johannes Kepler Universität Linz
A 4040, Linz, Austria

[2] Sección de Estudios de Graduados e Investigación
U.P.I.I.C.S.A., Instituto Politécnico Nacional
Calle Té # 950, Iztacalco D.F. 08400, México

THE GLOBAL MARGINAL EFFECTS ARE IN FACT PARADOXICAL SITUATIONS

Automation technology, as one of the culminating aspects of an assumed development of the technological concern of the human species has made possible to organize, in the time and in the space, the very **complex co-ordination of many** factors involved in tedious, wearisome, fatiguing, bothersome, dangerous,... and increasingly difficult and complicated operations, but it has been also, until now, the source of more and more serious **global marginal effects (degradation** of cultural values, **retrogression** of human relations **deterioration** of all kinds of natural environments) which make more and more **paradoxical** to deal with the evolvement of human affairs .

To all appearances, at first the emergence and afterwards the development of the (human) civilization on the Earth have been in a large measure the unavoidable outcome of ideas conceived and actions implemented by some (not many) human beings **technologically and scientifically minded.**

Most of these innovators seem to be almost always quite successful in finding out how to develop certain capabilities and skills, which enable them:

1) to get a good comprehension of **"causes and effects"** and their inherent **"developmental possibilities"** in natural and man-made **real** things, facts and phenomena; questions perceived quite distinctly under the assumption that they might become **"resources"**

Systems Science, Edited by F.A. Stowell *et al.*,
Plenum Press, New York, 1993

useful, and even indispensable, at least for the immediate survival of individuals and communities, but also for the progressive evolvement of human life.

2) to search out how to make use of the knowledge accumulated gradually by means of such kind of comprehensive views, while determining and judging what the relative "usefulness" of such things, facts and phenomena might be.

Gradually but always faster, all along the historical path followed by mankind, these innovators have invented and built, quite **consciously,** tools, devices, instruments, appliances, apparatusses, processes, systems,... "intending to improve" one or other aspect of the human way of life established in the community they belong to.

They have certainly learnt quite consistently to "handle technologically" the evolvement of the knowledge acquired

They have found by themselves what, where, when, why and how:

* to reproduce purposefully certain aspects of things, facts and phenomena perceived or
* to invent something using some parts of such phenomena, things and facts or
* to modify the causal sequence of original phenomena perceived in order to get improved predetermined results.

An artificial world has been built gradually through the efforts of people scientifically and technologically minded.

This artificial world, which is at present the support of the actual civilization, has become an ensemble larger and larger of many diverse technological means that are in fact **extensions** of human hands, fingers, arms. feet,... brain presumably needed for making stronger, more flexible, more precise or fuzzier, heavier or lighter,... the performance of human beings as required for "organizing better" their survival, for searching more easily how to "progress", for taking advantage of resources, facts, situations,... after realizing that they have to face natural forces and also are obliged to confront many other artificial forces generated through human endeavours.

However, very often most innovators do not care whether their tecnological "contribution" will spoil something else that had not been taken into account or will cause harm, damages and troubles to members of other communities or will reduce the surviving possibilities of any of the species (comprised the human one) on the Earth. Usually they take into account, if they do, no more than their own interests and those of their colleagues or accomplices.

It means: (1) that the "new technological possibilities" inserted in the human society are in fact innovations introduced rather **"blindly"**; (2) that on the Earth there is an artificial world, which so far has been built quite **unconsciously;** (3) that the World appears to be more and more a **chaotic "array" of stochastic actions.**

The very large global marginal effects, which constitute sources of the actual crisis in the World at present, are in fact undeniable evidences of consequences provoked in a large measure by one or the other part of that artificial world or even the out-growth of several of them intermingled (Table 1).

THE SOURCES OF THE PARADOXICAL CIVILIZATION

Every human culture seems to be essentially an ensemble of **"intellectual frames of reference"** that allows a person, an institution, a community, a nation,... in practice any kind of human entity, to determine its particular ways for "LEARNING":

* how to perceive the physical and the social features of its "environment";
* how to **determine** certain **aims"** that would be desirable to accomplish, accordingly to criteria arising from the evolvement of such framework;
* how to **organize** successfully its particular **performance";...** aiming to survive more easily, to progress faster, to develop more effectively its capabilities.

Table 1. Some of the actual **global marginal effects** at present

SOCIAL DISTURBANCES

* New illnesses provoked by physical and mental troubles: Aids, Alzheiner's disease, new kinds of cancer,...)
* Aggravation of ancient diseases: Drug-addiction, neurosis,...
* Weakening of human motivation: Lack of ethics, poor social solidarity,...

TECHNOLOGICAL SHORTCOMINGS

* Waste of valuable material resources.
* Exhaustion of non-renewable mineral reserves.
* Exhaustion of energy reserves.
* Wastage and underemployment of human resources.
* Disappearance of species.
* Attacks on man's genetic heritage.

DEGRADATION OF NATURAL ENVIRONMENTS

* Damages to vital properties of natural elements: water,air, soils, forests,...
* Destruction of natural beauties.

METEOROLOGICAL TROUBLES

* Climate's changes due to green house effects.
* Holes in ozone layers in boreal areas.

GROWTH OF INSTITUTIONS INEFFECTIVENESS

* Mismanagement.
* Unsuitable integration of large organizations: regional, national and international

INTERNATIONAL DISORDER

* Increasingly hidden arms' traffic, used for destruction of cultural accomplishments.
* Export of inflation and recession.
* Export of capitals from less developed countries to highly industrialized ones.

Such learning usually takes place without realizing thoroughly that, at the same time, many other "human entities" are trying as well to organize, usually each of them on its own, their respective particular performances in the same environment.

Most of these "entities" are facing nowadays a larger number of difficulties while they intend to assure at least, their physical, emotional, and intellectual survival.

Apparently, these are troubles that arise due to the fact that most of them have been obliged to be increasingly immersed in the **"cultural system"** that predominates at present all over the world. In this system everyone becomes individually **entangled in a mesh of "socio-technical" networks** which provoke the emergence of unfriendly "social forces", that grow and become blindly inhuman and even antihuman, because they have been conceived without taking into account the features of human beings.

Instead they appear as if they were good enough and the right ones, only because they are increasingly supported by a greater number of **automatic "artifacts"** (devices and processes) that have been conceived, designed, implemented and built by human beings and are certainly maintained in operation by them.

It is the largest paradoxical situation faced by the human civilization, getting up from the unsolved (unattended) contradictory character of such prevailing culture:
* Which pushes ahead the individualism without allowing the development of **individualities** increasingly aware of their **social commitment** (ethically interpreted) and their earthly commitment (ecologically interpreted).
* Which assumes that humanities on one side, and science and technology on the other are cultural spheres that exclude one to each other, without realizing either on purpose or due to lack of knowledge:

1) that science and technology are both unavoidably human concerns, human actions and consequently human products;

2) that the development of our civilization, where humanities has found the possibility to emerge, has been so far and apparently will be always necessarily the outcome of actions of men and women scientifically and technologically minded;

3) that in fact none of these spheres could never be developed independently.

* Which provokes in the mind of most people passive and unconcerned attitudes towards the community's progress, while technological innovations are introduced in accordance to criteria formulated by just a few, compeling the society to move rather blindly in the time and in the space.
* Which still pretends to ignore the **enormous possibilities of intrinsic features** of practically **any human being,** as a SUBJECT capable of developing his (her) capabilities on his (her) own and also of contributing to push ahead the development of his (her) environment, when suitable conditions are created and maintained.
* Which makes more and more scientists and engineers to believe in technolatry, to behave as technolators, and to work for **an utopian progress** by means of an **increasing automation** which so far, appears as if it were leading towards a "**cybernetic leisure society** which doesn't know exactly what to do with itself' (Bertalanffy), because quite a lot of these gentlemen work without realizing:

1) that the role of man on the Earth must be deeply and continuosly examined and critized, while facing the outcome of his perfomance;

2) that human needs, possibilities and limitations should be systematically updated, bearing in mind the state of resources available;

3) that such role should be consequently redefined and implemented, bearing in mind that his historical path and his teleological possibilities are changeable in the time because both are intellectual phenomena dependant from information perceived;

4) that such role such be taken into account when decisions related with the development of automation technologies are made;

5) that the four previous questions are unavoidably his concern, if and certainly if, they accept to work for the progress of the community and the development of the species they belong to.

* Which makes many scientists and engineers to assume that human beings are a peculiar kind of **nuisance** when they are allowed to be involved in a scientific, tecnological and/or technical environment, because they cannot avoid to behave as **ignorant foreigners** which means that they:

1) neither recognize the raison d'etre of artifacts that engineers conceived and built, presumably for the benefit of them (the ignorants);

2) nor are capable to understand and appreciate the theoretical explanations of facts and phenomena provided by the scientists.

* Which pushes forward the emergence and development of quite a number of ''artificial **intelligences",** but prevents de facto further development of millions of **natural intelligences.**

AUTOMATION TECHNOLOGY CONVEYS A PURPOSEFUL CULTURAL FEEDBACK

Due to inertial forces created by the presence of technological things, facts and phenomena in the human environment, it seems proper to assume that the technological thinking (comprised automation: its more sophisticated manifestation) can be, from now onward, the main resource for searching out how to overcome the worldwide crisis and then to push ahead the progress of communities, nations and the mankind and consequently to push forward the development of the human species.

However such possibility can become a real option, if and only if, the human beings involved develop soon enough their capabilities for learning how to deal properly with the complexity implicit in the society's development more and more implemented today by means of "automatic things". It is already a potential alternative, though very often it is ignored or twist on purpose in order to satisfy selfish interests.

In the history of mankind the insertion, from time to time, in every society of "automatic things" and "automatized processes" is undoubtedly the outcome of human capabilities such as reasoning, intuition, abstraction,... employed, accordingly to prevailing cultural frames of mind, in order to conceive, design, build and install self-moving, self-acting, self-regulating, self-governing,... devices, machines and systems which "ought to behave" as self-determining "subjects" while being located in very particular environments which necessarily must be specifically organized accordingly to previously specified purposes of such "automatic entities".

Also in the history of mankind, larger number of human beings have been more and more involved in searching how to increase their autonomy [freedom] as individuals and groups in order to increase their self-moving, self-acting, self-regulating, self-governing,... possibilities in every cultural environment.

The question now is how to lead reasonably these two human concerns ?

At present still few men recognize consciously that the only question of the reality that has remained always constant, that still remains constant at present, that certainly (?) will remain constant forever is the CHANGE, because it is (the change) the **unchanging** feature of every aspect of the whole reality.

If lately the way of life is changing faster than ever before it is due to an extensive automation, which has arisen and has been developed after discovering the many possibilities offered by **feedback** relations established **on purpose** among certain parts of preconceived devices, machines, apparatusses and systems.

However, changes are always seemingly caused everywhere in accordance to rather particular **involutionary and evolutionary** possibilities and restrictions, intrinsic to the nature of feedback relations. Then, it appears possible to interpret cybernetically the evolvement of everything in Nature and also the manifestations of the three main possible social changes: to **progress, to retrogress** and to **degrade or to deteriorate,** which are determined necessarily by involutionary and evolutionary processes. It means that it is indispensable to find out how to manage reasonably discovered and invented [+] and [-] **feedback relations**

The global marginal effects seem to arise without possible solutions essentially because most scientists and engineers have not yet realized that more often than not it is possible and necessary to find out how to complement: automation with the development of manual capabilities; centralized power with decentralized decisions; centralized economic decisions with free market requirements and possibilities; productivity's increases with a respectable income for everybody; econornic requirements with good quality needs,... instead of believing that the assumed conflicting situation in each case ought to lead unavoidably to a dilemma, a trilemma or...a multilemma, that **"cannot be solved"** until all the "participants": individuals, communities, ethnical groups, parties, nations,... and even books and thoughts

-except one- are eliminated; in order to allow the winner of the competition to be the "right", the indisputable, the unquestionable one.

This complementarity is closely related to feedback relations intrinsic to the dynamical features of every aspect of the whole reality. This assertion is still an assumption that obliges to find out how to deal reasonably with:

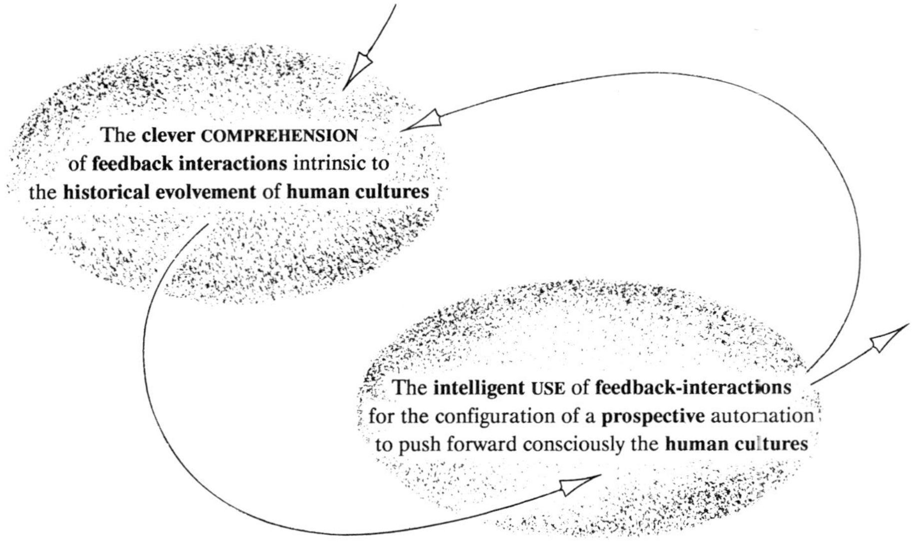

Figure 1.

This paper is related with
the evolvement of a **Research Programme** being outlined at
"(die) Wissenschaftliche Landesakademie für Niederösterreich"
with the support of
(die) Johannes Kepler Universität
for finding out in general
how to deal with **interactions**
that could and/or should be established between two of the more **significant** human concerns:
Automation Technology and Culture.
Through this programme it is intended, in particular,
to identify **possible** and **feasible answers** to a fundamental question:

> How might **automation** become a **human "tool"**
> for helping
> to GUIDE consciously the progress of mankind
> and
> to LEAD properly [reasonably] the human species' development?

Although automation technology has remained the **blind technological "tool"** used by
that **"man** (who) **makes himself"** (Gordon Childe)
who has not yet realized **what kind of man** he has been making;
who has ignored so far that assumed technological advances may cause **"automatically"**:
degradation of **cultural values**
retrogression of **human relations**
deterioration of **earthly environments**
in spite of this inconvenience,
this automation technology,
as one of the main factors of the **paradoxical panorama** that faces today the **civilization**,
is obliged as well,
to find out **ways** that contribute to overcome the actual **crisis**.

HEALTH FOR ALL AND COMMUNITY PARTICIPATION

Gilbert Mansell and Helen Pearson

University of Huddersfield
Queensgate
Huddersfield
HD1 3DH

HEALTH FOR ALL

The Health for All movement has its origins in the World Health Organisation's Global Strategy of Health for All by the Year 2000, which was accepted as WHO policy in 1981 by the thirty fourth World Health Assembly (Ashton & Seymour 1990). The objectives of the strategy were to reduce preventable ill-health, to encourage life-styles conducive to good health and to develop community based primary health care available to all. Health for All (HFA) related activity was seen to necessitate collaboration between agencies and sectors, for example between local government and health authorities and between the statutory and voluntary sectors. This was because the factors that impact on health are beyond the sphere of influence of any one agency or sector. The HFA philosophy also included an emphasis on community participation in decision-making and implementation of strategy. This was in tune with a "New Public Health" movement that stressed that health was a social rather than a medical concept, and criticised the elitism of the medical profession. The HFA strategy was global, but it was expected that each of the regions and nations of the world would develop strategies more specifically relevant to their own circumstances. The European member states (including the UK) agreed on a set of 38 targets whose achievement would mark progress on the way to health for all. The first target was that by the year 2000 differences in health status between countries, and between groups within countries, should be reduced by at least 25% by improving the level of health of disadvantaged nations and groups. The Black Report (Townsend & Davidson 1982) had shown that in the UK all the important determinants of health were linked to social class, sex, race and geography. For example cigarette smoking and physical exercise were class-related as were accidents and disability. The South-East of England was markedly healthier than the North. The principle of equity in health is fundamental to the HFA movement. Other important consequences of achieving the 38 targets would be an improved environment and a re-orientation of health care. Targets were set, for example, for controlling water and air pollution, improving food safety, protecting against hazardous wastes, improving housing conditions and protecting against work-related health risks. Provision of health care was to be community-based, target 29, for example,

Systems Science, Edited by F.A. Stowell *et al.*,
Plenum Press, New York, 1993

specifying that primary health care systems should be based on co-operation and team-work between health personnel, individuals, families and community groups. The target dates for improving the environment and re-orientating health-care were in the 1990s and little progress has been made towards them in the UK.

The HFA philosophy was affirmed by the 1986 Ottawa Conference on Health Promotion which produced the influential Ottawa Charter. (WHO 1986). This stressed the importance of public policy in health promotion (for example the practice of raising state revenue by taxing tobacco and alcohol consumption is "unhealthy" public policy), and argued for a socio-ecological approach to health that preserved the natural environment and organised work to help create a healthy society. In tune with a burgeoning global "green" movement the Charter called for action to counteract pressures towards harmful products, resource depletion, unhealthy living conditions and environments, and bad nutrition, and focussed attention on pollution, occupational hazards and homelessness. A WHO initiative, known as "Healthy Cities", started in 1986 involving collaboration among, initially, 21 European cities. The cities involved undertook to make plans to realise the 38 European targets, develop models of good practice, and monitor and evaluate their work. Ideas and experiences were to be shared among collaborating cities. The principles underpinning the project are defined as follows (Tsouros 1990).

-Health for All implies *equity*.
-The main emphasis should be on *health promotion* and the prevention of disease.
-A well informed, well-motivated and actively *participating community* is essential for health for all.
-*Multi-sectoral co-operation* is necessary for ensuring the pre-requisites for health and promoting healthy policies.
-The focus of the health care system should be on primary health care, meeting health needs as close as possible to where people live and work.
-Health problems transcend national frontiers.

HEALTH FOR ALL IN KIRKLEES

The Healthy Cities movement in the UK spread beyond the original participating cities and in 1988 a Kirklees HFA initiative was launched. Kirklees is a local authority in West Yorkshire that includes the towns of Huddersfield, Dewsbury, and Batley. The initiative was multi-agency because Kirklees Metropolitan Council, two Health Authorities, Kirklees Family Health Services Authority and the Polytechnic of Huddersfield all agreed to collaborate. Research conducted by H. Pearson in 1992 found some degree of uncertainty amongst stakeholders as to what counted as an HFA activity and what did not. A core set of activities (defined in Table 1) were generally accepted as "official" HFA initiatives. HFA Events are day or evening sessions, open to the public, where issues can be discussed and health promotion messages promulgated. For example at health fairs members of the public are given the opportunity to have a personalised test of their fitness and discuss their life-style. Groups and projects represent ongoing activities that tackle health related issues, for example child safety. The Deighton Health Project is an example of community development, with a health focus, staffed by a full-time worker. Deighton was targetted because the area has the highest level of multiple deprivation in the locality reflected in high levels of mortality and morbidity. Policy development is represented by award schemes that encourage health policy in, for example, schools, canteens and restaurants and retail outlets. Provision of healthy food is an issue because of the widespread prevalence of poor diet, and the links between diet and heart disease and cancer.

Table 1. Activities identified as specifically related to Health for All

Nature of Activity	
1.	**Health for All Events:** Health for All Forums Health Fairs "Sight for All" Conference
2.	**Groups and Projects** "Keeping Kirklees Children Safe" Group Deighton Health Project
3.	**Policy Development in the Public and Private Sector** Healthy School Award Heartbeat Award Scheme Kirklees Healthy Choice Award

The research revealed that in addition to HFA-specific activity, there was a great deal of other activity that contributed to health for all principles that did not operate under the HFA banner. Examples are road and home safety campaigns, quit smoking groups, HIV/AIDS initiatives and work with prostitutes.

Whilst great enthusiasm and commitment to the HFA principles were shown by particular individuals, not all agencies contributed equally (for example there had been little contribution made by the FHSA and little emphasis on the primary health care sector). The voluntary sector had not been involved in HFA despite the fact that many community-based voluntary groups were serving health-related needs (for example breast-feeding support groups). All the initiatives had been agency-led by "professionals" and true community participation was limited. Unlike most towns and cities Kirklees had not appointed an overall HFA co-ordinator (although such an appointment was made in 1992). The involvement of the local authority although valuable, was largely limited to Environmental Services, the Leisure and Recreation Service and the Education Service. Other functions of a local authority also impact on health, however, for example housing, social services and highways.

USE OF A SYSTEMS APPROACH

It seems evident that a systems approach is relevant to studying the phenomenon of Health for All. The health of communities is determined by the complex interaction of natural and social phenomena. No one human agency is responsible for bringing about good health and many of the consequences of human activity in many areas of the economy impair health. An investigation of health necessitates the investigation of a complex system with uncertain boundaries. A strong element of subjectivity is present in the area, however. The definition of health is itself problematic, ranging from the absence of disease to a feeling of complete physical and mental well-being. Attitudes to health policy are determined by value systems. For example ill-health might be regarded as a just punishment for an unhealthy life style (perhaps in the case of smoking-related diseases). On the other hand life-style might be seen as a consequence of social status rather than free choice. Unemployment and poor housing gives rise to poor health. The UK Government white paper "The Health of the Nation" (HMSO 1992) concentrates on a medical rather than a social model of health and ignores the problem of class-based inequality. The policy is not compatible with the principles of the HFA movement. Because of the presence of subjectivity, the importance of culture and the relevance of politics, Soft Systems Methodology (Checkland 1980, Checkland & Scholes 1991) will be used to investigate HFA in Kirklees. The status of SSM as an action-research methodology is also relevant to the choice because it is intended to address the issue of community participation in the provision of HFA. This will involve work with statutory agencies to develop ways of using participation as a strategy to improve health, as well as work with grass-roots associations and groups in the community. The work will be performed in collaboration with the Community OR Unit at Northern College which has some experience of work in the community health area (Townley, Thunhurst and Ritchie 1991).

Community OR in the UK has been characterised by its application area (work with non-hierarchical organisations) rather than any distinctive methodology. The area of methodology is addressed by Rosenhead (1989), however, in a book which presents a number of methods and techniques representing an alternative OR paradigm. This paradigm is characterised by simplicity and transparency, conceptualises people as active subjects, and facilitates planning from the bottom-up (Rosenhead 1989). These characteristics seem relevant to facilitating community participation in Health for All, and SSM is included as an exemplar of the alternative paradigm. There is need for caution in using SSM in community settings, however. Community groups may lack communicative competence in comparison with health professionals and local government officials and the outcome of using a "reformist" methodology such as SSM may be manipulative rather than emancipatory. Jackson's critique (Jackson 1982) of soft system approaches is salutary, but no more radical approach as yet commends itself. It is hoped that the research will contribute both to improving community participation in health but also to use of SSM in community settings.

REFERENCES

Ashton, J. and Seymour, H., 1990, "The New Public Health", Open University Press, Buckingham.

Checkland, P.B., 1981, "Systems Thinking, Systems Practice", Wiley, Chichester.

Checkland, P.B. and Scholes, J., 1991, "Soft Systems Methodology in Action", Wiley, Chichester.

HMSO, 1992, "The Health of the Nation: A Strategy for Health in England", HMSO Cm. 1986.

Jackson, M.C., 1982, "The Nature of Soft Systems Thinking: The Work of Churchman, Ackoff and Checkland", *Jnl. Appl.Sys.Anal.* 9, pp17-28.

Rosenhead, J., ed., 1989, "Rational Analysis for a Problematic World", Wiley, Chichester.

Tsouros, A.D., ed., 1991, "World Health Organisation Healthy Cities Project: A Project Becomes a Movement", Sogess, Milan.

Townley, C., Thunhurst, C., and Ritchie, C., 1991, Community OR and The New Public Health Movement, *in*: "Systems Thinking in Europe", M.C. Jackson, G.J. Mansell, R.L. Flood, R.B. Blackham and S.V.E. Probert, ed., Plenum, New York.

Townsend, P. and Davidson, N., 1982, "Inequalities in Health - The Black Report", Penguin, Harmondsworth.

WHO, 1986, "Ottawa Charter for Health Promotion", WHO, Copenhagen.

INTRODUCTION TO INFORMATION SYSTEMS STREAM

From an historical point of view it may be appropriate that the Information Stream attracted the greatest number of papers for the conference because for many commentators we are in the midst of an information revolution. However, it is unlikely that this assertion was the motivation for any of the writers represented here but the popularity of this theme may add weight to the observation itself.

Since the mid 1980's there has been a growing interest in the use of Systems ideas in information systems definition and latterly in the use of these ideas in defining appropriate technological support. As we move out of the data processing age into the information systems age researchers and practitioners have looked for methods of Information Technology (I.T.) development which are more in keeping with the satisfaction of client needs. There has been a shift away from technology driven design approaches towards methods which enable the client to control the design and development process itself. Necessarily this has meant radical re-thinking about information systems and the way in which we employ technology to support them. Lessons learnt from the past suggest that technology based I.T. development methodologies do not always produce the results that the client anticipated nor the required improvement in organisational performance.

The recent developments in technology have enabled I.T. professionals to contemplate novel ways of supporting information system provision which were hitherto unattainable. Technology has become easier to use, faster to respond, cheap to provide and freely available. The technology has become, to some degree, transparent to the user and consequently considered by many as a management tool in much the same way that the telephone is regarded. The concentration of effort for the Information Systems professional is now upon defining the support that I.T. can offer to management activities and not upon the marvel of the technology itself.

This stream includes papers whose authors are concerned with the use of systems ideas in the technological problems of I.T. provision and authors whose interests lie in the application end of the I.T. spectrum. Systems thinking is described as being of value in the development of Data-base and Knowledge Based System design, Requirements Analysis, Knowledge Elicitation, as the basis for the development of Information Systems Design methodologies and in Strategic Information Technology specification. This section includes a number of papers from outside of the United Kingdom and highlights the International appeal of Systems Science to this domain.

Frank Stowell
University of Paisley

ON THE INCOMMENSURABILITY OF HARD AND SOFT SYSTEMS APPROACHES TO INFORMATION SYSTEMS PROVISION

Richard B. Beeby

Information Systems
Department of Computing Science
University of Paisley
Paisley, PA1 2BE

INTRODUCTION

Hard engineering approaches to information systems development seem to be having a hard time recently. The methods are accused of a confusion between the computer based elements of an information system and the system itself (for example, Stowell, 1991) and of a (naive?) assumption that the system objectively exists "out there" (for discussions, see Klein and Hirscheim, 1987; Hirscheim and Klein, 1989; Stowell, 1991; Jayaratna, 1992; Lewis, 1992). In contrast, there is in soft approaches an emphasis that information systems are human activity systems, that they have no independent existence but are socially constructed by the individuals or groups that use them (a sample of discussions in the literature are Lyttinen and Klein, 1985; Liebenau and Backhouse, 1990; Stephens and Wood, 1991). So information systems, it is argued, cannot be investigated or developed using methods that assume they "objectively exist", but must be interpreted using methods that recognise their subjective and social nature (Checkland, 1981; Boland, 1985; Winograd and Flores, 1987; Oliga, 1988; Jayaratna, 1988).

INFORMATION SYSTEMS VERSUS DATA PROCESSING SYSTEMS

Some of the issues above can be, and sometimes are, summarised by saying that information systems are not just data processing systems. However, it is not clear whether this amounts to a strong ontological claim, that there are no systems that are just data processing systems, or simply a definition or criterion to be applied in deciding whether a system may be called an "information system" or not. What it does do is emphasise that approaches to the development of information systems must recognise the difference between information and data.

According to the glossary entry for "information" in Checkland (1981), "Outside information theory, the word 'information' is best used to denote a combination of fact plus the meaning an observer attributes to it." We shall return to the words "fact" and "observer"

Systems Science, Edited by F.A. Stowell *et al.*,
Plenum Press, New York, 1993

later, but note at this point that they do not sit well with the philosophy adopted by a number of todays practitioners of soft approaches. The more recent discussion on the nature of information systems on pages 54 to 56 of Checkland and Scholes (1990) defines information simply as "data with attributed meaning in context" and emphasises that attributing meaning is a "uniquely human act." The discussion informs us that, "An information system, in the exact sense of the phrase, will consist of both data manipulation, which machines can do, and the transformation of the data into information by the attribution of meaning" (page 55). This discussion has been cited with approval by a number of authors (e.g. Stephens and Wood, 1991; Gammack and Stephens, 1991; Galliers, 1992; Stowell and West, 1992). On this basis one can consider whether there is a possibility of successfully factoring information systems provision into the use of information by humans and the manipulation of data by machines. If so, the temptation is to pursue the use of soft approaches for the information aspects and hard approaches for the data manipulation aspects of the system.

Considering these issues has resulted in information systems practitioners participating extensively in philosophical debate about the nature of information systems and the assumptions used in their development. As information systems are seen as social constructions it is not surprising that much of the philosophical literature cited concerns the philosophy of the social sciences. Indeed social scientists have taken a much more active interest in philosophical issues than practitioners in the natural sciences. Gasper (1991) accounts for this by positing that social scientists perceive their disciplines as "relatively unsuccessful." The absence of any "overarching predictive or explanatory social theories ...has prompted methodological reflection: social scientists have attempted to root out mistakes in method that may be responsible for holding back their disciplines" (page 713). This motivation is certainly felt by information systems practitioners, who point to a widespread dissatisfaction with information systems provision (e.g. Hirscheim and Newman, 1988; Curtis, Krasner and Iscoe, 1988).

CONSTRUCTIVISM IN THE PHILOSOPHY OF SCIENCE

An influential framework for this philosophical discussion is that of Burrell and Morgan (1979), who examine the philosophy of the social sciences with respect to assumptions made in ontology, epistemology, human nature, and methodology. In each of these four areas they characterise positions as to whether they are subjectivist or objectivist in approach (this is interpreted in the Anglo-American tradition and largely ignores the poststructuralist European tradition deriving from objective idealism, for a discussion of which see Harland, 1987). They give labels for positions at each extreme in each of the four areas. This framework has been applied to information systems development by Hirscheim and Klein (1989), who provide a basis for discussion by others (e.g. Avison and Wood-Harper, 1990; Stowell, 1991).

Some of the labels chosen for objectivist or subjectivist positions by Burrell and Morgan are used more narrowly than is traditional in the philosophical literature. For example, under epistemology they contrast positivism (objectivist) and anti-positivism (subjectivist), while under ontology they contrast realism (objectivist) with nominalism (subjectivist). But one of the most significant distinguishing characteristics of positivism is its rejection of metaphysics (and hence, of course, ontology). So realists are anti-positivist in their ontological commitments, and technical literature on the philosophy of science (exemplified, for example, by the anthology edited by Boyd, Gasper and Trout, 1991) regard realism and positivism as very different competing positions. That the realist/positivist distinction also has methodological significance is argued by many social scientists (for a recent review, see Romm, 1991).

The word "paradigm" occurs in the title of both Burrell and Morgan (1979) and Hirscheim and Klein (1989). This term was introduced into the philosophy of science in a famous book by Thomas Kuhn (1962, 1970). Chapter 10 of Kuhn's book is entitled "Revolutions as Changes of World View" and describes several senses in which scientists adopting different paradigms "work in different worlds." This equating of "world view" with "world" is one of the central characteristics of the movement, generally called constructivism, that Kuhn helped to create.

Constructivism is defined in the glossary of Boyd, Gasper and Trout (1991) as follows:

> The view that the subject matter of scientific research is wholly or partly constructed by the background theoretical assumptions of the scientific community and thus is not, as realists claim, largely independent of our thoughts and theoretical commitments.

Probably the best argument that the relativist (or constructivist) offers for the assertion that the worlds of groups differ is that the meaning of statements in a language can only be understood holistically. To see the implications of this for the philosophy of science, consider the following extract from the book by Fodor and Lepore (1992) on semantic holism. They consider a semantic property R^*:

> An expression has R^* if it refers to something or other that currently accepted astronomical theories refer to. Suppose that R^* is anatomic, hence holistic. Then it might turn out that no theory could refer to (for example) stars unless it could also refer to (as it might be) planets, nebulas, black holes, the centre of the galaxy, the speed of propagation of light, and the location of the nearest quasar. It would follow that Greek astronomy (hence, Greek astronomers) couldn't ever haye referred to stars. And it would follow from *that* (what one had naively assumed to be) the Greek view that stars are very nearby and ride around the heavens on glass spheres is actually *not contested* by our view that the stars are very far away and don't ride around the heavens at all. In fact, strictly speaking, it would follow that the Greeks didn't have any views on stars; we can't in the vocabulary of modern astronomy say what, if anything, Greek astronomy was about.

As we have seen, the issue of shifting between paradigms is a live one in the IS domain, where there is considerable interest in whether hard and soft approaches can be combined in some way (for example, along the lines suggested above). To say that these approaches are incommensurable, that there is no way to translate any of the concepts of one approach into the language of the other, is to say that there is no possibility of such combination.

CONSTRUCTIVISM, HERMENEUTICS AND THE NATURE OF INFORMATION

One of the most influential texts for the IS community on the philosophy of design is that of Winograd and Flores (1987), who look at competing design philosophies. They place traditional design approaches within what they call "the rationalistic tradition". Their clear statement of what they take the assumptions of this tradition to be, and their coining a new term for it, avoids the problem of their critique being misunderstood as an attack, for example, on the straw man of positivism. Of course, one may still wish to argue whether the assumptions they claim the traditional approach to have do indeed hold.

Against the rationalistic tradition Winograd and Flores set out what they describe as a hermeneutical approach. This approach is largely influenced by Heidegger and by the theory of autopoiesis described by Maturana and Varela (1980, 1988). Similarly, hermeneutics in the tradition of Dilthey and Habermas in the social sciences, is an acknowledged influence on soft systems thinking (Checkland, 1981) and a convergence of ideas seems to have occurred.

Just as, for Maturana and Varela, perception cannot be claimed to be representational so, for Winograd and Flores, language cannot be claimed to be representational either. According to Maturana and Varela (1988; quoted by Stephens and Wood, 1991) "all knowing is doing and all doing is knowing." *All* information, in this view, is "information-for" (rather than "information-about"). Rather counterintuitively this causal aspect of information is what Kampis (1987, 1991) calls *referential information* (counterintuitively because what is distinctive in these approaches is their denial of reference: part of the reason for the choice of term is that what Kampis is interested in is system self-reference). The denial of epistemic information is distinctly constructivist and so places constructivism as a firm contender for anyone searching for a hermeneutical philosophy of information systems (see Stephens and Wood, 1991). It is to the constructivist that the words "fact" and "observer" in Checkland's (1981) glossary entry for "Information" are particularly likely to seem inappropriate.

Yet if information-as-action is the only notion of information to which a constructivist approach can be ontologically committed then it is hard to see how any existing approaches to IS provision even remotely hit the mark, for none have found any way to dispense with an epistemic (rather than causal) view of information. To do so would be to deny that an observer (or even a participant) can construct a model of the system, and so obtain information *about* it (what, for Kampis, is *nonreferential information*). Kampis makes the case that self-modifying complex systems use both referential *and* nonreferential information. He observes that the theory of autopoiesis is incomplete and is unable to explain the emergent properties of the systems with which it is concerned.

Given this critique, we need to ask whether, in recognising subjectivity as an essential feature in human activity systems, it really is necessary to commit ourselves to a fully-fledged constructivism. Indeed, Stephens and Wood (1991), despite their claim to take a constructivist approach, seem to draw back from this when they speak (page 473) of "the dialectical interpenetration of subject and object in which neither has primacy."

CONCLUDING REMARKS: INCOMMENSURABILITY

In addressing the question whether hard and soft approaches to information systems are incommensurable it is helpful to look at some examples from biology. Scientific theories are often expressed in two langauges, one natural and the other mathematical. Kampis (1991) gives two significant examples where competing approaches to a problem domain result in equivalent mathematical models. Firstly, behaviourism and cognitive science are diametrically opposed approaches to behaviour, but mathematical models arising from the behaviourist approach of relating stimuli to their responses (without reference to mental states) can be shown to be equivalent to models which explicitly set out to describe cognitive states (this argument stems from Casti, 1985). Secondly, Neo-Darwinian approaches to evolution assume that populations evolve in response to selection pressures imposed by an independent environment, while coevolutionary theories postulate a causal loop where population and environment select each other: the mathematical models arising in each case can be transformed into one another. A significant part of Kampis' thesis is that formal dynamical models are not capable of capturing the future behaviour of complex systems in biology. Kampis argues that in these cases attempts to formalize competing theories result in equivalent, impoverished, models.

It is not that the natural and formal languages cannot be translated into each other at all (if this were so they would express incommensurable world-views). Indeed, Kampis argues that the known past behaviour of complex systems can always be formalized, which partially accounts for the widespread belief in the scientific community that all physical systems behave according to mathematically formalizable rules (see Conrad, 1985 and Deutsch, 1985 for arguments supporting the computability of any finite physical system).

What makes hard systems approaches hard is their commitment to formal system models. If Kampis is right, human activity systems cannot be formalized or computed. One might say that soft approaches, like natural language descriptions of complex systems, capture aspects of systems that formal (that is, computable) system models cannot. Traditional hard approaches to information system provision aim to produce an unambiguous implementable statement of requirements: conventional widsom is that this requires formal notations, and certainly those parts of the requirements that will be implemented using computers *must* be formalized. So perhaps this is the way the world is: for a sufficiently complex system something will always be lost when going to a hard formal system model, and this is an argument for using soft systems approaches. If we wish to use computers to implement aspects of information systems we need hard approaches to address those aspects, but the soft systems community may be right in trying to find some way to carry the soft approach forward into the provision of the information system as a whole. This leaves open the question of whether there are techniques currently used in hard approaches that might be imported into or blended with a soft framework for the system *qua* human activity system, and this can only be answered by looking at the techniques in question.

The possibility of using data flow diagrams or data models as bridging techniques between them depends on more than "data" appearing in the ontology of both approaches. For one might concede this but argue that "data" does not refer to the same thing in each case: like the "stars" in ancient Greek astronomy that, on a holistic theory of meaning, might have no relation to the "stars" of modern astronomy. A view that hard approaches are ontologically committed to the objective existence of data while soft approaches deny any such existence would support this argument, but is this really a view to which anyone assents?

If we distinguish information and data as Checkland and Scholes suggest, and agree that machines can manipulate data, then data must be something with formal properties that can be described by hard systems science. The question of whether these formal entities exist "out there", independently of our thoughts and theoretical commitments would seem to be an issue best addressed from the philosophy of mathematics rather than the philosophy of science. Field (1980, 1989) argues ingeniously that mathematical entities are fictions without undermining the value of what mathematicians do, and it seems unlikely that hard approaches would be undermined if it turns out that data can be fictionalised in the same kind of way.

There may still be substantive philosophical disputes to resolve about whether an account of information can be formalised in a way that makes it, too, amenable to hard approaches (using, say, the kind of situation theory described by Barwise and Perry, 1983, and Devlin, 1991), or of the possibility of extending soft approaches to the computer-based aspects of provision, but only by adopting a thoroughgoing constructivism can one really make a claim of incommensurability convincing. And while a thoroughgoing constructivism, which abandons reference and epistemology, might be of interest to a sociologist of information systems practice (just as it is to some sociologists of science) it is hard to see how it could ever give us a foundation that retains any acknowledgement that information systems provision might involve systematic rational activity.

REFERENCES

Avison, D. E. and Wood-Harper, A. T., 1990, "Multiview - An Exploration of Information Systems Development," Blackwell, Oxford.

Barwise, J. and Perry, J., 1983, "Situations and Attitudes," MIT Press, Bradford.

Boland, R. J., Jr., 1985, Phenomonology: a preferred approach to research in information systems, *in*: "Research Methods in Information Systems," Mumford, E., Hirscheim, R., Fitzgerald, G. and Wood-Harper, A. T. (editors). North-Holland, Amsterdam.

Boyd, R., Gasper, P. and Trout, J. D., 1991, "The Philosophy of Science," MIT Press, Bradford.

Burrell, G. and Morgan, G., 1979, "Sociological Paradigms and Organisational Analysis," Heinemann, London.

Casti, J. L., 1985, Behaviourism to cognition: A system-theoretic inquiry into brains, minds, and mechnanism, *in*: "Real Brains, Artificial Minds," Casti, J. L. and Karlquist, A., editors. North-Holland, New York.

Checkland, P., 1981, "Systems Thinking, Systems Practice," Wiley, Chichester.

Checkland, P. and Scholes, J., 1990, "Soft Systems Methodology in Action," Wiley, Chichester.

Conrad, M., 1985, On diesign principles for a molecular computer, *Communications of the ACM* 28:464.

Curtis, B., Krasner, H. and Iscoe, N., 1988, A field study of the software design process for large systems, *Communications of the ACM* 31:1268.

Deutsch, D., 1985, Quantum theory, the Church-Turing principle and the universal quantum computer, *Proceedings of the Royal Society London* A 400:97.

Devlin, K., 1991, "Logic and Information," Cambridge University Press, Cambridge.

Field, H., 1980, "Science Without Numbers: A Defense of Nominalism," Blackwell, Oxford.

Field, H., 1989, "Realism, Mathematics and Modality," Blackwell, Oxford.

Fodor, J. and Lepore, E., 1992, "Holism: A Shopper's Guide," Blackwell, Oxford.

Galliers, R. D., 1992, Soft systems, scenarios and the planning and development of information systems. *Systemist* 14:146.

Gammack, J. G. and Stephens, R. A., 1991, The human-centred paradigm in KBS design theory, *in*: "Systems Thinking in Europe," Jackson, M.C., Mansell, G. J., Flood, R. L., Blackham, R. B. and Probert, S. V. E., editors, Plenum, New York.

Gasper, P., 1991. The philosophy of social science, *in*: "The Philosophy of Science," Boyd, R., Gasper, P. and Trout, J. D. editors. MIT Press, Cambridge.

Harland, R., 1987, "Superstructuralism: The Philosophy of Structuralism and Post-Structuralism," Routledge, London.

Hirscheim, R. and Klein, H. K., 1989, Four paradigms of information system development, *Communications of the ACM* 32:1199.

Hirscheim, R. and Newman, A., 1988, Information systems and user resistance: theory and practice, *Computer Journal* 31:398.

Hollis, M. and Lukes, S., 1982, "Rationality and Relativism," Blackwell, Oxford.

Jayaratna, N., 1988, Guide to methodology understanding and information systems practice, *International Journal of Information Management* 8(1).

Jayaratna, N., 1992, Should we link SSM with information systems! *Systemist* 14:108.

Kampis, G., 1987, Some problems of system description II: Information, *International Journal of General Systems* 13:157.

Kampis, G., 1991, "Self-Modifying Systems in Biology and Cognitive Science: A New Framework for Dynamics, Information and Complexity," Pergamon Press, Oxford.

Klein, R. and Hirscheim, R., 1987, A comparative framework of data modelling paradigms and approaches, *Computer Journal* 30(1):8.

Kuhn, T. S., 1970, "The Structure of Scientific Revolutions," 2nd edition, University of Chicago Press, Chicago. (1st edition, 1962).

Lewis, P. J., 1992, The feasibility and desirability of a closer linking of SSM with data focussed information systems development, *Systemist* 14:168.

Liebenau, J. and Backhouse, J., 1990, "Understanding Information: An Introduction," Macmillan, Basingstoke.

Lyttinen, K. J., and Klein, H. K., 1985, The critical theory of Jurgen Habermas as a basis for a theory of information systems, *in*: "Research Methods in Information Systems," Mumford, E., Hirscheim, R., Fitzgerald, G., and Wood-Harper, T. (editors). North-Holland, Amsterdam.

Maturana, H. R. and Varela, F. J., 1980, "Autopoiesis and Cognition: The Realization of the Living," Reidel, Dortrecht.

Maturana, H. R. and Varela, F. J., 1988, "The Tree of Knowledge," Shambala, Boston.

Oliga, J., 1988, Methodological foundations of systems methodologies, *Systems Practice* 1(1).

Romm, N. R. A., 1991, "The Methodologies of Positivism and Marxism: A Sociological Debate," Macmillan, Basingstoke.

Stephens, R. A. and Wood, J. R. G., 1991, Information systems as linguistic systems: A constructivist approach, *in:* "Systems Thinking in Europe," Jackson, M. C., Mansell, G. J., Flood, R. L., Blackham, R. B. and Probert, S. V. E. (editors). Plenum Press, New York.

Stowell, F. A. 1991. Towards client-led development of information systems. *Journal of Information Systems* 1:175.

Stowell, F. A. and West, D. 1992, SSM as a vehicle for client-led development of information systems: Utilising "ideal-type" mode 2, *Systemist* 14:168.

Winograd, T. and Flores, F., 1987, "Understanding Computers and Cognition: A New Foundation for Design," Addison-Wesley, Wokingham.

KNOWLEDGE ELICITATION: FIRST STEP TOWARDS MANAGING FLOATING INFORMATION IN FINANCIAL SECTORS

Dr A. Behrooz & Ms G. Roushanbakhti

Department of Management Systems
Bournemouth University
Bournemouth BH12 5BB U.K.

INTRODUCTION

Banks have been faced with intense competition largely as a result of environmental changes, which includes the growth of non-bank institutions and Government deregulation, as well as changes in the social trends which have influenced customer demand. Marsh (1988) agrees that with respect to all these changes banks have had to develop their marketing skills to at least, maintain their market share and profitability levels.

Banks could not afford not to respond to these environmental pressures hence, they have been forced to review their objectives and goals to adopt corporate strategies that would help them gain and maintain sustainable competitive advantage in the financial market. Kenneth (1990) sees the introduction of marketing as an answer to an increasing competitive marketplace. In another article Joe Carter of Anderson's Consultancy has been quoted mentioning that the reason for the growth of AI is the firm's increasing emphasis on knowledge intensive strategies that rely on unique proprietary information (LaPlante, 1991).

It is not however, a new approach for organisations to concern themselves with developing strategies which would help them achieve a competitive edge. In the process of planning and implementation of corporate strategy formulation, banks have discovered that customer satisfaction should be the focus around which corporate decision making is to be based. At the same time banks are finally realising that to serve their customers better, they need to identify who their customers are together with the type of services they require. In order to avoid mistakes made in the past Read (1991) directs the attention of banks to a number of factors. One of these factors emphasises on banks identifying the markets they serve more clearly.

This in turn has meant that banks have had to analyse their micro-environment to measure their strengths and weaknesses and match these to the opportunities and threats offered and imposed on them by the macro-environment factors. Analysing one's position in the market place, and comparing the results with the competitors' performance, are only

Systems Science, Edited by F.A. Stowell *et al.*,
Plenum Press, New York, 1993

primary steps in the right direction, whereas determining solutions and achieving them are the answers to corporate success.

In discovering the role of customers, banks are slowly becoming familiar with the concept of marketing and they are beginning to recognise the importance of integrating marketing with their existing business activities.

To examine the development and use of the marketing information resource in the UK financial services, amongst other purposes, over 370 of the largest banks, building societies, and insurance companies were chosen to take part in a survey. It was depicted that only half of the organisations had formally organised marketing research units with a significant figure ageing under one year (Morgan, 1989).

Banks need to segment the market in order to recognise and satisfy the distinct requirements of the selected groups of customers. This would then enable them to differentiate their services from those of their competitors. Median (1984) refers to differentiation as a tactic used in selection of a marketing strategy to achieve long-term corporate objectives.

Having justified the essentiality of marketing in the banking sector leads to an equally important issue of collation and usage of information. This day in age one does not have to be an expert to be familiar with such statements as *"... information is the lifeblood of any business activity"*.

There has been a revolution in the acquisition, processing and retrieval of information over the years and the industry is undoubtedly indebted to the advent of Information Technology (IT).

Response to Environmental Changes:

In order to respond to environmental changes effectively IT has been adopted to support corporate strategies. Nevertheless, the only way to make IT effective is to increase efficiency in information handling. This argument is supported by Duffy (1990) where she emphasises that financial institutions will have to process the information needed for their strategies in an effective manner.

Until recently, IT has been treated more as a state-of-art to enhance the image of banks, but that is changing and with the advancement in the development of specialised software, banks are encouraged to use applications that would meet the different needs of their corporate clients. Although competition is intensifying more rapidly than banks are able to respond, banks have managed to respond to the rapidly changing environment by the introduction of *'off-the-shelf'* packages.

Technology has heralded new services and ways of delivering services to the corporate clients; some of these include such services as Tele-Banking which is similar in concept to Home-Banking in that the customer is able to access a number of banking services online.

The introduction of a new generation of databases however, has made way for a more detailed and up-to-date customer records.

This brought about a whole new way of dealing with customers in that a more friendly approach to the customer's enquiries was taken on the phone, *'Phone Banking'*, and the adaption of a *'Personal Touch'* to gain competitive advantage.

Technology has also made *'Direct Marketing '* possible which means when a customer calls and gives his/her account number, a detailed record of the customer is displayed on the screen. This enables staff to form a pattern on the customers' purchase behaviour hence, maximising efficiency and effectiveness of customer targeting by offering them appropriate services.

A more recent development is the arrival of EDI (Electronic Data Interchange) where the system allows for information to flow freely from the customer to the bank's internal processing unit and from there to the intended banks.

However, the most elaborated, but not so profitable, electronic service offered to corporate customers is known to be *'Cash Management'*. In addition to the speedy collection and concentration of funds to companies or individuals who wish to exercise a tighter control over their financial affairs, a tailor-made installation is capable of incorporating a variety of special features; such as timely forecasting of receipts and disbursement; transaction and balance information etc. Merrill Lych is one of the pioneers in the introduction of such a system, known as the "Cash Management Account" (CMA).

Home Banking (HB), on the other hand, is one of the significant signs to reflect how technology has been incorporated into the customer's changing lifestyle and expectations. HB services were initially targeted to attract affluent consumers who required the sophistication of such services and were able to afford it. Due to lack of a well thought out approach, not surprisingly, the system was not successful. Banks, therefore, had to re-direct their target market and concentrate on the needs of a new group of users, namely the small corporate customers, who were thought to benefit more from the services offered by the system.

The new generation of software applications have contributed to the arrival of Decision Support Tools. This is where a computer system is programmed to follow a set of predetermined rules in order to reach a certain conclusion. Banks may use this to speed up the process of credit checking and loan approvals, at the same time customers would benefit from added value offered by a speedy reply. However, it is ignored, in most cases, that this system is to be used as a support tool to the decision maker and not a replacement to his/her judgment.

The system is up-dated regularly so a customer making an enquiry is provided with up-to-date information, as well as allowing banks to save human resource and time in dealing with such enquiries.

This concept coincides with the development of Customer Information File (CIF). A properly designed CIF has given banks the opportunity of selecting groups of small business customers enabling them to create a marketing tool used to target potential customers within these groups. Alper (1988) referred to CIF as a system that allows marketing representatives to obtain a unified view of a customer's relationships. Some believe that CIF helps financial institutions in studying their total business strategy. Karush (1990) however, believes that CIF can assist in resolving a wide range of marketing problems, in particular development of better marketing strategies.

The development and use of IT in the provision of services is spreading in the financial sector, except for the fact that currently banks are concentrating more towards the adaptation of technology to their needs, rather than continuing an irrational spending on expensive equipment just for the sake of having such equipment.

Identification of Floating Information:

It is only recently that banks have turned to Expert Systems where human expertise and knowledge is used to solve problems. This is achieved by following a set of facts and rules. Recent emphasis on the application of AI in the financial sector indicates its impact in the sector, particularly its influence on strategy formulation as a key factor to gain advantage over competitors (Keyes, 1992).

Successful examples are: Arthur Anderson's "Mortgage Loan Analyser" (MLA), Masterplan- a planning system developed for planning, constructing, and controlling an internal audit (Brown and Phillips, 1991), Authorised Assistant System of American Express designed to help determine whether to authorise credit card transactions that exceed current credit limits (Mott, 1988). AI has also contributed to the insurance world (Gluch-Rucys and Walker, 1991).

The state-of-the-art that artificial intelligence has promised the users has however proved to be immatured. Many have not been able to deliver what they had been promising for decades, simply for two basic reasons:

Firstly, AI (and its public relation face - expert systems) initially pretended to know all the answers. The practioners in natural language processing, knowledge representation techniques, robotics and other sub-fields of AI spoke too soon, so far as the users were concerned.

Second, the knowledge that was formalised and codified by the programmers (knowledge engineers) though produced futuristic results, were far from realistic imitations **of the human** expert's manipulation of heuristic and factual knowledge. This was because of a limited investigation into knowledge elicitation and knowledge validation techniques. This does not mean that the techniques did not produce results, but to say that the results, which were ended up as formalised knowledge, were incomplete. Nor does it imply that the knowledge that was entered into the KBS was not validated, but it suggests that the formalised knowledge was not validated during and after the elicitation phase.

Limitations (human and technological factors) forces elicitors to reduce the domain knowledge so that a quick and commercial result can be achieved. Behrooz (1986), having discussed his experience as a knowledge elicitor in full detail, warns the society of AI researchers that:

> *" A very important factor that must not be ignored is that knowledge elicitation is an activity in its own right The methodology of this newly discovered domain is obviously not well known and may be unfamiliar to many scientists and experts. This first step is time consuming, painstaking and needs the full co-operation of both the contributing members of the knowledge elicitation interaction.Both the elicitor and the expert must understand that they do not deal with machines and computers, but with human being and their specific ways of interactions."*

It is however acknowledged that domains must be divided into sub- domains and an appropriate sub-domain must eventually be selected, but he emphasises that such an inevitable action should not affect the overall process of knowledge elicitation:

> *"It should not reduce the quality of the elicited knowledge and its formalisation*

It is therefore the author's view that we should begin to examine our own judgement on views regarding knowledge formalisation and representation techniques. And the first step is to understand and produce a practicable and realistic knowledge elicitation methodology.

There are however many techniques available which have proven to be pragmatic and (Behrooz 1987a;1987b; 1992), (Bond, 1990), (Welbank, 1987), and (Boose and Gaine, 1989). They have demonstrated that a practically acceptable expert system may be developed if the necessary steps towards formalising a given knowledge domain are taken. Behrooz defines these steps as:

- **1-Identification of all the contributing elements in the domain;**
- **2-Identification of all the relationships (internal/external) between every two contributing elements;**
- **3-Identification of knowledge domain as an element within a more complex meta-structure.**

If the above steps are fully implemented then the possibility of manipulating the floating information in an organisation would undoubtedly be easier to control.

REFERENCES

Alper, A., 1988, MIS in Banking: Banks Seek Higher Yield from Information Systems Investment, *Computerworld*, Vol. 22, Iss.33, pp65-68.

Behrooz, A., 1986, Meta Description of Experimental Identification of Medical Knowledge: PhD Thesis, Brunel University.

Behrooz, A., Anderson, J., and Kohout, L., 1987a, Identifying the Expertise of the Expert, *in*: AAMSI Congress Proceedings, Washington.

Behrooz, A. and Kohout, L., 1987b, Experimental Identification & Formalisation of Knowledge Structures Representing Uncertainty, *in*: "Proceeding of Management of Congress Cybernetics & Systems", Imperial College, London.

Behrooz, A., 1992, A New Knowledge Elicitation Method, *in*: "KBSs for Multiple Environments", L. Kohout, J. Anderson and W. Bandler, eds., Ashgate Publishing Ltd. Gower House, Aldershot, U.K..

Bond, A., 1990, The Co-operation of Experts in Engineering Design, *in*: "DAI", Vol.2, L. Gasser and M.N. Huhns, eds., Morgan Kaufmann Publishers, Los Altos.

Boose, J. and Gaines, B., 1989, Knowledge Acquisition of KBS: Notes on the State-of-the-art, *Mach. Learn.*, 3, 4, pp377-394.

Brown, C.E. and Phillips, M.E., 1991, Expert Systems for Internal Auditing, *Internal Auditor*, Vol. 48, Iss. 4, pp23-28.

Duffy, C., 1990, I.S. in Financial Services: Staying in Step with Times, *Computerworld*, Vol. 24, Iss. 20, pp83.

Gluch-Rucys, M. and Walker, D., 1991, Underwriting Intelligence, *Mortgage Banking*, Vol. 52, Iss. 3, pp60-66.

Karush, G., 1990, The CIF and the Market Planning Process, *Bank Marketing*, Vol. 22, Iss. 5, pp32-36.

Kenneth, A., 1990, Bank Marketing in a Changing World, *Int. J. Bank Marketing*, (U.K.), Vol. 8, Iss. 5, pp2-56.

Keyes, J., 1992, Winning Back Investors' Confidence, *Information Strategy: The Executive J.* Vol. 8, Iss. 2, pp42-44.

Laplante, A., 1991, Using Your Smarts, *CIO*, Vol. 5, Iss. 5, pp54-58.

Marsh, J.R., 1988, "Managing Financial Services Marketing", Pitman, London.

Median, A., 1984, "Bank Marketing Management", Macmillan, London.

Morgan, N.A., 1989, Developing Information Strategies in the U.K. Financial Sector, *Marketing Intelligence and Planning*, (U.K.), Vol. 7, Iss. 7/8, pp24-28.

Mott, S., 1988, The Expert Back Office, *Institutional Investor*, Vol. 22, Iss. 14, pp7-9.

Read, C., 1991, What Did We Learn from the 1970's in Preparing for the 1990's?, *World of Banking*, Vol. 10, Iss. 2, pp24-27.

Rosenthal, B.E., 1990, Structuring Your Strategy with Computer Information Files, *Banking Software Review*, Vol. 15, Iss. 3, pp18-22, 27-29.

Schmidt, G., 1990, I.S. in Financial Services: Future Hinges Hinge on Present I.S. Choice, *Computerworld*, Vol. 24, Iss. 20, pp92-93.

Schoeck, M., 1989, Gaining Strategic Advantage Through Information Systems, *Ohio CPA J.* Vol. 48, No. 1, pp45-46.

Welbank, M., 1987, Perspective on Knowledge Acquisition, *in*: Proceedings of the SERC Workshop on KA for engineering applications, C.J. Pavel and M.D. Wilson, eds., Rutherford Appleton Laboratory Report, RAL 87055, pp14-20.

INFORMATION REQUIREMENTS:
A SYSTEMS AND KNOWLEDGE BASED APPROACH

Louis C-Y Cheung, Tony Holden

Cambridge University Engineering Department
Trumpington Street, Cambridge CB2 1PZ, UK

INTRODUCTION

Our research on information systems (IS) focuses on the upstream design activities of analysing the problem situation and defining the requirements. To this end, we propose a *systems and knowledge* paradigm (Cheung, 1992) as the theoretical basis. This paradigm recognizes the ill-defined nature of real-world situations, as well as the complexity and uncertainty which characterize the IS design process. Soft systems practice provides the epistemology to intervene in the concerned situation. The explicit role of knowledge helps us to understand the cognitive aspects of design, and the support required by the designer.

SYSTEMS AND KNOWLEDGE PARADIGM

Relevant lessons from systems thinking, AI (artificial intelligence) and theory of knowledge together lead to a coherent *systems and knowledge* paradigm to understand IS design. The *systems* component is characterized by soft systems thinking, while the *knowledge* component is concerned with suitable representation to aid understanding and exploration.

Systems Thinking

Modern systems researchers have extended systems ideas to real-world problem solving practice. To deal with problems of organized complexity, a holistic (rather than reductionist) approach is necessary.

Checkland (1981) proposes a systems typology which identifies the minimum four systems classes as natural systems, designed physical systems, designed abstract systems and human activity systems. Regarding designed systems, man as designer can be considered a teleological being who is capable of creating means for ends to be pursued. In contrast to natural systems, human activity systems can be adequately described only with respect to a particular *Weltanschauung* (worldview).

Systems Science, Edited by F.A. Stowell *et al.*,
Plenum Press, New York, 1993

IS are basically designed abstract systems, being manifested in designed physical systems (i.e. the computer hardware and software). However, they serve the information needs of the organization which belongs to the class of human activity systems. The human act of design itself, which operates in real-world situations, is also a kind of human activity system.

Checkland (1981) proposes the Soft Systems Methodology (SSM) as a cyclic process for enquiring and learning. It makes use of systems models (holons, or constructed abstract wholes) to study the purposeful human activity systems.

The SSM process starts with a situation which is considered problematical. The logic-driven stream of enquiry names (by root definitions) and models the relevant human activity systems. Selected models (as purposeful holons) are compared with the perceived part of the real world to structure a debate about change, i.e. to seek an accommodation among different interests in order to improve the initial problem situation. Cultural knowledge is required in reaching the accommodation and the motivation to action. The cultural stream consists of three examinations: the intervention itself, the situation as a social system, and the situation as a political system. The general models used relate respectively to problem solving, the social process and the power-based aspects of human affairs.

Checkland (1988) argues that the work on IS has so far followed the hard systems tradition, which assumes the existence of well-defined objectives and seeks to evaluate alternatives to meet the objectives. However, current problems about IS provision in organizations need to be addressed by soft systems thinking. The *project* approach needs to be complemented by the *process* approach which generates continuous rethinking of the organizational tasks and information flows in a changing environment.

AI and Theory of Knowledge

One major outcome of AI research on problem solving is the search model which corresponds well with the human problem solving behaviour in well-structured task environments such as logic and chess. A search problem generally consists of a complete and consistent goal state, and well-defined heuristics to search the known problem (and solution) space.

In contrast, design cannot usually be approached by applying a straightforward search. It is an ill-structured problem, though some of the tasks within each particular design process may be well-structured.

Therefore the latest AI-based research on design focuses on the nature of real-life design as a knowledge-based exploration task. To cope with imperfect knowledge, the designer has to explore the problem domain and go through numerous iterations and refinements. The design objectives need to be reviewed during the process. To manage design complexity, the designer has to reason with the designed artefact at different levels of abstraction.

Progress in the uncertain design process depends on the availability of both domain and design knowledge. The quality of design depends critically on the representation and analysis of such knowledge. Thus a representational scheme with sufficient expressive power is essential in supporting the designer to organize the knowledge and complete the design tasks.

Williams (1990) proposes an integrated model of design reasoning, which agrees well with other theories of knowledge, to describe how designers overcome the generic problems of complexity and uncertainty. The model features the following elements:

Problem situation: appreciation of a situation which the designer wants to affect.

Generalized models: idealized mental constructs which organize human experience and expertise, both theoretical and practical.

Specific model: activation of proposals from generalized models, while forming an explanation of the problem situation.

Proposal: selection among proposals for actions towards the intended goal.
Understanding: use generic knowledge to form a specific model of the problem situation.
Conjecturing: put forward a proposal based on the specific model.
Action: proposal guides actions on the problem situation.
Learning: adjust the generic knowledge if the actual results are not consistent with those expected.

Implications for IS Design

With rapid technological development, there is an ever-increasing demand for sophisticated IS which penetrate deeply into various organizational functions, and affect multiple levels in the management hierarchy. This inevitably brings IS designers into broader organizational concerns.

Firstly, the boundary of IS must include the attribution of meaning by the concerned users. In this enlarged problem situation, the formulation of desirable ends becomes itself problematical. Secondly, the dynamic nature of modern business environments presents challenges of equally dynamic IS support. The structured analysis and design techniques, which are dominated by logical and technical concerns, provide only a partial solution (Vitalari and Dickson, 1983). The lack of concern for social and political issues often results in pre-mature judgement by the IS designer.

SSM as an enquiring system facilitates *short-term* iterative learning in the IS study, and helps to bridge the gap largely neglected by the hard tradition. While models (purposeful holons) are purely intellectual constructions for further investigation, we must recognize that the world is always richer and messier than the pure concept. Therefore, the conceptual models being constructed are not strictly models of real-world activities; rather they are models relevant to debating about them.

The concept of modelling in traditional IS work has its major emphasis on representation and abstraction. Representation is the essential step to express the designer's understanding and interpretation of the real-world enterprise. This is where AI comes in (Cheung et al., 1991a) and supplies a whole variety of knowledge representation schemes for capturing the rich semantics of the world. Abstraction is the cognitive aid to help the designer to manage complexity and concentrate only on the relevant details at one time.

However, this trend of IS modelling (seeking to employ models to map onto aspects of perceived reality) risks the danger of equating an abstract epistemological device to an assumed ontological entity. The choice of relevant systems to describe human activities is essentially subjective. This leads to another dimension of design uncertainty, in parallel with the process uncertainty arising from the lack of perfect knowledge. Also, continuous learning in the process is bound to affect what turns out to be feasible at the end.

With a soft framework to embrace IS modelling, the designer tries to understand the perceived world using defined concepts, but recognizes that the world is also the source of all concepts. Thus each creates the other in a never-ending process, which is fundamental to learning. This emphasis of SSM on *learning* bears some basic similarity to *exploration* in the AI-based design models, although the role of knowledge is not explicitly stated. The similar characteristics include:

- No complete design goals or requirements available from start.
- Often imperfect knowledge about the problem space.
- Emphasis on each particular problem situation and design context.

In general, the AI-based approach provides more insights on the cognitive aspects of design, especially in its attempt to understand how knowledge is organized and manipulated in ill-structured design problems. There is also the attempt to represent knowledge in computational terms. Three elements of knowledge are readily identified in Williams's reasoning cycle, which agree well with the AI description of knowledge (Brachman, 1979) in terms of generic concept, specific concept and actual reality.

One long-term goal of AI is recognized as building a knowledge medium to make knowledge widely accessible. Given the present state of technology, it is not yet possible to construct fully automatic tools to replace the human designer. An alternative is the *assistant* approach (Cheung et al., 1991b; Holden et al., 1990) which aims at a human-computer partnership to tackle the complex and uncertain design tasks. A design support tool must be flexible enough to allow the designer to fully exercise his own reasoning power.

INFORMATION REQUIREMENTS

Following the systems and knowledge paradigm, we propose an intellectual framework to guide the design activities related to requirements elicitation. The designer interacts with the real-world situation to specify IS requirements, which are regarded as the means for improving the situation. To tackle the problem situation, the designer is engaged in a process of continuous negotiation with the user, until both parties become interested in and committed to a common problem, as well as a mutually acceptable solution. This process is assisted by a set of model building and knowledge representation techniques.

Our framework incorporates a model of design which is based on recursive application of the reasoning cycle augmented with soft systems thinking, as shown in Figure 1. Systemic enquiry into the situation achieves short-term iterative learning by means of abstract systems models, in order to propose desirable and feasible changes. Design reasoning, as *informed* by the systemic enquiry, applies generic and specific design knowledge (with feedback and goal

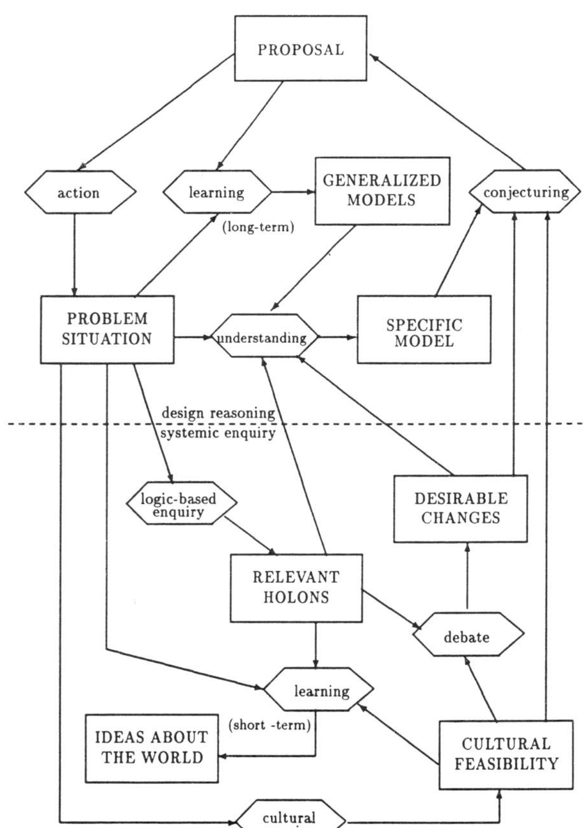

Figure 1. The design reasoning cycle augmented with soft systems thinking.

refinement) to produce the designed artefact. This model of IS design agrees well with other empirical research results (Guindon et al., 1987) and our own experience in several case studies.

Relevant parts of the generalized knowledge (i.e. the generic design concepts and techniques) is activated whenever there is the need to capture IS requirements according to an agreed *Weltanschauung* for the perceived situation. The specific model thus formed is effectively a collection of requirements captured in some IS models. Proposals raised according to each specific model can invoke another soft systems enquiry cycle, and are subject to debate about systemic desirability and cultural feasibility. Also, long-term learning is achieved by enriching the generalized models which are applicable across a range of problem situations.

In requirements elicitation, model building is especially important in organizing knowledge which is often complex, imprecise and context dependent. Better representation results in better *cognitive fit* to improve problem solving performance. Also, the explicit representation of mental models serves to stimulate thinking and facilitate communication.

The generalized models and specific models of design, as described in the reasoning cycle, form the two fundamental layers of models. The layer of generalized models contain the structured conceptualizations, as meta-models, of both design states and design processes (i.e. transitions between states). They represent the idealized mental constructs to organize generic design knowledge. According to different design purposes, the layer of specific models is further classified into four types:

Activity model: capturing the minimum necessary activities to satisfy the root definition, thus setting the problem scope for any potential IS study.

Information model: providing a high-level view of the minimum information resources to serve the set of activities described in the activity model.

Requirements model: bridging from a model of the relevant human activity system to a requirements model of the designed system.

IS model: representing the structure and behaviour of the IS to be realized.

The various model types are not to be taken as strictly consecutive phases. They can be developed concurrently, with interdependent concerns. Model building encourages rigorous thinking, and the models actually shape the thinking process. The commitment to a certain model implies that the designer is consciously filtering anything that does not fit the model structure. This can be compared to the linguistic principle that how we perceive reality (i.e. our habitual thought and behaviour) is heavily influenced by the language we use.

The information model links the relevant human activity system (represented by the activity model) to the potential IS application, so that both can be subject to debate. Between the information and requirements models, there is a conscious transition from human activity systems to designed systems. The requirements model also provides a focus for high-level design decisions to be structured to support construction of the detailed IS model.

The debate about change involves subjective reasoning with both logical and cultural aspects. Also, the design study raises various issues which require decision making. During the design exploration, multiple scenarios are constructed. The reconciliation between separate scenarios lead to the recognition of further design issues which must be resolved.

To cater for such needs, there is a third layer of models to evolve in parallel with the development of the specific models. This layer represents the subjective reasoning to facilitate the negotiation process. It also captures the pertinent design issues and their deliberation, as an important part of the design documentation.

Knowledge Based Support

Knowledge sharing becomes possible if the computer can communicate mental models similar to those of the human designer. So far there is still no unified model in psychology about concept storage and development in human learning. Our research seeks to explore the usefulness of symbolic representation of knowledge with the latest AI technology.

To manage the inherent complexity, the modelling process is supported by various knowledge representation techniques, in particular object orientation and conceptual graph. Specific models are formed as a result of successfully understanding the situation. The different layers of models together constitute an overall *concern net* which is explicitly maintained throughout the design process. This makes it easy to organize and re-organise the design documentation.

We advocate cognitive support for the designer in three ways. The support for design exploration guides the acquisition and organization of systems concepts about the problem situation. The support for problem negotiation enables the designer to formulate design issues and structure the deliberation process. The support for generic understanding helps inexperienced designers to build up generic knowledge required to complete the design tasks.

Following the assistant approach, a comprehensive support environment can be realized by the latest knowledge-based technology. Our prototype support system (Ip et al., 1992) is implemented on a LISP machine with object-oriented and rule-based software.

CONCLUSION

In the above discussion, we have shown how an abstract understanding of the early, upstream design process leads to a coherent systems and knowledge paradigm, which in turn provides an intellectual framework to guide the complex and uncertain design activities. The framework also incorporates the necessary techniques and tools to provide cognitive support for the IS designer.

REFERENCES

Brachman, R.J., 1979, On the Epistemological Status of Semantic Networks, *in:* "Associative Networks," N.V. Findler, ed., Academic Press, New York.

Checkland, P.B., 1981, "Systems Thinking, Systems Practice," Wiley, Chichester.

Checkland, P.B., 1988, Information systems and systems thinking: time to unite? *International Journal of Information Management,* 8:239-248.

Checkland, P., and Scholes, J., 1990, "Soft Systems Methodology in Action," Wiley, Chichester.

Cheung, L.C.Y., 1992, Towards a systems and knowledge paradigm for information systems design, *Research Report, Cambridge University Engineering Department.*

Cheung, L.C.Y., Ip, S., and Holden, T., 1991a, Survey of artificial intelligence impacts on information systems engineering, *Information and Software Technology,* 33(7):499-508.

Cheung, L.C.Y., Ip, S., and Holden, T., 1991b, Knowledge representation to aid the information systems design process, *in:* "Proceedings of the 4th International Symposium on Artificial Intelligence," F.J. Cantu-Ortiz and H. Terashima-Marin, eds., Editorial Limusa, Mexico.

Guindon, R., Krasner, H., and Curtis, B., 1987, Cognitive processes in software design: activities in early, upstream design, *in:* "Human-Computer Interaction, Proceedings of INTERACT'87," H.J. Bullinger and B. Shackel, eds., North-Holland, Amsterdam.

Holden, T., Cheung, L., and Ip, S., 1990, Intelligent support for the information system design process, *European ART User-Group Conference*, Rome, Italy.

Ip, S., Cheung, L.C.Y., and Holden, T., 1992, A knowledge-based requirement engineering assistant, *in:* "CASE: Current Practice, Future Prospects," K. Spurr and P. Layzell, eds., Wiley, Chichester.

Vitalari, N.P., and Dickson, G.W., 1983, Problem solving for effective systems analysis: an experimental exploration, *Communications of the ACM,* 26(11):948-956.

Williams, D.O., 1990, Developing systems for supporting design understanding, *PhD Thesis, Cambridge University Engineering Department.*

THEY SHOOT WEREWOLVES, DON'T THEY ?

Kevin G. Doyle

Department of Computer Studies
The University of the West of England (Bristol)
Coldharbour Lane
Frenchay
Bristol
BSI6 lQY

INTRODUCTION

This paper discusses the way that systems development methods have emerged in an attempt to solve the problem of the "software crisis", and questions whether information systems developers really understand the nature of the problem that they are trying to solve.

Using the current debate over whether object oriented design is a "silver bullet" with which to slay the beast that plagues the software industry, it seeks to show that in many instances not only are claims regarding improvement in the development and delivery of information systems exaggerated, they also indicate fundamental misunderstanding of the nature and role of information systems in contemporary business and commerce.

The paper concludes that the next great revolution in the design of information systems will come, not from the move from data orientation to object orientation, but from a paradigm shift involving a move from the objective view, typically embedded within an information system, to the subjective. Examples of how this might be achieved in practice are given, with reference to current action research based at Bristol.

THE SOFTWARE CRISIS

Planet Earth is over four thousand and six hundred million years old. Almost nothing is known about the first seven hundred million years of Earth's existence, and no coherent body of knowledge exists in relation to the next thirty five hundred million. We believe that the planet began to flower about four hundred million years ago and that dinosaurs and the great reptiles appeared around two hundred million years later. Mammals arrived sixty million years ago and Homo Sapiens have been around for barely a quarter of a million years.

Computers have been around for about forty five years, the "information economy" began in the fifties, it was noticed in the sixties and widely accepted in the eighties. The next

Systems Science, Edited by F.A. Stowell *et al.*,
Plenum Press, New York, 1993

economy, the so-called "bio-economy" is in sight (Davis, 1991). During the age of the information economy the application of computers and related technology has had a wider, and more profound, effect on mankind and, through mankind, the planet Earth, than anything else in history.

The term *"software crisis"* was coined during the NATO Software engineering conference of 1968 to indicate that software projects often ran late, cost far more than expected and frequently did not meet the needs of the client. Some seven years later in "The Mythical Man-Month" (Brookes, 1975) a book written in his own blood, Fred Brookes reminded us that the only unforgivable failure is the failure to learn from our previous mistakes, and yet the software crisis persists. Software is still scarce and often fails to meet user expectations. Projects continue to use much more in the way of resources than expected, and are notoriously difficult to manage. Recent research by the Kobler unit at City University discovered that very few chief executives felt that they received value for money from their investment in technology and that total failure to meet the requirements of the client is still common (UNICOM, 1992).

ESSENCE AND ACCIDENTS OF SOFTWARE ENGINEERING

In his "No silver bullet" paper, Brooks examines the difficulties of software systems development. Following Aristotle, he divides them "into *essence,* the difficulties inherent in the nature of software, and *accidents,* those difficulties which today attend it's production, but which are not inherent." (Brookes, 1987).

Brookes sees the essential difficulties as complexity, conformity, changeability and invisibility, and argues that the hard part of building software is the specification, design and testing of the conceptual structures that form the software entity, and not the task of representing them or testing the fidelity of the representation.

Brookes uses the werewolf as a metaphor for the software project which changes without warning from innocent and straightforward to a nightmarish monster of "missed schedules, blown budgets, and flawed products", and thus as a metaphor for the software crisis itself.

In seeking silver from which to fashion a bullet with which to slay the beast, Brookes concludes that there is no silver bullet, but that we can make real progress towards handling the essential complexity of software development by, where possible, buying rather than building, using prototyping to establish the requirements of the client, growing software organically, and by identifying, nurturing and developing great conceptual designers.

THE SEARCH FOR SILVER (buddy can you spare a paradigm)?

Despite the assertion by Brookes that there is no silver bullet, the search for silver continues. The evolution of structured development methodologies, combining methods, techniques and tools, within the framework of an approach, is an attempt, through the focus on process, to solve the problems of the software crisis. An abundance of such methodologies is available, SSADM, STRADIS, JSD, Information Engineering and MERISE, to name but a few. Structured Development Methodologies have been compared (Olle, 1982) analysed (Olle, 1983) evaluated (Conner, 1985) front-ended, back-ended, merged and blended, but still the crisis persists.

Almost two years ago Byte published an article by Brad Cox entitled "There *Is* a Silver Bullet" (Cox, 1990). In this article Cox explains that the silver bullet in question is a cultural change, a *'paradigm shift",* which moves the focus from the process of software development to the products that are required by the marketplace, using object-orientation.

Under this new paradigm, Cox argues, building software applications will be like plumbing a house. Reusable software components, based on object-oriented technologies like Objective-C, will be available at a number of levels, from "gate and block" to whole applications. Cox does, however, admit that his studies have revealed how difficult it is to design and build components that are genuinely reusable.

The author believes that there is a need for a paradigm shift, and that the move that Cox seeks, from processes to product will occur. That technology will become ubiquitous and "ready to hand" or even *"ready to mind"*, (Banathy, 1990) and that the **market perspective** will move from process to product is as inevitable as was the move from cottage industry to industrialisation. This is not a paradigm shift, it is merely a shift in emphasis, focus and practice. As Brookes has said the hard part of building software is the specification, design and testing of the conceptual structures that form the software entity, and not the task of representing them or testing the fidelity of the representation. The "industrialisation" of software production and the assembly of common building blocks to suit particular requirements is, therefore, not the answer to the problem of the software crisis.

Whilst object-orientation may be the silver bullet that will slay the **software project** werewolf, it will not end the software crisis. The software crisis will be with us as long as we continue to develop software systems as though this process were equivalent to the engineering of some physical artefact. If a paradigm shift is required, then it is one which replaces science, objectivity and engineering with systems thinking, subjectivity and evolution. Cox is right to assert that an emphasis on the process of systems construction will not solve the software crisis, but neither will the development of an infinite number of re-usable software components. At the heart of the crisis lies the fundamental *(essential)* dichotomy between the subjective nature of perception, and the objective models used to construct software systems. Until computer based information systems are able to support multiple views of reality they will continue to fail to meet user requirements. (Doyle, 1990)

OBJECTS AND SUBJECTS

When we design and construct an information system, it is necessary to embed and support models of the world that exists outside that information system. Indeed, it is the subject matter of the information system that is its raison-d'étre. In some circumstances this subject matter is tangible in nature and exhibits patterns of behaviour and qualities that are universally accepted and may be defined in absolute terms with precision and certainty. Such subject matter may easily be conceptualised as a series of objects. Such objects may be conceived of as a series of classification schemes, either hierarchical or overlapping, because they exhibit behaviour or qualities that may be commonly attributed to a definitive class. Other subject matter is intangible but may still be usefully considered as a collection of objects because of the universal accord over the meanings and structures associated with

A substantial amount of the subject matter that lies at the heart of many information systems does not, however, relate to universally accepted and consistently defined objects. On the contrary, it relates to material that only has meaning within the context of the wider human activity of which it forms part, and the purposes of which it is intended to serve. The subjective nature of "social reality" demands that information systems are developed to fit their social context, with all the inherent difficulties that this implies. Naturally users of Information Technology will seek simple and cost effective solutions to their particular requirements, and where this can be accomplished through the acquisition of useful hardware/software solutions, which are in existence either as modules or complete systems, such acquisitions will take place. Application packages, systems shells and business pattern software are commonly available, and widely used. It is not uncommon for vendors to make

4GL source code available to the purchaser of such software, along with systems documentation and, if required, the development staff to extend and enhance the base product.

The concept of software re-use is not new but an exhaustive and definitive set of software components as a panacea is both unattainable and undesirable.
If one takes a subjectivist stance, clearly the subject matter of "reality" can not be adequately represented in richness and complexity by a single objective model. The use of existing hardware/software environments forces the developer to rationalise and condense, and usually to impoverish, the richness of the world. The tools and techniques of structured analysis and design encourage, and enable, this transformation to take place during modelling by ignoring all but that which is definable by the tools used.

MULTI-PERSPECTIVE AND CONSENSUS FRAMEWORKS

There has been much interest recently in the theoretical and practical implications of grafting or embedding hard and soft approaches to the development of Information Systems. The desire to "Do the Right Thing Right" (Wood, 1989) is a powerful one, which has resulted in a number of multi-perspective and consensus frameworks. (Wood-Harper et al., 1985), (Wilson, 1984). The advocates of these frameworks can be criticised for attempting to merge methods that have different ontological underpinnings (Doyle, 1991a) and ignoring epistemological differences (Sawyer, 1991) in an attempt to bridge the gap between systems thinking and engineering design (Doyle, 1991b).

These frameworks generally require consensus or agreement over the models that the system under development will support: this is not the same thing as correctness. For the development of transaction processing and other operational level systems this approach is likely to be adequate. For management information systems involving support for non-programmable activities such as exploration and sense-making at the tactical and strategic level it is likely that consensus between stakeholders over a single set of objective, sufficient and correct models will be much harder, perhaps impossible, to achieve.

CAPTURING AND RETAINING SUBJECTIVITY

The authors current research is founded on the notion that in attempting to capture, represent and reflect the subjective nature of our views of the organisational reality in which we exist it is not sufficient to use methodologies, tools and techniques for systems specification that are rooted in subjectivity and/or the world of systems thinking, and then to channel and condense these subjective models into objective models that can then form the basis of the computer based information system. What is necessary is not a bridge between subjective human perception and objective computer software systems, but a set of technologies that are able to capture and retain subjective perceptions and requirements. (Doyle, 1992).

The research, which involves the specification of an integrated clinical workstation for the neurosurgery department of Bristol's Frenchay Hospital, will explore the means by which technology may be more effectively matched to human perceptions and requirements by consideration of temporal, spatial, application, and interpretative requirements.

The underlying information architecture will be based upon three dimensions. The first dimension deals with form, and consists of data, text, sound and image. The second dimension deals with function and considers generation, processing, storage and transmission. The third dimension concerns time, activity, space, and interpretation.

			Interpretation	**Generation**
		Space	Generation	Processing
	Activity	Generation	Processing	Storage
Time	Generation	Processing	Storage	Transmission
Data				
Text				
Sound				
Image				

The three dimensions considered above provide the beginnings of a framework in which to consider multiple perspectives. For instance consideration of the data/storage cell might suggest the following requirements.

Time — The use of event-bases rather than data-bases, in order to track changes in entities, attributes and relationships. (Waters, 1989).

Activity — The ability to support different modes of activity e.g. processing, audit, enquiry, exploration and proofing, as well as operations.

Space — Support for geographical considerations like national laws or local policies.

Interpretation — Support for a range of applications with different, perhaps even conflicting objectives and requirements.

CONCLUSION

In the film "Love at First Bite" there is a scene in which the vampire is shot through the heart with a silver bullet. Shrugging its shoulders at its adversary, the monster tells him *"silver bullets are for werewolves stupid!* If we expect object-orientation to end the software crisis we are likely to be similarly disappointed.

REFERENCES

Banathy, B.A., 1990, "A Perspective On The Design Of Evolutionary Information Systems", ISSS Conference, Portland, Oregon, pp392-400.

Brookes, F., 1975, "The Mythical Man Month" Addison Wesley, Reading, MA.

Brookes, F., 1987, "No Silver Bullet: Essence and Accidents of Software Engineering" IEEE *Computer,* April 1987, pp1069-1076.

Connor, D., 1985, "Information Systems Specification and Design Roadmap" Prentice Hall, Englewood Cliffs, NJ.

Cox, B., 1990, "There Is a Silver Bullet": *Byte,* October 1990, pp209-216.

Davis, S. and Davidson, B., 1991, "2020 Vision", Business Books Ltd, London.

Doyle, K.G., 1990, "Can Computer-Based Information Systems Support Multiple Views of Reality", ISSS Conference, Portland, Oregon, pp1172-1178.

Doyle, K.G. and J.R.G.Wood, 1991a, "Systems Thinking, Systems Practice: Dangerous Liaisons" Systemist, Vol. 13, pp 28-30

Doyle, K.G. and J.R.G. Wood, 1991b, "The Paradigm Shift & The Protozoon", *Systemist,* Vol 14, pp131-134.

Doyle, K.G., 1992, "Capturing and retaining subjectivity in systems development", ISSS Conference, Denver, Colorado, pp937-945.
Olle, T. W., ed., 1982, "Information Systems Design Methodologies: A Comparative review". North Holland, Amsterdam.
Olle, T. W., ed., 1983, "Information Systems Design Methodologies: A Feature Analysis". North Holland, Amsterdam.
Sawyer, K., 1991, "Linking SSM to DFD's: The Two Epistemological Differences" Systemist, Vol 13, pp76-80.
UNICOM, 1992, "Managing and Evaluating the IT Investment", UNICOM seminar, London.
Waters, S.J., 1989, "From Databases Towards Knowledge-Bases" in Knowledge-Based Management Support Systems, Ellis-Horwood, London.
Wilson, B., 1984, "Systems Concepts, Methodologies and Applications" John Wiley, Chichester.
Wood, J.R.G. and Doyle, K.G., 1989, "Doing The Right Thing Right: An Exploration Between Soft Systems Methodology and Jackson System Development". ISGSR Conference, Edinburgh, Vol. IV, pp146-155.
Wood-Harper, T., Antill, L., and Avison, D.E., 1985, "Information Systems Definition: The Multiview Approach", Blackwell Scientific Publications, Oxford.

FUNCTION-ORIENTED DATA MODELLING

Jun-Kang Feng

Information Systems
Department of Computing Science
University of Paisley
Paisley, PA1 2BE, Scotland

INTRODUCTION

This paper presents a method that is aimed at overcoming the following problems with contemporary data modelling methods:

• Lack of concepts for analysing *a system* being modelled and *data* used within it. This hampers the identification of data.

• Lack of mechanisms for revealing the relationship between *data* and real world *objects*. This results in the interpretation of data being conducted intuitively.

• Lack of systematic modelling procedures. This leads to data modelling being a 'black art'.[1]

This work is inspired by Rund's work,[2] and is developed by using both the *hard systems thinking* and the *reductionist analytical thinking*.

As argued in another paper of the author,[3] a *complete* modelling method must consist of a fundamental view of a universe of discourse, a modelling formalism and associated modelling procedures. This method takes two basic viewpoints. First, the system being modelled must be systematically analysed. Database systems are usually developed to meet the data needs of a human organisation. In most cases, such a human organisation can be viewed as a procedural system, therefore hard systems thinking is applicable. Second, in order to map the reality onto the constructs of a data modelling formalism, the relationships between a system being modelled, the data being used in the system and real world objects involved in the system must be revealed.

According to these basic viewpoints, a modelling formalism and a modelling procedure are developed, all of which put together constitute a complete data modelling method.

A MODELLING FORMALISM COMPOSED OF THREE LAYERS

Strictly-typed data modelling formalisms employ a set of basic concepts, namely *entity, attribute* and *relationship*[4]. The idea is that categories of data, namely *sets* and *relations* of data should be interpreted in terms of objects, their properties and associations. However, these concepts are general and ambiguous. They help neither analyse the system

Systems Science, Edited by F.A. Stowell *et al.*,
Plenum Press, New York, 1993

being modelled nor reveal the relationship between the functionality of the system being modelled, data used within the system and the real world objects. Therefore they alone do not constitute an adequate modelling formalism.

Our solution is to have developed an enhanced modelling formalism composed of three layers. The basic semantic concepts are now treated as the low layer, which are concerned with the building blocks of a data model. And two more higher layers of concepts are developed to fill the gap between the system being modelled and the data model.

The high layer of concepts is concerned with the functional analysis of the system being modelled, which consists of the following concepts: *system being modelled, root definition of a system, functionality-decomposition, task,* and *task-data pair.*

The middle layer of concepts is concerned with the revealing of the relationship between data and objects, which consists of the following concepts: *primary data, auxiliary data, parameters of data, empirical distributions of data parameters,* and *criteria for data classification.*

Using this modelling formalism, a systematic modelling procedure is developed. It consists of three stages, namely *data identification, data interpretation and structuring,* and the *implementation design* of a data model.

DATA IDENTIFICATION

With this method, a data modelling process starts with data identification via an analysis of the functionality of the system being modelled using the high layer of concepts. It creates a preliminary model of an organisation.

A system is composed of a hierarchy of sub-systems and can be described at different abstract levels.[5] We use *data flow diagram* to exercise the functional analysis. First, a root definition of the system being modelled is specified, which results in a function at the top level. Then this top level function is decomposed in a top-down manner till the lowest level, namely tasks. A task is defined as: A unique unit of work; Consisting of a set of serially performed steps; All steps directed towards a common goal; All steps utilising or creating a common set of data.

Essentially, what a task does is:

i. To identify a particular object or a particular set of objects on which some particular work needs to be done.

ii. To do something about them in order to achieve a particular goal, during which some extra information may be needed or created.

Note that first of all, tasks must be defined as being mutually exclusive; secondly, a task must be aimed at a single minor goal; and finally and most importantly, it must use or create a set of data from the very beginning to the very end of the task.

To reflect the relatedness between data and functions, and between data items themselves, a particular data structure called *task-data pair is* used. It is defined as the unique relationship created between data items when they are used together to fulfil a specific task.

When data within a task are identified, only the *necessary* and *sufficient* data for the task qualify. These data are of one of two kinds: either those that are used tor determining *objects* to work on, or those that are used for achieving a particular goal other than determining the objects. The result is the largest set of necessary data and the smallest set of sufficient data for the fulfilment of a task. No more, no fewer.

Note that first, we are dealing with *data types,* rather than *data tokens*. Second, data can be either *elementary data* or *aggregate data.* Whether a group of data should be defined as an aggregate datum depends on whether they are used as a unit when functions are performed. Finally, data should be defined to be mutually exclusive, because otherwise it would make little sense to compare them with each other.

Here a distinction must be made clear between an object, data that are a description of the object, and data that are only part of the description but that represent the object when we deal with it. For example, a particular hall, say, hall A is an object in reality. Here *hall* A is a description of the real hall A and it is a data token. The generalisation of individual hall data tokens is data type *hall,* which is a description of a class of real objects -- halls. A hall is composed of components, so *hall is* also an aggregate of a number of data types such as *hall-id, hall-address* and *hall-details.* When we refer to objects halls, we often use *hall-id,* which is, however, merely part (though a representative part) of the description *hall,* rather than the aggregate data type *hall* itself.

Understanding this distinction is vital, because the idea of data modelling via functional analysis is to reveal hidden objects and properly define aggregate data types for them by looking at the data used in functioning. These data are most often merely representations of these objects. We shall say more about this later.

DATA INTERPRETATION AND STRUCTURING

Have been identified, data need to be structured into a data model. The nature of this stage is to obtain a *sensible interpretation* of data captured. This is achieved using the middle layer and the low layer of the modelling formalism. At this stage, a bottom-up approach is adopted. The idea is to describe *usage* of data mathematically by using reductionist thinking.

Through functional analysis, tasks, data, and the link between them have been identified. We term this link *usage of* data, and it has two important aspects. One is how a datum is used by tasks. The other is how a datum is used with other data. Obviously, 'usage of data' is not a 'random event' in mathematical sense. However, according to Rund[6] and in our own experiments, some underlying trends concerning usage of data were revealed. These trends can be well explained using systems theory and data modelling theory. These trends define a set of *probabilistic* concepts[7] about data, namely *importance, independence* and *closeness.*

Parameters Of Data

From the point of view of the relationship between data and objects, as aforementioned, a datum is either a complete description of an object, a representative part of the description, or an ordinary part of the description.

From the point of view of data being used by a task, however, there are only two categories of data, one is those used in determining a set of objects to work on, the other is those used as auxiliary material for achieving a particular goal of a task.

These two viewpoints fit together nicely in data modelling, because a datum used in determining a particular set of objects within a task is always a datum which is a representative part of the description of an object, and a datum used as auxiliary material in achieving a particular goal of a task is always a datum which is an ordinary part of the description of an object. From now on, we will term the first category of data *primary data,* the second *auxiliary data.*

Furthermore, the functioning of an organisation can be regarded as taking a series of different actions on a set of objects, and these actions are organised in various *tasks.*

The set of basic objects is stable, while tasks vary. This means that the same set of objects (though often in different combinations) are worked on by all tasks. Given that a primary datum is used for determining an object, it is expected to be involved in many tasks.

Tasks must be different because of their distinct goals. To achieve their different goals, it is highly likely that tasks use very different auxiliary data although they possibly use different combinations of primary data as well. The consequence of this is that a primary datum is used with many other data; And moreover, because usually it is used with a group of different data in different tasks, it does not appear to be closely related to any group of them, therefore it is independent.

In the case of an auxiliary datum, the situation is just the opposite. Because it represents a particular aspect of an object, it is usually used only by a small number of tasks, and therefore used with a small number of other data and appears to be closely related to them, which means it is less independent than a primary datum.

All these show that we can classify a datum to be either a *primary datum* or an *auxiliary datum* by looking at how widely it is used and how independent it is. In fact, three parameters were used to describe these two points, and in addition, a fourth parameter is defined to describe how close a datum is to another datum or a group of other data. The main use of this parameter is to describe how close an auxiliary datum is to a primary datum or a group of primary data.

The first parameter is *usage across tasks* denoted as TU and defined as *'The number of tasks by which a datum is used'*. The second parameter is *joint usage* denoted as JU and defined as *'The number of other data elements with which a datum is used'*. The third parameter *usage ratio* denoted by UR is defined as *'The ratio between the number of other data with which a datum is used in most tasks and the total number of other data with which the datum is used'*.

When the set of most important and independent data have been selected and classified as primary data, how close they themselves are to each other is of interest. Furthermore, since the rest of the data (i.e. auxiliary data) must be attached to the primary data, their relationship with primary data is a key feature. A fourth parameter -- *joint usage ratio* denoted by JUR has been defined to describe this feature.

In general, datum A's JUR with data group B is defined as

(The number of the tasks where datum A and data group B are used together)
/ (The number of tasks where data A is used).

A special case of this parameter is the *joint usage ratio with primary data*. It reflects how closely a datum is related to a primary datum or a group of primary data. The relationship between JUR and UR is: For any datum, say, Datum

$$UR_i = Total\ Number\ of\ JUR_{high\_ix},$$

where UR_i is the UR of $Datum_i$, and JUR_{high_ix} means a high JUR of $Datum_i$ with another $datum_x$.

Now we are in the position to have a further look at the third parameter UR. A datum, no matter whether or not it is a primary datum, usually has a high JUR with a primary datum because the latter is often involved in many tasks. However, with the auxiliary data with which it is used, a primary datum usually has low JURs, because it is used with many of them in many different tasks, while an auxiliary datum has high JURs, because it is usually used with a small number of them in a small number of tasks, therefore it has a high probability of being used with them. Consequently, a primary datum has a low UR, whereas an auxiliary datum has a high UR. As a result of this, the UR is a powerful parameter to indicate the nature of a datum.

Empirical Distributions Of Data Parameters

Through a series of steps, all these parameters of data can be calculated. Normally, a primary datum has a high TU, a high JU, and a low UR. An auxiliary datum, depending on the nature of the tasks, may have different parameters in terms of being high, average, or

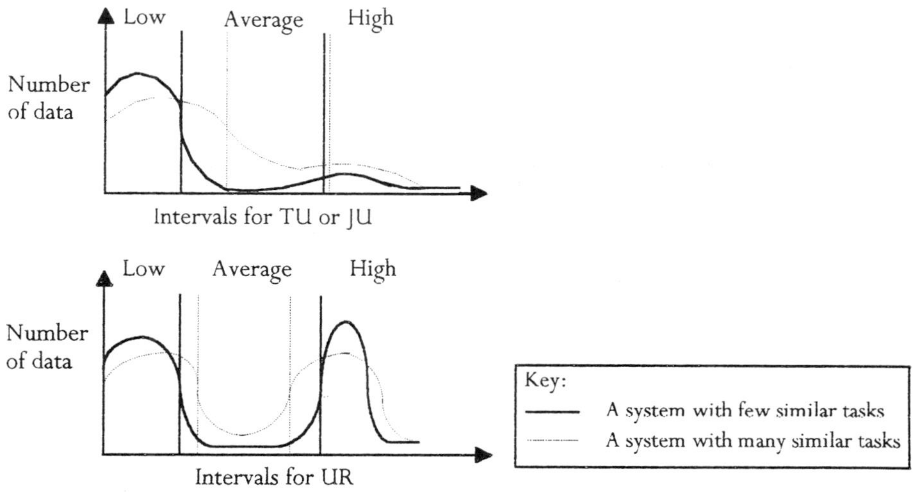

Figure 1. Empirical Distributions Of TU, JU And UR.

low. Where there are few similar tasks, an auxiliary datum will have a low TU, a low to average JU and a high UR. However, where there are many similar tasks, both its TU and JU may be average though its UR may still be high. Therefore the curves for the first three parameters may be slightly different. Figure 1 illustrates this situation.

With obvious turning points with high differential values, these curves can be divided into three scopes: high, average and low. These scopes form a set of criteria for the classification of primary data or auxiliary data: a primary datum should have a high TU, a high JU, and a low UR, otherwise, it is an auxiliary datum. With the criteria and parameters of each datum at hand, to classify them is trivial.

These curves also show that the separation of data into two groups should be obvious. If not, it is either because the tasks and data were not defined properly, in which case they should be re-defined; or because the tasks identified do not deal with the same set of objects, in which case the organisation should be divided into smaller sections according to functions at a high abstract level, and the process carried out for each of the sections.

Identification Of Entities

When the primary data have been identified, the underlying object classes that they represent can be revealed easily. The procedure for this stage is as follows:

First, check the primary data each a time to see whether they represent an object class. The idea here is to see whether an instance of a primary datum determines a unique object. If so, the object class is defined as an *entity*.

Second, pair each of the primary data that does not result in an entity to another primary datum or primary data group in descending order of JUR, and see whether the two together determine an object class. If so, the object class is defined as an entity too.

Identification Of Relationships

Whether entity A and entity B have a *relationship* depends on whether they are ever used together, namely, neither of their JURs with each other is zero, and also whether it makes sense semantically. The procedure for this stage is:

First, identify combinations of individual primary data or primary data groups that represent entities and that have non-zero JUR with one another. These combinations are candidates for relationships. When a large number of entities is encountered, we can identify relationship candidates via tasks.

Second, check semantics of these candidates, if it makes sense, then there is a relationship. Note that a group of primary data may determine both an entity and a relationship, for example, *hall-id* and *hall-room-no* determine entity *hall-room* and a relationship between *hall* and *hall-room*.

Assignment of Auxiliary Data To Entities Or Relationships

The basic idea here is that an auxiliary datum is an attribute of an entity or a relationship, and its assignment can be done by means of the following rule:

1. Always assign an auxiliary datum to the primary datum or primary data group that represent an entity or a relationship with which it has the highest JUR;

2. If there are more than one of such primary data that qualify, then always assign the auxiliary datum to the one which consists of the largest number of primary data;

3. Then use semantics to verify the assignment.

The reason behind the first part of the rule is that an auxiliary datum is an aspect of an object or a relationship between objects, so it can only appear with the primary data which represent its 'master'. This means that its JUR with the primary data must be the highest among all its JURs. The reasoning tor the second part of the rule is that, say datum A's JUR with datum B and datum C as a group is j, then both A's JUR with B and A's JUR with C must be greater than or equal to j. If datum A belonged to either B or C, A's JUR with B or C would normally be higher than j, according to the above reason.

The third stage of the modelling procedure transforms an entity-relationship data model to a relational data model. The relevant rules are well established so omitted here.

SUMMARY

This paper presents a data modelling method developed by virtue of both analytical thinking and hard systems thinking. Its cornerstone is the functional analysis of the system being modelled. Data are treated as a resource for the fulfilment of the functionality, thus data are identified systematically. Around the functionality, data are described by a set of parameters. These parameters are used as a collection of features which define such probabilistic concepts as *importance, independence,* and *closeness* of data whereby *primary data* and *auxiliary data* are classified, and auxiliary data's attachment to primary data are revealed. And furthermore, the relationship between data and real world objects is revealed, therefore data are interpreted and structured mathematically.

REFERENCES

1. M. Flavin. "Fundamental Concepts of Information Modelling," Yourdon Press, A Prentice-Hall Company, Englewood Cliffs (1981).
2. T. Teorey and J. Fry. "Design of Database Structures," Prentice-Hall Inc., Englewood Cliffs (1982).
3. J.K. Feng. Conceptual modelling and DBS/KBS design in "Systems Science: Addressing Global Issues,"F. Stowell, D. West and J. Howell ed., Plenum, New York (1993).
4. D. Tsichritzis and F. Lochovaky. "Data Models," Prentice-Hall Inc., Englewood Cliffs (1982).
5. Schoderbek, Schoderbek and Kefalas. "Management Systems," Business Publications Inc., Homewood, I.L. (1990).
6. T. Teorey and J. Fry. (1982) (OP.CIT.)
7. J.F. Sowa. "Conceptual Structures," Addison-Wesley, Massachusetts (1984).

CONCEPTUAL MODELLING AND DBS/KBS DESIGN

Jun-Kang Feng

Loreto College
Chichester Road
Manchester, M15 5PB, UK

Conceptual modelling is an essential issue of information systems design, particularly related to requirements specification. However, in the literature, its nature and principles for an adequate conceptual modelling method are not well defined, no systematic conceptual modelling method for the design of database systems (DBS) and knowledgebased systems (KBS) is specified, so it remains as a widely-used but vague jargon. This paper investigates these fundamental issues and presents the foundations of an approach that is based on the whole paradigm of scientific thinking, i.e., both the reductionist thinking and systems thinking, 'soft' as well as 'hard'.

NATURE OF CONCEPTUAL MODELLING

Data And Expertise

DBS and KBS are technologies, which are ideally based on a science of data and expertise. A particular DBS or KBS may appear to be a random sub-set of data or expertise, but a method for their design must offer common forms of data or expertise. These common forms are also best derived from such a science. This science should embody laws or trends that govern the regularity of data or expertise. However, there does not exist such a science, and due to the nature of data and expertise, such a science may not be possible.

Data and expertise are not a restricted scientific problem, but are in most cases results of human activities. Data are recorded facts or concepts about phenomena from which humans gain information about the world. So a datum is 'a fact' or 'something upon which an inference or an intellectual system of a sort is based'.[1] In the case of expertise, it is more obvious that any expertise is the product of human expert's long term experience in doing some particular tasks.

Humans have genuine freedom to attribute *meanings* to social reality and then to take actions which make contributions to the change of the social reality, so the social reality is a mixture of objective and subjective phenomena. The social reality is also ever-changing because humans in the social process are constantly creating the social world in reaction with others. They negotiate with others their perceptions and interpretations of the social reality, and at the same time constitute the reality itself. Therefore there is not a pre-given universe

Systems Science, Edited by F.A. Stowell *et al.*,
Plenum Press, New York, 1993

of objects but one that is produced by the active doing of subjects. While it is reasonable to assume that there is an unchanging physical, chemical, and biological world out there and that we can find out about it by using the method of natural sciences, there is no social reality to set alongside what appears to be well-tested physical regularities of the universe so no once-and-for-all substantial account of social reality is possible. The social reality is also very complex, a large number of unpredictable or unidentifiable variables may be involved in a seemly simple social event. Data and expertise are of social reality. They are a mixture of subjective phenomena and objective phenomena, ever-changing and complex.

Basic Ways Of Handling Data And Expertise

Data or expertise is a mixture of subjective phenomena and objective phenomena, so publicly testable knowledge of them can hardly be acquired, a match between reality and model that natural science seeks should not be expected. We are only aimed at sensible epistemological statements, rather than ontological statements of them.

Data or expertise is ever-changing. While we do not expect to reveal the laws within data and expertise like those that govern the regularity of restricted scientific problems, we hope to find some trends within them, Moreover, the ways one finds out about these trends may well be stable, so we should find these ways.

Data or expertise is complex, we can conceive some intellectual constructs to model them. In general, to tackle complexity, humans make intellectual constructs pertaining to complex reality by creating models. Then they use the models to test hypotheses and reveal laws or trends which govern the regularities and continuities of the real world, and therefore to understand the complex reality. The point of so doing is to reduce the multiformality of the world to some common forms so that things can be brought into a logical and conceptual relationship with each other. These intellectual constructs must be sensible, and is better to be validable. Natural sciences have been the most powerful means of making valid some of intellectual constructs used to handle the natural phenomena. In the case of data and expertise, our goal is also to find common forms of data or expertise, so that data or expertise created in deferent situations can become compatible.

In summary, to develop intellectual constructs and procedures of pertaining them to complex reality may create sensible interpretations of data or expertise, and is the basic and probably the only way of handling them.

Conceptual Modelling Approach To DBS And KBS Design

The intellectual constructs used to handle data or expertise help us to obtain some *interpretations* of them. Quantitative models or mathematical models of them do not seem possible. Moreover, it is all but impossible to obtain *one* overall model as an general purpose interpretation of data and expertise. So we look at the issue of data and expertise in a particular context, namely database structure design and knowledge base design. The purpose is twofold: To devise a set of sensible intellectual constructs for handling data and expertise; And to develop better methods for the design of DBSs and KBSs.

In the development of complex software systems, fashioning their conceptual constructs is the essence.[2] The main part of the conceptual construct of a DBS is its data schema, which is a conceptual model of the passive component of a universe of discourse. And the main part of a KBS is its knowledge base, which is a representation of some knowledge about a problem domain. Therefore, fashioning the conceptual construct of a DBS or a KBS is substantially that of modelling a universe of discourse. Many empirical studies have shown that the most critical activity in the development of a large system is that of requirement modelling, and the effectiveness of such a specification depends largely on the ability of the conceptual model developed to represent the problem domain.[3]

Thus understanding a universe of discourse is the most critical activity of DBS and KBS development, and it is a difficult job too. The author believes that such an understanding can and is better achieved by doing the following: First, developing a set of sensible intellectual constructs in terms of modelling formalisms or languages; Second, mapping real world phenomena onto basic concepts of the modelling formalisms or languages through a series of transformations, resulting in a conceptual model of the universe of discourse. Here, 'modelling a universe of discourse' is the focus and the joint point of the issue of handling data and expertise and the issue of DBS and KBS design. It implies that, first, from the point of view of handling data and expertise, a set of intellectual constructs, as common forms of data or expertise, is developed for the sake of modelling the universe of discourse; Second, from the point of view of DBS and KBS design, the use of these constructs creates blueprints of database structures or knowledge bases. This approach may be called *conceptual modelling approach to DBS and KBS design,* and the author believes that this is the right one. Our task is therefore to develop such an approach the most scientifically possible.

ADEQUATE CONCEPTUAL MODELLING METHODS

The modelling approach is widely adopted[4] and should be studied as an independent discipline.[5] This importance is reflected in the baseline for conceptual modelling established by the ISO,[6] which consists of a definition of conceptual model and a number of principles. A conceptual model is defined as

'The description of the possible states of affairs of the universe of discourse including the classification, rules, laws, etc. of the universe of discourse.'

And the three principles that a conceptual modelling process must conform with are: the 100% Principle, the conceptualisation's Principle and the Explicitness Principle.

The first principle effectively states that all the rules that govern the behaviour of the system should be considered, and neither they should become part of the implementation nor their definition should be distributed among different application programs. The second principle emphasises the clear separation of the conceptual model from any implementation related details such as physical data organisation. The third principle demands that all the rules of the universe of discourse must be clearly visible. The essence of conceptual modelling can be summarised as: Being independent of implementation; Being at a high level of abstraction; Being visible; And being necessary and sufficient.

The ISO baseline sets the target for conceptual modelling. To reach this target is non-trivial. It calls for adequate methods. A method may be regarded *adequate* if its proper use results in a model that meet the ISO principles. The author believes that such a method should consist of three components. First, a fundamental view of a universe of discourse. It is a philosophy invoked that determines the orientation or paradigm of the method. Second, a modelling formalism. It embodies a set of intellectual constructs, namely a collection of concepts and a modelling mechanism based upon these concepts. Third, modelling procedures. They are coherent sets of steps and transformations for mapping real world phenomena onto the concepts of the modelling formalism. A method composed of these components may be regarded *complete*. A complete method is not necessarily adequate, but stands better chance than incomplete methods as most of the contemporary methods are.

We used above principles and views to have analysed contemporary approaches to DBS and KBS development, which covers Function decomposition,[7] Data flow approach,[8] JSD,[9] Entity-relationship data model, Relational data model, Models of expertise,[10] Models of problem-solving activities,[11] and many more. This analysis demonstrates that the state of the art in this area is immature and confused. In the aspect of fundamental views of a universe of discourse, a huge variety exists in terms of different orientations and paradigms. In the aspect of modelling formalisms, concepts are usually general and ambiguous with neither syntax nor deduction apparatus, therefore they do not constitute formal systems and are difficult to use or to build upon. Their establishment seems based merely on common sense and therefore rather intuitively and arbitrarily. The modelling procedure aspect is often missing, no explicit and quantitative relationship between the reality being modelled and the concepts of the modelling formalisms has been revealed.

This conclusion explains why in information system development. projects that meets schedule, cost and performance objectives are still the exception, rather than the rule,[12] the cause is probably the lack of adequate methods, which in turn is caused by the lack of understanding and using appropriate paradigms of thinking.

FOUNDATIONS OF A PROPOSED APPROACH

Paradigms Of Thinking

It seems sensible to classify paradigms of thinking with the following hierarchy: Rational thinking and irrational thinking; Rational thinking is further divided into scientific thinking and common sense thinking. For example, modelling a universe of discourse by means of arbitrarily chosen concepts such as *data flow, object, function* and so on is of common sense thinking, most contemporary approaches to DBS or KBS design appear based on this kind of thinking; Scientific thinking is in turn divided into reductionist analytical thinking and systems thinking. Systems thinking consists of hard systems thinking and soft systems thinking, the distinction between them is whether one regards a system existing ontologically.

Due to its constructs being publicly testable, scientific analytical thinking is the most powerful means of viewing the naturally existing world. The problems that can be resolved by this paradigm of thinking must be reducible, repeatable and publicly falsifiable. However, in the area of information systems, many factors may be unknown or cannot be identified scientifically or controlled, therefore a sensible reduction is difficult to achieve.

Systems thinking[13,14] constitutes another set of intellectual constructs. Its basic view is that the existence at certain levels of complexity of properties which are *emergent* at that level and which cannot be reduced in explanation to lower levels is the illustration of the 'systems' paradigm.

Within the category of systems thinking, hard systems thinking has a built-in positivist ontology:[15] The world contains systems, which can be named and manipulated in the interests of efficiency. A particular implementation of hard systems thinking is *functionalism,* and functional analysis is a specific form of systemic analysis.[16] Functionalism has been a major concern in social theory in the last fifty years.[17] Soft systems thinking, on the other hand, takes the view that systems is epistemology before it is an ontology, and in the case of work whose concern is social reality, it may never be possible to make ontological statements in systems terms. Soft system thinking provides a structure for thinking and helps to gain insight of the problem or situation that are not necessarily ontologically existing systems as such.

Due to the nature of data and expertise, our solution to the problem of data and expertise modelling is to adopt both reductionist analytical thinking and systems thinking, soft as well as hard, and we have developed an approach with three *complete* modelling methods, namely a function-oriented data modelling method,[18] an object-oriented information modelling method, and a systems-based knowledge modelling method.[19] In the following section, the use of these paradigms of thinking in these three methods is briefly described.

Use Of Analytical Thinking

Analytical thinking is used in the function-oriented method. Any creditable method for data modelling must be based on the laws or trends that govern the regularity of data.

However, data, objects, and tasks are neither physical phenomena, nor variables in mathematical sense, so the laws or trends about them are difficult to discover. But by means of the reductionist analytical thinking, under a certain conditions, i.e., by discarding their rich variety of semantics, and looking at them abstractly, we can treat *data, object* and *task* as if they were mathematical variables and therefore describe their relationships mathematically. A particular characteristic of data, namely their relationships with tasks and objects is selected and analysed. This way, a means of measurement is achieved and expressed mathematically. This makes it possible to obtain a quantitative models of data and to reveal relevant regularity of data mathematically. Note that measurement is of particularly importance for developing a scientific approach to data modelling, because measurement is necessary for achieving the repeatability that a science must possess.

Analytical thinking is also used in the object-oriented method. The idea of object-oriented *per se* can be regarded as a result of reductionist analytical thinking in that anything and everything of interest is reduced to the simple concept *object* with most of their semantics discarded. This way, the ambiguous semantic concepts of conventional data modelling formalisms, namely *entity, attribute* and *relationship* are avoided, and the basic concepts used are only *object* and *information of interest about object.* Because they have simple, intuitive and clearly different implications, the semantic relativeness[20] is reduces, and the scope of design possibilities narrowed. These two concepts are further classified according to the basic methods of human organisation,[21] as a result, a powerful modelling mechanism is established.

Use Of Hard Systems Thinking

By and large, the function-oriented data modelling method is based on hard systems thinking. DBSs are developed to meet data needs of a human organisation, so the organisation must be systematically analysed first. With hard system thinking, an organisation is regarded as an ontologically existing system and such a system can be looked at in terms of its functionality. In the case of procedural systems, the functionalism can be well justified,[22] and most of the human organisation can be safely deemed as procedural systems. The use of hard systems thinking bring us the focus of attention in data modelling to be the relation of data to the more fundamental factor, namely the functionality of the system. And this way, data with variegated nature is simplified to a resource of the functioning of the organisation and parameters of data are arrived at by an analytical accentuation of this relationship.

This view is also the standing point for revealing the relationships between the functionality of a system, data being used within the system, and the real world objects.

Use Of Soft Systems Thinking

The knowledge modelling method is based on soft systems thinking. KBSs are concerned with human expertise, and human expertise is an *interpretation* of human expert's purposeful activities in problem - solving. This idea is not an arbitrary assertion but based on the basic concepts of Weber's *interpretative* social science.[23] Its relevant concepts include: A single deliberate action by an individual is directed to affect the behaviour of others; Such an action has *meanings* for the individual who acts; Society is the result of the totality of such intentional actions; The key issue is to interpret meaningful actions by the observer in terms of a means-end scheme of rationality; To obtain a sensible interpretation is possible, because actions of individuals are not random but fall into structures or patterns; And the interpretation would be made using generalisation about typical processes of actions, which are termed 'ideal types of actions'.

Human expert's behaviour may be studied using this idea. The purposeful actions of the expert are observed, generalised, and interpreted in terms of structures and patterns. And this interpretation can be regarded the relevant expertise. The SSM[24] that embodies the soft systems thinking is an appropriate tool for obtaining such a sensible interpretation. Its main idea is that a set of human purposeful activities is modelled in terms of a 'human activity system' which is then analysed using the formal systems model. Using soft systems thinking, a human expert's purposeful activity is viewed as a *problem-content-and-solving system.* The expertise relevant to this problem-solving activity therefore is viewed as the functionality of this system. And modelling expertise is to reveal this functionality.

SUMMARY

This paper looks at conceptual modelling in the context of DBS and KBS design. First, a number of fundamental issues are investigated, including the nature of data and expertise, the basic ways of handling data and expertise, and the conceptual modelling approach to DBS and KBS design. Then, the criteria for adequate and complete conceptual modelling methods are proposed. And finally, the foundations of a conceptual modelling approach developed by the author are described. This approach consists of three methods and offers a solution to the problems in this immature field.

REFERENCES

1. D.B. Guralnik ed. 'Webster;s New Word Dictionary of the American Language," William Collins and World Publishing Co. New York (1974).
2. F. P. Brooks. No silver bullet: Essence and accidents of software engineering, COMPUTER. 20:4 (1987).
3. C. Theodoulidis. "A Declarative Specification Language for Temporal Database Applications," PhD Thesis, UMIST, Manchester (1990).
4. N. Kheir. "Systems Modeling and Computer Simulation," Marcel Dekker Inc., New York (1988).
5. H. Kangassalo, S. Ohcuga and H. Jaakkla. "Information Modelling and Knowledge Bases," ISO, Amsterdam (1990).
6. C. Theodoulidis. (1990) (OP.CIT.)
7. P. Coad and E. Yourdon. "Object-Oriented Analysis," Yourdon Press, Prentice-hall, Englewood Cliffs (1990).
8. C. Gane and T. Sarson. "Structured Systems Analysis: Tools and Techniques," Improved System Technologies INC., New York (1979).
9. M. Jackson. "System Development," Prentice-hall, Englewood Cliffs (1983).
10. P.E. Slatter. "Knowledge Engineering," McGraw-Hill Inc, Maidenhead (1990).
11. W. Karbach, M. Linster and A. Vob. Models of problem-solving: one label - one idea? in "Current Trends in Knowledge Acquisition," B. Wielinga, J. Boose, B. Gbines et al ed.,ISO, Amsterdam (1990).
12. C. Theodoulidis. (1990) (OP.CIT.)
13. P. Checkland. "Systems Thinking, Systems Practice," John Wiley & Sons, Chichester (1981).
14. Schoderbek, Schoderbek and Kefalas. "Management Systems," Business Publications Inc., Homewood, I.L. (1950).
15. R. Aron. "Main Currents in Sociological Thought: Vol II," Weidenand Nicholson, London (1968).
16. P. Sztompka. "System and Function: Toward a Theory of Society," Academic Press, New York (1974).
17. P. Checkland. (1981) (OP.CIT.)
18. J.K. Feng. Function-oriented data modelling in "Systems Science: Addressing Global Issues," F. Stowell, D. West and J. Howell, ed., Plenum, New York (1993).
19. J.K. Feng and H.W. Rooms A systems based modelling technique for KBS construction in "Systems Thinking in Europe," M.C. Jackson, G.J. Mansell, R.L. Flood, et al ed., Plenum, New York (1991).
20. H. Kangassalo, S. Ohcuga and H. Jaakkla. (1990) (OP.CIT.)
21. P. Coad and E. Yourdon. (1990) (OP.CIT.)
22. P. Checkland. (1981) (OP.CIT.)
23. P. Checkland. (1981) (OP.CIT.)
24. P. Checkland. (1981) (OP.CIT.)

THE ALGORITHMIC NATURE OF PROBLEM SOLVING

F.J. Garlick and G.L. Leonard

Department of Information Science
The University of Portsmouth
Portsmouth
Great Britain

INTRODUCTION

The work on this paper arose from some studies into how practitioners solved the problem of knowledge elicitation. These studies were illuminating since the most common situation seemed to be that no strategy was employed and that the activity in essence was based on chance. In other words when practitioners were asked what strategy they were using the most common answer was that they *did not know but they could do it anyway.*

In this paper we will explore five mechanisms or approaches to generalised problem solving and our contention is that better problem solving ability results by explicitly using these ideas. Our particular emphasis will be on the problem solving process rather than on knowledge structures that support the process.

ALGORITHMS

Harel[1] in his book Algorithmics, made a very perceptive comment, paraphrasing, *we are a bit short on generalised algorithms.* Thus Harel is saying we have come along way and have many algorithms at our disposal but we do not as yet have an algorithm for finding other algorithms. Perhaps this is as it should be otherwise most computer scientists would be without jobs!

In this paper we will discuss five (we contend that there are only five) underlying problem solving strategies that have algorithmic-like properties. In this context we shall demonstrate the algorithmic properties that make these techniques useful in any problem solving situation using a metric proposed by Knuth[2] as long ago as 1973, which he called *finiteness, definiteness, input, output* and *effectiveness.* Briefly:

- □ **Finiteness** implies that the algorithm terminates.

- □ **Definiteness** implies that each step in the procedure is unambiguously defined.

Systems Science, Edited by F.A. Stowell *et al.*,
Plenum Press, New York, 1993

- ☐ **Input/Output** means simply that we need zero or more inputs and we must have at least one output.

- ☐ **Effectiveness** implies that each step in the algorithm can, in principle be done exactly and in a finite time.

Now these ideas can be formulated in other ways and research on algorithms has come quite a long way since 1973, however, there is still general agreement on those four principles. It can be argued that the most important property of an algorithm is its procedural nature. By this we mean that we can tell someone how to do each step unambiguously and that the steps themselves are sufficiently basic that they can actually be done. It is legitimate therefore to speculate on whether these metrics hold for more general problem solving methods. If they do, then the implication is that we can become better problem solvers, and teach others to solve problems by explicitly applying the algorithms inherent in our five strategies.

PROBLEM SOLVING

Analysis of the solutions to a number of problems leads us to the conclusion that there are five ways in which one solves problems, some of which correspond to a surface approach[3] whilst others, are representative of a deeper strategy. As you look at the five strategies you will see that they are comprehensive and in this sense they represent an holistic approach to problem solving. Indeed when using all five levels we might easily concur with Rodrigues-Ulloa[4] in his contention that we must see *the problem solving system* as part of the problem content itself. Briefly the five levels are:

- ☐ **Stochastic** where we imply that the process is akin to guessing and in essence chance is at work and therefore problem solving may or may not take place.

- ☐ **Resolution** which uses the classical idea of divide and conquer. This is a well understood mechanism where a large problem is resolved into smaller and therefore simpler problems.

- ☐ **Propagation** concentrates on ideas associated with previous experience. This is almost always the most productive mechanism for solving problems since we ask *have we seen something like this before.*

- ☐ **Contextualization** implies a deeper level of understanding since a clear idea of what view we have of the system is required. The point here is that solutions may come if we change our viewpoint so that we see the same problem in another light.

- ☐ **Symmetry** requires that we consider how elements affect and are affected by other elements. It is likely that this is the deepest level of solution for we try to consider how one system (or part of a system) is affected by another.

You may well be thinking at this point what about other strategies such as bottom-up or simulation. However, if you think carefully through these two examples then you will see that bottom-up is either stochastic or resolution whilst simulation is a means of testing a given solution rather than finding one. Other possibilities such as inspiration or intuition are not legitimate since we cannot get near to explaining how these are actually done.

SOME EXAMPLES

Here we consider some examples which serve to illustrate the five levels of the problem solving strategies outlined above. In particular we begin with a look at repertory grids as a complete example, then consider learning as problem-solving and finally some generalised examples illustrating all five levels.

Repertory Grids

Consider for a moment the repertory[5] grid as a means of either discovering new facts or establishing which facts are more important. The strategies employed therefore will be:

- ☐ If we employ interviewing then it will be mainly *stochastic* at the start for we may not have sufficient knowledge to ask the right questions.

- ☐ As a study proceeds then we can gradually be more structured in our questions and then we most probably will be using *resolution.*

- ☐ Alternatively we could use Kelly's[6] idea where we look for other situations that resemble the one under investigation. The actual similarities might be functional, organic, social and so on. Clearly here we are using *propagation* to find similarities that we know about and *contextualisation* because we might ask why the situation chosen is similar to the problem setting.

- ☐ When we have a complete grid then it is usual to look for links between the various constructs and this implies using *symmetry.*

The grid therefore might be regarded as a powerful tool since it *forces* the user to work at all levels. In practice we have observed that students find it quite difficult to work at the deeper levels but nevertheless find that it stimulates thought and frees the mind to some extent from working on sensory tramlines. For example, a student, using repertory grids noted the relationship *Garage Equipment Installer/Saville Row tailor* to some surprise from other class members. However, when he pointed out that *made-to-measure* was the relationship a new and illuminating line of enquiry was opened up. This of course is not new and Checkland and Scholes[7] discuss the usefulness of what they call A-B relationships such as parasite/host, master/slave, judge/accused since thinking through these relationships makes one reflect on real situations and therefore stimulates thought.

Learning Models

Now in a given problem solving situation one might use any of the five levels or perhaps composites of them. However, it is true that we often tend to work at the *stochastic* and *resolution* levels and it is also unfortunately true that learning is also only done at these levels. For example, suppose the problem to be solved is the teaching of a programming language then we could adopt either of the following methods:

- ☐ Teach in detail appropriate language constructs and gradually show how complete programs might be built to provide a particular function within a system.

- ☐ Discuss whole programs and how program elements hang together learning appropriate language constructs as they are needed.

It is probable that most people will have difficulty identifying which problem solving technique is at work in the first method but less so with the second. The point is that in the first method the outcomes are in the teacher's mind and the average student may have no real idea what it is that they are supposed to be achieving. Clearly in the second method the aim must be explicit and it is likely that greater understanding and therefore learning is taking place. This example illustrates the holistic nature of good problem solving since here we have an example of sharing and therefore owning a problem, in this case teacher and student. Indeed it is a contention of the paper that the deepest level of understanding comes when all levels are used but in particular, the fourth and fifth levels. It follows that the best problem solving techniques should explicitly use these levels or composites of them so that a comprehensive approach is developed.

Stochastic

An example might be the typical maze-solving or route-finding program, where a simple exhaustive search is made through all the possibilities (in an unordered or random way), until a solution is found. However, there are a large number of real life problems which are too complex or too dangerous to be solved in practice by this method, for example, putting men on the moon.

Resolution

Recursion in problem solving is a prime example of this technique. The typical Towers of Hanoi problem of moving a stack of N graded discs from a start pole to a finish pole results in moving N-1 discs from the start pole to an intermediate one, moving the largest disc to the destination and then finally moving the N-1 discs to the destination pole. The close comparability between recursion and induction means that induction can also be viewed as an application of Resolution.

Propagation

This is almost always the most profitable method yet its value is often underestimated even by professionals. An excellent example is provided by the Boeing Aircraft Company during the design of the doors for the new 777 aircraft. What the door team did was to look back at how the doors had been designed on previous aircraft and essentially learn all the pitfall that had existed. It was estimated that this exercise saved approximately $64M dollars.

The current emphasis on planning[8] in Artificial Intelligence research is an example of this principle, whereby previously solved problems have their solution broken down into definable sub-goals and the relevant activities to resolve them. These components then become ingredients for the formulation of a plan to solve a new problem.

Contextualization

This approach is typified by considering the solution to the missionary and cannibal problem, where a solution can be achieved for getting all of them from one bank of the river to the other by shuffling people in the small boat, without the missionaries being eaten (using a stochastic approach). However it might have been more sensible to walk half a mile upstream and use the bridge! De Bono's[9] *lateral thinking* is another expression of this principle.

To take an HCI example, that of designing hierarchical menu structures. This could be done using a semantic model[10] which embodies a logical ordering of the levels in the menu structure and within each level. If we take another view we may instead decide to order the levels by use and this may fit more precisely the semantic model in a users mind.

Another excellent example of this principle is found in the ideas associated with *interpretive knowledge structures* where, in Johnson's[13] words *facts become embellished and contextualised* as a means of organising your knowledge and taking a particular view is in effect a means of interpreting that knowledge and therefore giving it meaning.

Symmetry

The use of semantic nets[11] where concepts are linked by binary relations, and the movement towards using Object Oriented[12] approaches to modelling systems are examples of this principle at work. In the latter, the system's behaviour is not only determined by the behaviour of each of the autonomous objects, but also by the message transferred between objects.

To use HCI again, we see the use of symmetry in *proposition networks*[13] where ideas are linked by the type of relation between them. This appears to be powerful since linking ideas gives one access point to the network (knowledge) but other access points are available by means of *cued* or *primed* questions.

CONCLUSIONS

Our main conclusions are that better problem solving skills result when the five strategies outlined above are used explicitly. Further, we conclude that any methodology that embodies all the above strategies will be inherently more stable and better adapted to real world problems than one that is based on only a partial set of the above five elements. In a similar vein we might say that looking for the above in a methodology would be a means of testing that methodology.

The examples used illustrate another facet of problem solving and that is that composite of the five levels may also be used. Indeed we feel that the problem solving process would be impoverished by simply trying to work only in one of the levels we therefore encourage explicit use of all levels and composites of them.

Finally we assert that the approach is holistic and therefore we conclude that our problem solving ability will be improved whether we are engaged in an interventionist or interactionist investigation.

REFERENCES

1. D. Harel, "Algorithmics - The Spirit of Computing", Addison-Wesley, London, (1987).
2. D. Knuth, "Fundamental Algorithms", 2ed, Addison-Wesley, Wokingham, (1979).
3. G. Gibbs, "Improving Student Learning", The Oxford Centre for Staff Development, Oxford, (1990).
4. R.A. Rodrigues-Ulloa, The Problem Solving System: Another Problem Content System, *Systems Practice*, 1(3), 243-257 (1988).
5. A. Hart, "Knowledge Acquisition for Expert Systems", Kogan Page, London, (1989).
6. A.J. Kelly, "A Theory of Personality", The Norton Library, New York, (1963).
   The Open University, "Complexity, Management and Change", T301 Block IV, Open University Press (1983).
7. P.B. Checkland and J. Scholes, "Soft Systems Methodology in Action", Wiley, Chichester, (1991)
8. D. Chapman, Planning for Conjunctive Goals, *Artificial Intelligence*, 32: 333-377, (1987).
9. E. De Bono, "The Mechanism of Mind", Simon & Schuster, New York, (1969).
10. B. Schneiderman, "Designing the User Interface", Addison-Wesley, New York, (1987).
11. R.J. Brachman and H.J. Levesque, ed., "Readings in Knowledge Representation", Morgan Kaufmann, New York, (1985).

12. M.J. Stific and D.G. Bobrow, Object-Oriented Programming - Themes and Variations, *The AI Magazine*, 2(4), 40-62, (1984).
13. P. Johnson, "Human Computer Interaction", McGraw-Hill, Maidenhead, (1992).

SYNTAX AND SEMANTIC ANALYSIS OF MISSION STATEMENTS

F.J. Garlick and I.R. Lightfoot

Department of Information Science
The University of Portsmouth
Portsmouth
Great Britain

INTRODUCTION

Over the last two years there has been a world-wide change taking place in Higher Education. Typically we have seen a move towards full university status for suitable institutions, an emphasis on quality systems, particularly TQM[1], and a demand for full accountability. Almost universally old and new Universities have started with a search for a mission and a corresponding Mission Statement. In most cases Vice-Chancellors and governing bodies seem to know that it (the Mission Statement) would be used to **test** any proposed course of action in order to see if that action supports the transformation suggested by the Mission Statement.

The paper is an examination of University Mission Statements and looks at their meaning and forms of construction. We therefore offer our analysis for the following reasons:

- If the Mission Statement is to be used as suggested then it must be absolutely clear and leave little if any room for misinterpretation.

- If we assume that TQM or TQL[2] will soon be on all University agendas then in this context the Mission Statement would seem to be crucial.

ANALYSIS OF MISSION

For our purposes we equated mission with what is known as Root Definition in Soft Systems Methodology. The similarity is not entirely exact since Root Definitions typically deal with conceptual systems rather than real. However, it does seem legitimate to think of the production of Mission Statements by governing bodies as a form of Soft Systems Mode 2 working, for one supposes it is produced in a very interactionist manner.

Systems Science, Edited by F.A. Stowell *et al.*,
Plenum Press, New York, 1993

Method of Analysis

The analysis for this paper was carried out in two parts; firstly in terms suggested by Checkland[3] and aided by the CATWOE criterion - customers, actors, transformation, Weltanschauung, owners and environment, and secondly we used simple English Language syntax. Although a full analysis was carried out we shall only cover those elements that exhibited the most obvious flaws. In particular our research has shown the necessity for debate about core purposes and their implications particularly with regard to resources. Many authors, notably Checkland and Scholes[3], have discussed this issue and our study serves to reinforce the view that looking for core purposes is a necessary and important task if meaningful statements are to be produced.

Core Purpose

When defining mission, which we will think of as defining the purpose or core purpose of a University, two possibilities exist. Firstly, we can define mission in terms of real world systems that exist; in Soft Systems terms this is a Primary Task based rationale. Secondly, we could base our search for a mission on perceived issues in the real world. In both cases our perceptions will be based on particular Weltanschauungen. Checkland and Scholes[3] suggest a very simple input/output model with a central transformation but note that both input and output entities must be of the same kind for the model to make sense. Briefly, at least three possibilities exist:

1. *input entity* ↦ *core transform* ↦ *output entity*

2. *input entity* ↦ *transform 1* ↦ *transform 2* .... ↦ *output entity*

3. *input entity*
   ↦ *transform 1* ↦
   ↦ *transform 2* ↦ *output entity*
   ↦ *transform 3* ↦
   ....

There is no time to analyse in detail these models except to say that model 2 can be effectively reduced to model 1 by finding a central transform (to borrow a term form software engineering). Model 3 however is complex and finding the core or central transform may be difficult. In general the recommendation is to look through each transform attempting to decide when the entity is recognisable as an output rather than an input; the transform where this occurs is the core or central transform.

In our study most Mission Statements were of type 3 and, like Checkland[3], we take the view that they *enrich the debate* but do so at the cost of increased complexity. In general therefore we took the view that a Mission Statement modelled on the form suggested by Checkland and Scholes[3] was both adequate and useful. The model, simply stated and corresponding to model 1 above, is:

a system to do X by Y in order to get Z

SYNTAX ANALYSIS

A number of the Mission Statements investigated consist of one sentence with its normal minimal requirement of one main verb. But not all retain the simplicity of meaning which that construction might imply. The clarity of the Mission Statement is often blurred by too many qualifications in the form of relative and other clauses which merely combine to strengthen the impression that the institution has no single-minded purpose, but is seeking to be a panacea for the shortcomings of the educational world by attempting to achieve a wide-ranging set of objectives. The view of a single purpose disappears in the mists of multiple ideals.

Furthermore, many of the statements consisting of one sentence are broken down by semi-colons into a number of clauses each containing its own verb. Technically this is a valid method of sentence construction, but in this context it only serves to create added complexity to the semantics of the statement and added confusion for the reader. All the resultant sub-goals are competing for prominence and attention, and no dominant theme emerges. It is almost as if the reader is led to believe that the sentence is taking a certain course which he afterwards finds, to his confusion, it does not take. This is an obvious folly and should be avoided at all costs[5]. The use of a string of infinitives is similarly not to be recommended, as it, too, leads to a multiplicity of competing themes.

It can be quite simply stated that a Mission Statement should consist of one main verb indicating directly the singular function of the institution. There are many statements which adhere to this rule but which nonetheless seek to confuse the reader by the introduction into the text of gerunds, verbal nouns and participles. Each of the latter constructs implies in its own way an action, the introduction of the reference to which, results in a pattern of confusion, which is similar in nature to the confusion alluded to earlier in the discussion of the use of several verbs in a sentence. Table 1 below shows the average number of occurrences of significant syntactic elements found in the Mission Statements studied. As can be seen there is too great a reliance on the use of many verbs in a single sentence to express an essentially simple idea.

Table 1. Main Syntactic Element Averages found in Mission Statements

Element	Average
Sentences	1.5
Verbs	5.0
Adjectives	4.0
Adverbs	1.0

Construction Guidelines

The most significant purpose of the Mission Statement is to state a transformation distinctly, and the most suitable way to achieve this is to express the relevant action by the use of a single verb. The use of adjectives to describe the nouns used within the statement does not assist in the elucidation of the stated transformation, but merely attempts to explain its input and outputs. This can be useful to qualify and describe more accurately the characteristics of the customer or the Weltanschauung and the environment of the owner. But so many Mission Statements include adjectives such as *special, appropriate, wider, various and liberal,* which carry so vague a meaning as to render the intended meaning beyond the comprehension of the reader. Analysis of the superior Mission Statements would suggest that careful choice of nouns preclude the use of all but the most pertinent adjectives.

It might be thought that adverbs or adverbial phrases would predominate as major syntactic units in a Mission Statement. The verb should be the most singularly important feature of the statement, and it would not seem unreasonable to expect to see its meaning being distilled and refined in an attempt to make the nature of the transformation as clear and intelligible as possible. There is a surprising paucity of adverbs in Mission Statements. This is made only marginally more credible by the slightly more frequent use of adverbial phrases. These are typically and most commonly introduced by the word *through* suggesting, usually by implication, how, rather than why, a certain transformation might take place.

For all the Mission Statements reviewed it is evident that the most successful in fulfilling their objectives are those which express a single transformation succinctly and concisely through the use of one main verb in the sole sentence. However, such a simple structure in itself is not entirely sufficient to render a Mission Statement ideal or to be beyond question. In an attempt to eliminate all doubt and to suppress all debate concerning its meaning a Statement should be furnished with a structure clothed with words which carry a precise and unambiguous meaning. The author of a Mission Statement needs to eschew unnecessary complex syntax construction and should exercise extreme care and caution in his choice of words.

TRANSFORMATION

The transformation implied by a main verb is perhaps the most important and difficult element to capture in Mission Statements. We support the contention of Wilson[6] who suggests looking for just a **single** primary purpose for the system being described. The reasons for this, in terms of a University organisation, are simple:

- It forces one to be clear about the central purpose for any system

- When several activities are suggested it is almost impossible for them to be weighted satisfactorily.

- When a statement contains several verbs, what is being communicated may be a very complex idea but, in most of the research we have carried out, such definitions tend to be meaningless.

- When several activities are suggested it is common to find that no consideration has been given to the question of if and how they inter-relate.

- When multiple activities are listed it is always possible to show that, whatever you do, can be supported from the definition.

- A corporate image implies a clear view as to purpose and it is hard to see how the multiplicity of activities suggested in many Mission Statements could be faithfully reflected in any corporate image.

- When there is no clearly understood purpose, the transformation that should be implied by the definition will ultimately be replaced by the bureaucracy that was put in place to effect the transformation.

- Lastly, when several activities are suggested, it follows that each separate activity resulted from a particular Weltanschauung or, more simply, motive. Mission Statements of this form therefore lack internal consistency since multiple motives will be involved.

We are not implying that several activities should not be pursued, but rather that they should be derived from the principal activity in such a way that the main transformation should be guaranteed to complete.

Transformations and TQM

To digress, it is interesting to note, that lack of clarity in mission may lead to a seriously flawed TQM model. When multiple activities are suggested (as is common) the following TQM model often emerges:

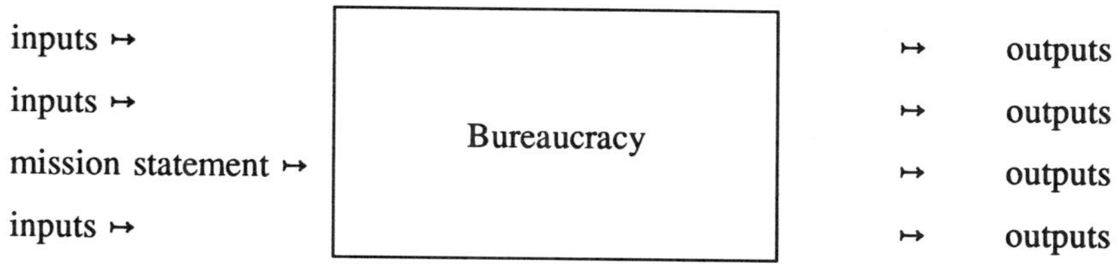

This is an appealing model but we think it is flawed. The Mission Statement should describe the transformation that the system performs on its inputs to derive its outputs. To summarise:

- The model suggests that structure (bureaucracy) determines what the transformation is. This makes little real sense for it is rather like saying we solve a problem before we know what it is.

- The model suggests that structure is all important and relegates processes to a secondary role. Such a view is inimical to TQM since we need to define processes and this can only realistically be done if we are clear of the purpose.

The point we are trying to make is that transformations and structure are symbiotically related but it is the transformation that defines the structure not the other way round.

If one looks at typical definitions one almost always sees that several activities are listed: *promoting, providing, pursuing, maintaining, teaching, researching, emphasising, ensuring, developing* and so on. Our analysis suggests that many, perhaps most, Mission Statements are of little value since it is probably impossible for anyone to say what a particular University is trying to achieve. In these cases the above TQM model will almost naturally arise and the emphasis is not on what is done but who is doing it. It is our contention that such a structure will tend to be self-preserving and sterile. We naturally understand that various people in the University are engaged in some of the above-mentioned activities, but it is not clear what their combined effort is supposed to achieve.

Motive & Weltanschauung

In Soft Systems we would expect to follow, at least to start with, a pattern similar to the following:

Motive - Idea - Process (transformation) - define and build

Weltanschauung is rather difficult to translate and define. However, it would appear useful to think of motive as one's immediate response to a given set of factors. Clearly in the real world the same motive in different people can lead to different conclusions. Hence we need to step back and ask why people act differently even when their motives are laudable; it depends entirely on their Weltanschauung.

The majority of Mission Statements contain an explicit Weltanschauung rather than a specification of the transformation itself. Such an approach, albeit unconscious, leads to fine-sounding missions, for example:

it strives to be the finest University in ...

for the particular benefit of .... and the good of the wider national and international community ...

contribute to the quality of life and ecologically sustainable development...

Mission Statement Formats

It would therefore appear that in the set of 20 Mission Statements which we examined, three possible transformation models emerged as shown below. It is clear that the first two do not engender explicit transformations, but they may lead the reader to form his own concept of what the transformation might be. Our research leads us to the conclusion that only the last model is able to deliver successfully a germane and lucid Mission Statement. In this context it is interesting to observe how the last model mirrors the conventional sentence structure of subject, verb and object.

input entity $\mapsto$ *management of ..* $\mapsto$ *output entity*

input entity $\mapsto$ *reason for ..* $\mapsto$ *output entity*

input entity $\mapsto$ *activity* $\mapsto$ *output entity*

CONCLUSIONS

Our analysis shows that the Mission Statements reviewed in this study are not sufficiently well-formed to be used as a mechanism for promoting change. We would therefore suggest that in order to achieve a successful Mission Statement the author should seek to state a single transformation process together with the Weltanschauung of the University. This latter idea should be central to the building of Mission Statements but we nevertheless would recommend that all authors test the complete definitions using the CATWOE criterion. Finally, Mission Statements are very much in vogue at present, and in this context it is helpful to recall a few apposite remarks.

if it requires more than five lines you are on the wrong track .... Dijkstra.

What can be said at all, can be said clearly; and on what we cannot talk about, we have to remain silent.... Wittgenstein

REFERENCES

1. L. Munroe-Faure and M. Munroe-Faure, "Implementing Total Quality Management", Pitman, London, (1992).
2. R. Lessem, "Total Quality Learning", Blackwell, Oxford, (1992).
3. P.B. Checkland and J. Scholes, "Soft Systems Methodology in Action", Wiley, Chichester, (1991).
4. The Open University, "Complexity Management and Change", T301 Block IV, Open University Press (1983).
5. H.W. Fowler, "Dictionary of Modern English Usage", Oxford University Press, Oxford, (1954).
6. B. Wilson, "Systems: Concepts, Methodologies and Applications", 2ed, Wiley, Chichester, (1990).

SUBLIMATING METHODOLOGIES:

THE FALLACY OF 'THE RIGHT THING RIGHT'?

James Gilligan

Department of Computer Studies
University of the West of England, Bristol
Coldharbour Lane
Bristol, BS16 1QY

INTRODUCTION

Flood, Jackson and Keys (1989), introducing the record of papers presented at the first conference of the UKSS, note that in attempting to create a well-founded pluralist approach to systems theory and practice: 'The difficulty remains for pluralism that once it accepts the existence of wholly different systems approaches resting upon apparently irreconcilable presuppositions (inhabiting different paradigms), how can the problem of paradigm incommensurability be overcome? Without privileging any one of the competing positions, how can a metatheory be created which respects the relative strength and weaknesses of each and oversees their 'correct' employment by systems practitioners?' One continuing response to this question has been the effort directed towards identifying strategies for the integration or synthesis of 'hard' and 'soft' systems methodologies.

A considerable volume of recent research, for example: Avison and Wood-Harper (1991); Wood-Harper (1985); Miles (1988); Doyle and Wood (1991b); Sawyer and Trahern (1991); Prior (1991); Keys and Roberts (1991), Merali (1992), seeks to address recognised methodological incompleteness by the selective accumulation of techniques drawn from epistemologically diverse paradigms. In other related work, for example Mingers (1992); Checkland (1992); Holwell (1992); Jayaratna (1992); Paton (1992); Jones (1992); Fitzgerald (1992); Prior (1992); Lewis (1992); Gregory (1992); and Sawyer (1992), researchers have reviewed and debated the intellectual underpinning of such an enterprise.

The central problem of plurality, however, is not that of integration or synthesis of epistemologically diverse systems methodologies. It is the creation of a metatheory *beyond* 'hard' and 'soft' paradigms, which may be used to provide, from a variety of epistemologies, an appropriate and sufficient response to real world systems activity.

This paper, through an exposition of the underlying ontological *similarity* between 'hard' and 'soft' paradigms, offers a test of the validity of the assumption that more appropriate and sufficient methodologies may be based on the integration or synthesis of such

Systems Science, Edited by F.A. Stowell *et al.*,
Plenum Press, New York, 1993

paradigms. It is believed that such a test provides evidence to sustain the argument that while the synthesis or integration of hard and soft systems methodologies may create functional benefits for the already committed practitioner, the theoretical foundation of any such approach is incomplete, and in failing to address the requirement for a metatheory supporting plurality without privilege, it thus fails to meet a principle objective of current systems theory.

EXPOSITION

Hirschheim and Klein (1989) describe by way of narrative the positivist epistemology and realist ontology of 'The Analyst as Systems Expert'; and the 'antipositivist' *(sic)*, sense-making epistemology and nominalist ontology of 'The Analyst as Facilitator'. In so far as these narratives describe 'hard' and 'soft' systems paradigms (rather than simply as narratives in their own right), we have a summation of the literature's epistemological and ontological classification of 'hard' and 'soft' systems. The positivism adopted by 'hard' systems methodologies reflects a reductionist view which allows for an external reality to be described and defined by cause and effect. The sense-making subjectivity of 'soft' systems replaces such well-bounded causality with what Husserl regarded as 'objects under aspect'. (Dreyfus quoted in Magee, 1987, p256). How then should we regard the recent history of Information Systems studies which contains a wide range of efforts to integrate or synthesise 'hard' and 'soft' methodologies despite their evidently divergent epistemologies?

In Miles (1988), for example, the possibilities of 'Grafting' or 'Embedding' are discussed as means of combining HSM and SSM. Miles makes the pragmatic nature of the exercise clear when he states that 'there are those who believe that these two schools of systems thinking are, in practical terms, incompatible. Efforts to combine them are therefore worthy of further examination.' 'Grafting' and 'embedding' follow the simplest approach to epistemological union by taking 'soft' and 'hard' methodologies in sequence or iteration. The boundary level, where 'soft' meets or gives way to 'hard', is not well defined. The epistemological gap is recognised, but not bridged.

Doyle and Wood (1991a) claim that there is a need to address this different 'epistemological underpinning', and one response by Sawyer (1991) is a stepping mechanism, described as two 'epistemological shifts to achieve compatibility between SSM and DFDs'. But can one 'shift' an epistemology? Or is it, as Jayratna (1992, p116-117) suggests, the abandonment of one epistemology for another? Doyle and Wood (1991) seem unwilling to accept the pragmatism of a 'shift' between SSM and HSM, but go on to accept the possibility of an epistemological bridge between SSM and *logical* forms of HSM descriptions, referring to the epistemological diversity of HSM and SSM as 'the Great Divide', a metaphor also used by Sawyer and Trahern (1991).

The problem addressed in each case mentioned above, exemplifying recent literature, is explicitly that of epistemological compatibility. The solutions offered may well provide a sense of epistemological reconciliation to those engaged in a grafted or embedded 'soft' systems study, and who wish to create related 'hard' systems descriptions by way of a formal methodological bridge between the incontrovertibly divergent paradigms. Indeed, as noted above, we can find much in the literature which attempts to show that HSM and SSM are by no means incompatible 'in practical terms'. The danger is that apparent compatibility may well be that of an unstable mix of techniques and arbitrary interfaces which provide temporary relief to the systems practitioner who feels the territory opened up by SSM must be assimilated into everyday practice as soon as possible. In some instances, the approach taken offers meaningful insights into the methodologies and their implementation. In others, there appears to be little in the way of sound methodological guidance, while affording far more opportunities for questioning the theoretical basis of any such move. Fitzgerald (1992, p127) is particularly forthright: 'The only appropriate linking of SSM and IS approaches is by performing SSM first and then the IS methodology with the only formal linking being the enhanced education and understanding of the actors....Any more formal linking of SSM outputs to IS inputs is impossible without fundamentally changing SSM or more probably both SSM and the linked IS...'

Wood and Doyle (1989) have provided an useful aphorism to describe the entire enterprise: 'Doing the Right Thing Right'. The 'Right Thing' decided through the use of 'soft' systems, with 'hard' systems engineering to do the thing 'Right'. If we take this as a statement of intent for integration or synthesis, we must not forget that within the statement 'Doing the Right Thing Right' is the implicit assumption that recognising 'The Right Thing' (SSM) and doing it 'Right' (HSM) are compatible and complementary. It seems natural to agree with the authors when they say: 'The desirability of developing methods which do both 'the right thing' and 'the thing right' is also clear, as to do one without the other will surely produce a perfectly engineered mess or a poorly implemented good idea.' Jayaratna (1992, p112) agrees: 'Attempts to link SSM with structured methodologies should be encouraged if they help the designers to develop better solutions [or better perceived solutions] to the problems they perceive.' While not everyone is convinced of this, the opposing arguments which claim that SSM cannot be combined with HSM because 'The paradigm, the philosophy, the objectives, etc. are totally different.' (Fitzgerald, 1992, p127); or that HSM and SSM represent 'opposing philosophies' (Galliers, 1992, p146), appear to confuse a diversity of epistemology with conflict of a more fundamental philosophical nature which is not justified by the above epistemological debate.

If the diverse epistemologies do indeed co-exist and complement each other at some level, how might this level be described? It is the author's view that the widespread belief in the possibility, at some level, of an intuitive complement of 'hard' and 'soft' paradigms, despite their epistemological divergence, is as a result of an underlying *ontological* similarity; and that the implications of 'realism' and 'nominalism' as understanding of the nature of reality, is of greater significance than is the evident epistemological conflict.

THE COMMON GROUND

The ontology which supports HSM is one which allows for empirical testing of a given external reality, and as such reflects one form of the Cartesian tradition. Reality is regarded as external and unchanged by the epistemology of the observer.

Soft systems theory, and in particular Checkland's philosophical underpinning for SSM, is described in the context of Husserl's phenomenology (Checkland 1981, Checkland and Scholes, 1990) as participating in a world of subjects among objects, of 'objects under aspect'. However, this tradition is *also* Cartesian in origin, differing from realism not by questioning the existence of the external world, but by recognising the mental constructs we place upon it.[1]

Epistemologically, HSM and SSM deal with objective knowledge and subjective interpretation: analysis of the world according to external principles and criteria (HSM); and according to internal principles and criteria (SSM). HSM selects from complexity in the real world by objective simplification and by assuming a single set of initial conditions. This approach is recognised throughout engineering and mathematics where linearity is used to approximate to the unmanagability of nonlinearity in the real world. SSM selects from complexity through subjective interpretation within a restricted set of initial conditions.

---

[1] Indeed, the *philosophical* debate between HSM and SSM may appear to some to resemble the debate between empiricists and rationalists in the 17th and 18th centuries, reconciled by Kant.

The differing epistemologies of hard and soft paradigms primarily affect how the choice of system description and options for change should be abstracted and presented. Significantly, both HSM and SSM require human intervention which is external to the methodology: in the choice of options for HSM; and by debate and through unspecified means in SSM. Indeed, *in practice,* HSM is forced to recognise the subjective nature of systems activity (Lewis, 1992, p173; Jones, 1992, p124; Stowell, 1992, p107), with the conclusion drawn that the IS development process is 'effectively a research act in which the chosen methodology itself is interpreted within the context of the problem situation' (Mingers, 1992, p107). Ontologically, HSM and SSM share the belief that through knowledge, abstraction, and analysis of appropriate, relevant, selections or interpretations *from* the world, purposeful and persistent change can be introduced *to* the world.

A TEST OF SUFFICIENCY

If we come to accept the possibility of purposeful, persistent change based upon knowledge of on external world or our interpretations of it, then (as Wood and Doyle *et al.* suggest), HSM may tell us how to do the 'right thing' (by deterministic, empirical means), while SSM may tell us (though debate) what the 'right thing' is. However, we must ask whether the complement of HSM and SSM does, in fact, address the recognised methodological insufficiencies of either or both paradigms? What, in each case, is felt to be missing and so provides the impetus for the many attempts at integration or synthesis?

HSM may be regarded as insufficient, not for failing to recognise the 'Right Thing', but for its inability to do anything 'Right' at all. HSM regards the world as deterministic, predictable and bounded. It is the paradigm inhabited by classical physics and engineering. As such, nothing is 'right' or 'wrong'. Solutions either 'function' or 'fail' and are of a mathematical and engineering kind. As such they can neither be right nor wrong except in a sense which is *external to the methodology,* that is, in *perception or expectation* (Lyytinen, 1988).

SSM may be regarded as insufficient in that it does not tell us what is to be done, or how to do that which is to be done. It offers the prospect of 'culturally feasible, systemically desirable change' (Checkland, 1981). It does not offer solutions of the mathematical or engineering kind. There are no such solutions, only processes of learning about the constructs we place upon the external world. Whether and how a solution may be implemented is not addressed by SSM except in a sense which is *external to the methodology*, that is, through a process of reconstruction of individual or social perception, or through engineering solutions such as are provided by HSM.

In the case of both paradigms, the insufficiency can be seen to rest on the evaluation of the solution, on the meaning of the word 'right'. In the case of HSM there is no meaningful use of the word 'right' in relation to methodological solutions. The word applies only to the external judgement made of the resulting systemic change. In the case of SSM, the matter is less clear. However, Wittgenstein provides direction in his Philosophical Investigations (1953, 270, p95e) through discussion of the meaning of the word 'right' in the context of an internal, personal language. For Wittgenstein, a system which provides a result from a given subjective sensation which has no *external* verification is always 'right', or rather: '...it seems quite indifferent whether I have recognised the sensation right or not. Let us suppose I regularly identify it wrong, it does not matter in the least. And that alone shews that the hypothesis that I make a mistake is mere show.'

Searle (Magee, 1987 p338) concludes that for Wittgenstein: 'if there is no distinction between getting it right and believing that we have got it right, then we can't talk about right at all'.

This may be taken to say that for SSM the belief that a solution is 'right' (that is, culturally feasible and systemically desirable), is meaningless except where such a belief is sustained in some way *external to the methodology*. That for the word 'right' to have any meaning at all, an 'inner process stands in need of outward criteria.' (Wittgenstein, 1953, 580, p153e).

Here then is evidence that the integration or synthesis of hard and soft systems cannot address the issue of methodological insufficiency within either paradigm, but rather that a closer examination of language and meaning shows the paradigms to diverge, creating the perceived epistemological misunderstanding found in attempts at integration and synthesis, rather than the intended convergence towards a common epistemological base. The epistemologies diverge: the *philosophies* are based upon a common ontological, and specifically Cartesian, foundation.

CONCLUSION

The insufficiency of both 'hard' and soft' paradigms may be recognised as the common belief that knowledge (gained empirically or hermeneutically), is capable of defining and enabling externally justifiable requirements for action in the real world; whereas in practice both paradigms critically depend upon the need for external human intervention in order to progress from methodological description into the solution space, and from there to the world. The mere accumulation of technique cannot meet the requirement for more appropriate and sufficient methodologies. Through integration and synthesis we are building bridges, not across the 'Great Divide', but across already solid ground. The epistemological diversity is intuitively bridgeable: the ontological underpinning is already in place. If the metaphor may be expanded: it is the rivers of Selection (from reality) and Decision (at which point to jump into the solution space) which are the true divisions between internal methodological description and solutions sustainable through external criteria, and which continue to flow with the dynamic complexity of the real world.

If what we can say that is meaningful about a system depends on what Wittgenstein (1953, 7, p5e) called the 'language game' we are playing, then in using specific epistemological bases we acquire something of a vocabulary. In attempting to mix our epistemological base, we may make assumptions analogous to literal mistranslation. First we must establish what can be said that is meaningful about a situation. An appropriate and sufficient epistemology follows from that, and does not precede it.

Current systems theory, expounded through epistemological debate, does not provide the foundation required to define the boundary conditions of paradigms created from such debate. External criteria are required. There already exists a body of work, for example in logico-linguistics, which takes relevant ideas such as Wittgenstein's concept of language and meaning, to *within* existing epistemologies. We must develop levels of understanding which *precede* specific epistemological enquiry, and among the first such studies must be those of language. Initial indicators from the author's present work is that such research is fruitful, and may well provide us with a means towards plurality without prejudice.

REFERENCES

Avison and Wood-Harper, 1991, Conclusions from Action Research: The Multiview Experience, in "Systems Thinking In Europe", Jackson et al., ed.,Plenum, New York.

Checkland, 1981, "Systems Thinking, Systems Practice.", Wiley, Chichester.

Checkland, 1992, SSM in Information Systems Design: History and some Current Issues, *Systemist*, Vol.14(3), pp90-92.

Checkland and Scholes, 1990, "Soft Systems Methodology in Action.", Wiley, Chichester.
Doyle and Wood, 1991a, The paradigm shift and the protozoon, *Systemist*, Vol.13(3), 131-134.
Doyle and Wood, 1991b, Soft Systems and Systems Engineering: across the great divide., Addendum to proceedings of UKSS conference 1991, Plenum, New York.
Fitzgerald, 1992, On linking Soft Systems methods and Information Systems methods, *Systemist*, Vol.14(3). pp126-127.
Flood, 1990, "Liberating Systems Theory.", Plenum, New York.
Flood, Jackson, and Keys, 1989, "Systems Prospects: The Next Ten Years of Systems Research.", Plenum, New York.
Galliers, 1992, Soft Systems scenarios, and the planning and development of Information Systems, *Systemist*, Vol.14(3), pp146-159.
Gregory, 1992, SSM to Information Systems: a logical account, *Systemist*, Vol.14(3), pp180-189.
Hirschheim and Klein, 1989, Four paradigms of information systems development, *Communications of the ACM*, Vol.32, No.10, pp1199-1215.
Holwell, 1992, SSM Information Approach: A Critique, *Systemist*, Vol.14(3), pp93-98.
Jackson, Mansell, Flood, Blackham, Probert (ed.), 1991, "Systems Thinking in Europe.", Plenum, New York.
Jayaratna, 1992, Should we link SSM with Information Systems, *Systemist*, Vol.14(3), pp108-119.
Jones, 1992, SSM and Information Systems, *Systemist* Vol.14(3), pp123-125.
Keys and Roberts, 1991, Information Systems Development and Soft Systems Thinking, in "Systems Thinking In Europe", Jackson et al., ed.,Plenum, New York.
Lewis, 1992, The feasibility and desirability of a closer linking of SSM with data-focused information systems development, *Systemist*, Vol.14(3), pp168-179.
Lyytinen, 1988, Stakeholders, information systems failures and Soft Systems Methodology: an assessment., *Journal of Applied Systems Analysis*, Vol.15, pp61-81.
Magee, 1987, "The Great Philosophers.", BBC Publications, London.
Merali, 1992, Analytic Data Flow Diagrams.,*Systemist*, Vol.14(3), pp190-198.
Miles, 1988, Combining soft and hard systems practice: grafting or embedding?, *Journal of Applied Systems Analysis*, Vol.15, pp 55-60.
Mingers, 1992, SSM and Information Systems: An Overview, *Systemist*, Vol.14(3), pp32-89.
Paton, 1992, Using SSM for Information Systems, *Systemist*, Vol.14(3), pp120-122.
Prior, 1991, Dangerous liaisons: a reply to Doyle and Wood., *Systemist*, Vol.13(2), pp81-85.
Prior, 1992, Linking SSM and IS Development, *Systemist*, Vol.14(3), pp128-132.
Sawyer, 1991, Linking SSM to DFDs: the two epistemological differences., *Systemist*, Vol.13(2), pp76-80.
Sawyer, 1992, A contribution towards the debate on linking SSM to IS, *Systemist*, Vol.14(3), pp199-201.
Sawyer and Trahern, 1991, Integrating Hard and Soft Systems Thinking: The Use of Current Activities Description in OPIUM, in "Systems Thinking In Europe", Jackson et al., ed., Plenum, New York.
Stowell, 1992, SSM as a vehicle for client-led design of Information Systems: Using 'ideal-type' mode 2, *Systemist* Vol.14(3), pp 99-107.
Wittgenstein, 1953, "Philosophical Investigations.", Basil Blackwell, Oxford.
Wood and Doyle, 1989, Doing the right thing right: an exploration between Soft Systems Methodology and Jackson System Development., ISGSR Conference 1989, Vol.IV, pp146-155.
Wood-Harper, Antill, Avison, 1985, "Information Systems Definition: the Multiview Approach.", Blackwell Scientific Publications, Oxford.

APPLICATION-ORIENTED TOOLS FOR SOFTWARE DEVELOPMENT

Donald Millington[1], Edwin Gray[2], and Iain M. Tulloch[1]

[1]Department of Computer Science
University of Strathclyde
Glasgow G1 1XH
Scotland

[2]Department of Computer Studies
Glasgow Polytechnic
Glasgow G4 0BA
Scotland

INTRODUCTION

Boehm[1] has drawn attention to the potential of domain specific programming languages for improving software productivity. This is in line with the view that productivity in software development is improved through the use of higher-level languages and that higher-level languages inevitably become more domain specific because they approach the language of the software end-user.

In the area of corporate planning applications, Tillman and Watson[2] discuss a survey of the use of 'planning languages' and general purpose programming languages in corporate modelling. The survey covered 10 companies in the vicinity of Atlanta, Georgia, U.S.A. Of these, two used general purpose procedural languages - APL and FORTRAN - and the remainder used non-procedural planning languages ("modelling software") - products identified as IFPS (Interactive Financial Planning System), PROFIT II, and FCS (Financial and Corporate Modeling System). As well as finding differences in the ways software was developed in the two groups, the study concluded that "The use of nonprocedural software also seems to be associated with projects that can be expected to consume less time and money." Although guidelines are quoted for the characteristics of the corporate models that were included in the study, it is not clear that they provide a sufficiently tight definition to ensure that like was compared with like, and therefore there may be an element of ambiguity in the conclusion given. However, the authors appear to support the view that specialist software development tools/languages do give advantages over general purpose facilities.

Moving beyond mere programming languages, CASE Outlook[3] reports on the availability of an 'application-specific CASE tool' aimed at the area of banking applications. MFS (Master Financial System) includes an active integrated data dictionary, a core system

Systems Science, Edited by F.A. Stowell *et al.*,
Plenum Press, New York, 1993

of commonly required banking functions and the facility for these to be tailored to the needs of specific users. Described in this way, the distinction between this and a software package with parameters may not be clear. The essential distinction is that the package is always the same code in operation, whereas application-oriented tools are concerned with building appropriate code for a specific environment.

Hrushka[4] identifies a parting of the ways for technically oriented and commercially oriented methods and tools, and expresses the view that "major application domains, like car manufacturing, insurance, telecommunications, etc. will have specialized methods and tools".

The current 'big' development of application-oriented software tools is that of the ATMOSPHERE project (Obbink[5]). ATMOSPHERE (Advanced Tools and Methods for System Production in Heterogeneous, Extensible, Real Environments) is a mega-project supported by the CEC under the Esprit II programme, and encompassing both hardware and software development. The work of ATMOSPHERE includes a major workpackage constructing system engineering environments in parallel in each of four major partners. Each partner is carrying through similar work, but with some variation of approach, for a specific application domain. These domains are:

i) Aerospace systems and distributed information processing systems.
ii) Digital computer network systems.
iii) Communication systems.
iv) Process control systems

The reason for following this route of development rather than creating a single monolithic general purpose System Engineering Environment was "the will of the ATMOSPHERE partners to have the project producing practical and tangible results and to exploit them."

There is thus a wide measure of support for the approach of developing application-oriented software tools, even though there are few available in the market place.

SOFTWARE DEVELOPMENT FOR HEALTH CARE APPLICATIONS

The need for health-care systems is huge. One has only to count the number of hospitals in any country, and the number of different activities which take place within any hospital to realise that the potential requirement for (high-quality) software is one which is difficult to meet (even if the needs of community medicine are totally ignored - which is unallowable). From early days of health service computing some creation of software tools has taken place. The MUMPS programming language (National Bureau of Standards[6]) had its origins in health computing in the 1960s, although few would regard it as domain-specific in its character.

The outline architecture for a software tool presented in this paper has emerged from work carried out in a project (Healthbench) supported under the CEC's AIM (Advanced Informatics in Medicine) Programme. The partners in Healthbench Project were INESC (Lisbon) and the information systems division of the Ministry of Health from Portugal, BAZIS (Leiden) from The Netherlands, ITS (Madrid) from Spain, and Strathclyde University (Glasgow) from the U.K. The project surveyed the current use of software tools in developing health-care applications (Healthbench[7]), and developed a perspective for a software support environment for the whole development life-cycle of such applications (Healthbench[8], Hooymans[9]). The proposals made subsequently in this paper concentrate on the implementation phase of the development life-cycle, and point towards a wider perspective on domain specific application generators.

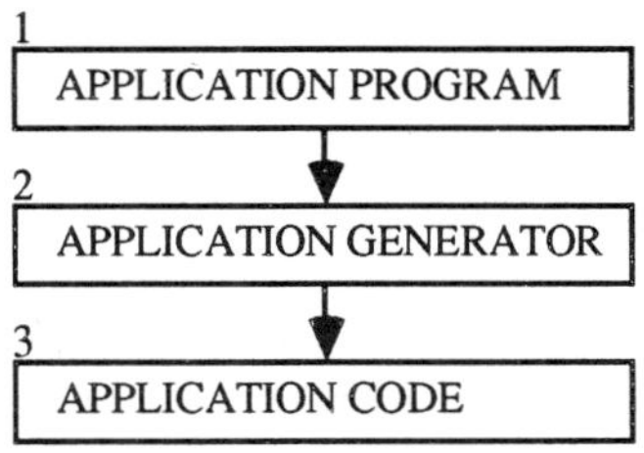

Figure 1 . Simple model of software implementation.

A MODEL OF SOFTWARE IMPLEMENTATION

A simple model of software implementation is given in Figure 1, in which general terms have been used capable of a range of specific interpretations. In one perspective the application "program" is presented in a programming language such as COBOL or PASCAL, the 'application generator' is a compiler and the 'application code' is object-code. An alternative view could be that the application "program" is expressed using fourth-generation language (4GL) facilities, the 'application generator' is the 4GL environment and the 'application code' is the resulting executable application.

If the applications to be implemented are restricted to a particular domain, e.g. medicine and health, the language in which the specification is presented might refer directly to commonly occurring processes and data in that domain, e.g. 'make an x-ray appointment' and the 'patient record', where these are known to the system. It may also be possible to identify processing capabilities required in many if not all applications in the chosen domain, e.g. costing of any service used by a patient.

Considerations such as those above suggest a more complex model of software implementation such as that of Figure 2. In this figure the symbols down the centre numbered 1, 2, 3 are effectively those of Figure 1. Symbol 4, the "Application Data Base" is 'organised' by the Application Processor for use by the "Operational Application". The symbols on the left-hand side of the diagram (numbered 5-8) represent domain reference data - or application knowledge - which may be used by both the generator ("Application Processor") and by the application when operational.

The symbols on the right-hand side - labelled as 'engines' in order to indicate their active nature - are envisaged as software tools called upon by the "Application Processor" to ensure that commonly required capabilities are incorporated into the implemented application. Often these tools may be enforcing standards - or at least encouraging standardisation - in their particular areas of concern. The roles of those shown might be:

Language Engine — to handle the tailoring of the application to a particular end-user (national) language.

Costing Engine — to ensure that appropriate applications include facilities for recording and reporting costs associated with the work of the application.

Security Engine — to incorporate appropriate security and data protection facilities in the application.

User Interface Engine — to handle in a consistent manner the user-interface of the application.

Quality of Care Engine — is specifically Health Service oriented and concerned that quality of care measures and quantities should be established in all appropriate applications.

At present the roles of these engines (if performed at all) are carried out manually but in future they could be, for example, automated, perhaps as expert systems. Such a development would require the initial "Application Program" to include some appropriate indicators (or parameters) which guide the engines in their work, or for the work to be carried out interactively by the designer.

Amongst these engines there are several which may be not application-oriented, and therefore built in a general manner, and interfaced to the application-oriented components. Neelamkavil and Mullarney[10] have, for example described such an approach in respect of the user interface.

"BACKWARD EXTENSION" - EARLIER LIFE-CYCLE TASKS

The first symbols on Figures 1 and 2, labelled 'Application Program', may be thought of as being a high-level representation, not necessarily procedural in form. With this perspective it may be recognised that a second area of study should be concerned with systems analysis and design and the creation of a specification from which the application may be programmed.

Trends are towards the automation of this step and the influence of the specific application domain might then be expected to appear formally in the specification at this earlier stage. This raises the questions as to how far any methodology for systems analysis and design (e.g. SSADM[11] or Yourdon/De Marco[12,13]) may be able to be application or domain

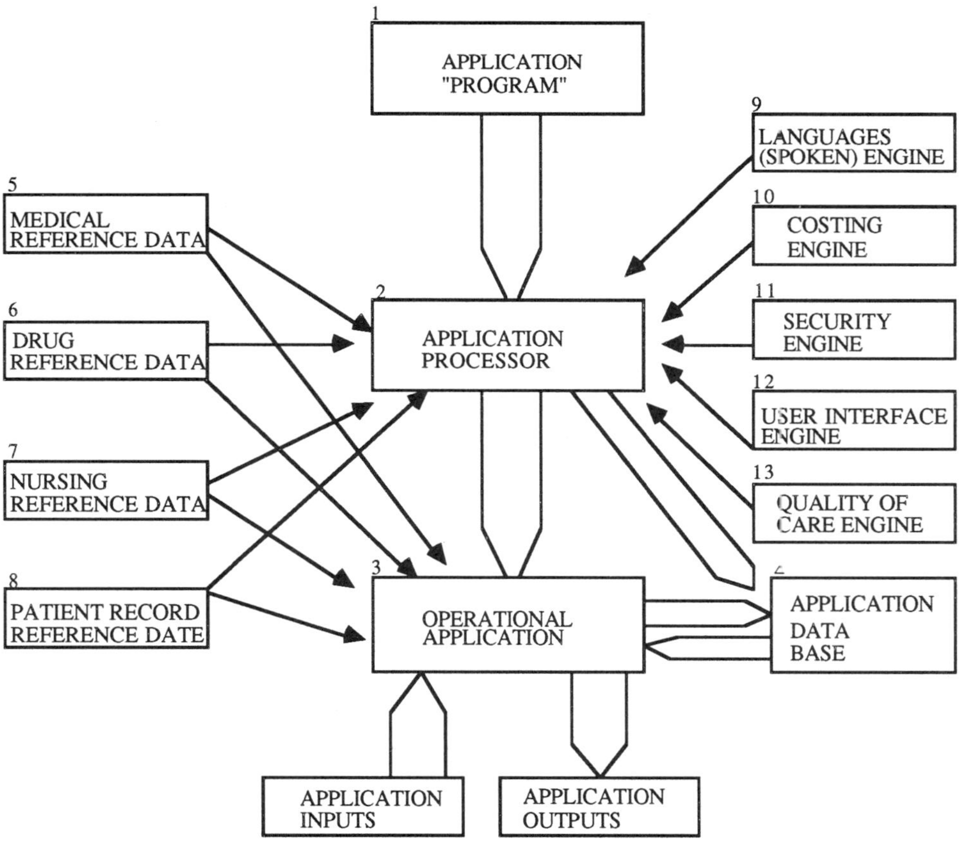

Figure 2. An architecture of software implementation.

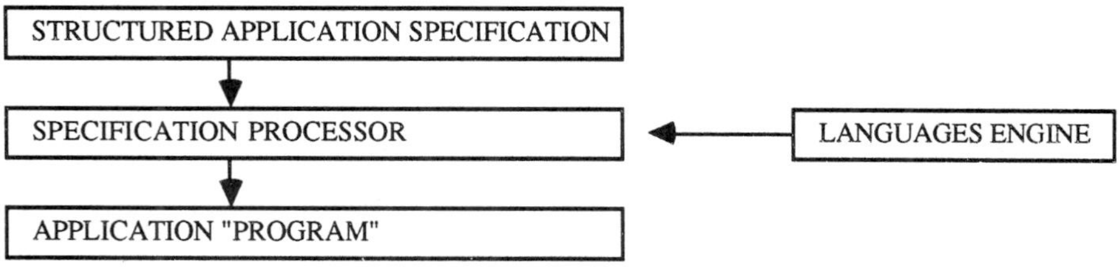

Figure 3. A model for a specification processsor.

oriented and what advantages would result. It also indicates the need for integration with CASE (Computer Aided Software Engineering) tools for systems analysis and design ('front-end' or 'upper case' tools in the language of some authors).

Figure 3 represents this situation and indicates a further role for a "Languages Engine", perhaps taking account of a European dimension and the fact that the Structured Application Specification may have been produced in any one of a range of languages. The view suggested is that the Specification Processor uses the Language Engine to create the Application Program in a chosen common base language.

DEVELOPMENT WORK

Most of the boxes in Figures 2 and 3 represent development ideas at the present time, particularly in respect of orientation to a particular application domain. Two important points for further development may be noted from Figure 2.

First, the symbols representing "engines" and those representing "reference data" will have their counterparts in any other application domain, so that Figure 2 is itself a particular example of a more general framework and therefore a model for a family of application-oriented application generators. Secondly the CEC's AIM programme has supported a number of projects working to define the various "Reference Data" and their uses, for example in expert system or decision support environments.

In Figure 3 the specification processor module is the repository of all syntactical knowledge about the specification language. The semantic knowledge is shared between this module and the application "program" module. Correct specifications, according to the specification language, are accepted and semantic routines are called to store them in a standard internal format which will be used eventually by the application "program" module. This internal format is domain dependent and can be any combination of tables, syntax trees, and lists of other data structures.

Cleaveland[14] suggests a seven step method for building application generators:

- Recognising domains.
- Defining domain boundaries.
- Defining an underlying model.
- Defining the specification input.
- Defining products.
- Implementing the generator.

Even though these steps may not be applied explicitly and in the sequence suggested, they nevertheless constitute a very useful abstraction for building application generators. The first six tasks are to be executed by the domain analyst and the seventh is a task for the domain designer. An example of a generator developed according to those steps is currently being undertaken by the authors.

CONCLUSIONS

This paper has presented a model for the development of a domain dependent application generator for medical and health applications.

It has done so with the objective of creating tools for improving software engineering productivity in that domain. Such tools are likely also to contribute to significant improvements in software quality.

The model presented is seen to be one of a family that could be created for different application domains and therefore points to a generic model for such software development systems.

ACKNOWLEDGEMENTS

This paper has come about largely through discussions with our Healthbench partners. Their involvement, and the support of the CEC AIM Programme are appreciated.

REFERENCES

1. B.W. Boehm, Improving software productivity, *Computer*, pp43-57, September (1987).
2. G.J. Tillman & H.J. Watson, The impact of planning languages on the development of corporate models, *Database*, 19:3/4, pp24-32, Fall/Winter (1988).
3. Case, Outlook 89, The CASE state-of-the-industry report : application specific CASE, *CASE Outlook*, 3, pp 46-47, (1989).
4. P. Hrushka, Should CASE be application dependent, *in*: "Proc. European Symposium on Software Development Environments and CASE Technology," pp 143-152, Springer-Verlag, Berlin (1991).
5. H. Obbink, Systems engineering environments of ATMOSPHERE, *in*: "Proc. European Symposium on Software Development Environments and CASE Technology," pp1-17, Springer-Verlag, Berlin (1991).
6. National bureau of standards MUMPS language standard, NBS, Washington (1976).
7. Healthbench Deliverable 1, The use of languages, tools and methodologies in health applications, Report to CEC AIM Programme, (1989).
8. Healthbench Deliverable 3, Description of the general architecture of healthbench, Report to CEC AIM Programme, (1990).
9. M. Hooymans, HEALTHBENCH - health information and decision support workbench. General architecture of healthbench, IMIA working conference on software engineering in medical informatics, Amsterdam, (8-10 October 1990).
10. F. Neelamkavil, & O. Mullarney, Separating graphics from application in the design of user interfaces, *The Computer Journal*, 33:5, pp 437-443, (1990).
11. National Computing Centre, SSADM manual, NCC, Manchester (1986).
12. E. Yourdon, "Modern Structured Analysis," Prentice-Hall, Englewood Cliffs (1989).
13. T. De Marco, "Structured Analysis and System Specification," Prentice-Hall, Englewood Cliffs (1979).
14. J.C. Cleaveland, Building application generators, *IEEE Software*, 5, pp25-33, (1988).

HOW TO DEAL WITH WICKED PROBLEMS USING A NEW TYPE OF INFORMATION SYSTEM

Severin Isenmann

Forschungsinstitut fur anwendungsorientierte
Wissensverarbeitung (FAW)
Helmholtzstraße 16
D - 7900 Ulm (Donau)
Germany
e-mail: isenmann@faw.uni-ulm.de

INTRODUCTION

Processes of problem solving involve more than just finding solutions to well described problems. They include identifying the problem, understanding it, looking for solutions, and weighing different solutions. For dealing with such problem solving processes, an approach known as IBIS will be introduced, which has been implemented in a system called HyperIBIS. Its operation is described through a discourse on the construction of a new railway track. Finally, extensions to the approach are proposed which will solve some difficulties that occur when using IBIS.

PROBLEM SOLVING

Problem solving is much more than selecting a method for solving a well defined problem. Problem solving is a process which consists of identifying that something is wrong, understanding the causes, defining objectives, developing possible approaches to reach these objectives, and comparing alternative approaches. But it should be clear that every approach considered may introduce new problems during implementation. Winograd and Flores[14] give many examples for such problems. Additionally, all participants in the problem solving process may have differing interests, opinions, and judgements on every aspect of this process. To contrast these problems with, for example, mathematical problems, which can often be solved by using formal procedures, they have been called "wicked" problems[9]. Typically, problems of this kind can be found in politics, e.g. when dealing with new laws or the construction of roads, but also in the field of strategic planning in the industrial sector. The solving of wicked problems requires the development of a structured view of the problem and its solutions. This requires a discourse among the participants, where issues are brought

Systems Science, Edited by F.A. Stowell *et al.*,
Plenum Press, New York, 1993

up and disputed, and argumentations are constructed in defense of or against the different positions. This process should lead to better founded solutions, where more aspects have been taken into consideration.

THE IBIS APPROACH

Dealing with wicked problems requires a new approach in the field of information systems that goes beyond the efficient storing, retrieving, and processing of well structured data. Such an approach has been suggested by Rittel and Kunz, known as the Issue-Based Information System (IBIS)[6]. The aim of IBIS is to support the whole process of problem solving by structuring and documenting the discourse around a problem. It is based on the fact that problem solving takes place by arguing. Basic elements of IBIS are not just records of data. Instead, this type of information system is based on issues emerging when dealing with a problem, differing assessments, alternative procedures, controversial opinions, and various lines of reasoning. The most important elements of IBIS are issues - questions emerging in the context of a problem solving process. Several types of issues are distinguished (see Table 1).

Table 1. Types of issues within the IBIS-concept

•	deontic issues	"should something be the case?" or "should a certain instrument be used to reach an end?"
•	factual issues	"what is the case?" or "what will be the case?"
•	instrumental issues	"which alternative actions are there that could lead to a desired goal?"
•	explanatory issues	"what are the causes for a certain state?" or "what are the expected consequences of an action?"
•	conceptual issues	"what is meant by a certain concept?"

Issues are the nuclei of discourse. Positions are taken, answers are given, and alternative measures are developed with relation to the issues. Positions and answers can be supplied with pros and cons. All of these elements may induce further issues, so that a network develops which contains all aspects regarded as relevant to the people involved in the problem-solving process.

HYPERIBIS

HyperIBIS is a computer-based tool that has been developed to support problem solving according to the IBIS approach. The tool has been realized using concepts of the hypertext-paradigm. Hypertext can be regarded as a special method for the representation of knowledge using a computer. The basic ideas are the partitioning of the knowledge into knowledge chunks that can be connected together, and the highly interactive handling of the knowledge. The knowledge-chunks - these can be textual units, but graphics are also possible - are stored as "hypertext-nodes" and the connections are realized by "hypertext-links". These links are also used for accessing the nodes. HyperIBIS provides a set of predefined types of nodes for the representation of different categories of issues, positions, arguments etc., and several link-types that represent relationships among the nodes. For a more detailed discussion of hypertext see an article by Conklin[1], or books written by Nielsen[8] or Kuhlen[5].

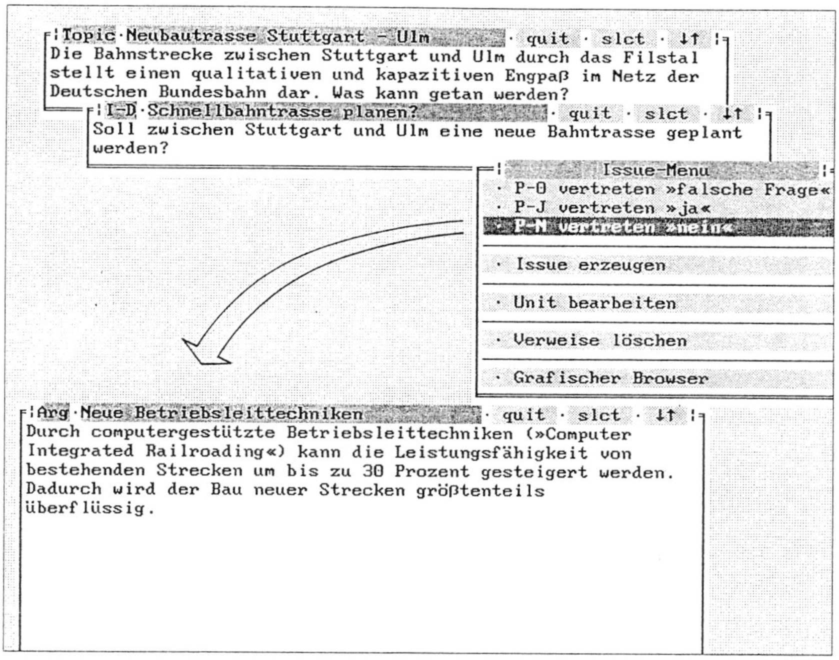

Figure 1 Inserting new arguments into the system. By selecting the highlighted item from the issue menu, a new argument-node will be created, which is connected to the deontic issue-node (I-D). The new argument-node will then be opened for insertion of the verbatim argument.

Working with HyperIBIS

This section will describe how HyperIBIS is used to structure and document a controversial debate. This will be done using some excerpts of the debate on the construction of a new high-speed railway track between the cities of Stuttgart and Ulm (see Figure 1).

One of the first issues that might arise is a deontic issue that asks *whether* to plan a new railway track. As any node, this deontic issue-node is identified by a significant "trigger phrase" ("Plan a high-speed track?"). A more detailed description of the item can be stored in a text field associated with every node. To a deontic issue, two contrary positions are possible: "yes" or "no". A third position, that can be taken with every type of issue, is called "wrong question". This position can be taken if someone does not agree with the question that is posed by the corresponding issue. This can be the case, for example, if the question is based on premises that are not accepted. All positions have to be reasoned by arguments, which are listed by their trigger phrases in Table 2. New arguments can be inserted at any

Table 2. Excerpt of the discourse whether to plan a high-speed railway track or not

Deontic issue:	Plan a high-speed railway track between Stuttgart and Ulm?
Yes:	**Argument 1**: Insufficient capacity of the existing track.
	Argument 2: Expected amount of traffic in the future.
	Argument 3: Insufficient flow of local traffic.
	Argument 4: Insufficient flow of long distance traffic.
No:	**Argument 1**: There are more important tasks.
	Argument 2: Problems can be solved by "Computer Integrated Railroading".

time. Figure 1 demonstrates how a new argument supporting the no-position on the current issue can be inserted into the knowledge-base. Each of these arguments can be the nucleus of a new issues. It could be discussed, for example, *whether* the flow of the long distance traffic should be improved. This may lead to another level of discussion, where the question would be *how* to improve the long distance traffic flow. At this level, the construction of a new railway track would be only one of a set of possible solutions.

Table 3. Instrumental issue on alternative tracks

Instrumental issue: What are the possible configurations for the high speed railway track?

Alternative 1: Track A-IV
Alternative 2: Track B-l
Alternative 3: Track C-I
Alternative 4: Track D-I
Alternative 5: Track H (a new track along an existing highway).
Alternative 6: Track K (a modification of the existing track).

Without having to decide at an early stage of the discussion *whether* to plan a new track, one can already ask, what a new track could look like. This is done by the raising of the instrumental issue shown in Table 3. In principle there are an infinite number of alternative tracks, but only few of them are realizable and can be taken into consideration. The considered alternatives will depend on the people who express their opinions on the problem. Typically, these are experts from different fields, but also people who will be affected by the implemetation of a track in some way.

Around each alternative, a separate discourse develops, as shown in Table 4. A factual issue initiates the description of the characteristics of track K. This issue is divided into several sub-issues concerning line layout, capacity, possible speeds, and costs. It should be noted that answers, even to these *factual* issues, are not necessarily undisputed. Two experts, for example, may calculate significantly different capacities for the same track. An explanatory issue asks for the effects that are expected, if this track were to be implemented. These effects can be classified into different categories. If a certain alternative is to work, it is obvious that there must be some desirable effects that correspond to the global objectives (e.g. improvement of long distance traffic flow or increase of capacity), which should have been discussed within deontic issues ("improve long distance traffic flow?" or "increase capacity?") on a global level. Otherwise, this alternative is likely to be rejected. But normally there will also be side effects, some of which will be unwelcome. This may lead to the emerging of deontic issues, such as whether to accept these effects. If they are not acceptable, the alternative can possibly be modified so that these effects can be removed, or this particular alternative may be rejected. Finally, a deontic issue asks whether track K ought to be implemented. The different positions are supported by arguments which might refer to characteristics of this track, to objectives, to supposed effects, or to other alternatives. As soon as no new issues are emerging and no new arguments are articulated, a decision can be made. For this, all aspects that have been articulated during the discourse should be taken into consideration. The system can support this by showing every aspect within its context. It is capable of creating maps that graphically show the relationships among different issues, positions, and arguments. This enables the decision makers to weigh different aspects and to find or to develop more satisfactory solutions.

RELATED WORK

Systems and models for argumentative problem solving processes have been drawing more attention during the past several years. Jarczyk, Löffler and Shipman[4] give a

Table 4. Excerpt of the discourse on track K

Alternative 6:		Track K (a modification of the existing track).	
Factual issue:		Characteristics of track K?	
	Factual issue:		Line layout of track K?
	Factual issue:		Capacity of track K?
	Factual issue:		Possible speeds on track K?
	Factual issue:		Costs of track K?
Explanatory issue:		Effects of track K if implemented?	
Deontic issue:		Implement track K?	
Yes:	**Argument 1:**	Better quality of long distance traffic.	
	Argument 2:	Higher capacity than track H.	
	. .		
No:	**Argument 1:**	Problems with groundwater.	
	Argument 2:	Many houses will have to be torn down.	

comprehensive overview of the best-known approaches. The characteristic of all approaches is a rigorous distinction between content information (information stored *within* a node) and structure information (membership of a certain category of nodes or relationships *among* different nodes). Probably the most well-known system is gIBIS[2], which is regarded as the first computer-based realisation of some basic concepts of IBIS. Special purpose systems include SEPIA[12] whose aim is to support collaborative authoring, and JANUS[3] which supports argumentative kitchen design. SEPIA combines ideas of IBIS with an approach suggested by Toulmin[13]. The systems EUCLID[11] and SIBYL[7] rest on special formalisms called ARL (Argument Representation Language) and DRL (Decision Representation Language) to represent argumentations.

The most important difference between HyperIBIS and other approaches is the explicit distinction among different categories of issues within HyperIBIS. This results in much clearer distictions among objectives, causes of problems, facts, and approaches to solve a problem, and leads to more structured and objective discourses.

EXTENSIONS TO IBIS

Working according to the IBIS-concept generates disputes which lead to the development of more structured views of problems and their solutions. Although the quality of decisions made should be better, this approach may make it more difficult to come to decisions. Another problem is that the users of IBIS can raise issues very freely. For example, an explanatory issue, "What are the environmental effects of track A?", can be raised corresponding to one alternative, while there is no obligation to raise an equivalent issue corresponding to all other alternatives. This may lead to unfair dealing with different alternatives.

Rittel has made some suggestions to enrich IBIS by some procedural rules[10]. He suggested raising fixed issues according to each alternative. These issues concern whether an alternative will work, whether the prerequisites for the implementation of an alternative will be available, whether there are side effects, and whether the advantages of an alternative will outweigh its disadvantages. The raising of such issues can easily be done automatically by the system whenever a new alternative is proposed. But for equal dealing with each alternative this might not be sufficient, because these issues are too general. In many discussions, very specific issues may arise which cannot be anticipated. It will therefore be possible to reject some alternatives on account of some unwelcome effects e.g. on the groundwater, which have not been raised within the context of other alternatives. For this reason, it should be possible for users to freely create "generic issues" associated with instrumental issues. For example, an explanatory generic issue might be "What are the effects on the groundwater?". Whenever a new alternative responding to the instrumental issue is articulated, an explanatory issue such as "What are the effects of this alternative on the groundwater?" will automatically be raised. This will lead to a more fair discussion of different alternatives.

When attempting to reach a decision, the decision makers will be confronted with many arguments which are documented within the system. This will make it very difficult to weigh up the alternatives. When considering arguments in the system, it may be discovered that there are arguments which occur in connection with several alternatives, e.g. "This alternative will improve the flow of long distance traffic.". Such arguments are not relevant to the process of comparing those alternatives to which they refer. In the future, mechanisms which hide such irrelevant arguments during the decision process, and make it easier to compare alternatives, are planned.

REFERENCES

1. J. Conklin, Hypertext: An Introduction and Survey, *IEEE Computer*, 20 (9), pp. 17-41, (1987).
2. J. Conklin and M.L. Begeman, gIBIS: A Hypertext Tool for Exploratory Policy Discussion, *ACM Transactions on Office Information Systems*, 6 (4), pp. 303-331, (1988).
3. G. Fischer and R. McCall, JANUS: Integrating Hypertext with a Knowledge-based Design Environment, *in*: "Proceedings of the Hypertext '89", pp. 105-117, New York, ACM, (1989).
4. A.P.J. Jarczyk, P. Löffler and F.M. Shipman, Design Rationale for Software Engineering: A Survey, in: "Proceedings of the 25th Hawaii International Conference on System Sciences", pp. 577-586, (1992).
5. R. Kuhlen, "Hypertext - ein nichtlineares Medium zur Darstellung von Wissen und Erarbeitung von Information", Springer-Verlag, Berlin, (1991).
6. W. Kunz and H.W.J. Rittel, Issues as Elements of Information Systems, *Working Paper No. 131*, Institute of Urban and Regional Development, University of California, Berkeley, California, (1970).
7. J. Lee, SIBYL: A Tool for Managing Group Decision Rationale, in: "Proceedings of the Conf. on Computer Supported Cooperative Work", pp.79-92, New York, ACM, (1990).
8. J. Nielsen, "Hypertext and Hypermedia", Academic Press, San Diego, (1990).
9. H.W.J. Rittel and M. Webber, Dilemmas in a general Theory of Planning, *Policy Sciences*, 4, pp. 155-169, (1973).
10. H.W.J. Rittel, The Reasoning of Designers, *Arbeitspapier A-88-4*, Institut für Grundlagen der Planung, Stuttgart, (1988).
11. P. Smolensky, B. Fox, R. King and C. Lewis, Compter-Aided Reasoned Discourse or, How to Argue with a Computer, in: "Cognitive Science and its Applications for Human-Computer Interaction", R. Guidon, ed., pp. 109-162, Hillsdale, Erlbaum, (1988).
12. N.A. Streitz, J. Hannemann and M. Thüring, From Ideas and Arguments to Hyperdocuments: Travelling through Activity Spaces, in: "Proceedings of the Hypertext '89", pp. 343-364, New York, ACM, (1989).
13. S. Toulmin, "The Uses of Argument", Cambridge University Press, (1958).
14. T. Winograd and F. Flores, "Understanding Computers and Cognition: A New Foundation for Design", Ablex, Norwood, (1986).

METAPHORICAL THINKING AND INFORMATION SYSTEMS: THE EXAMPLE OF THE MECHANISTIC METAPHOR

Richard Kamm

Department of Computer Studies
University of the West of England, Bristol
Coldharbour Lane
Bristol, BS16 1QY

INTRODUCTION

The purpose of this paper is to explain some ideas which are being developed about the application of organizational metaphors to the subject of information systems. The inspiration was provided by Flood and Jackson's (1991) use within a systems thinking context of Morgan's (1986) extended discussion of common images of human activity. The objective here is to show the relevance of organizational metaphors to the use of information and to discussion of the nature of information requirements within organizations. Specific reference is made to the mechanistic model of organizations because it is the metaphor which has exercised the greatest influence on information systems development as commonly practised. It is argued here, in an effort to extend discussion of the nature of metaphorical thinking, that principles relating to the management of information themselves reinforced the philosophy of the mechanistic approach to organizations and help to explain its importance to organizational theory in general.

MECHANISTIC THINKING AS METAPHOR

Almost every discipline has at some time been affected by the desire to "mold our world in accordance with mechanical principles." (Morgan 1986, 20) A mechanicist philosophy essentially involves separation and classification to achieve given objectives. Knowledge is divided into compartments with subject areas being restricted in their areas of competence. Humans are encouraged to specialise in particular kinds of work, and even to separate their roles as employees from their private lives. The authority of such ideas can be traced to the development of philosophical principles from at least the 17th Century which may be characterised, in Weber's terms, as "instrumental rationality", the progressive mastery by humanity of the world through an understanding of the general principles of nature and society (Mommsen, 1989, 161). While these ideas can be called mechanicist, the fact that

Systems Science, Edited by F.A. Stowell *et al.*,
Plenum Press, New York, 1993

they run much deeper than a fascination with modern technology is suggested by argument that the development of machinery was itself made possible by the embedding of these principles in western culture. The force of these ideas has been developed through assumptions about the nature of:

- the universe, operating through a set of fixed, discoverable laws which ensure regular patterns of behaviour (Leatherdale, 1974, 138-39, 230-31).
- society, in which similar regularities can be detected as a result of some unvarying aspect of human nature; aggressiveness in Hobbes, acquisitiveness in Adam Smith, the desire to avoid pain and maximise pleasure in Bentham (Deutsch 1951, 235). Society should tends towards equilibrium and its undesirable features are a result of the mechanism having been upset (Stark, 1962, 124).
- knowledge about society, which is based firmly on an atomistic understanding of individuals, rather than studying collectivities as a whole (Stark, 1962, 109-110). Social phenomena are the sum of the activities of individuals and no more.
- the language in which that knowledge is expressed. If facts are objective then they can only be described in literal, as opposed to metaphorical, terms. Scientific description is a straightforward conveyance of underlying meaning whereas literary or rhetorical language exists primarily for effect; "a perfect cheat" in Locke's view of metaphor (de Man, 1979).

Two points are relevant here to the use of metaphors in organizational analysis. One is that we do not necessarily require that every aspect of mechanical operations has its exact counterpart in the social institution to say that we are thinking of it mechanistically. A simple point-by-point analogy, a form of metaphor characterised as that of 'substitution' (Black 1962, 31), would miss some of the philosophical undertones outlined above which are essential to the mechanistic approach. Both Adam Smith (1921, 14) and Henri Fayol (1949, 20) compared the social with the biological to reinforce their arguments about the benefits of specialization, but retained underlying assumptions which were clearly mechanistic rather than organismic (Stark, 1962, 171; Morgan, 1986, 25). The force of an organizational metaphor goes beyond that of a simple 'likeness' (Flood and Jackson, 1991, 14), to its conjuring of qualities of orderliness and precision which can themselves shape our expectations of the essential nature of technology, without requiring other aspects of machines, their tendency to wear out for example, to be equally relevant. When we emphasise the mechanical nature of an organization we concentrate on its employees' ability to set targets and achieve them in a planned manner. This is closer to an 'interaction' view of metaphor (Black, 1962, 35-36) whereby the associations which the use of the metaphor provokes are part of a network of assumptions which can also affect the meaning ascribed to the metaphorical term itself (Black, 1979, 29).

The second important aspect of the mechanistic metaphor is that its assumptions represent one of the fundamental trends in intellectual history. Mechanicist principles have been enormously influential in the development of management theory (Morgan, 1986, 29) and machine analogies are commonly used by managers (Kendall and Kendall, 1992). But, to go further than a simple demonstration that the mechanistic metaphor is explicitly used, the coherence of these ideas suggests that they constitute a 'root metaphor', a fundamental structure which implicitly shapes many of our perceptions of the world around us (Black, 1962, 239; Booth, 1979, 64). Many of the individualistic, goal-oriented aspects of mechanistic thought are so familiar that they have come to be taken as common sense, or else developed into coherent ideologies with which to oppose more collectivist or organismic viewpoints.

The implications of the mechanistic metaphor for the study of organizations are that management itself can have general principles which exhibit this form of instrumental rationality. The prime goal of management becomes the determination of optimal solutions to problems in a form which can be specified in mathematical terms. The structure of the organization should encourage the work of specialists, partly on the principle of the division of labour (Morgan, 1986, 31) and partly because management itself is held to be a science with rational principles (Fayol, 1949, 16). Under these circumstances, formal hierarchies and unambiguous job descriptions become a necessary feature of the organization, working through an ethic which emphasises the correct performance of set tasks and loyalty to the institution; the principles of bureaucracy.

A constant interplay between views of technology and organization has been characteristic of the modern world, with the mechanistic metaphor the product of a variety of influences arising from both theory and the application of those ideas in organizations. One of the most important influences, it is argued here, is that of the use of information. This is illustrated to some extent by the adoption of the philosophical assumptions within conventional systems development approaches (Walsham, 1991, 84), but it extends back to the origin of principles of organization in government and industry which are now so familiar that we take them for granted. The organization of information was a significant problem well before Information Systems became a coherent discipline, and the answers given to the questions it raised played an important part in shaping the most prevalent approaches to management in the 20th Century.

THE MECHANISTIC METAPHOR AND INFORMATION MANAGEMENT

The characteristics of bureaucracy which allow it to be characterised as a mechanistic approach to organizations are the written definition of tasks and the specialization of administrative roles. One theme in the history of government in Europe from the 16th to the 20th Centuries is of a gradual extension of formal administration, with salaried career officials, taking the place of a structure in which positions in services such as tax collection and the military were farmed out to those willing to pay for the privilege (Brewer, 1989, 69-70; Dyson, 1980, 34-35). The principle of pre-bureaucratic administration was that large areas of public service were contracted out to private individuals who would then undertake the functions and hope to make a profit from their activities. Rather than separating office from individual and filling the post with the best-qualified candidate, positions were the property of those who could pay for them, and could be sub-contracted, sold and even, in some countries, passed on as inheritance (Elliot, 1984, 85). When private industry took on similar bureaucratic features it accompanied the separation of ownership from control as industrial corporations became too large for an individual to oversee. The development of management tasks effectively meant that new career structures were developed to allow specialised information workers, such as accountants, to acquire a critical role (Chandler, 1990, 32). This process arguably reached its apotheosis in the extension of scientific management, which required the measurement and calculations of workers' efforts and which relied on mechanistic assumptions about the relationship between financial rewards and penalties and the efficiency of production line employees (Morgan, 1986, 30).

If a bureaucratic organization separates individuals from the roles that they take, then it follows that the information which is used in the performance of that role is the property of the organization, not that of the individual who happens to be doing the job at a particular moment. This is such a common principle in 20th Century organizations that it is worth explaining the effects of administration without it. One of the earliest reforms of modern states was the establishment of central archives because a lack of control of information meant that the ambitions of governments would far outstrip their financial resources and administrative capacity (Shennan, 1974, 54). At worst, documents were unavailable because they had been taken by people who, perfectly within their rights, treated them as their personal property. Even as late as the 1880s it was quite legitimate for senior members of

British cabinets to lodge departmental records in their private papers rather than the Public Records Office, an indication of the length of time it took to establish formal procedures of information management.

The establishment of corporate ownership of information, through central archives and standardised procedures, was an important condition for the development of formal administration and an ethic of professional or public service. If authority was to be exercised by those who were entitled to it by their position in the bureaucratic hierarchy, the decisions which were made had to be derived from records of past practice and the precedents which they provided. Bureaucracy has been accused of having many undesirable effects, secrecy, inflexibility, refusal to acknowledge moral implications, but it has also represented a deliberate attempt to treat individuals equally. Whether it affects its own functionaries (opening careers to those talented enough to pass the entrance tests) or the general public (laying down rules by which all citizens are to be treated), written records and precedents are an important source of the consistency on which the benefits of the mechanistic approach depend. Central control of information by the organization has the disadvantage of inflexibility, as indicated by demands for information systems that are better able to adapt to the personal requirements of senior managers (Davis and Olson, 1985, 44; Anthony, 1965, 38), and may alienate the less senior information workers (Zuboff, 1988, 129). It does, however, allow the establishment of a stable framework for the application of rules through a corporate memory which allows the identity of the organization to be developed.

THE MECHANISTIC METAPHOR AND THE NATURE OF INFORMATION

The spread of formal bureaucracy has invariably been accompanied by an intense interest in statistical data. Periods of reform in British government in the late 17th and early 19th Centuries were nourished by a diet of numbers, usually on the uncritical assumption that they represented an accurate representation of social or economic conditions which could be directly affected by administrative action (Porter, 1986, 26). Figures of economic activity and trade, poverty and crime, were all taken to indicate the existence of problems which an extension of state institutions could alleviate (Brewer, 1989, 228; Abrams, 1968, 16; Gatrell, 1991, 251-52). Treating this 'political arithmetic' (Brewer, 1989, 223) as a concept of information as data plus meaning, it might be argued that the bureaucratic use of such statistics takes data as unambiguous and meaning as self-evident, as in the strong connections made between military strength and the level of population (Porter, 1986, 20).

The traditional view of information in systems analysis has it performing the precise function of reducing uncertainty (Davis and Olson, 1985, 200). It therefore becomes a resource for an organization which allows it to achieve targets by taking unambiguous decisions, which is possible where the meaning attached to data is unproblematic. Within the mechanistic approach to organizations, there are certain decisions which must be taken to achieve particular goals which must be satisfied if the organization is to continue to function, whatever the social or political considerations that affect them. If there is no room for debate about the objectives to be achieved, the tendency will be to define management information as data which deals with measurable qualities, notably output per unit of input, and develop the organization on the assumption that this aim is to be achieved through monitoring of output and provision of material incentives. Information can even be said to have a quantifiable value (MacDonough, 1973), supporting the qualities of unambiguity and precision which are so prized within a mechanistic view of organizations.

This emphasis on measurability and fixed responses has some similarity with the decision-based view of organizations developed by Simon and the cybernetic approaches which Morgan (1986, 81-87) groups under the brain rather than the machine metaphor. The connections can be seen in the importance which both give to the optimization of performance and the assumption that human motivation can be modelled as a set of choices which would allow individuals to maximise pleasure and minimise pain (Beer, 1979, 406). The difference between the two can be argued to lie in their conceptions of information. Just as mechanistic philosophy was related to the common image of clockwork in the 17th Century, cybernetics was inspired by the development of communications during and after World War II (Deutsch, 1951, 239), a form of technology which requires explicit definition of the information which forms its inputs and outputs. Within the decision approach to organizations, then, information has to be considered as an entity with a distinctive role and which requires a distinctive style of management. The mechanistic metaphor gives great importance to information, as the means by which commands are transmitted and decisions made, but treats it as a resource like any other. If, as Fayol asserted, there exists a set of management skills which are transferable between the organization of labour, finance, land and machinery, then the same would presumably apply to information, with automation of information work being the prime objective. As Zuboff (1988) has demonstrated, a traditional mechanical form of industrial production will tend to underestimate the mental skills which the use of information requires. The emphasis of Simon (1976, 155) on the importance of informal flows of information would also move it away from a more traditional bureaucratic emphasis on management through defined, vertical communication.

EVALUATING THE MECHANISTIC METAPHOR

As with all ideas which were once fashionable but whose enthusiasts have died off, the mechanistic approach to organizations has been vigorously criticised. From the perspective of organizational theory, it has been argued that the metaphor is appropriate only when objectives are well-understood, tasks are unambiguous and highly structured, the environment is stable and the employees will take orders without complaint. The mechanistic metaphor simply does not fit the demands for adaptability and creative organizational development which economic and political forces have placed on both the public and private sectors (Morgan, 1986, 33). In terms of information systems, the mechanistic viewpoint has been held to produce a stereotyped model of development as the automation of routine clerical work (Walsham, 1991, 84). It tends to ignore not only the social context of systems but also the possibilities for more flexible uses of information which technological advances have produced. Bearing these justifiable criticisms in mind, it is worth outlining some benefits which can arise from the mechanistic viewpoint.

If the growth of organizations has been an essential part of economic growth, as firms seek to take advantage of economies of scale, and of the redistribution of income, through social policies and institutions of welfare states, then the principles of bureaucracy have played an important part in the enhancement of the well-being of the individual. From an information systems viewpoint, the creation of archives and reliance on precedent can mean that measurable progress towards the achievement of objectives is made possible, as is the establishment of consistency in the application of regulations. For individual citizens, the power of corporate bureaucracies they deal with may appear overwhelming, but at least those organizations are subject to conventions which constrain their freedom of manoeuvre. If rules are to be rigidly followed, they must be written down and applied equally in all circumstances. Arbitrary law-enforcement, the classic criticism of pre-bureaucratic government, leaves the individual more vulnerable because there is no means of presenting a case against the authorities or of understanding why their decisions have been made.

Ultimately, the strength of an organization with mechanistic priorities may lie in its ability to retain information for its own use, under the kind of official secrets laws which were attached to governments as their powers developed. What is also evident, however, is that control of information is not a sufficient condition for the efficient development of organizations. Without a commitment to the equal and rational application of the laws, growing bureaucracies have tended to become bloated administrations with little purpose beyond the extension of their own authority (Shennan, 1974, 55). Bureaucratic information management and bureaucratic ethics are associated, but not always indissolubly intertwined.

The mechanistic metaphor is an excellent starting point for the study of organizational thought and its relevance for information systems because it is the origin of many practises which are taken for granted as the norm of managerial behaviour. Historical study is one possible source of the informed criticism which allows us to think creatively about the links between organizational structure and information and helps to make room for creativity by showing that, because views of organizations have changed in the past, there is nothing inherently permanent about the most common perspectives of today.

REFERENCES

Abrams, P., 1968, "Origins of British Sociology", Chicago University Press, Chicago.

Anthony, R. 1965, "Planning and Control Systems", Harvard University Press, Cambridge, MA.

Beer, S., 1979, "Heart of the Enterprise", Wiley, Chichester.

Black, M., 1962, "Models and Metaphors", Cornell University Press, Ithaca.

Black, M., 1979, More about metaphor, *in*: "Metaphor and Thought", A. Ortony, ed., Cambridge University Press, Cambridge.

Booth, W.C., 1979, Metaphor as rhetoric: the problem of evaluation, *in*: "On Metaphor", S. Sacks, ed., Chicago University Press, Chicago.

Brewer, J., 1989, "The Sinews of Power", Unwin Hyman, London.

Chandler, A.D., 1990, "Scale and Scope", Belknap, Cambridge MA.

Davis, G.B. and Olson, M.H., 1985, "Management Information Systems", 2nd ed., McGraw-Hill, New York.

Deutsch, K.W.D., 1951, Mechanism, organism and society, *Philosophy of Science*, 18:230.

Dyson, K., 1980, "The State Tradition in Western Europe", Martin Robertson, Oxford.

Elliot, J.H., 1984, "Richlieu and Olivares", Cambridge University Press, Cambridge.

Fayol, H., 1949, "General and Industrial Management", Pitman, London.

Flood, R.L., and Jackson, M.C., 1991, "Creative Problem Solving", Wiley, Chichester.

Gatrell, V.A.C., 1991, Crime, authority and the policeman-state, *in*: "The Cambridge Social History of Britain", F.M.L. Thompson, ed., Cambridge University Press, Cambridge.

Kendall, J.E. and Kendall, K.E., Existence of predominant organizational metaphors and their implications for information systems development, Unpublished paper.

Leatherdale, W.H., 1974, "The Role of Analogy, Model and Metaphor in Science", North-Holland, Amsterdam.

MacDonough, A., 1973, Information management, *in*: "Systems Analysis", S. Optner, ed. Penguin, London.

de Man, P., 1979, The Epistemology of Metaphor, *in*: "On Metaphor", S. Sacks, ed., Chicago University Press, Chicago.

Mommsen, W.J., 1989, "The Political and Social Theory of Max Weber", Polity, Cambridge.

Morgan, G., 1986, "Images of Organization", Sage, Beverley Hills.

Porter, T., 1986, "The Rise of Statistical Thinking", Princeton University Press, Princeton.

Shennan, J.H., 1974, "The Origins of the Modern European State", Hutchinson, London.

Simon, H.A., 1976, "Administrative Behaviour", 3rd ed., Free Press, New York.

Smith, A., 1921, "The Wealth of Nations", George Bell and Son, London.

Stark, W., 1962, "The Fundamental Forms of Social Thought", Routledge, London.

Walsham, G., 1991, Organizational Metaphors and information systems research, *European Journal of Information Systems*, 1:83.

Zuboff, S., 1988, "In The Age of the Smart Machine", Basic Books, New York.

IMPROVING THE IMPACT OF SYSTEMS THINKING ON INFORMATION SYSTEMS DEVELOPMENT

M. A. R. Kirby

School of Computing and Mathematics
University of Huddersfield
Huddersfield, HDl 3DH

ABSTRACT

Systems thinking has had little impact on the development of information systems. To improve the impact of systems thinking, there is a need to influence the awareness, education and training of the people involved in information systems development (ISD). This is difficult because information systems developers are not receptive to ideas which challenge the predominance of a technical orientation. There is increasing recognition that many technically good information systems fail because they are not acceptable to users or other stakeholders, yet this has not improved receptiveness to systems thinking. This paper reports on three cases which demonstrate the difficulties in challenging a technical orientation. Moving from the particular to the general, the paper looks at the lessons that the systems movement could learn from attempts by the Human-Computer Interaction (HCI) community to improve awareness, education and training of HCI methods. The HCI community, with extensive support from the Department of Trade and Industry, has undertaken major national programmes to influence the practice and teaching of information systems development. These programmes have promoted ideas which challenge the technical orientation of ISD and have had to address the problem of lack of receptiveness to such ideas. This paper discusses the impact of these programmes, and reviews the lessons that are relevant to the promotion of systems thinking as a means of addressing the high rate of information systems failures.

INTRODUCTION

The practice of information systems development takes place in a functionalist and technically oriented culture. It is easier to change the practice of information systems design if the methodological change does not require a significant cultural change. It is difficult for systems designers to adopt systemic and human oriented methods, which require fundamental changes in culture, when their education, training and work environment all reinforce the

Systems Science, Edited by F.A. Stowell *et al.*,
Plenum Press, New York, 1993

existing culture. The entrenchment of technical orientation is well documented, for example see Eason (1988) and Willcocks (1991). Systems thinkers, in general, would like the culture of ISD to change to a systemic and human oriented culture. Some see a need for a complete paradigm shift to systemic information systems development which would make the currently dominant 'hard' methods redundant. Others have a more modest aim to encourage the use of 'soft' methods for systemically defining the context within which 'hard' systems analysis is applied, for example using a systemic approach as a precursor to structured systems design. This paper does not explore the issue of how extensive the cultural change should be - for a debate on that issue, see the collection of papers in the 'Information Systems Special Edition' of the Systemist (1992). Whatever the extent of cultural change, there is an issue of how to effect the change. This is the issue addressed in this paper.

The issue of how to effect cultural change is one that has long been debated in the fields of social psychology and organisational development. The model that has been adopted here is based on Lewin's (1952) 'three step procedure'. Lewin argues that there are three steps to changing group standards: unfreezing, moving and freezing. In effecting methodological change, it is necessary to make the practitioners aware of problems with their existing methods and aware of an alternative method which has the potential to address those problems. They also need to be educated to understand why the alternative method is more appropriate for the purpose than the current method. Finally, they require training in how to apply the method. So, there are three steps. Awareness is needed to unfreeze an entrenched methodological view and make practitioners receptive to new methods. Education is required to establish a new philosophy or new way of thinking to ensure that practitioners will persevere with the move to a new method. Training is essential to ensure that new method can be applied and be frozen as the new standard. In keeping with Lewin's model, the three steps of awareness, education and training need to be targeted at changing group standards, rather than converting individuals who are likely to revert to the group standard when it proves difficult to use new ideas in the old environment. It should be noted that the model is not necessarily strictly linear as some education may precede awareness or follow training.

CHALLENGING THE TECHNICAL ORIENTATION: 3 CASES

For effective methodological change, there is always a need for awareness, education and training, but the relative criticality of the three steps may vary. It took decades for structured systems design methodologies to become the practitioners standard and replace unstructured approaches. This is not intended to imply that structured methods are rigorously followed or understood by systems developers. There is evidence to the contrary: for example, Avison et al. (1992) indicate that there is low use of ISD methodologies, and Steinberg & Baram (1992) argue that there is low use of CASE tools because practitioners do not have sufficient pre-requisite methodological skills. The assumption here is that there has been sufficient awareness and education in the structured approaches for them to have a dominant influence on the way practitioners think about ISD, even if there has not been sufficient training to ensure rigorous application. In spreading structured approaches, awareness and training were the critical steps which proved difficult and caused the rate of change to be slow. Education was less critical because the underlying functionalist philosophy and technical orientation of systems analysis were not challenged. To change the culture of information systems design to reflect a systemic philosophy, education is the critical and most difficult step, because systems thinking does challenge the technical orientation. The following three cases are indicative of some of the difficulties.

Case 1: A Naive Challenge To Entrenched Views

The first case is a simple example of how failure to address education needs can undo successful completion of the awareness step. The case involves an analyst/programmer in the shipbuilding industry. This analyst was trained in structured methods, which he applied with some rigour, was undoubtedly a highly skilled programmer and was able to apply his own choice methods with relatively little pressure-to conform to a group standard. He did not come into contact with users and worked from general specifications produced by an 'application support consultant', who understood the specialist languages of both users and analyst/programmers and acted as an intermediary between them. The author was the intermediary in this case and found that the analysts/programmer would deviate from the specification whenever he thought that there was a better way of processing a particular task. His innovations were usually more efficient in terms of machine resources but proved unusable for the users. The analyst/programmer's reaction was to regard the user as an idiot and be defensive about his programs. It seemed a good idea to take the analyst/programmer to meet a user. A user was chosen who could articulate his requirements well without excessive use of shipbuilding jargon. The analyst/programmer was reluctant and nervous before the meeting. Afterwards, he seemed delighted and enlightened. They clearly had a closer shared understanding of the requirements and the analyst/programmer began talking for the first time about systems in terms of their impact on users' jobs. He seemed fully committed to the new specification that they agreed. Later, a new system was delivered to the user which was considerably different from the specification and was not acceptable to the user. The analyst/programmer explained that it was not possible to meet all the requirements so he had changed the specification. It had not occurred to him to consult the user or intermediary again to check the acceptability of the changes. He was now convinced that talking to the user had created unnecessary work and that the user should accept whatever he provided as the best technical solution. It was perhaps naive to expect a lasting change in attitude from one meeting between the analyst/programmer and a user, but the case does indicate the strength of entrenchment that may exist as a result of traditional systems analysis education.

Case 2: A Gentle Introduction To Systems Thinking

In the second case, there was an attempt to educate as well as make aware, but the education was incomplete. The case concerns the specification of a medical information system for the obstetrics department of a large university hospital. The system was expected to contribute to improving patient care. The proposed system would integrate data from existing stand-alone systems to provide clinicians and nursing staff with improved bedside information in the wards. The integrated data would also be made more accessible to medical researchers seeking to improve preventative health care. The specifiers were classically trained system analysts who focused on the design of new data structures and the complexities of networking and communication between different systems. Our role was to validate the specification. The validation was based on the SURVE technique (Specification of User Requirements, Validation and Explication) (Kirby, 1992). The first step in the validation was a simple stakeholder analysis to identify who would be affected, directly or indirectly, by the introduction of such a system. A check was then made to see if the specifiers had considered the views and needs of these stakeholders in the relevant parts of the specification. The stakeholder analysis showed that the specifiers had consulted many of the stakeholders, including clinicians and nursing staff, with regard to data requirements, but not considered the impact of the system on their jobs. They had also neglected other stakeholders entirely, including the patients and community midwives. These deficiencies in the specification were analysed in more detail, raising a number of important issues about the acceptability of the proposed system. The issues included: would the system have a good or bad psychological effect on the patients, would it affect the autonomy of the clinicians, would it increase or decrease stress for the nursing staff, would the integrated data be sufficient to provide useful information without data from community midwives? We reported that the proposed system was potentially unacceptable and advised the specifiers to consider the views of all the stakeholders in order to ensure that they addressed all the issues that might affect the acceptability of the system. Both the hospital decision makers and the specifiers accepted that the issues were valid and initially expressed willingness to investigate the stakeholders requirements further.

It seemed that a valuable intervention had been made, encouraging the specifiers to start considering an information system as socio-technical and systemic. They were not yet aware of terms such as 'socio-technical' and 'systemic', but that would come when they sought to be able to do stakeholder analysis themselves at the start of a project rather than reacting to the results of our analysis. This first step in the conversion of some hard systems analysts had been achieved by a less intimidating method than expecting them to learn a whole new socio-technical or soft systems methodology. However, the initial willingness changed to reluctance in the face of political and financial pressure to move on to the next stage of development. The issues were still accepted as valid but the need to address them was rationalised away. Issues of psychological impact, stress and autonomy could, they argued, be addressed when the time came to install the system and would not affect the technical design of the system. The issue of data held by community midwives was more difficult to ignore, because many patients in an obstetrics ward are only in the ward for a day or overnight and their health care is mainly supported by the community midwives rather than hospital clinicians. The rationalisation for ignoring this aspect was that they could not think of a feasible technical solution so it was not worth investigating the issue any further! Unfortunately, the consequences of these decisions are not known because the project itself was stopped by cuts in health service funding before development work was completed.

Case 3: The Need For A Broader Curriculum

The first two cases indicate that the education step is not a trivial matter. This raises the question: how do you ensure that the education step will be successful? The third case suggests that the education step will be more successful if the general education of information system developers is broader, making them more flexible to new ideas and more likely to understand their significance. The case concerns the education of degree students. At the University of Huddersfield, there are, broadly speaking, two types of computing course. One is business oriented with emphasis on applications and business needs. The other is technology oriented with emphasis on the technical system. All students have a sandwich year in industrial training. There is only a limited exposure to behavioural aspects of computing in the early years of the technology oriented courses, while there is a far greater exposure to a broad range of human aspects throughout the business oriented courses. Currently, socio-technical approaches and soft systems methodology are introduced in the final year of each course. It is a much greater challenge to teach these subjects to students on the technology oriented courses, who have internalised a 'hard' philosophy during two years of teaching and a year as a programmer in industry. Although the business oriented students have also been trained in 'hard' approaches, they are more receptive to 'soft' approaches as a result of their greater exposure to human aspects, in teaching and in placement, where they are more likely to have had a job that brings them into contact with users and their problems. A possible further indicator of successful conversion is that a there is a high incidence of business oriented students choosing to use 'soft' approaches in their final year projects, and a low incidence for students on the technology oriented courses.

THE HCI EXPERIENCE

The difficulties of challenging the technical orientation of ISD are shared by the Human-Computer Interaction (HCI) community. The HCI community includes experts from many disciplines. At one extreme, it is concerned with the impact of computers on people's jobs, and the methods used include forms of task analysis underpinned by socio-technical principles. The other extreme involves a scientific approach to the ergonomics of the human-computer interface. The main difference between ISD and HCI methods is that the former focuses on utility and the latter on usability[1], as Grudin (1992) has clarified. Some of the methods challenge the functionalist philosophy. All of them challenge the technical orientation to some degree. It is argued, therefore, that the experiences of the HCI community have some relevance to the challenge faced by system thinkers.

Grudin (1992) reviews the integration of ISD and HCI in the USA and finds that there has been little integration. The main reasons given are the low awareness of socio-technical approaches and the problem that 'parallel consideration of utility and usability ... requires substantial restructuring of existing internal development practices'. Grudin concludes that market forces, particularly the increasing importance of usability to competitiveness, will force a merging of ISD and HCI.

In Britain, there has been a significant intervention to improve HCI awareness and training without leaving them to market forces. This is the 'Usability Now!' initiative by the Department of Trade and Industry, in conjunction with the country's leading HCI centres. The main device for improving awareness has been the wide circulation of case studies published by the DTI (1992). These are brief glossy accounts of successful cases of HCI development, designed to appeal to managers and practitioners in industry. They are based on the assumption that general receptiveness to HCI is dependent on two factors: clear proof that HCI techniques are successful and knowledge that the techniques are being used by leading companies. Success is shown to be related to usability but measured in terms of competitive advantage and profitability. The companies who are regarded as opinion leaders are household names such as Boots, the RAC and the Bank of Scotland. The focus of the 'Usability Now!' initiative has been to increase awareness. The need for training has been addressed by providing directories and newsletters which give information on universities and consultancies who provide training. In an (unpublished) report to the HCI'92 conference in York, the DTI expressed a feeling that the initiative had been successful in terms of requests for further information from industrialists and increased interest in training courses. They had no plans to follow up the initiative with a further programme or to formally evaluate the success of the initiative.

It seems an oversight to neither evaluate nor reinforce the 'Usability Now!' initiative, steps which would help to ensure long term success. More critically, the educational issues have been ignored. HCI education is being addressed by another project under the auspices of the British Computer Society (BCS), but this has longer term objectives and is not related to the DTI initiative. The BCS Human-Systems Interaction Technical Committee Education and Training Working Group are studying the design of HCI curricula and may advise the BCS on how to improve ISD curricula in order to fully incorporate HCI into ISD education. If this happens, there could be a knock on effect on ISD curricula in universities courses to meet BCS accreditation requirements. A similar project has been underway in the USA since 1988, under the auspices of the Association for Computing Machinery (ACM). They have defined the content of HCI, which is a contentious issue, and produced a variety of curricula

---

[1] The International Standards Organisation definition of usability is "the usability of a product is the degree to which specific users can achieve specific goals within a particular environment; effectively, efficiently comfortably, and in an acceptable manner".

for students in different disciplines, see ACM (1992), but acknowledge that they have not resolved the philosophical problems of integrating HCI and ISD. They have started to address these problems by outlining interdisciplinary curricula for a variety of computer courses. The main proposal is that computing students should be exposed to ideas from other disciplines, including natural sciences, behavioural sciences and social sciences to make them more receptive to HCI issues and techniques.

CONCLUSIONS

To improve the impact of systems thinking on ISD, there is a need to influence the awareness, education and training of the people involved in ISD.

Awareness is potentially the easiest step. There are two morals of the HCI experience. The first is that there is a need to understand the factors which will make developers receptive to the new ideas. The factors assumed by the 'Usability Now!' programme were: clear proof that HCI techniques are successful and knowledge that the techniques are being used by leading companies. These may also be relevant factors for receptiveness to systems thinking. There may be other critical factors. This requires further investigation. The HCI experience also indicates the potential value of an extensive awareness programme, made possible in that case by government involvement. However, it is essential that awareness is followed up with appropriate education and training.

Education is the most difficult step. The above cases confirm its importance rather than resolve how to make education effective. The idea of influencing the institutions which influence ISD curricula, which is being tried by the HCI community, might be effective in encouraging new practitioners to follow new approaches. This may only be successful if a way is found to educate the established practitioners to create a working environment in which the new approach can flourish.

Training may prove to be a difficult step. The above cases and HCI experience have not given clues to how to ensure training is effective. However, it has been mentioned that training in structured methodologies has been deficient, yet structured methodologies have a dominant influence on the way practitioners think about ISD. Although training is important, it is more critical to address the educational issues, because that is the way to influence thinking about information systems development.

REFERENCES

ACM, 1992, "ACM SIGCHI Curricula for Human-Computer Interaction," The Association for Computing Machinery, New York.

Avison, D., Shah, H., Powell, R. & Uppal, P., 1992, Applying methodologies for information systems development, *Journal of Information Technology*, 7, pp127-140.

DTI, 1992, "Usability Now! Case Studies," Department of Trade and Industry, London. Eason, K., 1988, "Information Technology and Organisational Change," Taylor & Francis, London.

Grudin, J., 1992, Utility and usability: research issues and development contexts, *Interacting with Computers*, vol. 4, no. 2., pp209-217.

Kirby, M.A.R., 1992, "The Development and Testing of SURVE: a Technique for Validating and Explicating Specifications of User Requirements," PhD Thesis (in preparation), University of Huddersfield.

Lewin, K., 1952, "Field Theory in Social Science," Tavistock Publications, London

Steinberg, G. & Baram, G., 1992, An investigation of CASE software integration, *Journal of Systems Management*, September, pp20-22.

Systemist, 1992, Information Systems Special Edition, *Systemist*, vol. 14, no. 3.

Willcocks, L., 1991, Introduction: human resource and organisational issues in the 1990s, *Journal of Information Technology*, 6, pp121-127.

TOWARDS A MULTIMEDIA BASED NEW INFORMATION CONCEPT

P. Dagorret, A. Gastambide, M. Larrasquet and J. Meyranx

I.U.T. de Bayonne - LICIAP
3, avenue Jean Darrigrand - 64115 Bayonne
Cedex, France

INTRODUCTION

Sound and image have an increasing and essential role to play in the modern world, not only as regards people's specific or general culture, but also as information and training support. Audio-visual products have established themselves as undeniably efficient vectors for the transmission of information and for their 'impact value'. The average rate of memorisation of information received through audio-visual channels are comparatively superior to the other media, i.e. image only, and text-reading only media (Bois, 1990).

The power of these objects is increased further by their universal character. The image, and to a lesser extent, sound (apart from structured spoken language) have an evocative dimension far beyond that of spoken or written language which confers greater universality upon them. It should be borne in mind, nonetheless, that the semantic accuracy of the information they convey is, therefore, *de facto* limited.

Another set of characteristics belonging to image, and almost certainly linked to our first set of considerations is concerned with the manner of apprehending or registering in an immediate or 'global' way (Halin, 1989); this is partially true for sound. These characteristics are essential considerations in cognitive psychology where the notion of a text to read or especially a text to write, is intimately linked to the notion of effort (Bart, 1991).

An aspect we regard as highly relevant is the latest development of image, and to a lesser extent, of sound as computer objects. Computer publications seem to show relatively little interest in this new phenomenon, whereas, in our view, it is of the highest importance for the future of the information system design world.

When we look at information handled by computers in particular, such impact, memorisation power values, user-friendliness and universality have little to do with the type of alphanumeric information already described, i.e. current information systems design . Needless to say, the sound and image objects will very soon have a much wider application in information system development within an irreversible and ongoing userfriendly and personalization context.

Naturally, this is why all interfaces produced over the last few years, the so-called 'user-friendly' interfaces, have given pride of place to the image and the use of 'visual metaphors' to refer to objects. It is only a matter of time before the same reasons that led to

Systems Science, Edited by F.A. Stowell *et al.*,
Plenum Press, New York, 1993

image and sound becoming the basis of modern interfaces, bring about a new situation whereby they become informational objects, sound objects and vision objects, to be handled, in their own right, beyond interfaces, as part and parcel of tomorrow's information systems.

However, it is also apparent, that their limited capacity for semantic accuracy, and thus for expressing the complexities of management, will not allow them to replace traditional computer alphanumeric data. The future lies with systems which combine traditional and these new types of information. We will be able to speak, thus, of generalised information, when referring to types of information. When referring to a system capable of operating on co-operating technology and/or different support types, we will need to speak of multimedia systems.

It would appear, unfortunately, that current usage is imposing the term 'multimedia' to refer, almost indifferently, to any and every system manipulating generalised information.

INFORMATION SYSTEMS DESIGN: A NEW APPROACH

Both at an organisational level and at a technological level (hardware and software) we seem to have arrived at a turning-point. On the one hand, the audio-videographic technology is advancing in different fields: text, image, sound analysis, synthesis and processing, integration of computing and TV, multimedia and transmission interconnection of multimedia heterogeneous systems. As for software technology, object-oriented design methods and sophisticated user-centered software tools are progressively developed. On the other hand, we foresee the rise of the number of sophisticated interconnected individual workstations being introduced and the aspirations of individuals and groups to become actors (active agents) in the informational processes.

All these factors are, each and all, so many vectors bearing the seeds of a profound mutation within business enterprise informatics. All this we know, but it may not necessarily be perceived in all its amplifications by the different actors or agents involved within the process.

GENERALISED INFORMATION AND ITS DESIGN

If the importance of these technological changes have been correctly gauged here we will see that the design revolution created by this explosion of generalised complex information (data, graphics, sound, static and animated images) has not yet been fully grasped, especially in the world of information systems design.

These information processes no longer involve mere symbolic objects (traditional alphanumeric data) of the real world as it is perceived to be modelled, but also multimedia objects as such which are going to be manipulated as "real objects" by the multimedia information system. These objects will be texts, graphics, fixed images, sounds, video sequences. We shall call them "documents".

These documents will have a number of inter-relations which will require identifying. They will also maintain external relations with the symbolic objects of perceived reality. Modelling of the information system will thus experience, all within one design format, objects of a different nature (symbolic objects / documents) and the relations potentially existing between all of these objects, independantly of their nature.

The innovatory characteristic of the appearance of these 'document' objects within multimedia systems lies in the fact that the system will have to manipulate these objects as 'real objects', and thus, qualification criteria for such objects will have to be defined (Rochfeld, 1990; Culet and Rieu, 1991).

Both with external symbolic objects, and in any relations possible with them, these objects will be identified, and even perhaps qualified by one or more classes of lexical attributes. Each multimedia object will thereafter be characterised in a plural manner .

One current research development will serve to provide a simple example. Sponsored by Su-Azia, a Zibero-Basque Cultural Association, and developed in cooperation between the Institut Universitaire de Technologie of Bayonne, The Computer Science Faculty of San Sebastian, and the Heriot-Watt University of Edinburgh, this research involves a highly interactive multimedia system dealing with traditional Basque songs (Larrasquet and Lopisteguy, 1991; Larrasquet et al., 1992).

The complexity of the information to be taken into account at the conceptual level has led to the following questions:

1- What is the 'type' of a mutimedia object (multi-type object) ?

- the information-type included in the document (interview, commentary, singing performance, instrumental, illustrative animation, etc.) as it has been grasped by the cultural analyst, drawing on a kind of 'cultural semantics'.
- the document-type (photography, sound recording, charcoal portrait, text or manuscript, video-clip etc.) and thus a type of definition which is external to computer science, drawn really on the type of original document as a real object (when it exists!), i.e. relating to the material form of the document, apart from any technical definition.
- the medium (static paper image, slide, TVHD sequence, floppy-disc, D.A.T. numerical sound, etc...) which characterises the document at a more technical level, but which also borders on the limits of design concerns, given that it still characterises the document in a manner that is essentially external to the multimedia information system going to acquire and manipulate it.

2- The complexity of the document.

A document may be composed of several documents, each of them may be itself composed of several documents. For such composed documents, hierarchies and recursivity will have to be managed for these complex documents.

The complexity may be more important in the case of handling objects of different types within these hierarchies.

3- Psycho-artistic aspects.

This is referring to the semiotic of the multimedia objects. The density of information is much more important in this kind of objects than in the case of alphanumeric classical information. The relationships of the user with these objects will often be based on affectivity and artistic sense. Colorimetry, composition, aesthetics for what concerns image, melody, diction and voice for the sound, style for the texts will be important elements related to the user's cognitive psychological characteristics.

The objects will often be considered by themselves, each item will have its own significance and will be an element linked with its own network of semantic relationships. Each document has to be considered more like an individual atomic object than like an occurrence of a homomorphous objects set possessing homomorphous linkages with other homomorphous object sets.

In highly interactive multimedia systems, a document, for instance, like a photograph will be known and be desired for itself, for the effectiveness it has in the user mind. It will have its own linkages with other items of the base: other photographs, sounds, or symbolic objects (for instance information about the person represented in the photo), but another photographic item of the base will be related to a video sequence (part of), and be linked to symbolic information about an artistic show.

The homomorphy paradigm we systematically use in information systems design and which seems reasonable with traditional alphanumeric information is no more acceptable for a system essentially structured around multimedia documents.

THE APPLICATIONS DOMAIN

Will these latest design categories be sufficient for generalising all highly interactive multimedia applications? This is a question for research. The solution would seem to depend very largely on the kind of multimedia system we consider and on the type(s) of approach, of perception, and on the type of manipulation of these objects which is being planned to be made available to the (classes of) user(s).

The previous systemic analysis will have to build clearly the requirements of the system and to define what kind of system is actually needed (Jayaratna, 1991).

A first kind of application consists of introducing some mutimedia objects (multimedia fields or multimedia entity types in the E-A modelling world) in systems of traditional types, essentially structured around alphanumeric information and whose accesses will be traditional set requesting (looking for a list of answers). This kind of applications can be considered as single extensions of traditional systems.

Another category is referring to systems whose whole logic is structured around multimedia objects, with a cultural or artistic dimension (applications in museography, in cultural or artistic collection diffusion systems, in interactive presentation systems...). In this kind of application, the user is not supposed to have a pre-defined request-type behaviour. S/he must be allowed to browse from one document item to the other, choosing his/her path, guided by the succession of his ideas and his sensibility. This new type of relationship to the system sets the kind of modelling and tools used for the specification of these systems on a new base as well for generalised information management for user's interfaces.

MODELLING APPROACHES

Multimedia documents entering the scene will carry new problems to the system modelling activity because of their special characteristics as discussed above. Propositions will be made around the extension of E-A and relational models. On the other hand, one development of great importance recently in the computer world has been the concept of object. It seems to open some interesting new ways for multimedia systems modelling. Let us discuss these topics, and underline what kinds of limits they have for what concerns multimedia problems, before trying to propose some directions for these particular cases.

The E-A Approach

Traditional E-A and relational models deal with little more than alphanumeric information relating to symbolic objects which are external to the information system. The 'customer's name', for example, is one attribute-class for the entity-class of 'customer' which is likely to receive string-type values. 'Customer's date of birth' is another attribute class from the same entity class and is a 'date-type' class.

We can see, in fact, that such information systems manipulate alphanumeric data relating to an object from the real world, perceived and "name-able' by one or several of these data values which, in turn, belong to an attribute-class associated with an entity-class in the meta-database (customer number, customer name).

Once the advent of these new multimedia documents into information systems has been accepted as inevitable, it is only too apparent, in our view, taking the present situation into account, that traditional information system design methodologies are simply not appropriately adapted to this new problem.

The Object Oriented Approach

The concept of an object in direct contrast to the prevailing paradigms in the world of informatics, according to which an information system possesses a static aspect (or structure of the group of information elements, its attributes and relations) and, separately, a dynamic aspect (or the text-processing group running this information and making it develop).

The concept of an object, considered from this traditional viewpoint, integrates elementary or complex attributes (they are themselves objects) and systemizes certain relations between objects: notably the possibility of grouping them into classes which are, in turn, objects themselves, and thus, like all objects, possessing their own attributes. The member objects (instances) of one class inherit the attributes of that class. Furthermore, an object can possess methods describing its behaviour whenever a triggering message comes to it.

Two directions seem to be open today for object conceptual modelling. The data conceptual models of the ENTITY-ASSOCIATION family have not yet been adapted as such. A certain amount of extension jobs using these models is currently taking place. To the best of our knowledge, it merely involves integrating certain object approach concepts -as well as refining integration constraints in modelling- into Entity-Association type models and their derivatives: these concepts include those of specialization / generalization and 'is a' links. The dynamic aspects of the object philosophy are not, however, being taken into account. (Tabo, 1991).

Another more innovative approach, developed by Rolland (1991) and her team, is based directly on the object idea and integrates immediately both specialization and generalization mechanics, as well as those of inheritance and instantiation; it also defines different types of links between objects. We would add that the dynamic aspects of the object prove to play a direct role in its definition, with the notions of actors, methods and trigger events (Flor, 1990; Rolland, 1991).

General Evaluation of Approaches

The object approach has the advantage of being able to take into account and to manage complex documents, because of the organisation into hierarchical classes and the aggregation mechanism it provides. This property is very interesting for the management of multimedia documents, their structure being very often a complex one (Delobel et al., 1991; Valduriez, 1987).

These models (E-A, Relational, Object Oriented) have been mainly built for managing important quantities of alphanumeric data. As a consequence, all of them have been designed for the storage of data organized in sets and only known as homomorphous elements of these sets. In the same way, the manipulations the user is allowed, are merely set queries. Even the extensions allowing multimedia documents manipulation define them as document classes, therefore as sets of documents (Kim et al., 1986).

All these functionalities have an evident interest for all the bulk applications, including the multimedia ones. However, in this kind of approach, we are limited to a set-based view of the information in which objects will not have any individual existence. But for cultural or artistic applications, in which there are few objects, but each one has important semantic content, these functionalities are not very well adapted.

It would be interesting to be able to define and to store hypertext-like links between multimedia instances. These links between instances would be defined considering semantic aspects. The user's "butterflying' from document to document would then be supported by these links.

CONCLUSION

The future in such multimedia systems is certainly in mixing symbolic objects (alphanumeric data) and multimedia documents. The object paradigm is surely a good global model for the general architecture of these applications. But the problem of the links between instances, necessary to set up completely friendly 'butterflying' interfaces must be resolved.

REFERENCES

Barthet, M.F., Palanque, P. and Alquier, A.M., 1991, "L'Ergonomie et la Conception MERISE: un Complement a la Methode", Actes du Congres Autour et a L'Entour de Merise, SOPHIA ANTIPOLIS, Avril, 1991.

Bois, J. P., 1990, "Informatique et Telecommunications - Penser autrement les systemes d'Information', AFCET / INTERFACES N° 89, March 1990.

Culet, A. and Rieu, D., 1991, Classification et Representations d'Objets, Actes du congres Inforsid, PARIS, Juin, 1991.

Delobel, C., Lecluse, C. and Richard, P, 1991, Bases de Donnees: des Systemes relationnels aux systemes a objets, InterEditions.

Flory, A. and Rolland, C., 1990, "La Conception des Systemes d'Information: Etat de L'Art et nouvelles Perspectives", Nouvelles Perspectives des Systemes d'Information, Selection d'articles du Congres Inforsid, par Andre Flory et Colette Rolland, Eyrolles.

Halin, G., 1989, Apprentissage pour la Recherche interactive et progressive d'Images: Processus EXPRIM et Prototype RIVAGE. These, Universite de Nancy 1.

Jayaratna, N., 1991, Systems Analysis: The weak Link in the Systems Development Process! *Journal of Applied Systems Analysis*, Volume 18, pp61-68.

Kim, W., Luther, W., Woelk, D., 1986, An Object Oriented Approach to Multimedia Database, "Proceedings of the ACM SIGMOD International Conference on the Management of Data", Washington D.C., May 1986.

Larrasquet, J.M., Lopisteguy, P. and Jayaratna N., 1992, Multimedia Research - A New Direction for Information Systems Design, *Systemist*, Vol 14 (2), pp67-68.

Larrasquet, J.M. and Lopisteguy, P., 1991, Multimedia research - Potential implications for information systems, *International Journal of Information Management*, Volume 11, No. 4, pp308-310.

Rochfeld, A., 1990, L'Heritage: Type et Sous-Type, Journee AFCET du 15 Novembre 1990, Le Formalisme de donnees MERISE: Extensions du Pouvoir d'Expression, AFCET, PARIS 1990.

Tabourier, Y., 1991, Propositions pour un Changement de la Perspective de Conception des Systemes d'Infor mation Organisationnels, "Actes du Congres Autour et a L'Entour de Merise", SOPHIA ANTIPOLIS, Avril 1991

Valduriez, P., 1987, Objets complexes dans les Systemes de Bases de Donnees relationnelles, *TSI*, Vol. 6, No. 5, pp405-418.

TOWARDS AN INTERPRETATIVE FORM OF DATA ANALYSIS FOR THE SOFT SYSTEMS METHODOLOGY

P.J. Lewis

Department of Systems and Information Management
The Management School
Lancaster University
Bailrigg
Lancaster LA1 4YX, UK

INTRODUCTION

The soft systems methodology (SSM) is a proven means of addressing questions of change, particularly where there are multiple views of the present situation and uncertainty as to what changes might be desirable. An SSM analysis can, by highlighting issues in a problem situation and allowing different views in the situation to be explored through the modelling of a number of different relevant systems, help those involved to understand the possibilities and consequences of different changes that they might choose to make.

SSM does not though presently allow for any consideration of the data or information implications of such changes, for none of the techniques of SSM focus upon that important dimension. This is clearly a problem where SSM is to be used in information systems work or where some part of the problem situation involves the storage or provision of information.

It has been suggested (Lewis, 1992) that where some part of the problem situation concerns the provision and use of information then employing an interpretative form of data model as part of the SSM analysis may be advantageous. In this paper we discuss the nature of such interpretative data models. We suggest that conventional forms of data model are not suitable for use within SSM because they embody philosophical presuppositions that are different to, and match badly with, the phenomenology of SSM. We propose, instead, the creation of a rather different form of model, one that shows the cognitive categories that are meaningful in relation to a particular, conceptualised human activity system.

THE NEED FOR A NEW FORM OF DATA MODELLING

Many different data modelling approaches and diagramming conventions already exist, for example the relational model (Codd, 1970), the entity-relationship model (Chen, 1976), and the more recent object-oriented model (Schlaer and Mellor, 1988; Coad and Yourdon,

Systems Science, Edited by F.A. Stowell *et al.*,
Plenum Press, New York, 1993

1990; Rumbaugh *et al*, 1991). A natural first step towards providing SSM with techniques to investigate data is to consider whether any of these might be used.

Unfortunately none of the existing data modelling approaches appear to be suitable for use within SSM. First, all conventional forms of data modelling rely on there being a high level of agreement as to the nature of the present situation, as to what is 'the problem', what are the organisation's objectives and what any new information-systems should do. This makes them fundamentally unsuited to use in SSM at the early stages of problem analysis, where it is just such things that are undecided, and which form the major subject for investigation and debate. This problem is compounded by the tendency of conventional forms of data modelling to commence with an examination of current data usage and storage, and to adopt a datalogical perspective (Methlie, 1980), where existing information flows are regarded as satisfactory representations of the information needs of the organisation.

Secondly, there is a serious mismatch between the philosophies that underlie SSM and the established methods of data modelling. The paradigm of conventional data modelling has been described as both scientific (Wood-Harper and Fitzgerald, 1982) and objectivist (Klein and Hirschheim, 1987). Although there are some grounds for doubting whether, in *practice*, data models are genuinely 'neutral' or value-free (Lewis, 1992), the evidence that is available in the literature does support the contention that conventional data modelling follows an objectivist paradigm, characterised by ontological assumptions that are realist and epistemological assumptions that are positivist. It is not, then, surprising that the starting point for conventional data modelling is an identification of real-world things or events. Nor, given the assumption that the data model should reflect reality, is it surprising that when a data model has been created it is checked by comparison to known application requirements. An independent reality is thus accepted as both the ultimate source of knowledge for creating a data model, and the means by which the integrity and correctness of a data model may be validated. This contrasts sharply with the phenomenological stance of SSM. Whilst SSM is a flexible methodology whose use may take many forms (Checkland and Scholes, 1990) all uses of the methodology employ, in some way, the idea of 'system' as an epistemological device. The models produced within SSM are of notional human activity systems; these models are not representations of what does exist, of some individual's or group's perception of what exists or what should exist in the future, but are devices used to gain insight into the problem situation.

Data analysis though need not necessarily be carried out within an objectivist paradigm; alternative, interpretative forms of data analysis are possible based upon different pre-suppositions that are more compatible with the philosophy of SSM. The basic premise of any interpretative form of data analysis is that the only knowledge that we may have of the world is that which is acquired through distorting sets of values, beliefs and expectations. This leads to significant changes in the way that we regard the process of data analysis and our expectations of its possible outcomes. It implies that the primary source of knowledge about the things that are 'relevant' to a problem domain cannot be an external, objective reality and that when data analysis is used as a means of enquiry into problematical situations then it can only ever investigate the nature of some part of the 'real-world' as it is understood by those who perceive it as problematical. The data analyst must therefore abandon the ambition to create data models that are iconic, analogue or mathematical representations of real objects or situations. Instead, all that may be aspired to is the creation of a data model that is coherent, internally consistent and has meaning in the context of particular beliefs, values and expectations.

Abandoning the belief in an objective reality does though present the would-be interpretative data analyst with a major practical problem; if there is no single 'real' problem situation out there, but only multiple, possibly conflicting, views of what the problem situation might be, then how may these be handled and progress made towards enabling change? One way of resolving such issues is to employ interpretative data analysis only *within* the context of a methodology which has proven capable of dealing with such dilemmas. The soft systems methodology is such a methodology, and henceforth our discussion will assume that interpretative data analysis is to be used as a technique within the framework of a larger SSM analysis.

A BASIS FOR DATA MODELLING IN SSM

A SSM study involves two interacting streams of enquiry, logic-driven and cultural (Checkland and Scholes, 1990), and it is in the former that we propose data modelling may be valuable. Typically, an initial entry into the problem situation through fact-finding and issue identification is followed by the conceptualisation of a number of relevant systems. The subsequent use of these conceptualisations, for comparison with the present situation and as the basis for debate amongst interested parties, is possible because the notional systems are explicitly and precisely described in System Definitions consisting of Root Definitions, Activity Models and CATWOE declarations. These already include explicit statements of the weltanschauungen and assumptions within whose context the conceptualised system must be understood.

But, at an even more basic level, a System Definition can only be meaningfully understood by a reader if that reader is able to assign meaning to the terms used within it. For example, if the defined system is concerned with recording the costs of medical treatments provided to patients, without some understanding of what a 'patient' or a 'medical treatment' is, then we cannot possibly really understand the idea of that system. This rather obvious fact may provide us with a basis for data modelling within SSM. If we can identify the *cognitive categories* that must exist in order for comprehension of the idea of a given system then these may act as the basic components of an interpretative form of data model for that system. Now, instead of trying to identify a neutral, objective data structure for a problem situation, we may aim instead to create a data model that is relevant to an explicitly described system. This model will show the cognitive categories required to understand the nature of the system, together with any implied associations between those categories. To remind ourselves of this fundamental difference and the special nature of this model we shall refer to it as a *System Data Model.*

CONCERNING THE NATURE OF SYSTEM DATA MODELS

A System Data Model records the agreed meaning for each of the cognitive categories, and the different types of association perceived to exist between those categories, that are meaningful with respect to a given System Definition. It may include both textual and diagrammatic descriptions.

Thus a System Data Model is **not** a representation of any real-world situation, nor is it a model of any individual's view of a situation, nor is it a design for data storage mechanisms. It is a model that reflects a particular idea, that of a notional system, that is deemed to be relevant in some way to a problem situation. This system may have similarities with the way in which the real-world situation is appreciated; it may be radically different and have little in common with any person's appreciation of the real-world situation. The System Data Model will reflect these possibilities, perhaps including categories that coincide well with the things about which data is stored in real-world data bases, or showing categories for which there are no representatives in presently existing data storage mechanisms. The single, essential requirement imposed by interpretative data analysis is that there should be consistency between the System Data Model and the system to which it relates.

Each of us uses, and continually defines and redefines, thousands of cognitive categories in order to make sense of the world in our everyday life. Interpretative data modelling does not attempt to identify all (or perhaps even any) of the categories so used by the modellers. It has the far more modest aim of creating data models that show only those cognitive categories that are required in order for a particular notion of a system to be understood. Just as when creating Activity Models we have sought to show only the minimum set of necessary activities, so now we seek to identify the minimum set of cognitive categories necessary for comprehension of a given System Definition. The meaning that is to be assigned to those categories will be a subject for discussion and debate amongst the same individuals who are involved in the modelling of the notional system described by the System Definition.

Interaction with process of defining the system

The processes of defining the system and of creating the System Data Model are clearly inextricably linked. The definition, as precise and unambiguous as possible, of some notional but purposeful system provides the starting point for interpretative data analysis but this does not mean that system modelling is finished once data modelling begins. The System Definition and the System Data Model are both reflections of the same idea of a system. They both provide a means of communicating that idea, and each must change in response to changes in that idea. During the course of an SSM study the relevant systems may be challenged, amended or taken to higher resolution levels; each or any of these may lead to changes in our understanding of the nature of the system being discussed, and thence to changes in the cognitive categories through which it is understood. Also, the process of identifying and agreeing the meaning of categories may lead to a sharper idea of what, exactly, the system under consideration consists of. When this happens we must go back and amend the original System Definition so it more precisely reflects our new idea of the system. Thus development of the System Definition and the System Data Model proceed hand-in-hand (Figure 1).

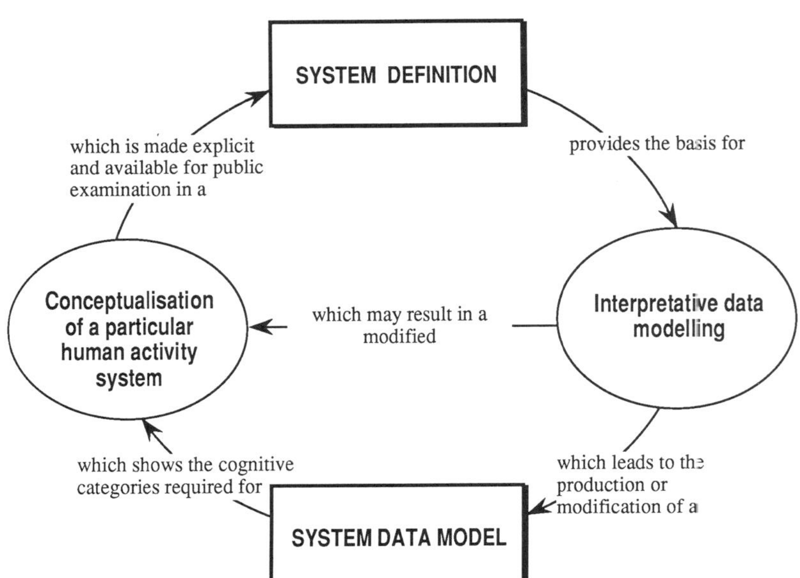

Figure 1. The interaction of the processes of system modelling and interpretative data modelling.

It should be noted though that whilst taking Activity Models to higher levels of resolution leads to a larger set of models, there will not be a corresponding multiplication of data models. So far as modelling to higher resolution levels leads to a better understanding of the original system concept, so it may allow us to make definitions of the cognitive categories more precise, to add new definitions and to modify the System Data Model; there will remain though only a single System Data Model, and a single definition of each cognitive category which will be applicable to Activity Models at all resolution levels.

Value of the System Data Model

Introducing an interpretative form of data modelling into an SSM analysis obviously requires additional time and effort, but in return offers several advantages. These include that the System Data Model:

- Establishes a vocabulary for discussion and debate in SSM, and ensures that such subsequent discussion and debate is organised around a shared understanding of the identified relevant systems. By ensuring that there are no differences in the meaning attributed to the terms of the System Definition we add rigour to the modelling and avoid the dangers of misdirected debate. If you and I attach different meaning to the terms of a System Definition then we cannot be understanding that System Definition in the same way. And if that is the case then we are unlikely to have a very useful discussion through it vis-a-vis the real situation.

  SSM does not already address this problem directly. This may be for two reasons. First, there may be an unconscious reliance that shared social conditioning (the participants in modelling generally coming from similar professional and cultural backgrounds), together with the interaction via discussion and debate will always lead to cognitive categorisations within an acceptable range of variation. Formal examination of the categories would in such cases be a relatively non-contentious process, but give some defence against the above assumptions being unfounded. A second possible reason is that ambiguity can sometimes be desirable, allowing for the temporary avoidance of confrontations that threaten the future of whole SSM analysis. In such situations then the flexibility of SSM allows that the detailed definition of categories may be postponed until politically feasible.
- Provides further insight into the nature of the system being considered. A central premise of SSM is that the discussion and debate that flows from modelling relevant human activity systems and comparing these with the real-world situation may lead to learning and thence to the identification of desirable and feasible change. That discussion and debate would be enriched if the human activity system models under debate were accompanied by corresponding data models. Considerations of feasibility might then include whether or not particular options for change are feasible in terms of their data implications. Such insights become more vital as strategic planning of the business activities becomes more closely linked with, and constrained by, the planning of technical support for the business activities.
- Complements the Activity Model of the System Definition. The Activity Model provides a dynamic, activity-focused view of the system; it may be thought of as showing the activities that any real-world expression of the system would logically be required to perform if it was to be an expression of that system. The System Data Model provides a structural view of the system, and shows those things about which a data storage mechanism to serve a real-world expression of the system would logically have to be capable of storing data.
- May provide the starting point for thinking about possible implementable data storage mechanisms. Just as we might say that if a particular relevant system, if it became realised, would need to do, in some way, the activities shown in the Activity Model, so we might now say that the system, if realised, would somehow need to store and use data about the cognitive categories shown in the System Data Model. The System Data Model therefore, whilst it is

not itself any kind of blueprint for data storage, might later provide the basis for more conventional entity-relationship, object-oriented or other types of design model.

CONCLUSIONS

This paper has suggested that if a data dimension is to be added to an SSM analysis then this should not be done by abandoning the phenomenological principles of SSM or by using conventional forms of data modelling. Instead, a different type of data model is proposed, one which is built around those cognitive categories essential for understanding the System Definitions used by SSM. The creation of such models is suggested as a technique that may be employed within the overall framework of SSM and not as an independent means of enquiry into problematical situations. When employed in this way several advantages may be offered to existing SSM practice, particularly where part of the problem situation involves the storage or provision of data.

REFERENCES

Checkland, P.B., and Scholes, J., 1990, "Soft Systems Methodology in Action", Wiley, Chichester.

Chen P.P-S., 1976, The entity-relationship model: toward a unified view of data, *ACM Transactions on Database Systems*, 1 (1), 9-36

Coad, P., and Yourdon, E., 1990, "Object-Oriented Analysis", Prentice-Hall, Englewood Cliffs, N.J.

Codd, E.F., 1970, A relational model of data for large shared data banks, *Communications of the ACM*, **13** (6), 377-387.

Klein, H.K., and Hirschheim, R., 1987, A comparative framework of data modelling paradigms and approaches, *Computer Journal*, **30**, (1), 8-15

Lewis, P.J., 1992, The feasibility and desirability of a closer linking of SSM with data-focused information systems development, *Systemist*, **14** (3), 168-179.

Methlie, L.B., 1980, Systems requirements analysis - methods and models, *in*: "The Information Systems Environment", H.C. Lucas Jr, F.F. Land, T.J. Lincoln and K. Supper, eds., North-Holland, Amsterdam.

Rumbaugh, J., Blaha, M., Premerlani, W., Eddy, F., and Lorensen, W., 1991, "Object-Oriented Modeling and Design", Prentice-Hall, Englewood Cliffs, NJ.

Schlaer, S., and Mellor, S.J., 1988, "Object-Oriented Systems Analysis: Modeling the World in Data", Yourdon, Englewood Cliffs, N.J.

Wood-Harper, A.T., and Fitzgerald, G., 1982, A taxonomy of current approaches to systems analysis, *The Computer Journal*, **25** (1), 12-16.

INTELLIGENT ASSISTANCE IN THE INFORMATION SYSTEMS DESIGN PROCESS

Michael Lloyd-Williams

Department of Computer Studies
University of Glamorgan
Pontypridd
CF37 1DL

INTRODUCTION

In their discussion on the commercial contribution made by the discipline of artificial intelligence (AI), Winston and Prendergast (1984) attempt to evaluate whether AI is indeed a field of exciting promise, or simply a white elephant doomed to eventual failure. This paper attempts to cut through some of the hype and jargon surrounding the subject, and examines the role that AI can and has played in the information systems design process. Specifically, it examines how AI technology has been used to provide intelligent tools for assistance in the design of information systems.

THE DESIGN TASK

A large amount of work has been undertaken previously on the general subject of design, including a number of researchers whose focus has been directed toward the definition of design and design knowledge as applied to the development of information systems. Coyne and Gero (1986) describe design as "*a process which often seems to elude formal representation*". They consider design tasks to be ill-structured because the goals of the process are either not clearly stated or not well understood. Furthermore, they argue that although human designers are generally undaunted by design problems, knowledge based design tools are likely to experience difficulty due to the complexity introduced by inter-dependencies and conflicts between individual design decisions. In their discussion of expert systems and engineering design knowledge, Rasdorf and Parks (1985) define a design task as being one to "*develop specifications to create objects that satisfy a set of constraints*". Gero and Maher (1987) suggest that the process of design can be broken down into a series of identifiable phases or subprocesses as shown in table 1.

Systems Science, Edited by F.A. Stowell *et al.*,
Plenum Press, New York, 1993

Table 1. Subprocesses associated with the design task

Design Phase	Function
Design Analysis	Identify the characteristics of the design problem
Design Synthesis	Exploration of alternate solutions to the design problem
Design Evaluation	Critical appraisal of the evolving design - applied implicitly
Design Detail	Augmentation of selected solution by addition of design detail

Gero and Maher (1987) also discuss the idea of innovative (or novel) design, which occurs when designs in the search space which have not been produced previously are generated or found. Creative design is said to occur when designs are discovered that did not exist conceptually before the system (that is, the extension of the boundaries of the search space). Given these definitions, information systems design is generally a process of innovative design, rather than creative.

INFORMATION SYSTEMS DESIGN

The complexity of contemporary information systems necessitates the need for a methodical approach to their development. A wide range of methodologies have been developed to support differing developmental approaches. Paradigms previously proposed include the waterfall model (Royce, 1970), the spiral model (Boehm, 1986), fourth generation techniques (Cobb, 1985), and formal approaches (Jones, 1981).

Despite the availability of such methodical approaches, the design of information systems remains largely a knowledge intensive activity beginning with an informal set of frequently vague requirements, and ending up with a systematically defined formal object. Lowry and Duran (1989) describe systems design as currently being a labour intensive process which is much prone to error. Furthermore, the end result of design is devoid of the design knowledge that led to its construction

Whichever design paradigm is employed, during the design process the design knowledge maintains close proximity with the evolving design (see figure 1). That is, design expertise is on hand as and when required if, for example, a design decision should be questioned or require explanation. The design and design knowledge can therefore be thought of as a single component during systems design.

However, this situation is not maintained. Once the design stage has been completed, the design knowledge (which is often held "in the head" of the systems designer) is effectively removed from the design. The designer moves on to another project and naturally as time progresses, becomes less familiar with the design in question. This combination of events creates what can be referred to as the "post design gap" as illustrated in figure 2. The design has become stripped of the knowledge that led to its initial construction. It is precisely this design knowledge that is needed to maintain and further develop the system. Note, the "post design gap" is not a geographical term. It has no relevance regarding the actual physical whereabouts of the design knowledge, but refers to the fact that the design knowledge is no longer readily available.

In certain circumstances this situation may not pose a serious problem (if for instance the system is relatively stable (requiring few fixes) and static (not to be further developed)). However, problems are likely to occur if design decisions are questioned in

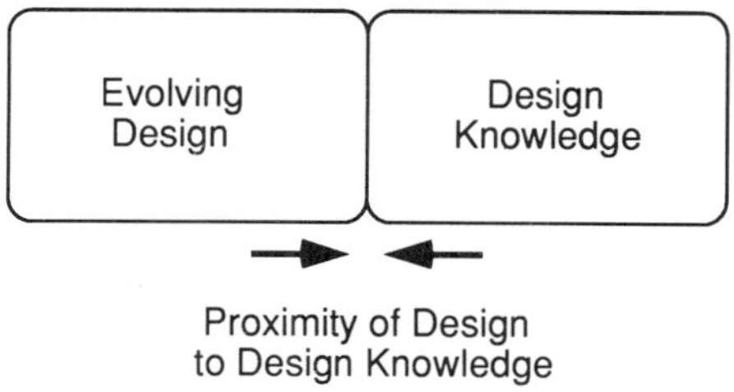

Figure 1. Proximity of the evolving design to the creating knowledge forming a single component.

retrospect and/or maintenance or further development of the system is required. The situation is exacerbated if for some reason it is not possible to reunite the designer with the design (for example, if a fixed term contract employee produced the design).

Contemporary computer aided software engineering (CASE) tools generally attempt to expedite the development process by increasing the productivity of system designers and programmers in a number of ways. For instance, contemporary design tools provide graphical interfaces (such as diagramming tools), cross checking and validation of systems, and automation of much of the task of system documentation. Although such tools provide assistance in carrying out many design tasks with improved efficiency, they are largely the results of the automation of well established structured design techniques. The facilities they provide do not solve the problem of the "post design gap". In general, the fundamental characteristic of design is not addressed by the existing CASE technology .

APPLYING ARTIFICIAL INTELLIGENCE

Oxman and Gero (1987) suggest that the two main approaches for the application of intelligent tools to the information systems design process are design synthesis and design diagnosis. During design synthesis, the tool is used to generate the actual systems design. During design diagnosis, the tool is used as a design critic, providing evaluation, criticism, and recommended changes. AI can thus be used to assist in the systems design process by developing design tools capable of providing the system designer with advice, proposing alternate solutions, and helping to investigate the consequences of design decisions. Such tools can be used to maintain the proximity of the design with the design knowledge by providing information should a design decision should be questioned or require explanation in retrospect. These tools understand the semantics of the design, and are an attempt therefore to close or at least narrow, the "post design gap".

Developments to Date

Given that artificial intelligence and conceptual modelling have developed similar knowledge representation formalisms (Hull and King, 1987), it is somewhat surprising

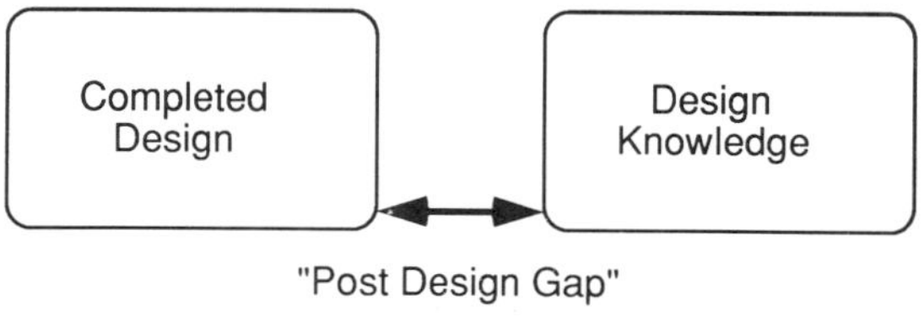

Figure 2. Creation of "post design gap" by the separation of design from the creating knowledge.

that relatively little activity has taken place in applying AI to develop systems design tools. However, a number of researchers have previously applied AI technology in the form of knowledge based systems to produce intelligent design assistants. Such tools have been developed to assist at all stages of the design process and are generally classed as those supporting specification acquisition (including conceptual, logical, and physical design) and those supporting program synthesis.

Tools for specifications acquisition are concerned with the generation of complete, consistent, and correct specifications and/or their validation and evaluation. Such specifications generally describe the user's requirements from the system, the structure of the system (in terms of software modules and flow of control), and the structure of the data to be used. Tools for program synthesis attempt to synthesise efficient programs from initial program specifications. This is generally achieved by the gradual refining of a high level specification until a program satisfying the original specification is obtained.

It is not the intention of this work to fully review intelligent design tools, or to provide a detailed comparison, however the following examples are of some interest as they illustrate points of interest, or are of historic value. As the focus of this paper is intelligent tools for systems design, no examples of tools for program synthesis are provided.

Examples of specifications acquisition tools include The Programmer's Apprentice (Rich and Waters, 1988), SECSI (Bouzeghoub et al., 1985), and Modeller (Tauzovich, 1990). The Programmer's Apprentice is designed to provide support at various stages of the development process. In this sense, its aims are similar to those of Modeller, an integral part of a family of intelligent products being developed by COGNOS inc. to assist in the process of systems design. The code specification module of both these projects can be classed as performing program synthesis. SECSI provides intelligent support for logical database design and is widely seen as the seminal work in the area of intelligent database design tools having influenced a number of successors and made a commercial breakthrough in being marketed by Infosys. A set of criteria for the evaluation of intelligent design tools is shown in table 3. These criteria are described more fully, along with a review of a series of intelligent tools for database design in (Lloyd-Williams and Beynon-Davies, 1992).

EVALUATION OF PROGRESS TO DATE

Overall, what would appear to be a potentially fruitful area for development has exhibited little commercial progress. with relatively few commercial examples of intelligent design tools available. Many intelligent design tools provide some support for the majority of the criteria described in table 3, however, few if any satisfy all the criteria. Perhaps the criteria with the greatest potential is also the most neglected - that of domain specific knowledge.

Domain specific (or "real world") knowledge and the ability to reason with this knowledge would be of obvious advantage to an intelligent design tool. There also appears little development in the area of transferring knowledge gained across design sessions. That is, the reuse of knowledge previously gained should it be applicable - such knowledge is generally restricted to within the original application.

Despite the apparent level of commercial inertia, the appearance of a limited number of commercial products such as Infosys' version of SECSI, and the forthcoming family of Modeller products from COGNOS indicates that a pool of interest for this approach does exist. Certainly there is no shortage of activity on the research front, with improvements to techniques and systems continually being developed. However, to date, relatively little of this research work has managed to successfully transfer to the commercial front.

Table 2. Criteria for the evaluation of intelligent design tools

Criteria	Description
Stage of Design Covered	Which stage(s) of the chosen design paradigm are supported
User Interface Employed	The method by which each tool receives information from the user
Method Used to Drive Design Process	The method used to provide required input to the tool
Domain Specific Knowledge	Whether the tool possesses predefined domain knowledge
Design Technique Used	The extent to which available design techniques are employed
Undo Facilities	Whether a system provides a facility to "undo" design decisions
Learning Ability	The extent to which design tools exhibit a learning ability

CONCLUSIONS AND FUTURE WORK

A proportion of the systems designed in commercial environments support broadly similar functions across organisations, for example, stock control, sales-order processing etc. The incorporation of domain specific knowledge representing generic models of these common systems within a design tool, and the ability to reason with this knowledge would be of obvious advantage. Such an approach would remove the need to ask trivial questions of the designer (thus enhancing the appearance of the intelligence of the tool), and would save time by removing the requirement to submit initial domain knowledge. The tool, having recognised an application domain, would present the generic model as an initial design attempt, and customise it to the designer's requirements during the design session. Some work has been carried out in this area (Mizoguchi et al., 1985, Kawaguchi et al., 1986), however, the tools developed, although possessing domain knowledge in the form of internal dictionaries, could not present an initial design suggestion for the designer to consider.

The majority of intelligent tools currently available are essentially used to support well established development methodologies. However, work is currently being carried out on the development of a new generation of knowledge based information systems design methodologies (KISM). Such methodologies incorporate knowledge, and explicitly state its role within the information systems design process. Ip et al. (1992) describe the Knowledge-based Requirement Engineering Assistant (KREA), supporting a knowledge based design environment providing a coherent set of techniques, tools, and procedures.

The author is currently involved in the development of a knowledge based tool which provides intelligent support for the design of object-oriented databases. Although the aim of the project is to produce a tool capable of designing object oriented databases, it is intended to examine the potential for the incorporation of domain specific knowledge in order to enhance the appearance of intelligence. However, it is recognised that this has the potential to form the basis of a research project in its own right.

REFERENCES

Boehm, B.W., 1986, A spiral model of software development and enhancement, *ACM SIGSOFT*. 11:22-42.

Bouzeghoub, M., Gardarin, G., and Metais, E., 1985, Database design tools: an expert system approach, *Proceedings of the 11th VLDB Conference*, pp82-95.

Cobb, R. H., 1985, In praise of 4GL's, *Datamation*, July, 1985, pp90-96.

Coyne, R.D. and Gero, J.S., 1986, Expert systems that design, *Proceedings of IAAIC'86*, pp1-11.

Gero, J.S., and Maher, M.L., 1987, A future role of knowledge-based systems in the design process, *Pre-proceedings of CAAD'87*, pp1-8.

Hull, R., and King, R., 1987, Semantic database modelling: survey, applications, and research issues. *ACM Computer Surveys*. 19:201-260.

Ip, S., Cheung, C-Y., and Holden, T., 1992, A knowledge-based requirement engineering assistant, *in*: "CASE: Current Practice, Future Prospects", K. Spurr and P. Layzell, eds., Wiley, Chichester.

Jones, C.B., 1981, "Software Development: A Rigorous Approach", Prentice Hall, London.

Kawaguchi A., Taoka N., Mizoguchi R., Yamaguchi T. and Kakusho O., 1986, An intelligent interview system for conceptual design of database, *Proc. of the 7th European Conference on AI*, pp39-45.

Lloyd-Williams, M., and Beynon-Davies, P., 1992, Expert systems for database design, Artificial Intelligence Review. 6:263-283.

Lowry, M. and Duran, R., 1989, Knowledge-based software engineering, *in*: "The Handbook of Artificial Intelligence: Volume IV", A. Barr, P.R. Cohen, E.A. Feigenbaum, ed., Addison Wesley, Reading, MA.

Mizoguchi, R., Kobayashi, H., Isomoto, Y., Nomura, Y., Toyoda, J., and Kakusho, O., Interactive synthesis of conceptual schema based on queries, *International Journal of Information Processing*. 8:207-216.

Oxman, R., and Gero, J.S., 1987, Using an expert system for design diagnosis and synthesis, *Expert Systems: The International Journal of Knowledge Engineering*. 4:4-14.

Rasdorf, W.J., and Parks, L.M., 1985, Expert systems and engineering design knowledge, *Working Paper, North Carolina State University*.

Rich, C., and Waters, R.C., 1988, The programmer's apprentice: a research overview, *IEEE Computer*, 21:10-25.

Royce, W.W., 1970, Managing the development of large software systems: concepts and techniques, *WESCON Technical Papers*. 14:328-338.

Tauzovich, B., An expert system for conceptual data modelling, *Proceedings of the 8th International Conference on the E-R Approach*, pp205-220.

Winston, P.H., and Prendergast, K.A., eds. 1984. "The AI Business: The Commercial Uses of Artificial Intelligence". MIT Press, Cambridge, MA.

INFORMATION SYSTEMS DEVELOPMENT SYSTEMS: WHAT GIVES?

Yasmin Merali

Warwick Business School
University of Warwick
Coventry CV4 7AL
United Kingdom

ABSTRACT

Increasingly organisations are investing in the introduction of methodologies for Information Systems (IS) development. The choices made often appear to be based on limited *weltanschauungen*, and some organisations have been frustrated in terms of their expectations of return on their investment in methodologies and related CASE tools. In this paper the *weltanschauungen* of various stakeholders are considered and some of the commonly held expectations of methodologies and CASE are analysed and questioned. In the final section, the impact of methodologies and CASE tools on the IS development system is reviewed in the context of Forbes and Checkland's criteria of systems efficiency, efficacy and effectiveness (Forbes & Checkland, 1987).

INTRODUCTION

As the role of IS within organisations has changed from being one of providing applications support to one of constituting the corporate information infrastructure, the engagement of stakeholders in the issue of IS effectiveness has increased.

The discussions in this paper are founded on observations, cases and anecdotal evidence accumulated over the past six years whilst dealing with organisations and management engaged in methodology and CASE tool adoption. The focus of this paper is on organisations where there is considerable IS development in-house. The stakeholders in the development process are defined as being managers of business areas, users of systems, the "manager" of the IS development activity and the systems development professionals themselves. It is possible to have a number of different conceptual definitions of the IS development system, some of which are given in Figure 1.

In the main, managers and users are more directly concerned with issues relating to the cost and performance of the IS developed. The IS professionals are concerned with issues relating to the quality of the development process and the acceptance of the developed product delivered. It is conjectured that an implicit super-ordinate goal for all

Systems Science, Edited by F.A. Stowell *et al.*,
Plenum Press, New York, 1993

stakeholders in the IS development process is the development, within resource constraints, of high "quality" systems which are very usable and which deliver high quality information.

Over the past two decades successive "classes" of tools, methods and methodologies have appeared, all directed at the enhancement of some facet of the Information Systems development process. Information Systems development methodologies and supporting CASE tools have been key areas for investment by organisations seeking to improve the quality of their IS and the productivity of the Information Systems developers.

MANAGEMENT EXPECTATIONS

Improved Productivity of Development Staff: The family of structured methodologies such as Yourdon (Yourdon, 1990), SSADM (Downs et al, 1988), Information Engineering (Martin and Palmer, 1991) and Merise (Quang et al, 1991), together with their supporting CASE tools can have a beneficial impact on the efficiency and efficacy criteria for the development process. These methodologies tend to divide the development process into stages, with well-defined inputs, tasks, outputs and documentation standards for each stage. If properly used, they promote consistency between and within stages of the development process. Productivity improvements may accrue from the "work organisation" imposed by these methodologies much as the Tayloristic methods "improved productivity" when applied to production systems (Taylor, 1911). Automation of documentation and completeness and consistency checking by CASE tools would also be expected to improve productivity. However, often these improvements do not manifest themselves as improved development times in the short term. This is because the thoroughness they impose is foreign to the prevalent development culture and thus represents an additional overhead on the development time. Also, methodologies and CASE tools both have a learning curve associated with them, which in the short term deflates the productivity figures. It is more realistic to expect productivity improvements

Stakeholder:	**System User**
Definition:	**System to build computer-based systems to automate boring manual tasks**
Weltanschauung:	**Users can devote more time to those elements of their job requiring human qualities and leave routine data processing to the machines**
Stakeholder:	**System User**
Definition:	**System to build computer-based systems to provide quality information & support for decision making**
Weltanschauung:	**Quality systems lead to quality decisions.**
Stakeholder:	**User management**
Definition:	**System to build systems to automate jobs**
Weltanschauung:	**Machines are cheaper than people, and more efficient too.**
Stakeholder:	**Developer**
Definition:	**System for developing a worthy career**
Weltanschauung:	**Organisations depend on IS and an up-to-date IS professional is worth investing in.**
Stakeholder:	**Enlightened Corporate Planners**
Definition:	**System to develop of a corporate information infrastructure**
Weltanschauung:	**A good infrastructure is essential for business success.**

Figure 1. Example *Weltanschauungen* and Definitions for IS Development Systems.

in the longer term, where the structure and internal consistency of developed systems together with comprehensive documentation and automated aids for documentation maintenance can be exploited for improving the efficiency of maintenance work. Here, the object orientated methods (Meyer, 1988) for software development which realise the "reuseabiltiy" concept do make a significant contribution to productivity gains in the overall development time. Object orientated "methodologies" for the entire systems development process are not mature, and developers are grafting object orientated methods onto their existing methodologies with successful outcomes.

Participative methodologies such as ETHICS (Mumford, 1983) and Soft Systems Methodology (SSM) (Checkland & Scholes, 1990) may impact badly on development times for new systems because of the overheads of user participation (Hirschheim, 1986). If used effectively, these methodologies can have a visible impact on the ease of implementation and user acceptance stages of the project.

Better Management of Development Projects: The well-defined tasks and standards and the staged approach of structured methodologies provide good check-points for monitoring the progress of the technical development process. The Tayloristic work organisation can facilitate the imposition of accountability and control mechanisms on development staff. The task-orientated approach and standardisation feature ease the problems associated with the management of projects in a climate of high staff mobility. The support provided by CASE tools for consistency checking and document maintenance improve the manageability of projects. The participative methodologies can contribute towards the management of user expectations and towards the resolution of political and cultural conflict. The requisite empowerment of the project management and human resource implications are key factors which influence the effective use of these methodologies.

Improved Flexibility/Maintainability of Systems: The top-down approach of structured methodologies and the clear definition of systems, sub-systems and the interfaces between sub-systems make it possible to modify relevant parts of the system in a controllable manner. However, the flexibility of the system in this context will depend on the "quality" of the initial partitioning. Object orientation and the recognition of multiple *weltanschauungen* in SSM, by contributing to the conceptual richness of systems design, will improve systems' ability to respond to changes in the "real" world.

Improved Quality of the Developed Systems: The concept of total information systems quality is not a universally defined one. In this instance it is useful to apply Forbes & Checkland's criteria of efficiency, effectiveness and efficacy (Forbes & Checkland, 1987) to the IS development process when attempting to evaluate the contribution that methodologies and CASE can make here.

The improvements that methodologies and CASE tools can promote in the internal consistency of systems design and development can improve the logical efficacy of systems. There is evidence to show that for certain types of applications, the use of 4GL code generators (which can be used in conjunction with CASE tools) may result in sub-optimised code with respect to systems performance (Institute of Data Processing Management, 1986).

The management of sub-systems integration supported by object orientation and structured methodologies may contribute to improvements in efficiency of design of the technical system, and the focus of the participative methodologies on human and organisational aspects may contribute to the efficient utilisation of the computerised system in its environment, and hence to the improved efficiency of the IS in its entirety.

The contribution of methodologies and CASE tools to improved effectiveness of systems is more difficult to construe. Because SSM addresses the issue of "relevant systems", and the stakeholders arrive at an explicit definition of what is systemically desirable and culturally feasible, it may be argued that the use of SSM contributes positively towards the definition of the effectiveness criteria for the system. However, as the stakeholders are also part of the realised system, the establishment of the objectiveness of criteria defined in this way is problematic.

TOWARDS THE SUCCESSFUL EXPLOITATION OF METHODOLOGIES & CASE TOOLS

In the previous section, management expectations of methodologies and CASE tools were reviewed in the light of existing capabilities. Methodologies have long since reached maturity and the current generation of CASE tools is well-defined. Why then do organisations have difficulty in realising the potential of their investments in methodologies and CASE tools? In this section the concerns of, and conflicting interactions between, stakeholders in the IS development system are examined and a set of "success factors" for the exploitation of methodologies and CASE tools is proposed.

Key Concerns

The conclusion from the previous section was that the effectiveness of the development process is decided by factors external to it, but that methodologies and CASE tools can be used to positively influence the efficiency and efficacy of the development process. Figure 2 illustrates the key concerns of stakeholders in the IS development system. It is suggested that the concerns and issues illustrated here serve to undermine the productive utilisation of methodologies and CASE tools. Many of the concerns can be dealt with by changing the attitudes of the stakeholders. However, in order to enable the attitudinal change to take place, some fundamental beliefs and perceptions need to be changed, and in many organisations this is only possible if preceded by a change in the organisational culture and philosophy. It is suggested that an organisation-wide ownership of the concept of "Information Systems as Information Infrastructure", a flexible team-based developer-user relationship, and a mixed-paradigm approach to human resource development are key "success factors" for the productive use of methodologies and CASE tools.

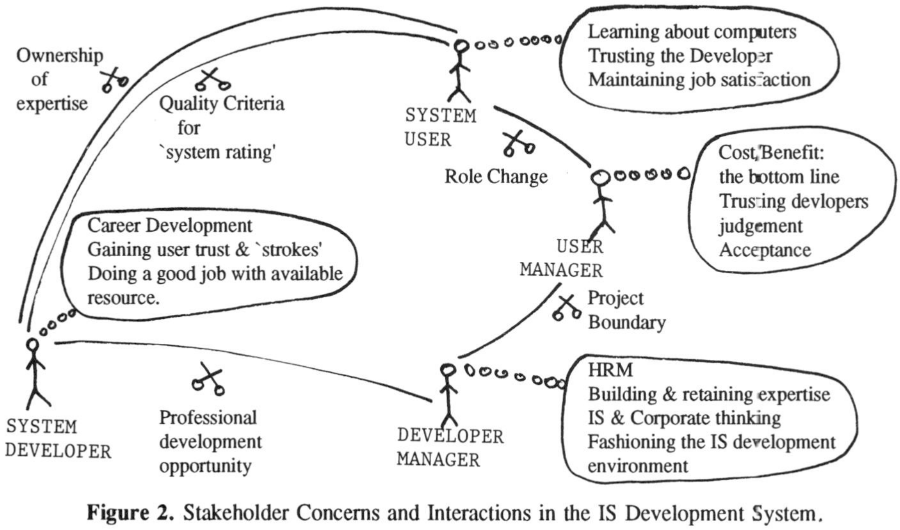

Figure 2. Stakeholder Concerns and Interactions in the IS Development System.

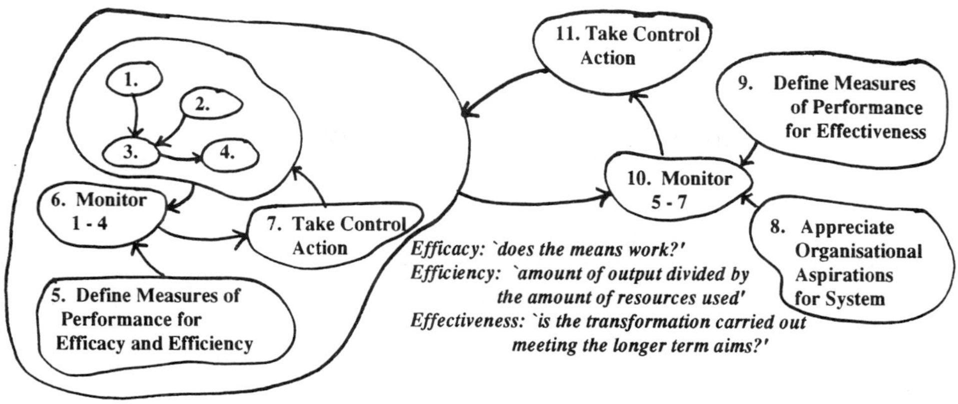

Figure 3. Efficacy, Efficiency and Effectiveness Measures for Systems (after Checkland & Scholes, 1991).

CONTRIBUTION TO EFFICIENCY, EFFICACY AND EFFECTIVENESS OF INFORMATION SYSTEMS DEVELOPMENT SYSTEMS

Efficiency and Efficacy

From the features discussed in previous sections, it can be concluded that providing they are exploited successfully, methodologies and CASE tools can both contribute positively to the performance of the IS development system. Besides providing opportunities to improve productivity and internal consistency within and between stages of the development process, these features make it easier to monitor, evaluate and improve the efficacy and efficiency of the development process. The explicitly defined outputs and the "staged" infrastructure of the methodology facilitate the definition of measures of performance in the context shown in Figure 3.

The argument here suggests that in an appropriate organisational context, the actual "engineering" or construction of computer software to meet a given specification is methodologically non-problematic, and that efficiency and efficacy criteria for both, the development process and the product of development can be objectively set if (and only if) we can assume the effectiveness of the specification as a "given".

Effectiveness

A Problem with Specification. As shown in Figure 3, the criteria for evaluating the effectiveness of a system must be set from outside of the system being evaluated. It was found that a common aspiration for IS development systems within organisations is captured in the definition of the IS development system as "A system to utilise available human and technical resources to develop effective, efficient and efficacious information systems", with definitions of efficiency, effectiveness and efficacy borrowed from the work done by Forbes & Checkland (Forbes & Checkland, 1987).

There is a fundamental problem here in that the effectiveness of the development process is defined in terms of the effectiveness of the product. Logically, the effectiveness criterion for a systems specification is taken to be satisfied by definition.

Objective evaluation of the effectiveness of systems at the specification stage is problematic. Of the currently popular IS development methodologies, SSM deals with this issue most explicitly. If the system to be developed is called "X", the SSM definition of the "effectiveness" criterion for "X" would be: "is the transformation to be carried out in "X" meeting the longer term aim?". In SSM the stakeholders will define "X" such that, in their opinion, the effectiveness criterion will be met in a way that is consistent with their perception of the valid "longer term aim". The dynamic nature of the real world and the transient nature of stakeholders detracts from the goodness of fit between the perceived and actual effectiveness of systems built to specification. Pragmatically, modelling tools and simulation can be employed to improve the approximation when defining the valid longer term aims upon which the effectiveness of "X" is predicated.

In most organisations management are striving for an improved development process in terms of efficacy and efficiency, are benchmarking systems in terms of efficiency and efficacy, but are failing to focus on the definition of criteria for effectiveness. Management continue to look towards investments in methodologies and CASE tools to ensure the development of effective information systems, although the notion that an effective IS development process will yield effective IS is basically tautological.

CONCLUSIONS

The contribution of IS development methodologies and their associated CASE tools to the effective performance of IS development systems was reviewed in the context of the effectiveness, efficiency and efficacy criteria for systems evaluation.

It is argued that appropriate stakeholder attitudes and organisational philosophy and culture need to be in place if methodologies and CASE tools are to be successfully exploited.

It is suggested that management expectations of methodologies and CASE tools are founded on the notion that an effective IS development process will yield effective IS. Although the efficacy and efficiency of the development system can be measured by objective criteria, the effectiveness criterion has to be defined externally by the stakeholders in relation to the business objectives and the environment. It is of fundamental importance that management focus on the effectiveness issue and define and communicate meaningful criteria for its evaluation. Without such criteria it is impossible to assess whether IS development systems will in fact deliver verifiably effective systems.

REFERENCES

Checkland, P.B. and Scholes, J., 1990, "Soft Systems Methodology in Action", Wiley, Chichester.

Downs, E., Claire, P. and Coe, I., 1988, "SSADM", Prentice Hall, Hemel Hemsptead.

Forbes, P. and Checkland, P.B., 1987, Monitoring and control in systems models, Internal Discussion Paper 3/87, Department of Systems and Information Management, University of Lancaster.

Hirschheim, R.A., 1986, Participative Systems Design: User Experiences, Evaluation and Conclusions, *Australian Computer J.* Vol.18, pp 166-172.

Institute of Data Processing Management, 1986, "Fourth Generation Languages (Volume I)", IDPM Publications, Sidcup, U.K.

Martin, J. and Palmer, I., 1991, "Introduction to Information Engineering Methodology", James Martin Co., Dublin.

Meyer, B., 1988, "Object Oriented Software Construction", Prentice Hall, Hemel Hempstead.

Mumford, E., 1983, "Designing Human Systems for New Technology", Manchester Business School.

Taylor, F.W., 1911 Principles of scientific management, reprinted *in:* "Scientific Management", Harper, New York, 1947.

Quang, P.T. and Chartier-Kastler, C., 1991, "MERISE in Practice", MacMillan Education, Basingstoke, U.K.

Yourdon, E., 1990, "Modern Structured Analysis", Prentice Hall, Englewood Cliffs, New Jersey.

A STRATEGY OF ENGAGEMENT: KNOWLEDGE ELICITATION FOR AUGMENTATIVE SYSTEMS DEVELOPMENT

Ian M. Neale

Department of Business Computing
The City University
Northampton Square
London ECIV OHB

INTRODUCTION

Clark (1987) reviews the difficulties of modelling human cognitive activity in an implementation-neutral style: ie, creating a description of cognitive processes which can be equally well realised on any of a number of different machines as in the human mind. The first problem is background knowledge: we cannot write a program which includes everything we know about a subject (which knowledge we acquire by being an integral part of the world). This has been well argued by Dreyfus (1979) and Collins (1990). The other principal difficulty is that some human problem-solving, such as planning a garden or solving a jigsaw puzzle, appears to involve manipulation of real objects in addition to our internal symbol manipulations. For future cognitive systems development, Clark recommends a strategy of "engaged cognitivism" based on a careful study of our embodied and embedded behaviour. This paper describes such a study.

Fault diagnosis is the paradigmatic task for expert system application. Knowledge elicitation from experts may proceed via classical hierarchical decomposition techniques such as card sorting, laddering, and structured interviewing (Cordingley, 1989). The result of analysis is frequently a set of production rules of the form IF (symptom) THEN (diagnosis); IF (diagnosis) THEN (action). The knowledge base may be refined to include measures of likelihood such as probability estimates, certainty factors, ranks, ratings, etc. These measures - or perhaps more accurately, analogues - of expert judgment may then be subjected to statistical analysis (Meyer and Booker, 1991).

As the output of the expert system eventually developed becomes more sophisticated and its recommendations more qualified, so does the knowledge required of the user to interpret the system increase. Insofar as real world events are embedded in a complex context, which we necessarily choose to describe in abstracted form in terms of a relatively few variables, an expert system alone can only deliver a 'rough cut' solution to a problem. Fully formalizable domains in which autonomous systems may be deployed

Systems Science, Edited by F.A. Stowell *et al.*,
Plenum Press, New York, 1993

are few (Winograd and Flores, 1986). Less tractable domains where it is necessary - or at least desirable - to retain a human in the loop are the rule. The events of a single day in October 1987, in which a wave of computer selling precipitated a stock market crash, demonstrated the error of omitting human insight and flexibility. Joint cognitive systems (Woods, 1986) on the other hand have a bright future in augmenting human performance, provided the potential of the machine is not overestimated in relation to the contribution of the person working with it.

In business and industry, a common problem is the need to distribute scarce expertise more widely, empowering less-skilled staff and improving service to clients and the efficiency and profitability of operations. Recent years have witnessed the dramatic decline of the apprentice system in Western economies: rare is the opportunity today to learn by 'sitting with Nellie'. In focussing on this problem, it is neither necessary, possible nor even desirable to attempt to elicit "everything" the expert knows. It is not necessary, because very rarely will the resultant system be intended for use by a completely naive person. Neither is it possible, because people do not have total access to their reasoning processes: much of our knowledge is tacit and as such largely beyond articulation (Johnson, 1983; Berry, 1987; Collins, 1990). The implementation-led approach to knowledge elicitation which has often prevailed has largely ignored the social character of a large part of what we know (Bloomfield, 1988). Finally, it is not desirable, because apparent success will lead to misplaced confidence in the system. The abstraction from the real world embodied by the system will inevitably, as it is only a partial representation, constitute an oversimplification (Neale, 1990). Reliance on this alone may prove dangerous (De Greene, 1991).

A valuable system will, far from mistakenly attempting to replace a true expert, enable a less experienced person to satisfactorily handle the majority of problems posed. Through a preliminary analysis of the problem, system and user together will typically be able to filter out the minority of cases which need to be referred to an expert. The remainder, which may be seen as conforming to a familiar pattern displaying no unusual features, or possessing a comfortable margin of error, can be satisfactorily resolved there and then.

A CASE STUDY: DOMAIN ORIENTATION

Screen printing is the most versatile of all printing processes: the surface of virtually any material can be screen printed, using a wide variety of types of ink deposited in a layer of controlled thickness. This is achieved by forcing the ink (paint-like in consistency) by means of a sharp-edged polyurethane or rubber squeegee through a finely-woven mesh, stretched over a frame to create the screen. The design to be printed is delineated by a stencil mounted on the underside of the screen.

Although in everyday life colour is expressed verbally, where colour matching is concerned, this is often inadequate: no two observers viewing an object in the same light are likely to propose precisely the same terminology for its colour. What is 'bright red' to one may be 'scarlet' to the other. Colours are more accurately (and completely) described in terms of three elements, hue or shade ('red', 'green', etc), value ('light', 'dark', etc) and chroma or saturation (degree of difference from a grey of the same value). These colour theory terms are not, it must be noted, often used explicitly by screen printers when matching colours. Any colour can in fact be obtained by mixing no more than three coloured pigments, plus white, or sometimes two plus white and black (Billmeyer and Saltzman, 1981).

Colour matching for screen printing is a skilled craft involving the production of a batch of ink in a colour which to the client's satisfaction prints an equivalent to an original sample. The sample is very rarely a wet sample of screen ink, as is preferable. More usually, it is a piece of paper or card which has been printed by a process such as litho which utilises quite different pigments and chemicals. It is important to note that unlike the domain of litho printing inks, where a commercially available library of some 520 colours provides a widely-accepted recipe system, there is no such comprehensive industry standard for screen printing inks. Several competing manufacturers have developed computerised colour matching systems linked to spectrophotometers, but this "high tech" approach is expensive for many printers working for small companies. Furthermore, it is arguable that the implicit deskilling which results from reliance upon such technology is socially and economically undesirable. We operate in a complex world which penalises unimaginative linear thinking and places an increasing premium on flexibility of response to changed circumstances.

Most if not all of the necessary conditions for expert system development, as stated by Waterman (1986, pp. 127-134), are met by the domain of colour matching screen inks. Many instances of colour matching are still performed by printers themselves, relying on experience but often working to some extent by trial-and-error. This may take a few minutes or a few hours: greys and pastel shades are particularly difficult. If too much colour is added to what is mostly white, a very large amount of white will have to be further added to return to the original shade, which means that the inevitable wastage has been increased.

The screen printing industry is increasingly subject to mechanisation and automation with concomitant deskilling; allied to the continuing decline in training in British industry, this is rendering skilled colour matching an ever-scarcer resource. The problem is important as clients are increasingly inclined to insist upon a match to their trademark colour (examples are 'Coca-cola red'; 'Marlboro red' and 'Rothmans blue'), by which their packaging and advertising can be instantly recognised. Prospective users of an expert system are printers who have a basic knowledge of colour and the tacit knowledge associated with the mixing and use of screen printing inks. An expert system in the modality of domain consultant would provide quick advice on what basic colours to mix and guidance on the proportions, to obtain a good commercial match to a sample.

KNOWLEDGE ELICITATION

Initial Interview

The expert is the proprietor of a small screen ink manufacturing company with over twenty years experience in the industry, as both printer and subsequently specialist colour matcher, but without any formal training. He relies entirely upon his own visual judgement rather than any instrumental assistance in measuring colour.

The language of screen ink colour matching is hard to locate in any literature and seems infused with a large amount of tacit knowledge, at first sight having all the hallmarks of a 'black art'. Fortunately the author had, through his background in the industry, become intuitively aware of this language. Otherwise, an extended period of engagement via participant observation (Spradley, 1972; 1979), as a helper/apprentice, would have been essential. It was decided, in view of the importance of using the expert's own terminology, to ask the expert to give a short informal tutorial interview, eschewing a tape recorder to make it as comfortable as possible for the expert. Notes were taken and transcribed the same evening. The purpose was to elicit domain concepts; as a focal point, the author selected from the commercial library a 'bright' or 'leaf' green, and asked the expert to describe how he would go about matching the colour. This was essentially an exercise in forward scenario simulation. The interjection of a single probing question by the knowledge engineer towards the end of his account led to some digression by the expert, composed nevertheless of a rich matrix of domain vocabulary and heuristics. The transcript also included many statements indicative of the kind of tacit knowledge involved; for example, "Once the shade is right, I would try to let it out with varnish". "To let out" means something like "to water down": to weaken the colour by diluting its strength with a clear additive. (An infinite regress is implicit in any attempt to codify such language in a set of programming rules; for illustrations of the problem, see Collins (1990)).

Verbal Protocol: The Next Step

Waldron (1985) emphasises two important points: first, the need to minimise interaction time with the expert and second, the communication and time-wasting problems which can occur if interviewing is used too early in the knowledge elicitation process. He discusses 'process tracing' more commonly known as verbal protocol analysis (Ericsson and Simon, 1984) - as a tool for streamlining knowledge elicitation and preparing the knowledge engineer for a later stage when more efficient and detailed interviewing can be deployed. The author agrees with this approach to the domain (in other circumstances it may be wiser to conduct interviews first; see Neale (1988) for discussion). Colour matching is very much a practical skill which does not easily lend itself to 'off the job' exposition.

The technique involves asking the expert to 'think aloud' all the time while solving a problem: this is tape-recorded, transcribed in every detail into a 'protocol' and then analysed for meaningful associations. The knowledge engineer intervenes during the task performance only with nondirective reminders to keep thinking aloud and should not ask the expert to attempt to explain his or her thoughts. The main advantage is that the unconstrained verbalisation of the expert may reveal the heuristic knowledge which he or she is unable to articulate in an interview. However, as expert knowledge is highly 'compiled', with many reasoning steps reduced to a single association, there is a risk the expert may simply not verbalise what is 'obvious' to him or her (Bainbridge, 1979).

In accordance with this description, the expert was asked to match an actual sample colour whilst thinking aloud, which the author tape recorded whilst observing and making notes of nonverbal actions. The tape recorder was left running after completion to collect retrospective comments. Some useful insights were obtained in this less constrained manner, including answers to questions carefully structured by the author to clarify particular data in the protocols.

A preliminary analysis of the protocol served two purposes. First, it identified the basic terminology, providing focal points for clarification in further interviews and an initial structuring of the knowledge base. Second, it demonstrated the particular way in which the problem was approached, which in turn suggested an appropriate KADS interpretation model (Breuker and Wielinga, 1989) for task analysis. Two alternative frameworks for further analysis of protocols are described by Waldron (1985) and Johnson, Zualkernan and Garber (1987). The KADS approach was selected as it offered an explicit integration of knowledge elicitation within the full process of expert system development.

Subsequent Work

The task of colour matching was viewed as assessment, an analytic task involving identification and classification and often sufficiently complex to involve other tasks such as diagnosis. The purpose of selecting the assessment model is to provide an organiser for the data in terms of the types of knowledge ingredients to look for (Wielinga and Breuker, 1986). The conceptual model of the domain which is eventually constructed acts as a bridge between the verbal data and the implementation and provides a functional specification for the system (Neale, 1990).

Colour matching is essentially a set of iterative procedures, for which decision trees are widely used as an intermediate representation and sometimes as the basis of an implementation formalism (eg Alvey and Greaves, 1987). As a cognitively accessible representation, on a single sheet of paper, the decision tree was also found to be a suitable basis for further knowledge elicitation work with the expert. The objective was to complete the analysis of the domain, providing a lexicon and a refined paper knowledge base. As the target knowledge and areas of ambiguity were more clearly defined by now, straightforward structured interviewing was the primary route taken.

At this stage the project terminated for lack of further resources. The direction further work should take could be discerned nonetheless. The first stage in colour matching essentially concentrates on matching the hue: broadly speaking, what is intuitively understood as the 'colour' by lay observers. The process by which the expert makes adjustments to the lightness and saturation is less easy to determine. Multidimensional techniques of knowledge elicitation, starting with card sorting as the quickest and most transparent to a domain expert, might be appropriate; the repertory grid technique is particularly suitable for accessing implicit knowledge (Gammack, 1987). This is also appropriate where, as in colour matching, there exist closely related concepts (for example, a "strong" colour and a "deep" colour), not well differentiated by novices but apparently susceptible to discrimination by experts.

DISCUSSION

No expert system could exactly replicate the performance of the expert in any new colour matching problem. The crucial reason is that human expertise is not a summation of parts which can be replicated unerringly on demand. It is always context-dependent, and it is unlikely that all the contextual parameters which might be relevant could be elicited. For example, there is the problem of metamerism: some colours - notably certain greens, for example - look radically different under different lighting. Do the experimental conditions match the light in which the results are to be viewed by the client? Daylight itself varies, as do fluorescent tube lights. Allowing for such continuous variation is a component of an expert's judgment which cannot be verbalised. Each new instance of the problem will differ in some way, in the contextual elements drawn upon by the expert; new insight is always at work (Compton and Jansen, 1990). Human performance cannot be totally consistent.

It is evident that some instrumental adjunct to the computer is desirable. Digital output from a spectrophotometer which compared the colour coordinates of the original sample and the colour match might be linked to a knowledge base of facts and heuristics elicited from the expert. The user must still contribute significantly to the problem solution, however, through the cultural knowledge of screen printing. For example, no expert system can judge the point at which the consistency of the mix is exactly right for printing. The task is very similar to a cook following a recipe: to manage the instruction: "Beat the egg whites until stiff and then fold in" one must contribute a great deal of knowledge of what the words mean, which it would be futile to attempt to program (Collins, 1990, pp 99-100). The way to gain this experiential knowledge is through a period of apprenticeship, formal or informal, which may be the route a knowledge engineer has to take.

CONCLUSION

A possible expert system for colour matching of inks for screen printing has been discussed in respect of the suitability of the domain and some methods for elicitation of the expertise involved. Much work remains before the project can be proven commercially viable. Success is more likely if the aims are comparatively modest, rather than seeking to replace an expert which it has been argued is inappropriate. The goal should be an augmentative computer system which empowers the person working with it. This is achieved by embodying the expert knowledge the person lacks, without denying them agency and initiative.

REFERENCES

Alvey, P.L and Greaves, M.F., 1987, Observations on the development of a high performance system of leukaemia diagnosis, *in:* "Research and Development in Expert Systems III", M.A. Bramer, ed., Cambridge University Press, Cambridge.

Bainbridge, L., 1979, Asking questions and accessing knowledge, *Future Computing Systems,* 1:143-149.

Berry, D.C., 1987, The problem of implicit knowledge, *Expert Systems* 4:144-151.

Billmeyer, F.W. and Saltzman, M., 1981, "Principles of Colour Technology", John Wiley, New York.

Bloomfield, B.P., 1988, Expert systems and human knowledge: a view from the sociology of science, *AI & Society* 2:17-29.

Breuker, J.A. and Wielinga, B.J., 1989, Models of expertise, *in:* "Topics in Expert System Design: Methodologies and Tools", G. Guida and C. Tasso, eds, Elsevier, Amsterdam.

Clark, A., 1987, Being there: why implementation matters to cognitive science, *Artificial Intelligence Review,* 1:231-244.

Collins, H.M., 1990, "Artificial Experts: Social Knowledge and Intelligent Machines", MIT Press. Cambridge.

Compton, P. and Jansen, R., 1990, A philosophical basis for knowledge acquisition, *Knowledge Acquisition,* 2:179-278.

Cordingley, E.S., 1989, Knowledge elicitation techniques for knowledge-based systems, *in:* "Knowledge Elicitation: Principles, Techniques and Applications", D. Diaper, ed., Ellis Horwood, Chichester.

De Greene, K.B., 1991, Emergent complexity and person-machine systems, *International Journal of Man-Machine Studies* 35:219-234.

Dreyfus, H.L., 1979, "What Computers Can't Do", Harper and Row, New York.

Ericsson, K.A. and Simon, H.A., 1984, "Protocol Analysis: Verbal Reports As Data", MIT Press, Cambridge.

Gammack, J.G., 1987, Formalising implicit domain structure, *in: C J.* Pavelin and M.D. Wilson, eds, "Knowledge Acquisition For Engineering Applications", Rutherford Appleton Laboratory, Didcot.

Johnson, P.E., 1983, What kind of expert should a system be?, *The Journal of Medicine and Philosophy* 8:77-99.

Johnson, P.E., Zualkernan, I. and Garber, S., 1987, Specification of expertise, *International Journal of Man-Machine Studies* 26:161-181.

Meyer, M. and Booker, J., 1991, "Eliciting and Analysing Expert Judgment: A Practical Guide", Academic Press, London.

Neale, I.M., 1988, First generation expert systems: a review of knowledge acquisition methodologies, *The Knowledge Engineering Review* 3:105-145.

Neale, I.M., 1990, Modelling expertise for KBS development, *Journal of the Operational Research Society,* 41:447-458.

Spradley, J.P., ed., 1972, "Culture and Cognition: Rules, Maps and Plans", Waveland Press, Prospect Heights.

Spradley, J.P., 1979, "The Ethnographic Interview", Holt, Rinehart and Winston, New York.

Waldron, V.R., 1985, Process tracing as a method for initial knowledge acquisition, *in:* "Proceedings of the Second IEEE Conference on Artificial Intelligence Applications: The Engineering of Knowledge-Based Systems", IEEE Computer Society Press, Washington DC.

Waterman, D.A., 1986, "A Guide to Expert Systems", Addison-Wesley, Reading.

Wielinga, B.J., and Breuker, J.A., 1986, Models of expertise, *in:* "Advances in Artificial Intelligence", B. du Boulay, D. Hoggs and L. Steels, eds, Elsevier, Amsterdam.

Winograd, T. and Flores, F., 1986, "Understanding Computers and Cognition", Ablex. Norwood.

Woods, D.D., 1986, Cognitive technologies: the design of joint human-machine cognitive systems, *AI Magazine* 6:86-92.

MANAGING IS IN GLOBAL COMPANIES

Richard J. Ormerod

Warwick University
Coventry CV4 7AL

ABSTRACT

Increasingly businesses are having to compete on a global scale resulting in organisational structures which exhibit varying degrees of connectivity between the local business units. The management of such information can at one extreme be centralised in a dirigiste fashion or at the other extreme it can be left to the local units in a hands-off mode. The paper will argue that if the balance struck between the roles of the corporate centre and the business units is consistent across policy areas the IS role will be better understood by both IS and non-IS managers. Further, drawing on empirical evidence, it will be argued that the balance struck must be consistent with the corporation's overall management style. As most corporations are continually adjusting their style, the management of IS must design their management strategy dynamically to avoid conflicting with corporate culture.

INTRODUCTION

With competition increasingly conducted on a global basis many companies have been seeking effective organisational forms. Early 'exporting', 'international' and 'multinational' models of organisation are seen as unbalanced and inadequate in the current climate. New models have been suggested and companies aspire to perfect their 'global', 'networked' or 'transnational' organisations. The coordination and control of these large, complex, far flung organisations raises many issues not least those of managing information systems and technology (see Deans and Kane, 1992).

As the technical characteristics of IT have evolved and the use of IT has grown in all organisations, the perception of IT's importance has changed. Organisations have struggled with the question of how to organise IS resources and how to coordinate their activities with the rest of the organisation. Companies organised across regions on a divisionalised basis face the problem of determining the relative roles of the different organisational units. Where should IS be sited in such organisations and how should the activities be coordinated?

Systems Science, Edited by F.A. Stowell *et al.*,
Plenum Press, New York, 1993

In 1991 a survey was conducted of 6 multidivisional companies to provide insight into the relationship between corporate centres and their divisions with respect to IS. What are the determinants, what are the dynamics? If the CEOs of these businesses are in a game of global chess, then the CIOs are engaging in a game within the game. What are the options? What are the rules? What are winning strategies?

RESEARCH BACKGROUND

The management of global companies has been the subject of extensive commentary and research. Bartlett and Ghoshal (1989) concluded that the conflicting aims of achieving operational efficiency, market responsiveness and organisational learning could be reconciled in what they termed the 'transnational organisation'. They also identified the 'administrative heritage' as an important contextual factor in determining how to manage a worldwide organisation. The term 'administrative heritage' referred to organisational capabilities as shaped by various historical and structural factors.

The management of IS within large and complex companies with global reach has been the subject of research at the Oxford Institute of Information Management (Feeney et al, 1987). The research concluded that two balances need to be struck (i) the division of IS responsibilities and resources between corporate and business unit levels (ii) the balance of roles between Users of IS and IS specialists.

These themes have been examined in a series of papers (Feeney et al, 1989b; Edwards et al, 1989; Feeney et al, 1989a). On the first question, the division of IS responsibilities and resources, Feeney et al concluded that four contingent variables should affect IS organisational design (i) the host organisational characteristics (ii) the IS heritage (iii) perceived IT strategic impact and (iv) technology assimilation. They suggest that the host characteristics and IS heritage should be the basis for determining the evolution of IS structure, which determines the balance between centralisation and decentralisation of IS resources. Hodgkinson (1990) has conducted a survey to explore the role the corporate centre in IS management. On the issue of organisational structure the federal model is advocated except for holding companies. Drawn widely enough the federal organisational models can encompass any balance between the centre and the divisions. However such a definition throws little light on where the balance should lie. In this paper a distinction is drawn between a market model of coordination and a federal model.

THE STRUCTURE OF THE SURVEY

The survey was conducted by interviewing a senior manager from within the corporate IT function or someone with an intimate knowledge of it. Attention was centred on the loose-tight grid shown in Figure 1 devised for the purpose. On the left hand side are various topics that the centre (the CIO, IT director or corporate strategy cell) could influence or control. The list is far from exhaustive and I would advise users of the grid to generate their own topics.

The loose-tight dimension is divided into four with an indication in each cell of characteristic behaviour. The four columns are labelled for convenience, hands-off, market, federal and dirigiste according to the coordinating style adopted. The two extremes are self-explanatory. 'market' envisages the centre offering certain services which the divisions can buy or not as they think fit. Policies and standards are laid down only so far as they assist the working of the market and may not be mandatory. 'Federal' envisages a more active role

	Light Touch			Heavy Touch
Strategy Formulation	None	Guidelines	Coordinate	Determine
People	None	Job Market	Core Competence	Career Planning
Common Systems	None	Spread Best Practice	Choose Leaders	Mandatory Systems
Data	Independent	Messaging	Agree Standards	Common Dictionary
Purchasing	None	Approved Suppliers	Corporate contracts	Central Purchasing
Technology Introduction	Nothing	Demonstrate	Pilot	Central Development
Standards	None	Comms	Agree Standards	Set Standards
Capital Approval	None	Monitor	Approval Hurdle	Budget Holder
	Hands Off	**Market**	**Federal**	**Dirigiste**

Figure 1. The loose-tight grid.

for the centre, but under the control of the divisions. Thus standards may be proposed and administered from the centre but divisional agreement is required to implement them.

The grid is used to position businesses along the loose-tight dimension for coordinating style. The grid can be used to indicate the evolutionary path of a company up to the present and forward into the future. Respondents were also asked to plot the course of the host organisation on a similar loose-tight dimension. Thus the contingent factors, host characteristics and IS heritage were examined.

THE RESULTS OF THE SURVEY

The interviewees were able to relate to the grid and use it to describe the evolution of management style and IT responsibilities within their own organisation.

Company A : A Decentralised Oil Company Business

The general management style of the organisation is to minimise the role of the centre except for certain reserved areas such as the global brand. This style represents a departure from the earlier more integrated approach. The business is organised into divisions where the IS resources are located. At HQ there is an IS strategy cell of about eight people which is establishing its role within the new organisational style. They currently find themselves in the 'market' column but the pendulum is starting to swing back and they are exploring the possibilities of achieving value added for the divisions. A recent strategy exercise concentrated on enabling infrastructure, leaving the divisions to determine their own forward development programme. The centre has no role in developing common systems, encouraging instead spread of best practice. Capital projects are monitored rather than approved and limited standards are in place. The organisation at present is 'market' in most activities. The possibility of a more 'federal' approach is being tested on several fronts.

Company B: An Information Company

The company has four organisational tiers, corporate, time zone, region and country or state. Each country or state is a profit centre with country businesses in Europe and states in America grouped into regions. The company's products are sold globally but control of product development is held by the corporate centre which allocates responsibility for each product to a time zone. Product sales and delivery are the responsibility of each country or state operation. During the last decade the globalisation of products had resulted in the shift from a national based organisation to the current zonal structure. Within the current structure there has been a shift in power from the centre to the zones. At the beginning of the decade the IS systems matched the overall organisation at the country level. By the end of the period most IS resources were located at the time zone level with the balance at the centre. Bespoke systems were designed and developed at the zonal level, and run in regional processing centres. At the corporate level the IS group developed and operated systems to support corporate activities and sought to define the role with the zones. An attempt by the corporate group to assert the European IS strategy and standards in other zones failed. A more decentralised approach is now being introduced with the central group in a coordinating and enabling role.

Company C: A Manufacturing Company

The company was created by merging two national companies. The cultures and corresponding IT structures were completely different. One company was centralised while the other operated as an informal federation, with the headquarters of the individual companies being in the same building. Since the merger the group has been seeking to effect some wide ranging changes. There is a drive to reduce central overheads and increase the autonomy of operating units. The first pan-company IS project has been initiated to develop an MIS system to allow the new structure to operate. One of the original companies ran their IT services from a single site with localised MRP and factory automation supported by central consultants. The other company had decentralised systems which, because of the loose federation, used broadly the same equipment. After merger the central systems were broken up and the resources distributed to the operating units. The intention was to devolve strategy formulation but some initiatives have run counter to this.

Company D: A Centralised Oil Company

The company has four levels, corporate, divisional, regional and country affiliates run as a highly integrated international corporation. The Divisional headquarters have functions that mirror the corporate functions with dotted line relations to them. Each affiliate has a similar structure. The delegated levels of authority at the affiliate level are very low with most investment decision being resolved at the Divisional level. The IS services are part of the finance and administration function at the corporate, divisional and affiliate levels. Data centres are located worldwide but controlled from the centre. They are in a position to ensure the effective implementation of standards. At the divisional level the IS function is concerned with common systems and technology transfer. At the affiliate level the operations, systems and planning managers are planned by the division, the remainder of personnel being locally employed. Most strategy decisions are taken centrally. In practice the divisions are able to exert some power.

Company E: A Consumer Goods Manufacturer

The group comprises three regional businesses. The very small corporate IT service supports the HQ and offers no services to group companies. The European business has grown from its UK origins by establishing sales organisations in most European countries to provide outlets for products made by the UK and continental manufacturing sites. There is a high degree of vertical integration. Towards the end of the eighties a new managing director was appointed to integrate the national profit centres into a single, pan-European business. A strategy was developed to develop a European wide approach to data and systems to support Pan-European product marketing, manufacturing and distribution management. A small IT resource has been set up at the European Business level to complement the existing resources in the companies.

Company F: A Construction and Shipping Conglomerate

The Group consists of seven divisions, each of which is made up of a large number of companies. Each company is run as a profit centre. A very small corporate office sets financial targets for the divisions. Divisional managing directors have operational discretion within the targets, including acquisitions and disposal of companies. Each division has a lean corporate centre where IT is championed by an IT coordinator. A large central IS resource group is in the process of being run down. The larger divisions have set up their own processing resources. Division X, for instance, has six development and operational centres in different countries. Each of the autonomous divisions defines its own strategy under the guidance of its IT coordinator. In division X the companies define their requirements which are brought together by the coordinator to exploit synergies and common requirements. The development centres then compete for the development work in an internal market. The coordinators have a regular fortnightly meeting to exchange ideas and to determine any demands on the corporate function.

GLOBAL CHESS

The discussions with the interviewees centred on the IS and host grid dynamics, giving the opportunity to consider possible strategies and outcomes. Where should the corporate IS function be positioned and in which direction should it be trying to move?

Most of the organisations surveyed were in a state of flux. The exceptions were those companies close to the extremes of 'hands-off' or 'dirigiste'. At these extremes long periods of stability could exist for both the host and the IS organisation, with a minimum of organisational tension. Of those in flux most tension seemed to exist when the IS and host organisations were not aligned (in the sense of lying in the same column of the grid). There are powerful reasons why this should be the case. Functions such as personnel and finance prefer to apply the same rules and procedures across all business areas. Senior managers strongly prefer their chosen management style, underpinned by core values, to apply throughout the organisation. The CIO or IT director is (or ought to be) uncomfortable when her role is out of step with those of her peers.

A question frequently raised by the interviewees was whether it was better if all their responses fell in one column. On the whole there was considerable consistency with typically 5 or 6 of the 8 categories lining up. One would expect a consistent approach to be easier for both the IS users and the IS specialists to understand and manage. With an inconsistent position there is good chance that the role of IS at the centre will be misunderstood. Organisations should find that the risks are minimised if they determine which column they intend to occupy and attempt to engage in activities consistent with that position.

Which column should a corporate IS group try and occupy? The simple answer is to occupy the same column as the host organisation in order to ensure its role is understood and to avoid the tensions of being a special case. On this the empirical evidence was strong: misalignment leads to tensions. CEOs are prepared to change their IT director if the tensions are felt to be too great. However, the starting position (or heritage) and the time it takes to effect change need to be considered. Because IT has only recently come on the organisational scene and its characteristics have been changing, the current position may not be aligned with the host organisation. For instance, the central role for IS may be newly established and must by definition enter the game from the left or 'hands-off' position whilst the host organisation may occupy the 'federal' column. If the two are not aligned a view has to be taken about which direction the host direction is moving so that IS can try to intercept it. One approach is to consider risks and preferences. Is it preferable for the corporate IS group to err on one side or the other of the host organisation? My own preference would be to err on the side of lightness. This is because there is a built in tendency in most functional organisations to try to gain more influence and resources (empire building). CEOs see it as their role to hold the ring and resist such tendencies. They are suspicious of attempts by any function to seek a greater role than implied by the company's management requirements and style.

CONCLUSIONS

The survey confirmed the findings of Feeny, Edwards and Earl that host characteristics and IS heritage should be the basis for determining the evolution of IS structure. Lack of alignment between the IS structure and the host management style was found to be a major source of tension. Tensions are minimised when: the IS and host characteristics are aligned at one extreme or the other on the loose-tight dimension; IS is managed across topics with a single consistent coordinating style; and when the IS coordinating style is consistent with that of the host organisation. Host organisations that are not occupying stable positions at the extremes of loose or tight coordinating style are usually in the process of changing their position. The corporate IT director must think about the role of the central IS group in the light of the dynamic process of change in the host organisation and risk preferences.

REFERENCES

Bartlett, C.A. and Ghoshal, S., 1989. "Managing Across Borders: The Transnational Solution", HBS Press, Boston.

Deans, P.C. and Kane, M.J., 1992, "International Dimensions of Information Systems and Technology", PWS-Kent Publishing Company, Boston

Edwards, B.R., Earl, M.J. and Feeny, D.F., 1989, Any way out of the labyrinth for managing IS?, "OXIIM Research and Discussion Papers", RDP 89/3, Templeton College, Oxford

Feeny, D.F., Edwards, B.R. and Earl, M.J., 1987, Complex organisations and the information systems function, "OXIIM Research and Discussion Papers", RDP 87/7, Templeton College, Oxford.

Feeny, D.F., Earl, M.J.and Edwards, B.R., 1989a, IS arrangements to suit complex organisations 2. integrating the efforts of users and specialists, "OXIIM Research and Discussion Papers", RDP 89/5, Templeton College, Oxford.

Feeny, D.F., Edwards, B.R. and Earl, M.J., 1989b, IS arrangements to suit complex organisations 1. an effective structure, "OXIIM Research and Discussion Papers", RDP 89/4, Templeton College, Oxford.

Hodgkinson, S.L., 1990, The role of the corporate IT function in the large multi-business company, "A Doctoral Research Project Report", OXIIM, Templeton College, Oxford.

KAIZEN: THE JAPANESE APPROACH TO I.T.

Peter Linecar and David Preston

South Bank University
Borough Road
London SE1 OAA, England

INTRODUCTION

Japanese culture is groupist in nature, Western culture individualist. This cultural difference explains why approaches to Information Technology vary so greatly between Japan and the West.

Kaizen, the Japanese approach, does not lend itself easily to examination; Japanese being non-proselytic means that traditional sources are often unreliable. The work here is based on extensive interviews with Japanese I.T. managers and software developers, and personal experience.

Whilst much of Kaizen is irrelevant in terms of Western application, there is much to admire, learn from and apply. The leitmotif of Kaizen is often a shadowed ball capable of movement in many directions; an apt reflection of Japanese management.

Attempting to identify how the West could more effectively utilise the Japanese experience should prove rewarding; in just over four decades Japan has grown from being an agrarian economy to the second largest economy and third largest trader in the world.

Soft systems are increasing in use in Europe. We highlight how Japan is sceptical about such use as they see the user and worker as given insufficient access. Dean (1991) called Kaizen 'the ultimate soft system'; we discuss this notion and examine the differing roles of standards.

Finally in an era of extended markets, where many predict that large organisations that do not compete globally will struggle, we conjecture how the course of I.T. approaches will develop.

JAPANESE AND WESTERN CULTURE

Japanese culture is groupist in nature, formed through centuries of Buddhist, Shintoist and Confucian tradition. Western culture is individualist, created by a past of Roman, Greek and Christian influence. In an oft repeated experiment Dean (1991), when asked to prove their existence western groups consistently reveal individualist response ('I think therefore I am') whereas Japanese return groupist replies ('You can see me'). This cultural difference explains why approaches to business generally, and information technology (I.T.) in

Systems Science, Edited by F.A. Stowell *et al.*,
Plenum Press, New York, 1993

particular, vary so greatly between Japan and the West. Business objectives vary enormously as shown in Table 1.

Table 1. Ranking of Corporate Objectives

Factor	U.S.A	Japan
Return on Investment	8.1	4.1
Share Price Increase	3.8	0.1
Market Share	2.4	4.8
Improve Product Portfolio	1.7	2.3
Rationalisation of Production and Distribution	1.5	2.4
Increase Equity Ratio	1.3	2.0
Ratio of New Products	0.7	3.5
Improve Company's Image	0.2	0.7
Improve Working Conditions	0.1	0.3

* Source: Economic Planning Agency. Japanese Government. Economic Survey of Japan 1980/1 p186

Large Japanese companies generally consider employing more graduates, improved working conditions for employees, providing improved customer service to be of significantly greater importance than their western counterparts. Other revelatory evidence of the groupist heritage lies in the *Fuyokai* or *industrial network;* groups of companies providing services, in particular information, for each other. Japanese business aims are, more often than in the west, long term in nature. Table 2 lists a major *Fuyokai:*

Table 2. Fuyokai - Fuji Bank Group

Fuji Bank	Marubeni Iida (Trading)	Nihon Cement
Yasuda Trust	Kokusaku Pulp	Oki Electric
Yasuda Life Insurance	Nehin Reizo(Food)	Showa Denko (Chemical)
Yasuda Fire Insurance	Nisshin Flour Mill	Showa Kaium (Shipping)
Tokyo Tatemono (Real Estate)	Sapporo Beer	Nippon Motors
	Nisshin Spinning	Nippon Seiko (Machinery)
Keihin Kyuko (Railroad)	Taisei Construction	Nihon Yushi (Oil)
Tobu Tetsudo (Railroad)	Toa Nenryo(Petrol)	Hitachi, Ltd (Heavy Mach)
Tobo Rayon		Nippon Kokan (Steel)

On examining the company list we can see the logic behind such relationships. 'Japan Inc.' is an overstatement but the integration of system from kindergarten to industry is enviable.

In just over four decades Japan has grown from being an agrarian economy to the second largest economy, the third largest trader and the largest capital exporter in the world (Drysdale 1989). Much Imported knowledge and the will to change have been the driving force, but commitment to improve from the nation as a whole has been no less important. Central to the issue of change is the philosophy of Kaizen; the intent is to stabilise change whereas in the west the ethos is of managing change. The western view point that innovation, where change takes place rapidly, is necessary is disputed by the Japanese, who prefer gradual continuous improvement of their products.

AN EXPLORATION OF KAIZEN

Kaizen attempts to produce gradual on-going continuous improvement; increments are small enough not to be disruptive to the environment, but simultaneously the change is enough to be noticeable. Kaizen is really a philosophy and defies rigid definition; rather it is an amalgamation of interrelated principles which singly are inconsequential but combined become a powerful method of initiating improvement. Kaizen is a holistic approach to problem solving and its difference lies in being people centred rather than system centred. Kaizen should be implemented as a policy that involves all company personnel. Discussions with Japanese software developers certainly support Professor Dean's description as 'the ultimate soft system'.

The Japanese way of implementing Kaizen has been initially in the maintenance of a set of current relevant standards but at the same time seeking ways of upgrading those standards, through worker participation. Indeed it is an astonishing truth that the Lexus motor car, generally recognised as the finest executive car under 80000 pounds sterling, is primarily the product of thousands of worker Kaizen groups discussing topics such as lowering error rate on wheel bearing production. The underlying principle of Kaizen is a recognition that any product involves human interaction. Much has been made of Japanese use of *total quality control* (TQC), *total quality management* (TQM) and statistics but these should be seen as outputs of their human activity studies. Kaizen is as much about altering behaviour patterns and changing ideals as it is about process control, because momentum for change is built and maintained only in the minds of the people (Huda 1992).

HUMAN ACTIVITY

Cross communication occurs both horizontally and vertically within the organisation; the former via management directives; the latter through *quality control circles* (QCs). QCs are voluntary groups within the organisation who meet, either in or out of company time, in order to discuss ways of initiating change and then of introducing it within the context of their work. Methods include using the *plan-do check-action (PDCA) cycle* or *Deming wheel;* a more realistic representation was given by Imai (1986). QCs are concerned with trivial matters, their primary concern is not saving money, although this does happen. Each member is versed in the use of statistical tools such as pareto diagrams, control charts and cause and effect diagrams, because fundamental to Kaizen is its solid basis on objective data as well as subjective evaluations. Another important concept behind Kaizen is the reduction of waste, particularly relevant to improvements in software.

Kaizen demands much of management. Japanese managers, having spent the first twelve years of their corporate career rotating through the organisation, are more multi-specialist than their western counterparts (Whitehill 1991). Further qualities of Japanese managers include an holistic approach to problem solving, the nurturing of relationships, and most crucially, a whole hearted commitment to the concept of groupism.

Tethered to the formal structure of the hierarchy is the informal horizontal structure; loyalties are built up through a network of contacts developed over years, lessening areas of conflict by approaching problems through indirect channels which reinforce company objectives. The tenets of Japanese corporate management: seniority based wages; lifetime employment; and smooth labour relations, also aid in reinforcing long term goals. Two other areas in which the Japanese excel are in consensus decision making and in allowing workers to use their initiative. Usually an outline structure is given but individuals are allowed to use their own inventiveness in the work, whilst adhering to company standards. At the same time staff are actively encouraged to improve standards, through the suggestion scheme system.

Realising the difficulties of transplanting the functional counterparts of Japanese management to fit comfortably into current Western business practices William Ouchi proposed a theory called *Theory Z* which adopted the best features of the Japanese system to make it more palatable to the West. The main features of Theory Z are that trust should be instilled in people and amongst groups within the organisation; a fostering of intimacy by nurturing subtle relationships should be given priority. Ways in which Theory Z differs are: both assessment and promotion are quick; both implicit and explicit means of control are used; the individual bears final responsibility for group decisions; and finally the holistic concern is not a hierarchical but an egalitarian relationship.

Pascale and Athos (1986) have determined that the main objectives to be tackled by the West are threefold: managerial practice should be examined wherein current strategy yields only diminishing returns; consideration should be given to shifting values which are socially and culturally determined; organisational effectiveness in terms of goal setting, co-ordination and motivation has to be re evaluated.

INFORMATION TECHNOLOGY DEVELOPMENT USING KAIZEN

Problems within the current models used for developing software abound (Huda 1992). Software engineering is an extremely human interactive process: it is somewhat surprising the extent to which this human side of the problem is ignored. Experience has shown that with due consideration for human issues the inherent difficulties of software production become more constrained in their unpredictability.

Consistent with the philosophy of Kaizen, an apt starting point for recognising areas of improvement would be to consider the people aspect of software development. Japanese management owes much to the work of Hackman and Oldham (1976), Kulik, and Herzberg (see Huda 1992 for a detailed analysis).

At the group level there should be more interaction between team members via formal and informal channels; the former is effected through comprehensive and comprehensible documentation, the latter through the use of software quality circles. Poore (1990) has conducted various field experiments using the concept of software quality circles to explore their relevance in introducing statistical quality control to the development process. He found them to be an effective means of elucidating a working definition for quality which is usable within the appropriate setting. In this way adherence to the standard will become more marked because it will have been pre-set by the group itself, allowing them access to autonomous decision making and control. Furthermore enhanced communication and dialogue will result in a more objective viewpoint because operational views can be analysed within the field setting.

Freeman (1987) cites a series of factors which are basic to software quality and others which can be thought of as extra. Basic quality pertains to the following: functionality; reliability; ease of use; and in certain aspects safety. Extra quality may be obtained through: flexibility; reparability; adaptability; lucidity; documentation; and enhancability of the software.

AN IMMEDIATE FUTURE: AUTOMATION-BASED PARADIGM

Balzer (1986) suggested that a new paradigm based on automated processes would lead to an evolutionary change of traditional methods. Automation would augment the effectiveness of using people to produce and maintain software through correction of 'bugs', and also enhancement of the released system. The two fundamental flaws of the existing paradigm are in the spread of optimisation information and in the undocumented development processes.

Balzer's new paradigm would perform maintenance procedures on the specification rather than the implementation, thus keeping the changes at the conceptual level understood by the user. Furthermore the user would be able to generate updated requirements and ideally be able to perform, given adequate training, the maintenance themselves.

Other major objectives of this paradigm are: rapid delivery through operational specification of a system matching the user's intent; and the provision of a fast, reliable, inexpensive automated implementation process which permits revision. Automated implementation support is based on formalisation and is knowledge-based in order to co-ordinate individual activities to support overall development. The advantages of the paradigm are: reduction in clerical errors; improved documentation; and reusability of software.

Automated assistance should radically enhance system maintenance and will eradicate portability problems. Standard packages can be manipulated for a more personalised manageable creation.

THE DESIGN PROCESS

Design covers the conception of a product or service to meet a perceived market need (Webb 1991). Good design is cost effective because it eliminates the need for rework and contributes to towards later maintainability.

Turner (1984) lists the five most important principles of good design as: modularity; abstraction; hiding; understanding; uniformity. Design reviews range from formal documented planned meetings to impromptu informal ones. Walkthroughs are a popular means of uncovering the majority of defects prior to coding testing and release. One form of review, known as the *inspection,* has become common since its introduction by Fagan in 1976. The inspection is a form of process improvement which combines the concepts of the PDCA cycle and the quality circle into an effective means of error detection. Each person involved in the circle has a role and a defined workload. Finally Booth (1987), disenchanted with the limitations of structured analysis, proposed a system based on the *usability approach* to design; the most radical form of user involvement, where the user designs the system with experts on hand to advise. Alavi and Wetherbe (1991) enhanced the prototyping paradigm using data modelling, resulting in fewer iterations and improved designs. Agrestri (1986) suggested improved use of operational specifications which, because it is capable of exhibiting behaviour, is attractive to both developer and user.

These new approaches to software development are really utilising existing methods most effectively: process control; striving to implement improvement; and exploring standards of quality and productivity. The presence of standards across the board provides a vehicle available for reuse as well as providing confidence in software packages.

CONCLUSION

There are fundamental differences between the cultures of Japan and the West, which have a profound effect on the philosophy underlying both business and the environment in which systems evolve and function. Japanese culture is essentially groupist in nature, while that of the West is individualist. Japanese culture encourages the stabilisation of change, and the process of gradual evolution, while that of the West encourages more innovative change, with all the implications for the management of such change. Kaizen is a concept which underlies this Japanese philosophy, and provides an environment in which it flourishes, by exploiting the potential of all personnel within and beyond the corporate boundary.

We find elements in the Western approach to systems and their development which have resemblance to the Kaizen philosophy: quality control circles; the PDCA cycle; total quality management, but we recognise the difficulty of translating Kaizen in all its manifestations to Western enterprise. William Ouchi's Theory Z presents perhaps a workable combination. We see the potential for greater use of consensus decision making in the West, and the encouragement of stabilisation of labour movement.

In software engineering it is surprising that user involvement been ignored for so long, in an environment where human issues are paramount. We see the Kaizen approach offering quality in this area. Perhaps the usability approach, where the user takes the initiative in systems design, represents a fairly radical move in this direction. Such approaches are not new, but rather make the best use of methods which have existed for a number of years.

REFERENCES

Agrestri, W., 1986, "New Paradigm for Software Development", IEEE Computer Society, New York.

Alavi, M. and Wetherbe, J.C., May 1991, Mixing prototyping and data modelling for information system design, *IEEE Software*, pp86-91.

Balzer, J., Nov 1983, Software technology in the 1990s using a new paradigm, *Computer*, pp39-45

Booth, P., 1987, "The Design and Development Process", Wiley, London.

Dean,W., July 1991, Japanese issues of management, *at:* South Bank University Invited Seminar

Freeman, P., 1987, "Software Perspectives", Addison-Wesley, London.

Hackman and Oldman, 1976, "Motivation through the Design of Work Behaviour and Human Performance", McLeish Publishing, Pomona, CA.

Huda, F. and Preston, D., March 1992, Kaizen: the applicability of Japanese techniques to I.T. *Software Quality Journal*, Vol. 1, No.l, pp9-26.

Imai, M., 1986, "Key to Japan's Competitive Success", McGraw-Hill, New York.

Pascale, R.T. and Athos, A.G., 1986, "The Art of Japanese Management", Cox and Wyman, London.

Poore, J.H., Nov 1988, Derivation of local software quality metrics, *Software Practice and Experience*, Vol. 18, (1), pp1017-1027.

Turner, T., 1984, "Software Engineering Methodology", Keston Publ Co., Birmingham.

Webb, J., 1991, "Quest for Quality", The Industrial Society, Cambridge.

Whitehill, A.M., 1991, "Japanese Management: Tradition and Transition", Routledge, London.

INTERPRETIVE ANALYTICS AND CRITICAL INFORMATION SYSTEMS: A FRAMEWORK FOR ANALYSIS

Stephen K. Probert

Department of Computer Studies
University of the West of England, Bristol
Frenchay Campus
Bristol, BS16 1QY

INTRODUCTION

It has been argued that in order to develop more effective information systems methodologies a better understanding of the power effects of using existing information systems methodologies needs to be gained, and a methodology based on Foucault's "genealogical" method has been advocated as one basis upon which to conduct this task:

> Genealogy, deriving from Foucault's writings, puts emphasis on the effect power at the micro-level can have on the formation and development of knowledges... Obviously a properly conducted genealogical study of the systems field could contribute significantly to the 'social awareness' of critical systems thinking. (Jackson, 1992, p. 86)

'Interpretive analytics' is the name given by Dreyfus and Rabinov to the abstract general principles of Foucault's "genealogical" methodology (Dreyfus and Rabinov, 1982). This paper is based on actual research that used a research methodology based on interpretive analytics to critically study the National Computing Centre's (NCC) Systems Analysis and Design Methodology and Soft Systems Methodology (SSM) (Probert, 1992).

What are the effects of the use of information systems methodologies on the members of organisations? Analyzing some of the "local" power effects of information systems practices might well be best achieved by empirical studies or action research(e.g. Zuboff, 1988), but analyzing the *will to power* which brought the prescriptive methodological texts into being may be achieved by analyzing the texts themselves. Foucault himself never wrote a detailed account of how to use his unique methodology; Jones makes this point succinctly:

> There is no shortage of commentaries on Foucault's later genealogical work but there is a shortage of studies that actually apply this work. One of the main reasons for this

Systems Science, Edited by F.A. Stowell *et al.*,
Plenum Press, New York, 1993

state of affairs is the absence of a comprehensive, programmatic statement of what the genealogical approach involves. The latter must be pieced together by combining Foucault's scattered theoretical pronouncements with his genealogical studies of the prison and sexuality. (Jones, 1990, p. 79)

Interpretive analytics does not square easily with the categorisations of social science methodologies normally propounded (eg Burrell and Morgan, 1979):

> We argue that Foucault's work during the seventies has been a sustained and largely successful attempt to develop a new method. This new method combines a type of ... analysis which preserves the distancing effect of structuralism, and an interpretive dimension which develops the hermeneutic insight that the investigator is always situated and must understand the meaning of his cultural practices from within them. (Dreyfus and Rabinov, 1982, p. viii)

The viewpoint from which an interpretive analytic analysis can proceed is that social practices have a history which can be analyzed. What makes such a history interesting is that we are situated at a particular moment in time, and that one of the features that makes this moment in time recognisable from the past or the future are the social practices of our age. Studying the (genealogical) history of these social practices is one of the key "tools" used in an interpretive analytical study.

THE RESEARCH DOMAIN: QUASI-SCIENTIFIC SOCIAL PRACTICES

Many of our concerns are connected with the social practices of our age; these practices may also be bound up with the technological developments of our age. But what makes genealogical history different from traditional social history is its focus:

> Genealogy opposes itself to traditional historical methods ... For the genealogist there are no fixed essences, no underlying laws, no metaphysical finalities. Genealogy seeks out discontinuity where others found continuous development. It finds recurrences and play where others found progress and seriousness. It records the past of mankind to unmask the solemn hymns of progress. Genealogy avoids the search for depth. Instead it seeks the surfaces of events, small details, minor shifts, and subtle contours. It shuns the profundity of the great thinkers our tradition has produced and revered... (Dreyfus and Rabinov, 1982, p. 106)

These are, perhaps, the most significant features of interpretive analytics. Interpretive analytics has as its subject matter social practices that can be regarded as dubious sciences - if indeed they are sciences at all. However a genealogical analysis does not concern itself with the complex theoretical issues that practitioners of such social practices may indeed discuss and dispute; it is concerned with the concrete functioning of power; the functioning of knowledge, and the ethical positions generated specifically by such practices.

SOCIAL PRACTICES AND THEIR RELATIONS WITH THEORIES

The kind of source material most relevant for an interpretive analytic study of systems practices are, in addition to the standard texts, items such as training manuals and procedural guidelines:

> The interpreter as analyst ... finds that the questions which were traditionally held to be the deepest and murkiest are truly and literally the most superficial. This certainly does not mean that they are either trivial or lacking in importance, only that their meaning is to be discovered in surface practices, not in mysterious depths. (Dreyfus and Rabinov, 1982, p. 107)

If we wish to explain the theories we must first understand how the practices are actually supposed to operate. This is an inversion of a popular commonsense intuition: that practices are based on a solid (or "semi-solid") body of theory. It is this inversion at the methodological level that makes interpretive analytics both unique and relevant to a study of systems methodologies:

> The genealogist sees that cultural practices are more basic than any theory and that the seriousness of theory can only be understood as part of a society's ongoing history. (Dreyfus and Rabinov, 1986, p. 115)

THE THREE AXES OF ANALYSIS

The actual effects of social practices may be abstracted thus:

> This leads to the study of what could be called "practical systems"... These practical systems stem from three broad areas: relations of control over things, relations of actions upon others, relations with oneself... we have three axes whose specificity and whose interconnections have to be analyzed: the axis of knowledge, the axis of power, the axis of ethics. (Foucault, 1984, p. 48)

Foucault himself did not analyze practices in the way that has been suggested here (ie explicitly treating "knowledge", "power" and "ethics" separately). However it proved both beneficial and worthwhile to separate my research into these three discrete sections. It is not, however, suggested that each analysis stands alone; all three form an integrated whole:

> It seems to me that in *Madness and Civilisation*, *The Order of Things* and also in *Discipline and Punish* ... I tried to locate three major types of problems: the problem of truth, the problem of power, and the problem of individual conduct. These three domains of experience can only be understood in relation to each other, not independently. (Foucault, 1988, p. 243)

It is also important to note that the ethical axis of analysis is very important, although this might not be evident from a first-hand study of some recent "Foucauldian" texts. I shall now characterise the nature of the three axes of analysis, taking power and knowledge together, before characterising the axis of ethics.

THE THREE AXES: POWER AND KNOWLEDGE

An interpretive analytic study is concerned primarily with the relations between the knowledge and power found in (or generated by) a particular practice. It is important to note that Foucault did not consider power and knowledge to be identical, however power and knowledge are closely related - in the sense that particular forms of power can be seen as closely linked (historically) with particular types of "knowledge" - particularly with respect to institutional social practices. "Knowledge" should be understood as what counts as valid knowledge in the particular practice under study, eg. that "person x has paranoia" could be a valid piece of psychiatric knowledge in this sense. In this example it could be argued that this type of knowledge is not strictly scientific, but it is obviously not commonsense knowledge either. This knowledge that "x has paranoia" could cause a doctor to invoke some of the power that he/she possesses, eg. person x could be held in hospital under the Mental Health Act. This would be an example of a power/knowledge relation. An interpretive analytic study of quasi-scientific practices will focus attention on the historical relationships between power and knowledge that can be discerned in those practices.

Foucault does not see power purely as an instrument of repression - as sometimes occurs in political theory. Power can function productively in the sense that it can lead us to *do* things to ourselves or to others. For example a power/knowledge relation might be that medical "knowledge" indicates that smoking damages your health (which *may be true*), and stopping (or resisting the temptation to take up) smoking *could* be seen as an example of medical power as "repression" (this is not, however, to attempt to deny the truth of such medical assertions). But we might also find people taking up jogging, taking vitamins etc, for broadly similar medical reasons. This would be an example of the productive dimension of power. The idea of a "productive" dimension to power is not just intended to be interpreted metaphorically. Power is seen as functioning to serve the productive requirements of western societies:

> I believe one must keep in view the fact that along with all the fundamental technical inventions and discoveries of the seventeenth and eighteenth centuries, a new technology of the exercise of power ... emerged, which was probably even more important than the constitutional reforms and new forms of government established at the end of the eighteenth century... a form of power comes into being that begins to exercise itself through social production and social service. It becomes a matter of obtaining productive service from individuals in their concrete lives... power had to be able to gain access to the bodies of individuals, to their acts, attitudes, and modes of everyday behaviour. (Foucault, 1984, pp. 66-67)

It is also important to note that Foucault did not equate power with domination. Where power is exercised one should expect to find resistance.

So the *axis of knowledge* examines the knowledge of a person that must be produced before that person can be made the object (or subject) of power, and the *axis of power* examines the power relationships that must operate if the production of such knowledge is to be made possible. Power and knowledge are thus seen as being systemically related.

THE THREE AXES: ETHICS

Along with a tendency to examine highly theoretically abstract texts, another dimension is beginning to emerge in management science / operational research etc. This tendency could be described as a preference for treating people as individuals; subjects etc. The advocates of such views sometimes berate other academics and practitioners for treating people as if they were (components of) machines; objects etc. Foucault was concerned with both the treatment of people as objects and as subjects:

> Foucault has attempted to isolate two trends: first, the genealogy of the objectifying trends in our culture; and second, that of the subjectifying practices which have been given increasing prominence and importance in recent years. In sum, Foucault is seeking to construct a mode of analysis of those cultural practices in our culture

> which have been instrumental in forming the modern individual as both object and subject. (Dreyfus and Rabinov, 1982, p. 120)

Whether or not people *should* be treated as subjects is not a question which an interpretive analytic study can properly address. The question that *can* be addressed is 'What are the *power effects* of treating people as "meaning-endowing" subjects etc? To address this question it is necessary to gain an understanding of what the practices mean by the phrase 'a subject':

> I don't believe the problem can be solved by historicizing the subject as posited by the phenomenologists, fabricating a subject that evolves through the course of history. One has to dispense with the constituent subject, to get rid of the subject itself, that's to say, arrive at an analysis which can account for the constitution of the subject within a historical framework. And this is what I would call genealogy, that is a form of history which can account for the constitution of knowledges, discourses, domains of objects, etc without having to make reference to a subject which is either transcendental in relation to the field of events or runs in its empty sameness throughout the course of history. (Foucault, 1984, p. 59)

THE INTERPRETIVE ANALYTICAL STUDY OF SYSTEMS PRACTICES

Foucault himself thought that his genealogical method could be adopted by other academics wishing to pursue an inquiry into their own subject areas:

> ... what you should be saying to me is, "You haven't occupied yourself with this matter which isn't particularly your affair anyway and which you don't know much about". And I would say in reply, "If one or two of these "gadgets" of approach or method that I've tried to employ with psychiatry, the penal system or natural history can be of service to you, then I shall be delighted. If you find the need to transform my tools or use others then show me what they are, because it may be of benefit to me". (Foucault, 1980, p. 65)

In my research hardly any "transformation" was necessary, although the inclusion of an analysis based on the philosophy of logic proved to be beneficial in the study of the role of knowledge in SSM; also some of Jean-Francois Lyotard's work on (Sigmund) Freud proved to be a useful analytical tool (Lyotard, 1989).

CONCLUSION

In my research I concluded that the two methodologies chosen for further study corresponded to Foucault's two conceptions: of an objectifying practice (the NCC methodology), and a subjectifying practice (SSM). The history of these systems practices is relatively brief (although their genealogical history is much older), but both methodologies examined are, most likely, currently being used in organizations. However, it is important to note that the question asked was not the same question as 'are they (in some sense) "successful" methodologies?'. One of Foucault's genealogical histories concerned the practices of the Victorian prison system (or penitentiary). Almost as soon as these institutions were introduced the criticisms (with which we are all too familiar with today) began to emerge: that they were "universities of crime" and so on. Genealogy does not focus its attention purely on "successful" institutions and their practices - but it does focus its attention on those that survive the test of time; those that are clearly "readily acceptable" to a significant sector of the society in which we live.

REFERENCES

Burrell, G. and Morgan, G., 1979, "Sociological Paradigms and Organizational Analysis," Heinemann, London.

Dreyfus, H. L. and Rabinov, P., 1982, "Michel Foucault: Beyond Structuralism and Hermeneutics," Harvester, Brighton.

Dreyfus, H. L. and Rabinov, P., 1986, What is Maturity? Habermas and Foucault on 'What is Enlightenment?', *in*: "Foucault: A Critical Reader," D. Couzens-Hoy, ed., Blackwell, Oxford.

Foucault, M., 1980, "Power/knowledge," Harvester, Hemel Hempstead.

Foucault, M., 1984, "The Foucault Reader," Penguin, Harmondsworth.

Foucault, M., 1988, "Politics, Philosophy, Culture," Routledge, London.

Jackson, M. C., 1992, An integrated programme for critical thinking in information systems research, *J. Inf. Sys.* 2:83-95

Jones, R., 1990, Educational Practices and Scientific Knowledge, *in*: "Foucault and Education," S. J. Ball, ed., Routledge, London.

Lyotard, J., 1989, "The Lyotard Reader," Blackwell, Oxford.

Probert, S. K., 1992, "A Critical Study of the National Computing Centre's Systems Analysis and Design Methodology and Soft Systems Methodology," Unpublished M.Phil thesis, Newcastle Upon Tyne Polytechnic, March.

Zuboff, S., 1988, "In the Age of the Smart Machine," Heinemann, Oxford.

SOFTWARE QUALITY - CAN WE REALLY AFFORD IT?

Neil Richards

Hobbits
Danesbury Park Road
Welwyn, Herts., AL6 9SS

INTRODUCTION

In the current economic climate, there is a tremendous pressure for company management to control costs, both direct and indirect. To protect the future for their shareholders and workforces, it is their duty to avoid spending unnecessary money. They may well decide to follow the traditional management model, described by Collard[1] as being the following sequence of events: reduce input costs (i.e. people, methods, equipment, environment); this will lower unit costs; which will increase profits; which will improve the return on investment; and enable the company to stay in business.

However these present-day financial pressures have to be seen in relation to the conflicting pressures from the market place. The world is experiencing a slow-down in trade. Consequently, competition is greatly increasing among the major companies who are striving to retain their share of the reduced market place in which their industries operate. This wider competition just described is a great opportunity for suppliers of application software, both software houses and in-house IT departments. This is because end users will need to employ Information Technology to gain this competitive edge. However the demand for software and IT services also puts heavy pressure on the smaller software companies, especially those who do not have a definite market niche in which to differentiate themselves from their market leaders.

The market place itself is also changing fast in almost every industry sector, from manufacturing to financial services. In the area of software supply specifically, there have been radical changes in the past 5 years. On the hardware front there has been a move away from dumb terminals attached to centralised computers toward workstations, graphics, and local area networks. On the software management side, new methods have begun to be proven. Structured methods, CASE and CAST tools, configuration management techniques are all forcing a new approach to software services. Therefore there is an essential need to spend money on research and development to keep up the momentum in the crucial task of finding the company's future products or keeping the current offerings up-to-date. These products of tomorrow need to be aggressively priced, be innovative and have, if possible, a sales appeal which extends wider than domestic borders. Such is the speed of technological change that software packages currently have a much-reduced life expectancy. This means

Systems Science, Edited by F.A. Stowell *et al.*,
Plenum Press, New York, 1993

that any new products these days need to be brought to market (or delivered to the in-house end user) rapidly or else they will be out of date before they appear. Production of computer software is therefore becoming more and more complex. This obviously means that direct costs are tending to be pushed upwards.

Other pressures being exerted on suppliers are those being applied by buyers, who are much more discerning these days. They too have financial pressures on them to keep their costs down so they are demanding good value - a quality product. There is also an increasing movement among companies to be certified under the international standard ISO 9000 series which requires that they ensure that their suppliers produce quality products.

WHAT IS QUALITY?

It is necessary to define the context in which the word "quality" is used here. The quality label for the purpose of this paper can be given to a wide range of products. For instance if examples were taken from the motor industry, the word quality could equally be used to describe the luxurious (say a Rolls Royce limousine) or a more humble car (say a Ford Escort) because quality has nothing to do with excellence. The quality label is rather given to products that do exactly what is expected of them. They conform to their specifications, not only at the time of their purchase but throughout their use, and this at the best cost. The official definition of quality as stated in ISO 8402 is: The totality of features or characteristics of a product or service that bear on its ability to satisfy stated or implied needs.

Of course software companies and IT departments have always strived to deliver quality products. It would be slanderous to say anything else. However the frequency with which projects come in late and over budget is well known. This highlights the fact that there is such a thing as poor quality. Poor quality unfortunately abounds.

A cost can be attached to poor quality. This is known as the cost of quality [COQ]. James Teboul[2] describes this cost as being of considerable proportions in industry, enough to fund a "hidden factory". He states that the cost can reach as much as 20-30% of sales turnover for companies at the first stage of awareness. This high per-centage is due to the incidence of multiple tests, inspections, modifications, rework, rejects, losses of capacity, stocks, irregular production, returns, claims, crises etc..

A similar order of magnitude is found in the software industry. Two reports commissioned by the DTI[3,4] and carried out by Price Waterhouse and CSC Index respectively each showed that the industry wastes approximately 20% of its investment in IT. This has been estimated as amounting to an industry-wide £2billion per year or £1M per working hour. A major benefit can therefore be gained by software developers by reducing these costs.

WHAT IS COST OF QUALITY?

There are two ways of looking at COQ - one is from the point of view of a user of a product and the other is from that of a supplier. The COQ from the user view is based on the cost of poor quality but with the objective of minimizing the cost of ownership through the life cycle. These costs include the price of acquiring the system, any maintenance charges, the costs of running the system (including costs of recovery following any incidents), the cost of obsolescence and any consequential loss.

The supplier of software would define COQ as those expenses which could have been avoided by getting the product or service right first time. From this point of view it is made up of four costs.

Prevention Costs

These are those costs which are incurred in trying to avoid defects from ever arising. They would include the cost of project reviews, design reviews, structured walkthroughs, training, generating and maintaining operating procedures, quality system audits etc.

Appraisal Costs

These are those costs which are incurred in trying to identify any defects which do occur. They would include costs for tests, inspections, audits, surveys, the gathering and processing of control data etc.

Failure Costs

These can be classified as internal and external.

Internal: These are those costs which are incurred when a defect is detected while the product or service is still in development or under the company's control. It the cost of scrap, rework, lost time, reruns, unused capacity, design change requests etc.

External: These are those costs which are incurred when a defect is detected after a product or service has been handed over to the purchaser. It includes the cost of returns, recalls, complaints, replacements, compensation, field repairs, repairs under warranty, product liabilities etc.

Opportunity Costs

These are the costs of lost market opportunity or loss of alternative use of resources. If one considers these categories from a Software industry point of view, one is forced to agree that there is plenty of scope for improvement. Prevention is totally inadequate. Too many defects are introduced very early into the development life cycle and then not removed until very late. Although software managers know that investment in appraisal is necessary and they emphasise it in most plans (in the form of reviews and walkthroughs), often the realities do not match the intention. Certainly testing is rushed and usually inadequate. As far as failure is concerned, the computer trade press is constantly reporting the painful instances of major failures. Yet for every one which makes the headlines, there are thousands of instances of major failures which never get reported.

So, what would be the profile of a company that tried to assess its COQ for the first time? If it were one that was unaware of its non-quality problems, it would probably have high failure costs, and low appraisal and prevention costs. Once a study had been made and the company became aware of its position, it would most likely develop a quality control system and attempt to reduce the failure costs by increasing the appraisal costs. This would have the effect of reducing the overall COQ, but only by a modest per-centage.

True savings would begin when the company understands the importance of investing in prevention costs. This has a dramatic effect on the COQ. To understand why this is so, it is necessary to appreciate the progressive nature that correcting a defect costs during the project life cycle. For instance, if the introduction of an error during the requirements study phase of a project is arbitrarily set at 1 unit, and if this error were to go through to the design phase, the cost of detection may well be 20 times greater. If this error were to be designed into the application system and consequentially moves into program build stage, detection may be multiplied by another 2. Maybe the detection cost can be increased by a further 5 to allow for the various testing processes,and when the system is finally in the hands of the user and breaks down, the cost to the supplier of correcting the defect may well be 5 times greater again. This means a total multiplication of 20x2x5x5=1000. Thus, it can be seen that, by increasing the cost of prevention by 1, it is possible to decrease the appraisal and failure costs by up to 1000. This ratio could be even higher where errors in certain pieces of software are so difficult to trace that they have to be rewritten entirely.

COQ is a fundamental concept of quality management. However it has to be admitted that some companies, particularly those who already fully embrace the idea of quality management - most notably the Japanese, see no need to calculate the COQ. But for the benefit of the doubting "scientific manager" it is necessary to develop the subject further.

COQ - MEASUREMENT AND METRICS

The reason why management measures anything is so that the business processes involved can be improved. There is a hackneyed expression that you cannot control what you cannot measure. Certainly you cannot improve the performance of any job if you do not have the measure of how it is done at the moment. Norman E. Fenton[5] recommends that, even if it seems difficult to work out how to measure certain attributes, we should always attempt to develop the criteria to measure the unmeasurable. Even characteristics such as 'dependability', 'quality', 'usability' or 'maintainability' should be tackled.

For companies new to COQ, it is necessary, initially, to calculate the current failure costs. This can be done by determining the number of defects, the time spent on corrective work. From this it will be possible to put an actual cost to them. Costs can be assigned initially by calculating a standard cost for labour and adding a rough overhead rate for all the indirect costs. If there develops a need for more precise COQ measurements, possibly as the company begins to spend more on the other cost categories, it may be necessary to subdivide the rough standard costs used to start with and have separate costs for direct development labour, direct maintenance labour, development platform, machine time, and overheads. Financial managers know that revenue from the sales of systems plus their maintenance must cover all the various costs for there to be a profit, it may be that much of this COQ information can be gleaned from the company's accounting system.

However, if software metrics data is not available via the company accounting system, it should be collected by using an in-line system. If the latter is used, data collection should be automated or, at the very least gathered on standard forms. It is necessary for any data collected to be objective at all times. In fact there is a lot to be said for the metrics gatherers to be independent of the project teams. The data, once collected, should be stored in a data base that will allow multiple access paths.

Norman E. Fenton[5] suggests that care has to be taken with software measurement data. In statistical terms, if it is not normally distributed, the data may need to be analyzed using other techniques than just averaging.

Two aspects of taking measurements should always be remembered during the time of first setting up a measurement programme. One is that metrics data should not be used to assess individuals unless it is done in a sympathetic manner and they agree to it in advance. It is imperative that the change process that accompanies the introduction of a quality programme is conducted in a spirit of goodwill for it to succeed. The second aspect is that any company that becomes involved in transferring to a quality programme must beware of the "Hawthorn" effect. This is the motivation effect caused through special attention by management. Hawthorn was the Western Electric plant in the USA where, in 1934, a series of experiments on the effect of working conditions on the productivity of the machine operators were carried out. It was noted that productivity continued to improve despite working conditions growing worse. This astonishing situation was due simply to the attention devoted to the subject by management and the research team. Management has to accept that real pay-back from a quality programme cannot be expected for at least two years.

As a post script to this discussion on COQ, it should be stated that in the shorter term, there will be a need for actual investment. Experience indicates that the establishment of an effective quality system involves substantial initial costs. For a company of up to 100 people, these costs may total perhaps £100,000 to £250,000, depending on both the scope and depth of the system that is introduced. These costs cover the definition of standards and procedures, the provision of a quality infrastructure, and the initial staff training. Subsequently there are recurrent costs of perhaps £50,000 or more per annum. The sales turnover for a company of this size is likely to be in the order of £6M. The COQ, without any change to the company quality approach and using the rule of thumb described previously, would therefore be approximately £1.2M per year. This figure is the amount the company has to think of as its challenge. It can be reduced substantially.

INSTALLING QUALITY

Once the COQ is calculated, decisions can be made by management on whether a quality programme should be implemented in the company, and what its scope should be. Objectives need be set and measurement criteria determined. It is also possible to begin reassigning costs between the different COQ categories.

If these decisions involve installing a quality programme, a major management project has to be set up. It requires the company going through a complete cultural change. Collard[1] describes the management model for a company involved in installing quality as being the following: install a quality improvement programme; which will increase productivity; which will bring costs down; which will allow prices to be brought down; which will allow market penetration to increase; which will enable the company to stay in business; which will create more jobs and improve the return on investment.

Consequently, it is essential that all levels of staff are fully committed to the project. Top Management have to take a leading role and show dedicated commitment. All other staff need to understand the purpose behind the quality programme and be ready for full involvement.

A significant amount of effort will need to be spent in producing a quality manual. This document is necessary to declare to its staff the company policy on quality. It will also communicate directives about various practises and procedures. The kind of subjects found in a quality manual would include some or all of the following:

- Quality policy statement
- Organisation structure
- Quality Responsibilities
- Procedures for:
    - development planning and control
    - reviews
    - testing and acceptance
    - document control
    - archiving, product releases, configuration management
    - change control, product & service problems
    - purchasing and sub-contractor control
- Policy on training

BENEFITS OF INSTALLING A QUALITY PROGRAMME

There are enormous benefits to be gained from installing a quality programme. Here is a list of some:

- massive reduction in COQ
- improved customer/supplier relationship due to the fact that the objective of the quality programme is customer satisfaction - improved internal control and visibility of work due to the improvement in standards and procedures
- confidence in work practise due to the cumulative effect of training, reviews, audits etc
- reduction in maintenance and life cycle costs because of the reduction in failure costs
- internal staff productivity benefits due to the higher motivation that comes from producing quality products for satisfied customers
- more predictability of work output because of the improved forecasting which comes from better measurement data
- better management because management, through its commitment to the quality programme, is involved in the 'front line'.
- improved software that can be independently audited - improved traceability due to the disciplines imposed by such techniques as configuration management
- increased quality awareness among staff which enables responsibility for quality to be delegated downwards and makes staff supervision easier
- marketing advantage gained from higher quality and lower prices
- reduction in assessments by customers (if an industry standard such as ISO 9000 is adopted and certification gained

SUMMARY

There is no denying that company management has to avoid spending unnecessary money. However, if the concept of COQ is understood and accepted, there is currently a large amount of money being spent in many companies on delivering poor quality. This can amount to 20-30% of sales turnover. This money, termed Cost of Quality (COQ), once identified, can be more advantageously spent. If a quality programme were to be introduced, large savings could be made on the COQ. Some of the money currently spent on repairing defects could be reassigned to preventing defects from happening and appraising whether any are slipping through the net. This, it is argued, is a much more cost effective investment and would bring major trading benefits to the company, over and above the COQ reduction. Implementing a quality programme is an improved way of cost cutting than arbitrarily to decimate the company's resources, especially as management still needs the company's products to be produced at the original quality level. The COQ concept can help company management come to this decision.

SELECTED BIBLIOGRAPHY

1. R. Collard, "Total Quality", Institute of Personnel Management, Wimbledon, (1989).
2. J. Teboul, "Managing Quality Dynamics", Prentice Hall, Hemel Hempstead, (1991).
3. Price Waterhouse, "Software Quality Standards: The Costs and Benefits", DTI, London, (1988).
4. CSC Index, "Key Issues Affecting Quality in Information Systems", DTI, London, (1992).
5. N.E. Fenton, "Software Metrics", Chapman & Hall, London, (1991).

ACCESSIBILITY AND UTILISATION OF ORGANISATIONS NEWEST ASSET: INFORMATION

Richard Rolfe

Department of Management Systems
Bournemouth University
EMail: rwrolfe@uk.ac.bournemouth

INTRODUCTION

As computerised information applications increase in size and detail, reaching an ever larger spread of the organisation population, the role of gatekeeper has for many organisations been the answer to providing workable information provision. With the movement towards decentralised end user computing, workstation provision will signal the end of the gatekeeper function, with the information user having direct, hands on access. However as application data model structures increase in size and detail, mounting pressure will be placed on the user interface to present the complete knowledge available in the information resource. This paper places this issue in a wider context, providing a model of user awareness whereby the user interface environment is challenged to move from a transparent medium towards a proactive knowledge presentation and analysis environment.

BACKGROUND

During the 1980's there has been a continual shift away from managing the computer technology itself, towards the explicit management of the data stored within the technology[1]. It is commonly believed that the value of such an asset is based on interpretation or capability of use in decision making; it is therefore, contextual, subjective and unquantifiable in nature[2]. The resource is based on human capacity for adding value -the management of the information resource is therefore dependent on managing the human potential for adding value[3]. Such a definition leads to problems of establishing the efficiency and effectiveness of this asset. Several surrogate measures have been suggested, ranging from incremental change in decision making[4], incremental change in user satisfaction[5] or increased system utilisation[6]. The underlying basis of this paper is that improved user accessibility will provide an increase in resource utilisation, and hence contribute towards the success of the information asset.

Systems Science, Edited by F.A. Stowell *et al.*,
Plenum Press, New York, 1993

THE IMPORTANCE OF USER ACCESSIBILITY

Accessibility to information can be examined on two levels. Accessing the physical terminal hardware to reach the available machine resources is the first level of accessibility that each user must accomplish. Rice and Shook[7] considered this level of accessibility, particularly with physical terminal location forming a determinant relationship with machine usage patterns. In this context accessibility is based on the physical accessibility to the hardware, rather than to the information resource itself. With increasing availability of hardware through the proliferation of personal computing coupled to networking technology, there has been a shift of emphasis away from this hardware accessibility towards what actively mediates between the user and the application. This second aspect of accessibility is increasing in importance, as the size and complexity of applications made available to the user community continues to rise.

Waguespack[8] identifies two layers of complexity: intrinsic and extrinsic. Intrinsic complexity refers to complexity inherent in the application itself, whereas extrinsic complexity refers to the application interface, for example the form of dialogue and interface environment provided. With continuing advances in processing power and storage capacity, increasingly large and detailed applications are being generated, for example enabling database modellers to capture growing semantic structures in database applications. Such advances fuel the rising levels of intrinsic complexity to which interfacing software must provide access to a widely diversifying audience. There is therefore, mounting pressure for an increase in extrinsic complexity to allow a user to access the functionality and depth provided in the structure of an application. However, increasing complexity of the user interface provides an increase in learning and usage costs, which have to be absorbed by the user during the accessibility process.

O'Reilly[9] investigated the costs implicit in the accessibility process, in particular focusing on why such resources were often left underutilised. O'Reilly argued that accessibility to the information resource was a key motivation for resource utilisation, those sources of information perceived as difficult to access were left underutilised. Bozeman & Shangraw[10] support this view, establishing that increasing the cost of any information resource negated accessibility, with subjects tending towards cheaper and more diverse sources of information. With the increasingly diverse user community coming into contact with computer applications and advancing complexity of application design, particularly in the database domain, the role of the interface becomes vital in ensuring accessibility through minimising user cost.

From a practical viewpoint accessibility to large database information systems has tended to be mechanistic, an afterthought of the application designer. This question of accessibility has often been tackled by providing hierarchies of standard reports, forms or menus which were set up at implementation time, supplemented by a catch all command line query generator. Providing such a structured range of access tools imposes an immediate limit to the value of the information resource, restricting ad hoc query generation and hence the key value of a database information system.

As organisations' continue to invest in ad hoc decision making capability, founded on database structured information systems, the issue of utilisation via user accessibility is increasingly important in MIS research. To investigate this in detail, fieldwork was carried out with the UK National Health Service across the south east of England[11].

FIELDWORK

The UK National Health Service is under acute resource constraint, requiring justification of current and future investment in information technology as a diversion of funds away from 'patient care'. The issue of information resource utilisation is therefore of growing practical importance as the provision of health care is resource intensive, both in terms of direct hospital provision and administration related processes. As a result of a UK government initiative, in the early 1980's each District Health Authority(DHA) was to install a Patient Administration System to provide an information resource against which users from different aspects of the Health Service could draw information: consultants for patient level detail, adminstration for operational data and central government for aggregated service reporting. The overall aim was to provide an information resource to promote improved management decision making at the lower level, whilst providing aggregated information for central government control.

The resource had been implemented approximately two years prior to the completion of the study on a standardised top down basis. The implementation processes, hardware, software and database datamodel used were standardised across all districts, but each district was left to develop its own information utilisation approach. Data was collected using indepth interviews and questionnaires supplemented by onsite observations of all staff who came into contact with the information resource.

RESULTS

Each study identified an overall low level of information resource utilisation, despite the significant capital investment made. In each District, utilisation of the resource was mediated via an intermediary gatekeeper role. This had the effect of installing two stages to the user interaction relationship: between the gatekeeper and the information resource, and secondly between the final information user and the gatekeeper. The gatekeeper was therefore viewed as part of the information resource by the final end user, providing a buffer to convert vague verbal information requests into syntaxically correct computer queries.

The level of information utilisation can be analysed corresponding to the model provided by Waguespack above. For the gatekeeper, utilisation of the information resource was identified as contingent upon an aggregation of the interface (extrinsic) and the application data model (intrinsic) complexity. The interface environment for ad hoc query generation provided a single rigid interface for all users regardless of computing skill and background, demands to be made of the system or individual knowledge. Those gatekeepers whose career had progressed up through the Health Service found this environment difficult to use, for example one gatekeeper deferred information users to paper based standard reports rather than tackle the user interface.

The fundamental disincentive to resource utilisation reported by all gatekeepers was the complexity of the application data model. The gatekeeper was required to manipulate the entire datamodel structure for the development of queries, mapping explicit paths and managing the structure directly. This was the key disincentive to utilisation even for those gatekeepers with a strong IT background, for example two gatekeepers were independently building a completely separate information resource on an alternative machine which would then be used for management reporting. These alternative information resources were fed via file dumps from the original information resource.

Interface and application data model complexity were identified as having a negative relationship with resource utilisation. In addition a third variable was identified: user awareness. For the gatekeeper, this became apparent in terms of the training provided. Minimal training in interaction detail was provided to all intermediary gatekeepers, for example in syntax, dialogue procedures and query generation. However, staff which interacted with the information resource had not received any contextual training, in particular how to manage and apply the value of the resource within the organisation. This was compounded further as the only staff available to promote final information awareness was the gatekeeper. In addition those gatekeepers who had entered the information function from outside the Service lacked a "common sense test", they were therefore, unable to identify the relevance or integrity of much of the information they were producing.

For the final information user, for example a consultant or unit manager, awareness of what information is available was identified as critical to information resource utilisation. Those gatekeepers who actively promoted an information culture, providing user awareness mechanisms, for example information shops providing face to face open access, information compendiums and shopping lists of available information, consistently showed a higher level of utilisation of the information resource.

The role of the gatekeeper has for many years been the answer to increasing the utilisation of a large information system. The human element of interaction masks the final information user from the application interface and data model complexities inherent in such large database systems. However, this introduces an awareness gap between the hands-on user and final information user. With the movement towards desktop workstations, the role of gatekeeper will diminish as complete access to the information resource is available to the decision maker from the desk top. The requirement of an established information culture to maintain information utilisation will then increase in dominance as the information user will then have to manipulate the structure of the information resource directly via a transparent user interface. As applications increase in size and detail, new ways will have to be found to present available knowledge to the decision maker based on improving user awareness.

A USER AWARENESS PERSPECTIVE TO INTERACTION

Based upon this fieldwork, a model has been developed which regards user interaction with an information resource as a process comprising both mental and physical user activity. It has been a common approach to evaluate interaction from a technical perspective, for example command line grammars and knowledge formalism, fitting the interaction processes into a rigid mathematical notation. This model takes an alternative perspective, viewing interaction as a layered discovery process, originating from the user and mediated by accessibility.

The model presented is comprised of four layers, where the user graduates from the abstract to a system specific user awareness in the interaction process. The outer layer

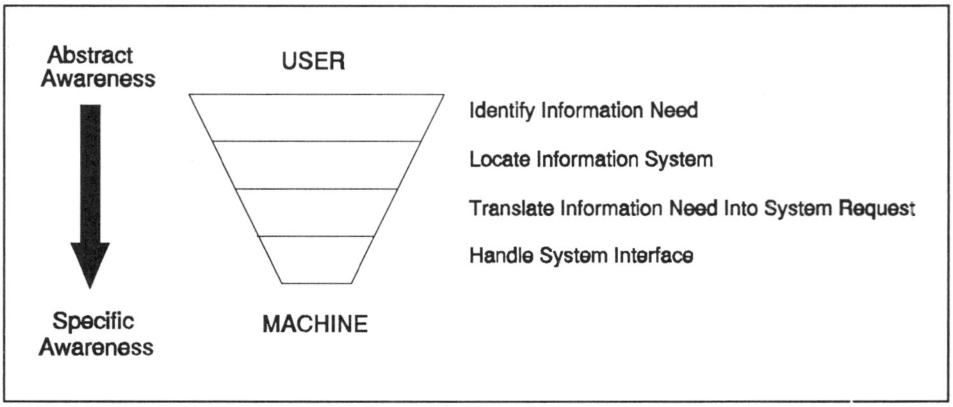

Figure 1. User Awareness Model.

depicts the information user arriving at a stimulated knowledge need, for example those organisations who actively promote an information culture posses a higher abstract awareness of what information is available from the resource. The initial process depicted is for the user to map this knowledge need into a more detailed information requirement, for example into a requirement to manipulate waiting list date, speciality and location information. This "shopping list" of information required will be necessarily abstract, likely to contain inconsistencies, error and may be incomplete.

On identification of an information requirement, the next layer presents the information user with two hurdles: whether the information is in existence within the organisation, for example is it collected and stored at all? and secondly the existence of any information and storage processing systems within the organisation which may hold the required information. This layer depicts a required awareness of the information user of what information is stored and processed by the organisation as a whole, for example identifying whether it is paper or computer based, departmental location and ownership. As applications are brought to the organisation which integrate data from distributed sources, this level of user awareness will be crucial to the effective utilisation of the information asset.

Once the user has established an information requirement, aggregated from one or more information processing systems, an explicit conversion from a vague unstructured request into a precise computer acceptable form is required. This requires an awareness of what is explicitly requested versus what is explicitly available. In particular the user is required to have an awareness of the data structures of each of the various sources, and the functionality and syntax of the required query languages.

The final layer of user awareness is systems specific, requiring the user to manipulate the functionality provided by the application through the user interface. This will vary according to the provided environment, for example command line vs WIMPS orientation. This final layer has drawn considerable attention as researchers pursue the goal of interface transparency. Recent approaches are moving beyond this interface layer, out towards the user, for example integrating the conceptual representation of application structures with database interrogation and user interface design[12,13]. For the future this movement must increase, in its ultimate form providing a user advice and awareness environment, a combination of decision support system, and intelligent knowledge based systems tailored to the information user and his/her task set.

CONCLUSION

As information applications increase in size and detail, reaching an ever larger spread of the organisation population, the role of the gatekeeper has for many organisations been the answer to providing workable information provision. The issue of accessibility has been split into two stages. For the gatekeeper, accessibility has been identified as a combination of interface and application data model complexity whereas for the final information user the issue of user awareness has been identified.

With the movement towards decentralised end user computing, workstation provision on the decision maker's desktop will signal the end of the gatekeeper function. This role will be replaced by interrogation software and knowledge representation applications. For this movement to be successful, it has been suggested that the gatekeeper must be supplanted by an established information culture. With this structure of accessibility supported by an information culture, the people who have the understanding of the context and knowledge have direct hands on access. However as the application datamodel structures increase in size and detail, mounting pressure will be placed on the user interface environment to present the complete knowledge available in the information resource.

As a means to placing this in a wider context, a model of user awareness was presented providing a user oriented process view to user interaction. With this increasing level of complexity, for the information resource to be successfully utilised, it has been argued that the user interface must undergo a change in role definition, moving from the transparent, passive medium towards a proactive knowledge presentation and analysis environment. The challenge therefore, is to combine the components of accessibility: interface and application data model complexity, user awareness and information culture mechanisms to promote information system utilisation and value.

REFERENCES

1. D.P. Best, The future of information Management, *International Journal of Information Management*, Vol 8, No 1, pp13-19, (1988).
2. P. Molholt and P.M. Hohenberg, What price the information age? A model for fair payment for the use of information, *in:* P. Zunde and J. Agrawal, ed., "Empirical Foundations of Information & Software Science IV: Empirical Methods of Man Machine Interfaces", North Holland, Amsterdam, pp479-486, (1988).
3. B.K. Brussaard, IRM in the Public Sector, *Information and Management*, Vol. 15, pp85-92, (1988).
4. L. Raymond, Validating and applying user satisfaction as a measure of success in small business organisations, *Information and Management*, Vol. 12, pp173-179, (1987).
5. B. Ives, M. Olson, and J. Baroudi, The measurement of user information satisfaction, *Communications of the ACM*, Vol. 26, No. 10, pp785-793, (1983).
6. R. Zmud, Individual differences in MIS success: A review of the empirical literature, *Management Science*, Vol. 25, pp966-979, (Oct. 1979).
7. R. Rice and D. Shook, End user computing: access, usage and benefits, *Proceedings of the American Society For Information Science*, Vol. 23, pp265-270, (1986).
8. L.J. Waguespack, Software complexity assessment and human machine evaluation, *in:* P. Zunde and J. Agrawal, ed., "Empirical Foundations of Information & Software Science IV" in: Empirical Methods of Evaluation of Man Machine Interfaces, North Holland, Amsterdam, pp187-204, (1988).
9. C. O'Reilly, Variations in decision makers use of information sources: the impact of quality & accessibility of information, *Academy of Management Journal*, Vol. 25, No. 4, pp756-771, (1982).
10. B. Bozeman and R. Shangraw, Computers as a public management decision tool, *Knowledge: Creation, Diffusion, Utilisation*, Vol 10 No 2 pp111-139, (Dec. 1988).
11. R.W. Rolfe, "Information Resource Utilisation: Query By Concern", PhD Thesis, Kent University, (1991).
12. M. Angelaccio, T. Catarci, G. Santucci, QBD: A Graphical Query Language with Recursion *in*: "Designing and Using Human Computer Interfaces and Knowledge Based Systems", Proceedings of Third International Conference on Human Computer Interaction, Vol II, Boston MA, USA, (1989) Amsterdam, Netherlands, Elsevier, pp384-391.
13. S.A. Goldman, P. Kanellakis, and K. Zdonik, ISIS: Interface for a semantic information system, *in:* "Proceedings of ACM SIGMOD International Conference on the Management of Data", pp328-342, (1985).

APPLYING SOFT AND HARD SYSTEMS APPROACHES TO A TECHNICAL PROBLEM

Margaret Ross and Geoff Staples

Southampton Institute
Information Systems Division
East Park Terrace
Southampton
Hants SO9 4WW

INTRODUCTION

The need to examine and monitor the condition of engineering systems has increased in recent years in order to reduce environmental, safety and breakdown risks and as a result to improve profit. Over the last twenty years, manufacturing and other industries have become more automated, involving complex machines and the electronic and computer systems necessary to control them. Failure to maintain adequate maintenance monitoring can result in breakdown or malfunction, resulting in some cases in serious accidents. Maintenance strategies have become more complex to cope with these new systems; simple corrective maintenance (i.e. when it breaks down, mend it) was replaced by planned maintenance (replace items at regular intervals) which is now being supplanted by preventive maintenance. This involves the continuous monitoring of a machine, to predict possible problems and hence to prevent unscheduled shutdowns due to unforeseen failures. Health indicators such as noise, vibration signature analysis, analysis of particles in oil and analysis of thermal data, are now accepted as reliable indications as to condition of a machine. This monitoring requires specialist data collection equipment and usually requires micro computing facilities to analyse the high volume of raw data.

The benefits of using condition monitoring have been recognised by Rao (1992). A survey by the Department of Trade and Industry indicated that the cost to the UK industry of poor maintenance was over one billion pounds per year, and that a five percent increase in machine availability could lead to a potential saving of one and a half billion pounds. More specific cost benefit examples can be examined, such as reduction over a four year period of stoppages in a paper plant from three hundred to thirty with an average cost of six thousand pounds per stoppage, or a saving of forty three percent over one year in a twenty five year old flour mill on the introduction of a condition based maintenance strategy.

Systems Science, Edited by F.A. Stowell *et al.*,
Plenum Press, New York, 1993

A breakdown of machinery can result in damage or death when considering robotics equipment or the aviation industry and the harm to the environment can be enormous if a breakdown is considered in say an oil tanker or a power station. The need to prevent or minimise malfunction should be a major consideration for industry today.

BENEFITS AND DISADVANTAGES

Currently, small, dedicated computer systems are used to capture data either at source, or from readings entered manually and to perform calculations on often very large quantities of data. Warning messages, and sometimes trend identification, are then generated. These systems are usually resident near the machinery. Additional desirable facilities were identified (Cross, 1990) including plant history, scheduling of maintenance, cost breakdowns against various cost centres together with various progress and financial reports.

The need to be able to integrate the monitoring system with the computer system of the company has been identified. This would

- utilise the existing hardware and software expertise
- maintain compatibility of hardware
- provide consistencies of analysis and design methods
- provide common documentation standards
- provide minimal additional purchase of software
- interface with Management Information System

This would enable company consistency and provide management visibility of the condition monitoring system. These and other objectives can be addressed by the use of a relational data base, such as Oracle, using the SQL language. This provides facilities for:

- report writing facilities of rdb
- graphical facilities
- spreadsheet facilities
- use of standard statistical packages, from languages such as SQL
- facilities for embedding SQL within third generation languages, such as Ada and C
- the existence of the ISO standard for SQL
- the use of case tools, integrated with the rdb
- the use of a common data dictionary
- the availability of larger record sizes, such as two gigabytes for the new Oracle release
- interface between rdb such as Oracle, and expert systems,the latter now being used in trials on dedicated machines, for monitoring purposes

The disadvantages include:

- the use of methods, software and hardware not specifically designed for the purpose
- the high volume of data collected, and hence the need for selective archiving
- the requirement for fast response in case of actual or potential problems
- the removal of control of the system away from appropriate sections

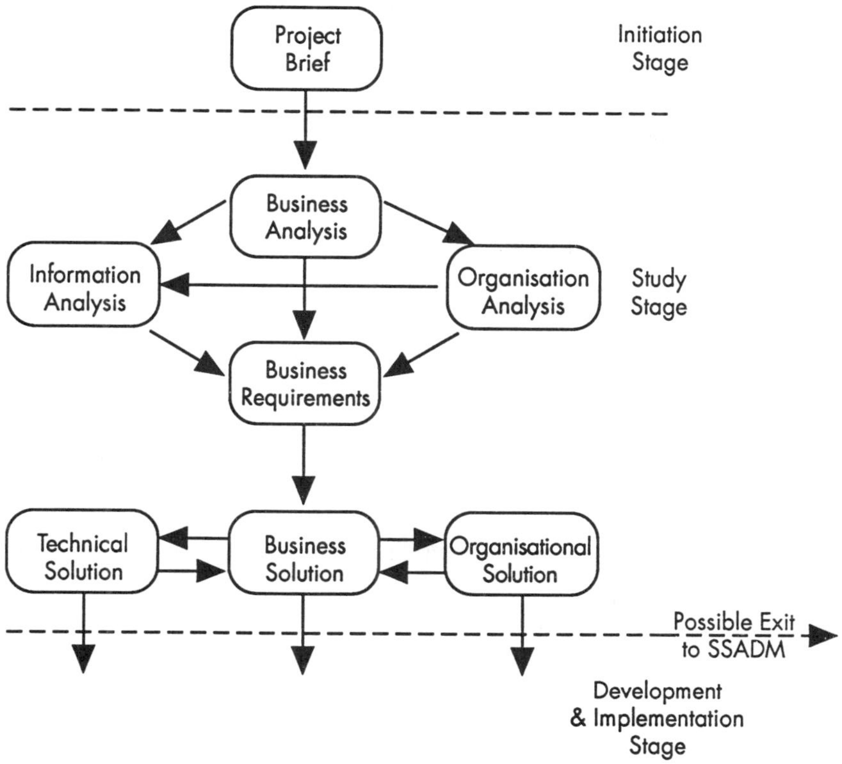

Figure 1. Relevant Stages of Compact

SOFT AND HARD SYSTEMS STRATEGIES

The analysis and design of the system for condition monitoring could be viewed as a continuous progression from the soft systems approach, via the Compact method, as devised by the Civil Service from the work of Checkland, to the structured approach of SSADM, as chosen by the CCTA for government use.
Both Compact (figure 1) and SSADM can be viewed as steps within stages. The initial stage of Compact is to obtain or clarify the briefing, asking questions concerning:

What:	to produce a system for condition monitoring
How:	eventually using a relational database
How Long:	a prototype within the next year
Resources:	the authors, together with a post graduate student

The Study Stage is concerned with first analysing the business. This involves consideration of the problems and the opportunities, both discussed above, and the whole system. This includes:

- task identification

- the people, such as operators, design and maintenance engineers, management (figure 2)

- the organisation, into which the condition monitoring system will become a more integrated element

- information, available from human and machine, and required, by both, as discussed earlier

- communications, between humans, between machines and the communication interfaces between humans and the machines

- technology, currently in use in separate standalone systems and in the MIS systems, together with the developments of relational databases

- Business Systems Definition, of the condition monitoring system, including the VOCATE (CATWOE) statement

- Viewpoint - Machine and Staff Performance can be improved by better integrated monitoring systems

- Owners - the management of the factory

- Actors

- Customers - the users of the system

- Transformation - improving and integrating the monitoring information, to provide better informed humans

- Environment - taking account of the working procedures and requirements, and constraints of resources, including cost, human and computing limitations

- Key Performance Factors - together with the implications

- maintaining normal working conditions of machines, computer systems and employees

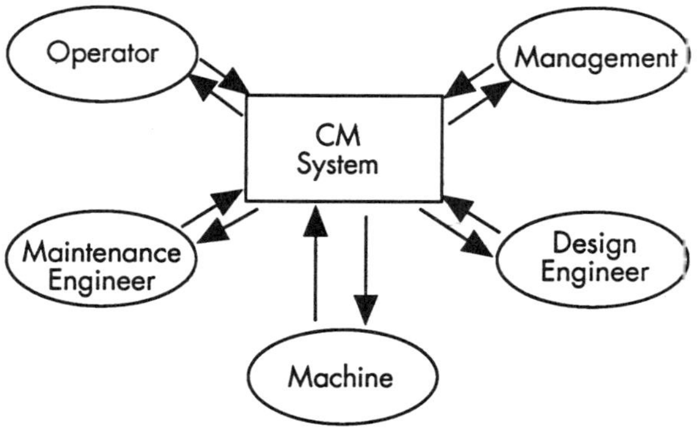

Figure 2. Context Diagram of Technical System

- providing reliable information within required time frame

- containing costs, and utilising existing computer systems and human expertise

- create common standards, in documentation, methods of analysis and design, development and testing and create storable data throughout the company

- Information Needs Table

The next phase would involve Analysing the Organisation, involving suitable structures, the heirachical organisation charts, job descriptions, organisation maps across the whole company. Particular care must be taken to identify duplication and locate any missing links.

The Analysis of the Information phase identified what was the adequate level of support needed for each activity, to make accurate and timely decisions, by humans or by the machine. The analysis was made of the information needed, what was currently available, and the variation of the latter from the former, together with examination of current documentation.

The phase to decide Business Requirements was concerned with identifying what changes are expected, their feasibility and expected effectiveness of the change. A Business Requirement List was raised with agreed changes to the Business Function, of monitoring machines, to the Information System for more effective monitoring and to the Organisation for improved visibility and consistency, identifying in all cases the benefits.

The Business Solution was then described, which incorporates the agreed business requirements, defines what has to be done from this point and defines performance standards. The degree of close fits, comparing the proposed solution with the previously documented needs, was created as was the Business Case for the condition monitoring system.

The Technical solution and the Organisational Solutions were then considered which, together with the Business Solution, are inter-dependent, resulting in changes potentially to all three solutions. The Technical Solution was concerned with the use of IT in the solution, in this case the questions - what technology, how feasible and how effective - were asked. The documented technical solution was then created, which was the relational data base, Oracle, using SQL, on a Vax system, as discussed earlier.

The Organisational Solution examined alternative models, if the solution was culturally acceptable, Effective and Efficient for the business, and job satisfaction for the staff. The documented organisational solution should then be completed. The completion of these three solutions provide an officially recognised bridge to SSADM at Stage two, the step 230, the Business Solution Option.

The SSADM methodology Version 3 was then used to provide a Required Data Flow Diagram, level 1 and 2, and a Required Logical Data Structure. This entity model, together with the entity descriptions, was validated with the DFD, and entity life histories were created. Following relational data analysis, the composite logical data design was created. This was used to form the tables for the tables for the relational data base.

CONCLUSIONS

The use of structured systems analysis and design methods and of soft systems approaches to the condition monitoring of engineering systems was felt to be of benefit. It has allowed the strengths of both methods to be harnessed to obtain a realistic solution. The wide range of interested parties, from operators, engineers, safety officers to financial managers, together with the different aims, was a factor in the choice of approach to this system. At the time of writing, the prototype of the system is being developed. It is felt that this approach to engineering problems, where appropriate, can provide company wide consistency, and enhance the management information system for the benefit of the company and its employees.

REFERENCES

Cross, P., 1990, Cost-effective computer software in maintenance management, in: "Proceedings of COMADEM 90", B.K.N. Rao, J. Au and B. Griffiths, eds., Chapman and Hall, London.

Rao B.K.N., 1992, Profitable condition monitoring and diagnostic engineering management, in "Proceedings of Profitable condition monitoring conference", Elsevier Applied Science, London.

Rao B.K.N., Ross, M., and Staples, G., 1992, Use of structured techniques for specification and implementation of a condition monitoring systems using a relational database, in: "Proceedings of COMADEM 92", P. Collette, B.K.N. Rao, and P. Souquet, eds., CETIM, Senlis, France.

Staples, G., 1991, Implementing a maintenance management system using a relational database management system, *in*: "Proceedings of COMADEM 91", B.K.N. Rao, and A. Hope, eds., Elsevier Applied Science, London.

LINGUISTIC CATEGORIES IN KNOWLEDGE BASED SYSTEMS

Robert A. Stephens[1] and John G. Gammack[2]

[1]Faculty of Computer Studies and Mathematics
University of the West of England, Bristol
Coldharbour Lane, Frenchay
Bristol BS16 1QY, UK

[2]Information Systems
Department of Computing Science
University of Paisley
High Street
Paisley PA1 2BE

INTRODUCTION

In the knowledge elicitation phase of system development, (which has affinities with general requirements analysis, and by extension almost any psychological investigation of someone's "knowledge"), normally an investigator with expertise in modelling or system building enters into communication with a practitioner or theoretician in a domain of interest. Their interaction is largely mediated through linguistic channels (i.e. talking) and will ideally result in a tokenised representation of the expert's utterances. This should be both computationally or systematically tractable, and interpretable by a (usually dissociated) third party using the representation for practical purposes.

Taken at face value, most practical knowledge elicitation is a formal exercise in language use. Language use pervades the process of knowledge acquisition from the initial interview or other technique, through the representation of linguistic tokens in a formal symbol system, to the informed use of the communicated expertise in the end user's operational environment. Viewed as a social communication, the emphasis in knowledge elicitation shifts from pinning down the objective terms as instructed by domain experts, towards supporting a process of shared language use. While both these aspects are required in successful knowledge elicitation, this paper considers the relation between KBS and linguistic categories and explores some of the consequences entailed by the inevitable use of language in KBS development methodology.

The process of knowledge elicitation involves the establishment of meaning upon symbols, which are eventually presented to the user. Normally this takes place in a dynamic social situation, where meaning is established dialectically, the product of negotiation. The importance of language and discourse in this activity becomes clear if we consider a typical analytic encounter: the 'subject' reports anything and produces further associations during

Systems Science, Edited by F.A. Stowell *et al.*,
Plenum Press, New York, 1993

the session; the knowledge engineer asks questions, re-narrates, re-associates, places constructions upon the subject's material and, by way of the conversational engagement, justifications and explanatory moves, (perhaps involving metaphor and metonymy), moves towards an interpretation. Although the product of this encounter will be a symbol set with logical operators defining their relations, all the crucial events here take place in language, of which logical implication is only a small and stylised operation.

Not only does language use pervade the knowledge elicitation stage: at the concomitant stage of representation development a communication leading to an agreed formulation is produced; and at the subsequent operational stage (system use) there is a requirement for third party interpretation of linguistic categories to proceed as intended.

KBS are usually constructed in symbols representing linguistic categories, but these have a more complex status than the 'objective' categories used in data processing systems, which does not make the properties of the representations strictly comparable. One aspect of this is the communicative transfer of highly specialised, and not commonly understood, concepts and categories. For effective this transfer, however, the symbols employed by a KBS must be considered for both their communicative and representational validity.

Successful communication implies a shared language between speaker and listener; correct interpretation requires understanding the concepts which the symbols represent. It is therefore important that a transfer from the original expert to the end-user be mediated through categories to which both can relate meaningfully. If the communicative properties of symbol systems are important then a major factor affecting uptake of KBS in practice will be the success of the match between the needs of the user and the properties of the symbol system. Freyd (1983), for example, points out the conventions used for concept representation, do not so much represent the concept faithfully, but instead allow the meaning to be shared via the communication artefact.

KBS and IS modelling and planning takes place in a social context where development is defined and understood via language, and the technologies, methods and tools employed promote novel language uses. Given this reliance on linguistic categories throughout all phases of the knowledge engineering endeavour, it is relevant to consider some of their general and systemic properties, and what their use implies for Information Systems, and especially Knowledge Based Systems development.

THE STABILITY OF SYMBOLS AND LINGUISTIC CATEGORIES

A fundamental concept for many systems thinkers concerns the act of distinction. Bateson (1980), for example, uses the word difference to denote the identification of entities, and according to Maturana and Varela (1980) it is the fundamental cognitive act. In Gestalt psychology, the separation of figure from ground is a primitive operation; more recently Winograd and Flores (1986) have drawn on Heidegger to show how distinctions emerge from a breakdown of an un-named process or tacit background. Indeed, the existence of language is predicated on such distinctions. The particular distinctions that emerge, however, are relative to the community of language users: i.e., they are contingent on the practice and concerns obtaining in that particular environment of language use, and further moderated by the common experience of communicants.[1]

The activity of symbolisation has considerable implications for the ways in which individual realities develop. Our experience encourages us to take for granted the interaction between objects and our senses and regard the existence and characteristics of objects as if

---

[1] Inasmuch as the concept lying behing word identification has a greater universality, the ideas represented may be communicated beyong the original environment. The issue of communication, however, is a separate and beyond the scope of this paper.

they were independent of our experience of them. However by assuming that the properties of objects are constant and reside in the objects themselves, we habitually overlook the fact that our understanding of objects is created through the evolving categories of meaning we inherit.

Descriptive categories are reified through linguistic practices, but as practice evolves or changes, the labels of these categories may become related to phenomena different to those they originally signified. For example, in recent years consumer borrowing, conventionally known as *debt*, came to be regarded as consumer *credit*, which has a positive and less inhibiting connotation. This semantic shift, when coupled with the evolution of the UK housing market into an asset market contributed to the failure of many macro-economic models to forecast the direction of the economy from the mid 1980s (see Chote, 1992). In extreme cases even quite basic words can, over time, actually reverse their meaning, as Williams (1983, p 308- 312) has shown for the words 'subject' and 'object'.

These examples illustrate a general instability of words and their tendency to decay, transmute, and shift their semantic content, all of which undermines the claim of a stable description of the original phenomena. However, rather than being an unfortunate property of language use, this imprecision of meaning is a necessary prerequisite for social co-ordination. Much communication is only effected because this ambiguity or imprecision allows different understandings to be accommodated by a common symbol. Common categories, such as bird, stabilised through frequent use, may lack defining criteria, but their communicative validity may be assumed by their resemblance to a prototype, a Wittgenstein notion developed by Eleanor Rosch (e.g. Rosch and Mervis 1975). For more on the representational stability of expert categories see Gammack (1990).

While the ambiguity of language is a prerequisite for social co- ordination it does create a tension for KBS because of the formal requirements of precise representation. There is an imperative for ambiguity reduction, yet symbols, by their nature, are open to non-prescriptive interpretation. Indeed, at a social level, and in sharp contradiction to this imperative, symbol systems must adapt and change with their environment to remain viable.

The ability of symbols to stably represent categories is, however, a *sine qua non* for formal representation in KBS. Knowledge representation typically relies on an assumed correspondence between words and objects in the world. Such correspondence is assumed by the realist ontology that informs many of the cognitive and informatic sciences, most explicitly articulated in the physical symbol system hypothesis (Newell and Simon 1976), in which symbolic tokens act as the building blocks of meaning, and thinking is essentially algorithmic computation of internal representations that correspond to an external reality. Corollaries of this view of the mind as a mirror of nature include the assumption that correct reason mirrors the logic of the external world, and that thought is logical in that it can be modelled accurately by systems of the type used in mathematical logic. These assumptions, which are common in the cognitive sciences are identified by Lakoff (1987) as the *objectivist* view, which his careful examination of anthropological, psychological and linguistic research on conceptual categories discredits. The evidence suggests a view in which linguistic categories are used in an imaginative and embodied sense, going beyond literal mirroring, in a non- atomistic and ecological structure: a view which he terms *experientialism*: an analytic distinction having affinities with other recent critiques of positivist notions of knowledge.

The limitations of representational hypotheses have been increasingly recognised in recent years as it has become realised that symbols can act only as *vehicles* for meaning, conveyed between (human) intelligences who interpret and impute meaning in specific contexts. The underdeterminancy of meaning by symbols is an inevitable problem for AI particularly when the system is explicitly concerned with the transfer of communicative intents via computer based media. Symbols encode regularities in the use of language that make information processing meaningful. But, as Winograd and Flores (1986, p 64) say, these regularities "grow out of mutual coupling among language users ... but this does not mean that formal accounts are useless ... computers can play a major role as devices for facilitating human communication in language".

Symbols, however, as we have seen, cannot ultimately stably represent concepts because of a) their propensity to shift semantic content over time; and b) what a symbol stands for is dependent on the interpreter of that symbol (the symbol itself cannot fully constrain its interpretation). Ultimately they gain their stability as part of a social process; they are stable objects not in the world, but in their conversational domain. We have seen that symbols themselves are not sufficient to determine their use; the stable use of symbolic representations is, or must be, embedded in the social fabric.

This social fabric itself however, is pluralistic, and understandings may be restricted to specialised communities, using words in localised senses. For example, in studying the exchanges between Bohr and Einstein, physicist David Bohm claimed their famous scientific disagreement to be founded on incompatible uses of quite elementary physics concepts such as process and event (Bohm and Peat, 1987, p 84- 87). A similar example is given by Kuhn (1970, p 50): "An investigator who hoped to learn something about what scientists took the atomic theory to be asked a distinguished physicist and an eminent chemist whether a single atom of helium was or was not a molecule. Both answered without hesitation, but their answers were not the same. For the chemist the atom of helium was a molecule because it behaved like one with respect to the kinetic theory of gases. For the physicist, on the other hand, the helium atom was not a molecule because it displayed no molecular spectrum."

Rejection of an objectivist ontology does not necessarily imply a loss of stability in language use; it relocates the basis of stable or recurrent use of linguistic categories from the symbols themselves to the social arena where representations are constructed and reproduced. If the stability of linguistic categories must be referenced to the conversational domain of use, the representation of knowledge in Information Systems must make provision for its interpretation by a third party. The apparently unproblematic transfer of knowledge in a simple Expert Systems disguises the fact that these tend to be developed only in situations where the expert and end user share enough experience in common to use the symbols consistently. As this assumption is generally only warranted for trivial domains however, there are explicit design consequences for realistic developments.

LINGUISTIC CATEGORIES AND DESIGN ISSUES

Information System developments may be differentiated by the way in which they manipulate conceptual categories. The differing attitudes adopted towards this point provide a dimension with distinct consequences for the end-user. At one extreme, categories are fully prescribed by unallied designers to which users must accommodate. At the other extreme are user-enhanceable systems whereby practitioners are encouraged to construct, augment or define their own systems applications, designed in terms meaningful within their own environment. Many traditional expert systems and database developments typify the former extreme: categories are chosen by the expert, with little regard to the context of their use or of the language familiar to the user. Although this may be entirely appropriate where there is a recognised need or will to direct the structure from above, such an approach is not always warranted. We now consider two cases where explicit attention is paid to linguistic categories emanating from users.

Gammack and Cui (1992) designed an environment for knowledge based decision support which was informed by cybernetic principles. This environment allowed the building of business models for prediction and classification. For instance good or bad risks for bank loans can be predicted from indicators such as "time at present address", and "nature of employer". Our approach to the problem of user participation was to allow the user design-level control over the detailed decision support model itself, and to define the categories in which the model would be built. The verification of these categories was not referenced to any notion of true or false, but instead to their viability in the user's context, and their meaningfulness to that user. This type of information system need not require formal validation, but can instead be judged pragmatically in terms of the usefulness of its predictions. One of our aims was to utilise both hard, publicly specifiable "objective" categories available from databases, with minimum semantic ambiguity; and soft, user-defined "subjective" categories intended to make a model more meaningful to its developer through the personal theories brought to understanding a complex domain. The predictive models developed can be (a) based purely on the linguistic categories provided by a database, or (b) based purely on linguistic categories provided by the user or (c) a hybrid.

To take one example, a financial analyst might have a database of good and bad bank loan decisions, along with customer profiles using "objective" indicators such as "months in credit". However, they may also have a personal theory that "wearing brogues" is the best predictor, and been recording this information in a private survey. This personally meaningful category may be incorporated into their formal decision model through utilities designed into our system, and can be formally tested against known data. A prejudice of experience in this way can be validated or weakened by a "reality test" in the data environment. Another example of this is the "Hemline Indicator" in economics: the theory that the length of skirts is correlated with the state of the economy. "Investors who (noticed that designers were thinking about long skirts several years ago) were well advised."[2] Although officially not factored into computer models of the economy, such theories (for which rationalisations can be produced) remain controversial.

Our second example concerns the automatic detection of defects on strip steel, in which quality inspectors' perceptions of faults was investigated. Taking images of defects as the elements, Kelly's repertory grid technique was used to estimate the commonality of categorisation over each member of the four inspection teams (18 individuals in total). The exercise revealed inspectors employed a wider category set than that officially recognised by the company, but also one that was organised somewhat differently to the received view. Specifically, inspectors' categories were constructed according to the type of production fault causing the defect: conversely, management tended to employ a less extensive category set, but one more consistent with the steel community at large. The image of one particular defect was both labelled and classed differently by the two parties. For the purposes of rectifying its cause the defect needed to belong to the 'roller' class of defects, but according to other criteria, such as further processing or merchandising, it needed to belong to the 'scale' class of defects.

This absence of uniformity may be seen as both a strength and weakness. A diversity of representation makes for an effective and viable interaction with a complex and changing environment and the lack of internal consistency leaves room for change in conceptual structures. However, problems arise where there is a need for fixing meanings, as for example in a computer based systems. Formalisation requires a commonality of understanding across users, but where this is absent, qualitative techniques such as Grounded Theory are appropriate for identifying and protecting contrasting viewpoints (see Pidgeon et al, (1991) for its use in knowledge elicitation). In our example, such differences were accommodated by using a common project reference document for developers and users to establish a minimal level of agreed categorisation.

These two examples have in two different ways shown how knowledge relevant to users can be incorporated into a KBS by paying explicit attention to linguistic categories

---

[2] Wall Street analyst David Fuller cited in 'Night of the Long Skirts', *The Independent on Sunday*, 18 October 1992.

used by end-users, firstly by having the users supply the categories themselves, and secondly, by grounding them in definitions understood by particular social communities.

REFERENCES

Bateson, G., 1980, "Mind and Nature: A Necessary Unity", E.P. Dutton, New York.

Bohm, D. and Peat, F.D., 1987, "Science, Order and Creativity", Routledge: London

Chote, R., 1992, Why the Chancellor is always wrong, *New Scientist*, 31 October 1992, pp26-31.

Freyd, J., 1983, Shareability: the social psychology of epistemology. *Cognitive Science*, 7, pp191-210.

Gammack, J.G., 1990, Expert conceptual structure: The stability of Pathfinder representations, *in*: "Pathfinder Associative Networks: Studies in Knowledge Organisation", R.W. Schvaneveldt, Ablex, N.J.

Gammack, J.G. and Cui, J., 1992, Automated decision support from very large databases, *Proc. Int. AMSE Conf. on Signals and Systems*, Geneva, June 17-19, pp3-12.

Kuhn, T.S., 1970, "The Structure of Scientific Revolutions", University of Chicago Press, Chicago.

Lakoff, G., 1987, "Women, Fire And Dangerous Things: What Categories Reveal About The Mind", University of Chicago Press, Chicago.

Maturana, H.R. and Varela, F.G., 1980, "Autopoiesis and Cognition", Reidel, Dordrecht.

Newell, A., and Simon, H.A., 1976, Computer science as empirical inquiry: symbols and search, *Communications of the ACM*, vol. 19, No. 3, March, pp113-126.

Pidgeon, N.F., Turner, B.A. and Blockley, D.I., 1991, The use of grounded theory for conceptual analysis in knowledge elicitation, *Int. J. of Man-Machine Studies*, 35, pp151-173.

Rosch, E. and Mervis, C., 1975, Family resemblances: studies of categories, *Cognitive Psychology*, 7, pp573-605.

Williams, R., 1983, "Keywords", Flamingo, London.

Winograd, T. and Flores, F., 1986, "Understanding Computers and Cognition: A New Foundation for Design", Addison-Wesley, New York.

ENTERPRISE DESIGN - A TOTAL SYSTEMS APPROACH TO INFORMATION SYSTEMS DEVELOPMENT?

David C. Sutton

SYSTEM SIX LTD
PO Box 107
Macclesfield
Cheshire SK11 9SX
England, UK

INTRODUCTION

The ability of the profession of systems design to meet the demands placed upon it is increasingly being questioned. Cost over-runs and systems abandoned before completion are reported in the press almost everyday to the extent of being an academically verified phenomenon (e.g. Belotti, 1988; Stowell, 1991). In addition, the satisfaction of users with systems that do achieve completion is by no means universal, frequently executives and professionals find that computing and information technology (IT) either offers them nothing of personal relevance or creates new problems that outweigh the anticipated benefits (e.g. Zuboff, 1988; Mumford, 1983).

Success in this sphere requires far more than a regimented approach to the management of IT projects and far more than increasingly sophisticated tools. Indeed, the adoption of many new tools and methods can be counter-productive in that they allow the perpetuation of attitudes and habits which are themselves part of the problem. The fact is, to achieve the dramatic contributions to effectiveness that are possible, information systems have to be appropriately integrated into the operations and procedures of the entire enterprise (e.g. Kaye, 1989).

It is only now being recognised that this cannot be achieved without a significantly broader perspective upon the design and use of technology (e.g. Wood-Harper et al., 1985). If these new attitudes are not engendered, they guarantee that an organisation will continue to fail to enjoy the real benefits of information technology. New information and communication patterns that are critical to the eflective use of IT have been researched and piloted. This paper outlines the new project management and work styles that are equally critical to success in this sphere.

SALIENT CHARACTERISTICS OF OUR APPROACH:

The necessary changes are far-reaching and fundamental, they require changes in attitudes throughout the whole enterprise. Fundamentally there is greater focus upon the organisational and user aspects in combination with the maintenance of a "Systems" perspective. The approach we now call "Enterprise Design" emerged from a project for a major client to develop a new methodology for computing and information systems development. A key element of the brief was that the approach was to be founded on first principles and employ a 'systems' approach to develop the methodology and its contents.

The details of the history and foundations of the resulting "Coevolution Systems Development Methodology" are in press (Sutton 1993). In addition to specifying the key requirements for a systems development methodology, the coevolution project made it explicit that all technology development takes place in an organisational context which itself needs managing (e.g. Sutton and Sutton, 1990). This necessitates greater involvement of users and user management in specifically designed processes. These processes and their management are the suhject of the Enterprise Design approach which has a number of key features which may be outlined as follows:

Focus upon user role

All stakeholders must be directly involved in the process from start to finish. Active participation by individuals and groups is employed to make explicit what they need to deliver trom their work and what they need (equipment, information etc.) to deliver it.

Done well this is a challenging and stimulating process which generates energy of its own. The deliberations must question all procedures and avoid the temptation to "automate" procedures which were designed in the context of technological capability that has without doubt been superseded. The drive must always be to enhance the underlying purpose of the activity rather than its historically evolved pattern of execution.

This is a process which can raise uncomfortable basic questions such as "why do we do it this way?"or "why do we do it at all?" A variety of support techniques are used to achieve this involvement and to enable participants to construct well founded answers to the questions that arise.

Users as designers

The process above is intended to be "enterprise"driven and to converge upon specific operational activities. When it does so, examples of how technology could contribute are introduced into the conversations. The users can then say whether the examples would help and what changes would make the offering more acceptable. By interacting with simulations and prototypes, users (and their own customers and managers) say what they would find useful in a real system; features are not imposed that the users themselves have not found useful.In this way, the users become designers and their involvement at all phases is essential.

The term "enterprise driven" is used above to emphasise two points. The aim to relate all development to the over-riding mission of the organisation does not preclude alerting the policy makers to new opportunities afforded by technological advances, but it should ensure critical examination of all initiatives that come from the technical specialists to guard against "technology push".

Secondly the term "enterprise" is meant to embrace all forms of organisation whether commercial and profit driven, governmental or in the voluntary/non-profit "third sector" (Drucker, 1990).

Changed role for technology providers

The "user as designer" approach does not replace the need for the classical analyst and programmer roles but, by bringing the users and the analysts closer together, it changes the relationship to one of joint membership of a team rather than the "doctor-patient" model which has characterised the relationship in the past. The "expert" status of the analysts and programmers is recognised to be of equal status to the expertise of the users in respect of knowledge of their tasks and of the managers in respect of the organisational objectives the tasks are to support. All perspectives are crucial to success and the conduct of projects must seek to harmonise these perspectives rather than arbitrate between them.

This is itself a challenge to both technical staff and users who have become accustomed to the old style patterns of sharing out the work and the blame. However, there is potentially a significant benefit to technical personnel as they no longer have to bear such a large proportion of the workload and can contribute to a larger number of projects than under the traditional approach.. The implications of all of this must be appreciated at the highest levels if outdated structures, appraisal and reward systems are not to impede the new ways of working.

Coevolution is continuous

The most notable feature of the business environment is its turbulence. The environment of any enterprise is subject to rapid change and may cause the needs of users to change overnight. In the extreme, the strategies and objectives of the enterprise itself may be reappraised.

Technology must be able to reconfigure its offerings just as rapidly, fixed solutions or applications which institutionalise yesterday's ways of working must be abandoned. We therefore need systems design and procurement frameworks that accept coevolution and change as their normal environment and not as deplorable signs of user vacillation.

The design process needs to be well specified and well facilitated. The impact upon users work of the technology offered can have far reaching implications, mechanisms to appreciate and assimilate these must be provided. As the capabilities and potentialities of the technology are recognised, work patterns and information flows can change and lead to changes in the roles and responsibilities of individual users. Also to be abandoned are the institutionalised procurement procedures which signally fail to ensure quality whilst ensuring systems are out-of-date even before they are delivered.

Dynamic specifications

The coevolution dialogue approach to technology provision makes stringent demands upon the technology. It is necessary to assist users to appreciate:

a) what technology might do for them,
b) how having that level of technology support might change their work and that of others.
c) how having that level of technology support might necessitate changes to both organisational structures and procedures if the full benefits are to be enjoyed.

This requires technology that is:

a) able to demonstrate any work effects that users request,
b) able to reconfigure in real time as users' ideas change,
c) not constrained to whatever is easy, or customary, for the existing installation to provide.

Few organisations have technology that can provide the capability required. The solution that has found to work well is to simulate the effects desired using whatever is available from any source. The simulation acts as a working model of what the final system is to achieve when it is constructed. This approach generates an "animated specification" to demonstrate and agree the operational effects required, whilst the procurement specification is the, contract oriented, formalisation of the agreed statement of requirements. In the long-run the technology kit which is used to construct the animated specification will be able to formally record its own configured state and automatically translate the data on its state into formal specifications and, ultimately, into a working implementation.

Although the process has similarities to "rapid prototyping" it requires new orders of rapidity and a facilitated process which is at least as important as the visible technology which supports the interactions. The rate at which the technology must alter is intended to be virtually real-time. This places different demands upon the analysts, the implementors and upon the technology itself.

Technology stimulation

The simulation approach provides a clear target for the developers of the new system. This target is presented as a simulated system with properties, capacities and behaviour that must be matched by the working system. Being expressed in terms of user effects and unconstrained by current technical functionality, the desired performance can often only be achieved in full if new functionality is developed. This provides clear signals of need which influence the direction of the evolution of technological advance. We have already found this approach to expose fundamental flaws in the underlying architecture of both hardware and data architectures of major providers and research issues were identified and incorporated into those organisation's development programmes.

UNDERLYING VALUES

The features listed above can only achieve their full potential if they are underpinned by values which seek to support not only the reductionist perspective and methods required for detailed analysis of situations but also the synthetic or total systems perspective required for the planning of effective intervention and the management of real-world situations and interventions. Some key elements of the Enterprise Design outlook may be summarised as:

- Start in the real world,
- Systems have human dimensions as well as technical components,
- Nobody is perfect but a team can be,
- Focus on differences that make a difference,
- Avoid Arbitrary Boundaries.
- Top Down Planning, Bottom Up implementation at *every level*
- You cannot have Task without Process

These beliefs were distilled out of reflection upon the underlying values of the methodology developed. They shape the use of many of the tool employed and also impact significantly upon the overall project management. In addition, the underlying attidudes, influence the make-up and brief of a project team, highlight the need for a multi-departmental steering function and contribute to its terms of reference. The values imply:

Start in the real world

No attempt should be made to intervene in a situation without a clear definition of the purpose of the system and the purpose of the intervention. All exercises must be based upon direct contact with the situation aimed at securing a clear understanding of what the system really has to do.

Systems have human dimensions as well as technical components

It is increasingly heard that "the only real problems are people problems". Whilst we are not quite so dogmatic, it is nevertheless true that the backgrounds of technical specialists tend to lead them to see only the "objective" sides of situations. Organisations are primarily social systems and issues of disciplinary and departmental perspective, motivation, general resistance to change and a host of cultural and environmental issues can all contribute to the "logic" of any choice of action. Failure to appreciate the human dimension is generally the prime cause of system failure.

Nobody is perfect but a Team can be

The total scope of most important business situations is always larger than we can comprehend. Any one person's ability to understand the world is shaped and limited by their experience and perspective. It is well established that multi-skilled teams and those which involve all relevant stakeholders and perspectives have the potential to develop higher quality solutions in pluralistic contexts than do individuals or single perspective groups.

Focus on Differences that make a Difference

The features examined and questions asked must be "pragmatically discriminating". In other words it is no use minutely examining factors of marginal impact. The outcome of the exercise must enable someone to make a choice between significant alternatives and an impact that makes a visible and desirable contribution to the over-riding mission.

Avoid Arbitrary Boundaries

There is little point basing an intervention upon an exercise that has ignored major factors that could swamp all your efforts. 90% of problems have causes outside the direct control of those who experience them. It follows that the factors examined must include all the significant influences upon the situation of concern and the search for such factors must not be summarily constrained by a desire to protect turf or egos.

Top Down Planning, Bottom Up implementation, at every level

It is emphasised that all projects must review the tasks they are seeking to empower in the light of the overall enterprise's mission. However, it must also be recognised that missions must take account of operational realities and technical feasibilities. The ideal must be conducted as a dialogue between these two levels of perspective with the further recognition that there are likely to be intermediate levels of debate. Between "critical success factors" and "business data model" there will be many levels of interaction between requirement, alternative and feasibility that must not be pre-empted.

You cannot have Task without Process

Too often organisations focus on concrete objectives without consideration of the means whereby those objectives are to be reached. Many project management frameworks simply divide major goals into sub-goals of increasingly fine grain. No goal is achieved without a process to achieve it. Using a computer is a process, designing a system is a process, these processes need to be given as much design and management attention as the outputs the processes are expected to provide.

IN CONCLUSION

None of the above features discount or replace existing methods and tools currently employed in the systems design field. Instead Enterprise Design widens the project management scope and the context within which such tools are employed. However, it is to be expected that, in the fullness of time, the methods will themselves change in response to the widening of perspective entailed.

This approach has been used and, where those involved are prepared to recognise that the key to success lies in the attitudes of the participants at least as much as in their technical and operational skills, it works. In fact this paper is an enlarged version of a position paper prepared for a client in order to spread awareness of the approach within his own organisation.

Whilst specific mention has not been made of the tools used at the various stages, it should be appreciated from the principles listed that the approach is both systemic and contingent. Although the framework and methodology were developed in the context of practice and commerce, it is interesting to see that the principles and practice of Enterprise Design have a considerable overlap with the critical systems perspective and the concept of Total Systems Intervention advanced by Flood and Jackson (1991). Might there be mutual benefit is regarding Enterprise Design as one approach to Critical Systems Practice?

REFERENCES

Belotti,V.,1988, Implications for current design practice for the use of HCI techniques, *in:* "People and Computers IV," Jones, D. M.. and Winder, R., eds, Cambridge University Press, Cambridge.

Drucker, P. F., 1990, "The New Realities," Mandarin, London.

Flood, R. L. and Jackson, M. C., 1991, "Creative Problem Solving: Total Systems Intervention," Wiley, Chichester.

Kaye, D., 1989, "Game Change," Heinemann, London.

Mumford, E., 1983, "Designing Human Systems," Manchester Business School, Manchester.

Stowell, F. A., 1991, Client participation in information systems design, *in:* "Systems Thinking in Europe,"Jackson M. C. et al, ed., Plenum, NY.

Sutton, D. C. and Sutton, M. M., 1990, Wheels within wheels - a development of traditional socio-technical thinking, *Journal of Management Education and Development, 21, 2*

Sutton, D. C., 1993, Submission to the First European Conference on Information Systems, Henley, Oxfordshire, UK, March.

Wood-Harper, A.T., Antill, L. and Avison, D.E., 1985, "Information Systems Definition: The Multiview Approach", Blackwell, Oxford.

Zuboff, S., 1988, "In the Age of the Smart Machine", Heinemann, Oxford.

THE ROLE OF ACTIVE INFORMATION SYSTEMS IN CREATING AND SUSTAINING A LEARNING ORGANISATION

A.J. Walkden and J.R.G. Wood

Department of Mathematics and Computer Science
University of Salford
Salford M5 4WT

INTRODUCTION

Increasingly, organisations are seeing the ability to learn as an important factor in maintaining competitive advantage. A paradigm for building a learning organisation has been put forward by Peter Senge, Director of the Systems Thinking and Organisational Learning Program at MIT. In this paper we argue that current information systems are inadequate for developing learning organisations, and that a new type of information system - called an active information system is required. A basis for designing such systems is discussed, together with the need for systems thinking both generally and in the design process itself. Finally, the paper discusses future actions for developing systems thinking as well as active information systems.

DISCIPLINES OF THE LEARNING ORGANISATION

"The ability to learn faster than your competitors may be the only sustainable competitive advantage", says Arie de Geus (Senge 1990 p4). Senge (1990) defines a learning organisation to be one within which people are actively expanding their ability to both learn and create desired results together. He describes five disciplines which are required to build a learning organisation: systems thinking, personal mastery, mental models, shared vision and team learning.

Systems thinking is the cornerstone of the other four disciplines. It is holistic and views actions and events as being interconnected. Personal mastery is the discipline of learning about oneself and ones interconnectedness to the world. The discipline of mental models involves surfacing and questioning the appropriateness to the real situation of peoples' mental models. Shared vision is the discipline of building a vision for the organisation that is drawn from each member's personal vision while team learning enables a team to perform at a higher level than the individuals can.

Systems Science, Edited by F.A. Stowell *et al.*,
Plenum Press, New York, 1993

THE NEED FOR SYSTEMS THINKING

Our animal background has left us with the legacy of a survival mechanism which is "geared" to responding to sudden change. This affects our perception and limits our horizon for perceiving threats. This is graphically illustrated by the tale of the boiled frog. A frog placed in boiling water will instantly attempt to leap out. If the frog is placed in cool water which is gradually warmed to boiling, the frog gets drowsier and will fall asleep. It will let itself be boiled to death, because its survival mechanism is not "geared" to perceiving the gradual change as a threat. The inertia of US car manufacturers in the face of an increasing Japanese share of the US car market can be compared to the tale of the boiled frog.

Sources of Non Systems Thinking

Our survival mechanism is not a physical thing but a mental habit. Systems thinking is a different mental habit. But why don't more people develop this habit? Most parents will, at some time or other, be confronted with a question by a child that they cannot answer: "Why does this work Daddy?"; "Well it's because ..."; "Well why does that work?". Whether the answer becomes too complicated for the child or "out of bounds" for the child or simply unknown by the parent is not important. It is the answer given that shapes the child: "Don't ask silly questions"; "Because it is"; "Shut up and eat your dinner". Education also contributes: "Please sir, why are we doing this?"; "Because I tell you to, now be quiet or ...". Within education, the teaching of knowledge is split up into subject areas which are taught and practised separately. The interconnections between subjects are not made explicit and children are not called upon to use relevant knowledge to solve the interdisciplinary problems which reflect real life.

We argue here that children are natural systems thinkers, and that by asking "why?" they are trying to make sense of the systems to which they belong. Unconsciously however, this habit is discriminated against and thus discouraged by older people. This is not to place blame on them however since their own discriminatory habits were acquired from their parents who acquired them from theirs who acquired them ... and so on. Implicitly therefore, we discourage children from systems thinking and reductionist thinking thus becomes the dominant thinking habit.

Implications

The 20th Century has seen significant changes in society, technology and the economy. Much of this change has been unforseen and many changes have been undesirable. People feel remote from the changes, even those that affect their own life. They feel there is no link between their own actions and the changes around them. Even managers and government ministers who instigate change often feel that they had no other choice, that the restrictions and constraints to acting were outside of themselves. One consequence of this lack of systems thinking is that change becomes highly political, and undesirable change causes defensive behaviour in the form of shifting blame. There is no working together of the affected parties and nobody takes responsibility for the change. Everyone believes that authority to improve the situation lies elsewhere, that someone else is the problem owner.

Systems Thinking

Systems analysts might well argue that they practice systems thinking, but research (Jayaratna & Wood-Harper,1991) suggests that such people are in fact performing reductionist analysis. Systems thinking allows us to perceive patterns other than those simple cause-effect relationships which we are conditioned to respond to. It enables us to behave differently from the stimulus-response, knee-jerk type of reaction which we typically display. With systems thinking, we can perceive wider and farther-reaching implications of the actions of today, and make our current actions wiser.

THE NEED FOR BETTER INFORMATION SYSTEMS

People as Information Processors

As information processors, people have cognitive limitations which restrict the amount of information that can be processed. Yet the world in which people operate is extremely information rich. The imbalance between the amount of information and the capacity to assimilate it necessitates the use of a process of information management (Espejo & Watt, 1991). The concept of information management (IM) here refers to the skilful and efficient use of information to generate action.

Espejo & Watt have used the science of Cybernetics to gain insights into the process of IM. They consider information managers and their interaction with organisational tasks as a case of controlling the variety (a measure of complexity) of the situation. Given the limited capacity for processing information, and the richness of information in their environment, information managers need to attenuate the variety of information received. Also, given their limited capacity for generating actions and the range of possibilities, they need to amplify their own variety to affect the environment. Mental models aid the information management process by assigning meaning to received information. They naturally attenuate the received information since any information that does not "fit" the model is discarded. For example, a sales manager who is told that an irate customer with a large moustache is waiting to see him/her will fit "irate customer" into the model of his/her sales responsibilities and will discard "large moustache". Mental models are formed experientially and remain tacit most or all of the time. Since peoples' mental models are constructed from experiential data, such models are often incomplete and even faulty. They affect the way in which people perceive events, and ultimately peoples' actions, since people base their actions on the perceived situation.

For most of the time, these mental models need to be tacit in order for people to function efficiently. Yet when the requirements of the real world change, these models need to be surfaced, questioned and changed. Argyris calls this double loop learning and evidence suggests (Argyris,1991) that many people in organisations are accepting their perception about reality, their information requirements and so on without surfacing or questioning the underlying mental models. This leads to what Argyris calls single loop learning.

Designers, Decision Makers and Problems

Two models influenced the thinking of the designers of Computer Based Information Systems (CBIS) in the 70s/80s: first the model of a manager as a classical decision maker who makes rational decisions in accordance with organisational goals; second the model of managerial decisions being structured, semi-structured or unstructured (Wood & Wood-Harper 1992). It was believed that an analyst and designer could reduce all unstructured problems to structured decisions by the process of systems analysis and design.

Schumacher (Senge, 1990) views problems as being either convergent or divergent: convergent problems converge to a solution the more one thinks of them; divergent problems diverge to many solutions the more they are thought about. Today, analysts and designers are more aware of the divergent nature of the problems information managers face, and can see that developing CBIS by traditional techniques is unsatisfactory. Design of CBIS should thus move to being driven by users in that it is much more the users' perception of their information requirements rather than the analysts which should determine the final design.

Impact of Technological Development

Prototyping is one example of a user driven design methodology. By working closely with the designer/team, it is possible for information managers to get a working system which delivers the required information. Such a CBIS however still only delivers the information that the information managers want. The design process does not cater for questioning the need for the information, and so the design does not allow users to question why they want the information. This design method is therefore unlikely to stimulate double loop learning.

Zuboff (1989) has carried out a number of studies of the impact of computers and IT on the work patterns of people and on the knowledge distribution within organisations. She interviewed staff in a number of companies, both staff who were employed before the new technology came, and staff who came after the new technology. In the context of the use of CBIS, three of her findings are of particular significance.

The first finding concerns the change of perception of the real situation. In a paper mill plant, automation meant pre-technology staff had to learn to perceive the situation differently. Where previously they could judge the condition of a process or paper by touch, eye or sound, now they had to judge by a dial or chart on a display. They tended to feel more remote and less in control of the situation. Post-technology staff however were often quite happy to learn how to operate the technology with the charts or dials as the only information inputs. They often did not know what the real situation was or how variations of these devices affected the real situation. A similar event occurred in a bank where computer terminals replaced the physical paper based system. Although the accounts IS was abstract, bankers' perceptions of it were "anchored" in the physical handwritten ledgers, and computerisation brought about feelings of being less in control.

The second finding concerns the loss of cognitive skills. Before computerisation, clerks in a dental claims operation could easily recite vast quantities of information given simple cues such as a signature to aid recall. With the new technology, volumes of transactions processed increased and the job became more impersonal. The clerks rapidly lost their ability to recall these details and instead relied on the system.

The third discovery concerns the loss of situational knowledge. Before computerisation, employees at a stock and bond transfer operation had a great deal of situational knowledge about what other people in the office did through occupational interaction. After the technology, transaction volumes increased and the work patterns became more self contained. The need and opportunity for interaction was diminished. The situational knowledge was lost. One event that illustrated this was where one benefit analyst sent a letter by post to another analyst who sat next to her at work without realising who she was sending it to.

Implications for CBIS Users

The implications of both Zuboff's and Argyris's studies are that current CBIS act only as variety attenuators and thus become a "crutch" to support the mental activities of users. The disciplines of personal mastery and mental models are thus not being practised. As more organisations become learning orientated, current IS will work against the strategic aims of these organisations. What is required is a new kind of information system; one which amplifies the variety of the user by actively improving their cognitive performance. This we will call an Active Information System (AIS).

FOUNDATIONS FOR DESIGNING ACTIVE INFORMATION SYSTEMS

Development of AIS will need to be different to the current development methods for IS in that development focuses on learning. The design of such systems can thus be classified as learning driven design. The specifics of analysing and designing such systems have yet to be researched, but it is clear that they will need to be based on stimulating learning and cognitive performance.

Learning

People often consider being taught something or remembering facts as learning. This single loop learning is an opportunity for improving ones skills, but it is also an opportunity for improving ones awareness of the mental models with which we operate. It is when we learn some new aspect of the real situation that does not fit with our model that the opportunity to question and change the mental model arises. Reflecting on and challenging these mental models leads to double loop learning.

Cognitive Performance

Research recently carried out at Salford University suggests that cognitive performance is a systemic property of a small number of fundamental learning skills. In many respects, the findings at Salford mirror and augment the work of Argyris, Buzan, and Senge. Four learning skills have been identified through the Salford work. These learning skills have been termed memory, perception, self awareness and changing habits. The habit of practising these skills can improve cognitive performance.

What is commonly called remembering is actually recall. It is often the case that someone might not be able to "remember" a name but will recognise it when it is spoken or seen. There are people who can recall, word for word, everything that is spoken to them. This is a habit of mind and Zuboff's research shows how this skill can wane with the advent of new technology. Perception is defined as the ability to gain insight into something by becoming aware of it through ones senses and interpreting it in the mind. Mental models facilitate the process of perception by providing a framework for interpreting percepts. Systems and reductionist thinking can be considered to be habits of perception. Self awareness is the skill of being aware of oneself. It involves being aware of the mental models with which we operate, but also includes awareness of our feelings of stress or mental pain and awareness of how we are doing mental tasks such as memorising, problem solving or decision making. Acquiring systems thinking involves a change of habit from the current one. The skill of changing habits is an essential part of modifying our perception and of learning.

LOOKING TO THE FUTURE

Current problems such as recession, rising crime and pollution show there is a need to change the way we think about ourselves and the world. We argue that people capable of systems thinking are needed to create and sustain learning organisations. Two steps which we feel will contribute to the development of systems thinkers are establishing courses teaching systems thinking and developing AIS.

Teaching Systems Thinking

The teaching of systems thinking has been advocated by Vickers (1980) who suggested that it should become part of the school curriculum at both primary and secondary level. He believed that it was necessary to equip all people with the skills to deal with the "requiredness" of life. We believe there are potential problems with this approach due to the manner of children's upbringing. To implement systems thinking courses in schools is desirable since generally, it is easier to acquire new habits early in life than later. However the "requiredness" that is placed on the schools by the government, for example through the national curriculum, make this a difficult option.

An interim suggestion is for individual organisations, both universities and businesses, to run systems thinking training courses. How will we categorise these courses and who will run them? Systems thinking is a meta subject discipline that needs to draw on all knowledge, whether it be Maths or IT or Sociology or ....or whatever. By their inter-disciplinary nature, such courses will not fit into a traditional knowledge domain or academic department. There will be a need to assemble units of knowledge experts from a number of knowledge domains or departments in order to run such courses. In order to facilitate communication between experts in different knowledge domains, there will be a need for multiskilled staff who have a grasp of boundaries and constraints in each domain. AIS will support the development of these staff by providing a rich learning culture within which to develop the essential cognitive skills.

AIS Development

We believe that systems thinking skills will be required to support the development of AIS. Systems thought will enable both designers and users of AIS to clarify the underlying information needs of the system and of the user. In turn, the use of AIS can support the development of systems thinking by enhancing the latent cognitive skills of the users. Who will be the analysts and designers of future AIS and what skills will they require? Such people will need to have gained personal mastery. They will be skilful learners themselves. They will be trained to recognise the learning requirements of an organisation. This process will be a key activity in the analysis of AIS. Design of AIS will be based on the stimulation of learning skills. Designs should allow for information to be stored, processed or displayed in ways which aid recall rather than recognition. They should aid perception by allowing information to be displayed in different ways and provide opportunities for people to question their own thought processes. The most important design issue will be determining the balance between the total support offered by the AIS and user responsibility. Make the AIS too capable and users will again be dependent on the technology. Make it too simple and the burden of learning on top of the workload will ensure that little learning will take place.

An information system that can release the untapped capacity and allow natural learning acts as a variety amplifier. This paper was developed using a software package which allowed us to rapidly develop a mental model for the paper. We have the technology for AIS; all we lack is the methodology!

REFERENCES

Argyris, C., 1991, Teaching smart people how to learn, *Harvard Business Review*, 69:3.

Buzan, T., 1989, "Harnessing the ParaBrain", Colt Books Ltd, Cambridge.

Espejo, R. & Watt, J., 1990, Information management, organisation and managerial effectiveness, "Working paper", University of Aston Management Centre.

Jayaratna, N. & Wood-Harper, A.T., 1991, The implication of missing systemic analysis in the systems development process!, "Technical report for the Dept. of Comp. Science", Heriot-Watt University.

Senge, P., 1990, "The Fifth Discipline", Doubleday, New York.

Vickers, G., 1980, Education in systems thinking, *Journal of Applied Systems Analysis*, 7:5.

Wood, J.R.G. & Wood-Harper, A.T., 1992, Information technology in support of individual decision making, "Technical report MCS-92-22 Department of Computer Science", University of Salford.

Zuboff, S., 1989, "In the Age of the Smart Machine", Heinemann Professional Publishing, Oxford.

USER CONSIDERATIONS IN COMMUNICATING INFORMATION TECHNOLOGY SECURITY REQUIREMENTS

Dr. Adrian R. Warman

Department of Management Systems
Bournemouth University
Talbot Campus, Bournemouth
Dorset, BH12 5BB
email: awarman@uk.ac.bournemouth

ABSTRACT

This paper considers user aspects of information technology security. The significant problems for computer security consultants rarely concern the development of methods for protecting systems. Instead, they focus upon the difficulty of educating the staff that the methods apply to. Without this information, it is possible that staff could cause problems within the systems, simply by failing to consider all the issues.

Research described by Warman (1992) indicates that users will frequently transgress the organisational security procedures either as a result of poor understanding, or alternatively because of ignorance of the issues. Unfortunately, it will not be enough to encourage understanding of the problems. Further, the fact that appreciation of a problem has become wide-spread will be poor recompense for an organisation that has been subject to a computer security breakdown.

There are two key issues addressed within this paper. The first is that thoughtless or ill-considered publicity given to security matters will actually reduce the competitiveness of organisations. The second is that the same publicity intended to improve awareness of official policies and guidelines may, at the same time, indicate other security weaknesses that may be exploited.

In particular, an objective for this paper is to highlight the need for organisations to establish a point of equilibrium for providing awareness of security without endangering the quality of the system, issues of considerable importance which are also discussed in Timms (1990) and Wilkes (1990).

USER CONSIDERATIONS

Information systems may often be characterised by the level of consideration given to user criteria, identified during the first stages of system construction. Ideally, every user

Systems Science, Edited by F.A. Stowell *et al.*,
Plenum Press, New York, 1993

involved with applying the system should find their work enhanced in some fashion. The recommended trend, therefore, has been to develop systems that take account of organisational, departmental and user objectives.

However, once the system is in place, the way in which the users apply it to perform their work will alter. Inevitably, some resources within the system are not used, although Burnes (1990) indicates that this will not be a problem. Nevertheless, having become a standard tool, subsequent changes to, or unavailability of, the system may be a source of increasing frustration.

The cause will be that unexpected or inconsistent variations in system behaviour make it appear unreliable. Resentment may build up if users cannot rely upon a system to be available and to aid them in their tasks.

System availability is a crucial issue. However, the need for, and nature of, availability is not always clear to all those involved, and the result can be increasing difficulty in developing a satisfactory system.

The problem can be highlighted by reference to the difference between availability and reliability. System development processes may confuse the issues during implementation, and consequently fail to balance the requirements and effort while attempting to address such factors as correctness and robustness (Ghezzi, 1991). In fact, availability and reliability *are* different, particularly when considered from the user perspective.

System development must take account of the difference between, and importance assigned to, availability and reliability. This is because the two concepts will affect the security of the system according to the need to ensure accessibility and accuracy.

POLICING SYSTEMS

Problems can also occur if apparently less important objectives are given more importance than those objectives which require systems functions to contribute to user effort. This is particularly the case with security mechanisms, which can often appear to authorised users as a real hindrance to their work. Jackson (1991) highlights the importance of communicating the value of security, not least to senior personnel, with Tantam (1990) further supporting and extending the concept to all levels.

A specific instance of concern involves the 'policing' nature of some security mechanisms. Users may worry that the monitoring tools could be misapplied, for performance reasons rather than security. An example concerns the monitoring of key-presses per hour by data-entry clerks. Anyone entering data too slowly may be sacked. Here, a mechanism used to monitor if equipment is being used out of hours can be applied for another, less constructive purpose. Concerns of this kind can lead to dissent, which may become apparent in a variety of ways. By recognising the existence of such issues, preemptive preventative action may be taken.

THE NEED FOR TRAINING AND POLICY DISSEMINATION

The motivation for someone to attack a computer system may arise from a variety of circumstances, not all of which will be for technical reasons. Accordingly, it will be inappropriate to concentrate on technology alone to address the problems. Indeed, technology itself is often the cause of the problems. As an example, given the large quantity of manuals needed to learn many new packages today, it is not surprising if they will sometimes include information that can be misapplied.

Even in the absence of hard-copy documents, on-line help facilities can often be just as informative - not least owing to their purpose of providing information to the reader. As an observation, although users and systems may be classified into different security domains, the on-line help is unlikely to be so domain-sensitive, increasing the risk of inappropriate distribution of information. Using such techniques, it may be possible to get hold of information on how to bypass or compromise security measures. A good example concerns the recovery of deleted files, some of which may contain sensitive information.

Clearly, any new employee or user should receive training in the use of the computer resources. Ideally, it will be recognised as a progressive, continuing process. Yet this may turn out to be counter to the actual objectives of the enterprise, as it may suggest further computer security attack techniques. Nevertheless, the issues of computer security must be discussed. Rather than providing excessive detail about the mechanisms, it would be appropriate to provide an overview of the threats and how the mechanisms help to deal with them.

Having initiated an appreciation of the issues, that awareness must be maintained. There are a variety of techniques, but two significant errors must be avoided. The first is the risk of making a particular issue appear unimportant. Use of cartoon figures spouting cliches suggests a frivolous technique which contrasts and conflicts with the need for serious consideration of the issues. The second arises from the possibility that the target market, the staff, may become immune to the issue as a result of excessive saturation.

Very careful handling of the computer security issues is required. Simple provision of checklists of things to do or things to avoid are not really very helpful, as they can be taken too literally and so fail to address the actual problems.

There is no reason why considered discussion of security issues should not form part of the regular training and development for all staff members. In particular, any modifications or to earlier guidelines or procedures - no matter how small - should be emphasised.

As a consequence, the organisation must accept that a degree of trust in its staff is essential. Only then will it be possible to equip staff, not only with the means for preventing with threats, but with an understanding the problems themselves. This information can then be used to justify the presence and application of the security mechanisms, and further emphasise in a non-threatening fashion the implications of failure to conform.

A good example, that refers directly to computer-based systems, concerns simple protection of information by taking backups, and subsequently keeping them secure in their own right. The perspective of security should be extended to include broader issues, not least the general operational circumstances of the business. For example, strangers should always be challenged, rather than simply accepting them as visitors or new employees. In essence, a culture must be encouraged where problems should never be ignored in the hope that someone else will deal with them. The well-being of the organisation will depend to a primary extent upon the active contribution of all those involved within it.

Equipping the employee with details of security problems should not be detrimental to the business. By definition, many employees will only be in their current job if they are assigned a certain level of trust and responsibility. The briefing on security issues should reflect that level of trust. Furthermore, it must become a part of company culture to recognise that actions taken for justifiable reasons of security will not be punished, no matter how politically undesirable the act may be. In particular, it should be possible for junior members of staff to refuse a request or instruction by a superior to carry out an act that is clearly wrong. Examples include the photocopying of material in contravention of copyright, or short-term 'loan' of software for computers.

At the current time, many unlawful actions, such as the breaking of copyright or piracy of software is interpreted very 'loosely' by some surprisingly senior people. Until such a corporate culture develops, the overall security and in particular its integrity will be severely constrained.

SPREADING THE MESSAGE

Most organisations will already be acquainted with basic security issues, as a result of the need to prevent loss or damage. Standard mechanisms, such as secure locks on doors or windows, installing alarms and security lighting systems, and carrying out unpredictable but comprehensive checks by specially trained personnel are generally widely appreciated. Frequently, the implementation of these duties will be allocated to form part of the domain of a premises division within the organisation. However, the responsibility for computer security may not be so straightforward.

For most organisations, there would normally be two divisions or departments that could be expected to make a major contribution to the task of conveying the issues of computer system security to staff. These two departments are the computer or information systems services group, and the staffing office.

Information systems or computer services groups should, by the very nature of their resident expertise and perhaps even mission statements, be expected to have a substantial awareness of the threats to the information technology systems that they are responsible for; if for no other reason than they will have to deal with any problems that arise from those threats. In the majority of such departments, however, there will be a clearly observable priority given to the technical issues rather than to helping the staff and users of the technology. In many such departments, end-users are viewed as an occupational hazard to be avoided wherever and whenever possible. Training or 'help-desk' manning is a 'dirty job that has to be done'. The result of this is that not only can the relationship between the information systems group and the rest of the organisation be rather poor, but further that the information systems group has a very sparse understanding of the organisational requirements of users and business units.

The preference, particularly from within technical departments, will be to assign the 'social' issues to other departments. In particular, as Thompson (1991) suggests, staffing offices may be given the extended additional task of dealing with personnel-related security issues. A result of this might be that within some organisations - and especially those with notably sensitive material or technology dependencies - new approaches will be taken to assess the character and integrity of current and prospective employees. For example, an increased amount of effort may be applied to the development of measures that identify the way in which staff work-habits alter when they are constrained or influenced by organisational policies and guidelines. It is to be hoped that such a trend would not result in excessive coercion or forceful control or employees, but instead that all the staff would be better able to appreciate the reasons why the rules and regulations are necessary.

The extension of staffing office duties in this way may be viewed as being inefficient and ineffective, particularly if there is already some form of judgmental procedure established for assessing staff. However, in very many organisations, the appraisal process may be treated as something of a joke, performed to keep senior managers happy with lots of paperwork. The fact that the review is normally carried out by the performed by the reviewee's immediate superior further undermines the process, as the familiarity engendered by normal day-to-day contact makes a more formal assessment extremely unnatural and incongruous with the normal working relationship. At the very least, it is possible to miss potential problems at the time, until they subsequently become much more obvious and, in all probability, much more difficult to deal with. Consequently, it is just as inappropriate to allocate security problems as an appendage to the existing role of the staffing offices.

The highly simplistic technique which is all too often found in organisations, where the presence of a department or team of specialists is effectively seen as a form of dumping ground for the problem. Such an approach simply cannot work. Any allocation of responsibility to a specific function is certain to result in the responsibility being addressed from the perspective of that function, and probably to the exclusion of any other perspective. There is also an additional danger that the additional responsibility may not form a coherent match with the normal duties, which will be degraded as a result.

A variation on the theme is to create a new and dedicated team of people that will focus on the issues anew. This too, however, fails to address all the problems in that it will still allow other staff to perceive the problems as being someone else's responsibility.

Furthermore, any such team of this kind - and particularly one which focuses upon the strategically and operationally essential matter of information and information protection - is bound to be viewed with some disquiet within the rest of the organisation. It could rapidly be seen as some form of secret police, since the very nature of the work would require it to ignore operational or departmental boundaries. As such, it would be extremely difficult to build up the constructive and cooperative culture necessary to achieve higher levels of success in combating computer security problems.

In practice, therefore, it is impossible and meaningless to assign responsibility for the security of computer-based information systems to any one department. It is an implicit imperative that greater liaison should be developed throughout the organisation, so that a cooperative culture can be established. The complexity and subtlety of many computer security threats mean that they are more likely to be detected and dealt with at an earlier stage if there is a general and continual attention given to the issues throughout the organisation. Computer security analysis must be carried out frequently and in detail, but at all times the resulting mechanisms and defences must be applied fairly and with respect for members of staff who, in most situations, should be viewed as an invaluable addition to the defence measures.

There is no advantage for an organisation to attempt to assign allocate the responsibility for dealing with security problems, nor for spreading the message, to a single department or team. A more meaningful approach would appear to be to supplement the existing security mechanisms by the development of a cooperational and participative culture. The responsibility for, and ownership of, security issues is one that must be shared by everyone at all levels within the organisation.

SUMMARY

In this paper, we have looked at some of the problems of communicating the message of computer security. Unlike many other organisational issues, computer systems and information systems have an increasingly deep-rooted status. Failures in the systems can rapidly and dramatically affect the performance and success of the organisation.

Accordingly, the systems must be protected, not only by the wealth of technological methods, but also by the awareness and cooperation of all staff in the organisation. This in turn can best be achieved by greater publicity using methods that reflect the seriousness and unusual nature of the threats faced.

The boundaries of acceptable and unacceptable actions must be clear to all, along with the knowledge that transgression of the boundaries cannot be tolerated or excused. However, communicating the existence and nature of those boundaries is a management consideration that cannot easily be delegated.

REFERENCES

Burnes, B., 1990, Managerial competence and the new technology, *Behaviour and Information Technology*, Vol. 10, No. 2, pp 91-109.

Ghezzi, C. et al, 1991, "Fundamentals of Software Engineering", Prentice-Hall Inc., Englewood Cliffs, New Jersey.

Hafner, K. and Markoff, J., 1991, "Cyberpunk: Outlaws and Hackers on the Computer Frontier", Fourth Estate, London.

Jackson, C., 1991, How to sell top management on information security, "DataPro Reports on Information Security", January 1991.

Tantam, M., 1990, Turning the key on employees' IT Power, *Computing*, 12 July 1990.

Thompson, H., 1991, Managers urged to vet employees, *Computer Weekly*, 24 January 1991.

Timms, M., 1990, Integrated computer security', Paper presented at "Corporate Computer Security conference", London.

Warman, A., 1991, Computer security policies, "LSE Working Paper", London School of Economics.

Warman, A., 1992, Organisational computer security policy: the reality, *European Journal of Information Systems*, Vol. 1, No. 5, pp 305-310.

Wurman, R., 1991, "Information Anxiety", Pan Books, London.

Wilkes, M., 1990, Computer security in the business world, *Communications of the ACM*, Vol. 33, No. 4, pp 399-401.

RESOURCE OBJECT DATA MANAGER: STRUCTURED APPROACH TO SYSTEMS AND NETWORK MANAGEMENT

Wita Wojtkowski and W. Gregory Wojtkowski

Computer Information Systems And Production Management
Boise State University
1910 University Drive
Boise, Idaho 83725
USA

INTRODUCTION

Operational management of business computer systems and computer networks was not long ago performed by an operator or a team of operators who interacted with consoles displaying status messages. In such a set-up each system component independently resolved its own set of notices, identifying conditions needing attention.

As business computer systems and computer networks increase in size and complexity, their expedient management becomes critical to their user organizations: the operational control information must be made available to a variety of systems and network management applications. The amount and complexity of the status information needed for control and coordination of multifarious systems makes it unlikely that operators will be able to keep up with diverse needs unaided by automation.

TOWARD AUTOMATED OBJECT-ORIENTED SYSTEMS MANAGEMENT

Automation of the control data management commenced with aids for the textual message processing. Basic functions and limitations of such facilities are summarized in Figure 1.

Enabling Automated System Control and Coordination

In order to reduce the complications of automating the control and coordination of today's systems, object oriented paradigm may be employed (Korson and McGregor, 1990).[1]

[1] We refer here to object oriented conceptual view (OCV) as oposed to object oriented programming view (OPV). For details see Winkler, 1992.

Systems Science, Edited by F.A. Stowell *et al.*,
Plenum Press, New York, 1993

FACILITY	BASIC FUNCTIONS	LIMITATIONS
Message Processing Facility (**MPF**)	Monitor and control of computer system through text message processing;allows reduction of message trafic at the operators' console; messages can be marked as suitable for automated processing	Deals only with the handling of messages; cannot create automation applications
Network Communications and Control Facility (**NCCF**)	Allows responses to well defined, immediate conditions (as specified in a message) programmed through command lists; a particular command list program executes in its own self-contained environment, whenever a prticular category of message is encountered	Automation capabilities are limited. Command lists programs are triggered only by the arrival of particular message. Difficult to share information between command list programs executing in response to different messages. Dependencies between messages in space and time cannot be taken into account.

Figure 1. Basic functions and limitations of automated textual message processing facilities.

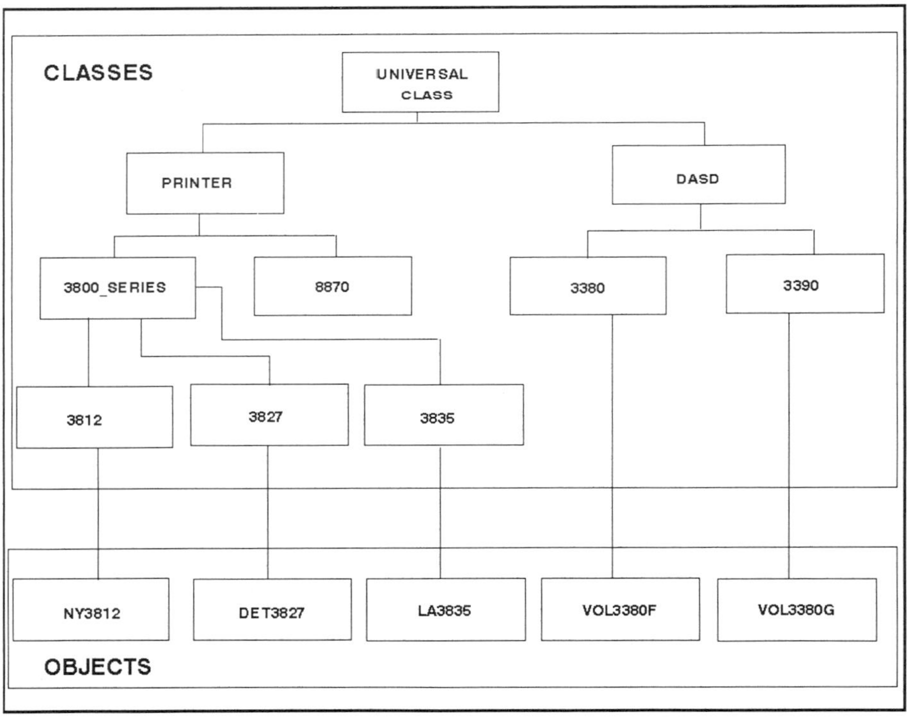

Figure 2. An example of hierarchy for the Printer and DASD classes and objects.

Class or Object Name	Locally Defined Fields	Inherited Fields
3800_Series	Stae Control_Unit Peers_in_String OwnerID	
3812	Seeks_in _Last_Hour Percent_of_Capacity Date_of_Last Service	State Control_Unit Peers_in_String OwnerID
3827	Date_of_Installation Number_of_Failures	State Control_Unit Peers_in_String OwnerID
NY3812		State Control_Unit Peers_in_String OwnerID Seeks_in _Last_Hour Percent_of_Capacity Date_of_Last Service

Figure 3. An example of the inheritance of structure.

This paradigm's design principles define four basic elements of an object:

- *Object class.* Describes objects as being related to one another and able to inherit attributes and data. A class can be a superclass of another class or an aggregate of other classes.
- *Object instance.* The manifestation of a class.
- *Method.* The behavior of operation supported by classes of objects.
- *Link.* That which connects two classes or instances of classes and defines relationships between them.

Applied to information systems, object orientation treats each resource type as object. That is, any real-world objects within a system that need to be controlled and coordinated are abstractions of real-world objects, such as printers, modems, terminals, DASDs, tape drives, and so on. The required behaviors of these objects are captured in their associated methods. Relationships and configurations of objects are maintained by object linkages. A collection of classes and associated objects forms structured hierarchy at the root of which is a class which has no parent. This class is referred to as Universal Class. An example of such hierarchy for Printer and DASD classes and objects is shown in Figure 2.

The structure of an object is completely determined by the definition of the class to which it belongs. Objects of similar structure need to be defined only once, at the class level. Figure 3 shows how this structure inheritance might work.

The inheritance functions as a reusability and information sharing mechanism: just as structure may be inherited by classes and objects, so may the values of subfields. Applications manipulate the objects in the data model as if these were real object. For instance, to query a status of a printer, an automation application might query the status field of the object data manager that corresponded to that printer. Programs stored with the object (methods), ensure that the status fields contain correct information. The application need not have contact with the actual printer. The methods associated with the object maintain contact with the real printer and assure the validity of object data.

Conceptually the situation is as follows: an active, object oriented model of the system is provided and maintained. The automation applications deal then exclusively with this model to effect management processes. To be able to automate the management of the resource object data requires:

- object data modeling capability
- active object management
- data sharing and support
- dynamic schema evolution

Data modeling

When commands are issued to physical devices, they respond with messages. Since the information needed for maintaining an accurate model of system resources might be contained in such status messages and alarms, special applications must be provided to recognize these messages, determine their relevancy, and use applications programming interface (API) to update appropriate objects. This special application is a status sender which converts messages to object data updates. By means of the API, classes can be defined, objects created or obliterated, information regarding objects manipulated. Instead of being concerned with messages, commands and the validity of data, API facilitates access to data objects contained in a structured data model. This is schematically shown in Figure 4.

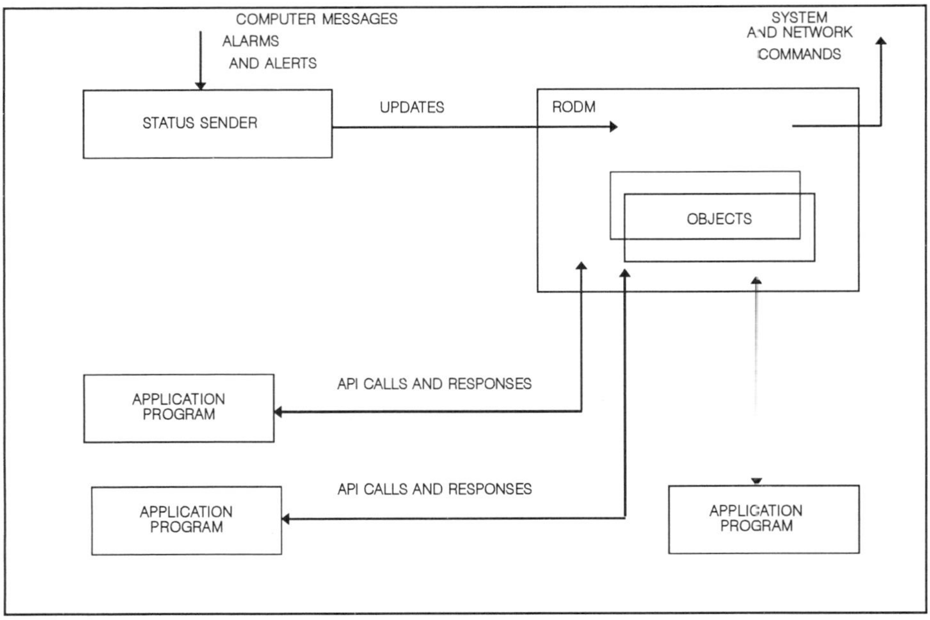

Figure 4. Relationship between objects in resource object data manager, API and status sender.

Active Object Management

Incorporation of methods into the information model allows the management of the active objects [2]. Facilities of this kind provide, in the object definition, programs that allow a given object to automatically maintain itself. This in contrast to conventional treatment of data, where data are passive, and external programs are required to specifically alter a data record in order for a value of that record to change.

Data Sharing

The data model should be shared among many users and applications. Numerous automation activities may take place at the same time. Thus each activity may be viewed as a different task (Cox, 1992).

Dynamic Schema Evolution

The object schema will change over time. New class definitions must be allowed, as well as creation and deletion of objects. Addition of new fields to classes ought to cause automatic propagation of the new fields to existing classes. All such changes ought to occur dynamically.

NETVIEW* RESOURCE OBJECT DATA MANAGER

The NetView* Resource Object Data Manager (Finkel and Calo, 1992; SystemView Overview, 1992) supports object data management as described above. It is a part of SystemView*, the IBM systems management strategy for coordinating and operating distributed and heterogenous enterprise systems. The goal of this product is to allow management of systems consisting of resources defined in terms of System Network Architecture (SNA) and resources that are not defined in terms of SNA (Szabat and Meyer, 1992).

One of the subsystems of the NetView* is a Multiple Virtual Storage (MVS) host that serves as a focal point for system status information. Its vital component is a new object oriented data cache manager (it runs as a subsystem on both MVS/XA and MVS/ESA). Figure 5 shows its basic architecture.

CONCLUSION

Resource object data manager is intended to store both operational data and control information. Fields in objects are to contain information about the status of the resources that the object represents. Applications query these fields to obtain information about system status. This status information is highly dynamic in nature and since it utilizes object-oriented paradigm it captures information on the interrelationships and dependencies as well.

Automation plays an essential role in the management of today's information systems. Many organizations have significant investment in existing systems management applications.

---

[2] Objects are said to be active if they can maintain their own state. The required behaviors of such objects are captured in their associated methods. Relationships and configurations of objects are maintained by objects linkages [Henderson-Sellers and Edwards, 1990].

* Trademark or registered trademark of the IBM

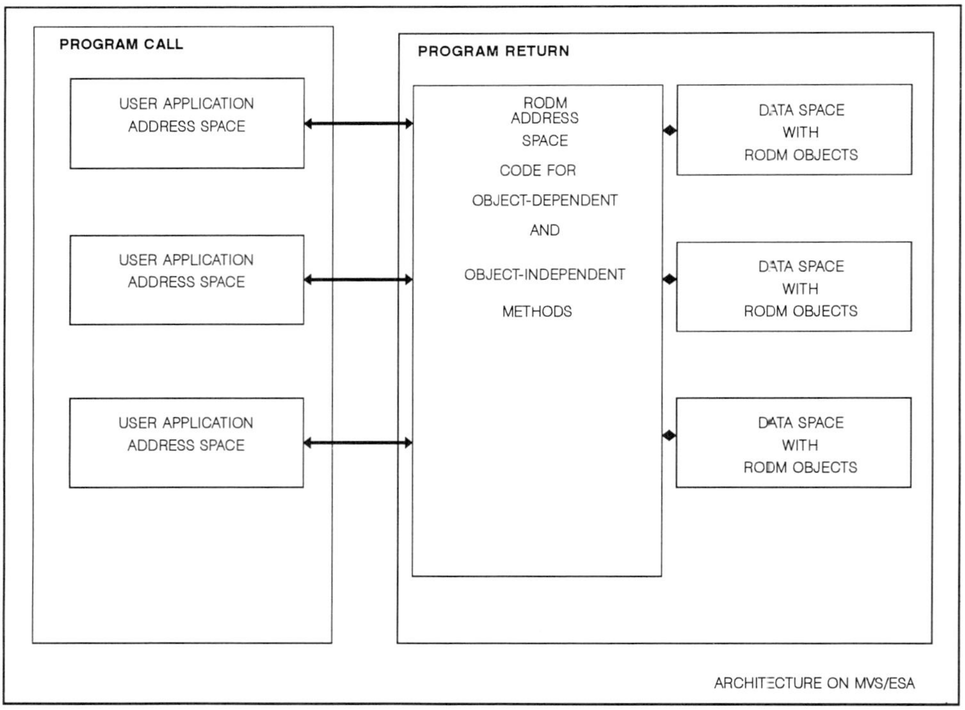

Figure 5. Basic resource object data manager architecture.

Some of these applications manage large amounts of data. Protecting investments in existing systems management applications is often of paramount importance. Resource object data manager data model capability should provide guidance for the data sharing among existing as well as new systems management applications.

REFERENCES

Cox, B., 1992, Superdistribution and electronic objects, *Dr. Dobbs Journal.* 44:48.

Customized Summary, 1992, SystemView overview, *IBM 292-578.*

Finkel, A., Calo, S., 1992, RODM: a control information base," *IBM Systems Journal.* 31:252.

Henderson-Sellers, B., Edwards, J., 1990, Object oriented systems life cycle, *Communications of the ACM.* 33:142.

Korson, T., McGregor, J., 1990, Understanding object-oriented: a unifying paradigm, *Communications of the ACM.* 33:40.

Szabat M. M., Meyer, G. E., 1992, IBM network management strategy, *IBM Systems Journal.* 31:154.

Winkler, J. F., 1992, Objectivism 'class' considered harmful, *Communications of the ACM.* 35:128.

Wirfs-Brock, R. J., Johnson, R., E., 1990, Surveying current research in object oriented design, *Communications of the ACM.* 33:9.

INFORMATION SYSTEMS DEVELOPMENT EDUCATION: ASSUMPTIONS AND PRACTICE

Stanislaw Wrycza and Witold Rybinski

University of Gdansk
Department of Information Systems
Armii Krajowej 119/121
81-824 Sopot
tel. (48 58) 5100 61 ext. 400
fax (48 58) 52 2212

ABSTRACT

The aim of this paper is to present the theoretical background of information systems development education at the university and to generalize the experiences connected with teaching of this subject. The teaching process is directed at combining both "hard" and "soft" approaches to systems development. It is concentrated around case solutions. The extensive use of CASE tool SDW (System Development Workbench) eliminates the rudimentary work connected with students project accomplishement. In the didactic process students use the following methods in the first three phases in the system lifecycle:
1. Planning: Critical Succes Factors (CSF), Meta Plan Session, Consistency Model of Broekstra (CMB), Control Paradigma, Information Cross, SWOT,
2. Analysis: ISAC, Data Flow Diagram (DFD), Entity Relationship Model (ERM), Decision Tables, Decision Trees,
3. Design: Data Flow Diagrams, Entity Relationship Model, Normalization, Bubble Charts, Structure Charts, Jackson Diagrams, Prototyping.
The paper starts with the assumptions of the proposed methods, then the practical experiences are extensively described. The paper is finalized by the conclusions.

INTRODUCTION

The courses of Information System Development (ISD) have been also included in the curricula of our Information Systems Faculty since 70's. Then, the commonly used methods of systems analysis and design were adapted to teaching purposes. The research work performed by department's scientists as well as contacts with foreign partners influenced the gradual change of the program of ISD. However, the radical change took

Systems Science, Edited by F.A. Stowell *et al.*,
Plenum Press, New York, 1993

place at the end of 80's. It was caused by the broadening impact of modern system development methodologies and CASE tools.

Finally, the growing demand for more sophisticated system development methods, techniques and tools has appeared. There were two reasons for this phenomenon:
a) newly introduced market economy needs more complex and active information systems facilitating the business strategy of many firms,
b) the economy becomes more open, i.e. many undertakings carried out with involvement of international institutions or business partners demanding internationaly recognized standards for system development.

The above-mentioned crucial factors first and foremost have stimulated the interest in learning modern approaches in system development. Primarily, managers, analysts and designers from privatized firms have expressed their interest in such courses. Of course, the experience gained then by university trainers was transferred to university courses.

In this paper the philosophy of ISD teaching at the university courses is presented. The roles of both trainer and student are described. The proposed didactic method concentrates around case solution in teams. During this teamwork students are supported by trainer's consultancy, literature, CASE tools and lectures. This method allows self-dependent application of modern system development workbench in practical case.

ASSUMPTIONS

The teaching team was aware of its need for radical style change of information systems development training for students.

The basic problem which arises when establishing new programme of information systems faculty studies is the question: what kind of professional skills will the graduate need to cope with technological, organizational and social challenges in course of his/her work? In our opinion, the modern systems analyst or designer, after graduating a course of information systems development should be able:
a) to develop ISs, supporting the goals of the business and administrative organizations,
b) to use efficiently versatile methods and techniques if IS development for solution of business problems,
c) to improve and train the social skills to cooperate in project teams.

The prerequisites concerning the University graduate abilities have the direct influence on assumptions of the didactic process. They can be divided into didactical and methodological ones. The former are concerned with the process of students' training, while the latter with essential approaches, methods and tools supporting didactic process. The main didactic assumptions are as follows:
a) the courses should be creative and active directed at problems solving,
b) the trainers play rather advisory role - the basic work is performed by students,
c) the system of student self-assesment should be introduced,
d) the course is centered around case analysis and solution.

On the other hand, the crucial methodological assumptions are following:
a) the students are equiped by means of lectures and literature with modern knowledge on methodological approaches to ISD,
b) the five-stage system lifecycle (SLC) was assumed (it comprises the following phases of systems development: IS Planning, System Analysis, System Design, Implementation, Operation and Control),
c) the CASE Tool SDW supports the use of specific techniques.
Information Systems predominantly are developed in teams. Therefore it is not a point to teach students exclusively the most modern methods and techniques, tools of ISDM, but to

teach them to do it in teams. This approach is more difficult for both students and teachers as creates the new challenges in didactic process. The above mentioned prerequisites cause the significant changes in the training process.

PRACTICE

The main ideas of the approach used are presented on the figures 1 . 5. The process of passing from case statement to system project by the above mentioned System Lifecycle (SLC) is shown on figure 1. It shows also the influence of consultancy, teamwork, methods & CASE on development process according to assumptions from chapter 2.

The discussed course deals with three initial phases. It was the intention of authors to propose the wide variety of methods for future systems developers. Therefore, the students were acquainted with the following methods:
1. Planning: Critical Succes Factors (CSF), Meta Plan Session, Consistency Model of Broekstra (CMB), Control Paradigma, Information Cross, SWOT,
2. Analysis: ISAC, Data Flow Diagram (DFD), Entity Relationship Model (ERM), Decision Tables, Decision Trees,
3. Design: Data Flow Diagrams, Entity Relationship Model, Normalization, Bubble Charts, Structure Charts, Jackson Diagrams, Prototyping.

Dependencies among the above mentioned three phases and connected with them CASE and methods, teamwork and consultancy are presented on figure 1.

The whole process starts with the group formation. The psychological test is used to indicate the proper composition of groups. There are four types of dominant behavioural styles in group: promoter, controller, analyst, supporter. The table 1 explains features of each style.

Table 1. Characteristics of four dominant behavioural styles in groups

PROMOTER		SUPPORTER	
+ Main Line Thinker	- Unsystematical	+ Helpful	- Unprincipled
+ Stimulating	- Manipulating	+ Intuitive	- Unstructured
CONTROLLER		ANALYST	
+ Natural Leader	- Lusting for Power	+ Facts gatherer	- Uncommunicative
+ Active and responsible	- Uninspiring	+ Very systematical	- Introvert

Of course, the composition of each group should be carefuly specified i.e. team group cannot consist of exclusively the same behavioural style people. The desired composition of team group is: one controller, one promoter, several analysts and supporters. There are following advantages and disadvantages of this type of group formation:
+ concrete way of group formation,
+ students get acquainted with the behavioural styles of themselves and co-workers,
+ possibility of avoiding many conflicts in course of team work,
-- the other important psycho- and sociological factors are ommited,
- it is not always possible to form groups combinig all four styles as the roles of promoter and controller are rare,
- students are reluctant of tests.

In the second stage students receive ready-made case and start to analyse it and askquestions connected with its context and task specified. There are still not enough

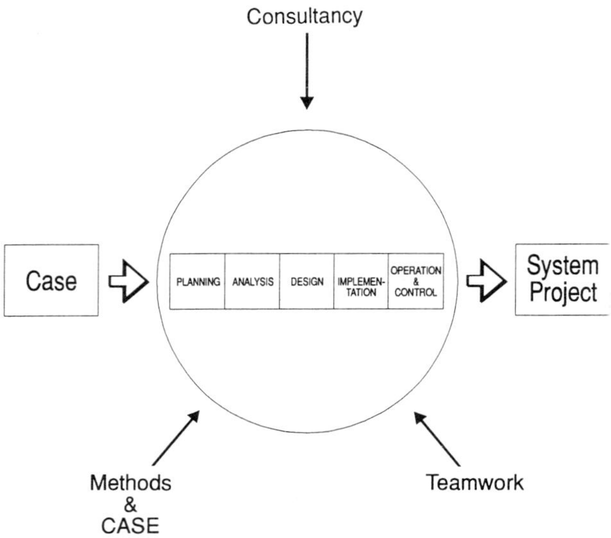

Figure 1. From case statement to project accomplishment.

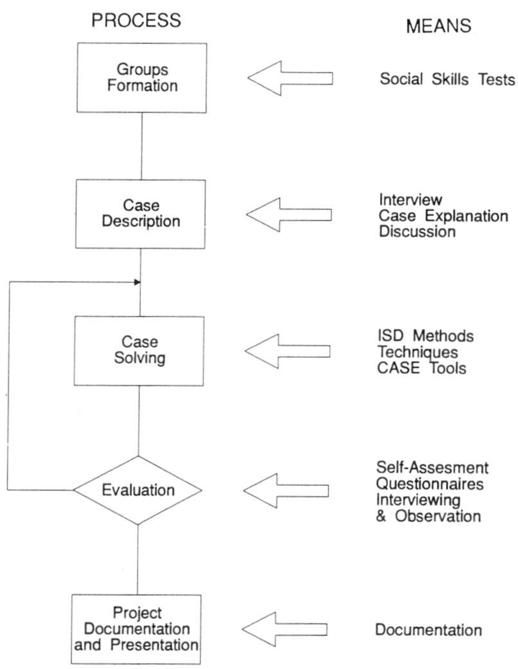

Figure 2. Teamwork of students' group in project realization.

didacticaly sound cases in the area of ISs. They should fulfil the following features:
a) cannot be simple i.e. should create the possibility of alternative solutions,
b) should be suitable for application of wide range of methods and techniques of ISD,
c) should be divident to individual specific tasks possible for solution by members of team,
d) should be only partly structured to stimulate students' creativity.

Students have already participated in lectures on system development process, methods and tools and are well equipped with this type of knowledge. The basic problem

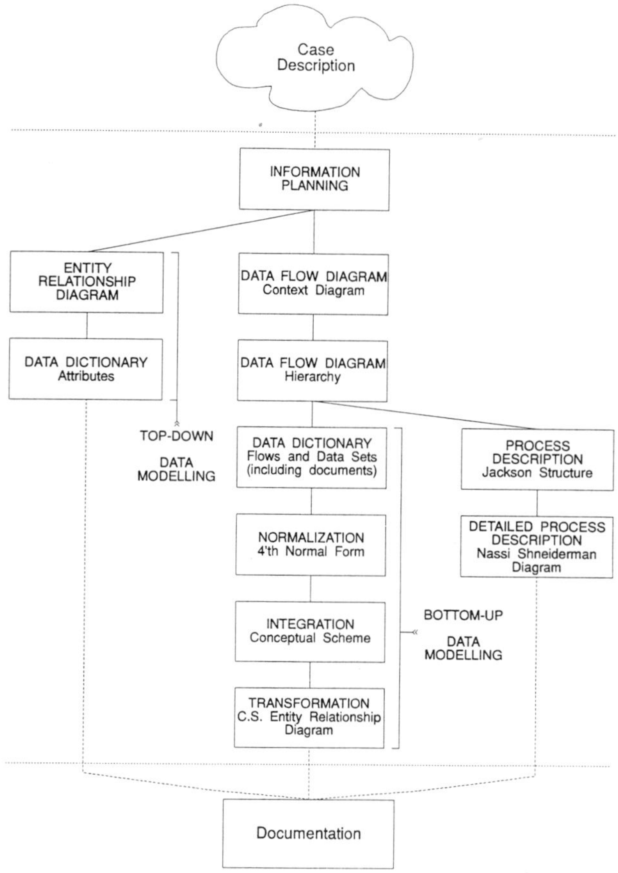

Figure 3. ISD methods & techniques in case solving.

arises around non structuring of the problem. The lack of precise description is a source of long discussions. The questions are asked in order of two succeeding waves.
1. How to deal with the understanding of the idea of the case 2. Analysis of significance of details of case description.

The discussion finally leads to understanding the creative nature of case. After discussion, as well as, question and answer session, the previously formed groups start to solve the case.

The process of case solution follows according to figure 2. Using the knowledge on methods and CASE as well as the trainers consultancy, students start to solve the case in teams. They use methods like Business System Planning (BSP), CSF, CMB, DFD, ERM, Normalization (4NF), Integration, Jackson Diagrams, Structure Charts. The diagram of specific problem solution is presented on figure 3. The phase starts with Information Planning enabling system objective statement by situation analysis techniques like SWOT, CSF, Information Cross i.e. dependencies among process, organizational functions, objects and information systems. This methods allow to specify system architecture and functional decomposition. The leading role is played by Data Flow Diagram. The whole hierarchy of them is starting from initial context diagram. They are the starting platform to both bottom-up data modelling and process description. The data modelling maybe solved in two ways: top-down - starting from general ERM perceived by system's user and bottom-up approach - starting from documents and data analysis (figure 3).

The final step is connected with completion of project documentation and oral presentation by team. After this presentation they are asked for explanation by trainer and

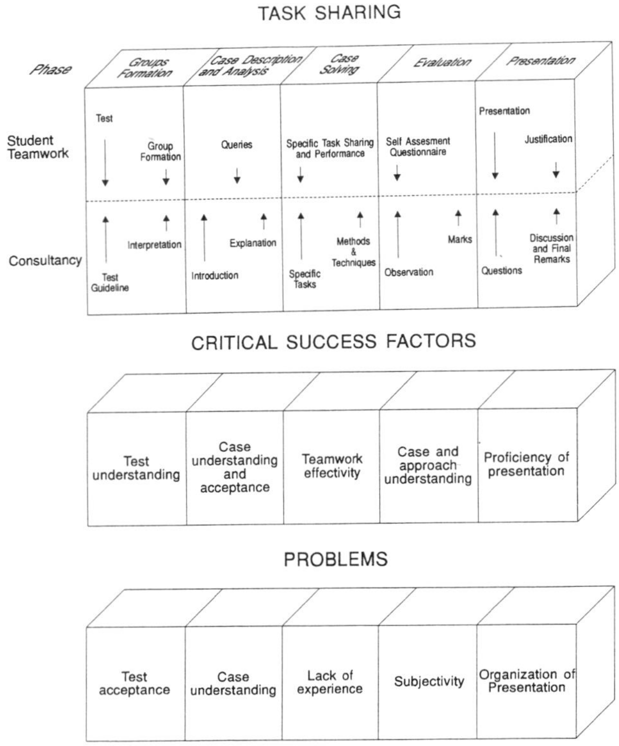

Figure 4. Critical Success Factors (CSF) in systems development education.

students. This presentation has the most significant influence on assesment of the group and individuals.

The synthesis of the main ideas of the approach is presented on figure 4 where in respect of basic steps of system development the Critical Success Factors and essential problems for each step are analysed. The progress of work is under continuous evaluation performed by both trainers and co-workers in teams.

CONCLUSIONS

The essential conclusions of method used are the following:

- the method stimulate students for active work in group and enhance their knowledge on ISD,
- the case gives the chance to work in conditions similar to practical tasks in future work,
- the style of work at the first sight looks as disordered and unstructured, but it finally allows the creative work accepted by students,
- the intensity of work of each individual is higher as everybody is controlled by team.

The presented proposal is under continuous development and improvements according to basic assumptions specified in this paper.

REFERENCES

1. Buckingham R.A. at al., "Information Systems Education: Recommendation and Implementation", Amsterdam, North Holland 1987.
2. Olle T. W. at al., "Information Systems Methodology: a Framework for Understanding", Addison-Wesley, Wokingham, England 1991.
3. Turner W. S. at al., "System Development Methodology", PANDATA, Amsterdam, North-Holland 1988.

USER SATISFACTION AS A MEASURE OF INFORMATION SYSTEM EFFECTIVENESS

James Y.L. Thong, Chee-Sing Yap, and K.S. Raman

Department of Information Systems and Computer Science
National University of Singapore
Republic of Singapore 0511

INTRODUCTION

The subject of information system (IS) effectiveness has been widely reviewed in the IS literature and its importance cannot be understated. There are theoretical, empirical, and practical reasons behind this. Theoretically, the construct of IS effectiveness lies at the centre of all IS models, i.e., all conceptualizations of the nature of IS have embedded in them notions of the nature of effective IS, and the difference that exists between effective and ineffective IS. Empirically, IS effectiveness is the ultimate dependent variable in IS implementation research. Practically, practitioners are always faced with the need to evaluate effectiveness of organizational IS.

In this paper, user satisfaction is examined for its suitability as a measure of IS effectiveness. It is noted that user satisfaction, commonly used as a measure of IS effectiveness in IS implementation research, has come under considerable criticisms in recent years. The criticisms include: (1) questionable operationalizations of the user satisfaction construct; (2) poor theoretical understanding of the user satisfaction construct; and (3) misapplication of user satisfaction instruments. Consequently, there have been mis-specifications of research models and mixed findings in IS research utilizing this variable. Because user satisfaction is widely understood to be an attitude held by the individual users, there is a need to examine the root of the user satisfaction construct, i.e., attitude. Although there is a consensual agreement on a definition of attitude, the definition allows for multiple interpretations. Thus, the user satisfaction construct suffers from the same ambiguity that afflicts the attitude construct. To clarify the attitude construct, we draw on Fishbein and Ajzen's (1975) conceptual framework for the prediction of specific intentions and behaviors. Finally, we examine two recent models of IS behavior grounded in the behavioral sciences and suggest some principles for measuring IS effectiveness.

Systems Science, Edited by F.A. Stowell *et al.*,
Plenum Press, New York, 1993

IS EFFECTIVENESS

Computer-based information systems are getting more pervasive in organizations; automating mundane tasks and aiding decision-making. Information systems are deemed to be effective only if they contribute to organizational effectiveness. But are they improving the effectiveness of the organization? To date, there is still no strong model of organizational effectiveness nor is there agreement on their measurement. Cameron and Whetten (1983) gave three reasons for the existence of multiple models of organizational effectiveness: (1) multiple models are products of multiple, arbitrary models of organizations; (2) the construct space of organizational effectiveness is unknown; and (3) the absence of a consensual criteria for assessing organization effectiveness. They noted that *universalistic propositions linking a set of variables to effectiveness can never be known because the meaning of the dependent variable continually changes.* The relationships among variables may become irrelevant, disappear, increase, or reverse, depending on the model of organizational effectiveness utilized.

Popularly, IS effectiveness is defined as the extent to which a given information system actually contributes to achieving organizational goals, i.e., its effect on organizational performance (Hamilton and Chervany, 1981a). However, there is no consensus among IS researchers on the conceptualization and operationalization of IS effectiveness (DeLone and McLean, 1992; Goodhue, 1992). The difficulties with measuring organizational effectiveness are reflected in the conceptualization and measurement of IS effectiveness. Similarly, it is pointless to search for a precise measure or set of measures of IS effectiveness that will be common across all organizations. In a single organization, the effectiveness criteria can vary with changing value structures of the people involved in evaluation, levels in the organization, and phases in organizational growth (Miller, 1989).

Ideally, the determination of IS effectiveness should utilize an objective measure such as cost-benefit analysis. The net value of the information system to the organization thus equates to the difference between the actual benefits (improved organizational effectiveness) and the cost of the information system development and operation. In practice, it is not that simple. Costs and benefits (especially benefits) of information systems are difficult to quantify. Further, objective assessment of benefits of information systems for decision support often cannot be demonstrated with any certainty. Even if objective data can be determined, they are generally not recorded and thus not available.

In view of the operationalization difficulties with economic analysis of the value of information systems, researchers turn to proxy measures of IS that are more easy to operationalize. One such proxy is system usage which measures the level of usage of the computer system and the computer generated reports. It reflects the degree of confidence the users have in the effectiveness of their information systems. Furthermore, IS objectives are typically stated in terms of the requirements definition which specifies the information to be provided by the system (Hamilton and Chervany, 1981a). These objectives affect organizational performance only through the use process. However, the relationship between system usage and IS effectiveness is complicated by the issue of voluntary versus mandatory use. Users may be dissatisfied with the systems but still use them extensively because of subjective motivations such as mandate from management, political motivation or self-protection for justifying 'poor' decisions (Ives, Olson, and Baroudi, 1983). Alternatively, a technically successful system may result in an unused and therefore unsuccessful system. Hence, the link between system usage and IS effectiveness is by no means simple, and perceptual measures, notably user satisfaction, gained prominence as surrogates.

USER SATISFACTION

Nolan and Seward (1974) argued that user satisfaction is the most important criterion in measuring IS success and failure. User satisfaction is based on the notion that if an information system meets the requirements of the users, the users' satisfaction with the system will increase. Conversely, if the system does not provide the needed information, the users will become dissatisfied. Thus, user satisfaction is a perceptual or subjective measure of IS effectiveness. It can also be an adequate substitute for objective measures of IS effectiveness. A 'good' system perceived by its users as a 'poor' system is a poor system. While it is easiest to examine system usage, success ultimately depends on how well the IS has supported decision-making. Hamilton and Chervany (1981a) believed that while formal evaluation approaches may provide objective measures, informal approaches to capture perceptions of system effectiveness are necessary and helpful to confirm results of formal evaluation approaches. Ives *et al.* (1983) noted that user satisfaction provides a meaningful surrogate for the critical but unmeasurable result of an information system, namely, changes in organizational effectiveness.

User satisfaction has come to be identified with the Bailey and Pearson's (1983) instrument and its derivative: the psychometrically sounder version developed by Ives *et al.* (IOB) (1983). The IOB instrument, especially, is the *de facto* user satisfaction instrument. Recently, a number of researchers have expressed reservation over this instrument and measurement of user satisfaction in general.

On the instrument, Treacy (1985) found that: (1) the variables identified through exploratory factor analysis are labelled in imprecise and ambiguous terms; (2) many of the questions used are poor operationalizations of their theoretical variables; and (3) the instrument fails to achieve discriminant validity. Galletta and Lederer (1986) questioned the test-retest reliability of the instrument while Doll and Torkzadeh (1988) labelled the instrument as outdated in an end-user computing environment. Goodhue (1986) argued that the IOB instrument leaves the IS field with a standard measure, but without a strong theoretical basis. Ives *et al.* have chosen the best of existing user satisfaction measures but as none of the measures chosen has a strong underlying theoretical basis, they did not consider theoretical bases in their evaluation.

Iivari's (1987) criticism of existing user satisfaction research is based on the finding that research thus far has neglected the information system as the antecedent and determinant of user satisfaction. Since different constructs for user satisfaction often include implicitly different assumptions about the above relationship, it is no wonder that there are interpretation problems. These different constructs may wrongly be assumed to commensurate leading to inconsistent findings in IS research. Kim (1989) showed that some IS researchers have misapplied the user satisfaction instruments. A common problem is the inclusion of the same variables as both dependent and independent variables in their studies. Two important consequences of incorrect application of the user satisfaction construct are: (1) mis-specification of research models; and (2) mixed findings in IS research utilizing this variable (Kim, 1989).

Based on a theoretical assessment, Melone (1990) concluded that in the strictest sense, user satisfaction cannot be a surrogate for IS effectiveness because satisfied users alone are not indicative of an effective information system. A weaker justification for employing user satisfaction assumes a strong correlation with IS effectiveness. As the individual's attitude toward an information system is more complex than previously conceptualized, much can be gained from a more theoretical understanding of the user satisfaction construct. Melone noted that there is a fundamental similarity between user satisfaction and the social and cognitive psychologists' notion of an attitude. Hence, there is a need to understand how attitude is related to behavior, to gain a clearer conceptualization of user satisfaction.

ATTITUDE

Attitude, the most distinctive and indispensable concept in IS effectiveness evaluation, is characterized by a great deal of ambiguity and confusion. Attitude can be described as a learned predisposition to respond in a consistently favourable or unfavourable manner with respect to a given object (Fishbein and Azjen, 1975). But this definition allows for multiple interpretations and gives no clear directions as to how attitudes are to be measured, resulting in arbitrary selection of measurement procedures. Fishbein and Ajzen (1975) reviewed the contemporary attitude theories and identified the basic constructs and interrelationships in terms of a conceptual framework of beliefs, attitudes, intentions, and behavior (theory of reasoned action). They found that the theories differed in terms of conceptualization and operationalization in empirical research. Many researchers fail to distinguish between the different conceptual variables at the operational level. Although a theory's conceptual variable may be attitude, the operationalization may be assessing beliefs, intentions, or even behavior. The situation is not improved when some theories do not specify exactly the cognitive elements of relevance and leave the decision to the researcher's intuition. Hence, the difficulties in operationalizing the theories in behavioral science carry over to the IS field.

The theory of reasoned action has been applied successfully in the field of psychology and marketing. In IS, Hill, Smith, and Mann (1987) confirmed the predicted path between beliefs, attitudes, intentions, and behaviors. Collins and Mann (1988) provided evidence that usage of a database is influenced by the individual beliefs and group norms. Robertson (1989) related the user's behavior (usage of the IS) directly to the user's attention to the social influences and demands of those around him or her, and indirectly to the social system's influence on how the user interprets the information systems available. Hence, there is evidence that the theory of reasoned action can clarify the ambiguity that afflicts the user satisfaction measure. Two recent models of IS behavior which utilized this theory are reviewed below.

RECENT MODELS OF IS BEHAVIOR

Using job satisfaction research as a theoretical base, Goodhue (1986) proposed a model of IS attitudes, beliefs, and individual performance or what he termed IS satisfactoriness. It differs from previous conceptions of IS attitudes or satisfaction in two ways. First, the model suggests that the correspondence or fit between task requirements and IS functionality determines individual performance. IS satisfaction is an attitude held by the individual about the subjective characteristics of the IS; it is measured by the individual's feelings about the subjective aspects of the IS. On the other hand, IS satisfactoriness is the individual's belief about the objective fit between task requirements and IS functionality, mediated by his or her abilities. In other words, IS satisfactoriness is the degree to which an IS environment assists the individual in his or her performance of job-related tasks. This emphasis on task implies that IS functionality only has value when applied to tasks. Second, a distinction between beliefs and attitudes is advocated. The model's focus on beliefs suggests we should ask individuals not whether the IS has accurate data as the individual may not be able to give an objective answer, but whether the IS provides data accurate enough for his or her work. Similarly, we should ask not whether the data are accessible, but whether they are accessible enough for the individual's work. Goodhue (1992) expanded the model to include social norms and habit in a system-to-performance chain.

Another model of IS behavior was proposed by Miller (1989). Miller's model suggests that user perceptions about the fit between business needs and IS capabilities shape attitudes toward IS and that these attitudes influence usage and subsequently value to the organization. The model focuses on the user and suggests that process and economic outcomes of IS result from the user's behaviors such as usage of IS facilities. These behaviors are outcomes of the user's intentions to behave. Intervening variables such as poor timing or changes in task requirements may exist to obstruct behaviors. The intention to behave is shaped by personal feelings toward the behavior and social pressures for or against that behavior. Personal attitudes are the result of cognitive beliefs (expectations) regarding the behaviors in question and subjective norms formed as a result of beliefs or expectations about the views of important others (managers, supervisors, work groups, IS department). The individual's beliefs about outcomes are influenced by his or her evaluation of the correspondence between task needs and IS capabilities. If a large discrepancy is perceived, expectation of success is lowered leading to negative attitude and a disinclination to perform the behavior. Conversely, perception of a close fit between task needs and IS capabilities will encourage the behavior. Likewise, notions of fit between business needs and IS functionality within the referent group will determine the performance or non-performance of the behavior.

CONCLUSION

From the discussion, it is noted that IS effectiveness is a complex variable with much theoretical and operationalization issues still to be resolved. This conceptual difficulty with IS effectiveness has been with us for a long time and will continue to be with us in the foreseeable future. Nonetheless, this review of the relevant literature indicates that there are several principles that should be considered in measurement of IS effectiveness.

First, user satisfaction can be a strong measure of IS effectiveness. However, there is a need to focus on fit between task requirements and IS functionality, and to make a distinction between beliefs and attitudes. Theorizing and empirical research into user satisfaction have led to a focus on attitudes. Previous conceptualization and measurement of user satisfaction via attitudes are imprecise and mix cognitive and affective mental processes. While satisfaction might be best determined by measuring strictly attitudes, the correspondence between task requirements and IS functionality is best determined by measuring beliefs. The fits between specific task needs and perceived IS capabilities by the individual, and between business needs and IS functionality of important others, shape the individual's attitudes which in turn determine intention to behave resulting in the intended behavior.

Second, evaluation of IS effectiveness will require multiple evaluation approaches, inclusive of both objective and subjective measures, to satisfy differing evaluation objectives and requirements. This is because a number of stakeholders possess vested interests in an implemented IS. Hamilton and Chervany (1981b) believed that validated user satisfaction instruments can be administered to the different stakeholders in an IS, to incorporate multiple evaluator viewpoints on IS effectiveness. To the extent that user satisfaction could be linked with output-oriented criteria (i.e. objectively measured user behaviors), measurement of IS effectiveness would be improved (Melone, 1990).

Third, researchers must make explicit the underlying assumptions of the IS effectiveness measure in their studies. The respondents and items chosen should match the purpose of the research, which may be an assessment of the total IS function, a type of IS, or a specific IS. Respondents could include users, managers, or IS staff while the items chosen must reflect the domain of interest. The domain may be end-user computing, microcomputers, mainframe systems or IS department. This principle is reflected in the in-context assessment approach and the multivalent approach (Bjorn-Anderson and Davis, 1988).

Finally, future IS effectiveness research should be based on stronger theories, possibly from the behavioral sciences, and hopefully more consistent empirical findings will result, allowing for an accumulation of knowledge. Goodhue's (1986; 1992) and Miller's (1989) models are possible starting points.

REFERENCES

Bailey, J.E. and Pearson, S.W., 1983, Development of a tool for measuring and analyzing computer user satisfaction, *Management Science*, 29(5), pp530-545.

Bjorn-Anderson, N. and Davis, G.B. (eds.), 1988, "Information Systems Assessment: Issues and Challenges," Elsevier Science Publishers B.V., Amsterdam.

Cameron, K.S. and Whetten, D.A., 1983, "Organizational Effectiveness: A Comparison of Multiple Models," Academic Press, New York.

Collins, F. and Mann, G.J., 1988, Change-related behavior and information systems, *OMEGA*, 16(5), pp369-381.

DeLone, W.H. and McLean, E.R., 1992, Information system success: the quest for the dependent variable, *Information Systems Research*, 3(1), pp60-95.

Doll, W.J. and Torkzadeh, G., 1988, The measurement of end-user computing satisfaction, *MIS Quarterly*, 12(2), pp259-273.

Fishbein, M. and Ajzen, I., 1975, "Belief, Attitude, Intention and Behaviour: An Introduction to Theory and Research," Addison-Wesley Publishing Co., Massachusetts.

Galletta, D.F. and Lederer, A.L., 1986, Some cautions on the measurement of user information satisfaction, Working Paper #643, Graduate School of Business, University of Pittsburgh.

Goodhue, D., 1986, IS attitudes: towards theoretical definition and measurement clarity, *Proceedings of Seventh International Conference on Information Systems*, San Diego, California, December 15-17, pp181-194.

Goodhue, D., 1992, User evaluations of MIS success: what are we really measuring? *Proceedings of Twenty-Fifth Hawaii International Conference on System Sciences*, Vol IV, pp303-314.

Hamilton, S. and Chervany, N.L., 1981a, Evaluating information system effectiveness part I: comparing evaluation approaches, *MIS Quarterly*, 5(3), pp55-69.

Hamilton, S. and Chervany, N.L., 1981b, Evaluating information system effectiveness part II: comparing evaluators viewpoints, *MIS Quarterly*, 5(4), pp79-86.

Hill, T., Smith, N.D., and Mann, M.F., 1987, Role of efficacy expectations in predicting the decision to use advanced technologies: the case of computers, *Journal of Applied Psychology*, 72(2), pp307-313.

Iivari, J., 1987, User information satisfaction (UIS) reconsidered: an information system as the antecedent of UIS, *Proceedings of Eighth International Conference on Information Systems*, Pittsburg, Pennslyvania, December 6-9, pp57-73.

Ives, B., Olson, M.H., and Baroudi, J.J., 1983, The measurement of user information satisfaction, *Communications of the ACM*, 26(10), pp785-793.

Kim, K.K., 1989, User satisfaction: a synthesis of three different perspectives, *Journal of Information Systems*, 4(1), pp1-12.

Melone, N.P., 1990, A theoretical assessment of the user satisfaction construct in information systems research, *Management Science*, 36(1), pp76-91.

Miller, J., 1989, Information systems effectiveness: the fit between business needs and system capabilities, *Proceedings of Tenth International Conference on Information Systems*, Boston, Massachusetts, December 4-6, pp273-288.

Nolan, R.L. and Seward, H., 1974, Measuring user satisfaction to evaluate information systems, *in*: "Managing the Data Resource Function," R.L. Nolan, ed., West Publishing Co., Los Angeles.

Robertson, D.C., 1989, Social determinants of information systems use, *Journal of Management Information Systems*, 5(4), pp55-71.

Treacy, M.E., 1985, An empirical examination of a causal model of user information satisfaction, *Proceedings of Sixth International Conference on Information Systems*, Indianapolis, December 16-18, pp285-287.

APPLICATION OF INFORMATION TECHNOLOGY IN SMALL BUSINESS IN SLOVENIA

Borut Verber, Jože Zupančič

University of Maribor, School of Organisational Sciences
64000 Kranj, Prešernova 11
Slovenia

INTRODUCTION

Small businesses, an essential part of the economy for United States, West European and newly industrialised Pacific countries, constitute majority of business organisations and contribute an ever increasing proportion of economic growth. For example, in the United States they account for about 97% of all businesses, contribute 45% of the Gross National Product and represent 57% of the employment (Ibrahim and Goodwin, 1986; Park, 1990). In Singapore they represent 90% of all registered companies and employ 44% of the workforce (Grant and Raman, 1992). In Slovenia, and in other East European countries, the proportion of small businesses is considerable smaller but is growing fast, because of corporate restructuring or founding of new firms. Small business research often defines small firms differently. Criteria depending on the nature of the business and industry, also vary. Most authors use the total number of employees as a criterion for research analysis. Sometimes this is combined with such financial indicators as total annual sales, operating capital or total assets. Employee criteria range from 15 (Ibrahim and Goodwin, 1986) to 50 (Schleich et al., 1990) for a service sector, to as much as 500 for manufacturing companies (Park, 1990; Yeh et al., 1990). In this study, small business is defined as a company employing 4 to 100 employees.

Remarkable progress in information technology (IT) with a decline in prices reduced the cost of computing systems allowing small businesses to use computers for information processing. This preliminary research analyses the use of IT by small businesses in Slovenia. Level of use, characteristics of computerisation and computer usage were surveyed. The survey showed that small businesses in Slovenia are relatively well equipped with hardware and software, and use various application packages. A prevalence of stand-alone PCs indicates that information systems are in an early stage of maturity for most of companies. It also indicates underutilisation of available technology to support business operations. Differences between IT use for manufacturing and service firms were analysed. Responses to types of applications showed that IT is usually used to automate routine tasks or to improve the availability of data. Chief executives direct involvement in data processing operations strongly impacts on the acceptance and use of IT. Findings about the relatively

Systems Science, Edited by F.A. Stowell *et al.*,
Plenum Press, New York, 1993

large amount of available technology, the widespread use of computers and end-user involvement in small business warrants more research. This paper offers some suggestions for further research topics.

SPECIFICS OF INFORMATION MANAGEMENT IN SMALL COMPANIES

The small firm characterised by a simple highly centralised structure, often is "resource poor" in terms of human, financial and material resources and faces greater environmental uncertainty since it has less control of external factors (Raymond, 1990). The small firm's decision-cycle or time-frame is generally short-term with a reactive rather than a proactive approach. There is also less use of formal management techniques such as capital budgeting, project management, inventory control or financial analysis. Decision-making for managers of small businesses - who may also be owners - is done more intuitively than by relying on formal information. Small businesses seldom use organisational information system (IS) as a communication medium between layers but use computers more as tools for specific tasks (Heikkila at el., 1991). Small business managers as well as other personnel tend to have less computer experience and training and are less demanding of the computing functions.

The IS function is in early stage of maturity in most small enterprises (Raymond, 1987). It generally supports accounting functions. Transaction processing applications such as: payroll, inventory control, billing, etc. which produce documents and reports of an operational nature dominate. These applications usually are purchased as pre-programmed software packages. Small corporate databases with few transactions require small, simple computer systems. Therefore, the IS in small firms is usually done on microcomputers.

Small companies usually cannot afford their own IS professionals, so they rely on technical and IS expertise from outside the organisation. They are frequently disappointed with the software because it is not adapted to the needs of the company and/or is difficult to use (Heikkila et al., 1991). On the other hand, user friendly software tools such as spreadsheets, database tools and 4GLs, provide greater opportunity for users in small firms to develop tailor-made applications based on interactive data storage, interrogation and analysis capability of these tools -- especially in areas where information needs are specific and packages are less available. This activity is known as end-user computing (EUC): the process in which computer users develop and operate their own applications instead of the traditional process where system development is controlled by computer personnel from the organisational computing function (Yeh et al., 1990).

Computerisation, if properly planned and managed, can contribute to small business operations but has high risk (DeLone, 1988). Lack of computer knowledge and experience on the part of the manager/owner can result in expensive mistakes for data processing. End users, usually not trained in system development methods, develop their own applications with little attention to formal system analysis or design rules (Schleich et al., 1990). Most users do not rigorously specify how data will be stored and manipulated; nor are they sophisticated in the design of data structures and user interfaces (Kettelhut, 1991). As a result, user-developed applications frequently do not function correctly and are difficult to maintain and upgrade.

RESEARCH APPROACH AND SUMMARY PROFILE OF RESPONDENTS

This study explores the level and characteristics of computerisation and computer use by small companies in Slovenia. As mentioned earlier, the definition of a small company for

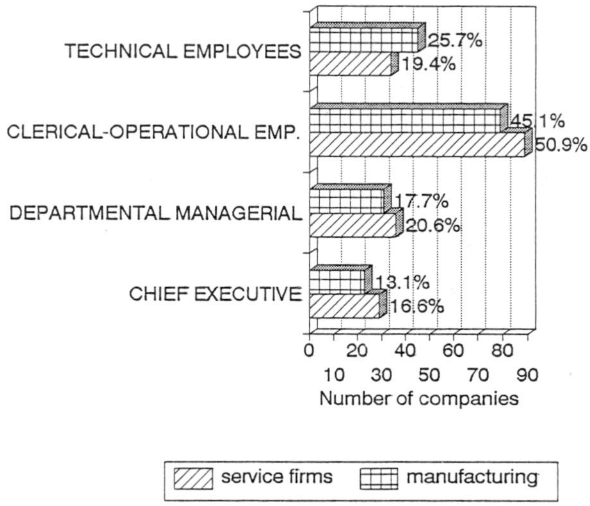

Figure 1. Who uses computers in small companies.

this research is 4-100 employees. A four-page questionnaire, which included five open-ended questions, was Mailed to 500 managers of randomly chosen small companies in Slovenia. Two hundred-twenty usable questionnaires were returned. From the total 220, 175 came from companies which use computers with 97 (55.4%) from manufacturing firms and 78 (44.9%) from service firms. Since the non-return of questionnaires was not investigated this percentage cannot represent computer-user percentages for entire small company population. Among the 45 (20.5%) respondents who do not use computer data processing 23 (51.1%) plan to implement it in one year. Average number of personnel using computers is 39.3 (47.6 in manufacturing and 31.9 in service firms). Average length of computer experience is 4.1 years (4.3 years in manufacturing and 3.9 in service firms). A breakdown of the survey sample by years of computer usage show that 27.4% of the firms have less than 1 year experience, 37.1% have 1 to 3 years experience, 27.4% have 4 to 10 years experience and 8% have more than 10 years experience. In firms with computer technology, 63.4% employ less workers then three years ago while firms 41.8% of firms without computer technology employ less employees than 3 years ago. However other factors such as redundant lay-off, retirement, etc., need to be considered in this finding.

FINDINGS

Mainframes or minicomputers, together with personal computers (PCs), are used in 29 (16.6%) of surveyed firms while 28 (16.0%) use rented terminals. Only one company (0.6%) has no PCs computers, but 144 (82.3%) use only PCs. PCs are connected into local area networks (LAN) in 38 companies (21.7%). The average number of employees per PC is 7.8 with 4.5 per PC or terminal. Companies where the chief executive (director) uses a computer have an average of one PC per 5.3 employees and one PC or terminal per 3.4 employees. In manufacturing companies the average number of employees per PC is 10.0 with 6.2 per PC or terminal, bat 5.9 per PC and 3.6 per PC or terminal in service companies. Figure 1 which presents usage of computers by different groups of employees indicates that only a small portion of chief executives use computers.

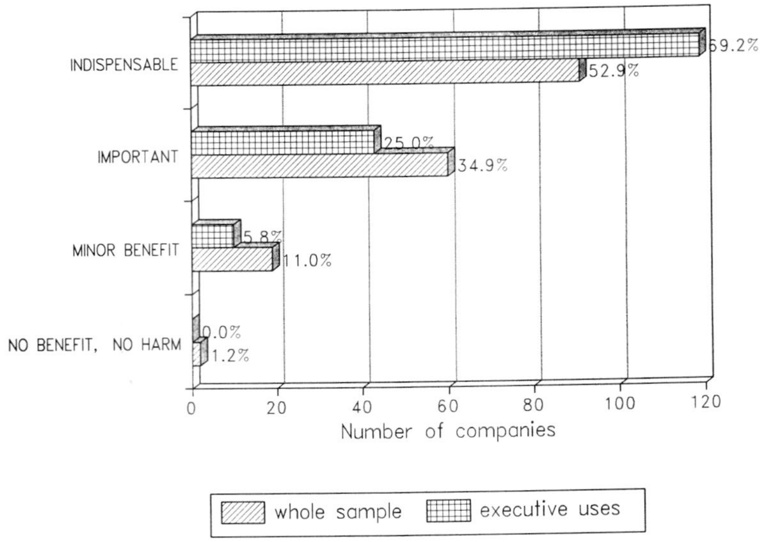

Figure 2. The role of data processing for the firm.

Companies where the chief executive personally uses computers rate its importance higher, but the duration of computer usage seems to have no significant impact on the perceived importance of computerised data processing. Managers estimate of the role of data processing is shown in Figure 2.

The most widely used tools are word processors and data base tools followed by spreadsheets. Only a relatively small portion of firms use additional tools.

Only 19 (10.9%) of firms use no software tools or did not respond this question, 73 (41.7%) use one or two tools, 3 or 4 tools are used by 68 (38.9%) firms, and 15 (8.6%) use more than 4 tools. Most companies have some experience with software tools for in-house development, which can be referred as end-user computing. Use of software tools by the surveyed companies is shown in Figure 3.

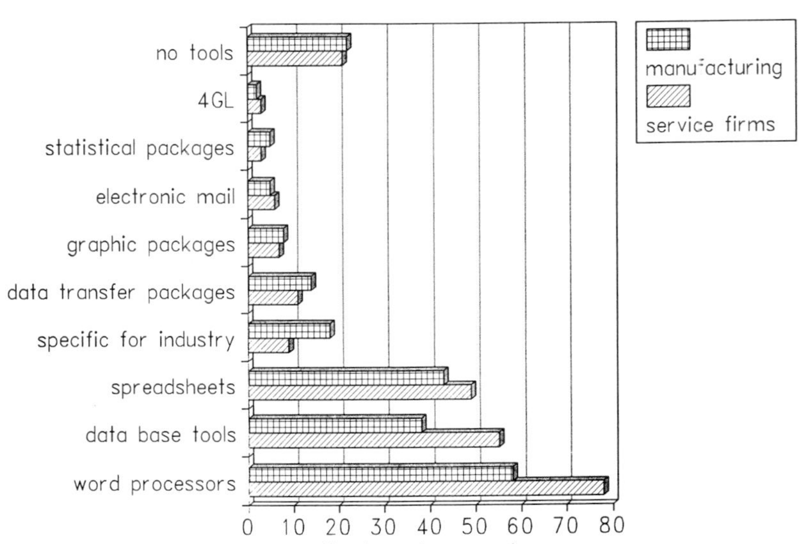

Figure 3. Use of software tools.

Table 1 presents application types and sources of software. Only 4 (2.3%) use no application programs or did not respond this question. One or two applications are used by 14 (8%) of respondents, three or four by 20 (11.4%) and 137 (78.3%) respondents use more than four applications. A relatively large portion of in-house developed applications indicates the presence of a considerable amount of end-user computing.

Table 1. Application types and sources of software

	NUMBER OF FIRMS				
Application types	Purchased package	In-house development	Customer program	Rented	Total
1. General ledger	85	9	54	4	152
2. Payable/receivable	69	13	53	3	138
3. Payroll	63	15	52	4	134
4. Assets	68	12	44	3	127
5. Billing	42	22	52	2	118
6. Specific for industry	44	32	30	0	106
7. Materials management	39	13	40	2	94
8. Inventory management	31	14	38	1	84
9. Sales	23	13	28	1	65
10. Personnel management	24	18	21	2	65
11. Purchasing	18	14	22	1	55
12. Work orders	10	15	19	1	45
13. Cash management	19	8	13	2	42

DISCUSSION

This study shows that small companies in Slovenia are relatively well equipped with information technology (IT). Findings show the average number of employees per PC or terminal is 4.8, while in larger companies it is 16 (Zupancic and Leskovar, 1991). Different research approaches and different definitions for a small company in other research make comparisons to small companies in other countries difficult. There appears to be no significant difference in the number of software tools and application packages used by small firms (DeLone, 1988) with the level of computerisation (Grant and Raman, 1990; Yeh et al., 1990). Results show that there is a difference in amount of computer use in manufacturing versus service firms. Prevalence of only stand-alone PCs in most firms indicates that the information system function is in early stages. It also indicates under-utilisation of available technology for supporting business operations. Types of applications (Table 1) and also responses to open ended questions show that IT is used usually to automate routine tasks, reduce the amount of manual work and improve the availability of data. In most cases, it is not used to support the decision-making process of managers. There are two groups of end-users: 52.6% are passive users who use only application packages and do no programming; while 47.4% are active users who develop their own programs and solutions. These two groups demand a different support environment.

Responses to open-ended questions show that small firms feel a need for external support in the following areas:

- maintenance of existing hardware and software
- custom development of application software
- education and training especially new IT application
- consulting support in planning and implementing corporate IS

Insufficient access to and quality of these services, as well as limited financial resources restrict IT usage in small companies. One possible solution for this need could be a program to support small companies in computerising and upgrading IS, similar to the Small Enterprise Computerisation Program in Singapore as described in Grant and Raman (1990). The primary goal is to encourage and to assist small firms by providing funds and appropriate technical support for determining computer requirements and designing computer and information infrastructure. The chief executive's knowledge and direct involvement in data processing operations has strong impact on the adoption and use of computer technology as shown in Figure 4. This agrees with the findings of other research (DeLone, 1988).

CONCLUSION

This preliminary investigation shows that small companies in Slovenia are relatively well equipped with computer technology. No similar research from other East or Central European countries are known to the authors. Therefore, it is uncertain if the findings can be generalised for these countries. The amount of available IT, the extent of computer usage and end-user computing in small companies encourage further and more detailed research in this area. Authors suggest the following issues for more detailed research and analysis:

-success factors for computerised IS in small companies
-training needs and outside consulting support
-success factors for implementing packaged software
-development of customer software
-implementation of end-user computing to increase number of active users
-justification and plan for a state-sponsored support program
 to encourage successful adoption of contemporary IT in small companies
-relationship of computer usage and decrease in number of employees

Further work will address some of these questions and results will be compared with similar research from other countries.

REFERENCES

DeLone, W.H., 1988, Determinants of success for computer usage in small business, *MIS Quarterly*, 12:51.

Grant, G.G. and Raman, K.S., 1992, Government initiatives for IT adoption in small business - experiences of the Singapore small enterprise computerisation program, *International Information Systems*, 1:69.

Heikkila, J., Saarinen T. and Saaksjarvi M., 1991, Success of software packages in small business: an exploratory study, *European Journal of Information Systems*, 1:159.

Ibrahim, A.B. and Goodwin, J.R., 1986, Perceived causes of success in small business. *American Journal of Small Business*, 11:41.

Kettelhut, M.C., 1991, Don't let users develop applications without systems analysis, *Journal of Systems Management*, July 1991:23.

Park, S.W.,1990, "The Characteristics and Usage of Computerised Information Systems in Small Apparel and Textile Industry", Dissertation, Georgia State University.

Raymond, L., 1987, The presence of end-user computing in small Businesses: an exploratory investigation of its distinguishing organisational and information systems context, *INFOR*, 25:198.

Raymond, L., 1990, End-user computing in the small business context: foundations and directions for research, *Data Base*, Winter 90:20.

Schleich, J.F., Corney, W.J. and Warren, J.B., 1990, Pitfalls in system implementation in small businesses, *Journal of Systems Management*, June 1990:7.

Yeh, H., Hsieh, C. and Lu, M., 1990, Implementing end-user computing in smaller firms, *Industrial Management and Data Systems*, 90,6:3.

Zupancic, J. and Leskovar, R., 1991, An analysis of key issues in information system development, Proceedings of the 13th International Conference "Information Technology Interface" ITI-1991, V. Ceric, V.Luzar, V.Dobric, and P.Ray, ed., University of Zagreb.

END-USER COMPUTING ENVIRONMENT: A CASE STUDY

Jože Zupančič

University of Maribor, School of Organisational Sciences
64000 Kranj, Slovenia

INTRODUCTION

End-user computing (EUC) is direct hands-on use and development of Information Systems (IS) by principal users of the system's outputs or by their immediate support staff (Werherbe, 1988). The rapid growth of EUC has dramatically affected the nature of information systems in many countries. Traditional IS development methods have been overwhelmed resulting in long lead times and application backlog. User needs far exceed what IS departments can deliver. Recent advances both in hardware and in software alter the economics of computing. Networks of powerful PCs gradually replace mainframes. Decentralisation of organisation business units causes a decentralisation of computing power. The MIS departments have less involvement and less control over purchasing decisions for individual departments. Many organisations implement a strategy that shifts more of the responsibility of system development away from the IS Department toward the end-user. On the other hand, the IS staff often must support an increasing range of hardware and software and offer appropriate training for end-users (Rogers, 1992).

This paper presents an overview of the end-user computing activities for companies in Slovenia and the role of EUC in corporate information systems. The EUC support environment for a large manufacturing company in Slovenia is investigated in some detail. An in depth analysis of end-user needs shows reasons for their involvement in IS development activities, ways they acquired the computing knowledge, sources of end-user support, and barriers to wider and more effective personal use of computers. The results are compared with trends in developed countries.

CHARACTERISTICS OF EUC IN SLOVENIA

The shift in emphasis from traditional computing practices toward EUC is not limited to highly advanced countries (Igbaria, 1992). Actually, the trend may be more evident in developing nations where the majority of organisations are relatively small to medium sized. These organisations cannot afford expensive hardware and supportive technical staff such as system analysts and programmers. Reality forces professionals and managers to satisfy their computational needs alone with inexpensive hardware (Abdul-Gader, 1990). Many

Systems Science, Edited by F.A. Stowell *et al.*,
Plenum Press, New York, 1993

organisations using computers do not have formal IS Departments depending on end-user computing and on microcomputer based application packages.

A survey of a randomly chosen sample of large and medium-sized companies in Slovenia as shown in Zupancic and Leskovar (1991) indicates information technology usages and the proliferation of EUC. The companies surveyed have an average of one terminal or PC per 16 employees; the estimated number of employees who are not IS professionals but use a computer or a terminal is 9% of total work force. This is relatively low in comparison to a United States study where Ein-Dor and Segev (1988) found that almost 40% of work force were end-users. The survey shows for end-users in Slovenia, 85% use a terminal or PC only on menu level (non-programming end-users), 9% use the command level for ad hoc query and reporting, and 6% use programming to develop applications for professional work, using such tools as spreadsheets, data base tools or 4GLs. Some kind of end-user support environment (information centre) exists in 38% of the organisations. If the IS Departments developed all the applications that were developed by end-users at the time of the survey their average workload would be increased 10.2%. The average application backlog, as estimated by IS Department managers in the companies surveyed, was 22.4 months.

Research of computing practices in small businesses in Slovenia shows a reasonable level of information technology with the average number of employees per PC or terminal at 4.8 (Verber and Zupancic, 1992). The study also shows that there is a substantial proliferation of end-user computing in small companies compared to large ones. On the other hand, small companies mostly use inexpensive hardware and packaged PC software to automate routine tasks and improve availability of data. Usually computers are not used to improve the decision-making process for managers.

A recent study of 18 companies in Slovenia shows the relationship between organisational factors and efficiency of personal use of computers (Nagel, 1992). The study shows that effectiveness of EUC correlates with organisational processes such as definition of EUC policy, planning, support and control. Centralisation of the purchasing decisions as well as formalisation and organisational complexity of personal use of computers seem not to be related to end-user satisfaction but positively correlate to quality and extent of computer usage. End user satisfaction associates strongly with the portion of employees in the firm that use computers.

CASE STUDY

The subject of this case study is a manufacturing firm with a total work force of 4200. The firm began computerised data processing in 1968. Current computing facilities include a mainframe with about 200 terminals and about 170 PCs with 150 of these available to end-users. A purchase of 70 new PCs is planned for the near future. Only about 30 PCs are linked to the mainframe, and 21 networked in four LANs.

The need for a support environment for end-users became obvious in 1987 by the purchase of PCs by the end-users. They quickly recognised that just purchasing of microcomputers didn't solve their information problems. They accepted the PCs as tools but then demanded various applications from the IS Department. Because the IS staff could not meet these demands quickly some users began to develop needed applications using diverse approaches and various tools. Others became discouraged and stopped using computers. The IS department quickly realised these problems and allocated one IS professional, who was also offered additional training in microcomputer software and hardware, to support end-user microcomputer problems. The support included designing and teaching an educational programme for end-users, assisting the users in planning and developing applications, selecting products, transferring data, etc.

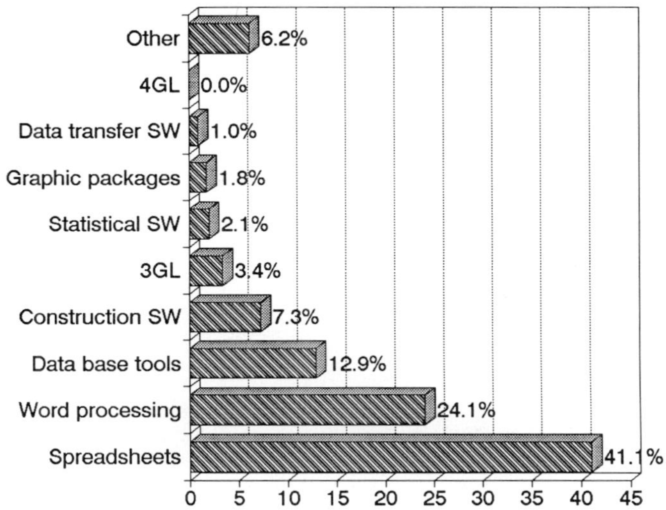

Figure 1. Percentage of time spent using individual tools.

In 1990, the IS Department proposed an end-user support organisation based on the information centre concept. In spite of the well-prepared plan for implementation of the information centre within the IS Department, the plan was only partially realised due to scarce financial resources. As the number of PCs continued to increase so did the demand for end-user support. Some of the most enthusiastic end-users were offered training courses and became responsible to organise support by departments; users who were given additional education, assisted other users by teaching short courses focusing on use of specific tools or application packages. The initial IS professional is responsible for organising internal courses, preparing or translating of manuals, selecting of information technology, advising end-users how to analyse their needs, etc. A second IS professional is responsible for technical problems, hardware installation and servicing.

RESULTS OF THE SURVEY

Managers and professionals from all business functions who had microcomputers on their desks or had easy access to a microcomputer in their work place, and used them to develop their own applications received a questionnaire. Rudez (1992) gives detailed results of the survey. From a total 80, 59 (74%) usable questionnaires were returned. The average age of respondents was 34.5 years, the average education level was relatively high with 73% having at least an associate degree. Respondents used computers to develop applications an average of 3.8 years (range 0.5 - 9 years). They spend an average of 16.7 hours a week using microcomputers. Figure 1 shows the frequency of use for individual software tools and programming languages. Results compare favourably to those presented in (Nagel, 1992; Igbaria, 1992).

Respondents were asked to rank reasons why they use PCs and major usage problems. Necessity to perform their job effectively and efficiency ranked first, followed by faster development of the needed applications. Among the major problems in using computers was lack of time for application development.

Next, respondents estimated the amount of contributing sources of knowledge, needed for efficient use of microcomputers. Figure 2 presents the results. Relatively low contribution

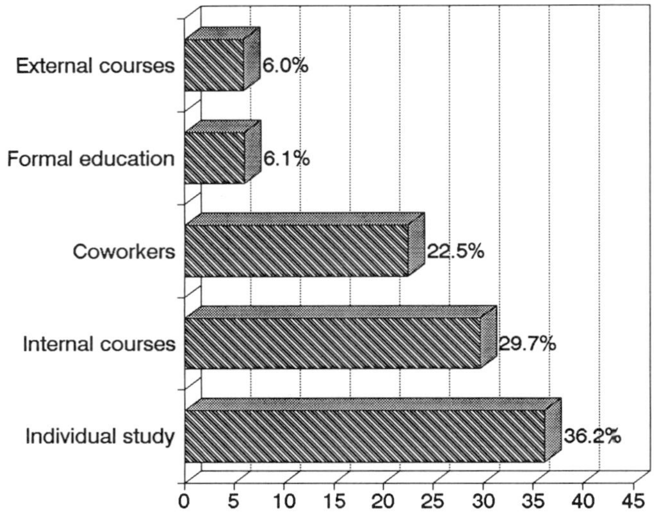

Figure 2. Estimated percentages of knowledge obtained from various sources.

of formal sources, such as internal and external courses indicates the need for more intensive and organised end-user training.

Figure 3 presents various support sources. Support from outside the company designates informal assistance by friends and acquaintances. The relatively small portion of end-user support service indicates the need for a better organised support environment available to wide range of users.

Respondents also rated the importance of particular services by choosing and ranking 7 among the 14 listed forms of end-user support. Table 1 presents the results. The number of points was derived by rating the first ranked choice by 7 points, the second ranked by 6 points, and so on.

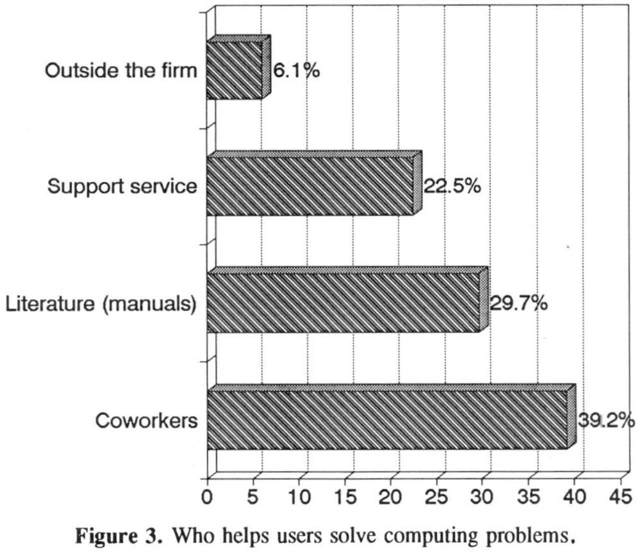

Figure 3. Who helps users solve computing problems.

Table 1. Ranking of end-user support services by importance

Rank	SERVICE	No.of points
1	Assistance in selecting application packages	182
2	End-user education and training	173
3	Evaluation and demonstration of software	149
4	Assistance in analysing and designing applications	148
5	Assistance in solving technical (hardware) problems	132
6	Hardware installation and maintenance	124
7	Assistance in developing applications	121
8	Assistance in selection of hardware	116
9	Assistance in choosing software for specific applcations	109
10	Evaluation and demonstration of hardware	101
11	Hot line/ technical support, debugging	90
12	Assistance in problem specification	75
13	Maintaining software library	48
14	Assisting application maintenance	46

To identify barriers to further progress of UEC, respondents rated 5 reasons, given in the questionnaire.

Figure 4 presents results and indicates the need for a better management of EUC activities. The first rank for scarce hardware resources contradicts with the results of recent investigation of key issues in information management by Dekleva and Zupancic (1992) who found financial resources were not considered by IS department managers as a major barrier for development of IS. Insufficient hardware resources may be related to the departmental attitude or lack of management support: when departmental managers do not believe in the benefits of EUC they cannot justify the purchases of information technology for end-users. In addition to insufficient basic and specific knowledge in the IS area poor understanding

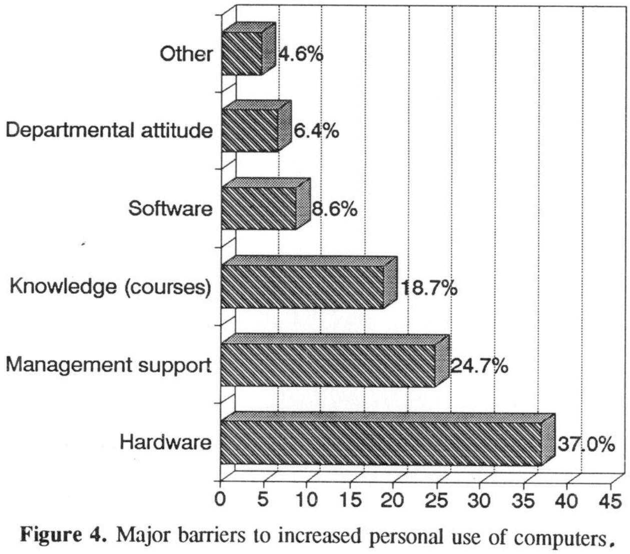

Figure 4. Major barriers to increased personal use of computers.

of the language of software tools and packages, usually English, makes it difficult for users to read manuals and related literature.

CONCLUSION

For the company studied end-user computing is no more a marginal activity but is becoming an increasingly important component in corporate IS development. In the future, further proliferation of personal use of computers through the organisation is expected. New employees learn to use computers during their formal education and accept computers as tools. Level of end-user computing knowledge and sophistication of user developed application continually increases. Thus, the demand for end-user support and consulting increases. Existing support based on a mutual assistance for end-users is not adequate any more. More and more advanced software tools require substantial learning and additional time. The company plans to connect more and more of the PCs and workstations to LANs which will enable some mission critical applications to move away from the mainframe and will shift more responsibility for systems development to the end-user. To remove the risk of the disintegration of corporate IS, a more sophisticated support environment based on the information centre concept is necessary.

The relatively large amount of computing skills that users obtained from informal sources, such as aid from their coworkers and individual study of literature and manuals, demonstrates a need for a more complete, better-organised more intensive training programme, which is available to a wider range of end-users. This will minimise time needed to learn to use new tools and methods. Need for more effective training is evident also from the following comment by one of the respondents: "Education and training don't stop when the course is finished; in contrary, active learning starts after the course. This requires, besides enthusiasm and knowledge of the foreign language (English), a lot of time. Preoccupation with regular work doesn't permit learning during work hours, and there is not sufficient motivation for individual learning after regular work time."

REFERENCES

Abdul-Gader, A.H., 1990, Computing success factors: further evidence from a developing nation, *Information Resources Management Journal*, Winter 1990:2.

Dekleva, S. and Zupancic, J., 1992, An empirical investigation of key issues in information system management, *in:* "Proceedings of the Third International Conference on Information System Developers Workbench", Stanislaw Wrycza, ed., University of Gdansk.

Ein-Dor, P. and Segev, E., 1988, Information resource management for end-user computing: an exploratory study, *Information Resource Management Journal*, 1:39.

Igbaria, M., 1992, Examination of microcomputer usage in Taiwan, *Information and Management*, 22:19.

Nagel, M., 1992, "Personal Use of Computers - Organisation and Effectiveness", Master Thesis, School for Organizational Sciences, Kranj 1992 (in Slovene).

Rogers, G.B., 1992, End-user computing in the 1990s, *in:* "Proceedings of the International Conference on Organization and Information Systems", Appendix, Z. Kaltnekar, ed., University of Maribor.

Rudez, S., 1992, "Organization and implementation of end-user computing", Graduation Assignment, School for Organizational Sciences, Kranj (in Slovene).

Verber, B. and Zupancic, J., 1992, End-user computing in small business in Slovenia. *in:* "Proceedings of the Third International Conference on Information System Developers Workbench", Stanislaw Wrycza, ed., University of Gdansk.

Wetherbe, C.J., 1988, "System Analysis and Design", West Publishing Company.

Zupancic, J. and Leskovar, R., 1991, An analysis of key issues in information system development, *in:* "Proceedings of the 13th International Conference Information Technology Interface ITI-1991", V. Ceric, V. Luzar, V. Dobric and P. Ray, eds., University of Zagreb.

INTRODUCTION TO MANUFACTURING SYSTEMS STREAM

Manufacturing industry has gone through a number of development phases in needing to respond effectively to constantly increasing customer demands for new, cheaper and 'better' products and products that can as far as possible reflect the purchaser's own identity and taste. Successful competition has been and is based upon the capability to handle and deliver a product mix which changes constantly according to customer preferences. The implications of these emerging customer demands have been profound throughout modern industrial history, and many 'well established' manufacturing companies have simply not been able to respond and have disappeared while new companies have emerged, the normal ebb and flow of economic activity. The present challenge of change demands fast and 'accurate' capture and definition of customer requirements, subsequent product development cycles which are short, accurate and 'correct first time', and the flexibility to handle small production batch sizes. These are formidable demands for industries oriented towards long production runs of products which the customer would happily purchase without demanding customised features subject to rapidly changing fashion demands. This requires a level of data and knowledge sharing that may be perceived to reject the cherished virtues of individualism and the autonomy of the individual in one section of the company while demanding unheard of flexibility and autonomy in another.

Computer based information systems have to some extent provided a useful tool in improving the efficiency and effectiveness of manufacturing industries. The development of computer based information systems has been a bottom-up development. This development has reflected the historic nature of the product life-cycle. The traditional product life-cycle is a clearly defined sequence of discrete events. The sequence progresses the development of the product from customer requirements specification and conceptual design to the series production, final delivery and customer feedback. Each event is very often perceived as being loosely coupled to its preceeding and following events and is therefore often operating with its own unique information base. The development of CAD, CAM, CAPP and various production control systems has strengthened the orthodox product life-cycle and promoted the discretisation of its events and procedures. The total information systems requirements of the manufacturing organisation has therefore not been achieved. To support this it is only necessary to examine the efforts that have been dedicated to the development of CAD data exchange standards and neutral data representation schemes. 'Islands of Automation' has become a recognised attribute of manufacturing oriented information systems and it is only recently that a holistic perspective has been given serious currency.

Relatively little effort has been channelled towards the development of open system architectures. It is tempting to draw an analogy with a computer programming environment where the low-level library routines have been developed and these routines are either used singly or as small sets through primitive parameter passing and manual activation. 'Islands of Automation' denotes the resultant 'rich tapestry' of incompatible, single function, computer based support systems which promote the continuation of the orthodox and highly ineffective product life-cycle. The common situation picture of the constant reformatting and

Systems Science, Edited by F.A. Stowell *et al.*,
Plenum Press, New York, 1993

interpretation of product information from one product development life-cycle event to the other is probably one of the main problems which European manufacturing industry will have to overcome.

The total information system requirements for manufacturing organisations are beginning to be formulated through the concepts of CIM and 'Concurrent Engineering'. The implications of this recognition is that the actual user requirements are the driving force for the formulation and development of information systems for the manufacturing environment. Effective information system support throughout the complete product life-cycle is dependent on the ability to access and process information relevant to the various stages of the product life-cycle, and the ability to allow the user complete freedom to use the available human based expertise and information. Information availability cannot be dependent on repeated re-entering or re-interpretation of information which is specific to preceding or subsequent life-cycle stages, but must be part of a complete product information model that is sourced only once. For instance the capture of a mechanical engineering product's physical attributes must necessarily be sourced only from the detailed design stages and thereafter be used for the following Production Planning and Machining Sequence and Tool Path generation. Any intervening interpretation or reentry of data can only introduce errors and reinforce the discrete, sequential nature of the orthodox product life cycle.

Once a comprehensive and globally valid product model exists product life-cycle events may access information in an incremental mode, so that once the product data is generated it automatically becomes meaningful across the whole product life-cycle. For a product model to be relevant throughout the whole product life-cycle, the modelling capabilities must go beyond the geometric attributes of the product. The information must incorporate not only the data immediately related to the product's physical attributes but must incorporate such attributes as are related to the product's expected behaviour during use, the implications of its design to the available manufacturing resources, and the likely implications to the eventual dismantling and possible reuse of its components and material.

Successful CIM is also dependent on the freedom given to the user to tailor the **processing** of the product model information to the specific processing and procedureal needs of the individual manufacturing organisation. Furthermore, the processing of the product information is dependent on both the product type as well as the manufacturing organisation and its customer domain. The latter of the two factors is at least as important as the former. The designers and manufacturers of identical products are quite likely to interact differently with its individual customer base, because customers will demand product conformance specific to the domain in which the customer himself is operating. Differences in manufacturing technology will cause one manufacturing organisation to process his product model uniquely. The implication of this is the need to create a systems reference structures, or reference architectures, so that individual system solutions can be created that are consistently applicable across enterprises. A reference architecture must be generic enough to apply to both inter and intra industry sector solutions. Therefore, the architecture has to provide facilities by which the specific system instance can be generated to reflect the user requirements accurately and efficiently. System design is a continuous affair and the reference architecture must supply enough support for updating, modifying and extending the instantiated system solutions in order to relect the dynamic nature of the user environment. The reference architecture naturally incorporates the required product modelling support as well as the processing aspects of the system solution. Enterprise wide CIM must therefore be seen as a four staged programme which incorporates the **technical, procedure** and **goal integration** as well as **architectural integration**.

Professor C. Irgens
University of Paisley

SYMBOLIC MANIPULATION CAD FOR PARAMETERIZED FAMILIES OF SYSTEMS

Octavian Păstrăvanu, Mihail Voicu, Laurenţiu Isopescu, and Radu Călinescu

Department of Automatic Control and Industrial Informatics
Polytechnic Institute of Iaşi
Str. Horia 7-9, 6600 Iaşi, Romania

INTRODUCTION

In spite of the vigorous development of modern control theory, its impact on the practice is still modest. Multi-model approaches represent a new promising trend which tries to cope with the shortcomings due to the inability to use accurate mathematical models. Such a new trend has been fully justified by the need for a more realistic investigation of system behaviour and has opened at the same time the perspectives to enrich the traditional concept of CAD in control.

The recent evolution of the symbolic manipulation software as for instance *muSIMP & muMATH* (Soft Warehouse, 1983), *REDUCE* (Rayna, 1987), *DERIVE* (Soft Warehouse, 1989), *PC-MACSYMA* (Symbolics Inc., 1989), *MATHEMATICA* (Wolfram Research, 1991) founded the premises for a global treatment manner, fundamentally different from the pointwise one specific to numerical computation. It is the purpose of this paper to emphasize the role played by the possibility to operate with symbols in control system analysis and synthesis.

Starting from some results reported in (Fliess et al., 1983), (Lecourtier et al., 1984), (Grossman, 1989), we have attempted to derive a full benefit from this experience in symbolic computation, developing a sotware tool *muSCCS* (micro Symbolic Computation in Control Systems) devoted to the key problems involved by the usage of the parameterized families of systems. After a short discussion (section 2) of the position presently hold by symbolic computation in the large area of CACSD, section 3 deals with the main functions implemented by *muSCCS* for linear and non-linear parameter dependent systems. Illustrative examples of the usage of *muSCCS* are given in section 4 and further research directions are mentionned in the last section.

BRIEF OVERVIEW ON THE RELATION NUMERIC-SYMBOLIC IN CACSD

Even though numerical applications currently dominate CACSD, formula manipulation software created a new framework able to transform many notable theoretical results (inadequate from the numerical computation point of view) in very useful algorithms for control software. Important is to say that such theoretical results requiring a cumbersome hand calculus even for parameterized families of low order systems can be successfully implemented by means of symbolic computation; in this respect, Leverier-Fadeeva algorithm, Kalman's criteria for controllability and observability, eigenproblems, pole placement represent just a few conclusive examples. Moreover it is worth noticing that the development of symbolic computation seen as a source of new capabilities contributed to the appearance of algorithms whose main ideas had been suggested even by these new capabilities, as for instance the remarkable correspondence introduced by Fliess (Fliess et al., 1983) between the Volterra series and the formal serries in non-commutative variables.

Unlike the great experience accumulated by the houses producing numerical software libraries, experience simultaneously supported by the requirements of a large number of users, the usage of symbolic computation is still limited to small groups of initiated persons. As a matter of fact the history of symbolic computation means about ten performant packages used along the last decade, among which the most recent are running only on very powerful machines. However the previously unimagined applications emerging today as a consequence of the elan of symbolic computation give the possibility to forsee an impact on CACSD similar to that produced in the mid eighties by incorporating the linear algebra packages *EISPACK* and *IINPACK* as irreplaceable tools in many stand-alone control packages. It is now simple to understand that the future of the academic and applied research in the field of CACSD will belong to a fruitful numeric-symbolic symbiosis, combining by complementarity their advantages.

muSCCS - A SYMBOLIC COMPUTATION APPROACH TO PARAMETERIZED FAMILIES OF DYNAMIC SYSTEMS

muSCCS was developed using the capabilities of *muSIMP & muMATH,* the most popular symbolic computation software running on all compatible IBM - PC machines. It implements functions devoted to linear and non-linear parameter dependent systems, which are to be detailed below.

The main functions available for the current user are listed in table 1.

For systems both state-space and input-output multi-models are considered. A continuous-time system can be described by the state-space representation:

$$\begin{gathered}\dot{x}(t) = A(\theta)x(t) + B(\theta)u(t), \quad y(t) = C(\theta)x(t) + D(\theta)u(t); \\ t \in \mathbf{R}_+, \quad x \in \mathbf{R}^n, \quad u \in \mathbf{R}^m, \quad y \in \mathbf{R}^p\end{gathered} \tag{1}$$

or by the transfer matrix $G_\theta(s)$, θ denoting the physical parameter vector. Similarly the models used for a discrete-time system are the state-space representation:

$$\begin{gathered}x(t+1) = E(\theta)x(t) + H(\theta)u(t), y(t) = C(\theta)x(t) + D(\theta)u(t); \\ t \in \mathbf{N}, \quad x \in \mathbf{R}^n, \quad u \in \mathbf{R}^m, \quad y \in \mathbf{R}^p\end{gathered} \tag{2}$$

or the transfer matrix $\tilde{G}_\theta(z)$, the notation θ preserving the same meaning as above.

Table 1. Functions available in *muSCCS* **for the current user**

Function name	Input Parameters	Output Parameters
AIEQSO	n, m, b_i, c_j in (3), N in (4)	coefficients of f in (4)
ALOC	A in (1) or E in (2), assigned closed loop poles	state feedback matrix
CHRPOL	A in (1) or E in (2)	characteristic polynomial
CLOSED	System transfer matrix, controller transfer matrix	closed–loop transfer function (monovariable case)
CONTR	A, B in (1) or E, H in (2)	controllability matrix
DECOMC	$G_\theta(s)$	Laplace transform of the step response
DECOMD	$\tilde{G}_\theta(z)$	Z transform of the step response
DISCRA	A in (1)	A discretized
DISCRB	A discretized, B in (1)	B discretized
EIGVAL	A in (1) or E in (2)	eigenvalues (when analytic form available)
EIGVEC	A in (1) or E in (2)	eigenvectors (when analytic form of eigenvalues available)
ESTIM1	A, C in (1) or E, C in (2), assigned eigenvalues	observer gain matrix
ESTIM2	A, C in (1) or E, C in (2), observer gain matrix	observer evolution matrix
FREQ	$G_\theta(s)$	frequency response (monovariable case)
GIEQSO	coefficients of f in (4), N,L-B transform of $u(t)$	coefficients of g
HURWM	open-loop polynomial	Hurwitz matrix
HURWD	Hurwitz matrix	stability test with Hurwitz
LEVF	A in (1) or E in (2)	adjoint matrix with Levérier-Fadeeva
REALCO	$G_\theta(s)$ or $\tilde{G}_\theta(z)$	controllable realization
REALOB	$G_\theta(s)$ or $\tilde{G}_\theta(z)$	observable realization
TRANSF	A, B, C, D in (1) or E, H, C, D in (2)	transfer matrix

For parameter dependent non-linear systems whose behaviour can be described by the differential equation:

$$Ly(t) + \sum_{i=2}^{m} b_i(\theta) y^i(t) = u(t), \quad L = \sum_{j=0}^{n} c_j(\theta) \mathrm{d}^j/\mathrm{d}t^j, c_n = 1 \tag{3}$$

(where *u(t), y(t)* are the system input and output respectively, and H denotes the physical parameter vector), an approximation of the time response can be computed. In what follows the mathematical background (Fliess et al.,1983) justifying the need for symbolic computation is briefly presented for a better understanding of the details in table 1. For an arbitrary signal *u(t),* a polynomial f in the non-commutative variables z_0 and z_1:

$$f = \sum_{k=1}^{N} f_k(1 + a_k^0 z_0)^{-1} z_k^1(1 + a_k^1 z_0)^{-1} \ldots z_k^{i_k}(1 + a_k^{i_k} z_0)^{-1}, \tag{4}$$
$$f_k \in \mathbf{R},\ \ z_k^j \in \{z_0, z_1\},\ a_k^j \in \mathbf{C},\ j = \overline{1, i_k},$$

is associated to a finite number of terms in the Volterra series of the output *y(t)*. From (4), for a given input signal *u(t)* one can obtain a polynomial *g* in one noncommutative variable z_0, which represents the Laplace-Borel (L-B) transform of the truncated series of the output *y(t)*; the time response results applying L-B inverse transform.

COMMENTS ON TWO CASE STUDIES

Both examples are referring to physical systems with variable parameters, whose behaviour can be globally described only in terms of the multi-models. Sequences of the user interface dialogue are reproduced below for demonstrating the ease to handle of this software.

Example 1: Consider the loading bridge in (Ackermann, 1985) with variable cable length *l* and load mass *m* described by a state space multi-model:

$$\begin{bmatrix} \dot{x}_1 \\ \dot{x}_2 \\ \dot{x}_3 \\ \dot{x}_4 \end{bmatrix} = \begin{bmatrix} 0 & 1 & 0 & 0 \\ 0 & 0 & f_{23} & 0 \\ 0 & 0 & 0 & 1 \\ 0 & 0 & f_{43} & 0 \end{bmatrix} \begin{bmatrix} x_1 \\ x_2 \\ x_3 \\ x_4 \end{bmatrix} + \begin{bmatrix} 0 \\ g_2 \\ 0 \\ g_4 \end{bmatrix},\quad y = \begin{bmatrix} 1 & 0 & l & 0 \end{bmatrix} \begin{bmatrix} x_1 \\ x_2 \\ x_3 \\ x_4 \end{bmatrix}$$

where $f_{23} = m/1000$, $f_{43} = -(m + 1000)/(100l)$, $g_2 = 1/1000$ and $g_4 = -1(1000l)$. The call sequences of the routines *CHRPOL, TRANSF, DECOMC* as well as their outputs are listed in figure 1. The analysis of the structural properties (controllability and observability) allows the design of a parameterized state feedback controller based on the estimated state using the routines *ALOC, ESTIMl,* and *ESTIM2* (figure 2).

Example 2: Consider a nonlinear electronic circuit (Fliess et al., 1983) whose functioning equation is:

$$\mathrm{d}y(t)/\mathrm{d}t + k_1 y(t) + k_2 y^2(t) = u(t)$$

with $k_1 = 1/(RC)$ $k_2 = \alpha/C$, where *R* and *C* denote a resistance and a capacitance respectively and <x is a variable parameter. The coefficients in *f*(3) for *N*=4 and an arbitrary input signal *u*(*t*) are computed by means of *AIEQSO* as shown in figure 3.

Calling *GIEQSO* and subsequently applying L-B inverse transform one could obtain the first 4 terms of the Volterra series of the unitary step response:

$$y(t) = (1 - e^{-k_1 t})/k_1 - k_2(1 - 2k_1 t e^{-k_1 t} - 2e^{-2k_1 t})/k_1^3 +$$
$$k_2^2[2 + (1 - 2k_1 t - 2k_1^2 t^2)e^{-k_1 t} - 2(1 + 2k_1 t)e^{-2k_1 t} - e^{-3k_1 t}]/k_1^5 + \ldots$$

```
? PC:CHRPOL(A);
@: Characteristic polynomial
-s^2 F43 + s^4
? FT:TRANSF(A,B,C);
@: Transfer function
(F23 G4 - F43 G2 + S^2 G2 + S^2 G4 L)/
(-S^2 F43 + SA4)
? RT:DECOMC(FT);
@: Unitary step response
-F23 G4/(S^3 F43) + G2/S^3 -
F23 G4/(S F43 2) - G4 L/(S F43) +
(S F23 G4/F43^2 + S G4 L/F43)/(-F43 + S^2)
```

Figure 1. Results of *CHRPOL, TRANSF, DECOMC*.

```
? R;
@: Closed loop assigned poles
{-1, -1, -141/1000 + 141/1000 #I, -141/1000- 141/1000 #I} ? AL:ALOC(A,B,R);
@: State feedback matrix
[19881/(500000 F23 G4 - 500000 F43 G2), 90381/(250000 F23 G4- 250000 F43 G2), (-19881/500000 G2 -
801881/500000 F43 G2 -F43^2 G2)/(F23 G4^2 - F43 G2 G1) + (801881/500000 F23 + F23 F43)/(F23 G4 -
F43 G2), 1141 F23/(500 F23 G4 - 500 F43 G2) + (-90381/250000 G2 - 1141/500 F43 G2)/(F23 G4^ 2 - F43
G2 G4)]
? R1;
@: Observer assigned eigenvalues
{-5, -5, -5, -5}
? ES1:ESTIM1(A,C,R1);
@: Observer gain matrix
[-20 F23/(-F23 - F43 L) +500/(-F43 - F23/L), (-150 F23 - F23 F43)/(-F23 - F43 L) + 625/(-F43 - F23/L), (-500
- 20 F43)/(-F23 - F43 L), (-625 - 150 F43 - F43^2)/(-F23 - F43 L)]
? ES2:ESTIM2(A,C,ES1);
@: Observer evolution matrix
{[20 F23/(-F23 - F43 L) - 500/(-F43 - F23/L), 1, 20 F23 L/(-F23 - F43 L) - 500 L/(-F43 - F23/L), O], [(150
F23 + F23 F43)/(-F23 - F43 L) - 625/(-F43 - F23/L), 0, F23 - 625 L/(-F43 - F23/L) + (F23 F43 L + 150 F23
L)/(-F23 -F43 L), 0], [(500 + 20 F43)/(-F23 - F43 L), 0, (500 L + 20 F43 L)/(-F23 - F43 L), 1], [(625 + 150
F43 + F43^2)/(-F23 - F43 L), 0, F43 + (625 L + 150 F43 L + F43^2 L)/(-F23- F43 L), 0]}
```

Figure 2. Results of *ALOC, ESTIMl, ESTIM2*.

and the first 3 terms of the Volterra series for the input $u(t) = \cos\omega t$:

$$y(t) = \left[-k_1 e^{-k_1 t} + k_1 \cos\omega t + \omega \sin\omega t\right] / (k_1^2 + \omega^2) +$$
$$+k_2 \left[e^{-2k_1 t} /\ k_1 + 2 \sin\omega t\ \ e^{-2k_1 t}/\omega - 1k_1\right] / \left[2(k_1^2 + \omega^2)\right] + \ldots \tag{5}$$

On the other hand the comparative study in [Voicu and co-workers,1991] intended to demonstrate that the differences between the values got by the numerical evaluation of the truncated serries listed above and the results obtained with the numerical differential equation solver *DGEAR* (IMSL, 1981) are quite small. This comparison states for a thorough guarantee that the analytic form of the time response calculated by symbolic computation using the first three - five terms of Volterra series usually represents a good approximation of the whole Volterra series of the output.

CONCLUSIONS

The software presented in section 3 and its facilities for studying parameterized families of systems, briefly illustrated in section 4, have to be seen as a first step of a more ambitious project aiming to elaborate a collection of powerful symbolic computation routines for control systems. As a general remark it must be noticed that the CPU time required by high order problems (seven or more) increases significantly, what represents a certain inconveniency of this version of *muSCCS* mainly due to the *muSIMP-based* implementation.

In this respect a more performant tool-box based on *MATHEMATICA* *is* already planned for the near future, supported by the generous resources offered by 386 and 486 environments.

```
? N:1;
@: 1
? L:{K1,1}
@: {K1,1}
? M:2;
@: 2
? P:{K2}
@:{K2}
? AIEQSO(N,L,P,4);
@: System output for an arbitrary input
1
    Z1
 K1   0

-2 K2
    Z0      Z1   Z1
 K1   2 K1   K1   0

12 K2^2
    Z0      Z0       Z1       Z1    Z1
 K1   2 K1   3 K1   2 K1    K1   0
```

```
4 K2^2
    Z0       Z1    Z0       Z1   Z1
 K1   2 K1   K1   2 K1   K1   0

-72 K2^3
    Z0       Z0       Z1       Z0       Z1       Z1
 K1   2 K1   3 K1   2 K1   3 K1   2 K1   K10
-24 K2^3
    Z0       Z0       Z1       Z1    Z0       Z1   Z1
 K1   2 K1   3 K1   2 K1   K1   2 K1   K1   0
-8 K2^3
    Z0       Z1    Z0       Z1    Z0       Z1   Z1
 K1   2 K1   K1   2 K1   K1   2 K1   K1   0
-144 K2^3
    Z0       Z0       Z0       Z1       Z1       Z1   Z1
 K1   2 K1   3 K1   4 K1   3 K1   2 K1   K1   0
-24 K2^3
    Z0       Z1    Z0       Z0       Z1       Z1   Z1
 K1   2 K1   K1   2 K1   3 K1   2 K1   K1   0
```

Figure 3. Results of *AIEQSO*.

REFERENCES

Ackermann, J., 1985, "Sampled-Data Control Systems", Springer Verlag, Berlin.

Fliess, M., and Lamnabhi-Lagarrigue, F.,1983, An algebraic approach to nonlinear functional expansion, *IEEE CAS* 30:554.

Grossman, R., ed., 1989, "Symbolic Computation. Applications to Scientific Computing", SIAM, Philadelphia.

IMSL, 1981, "Library Reference Manual. 9th edition", IMSL Inc., Houston.

Lecourtier, Y., Walter, E., and Raksanyi, A., 1984, Test of structural properties of state-space models through algebraic computation, *in:* "Prep. IFAC 9th World Congress", vol VIII, 250, Budapest.

Rayna, G., 1987, "*REDUCE*. Software for Algebraic Computation", Springer Verlag, Berlin.

Soft Warehouse, and Microsoft Corporation, 1983, "*muMATH* & *muSIMP*. Symbolic Mathematics Package", Soft Warehouse Inc., Honolulu.

Soft Warehouse, 1989, "*DERIVE*. Symbolic Mathematics Package", Soft Warehouse Inc., Honolulu.

Symbolics, 1989, "PC-MACSYMA", Symbolics Inc., Burlington.

Voicu, M., Pastravanu, O., and Calinescu, R., 1991, A comparison between numerical and symbolic simulation of nonlinear dynamic systems, *in:* "Prep. 8th Int. Conf. Contr. Syst. and Comp. Scie.", vol I, 90, Bucuresti.

Wolfram Research, 1991, "*MATHEMATICA*: A System for Doing Mathematica by Computer", Wolfram Research Inc., Champaign.

SUPPORT SYSTEMS FOR TOTAL QUALITY

Barbara M. Savage

Faculty of Engineering
University of the West of England
Coldharbour Lane, Frenchay
Bristol BS16 1QY

INTRODUCTION

The strategy of Total Quality Management (TQM) states that every organisation activity contributes to the final quality of the product as perceived by the customer, and thus requires that the quality of all elements of organisational activity be monitored and improved. It is not sufficient for the quality system to focus only on the production cycle, because quality, as evidenced by customer satisfaction, may be affected by a wide range of factors, over and above the mainstream of production activities. Thus it may be seen that all data available in an organisation may have potential for informing the quality improvement process. Hence, the quality system must be provided with the means of accessing any of the data held by the various information systems extant in the organisation. Indeed, a case may be made for significant data links between organisations throughout the supply chain.

Quality monitoring activities, however they may be focused, generate large quantities of raw data which must be analysed to determine trends and identify weak points in the process. They must be properly coordinated, with control being exercised over measurement methods and criteria. There is, therefore, a need for mechanisms to support the quality system by acquiring and storing data, and subsequently to provide a flexible access service to the data for analysis.

It is possible to apply a wide range of interpretations in the definition of a system for achieving quality in a manufacturing organisation; from inspection at the lowest end, through Quality Control and Quality Assurance, up to Total Quality Management. Given that the focus of this paper is to examine how best to utilise computers to support the necessary data handling activities of a quality system, it is desirable to state at this point the interpretation of this author. It is her contention that, in order to achieve continuous quality improvement and hence full customer satisfaction, a quality system must be all embracing in the manner proposed under TQM. As a corollary of this definition, the issue of what is "Quality Data" and what is not, can be resolved: all data extant in an organisation, or indeed, in a supply chain, has potential to help in the process of ascertaining organisational strength or weakness in quality terms. Hence, all data may be considered to be "Quality Data". It could be argued that this definition is somewhat extreme, since there is a strong school of thought which states that one should not seek to achieve product quality levels which exceed

Systems Science, Edited by F.A. Stowell *et al.*,
Plenum Press, New York, 1993

the customer's requirement, however, the teachings of Deming, now widely adopted in Japan, present another view; that only the best possible is good enough. It is this latter stance that this author supports.

Initially, it is necessary to investigate the problems posed by the inadequacies of traditional methods of data storage for the purpose of long-term quality improvement, and to consider the ways in which information technology may be used to address these problems. It is also necessary to consider the relationship between existing information systems (such as production planning and inventory) and the proposed quality information system. Whilst a case can be made for the quality system being the superior information system within an organisation, it is vital to realise the impact of such a solution upon companies wishing to adopt it. Such a dramatic change of direction in an organisation's information management system would be extremely costly to implement, and most definitely not without significant risks. It is therefore proposed that the quality information system be considered as the common ground upon which the integration of the presently separate systems which exist within an organisation may be based.

REQUIREMENTS OF THE QUALITY ACTIVITY

The demand for quality data collection, storage and analysis arises from the requirements associated with quality monitoring, quality improvement and documentation pertaining to product, process and quality system. These may be broken down into several subsections, so that quality monitoring requires such things as control over inspection procedures and calibration of measuring equipment. Similarly, quality improvement includes the strategic use of quality data to support and track the progress of a company in fulfilment of its business aims and strategy. This may be achieved in three areas:-

a) reactive quality improvement stimulated by identified failures found either in house (via inspection or SPC studies) or through customer complaints or reports from staff in the field.

b) proactive quality improvement designed to increase the robustness of the process, and thus reduce the likelihood of quality failure. This includes both strategic quality improvement planning by management, and local improvement. Greater customer satisfaction may be achieved by means of improved awareness of customer requirements, coupled with improvements in quality of design of both process and product.

c) focusing the combined efforts of supplier and customer in partnership to achieve improved quality throughout the supply chain.

The system must also address the provision of documentation for the following purposes:-

a) the need to produce information in some specified format to satisfy the documentary requirement of customers and standards. Additionally to demonstrate adherence to any relevant standards.

b) the need to be able to produce evidence that proper procedures have been followed in case of product liability prosecution.

c) the provision of a communications channel to allow the distribution of quality related information so that procedures may be monitored and areas where improvement may be needed can be identified. Also to monitor the effects of change on procedures and processes.

SUMMARY OF QUALITY DATA NEEDS

Clearly the data needs of these tasks vary considerably, so that a solution suitable for one may not satisfy another. It is therefore necessary to establish the requirements of each of these quality activities in order to harmonize them into an integrated information gathering and enquiry system.

Reactive problem solving relies upon the ability to identify and examine quality information on a historical basis. In some circumstances (e.g. aircraft parts), there may be an additional need to trace other products from the same batch in case they too exhibit the reported fault. In this situation, it should also be possible to establish a link back to the source materials or raw materials if they may be the root cause of the problem, so that other batches in which those materials were used can also be traced. Combined with this is the need for control of the Corrective Action process to ensure that problems are corrected and that the effectiveness of the corrections applied are formally reviewed so that further action can be taken if necessary.

However, whilst quality improvement techniques such as Statistical Process Control (SPC) and Quality Circles are an important part of a company wide quality improvement programme, there is a need for strategic quality planning by management. To be effective, this planning must be supported by a flexible, comprehensive information server which is able to supply data for investigative analysis and to keep track of the progress of strategic plans.

Quality of design is arguably more vital to the overall quality of the finished product than quality of manufacture. It is therefore necessary that designers have access to historical data pertaining to the performance of earlier similar products in the field, along with details of the capabilities of the processes used to produce them. Also of interest will be details of any available alternatives in process or product design. The work of ESPRIT project 2178 (RA-IQSE) has focused on this part of the manufacturing process and has resulted in the development of a prototype quality support system for design (RA-IQSE Consortium[1], Irgens[2] and MacArthur[3]).

With the increasing demand for manufacturers to satisfy company-specific, national and international standards, and the recent changes in Product Liability legislation, there is a need for an organisation to have documentary evidence of conformance to quality procedures both in all elements of manufacture and design.

As quality improvement programmes become established, the role of suppliers in the achievement of quality becomes more evident. There is a need for communication between customer and supplier as they begin to work together to achieve mutual satisfaction. There must be a clear understanding of the requirements of the customer by the supplier. This may be attained by direct transfer of specifications from the customer's database to the supplier's. Further, evidence of conformance to those requirements may be passed back in a similar way.

INTEGRATING QUALITY INFORMATION

Bringing together the needs of the various quality related tasks considered above leads to the proposal that an integrated database of quality information should be set up. This database should have the ability to answer not only a set of standard, pre-defined queries, but also unforeseen, ad-hoc enquiries. Furthermore, that there is a requirement for links with other systems within the organisation (for example, production planning, clerical, CAD/CAM) to allow the relationships between quality problems and other factors to be examined.

It is first necessary to consider the nature of present management information systems. Within a manufacturing organisation there exist a number of independent data resources, each assigned to some specific function, and normally isolated from one another. These data resources may be of several types; firstly, purely paper based (i.e. traditional filing systems) although it should be noted that these may incorporate space saving devices such as storage of data on microfilm; secondly, simple microcomputer based systems servicing the needs of well defined, but strictly limited functional areas; and finally, site or company wide databases to support given organisational functions (for example, payroll or stock control).

As has been stated previously, this author considers that all available data should be viewed as "quality data". Even so, the design of an integrated information system for quality improvement is somewhat problematic, because so much of the data to be included is primarily associated with other information systems throughout the organisation. Hence, issues of responsibility for obtaining and maintaining quality data, and control of access to that data must be addressed.

Since a given functional system (eg. production planning, design, procurement) has the most knowledge and experience of the operations which it controls, it is logical that it should also take responsibility for quality in those operations. This follows from the findings of those who have investigated quality control in manufacturing processes and have pointed out that the person doing a particular job is the one who usually knows most about it, and therefore that person should be in control of quality for that job. Thus quality information derived from given functional systems should remain under the local control of that system but that an access mechanism should be provided so that it may be utilised via the Integrated Quality Systems (IQS) to satisfy quality queries, possibly in combination with data gleaned from other sources for cross-functional quality investigations.

Quality is a strategic activity which requires focused management involvement. In practice, it comprises three basic functional elements; prevention, cure and documentation. Quality Control may be considered to be the "cure" side, while continuous quality improvement can be identified with "prevention". The needs of prevention are rather different from those of cure, it is therefore necessary to ensure that the quality information system can satisfy both. Thus it must not only provide the means of detecting and identifying the root cause of quality failures ("cure"), but must also supply information to the planning process to allow quality to be built in to both product and process ("prevention"). Finally, it must provide evidence that proper procedures have been followed in order to satisfy contractual obligations and international standards.

PROPOSED IQS STRUCTURE

The diagram at Figure 1 shows a schematic of a typical distributed quality information system. This structure reflects the need for flexibility and the ability to expand the configuration in response to future need are essential. The central core of the IQS is the central quality database (CQDB) which will not only be the repository of all summary quality information, but will co-ordinate data exchange with other information systems (any or all of which may be resident on a physically separate computer), such as those concerned with production planning, scheduling and commercial activities, via a local area network (LAN). Also linked to this LAN would be a number of microcomputers which would provide direct process/product monitoring and support process control activities such as SPC. These units are known as Quality Data Acquisition and Analysis Stations (QDAAS). There are no limitations as to the number or type of QDAAS, hence this approach may be used in plants of any size, with QDAAS situated in various parts of the plant; for example, incoming goods inspection, production areas and the desks of quality function staff and management. Communications with other automated elements of the manufacturing system, such as production scheduling and design are also necessary. As a long-term goal, external links to suppliers, customers, other company sites and sub-contractors might be considered desirable, although clearly, access to the database from these external sources must be strictly controlled.

This structure is based upon the functional dichotomy evident in quality activities. Firstly, there are operations which are closely associated with the actual manufacture of products, wherein quality control and certification are the major activities. Secondly, quality management activities are focused upon study and improvement of all operations in the

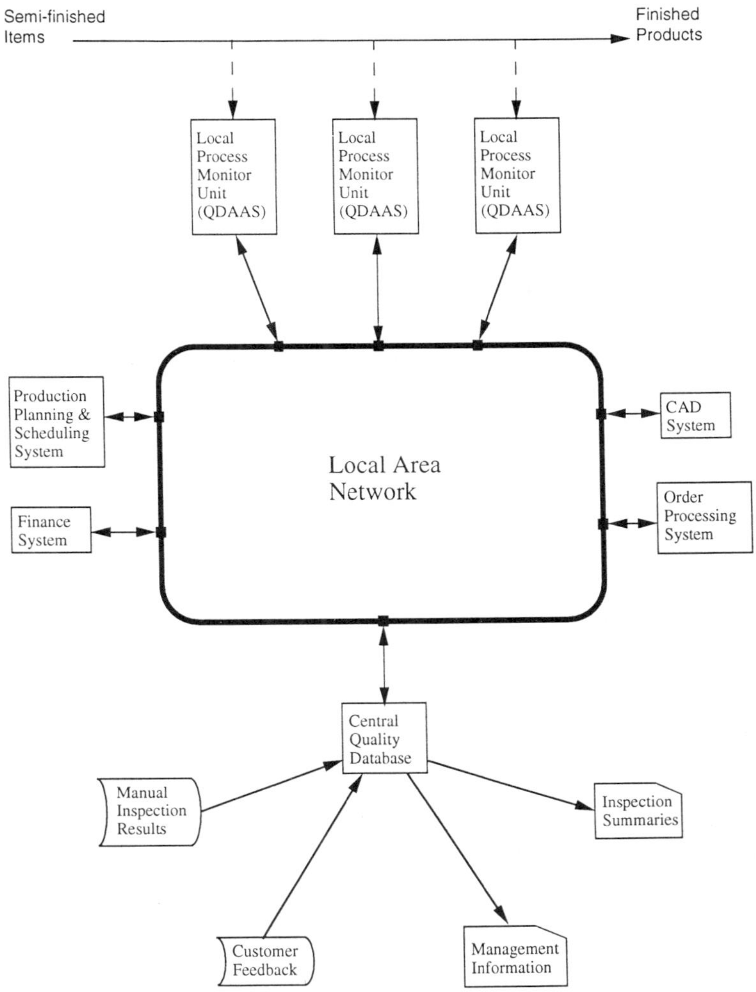

Figure 1. Distributed Quality Information System Schematic.

organisation with a view to overall quality improvement. This latter category includes product and process design activities which rely heavily on records of past experience, predefined product or process specifications or constraints, and legislative requirements and restrictions.

The functionality of the initial prototype was very simple; it gathered inspection specifications from their source (typically the CAD system), from the production schedule, it caused the appropriate inspection specifications to be issued to the QDAAS, and finally, it collected inspection results from the QDAAS at the conclusion of a batch. As the prototype was developed, this functional element became known as the Inspection Control and Records module (ICRM). However, this simple system was not sufficient for true TQM support because it only provided enough functionality for Quality Control, with perhaps limited support for some aspects of Quality Assurance. Thus it was necessary to extend the prototype in such a way that it might approach full support for TQM.

It is evident that there is a need for linkage between the ICRM and other modules, such as Equipment Calibration Monitoring and Training Records, so as to demonstrate that all inspection results collected are reliable, and have not been affected by the use of out of calibration measuring equipment or inadequately qualified staff. Hence, the extended prototype comprised a number of loosely linked functional modules, each designed to operate as free-standing, independent entities, but able to provide or request data from other modules as required. The modules defined in the extended prototype were Equipment Calibration Monitoring, Staff Training Records, Corrective Action, Customer Complaints and Vendor Monitoring. Several of these functional modules were fully contained within the CQDB as the necessary data was not already held in any external information system. This modular structure provides easy user access both to specially designed functions and to the access gateway for wider queries for which are provided sets of predefined reports, along with facilities for ad-hoc querying of the data and the definition of new reports which may then be linked into the system for future use if required. In addition to the these in built functions, the CQDB requires the ability to communicate information between the various physical parts of the integrated quality system without the necessity for direct human intervention. This has been effected by means of a set of continuously running processes to carry out a selection of data transfer and control tasks. These processes have been given the collective title the "Active Element".

THE IQS CONCEPT APPLIED TO FOOD PRODUCTION

The concept of an IQS has wider application than just within manufacturing operations. This idea is now being applied to the production and transportation of Citrus Fruit with the remit of ESPRIT project 5379 (FRUIT). The goal of this project is product life tracking for fruit in transit with a view to the provision of information on how changes in the conditions experienced by the fruit during transportation effect the quality of individual shipment of fruit. This will be achieved be the development of a cheap, re-useable sensing device which will travel with the fruit. Data collected by this device during the journey will be combined with data concerning both pre and post harvest conditions and treatments, and known characteristics of the relevant fruit variety. Analysis will then provide information concerning the aging of the fruit in the shipment and whether there is evidence to suggest that the fruit has been damaged, either physically or by disease. The results of this analysis will be of use to both the wholesaler and retailer and, in the longer term, might be used to provide the final customer with a detailed history of the fruit which they are purchasing. Further details of the FRUIT system are given in Knight et al[4].

REFERENCES

1. RA-IQSE Consortium, "Deliverable 10: RA-IQSE Project Evaluation - Achievements and Perspectives", *reprinted in*: "Proceedings of CIM-Europe Conference on Results of ESPRIT Projects, Bilbao", Commission of the European Communities, DG XIII, Brussels (1992).
2. C. Irgens, "A Feature Based KBS for Quality Prediction of Machined Parts and Products" *in*: "Proceedings of the 7th Annual Esprit CIM-Europe Conference", R. Vio and W. Van Puymbroeck ed., Springer-Verlag, Turin (1991).
3. E.W. MacArthur, "Quality Support in RA-IQSE", *in:* "Proceedings of CIM-Europe Conference on Results of ESPRIT Projects, Bilbao", Commission of the European Communities, DG XIII, Brussels (1992).
4. J. Knight, B.M. Savage, C.L. Honeybourne, G. Emmanoulopoulos, and J. Puig Gomez, "Fresh Fruit Product Life Tracking" *in:* "Proceedings of the 8th Annual Esprit CIM-Europe Conference", C. O'Brien, P. MacConaill and W. Van Puymbroeck, ed., Springer-Verlag, Birmingham (1991)

PRODUCTION AND INFORMATION IN SYSTEMS

Bernd Schiemenz

Universität Koblenz-Landau
Institut für Wirschaftsinformatik
Rheinau 3-4
D-W5400 Koblenz

INTRODUCTION AND DEFINITIONS

The purpose of this article is to reflect production from a cybernetic and systems view and in doing so to contribute to the question of information as a factor of production.

This appears promising, because business cybernetics has investigated and gained considerable insight into the three aspects: system, information and control (Schiemenz 1982, 1993).

A system is a general frame into which reality is modelled, when the systems view is used. It consists of elements (objects, parts, subsystems, systems of lower level) with attributes (properties, qualities) and the relations between them. Together with other objects (its environment) it is, on the other hand, part of a more comprehensive system (system of higher level, supersystem) with which it interacts (see e.g. Hall, 1989). Subsystems as well as a supersystem themselves can be understood as systems (aspect of hierarchical order resp. recursion). Fig. 1 shows a systems view of machines as an example.

With regard to information we can distinguish three semiotic levels: Syntactics, semantics and pragmatics. For the purpose of this article, we use the pragmatic level and define information as purposeful knowledge (e.g. Wittmann 1969). Information then can be comprehended as a relation between the three entities data, problem and problem solver (see fig. 2). They indicate relevant data (data that belong to a problem) and are (therefore) needed by the problem solver who has to solve that problem.

Information as a process should be understood as delivering relevant data unknown to the problem solver.

Control is the process of influencing a system by gaining and processing information from this system and its environment (see fig. 1).

By production I understand the transformation of goods (including information, services and 'ungoods', i.e. goods of negative value) into others. It takes place in trade, banking, insurance, education etc. However, in the following we concentrate on industrial production.

Systems Science, Edited by F.A. Stowell *et al.*,
Plenum Press, New York, 1993

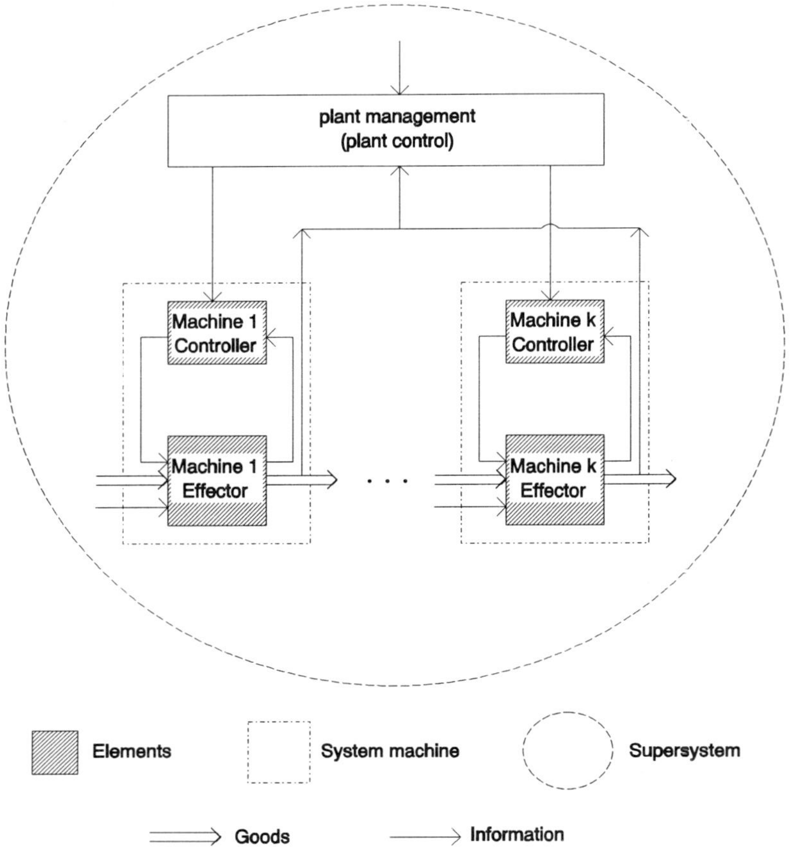

Figure 1. A systems view of machines and their control.

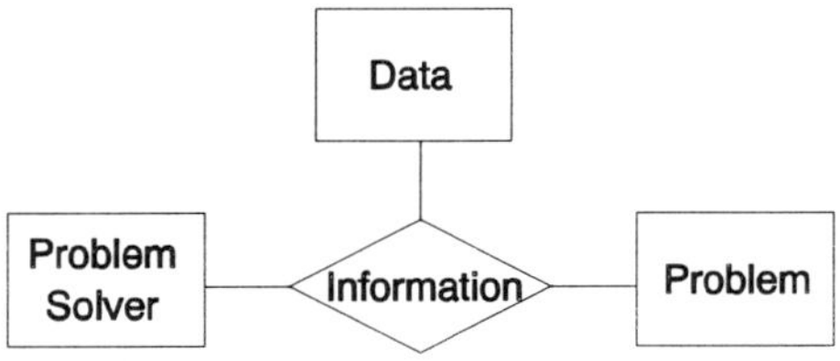

Figure 2. Information as a relation between Data, Problem and Problem Solver.

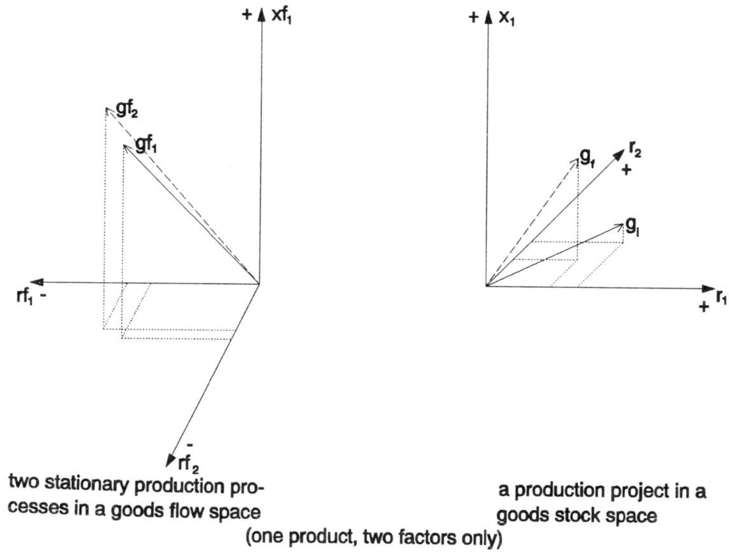

Figure 3. Models of production processes and a production project.

PRODUCTION PROCESSES AND PRODUCTION PROJECTS

Production can be described as a process (production process) or as a project (production project). Both are related. The time integral of a process may be seen as a project. When there is no need to distinguish between them we speak of production.

In production processes we have a continuous flow of input rf and output xf (of productive factors resp. products). Following the activity-analytical approach (e.g. Fandel, 1989), this fact can be described as a vector (or point) gf_1 in a n-dimensional space of flow of goods ('goods flow space'). The vector components with negative sign represent the factors of production. Components with a positive sign are products. In stationary processes gf_1 is independent of time, in nonstationary processes we have a time dependency $gf_1(t)$. In the latter case it may be preferable to model production as a project. We therefore concentrate on the first, stationary, case. Here a change of production (activity) can be described as a switch from gf_1 to another vector gf_2 (see the left diagram in fig. 3).

In production projects, a defined initial state is transferred into a defined final state. Initial and final state can be modelled as vectors (points) g_i and g_f in a n-dimensional space of stocks of goods ('goods stock space'). The production project itself can be described as a switch from g_i to g_f. Stocks that are reduced during this switch are production factors. Stocks that are increased are products. A change of the production (activity) would correspond to a change of the difference-vector $g_f - g_i$.

Production changes themselves may be described as production processes or production projects.

Also gaining knowledge about production (activities) may be described as production (process or project). Input are factors of production, output is knowledge.

DECOMPOSITION, AGGREGATION AND BACKWARD AND FORWARD CHAINING OF PRODUCTION

Production (processes and projects) are generally systems of production of lower order and parts of supersystems of production of higher order.

The factors of a production p are (with the exception of primary matter) as preproducts products of preceding productions pp. Backward chaining is the inclusion of some of the productions pp into p. Inclusion of production pp_1 corresponds to a substitution of vector gf_1 of p by another one that contains the factors of production of pp_1.

The products are (with the exception of consumption goods) factors of production of a following production. Forward integration corresponds to the displacement of gf_1 to a point that contains the products of this further production.

EVALUATION OF PRODUCTION

Production, on any level of aggregation, is positive when the value of output is higher than the value of input. This evaluation is a most difficult problem because of its systemic nature.

In business the value is an indicator of the utility (and disutility) of a good (resp. 'ungood') for its environment. The value therefore depends on the alternative uses in this environment. Indicators for these alternative uses are among others the alternative production activities (resp. consumption activities) of the environment. This would lead to an infinite recursion. The value of broken plastic articles, e.g., may be higher if they can be recycled or burned than if they need disposal.

Therefore, the evaluation of the goods has to be done differently. This may be in a subjective way by the person responsible for the respective production. More objective methods include descriptive or causal prediction. Market prices then can be helpful (for goods with offer and demand). But in many cases there seems to exist no functioning market giving the correct price information. Examples for this may be found in the exploitation of the sources and sinks of the earth like primary energy, ore, rain forests, emission of $C0_2$, disposal of (especially nuclear) waste etc. As a solution one discusses the installation of a workable or respectively an additional market (e.g for pollution rights) combined with appropriate property rights.

We recognize that any evaluation of production presupposes sophisticated information. The deeper contents of this information concerns the position of the production in its environment.

TECHNOLOGICAL INFORMATION, PRODUCTION OF INFORMATION AND INFORMATION IN PRODUCTION

A change from production (activity) gf_1 to gf_2 is positive, when the evaluation (in the sense of chapter 4) of gf_2 is higher than that of gf_1 (if one neglects the transition of the production change). Knowledge about positive production changes is purposeful and is therefore information for people in charge of this production.

Forward and backward chaining are positive, when their evaluation is positive. Knowledge about positive backward and forward chaining is positive and is therefore information.

Information is the result of production. It can in principle be described the same way as the production of material goods. That holds true especially for primary information, that is, for the first modelling of reality for the solver(s) of a specific problem.

The case of derived information is somewhat different and quite specific. There the emphasis lies on the allocation of primary data to other problem solvers and/or problems. The product of the production here lies in the recognition and implementation of new relations 'information' (problem solver, problem, data). Copying and communicating data is only one - and generally the cheapest - part of this production. It is more important to ascertain the fit between data and the other two entities (problem and problem solver). One aspect of this is to avoid what people call 'information overload' or even 'information pollution'. I prefer the term 'data overload' because of our positive definition of information as purposeful knowledge.

Technological knowledge can be integrated into production. This also leads to a change in production or even production technology and a change in the goods space. In automatic control, e.g., technical factors measurement, transmission, storage, processing and activation as well as the knowledge-factors use of models and procedures are added. The relative input of other factors is being reduced. Possibly, the products or their quantities are modified as well.

Production with automatic control is production for which all the above-mentioned statements are valid (recursively). That means that automatically controlled production (including the control process) can be described as a production process or project, that backward or forward chaining is possible and evaluation is necessary. One can also produce information about automatically controlled production processes and incorporate this information into a controller (of a higher level). Similar statements could be made for forward and backward chained processes.

Automation of control as well as forward and backward integration (of the flow of material and information) are important aspects of the tendency towards computer integrated manufacturing (CIM). CIM, too, can be used recursively in the sense of computer integration of computer integrated manufacturing.

DUAL PRODUCTION CONTROL

This shows that production (at a specific level) can be split into information and again production (now at a lower level). Especially, production control is of an informational nature, as well as production functions, models of production, product utilities etc. (see also fig. 1).

That corresponds to the thesis of Conant/Ashby (1970) that every good regulator of a system must be a model of that system. Production functions (which contain the efficient production activities or production points gf) are an important part of such a model for the control of production processes.

Therefore, there are two directions for an improvement of production: Improve the information (resp. control) or improve the (rest of the) production. The overall production is optimal if the utility per $ is the same into the two directions. This is the approach of dual control (Feldbaum 1960).

Dual control itself can again be seen as production for which the statements made for production are valid.

CONCLUSION

Production and Information in systems are recursively related.
The horizontal integration or an aggregation of production can be described as
$\text{production}(i) = \text{production}_1(i\text{-}l) + \ldots + \text{production}_k(i\text{-}l)$
(where i and i-l indicate the system level and 1, .., k the elements)

For the vertical control structure we can write
$\text{production}(i) = \text{control}(i\text{-}l) + \text{production}(i\text{-}l)$, where
$\text{control}(i\text{-}l) = f(\text{information})$

Input and output goods follow the equation
goods(i) = information (i-l) + goods(i-l)

Production functions can be described by
$output_k = f(input, output_{k-1})$, where input, output = goods

The value of goods of a system (production) depends of the value of these goods for the subsystems and the supersystem of this system:

$$value(i) = f(value_1(i-1)), ..., value_k(i-l), value(i+l))$$

REFERENCES

Conant, R. C. and Ashby, W. Ross, 1970, Every good regulator of a system must be a model of that system, *Int. J. of Systems Sci*, Vol. 1, No. 2, pp89-97.

Fandel, G., 1989, "Produktion I", 2. Aufl., Springer-Verlag, Berlin.

Fel'dbaum, A.A., 1960, Dual-Control Theory I-IV, *Automatika i Telemekhanika*, Vol. 21, pp1240-1249, 1453-1464, 1961, Vol. 22, pp3-16, 129-142 (in the English translation in *Automation and Remote Control*).

Hall, Arthur D., 1982, "III, Metasystems Methodology - A New Synthesis and Unification", Pergamon Press, Oxford.

Schiemenz, Bernd, 1982, "Betriebskybernetik", Poeschel, Stuttgart.

Schiemenz, Bernd, 1993, Systemtheorie, betriebswirtschaftliche, *in:* "Handwörterbuch der Betriebswirtschaft", 5. Auflage, Schäffer-Poeschel Verlag, Stuttgart.

Wittmann, Waldemar, 1973, Information, *in:* "Handwörterbuch der Organisation", hrsg. von E. Grochla, Poeschel, Nachdruck, Stuttgart, Sp.699 ff.

SYSTEMS DESIGN AND CONCURRENT ENGINEERING - AN ORGANISATIONAL PERSPECTIVE

Carys Siemieniuch and Murray Sinclair

HUSAT Research Institute
The Elms, Elms Grove
Loughborough, LE11 1RG

INTRODUCTION

This paper describes some implications of the move to Concurrent Engineering (CE) and the requirement for IT -based tools to support this philosophy. The emphasis of the paper is user-centered design, with an organisational perspective. The paper introduces the project on which this research is largely based, then discusses the organisational requirements for the design of IT systems to support a Concurrent Engineering environment.

The Car Project

CAR (CAD/CAM in the Automotive Industry in RACE) is a 4-year project in the European RACE (Research into Advanced Communications in Europe) initiative, focused on the identification of opportunities made possible by the use of IBC in CAD/CAM applications in the European automotive industry. The aim of this project is to improve the effectiveness of the European automotive industry by accelerating design-to-production timescales through the concept of 'concurrent engineering' (CE) or co-operative working, by providing applications based on low cost, high speed, public switched communication links and to identify any requirements for public network services that might emerge. The CAR project aims to show how a combination of high speed data transfer links and multi-media communication facilities can support this by making real-time cooperative working at widely separated sites a reality, both within and between user organisations. It is a crucial element in the project's philosophy that this can only happen if proper attention is given to the human and organisational aspects of the implementation of such technology.

In common with many other organisations, companies in the automotive industry have widely distributed design and manufacturing sites and close contacts with equally widely distributed clients, suppliers and subcontractors. At many sites heavy use is made of CAD modelling systems and other complex CAD/CAM tools. Electronic data transfer is common, but the capacities and speeds of current systems limit the usefulness of this channel. Current technology available to the automotive industry allows sequential integration of design and manufacturing activities, but CE is carried out only at the risk of

Systems Science, Edited by F.A. Stowell *et al.*,
Plenum Press, New York, 1993

prejudicing work elsewhere. Hence the overall integrity of the design is compromised, leading to costly and time consuming remedial work (Heskett, 1991). There is also an increasing emphasis on a requirement for much closer working with suppliers and sub-contractors, but little technological support exists for such activity. Therefore, because the industry has a wide distribution of its various activities, there is an important requirement for distributed computing enabling interaction between all the parties involved (often in different locations), from the initial design to the manufacture of the component concerned, ie sales, finance, maintenance and in particular between design, engineering and production personnel.

Collaborators in the project include three international automotive companies as users, a national telecommunications organisation, international organisations providing expertise in computing, graphics, video- and telecommunications technologies and leading academic institutions, specialising in the computer science and human factors domain respectively.

ORGANISATIONAL AND END USER REQUIREMENTS

One of the main principles underlying the CAR project was that of *user centred design.* Accordingly the project's first emphasis was to provide an in-depth user requirements specification (Siemieniuch, 1990) to be iteratively translated into a functional specification for the design and and implementation of Application Pilot Demonstrators (APDs) for the project. These APDs demonstrated the use of the CAR system in a real-life environment, as in 'Alpha-site' tests in the IT software industry. In parallel was an investigation of the user companies' current technical base and any emerging technology which might influence the design of the APDs. All work in the project was oriented to satisfying the collective needs of the project user organisations. These needs have also been validated by two other major european automotive manufacturers outside the project, thus representing the requirements of approximately 50% of the industry in Europe. The methodology is described in Powrie and Siemieniuch (1990). After a thorough analysis, it was eventually decided to concentrate on the following functional areas (all incorporating high speed file transfer aspects):

- Multimedia electronic user conferencing
- Multimedia messaging system
- Realtime interactive sketching facilities
- Full motion video transfer facilities

The requirements also generated a list of multi-user facilities for support by IBC systems:

- multi-media messaging tool
- interactive informal sketching tool
- formal interactive graphics facilities to allow immediate multi-user access to drawings, solid and surface models, related design data etc
- search and retrieval systems for distributed databases
- comprehensive change control and electronic signing-off systems
- real time multi-media conference systems (point to point and group to group)
- multi-media conference logging and minuting tools
- multi-media facilities for incorporating data not held on-line
- on-line supplier's catalogue/parts database
- non-intrusive, but effective, security mechanisms
- requirements for both specialist workstations with a full range of facilities, with perhaps sub-sets of facilities available on individual workstations.

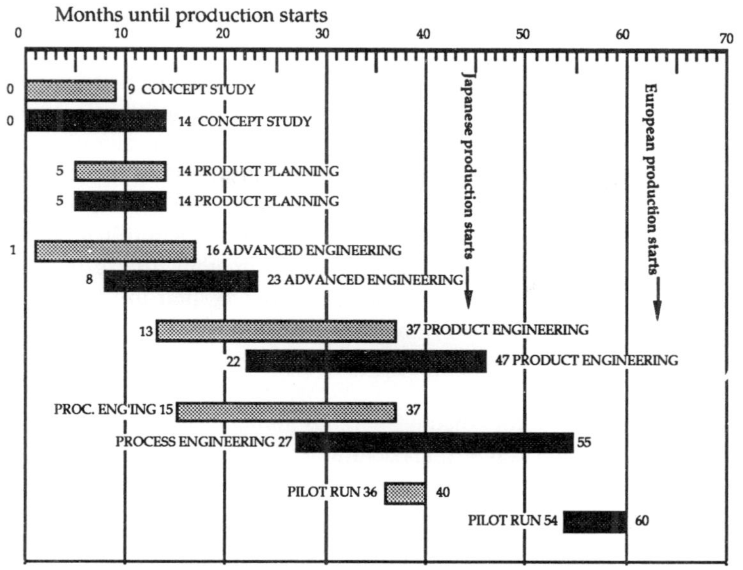

Figure 1. Comparison of average automobile product development times in European (dark bars) and Japanese companies (light bars). From Clark & Fujimoto, 1987a.

CONCURRENT ENGINEERING

To identify the impact on organisations of new applications based on of cooperative working, one must understand the CE concept and the differences between this concept and more conventional engineering philosophies, with respect to the automotive industry. Conventional engineering ('over the fence' engineering), can involve frequent backtracking leading to considerable reworking, which is costly in terms of time and money.

In contrast, CE implies almost-simultaneous design of a product, its development, and preparation for regular volume production. Many authors state that although there are great advantages from this strategy, it is also risky; for instance, Jurgen (1990) has pointed out that development costs can be ten times as much as research costs, and production costs can be 100 times as much. Hence, because in CE there is a rapid commitment of resources to any project, a mistake can be expensive if not caught early. If time is lost because of mistakes, the effects can be severe; a survey indicates that 'a product which is late to market by six months in a product life of five years can lose one third of its total recoverable profit. But if there is a 50% overspend in product development, the loss of profit is less than 4%.' (Nichols, 1990). Similarly, press reports show that in Europe Ford lost $1.5 billions profit to recover market share, by being one year behind General Motors in introducing a new model (Holberton, 1991). Fig. 1 (from Clark & Fujimoto 1987a), shows the reductions in product development time that Concurrent Engineering can produce in the automotive industry (about one third), and in a later paper (Clark & Fujimoto, 1987b), they state: '...the US and the Europeans use about twice as many engineering hours to complete a project of the same size class, with the same number of body types, selling for the same price, as the Japanese.'

ORGANISATIONAL REQUIREMENTS FOR I.T. SYSTEMS TO SUPPORT CONCURRENT ENGINEERING

The context outlined previously gives rise to a multitude of requirements for systems to support cooperative working. Some classes within these levels are discussed below:

- **The importance of creating appropriate organisations, roles and culture**. The advent of usable, efficient means for co-operative working will revolutionise organisations. Given the right technology and communications, it is feasible to uncouple the organisation intellectually, as well as geographically. For some time there have been trends to turn organisations into 'networks'; Drucker (1989), Porter (1990), and Roberts (1990) discuss this from a number of viewpoints. There are other aspects as well. There is more emphasis on the accretion and restructuring of knowledge within human heads, and companies must retain this high-cost expertise. Next, this knowledge must be applied together with other peoples' expertise to solve problems, requiring devolution of authority and responsibility, usually held by managers higher in organisation downwards to the relevant group of people who 'own' the problem that needs to be solved, together with the means to execute this authority. This demands redefinition of the roles of managers. In the workgroup there are conflicting demands for equality of status and for recognition of individual work (at least in the West). This demands careful assessment of the design of jobs, and of reward structures. These requirements have given rise to concepts such as 'skill-based pay', which offer alternative means of satisfying the aspirations of individuals. These, and other issues are discussed in Bainbridge & Quintanilla (1989), Majchrzak (1988), Rosow & Jager (1988). What these authors state is that any evolutionary process in organisations will require changes to the organisational structure; this cannot be avoided.

- **Identification of the competence requirements, and training needs, for Concurrent Engineering**. Today, it is enough for companies to group people together. In future, as more organisations do this, further improvements will be necessary for the organisation to retain its competitive edge. Design engineering knowledge will need restructuring and the old design culture in which people have operated will have to change. Especially, changes in interpersonal and technical communication skills, and understanding of other team-members' technical domains, will be required. These issues are discussed in Cole (1989), Roberts *et al.* (1990), Majchrzak (1988), Rasmussen (1988),and Sinclair (1988).

- **Variations in the design process as practised by individual stakeholders, workgroups, and the organisation**. The design styles of individuals (Dillon & Sweeney, 1988) and of the organisation must be considered to define requirements for design tools. Individuals must be able to tailor applications to suit their needs, and there is a need at the organisational level for tools to monitor the progress of the design process as a whole. These requirements conflict - there is a management-orientated 'official' design process (e.g. Pahl & Beitz, 1984), and the 'unofficial' design process, which is what happens when there is sweat on the individual designer's brow, sitting in front of the workstation. These design processes are not the same, but both must be supported, especially the latter with all its idiosyncracies. Some implications of this are discussed in Sinclair *et al.*, (1989).

- **Identifying requirements in collaborative working to accomplish and control work**. Arising from the comments above, a usable system to enable collaborative working both within an organisation and between organisations must include:
  - an information architecture conforming to Open Distributed Processing principles;
  - a multi-user shell for single and multiple-user applications ;
  - non-intrusive security systems, with security based on the 'ownership' of information, and with information bound tightly to its appropriate applications;

- comprehensive directory facilities linked to the security system;
- usable identification methods for users (who may flit between locations).
- comprehensive version management systems and signing-off protocols ;
- a multi-media messaging facility;
- an interactive, informal sketching tool;
- real-time, multi-media conferencing systems, including point-to-point and group-to-group, able to cope with heterogeneous equipment;
- search mechanisms able to handle multi-media databases in real-time operation;
- multi-media systems to capture data not held on-line;
- high-speed file transfer facilities;
- usable, consistent, robust, and user-configurable interfaces.

Many of these are discussed in Powrie et al. (1990); Palfrey (1991) addresses some user requirements and cultural aspects of security mechanisms. Most of the technical issues of security architectures and mechanisms are addressed in Bull (1991a, 1991b), Stallings (1990), and the series of conferences organised by the IEEE (Anon, 1990). These security and integrity issues are non-trivial; they go to the heart of the structures for distributed co-operative working, since they involve matters of security against outsiders, network administration, the problems of 'ownership' of data and information, response times, directory and naming issues, dynamic re-allocation of system resources to relocating users, absentee users, maintenance of long-term application sessions etc.

Identification of communication issues. This covers quality of service provided; many of these should be invisible to the user. From an ergonomics perspective, this includes:

- dynamic allocation of communications services and bandwidth according to user needs;
- network management issues;
- provision of planning information (costs, quality of service, information about destination facilities) to enable users to specify their needs;
- on-line usage information (cost, facilities used etc);
- assessment of the consequences of new data compression techniques;
- the specification of the minimum requirements upon heterogeneous networks and equipment in order to ensure a usable service to end users.

Brief consideration of these ergonomics-based requirements,will reveal the importance of standards in order to make IT-based co-operative working possible.

OPPORTUNITIES FOR THE FUTURE

Successful implementation of systems like that developed in the CAR project will give substantial benefits to companies by facilitating the introduction of CE concepts. Other benefits include the potential to remodel organisational structures and communication links to take advantage of regional development incentives, differential tax regimes, low labour costs, local availability of raw materials, etc. Organisations may be restructured flexibly, towards teams based on product groups and business activities rather than skill specialisms, with personnel moving between teams as necessary, unhampered by the need to change physical location. Much work might be sub-contracted to individuals or groups working at their own premises, thus reducing overhead costs. Working relationships within the organisation and with clients and suppliers will become closer, facilitating rapid development of products more closely tailored to customer needs. Such benefits will demand considerable attention to the process of integrating the system into its host organisation. These systems will affect user organisations at four levels: interorganisational, organisational, end-user, and task. While the essential demands of the design process itself stay the same, the nature of tasks, the jobs of individuals doing those tasks, how these tasks are completed and their organisational context will all be subject to change. These changes must be carefully managed to take full advantage of the powerful potential of IBC and CE-based systems.

REFERENCES

Anon, 1990, *Proc. conf. research in security and privacy*, May 7-9, Oakland, Ca., IEEE Computer Society Press, Los Alamitos.

Bainbridge, L.S. & Quintanilla, S.A.R., 1989, "Developing skills with information technology", Wiley, Chichester.

Bull, J.A., 1990a, ISA Project: security mechanisms. Report no. APM/RC.154.02, for ESPRIT programme. Architectural Projects Ltd. Cambridge, U.K.

Bull, J.A., 1990b, ISA Project: security architecture. Report no. APM/RC.188.01, for ESPRIT programme. Architectural Projects Ltd. Cambridge, U.K.

Clark, K.B. and Fujimoto, T., 1987a, Overlapping problem solving in product development, "Harvard Business School Working Paper", 87-048.

Clark, K.B. and Fujimoto, T., 1987b, Product development in the world auto industry: strategy, organisation and performance, Paper presented to the "Brookings Inst. Microeconomics Conf.", Dec.3, p22.

Dillon, A. and Sweeney, M., 1988, The application of cognitive psychology to CAD, *in*: "People and Computers IV", D.M. Jones and R. Winder, R. eds., Cambridge University Press, Cambridge.

Drucker, P.F, 1989, "The new realities", Heinemann, Oxford.

Heskett, I., 1991, Product Integrity, Keynote Address at "Product Integrity: a symposium to discuss the process of product development", Nov. 1, 1990, at IIT Centre Boston, Mass.

Hirschhorn, L., 1984, "Beyond mechanisation", MIT Press, Cambridge.

Holberton, S., 1991, R&D deserves a strategic position, *Financial Times*, 12 February.

Jurgen, R.K., 1990, R&D isn't what it used to be..., *IEEE Spectrum*, 27 (10), 34-39.

Majchrzak, A., 1988, "The human side of factory automation", Jossey-Bass, San Francisco.

Nichols, K., 1990, Getting engineering changes under control, *J. Eng. Des.*, 1(1), 5-16.

Pahl, G. and Beitz, W., 1984, "Engineering design", (English edition), Pitman Press, Bath, U.K.

Palfrey, M., 1991, CAD/CAM for the Automotive industry in RACE: Generic application model - intermediate version, Report No. 79/BT/CAR/DS/B/530/b1 of RACE project R1079.

Powrie, S.E., Siemieniuch, C.E. and Sinclair, M.A., 1990, Assimilating IBCN into CIM - some Human Factors aspects, *in*: "Proc. 2nd. Conf. Human Aspects of Advanced Manufacturing and Hybrid Automation", Aug. 12-16, 1990, Honolulu, Hawaii, 307-312.

Powrie, S.E. & Siemieniuch, C.E., 1990, An investigation of User Requirements for broadband communications in the Automotive Industry, *in*: "Human Computer Interaction: Interact '90", D. Diaper, D. Gilmore, G. Cockton, & B. Shackel, eds., Elsevier, Amsterdam.

Porter, M.E., 1990, The competitive advantage of nations. *Harvard Business Review*, 90(2), 73-90.

Rasmussen, J., 1988, A cognitive engineering approach to the modeling of decision making and its organisation in: process control, emergency management, CAD/CAM, office systems, and library systems, *in*: "Advances in man-machine systems research, Vol 4", pp165-243.

Roberts, T. et al., 1990, Manufacturing into the late 1990's. Report AMR/pm 4A&B, to Dept. of Trade & Industry, H.M. Goverment, U.K.

Rosow, J.M. & Zager, R., 1988, "Training - the competitive edge", Jossey-Bass, San Francisco.

Siemieniuch, C.E., 1990, Automotive Industry Computer Aided Engineering Communications Requirements for IBCN - Final Version, Report No. 79/HUSAT/CAR/DS/B/001/b1.

Sinclair, M.A., 1988, Future AMT and Ergonomics: knowledge, organisational issues and human roles, *Applied Ergonomics*, 19(1), 49-54.

Sinclair, M.A., Siemieniuch, C.E. & John, P.A., 1989, A user-centred approach to define high-level requirements for next-generation CAD systems for mechanical engineering, *IEEE Trans. Engineering and Management special Issue: Social impacts of Computer-Aided Design Systems* 36(4), 262-27.

Stallings, W., 1990, "Business data communications", Macmillan, New York.

INTRODUCTION TO STRATEGIC IS AND SYSTEMS METHODOLOGIES

The field of strategic planning and systems methodologies continues to progress through the results of research and the accumulation of practical experience. The papers in this stream reflect this progress in both areas. The stream itself results from the response to the Call for Papers which produced papers that recognised a common element of systems methodologies being employed in systems development, including the strategic planning exercise. This underlying trend recognises the need for an integrative approach to systems development and the use of accepted techniques. For example, data flow diagramming and data modelling being employed at the system and corporate modelling level with the analysis being committed to an organisational data dictionary.

There is some debate about what corporate strategy (and hence strategic IS) is

> " ... almost no consensus exists about what corporate strategy is, much less about how a company should formulate it"[1]

but it has been recognised that strategic planning of information systems should be part of corporate strategic planning:

> "strategic planning ... must be carried out as part of the overall business planning functions if the Information Systems to meet future organisational needs are to be available when required."[2]

It is worthwhile to provide a definition of Strategic Information Systems which can be considered in the light of the papers to follow; Strategic IS is the consideration of the policy and position of Information Systems organisation-wide for some years ahead.

A working definition of methodologies can be suggested to embrace a language, a philosophy and a synthesis of best practice rather than as "a more impressive-sounding synonym for methods" recognised by Blaug[3] some years ago.

Methodologies and methods are now being used more widely in systems development than ever before and their applicability considered in other aspects of systems development such as the management of change and human resource development. The papers that follow provide further insights to this changing field of strategic planning and systems methodologies and provide some evidence of the progress toward cohesive, integrated systems.

James G. Howell
Information Systems, Department of Computing Science,
University of Paisley

Systems Science, Edited by F.A. Stowell *et al.*,
Plenum Press, New York, 1993

REFERENCES

1. M.E. Porter, From Competitive Advantage to Corporate Strategy, *in*: "Readings in Strategic Management", D. Asch and C. Bowman, eds., Macmillan, London (1989).
2. L. Capper, "Information, Data and Information Systems", Open University text, P792, p16, Open University Press, (1988).
3. M. Blaug, "The Methodology of Economics", p xi, CUP, Cambridge, (1982).

SOME AREAS OF ECONOMIC THEORY VIEWED SYSTEMICALLY

Tony Brauer

School of Development Studies
University of East Anglia
University Plain
Norwich NR4 7TJ

AN OUTLINE OF THE ARGUMENT

In general it is unwise to use maps out of context. Economic theory is an approach to mapping that suffers from severe constraints, which are not always recognised. It may be inferred that if economic theory is to be practically useful it must be subordinate to at least one other mapping technique. The role of economics can be defended, but not adequately.

THE METHODOLOGY OF ECONOMICS

It is a notable characteristic of human beings that we make maps. These maps can be extremely useful in helping us to get where we're going, though it is clearly important to have the right map. At the same time, it is not necessary, and may not even be possible, for any map to be a perfect truth; Picasso may have helped us to understand ourselves and the context of our lives, but we wouldn't rely on a surgeon who had learned anatomy from a cubist painting. A map is only applicable in certain contexts.

This is not to say that a map cannot evolve. If the function of a map is to help us to understand the world in which we live, it may be necessary for it to evolve both in the range of its concerns and in its techniques of representation. The world is changing, and we change too, in what we are seeking to achieve, and in how we are seeking to achieve it. As we change, we learn, and our overall conceptual map changes; though, of course, few, if any, individuals have a completely integrated set of maps. We all suffer from cognitive dissonance to some extent: but where there are inconsistencies, explanations are required. These explanations may uncover internal inconsistencies or misrepresentations in a map, or may indicate that we are using it out of context. Nevertheless, maps which have been superceded at one level may continue to be useful at another. We probably use Newtonian physics and Euclidean geometry when playing billiards, even though we accept the ultimate superiority of Einsteinian relativity in a curved universe. Any map may be useful in certain circumstances, but it is an essential aspect of its use that it should be used in the right context.

Systems Science, Edited by F.A. Stowell *et al.*,
Plenum Press, New York, 1993

It is my contention that orthodox economic theory, and neo-classical economics in particular, are mapping systems which tend to be used grossly out of context. The basis of my complaint is probably familiar. Economics in general seeks to avoid the obvious assumption that it is a social science. It might be thought that a sufficient reason for economics to accept the methodological paradigm of social science would be that it is concerned with human motivation and behaviour. Economics, however, evades this inference by using Rational Economic Man as its model of a human being. Such a creature is assumed to have perfect knowledge of the present and the relevant future, and an inexhaustible drive to maximise 'his' satisfactions through the market. Theories based on this model can thus evade the problems of complex motivation and behaviour by denying their existence; but even though this allows economics to maintain its status as a non-social science, it cannot yet be described as a natural science, since that which it studies does not exist in nature. Conventional economics, in this analysis at least, appears to be an inapplicable abstraction.

A second reason for insisting that any socially applicable economic model should be translated into the social scientific paradigm is that it can have no other purpose than to enhance the welfare of human beings. (This, of course, incoporates the welfare of our life support system, planet earth, and, increasingly, other fragments of the universe.) Many people, myself included, would argue that welfare can only be measured in terms of the extent to which human needs are satisfied. The effectiveness of any social system is measured by criteria related to needs which are common to us all, despite the fact that the way in which they are satisfied is culturally, situationally, and personally specific. An applicable social science must therefore take into account the level of satisfaction of our needs for protection, sustenance, and participation; affection, understanding, identity and autonomy; creativity and play. (The list is adapted from Mallman's original.) (Max-Neef, 1986) Such judgements are far too nebulous to permit their use in a natural science, so economic methodology reproduces its familiar trick of declaring that that whatever cannot readily be explained is irrelevant. The welfare of human beings is declared to be whatever emerges from the market process, on the ground that we freely express our preferences through the market. Neo-classical market theory therefore depends for its relevance on the assumption that the market is perfectly free of constraints.

This is clearly not true. The market cannot be free in this sense, if for no other reason than that a precondition of the perfect market is perfect information, which would require illimitable time and understanding, and a degree of insight into the future which none of us has. It is the well-being of the market which is measured by economists, and this variable does not share an identity with the welfare of human beings and of our life-support system. Both rational economic man and the free market may be special cases, but it is more likely that they are purely theoretical abstractions. Economic theory can only maintain its status as a value-free science by disqualifying itself as an applicable social science.

THE OPERATION OF THE MARKET

It is as yet uncertain the extent to which a systems approach can supplant the conventional mapping techniques of economic methodology, but, in terms of its capacity to explain the outcomes of social policies, it would appear to have advantages. The application of systems techniques can certainly not be considered unreasonable, since the claim that the free market is an effective social mechanism depends on the assumption that it operates as a communication system. Milton Friedman's highly influential view, for example, is that the market is efficient because it automatically transmits incentives to produce at the same time as it transmits consumer choices (Friedman, 1984). However, this essentially traditional idea may easily be translated into systemic terms: *the market is a system to enable the distribution of entitlements to act as an incentive to useful activity.* It is described graphically in figure 1.

Figure 1. A first version of the market as a communication system.

It should be noted that unit price and effective demand form a negative feedback cycle. If an exogenous factor such as technical innovation or a change in tastes leads to temporary disequilibrium, the system is supposed to correct itself automatically, so that what is produced is consumed, and the efficient allocation of resources is maintained. This allocative efficiency is often perceived as the great virtue of the market, and in so far as that perception is true, market forces may be thought to lead to desirable outcomes; but there is much more to be said on the subject. Friedman, for example, has extended the claims to incorporate the idea that the majority of social motivation should take place through the market (Friedman, 1984). The system is efficient, he argues, because the incentive to produce is identical to the notification, through purchase, of what is desired.

However, a map is only useful within its range of applicability. In any communication system the paradigm suggests that we look for distortion, and once we do so, it swiftly becomes apparent that the model will have to be emended if it is to represent a fundamental mechanism of social organisation. Distortions arise in the system for a number of reasons, two of which may be thought to be of particular significance. We simply do not have the time to be aware of all the implications of our market decisions; nor is it in the interests of producers to make us aware of those implications which might deflect us from a purchase, particularly since production decisions in a mass society generally have to be made before consumption decisions. Also, in the general case, it is not in the interests of producers actually to satisfy consumers; we must be persuaded that satisfaction arises from continual consumption, whether or not this is in our real interests. The market is structurally incapable of transmitting our values as effective incentives, because of the transaction costs of perfect information, and because it is in the interests, and within the power, of large scale producers to distort our perceptions of our needs.

It is one of the alleged virtues of the market as a social mechanism that it can be assumed that voluntary exchanges express our values and our political preferences. We are

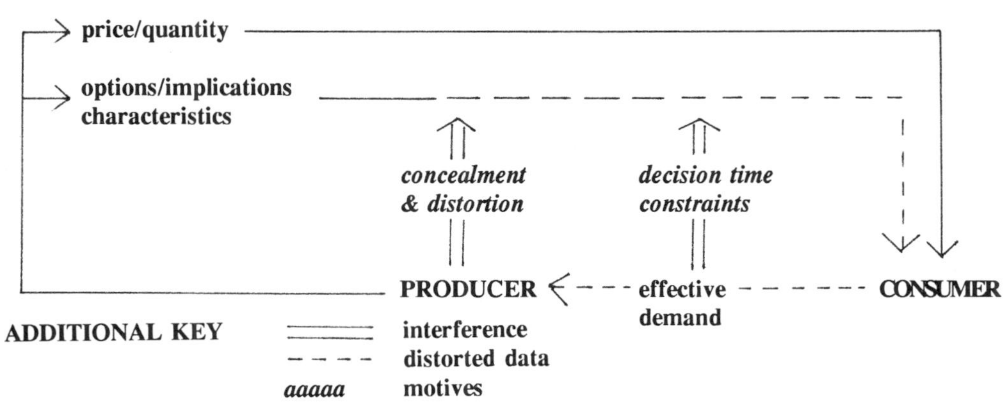

Figure 2. A second version of the market as a communication system.

voting with our money. It has been shown why it is doubtful in theory that this is an adequate description, and in practice we almost certainly do not trust the market. When we buy the cheaper of two similar products, we make assumptions about certain minimal conditions associated with the productive process; for example, we assume that the reason for its cheapness is not brutal and systematic exploitation in an oppressive society, but efficiency in production. However, if we make this assumption, it is not because we think we have provided an incentive to treat humans with respect by our purchase; it is because we assume that there is a superordinate power which deploys constraints on the context of production. There seems, therefore, to be neither a theoretical nor an empirical basis for the assumption that the market is adequate as a system for the communication of political decisions.

If such is the case, it is not surprising that the market does not necessarily lead to outcomes which most people would have chosen. In addition, in the words of Keynes, "the world is not so governed from above that private and social interest always coincide"; our individual decisions may in aggregate be against our individual interests, even though each decision in itself appears rational. On this reading, the invisible hand model is intrinsically deceptive. If we want our social mechanisms to reward socially desirable behaviour, and if we want the market to act as such a mechanism, it will clearly require continuous intervention to compensate for distortions in communications.

This, however, is not the end of the story. It must be asked what effects the incentives to distort or inhibit flows of information will have on the system: a system which, it will be recalled, is intended to enable the distribution of entitlements to act as an incentive to useful activity. It has been shown why the output of a market economy cannot be assumed to be intrinsically beneficial. In the heirarchy of maps, the market model is evidently subordinate to at least one other; to the map which describes what we actually want. Economic theory must be subordinate to other analytic approaches if it is to be beneficial to human welfare; and this becomes still more evident if we look at the system in its distributive role.

In a just society, entitlement levels are a function of contributions to the welfare of the society. This is pragmatically as well as morally desirable: unless a distribution is seen to be acceptably just, it will have a divisive effect on society.

Contributions are assessed in relation to a social evaluative system. Individuals will not share all the values, and one of the most significant influences on the ratification of social value systems is ideology. Ideology is accessible to influence, and that influence is largely determined by wealth. Thus someone with a high entitlement level can use some of their resources to influence the assessment of contributions to reinforce those differentials which

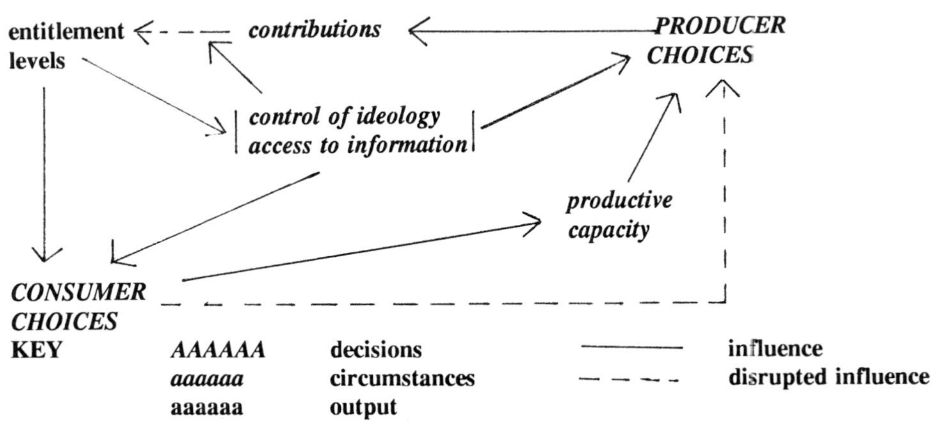

Figure 3. The market as a system to enable the distributions of entitlements to act as an incentive to useful activity.

are to their advantage; and there are other positive feedback loops within the system. A high entitlement level gives differential access to information, which allows production decisions to be made in one's favour. A high entitlement level will also provide the individual with a wider range of well-informed consumption decisions, which may in turn lead to differentials in individual productive capacity which will perpetuate advantages. It should also be noted that just as a position of advantage within this system will initiate a positive feedback loop, so too will disadvantage. Those among us who have least entitlement are also likely to have least health, least development of productive capacity, least information to guide our consumption and production decisions, and least influence over what is to be considered a valuable contribution to society.

The crucial point is that the market systematically leads to the reinforcement of entitlement status; the rich get richer and the poor poorer. This is evident within Britain after twelve years of neo-classical economic policy, and in international affairs, where the World Bank and the IMF have sought to impose free market policies on world trade. There are certainly sub-systems which allow individuals and nations to produce different outcomes from those which the free market would produce, but the overall tendency is clear. As a system to enable the distribution of entitlements to act as an incentive to useful activity, the market is severely constrained. It is a communication system with an explicit incentive for those involved to distort the flow of information, so that an advantage can be gained; and it also reinforces morally arbitrary differences in the opportunities available to individuals and communities through differential access to information, and authority in interpretation.

THE EPISTEMOLOGY OF ECONOMICS

The first step in this discussion was to point to a qualification to the evident usefulness of maps; that they should only be used within a recognisable scope of application. In this context, neo-classical economic theory appears to have a severely restricted scope. Because the characteristics which determine its epistemological status are also those which determine whether or not it is applicable as a social mechanism, it can be a value-free science, or it can be an applicable social science, but it evidently cannot be both.

Many economists would, of course, deny this. Friedman argues that economics is a positive science despite the remoteness of its models from our experience (Friedman, 1979). As long as economics is predictively successful, he claims, it has a practical function; but how predictively successful is it? The Treasury's computer model may on occasion make an approximately accurate forecast, but two questions must be borne in mind. Firstly, do econometric models provide evidence of an exceptional predictive capacity? As Morgernstern has shown, the problems of data collection of, for example, national income, imply an error of ± 10% to ± 15%: and, as we know, however skilfully nonsense is processed, it can never cease to be nonsense. The data themselves are fairly meaningless as an empirical verification. Furthermore, to take a simple example, the inaccuracy of Treasury predictions for GDP between the years 1985 and 1990 averaged 1.6 per cent. Had one instead simply assumed that for any year GDP could be predicted by multiplying the previous year's GDP by the same year's growth, the average level of inaccuracy would have been 2.4 per cent. It seems that the crudest extrapolation, ignoring all knowledge of other variables, would have been nearly as accurate as the Treasury's econometric model. It is true that the variation over the whole period was only between 6% and 10% growth, which would help extrapolative predictions; but if the variance was only 4%, the average Treasury error of 1.6% is singularly unimpressive. The second point has been made most thoroughly by Meadows and Robinson (1985):

> The mathematical requirements of estimation...impose restrictions on econometric models which cause them to represent a world quite different from either the world of economic theory

or the world seen by other modelling schools. Econometric models cannot, then, be used as evidence of the positivism of economics.

It would seem, therefore, both that the predictive success of economics is severely constrained, and that it is, in any case, not economic but econometric models which achieve the limited predictive success that might be used to support Friedman's claim. Certainly where his neo-classical approach has been most enthusiastically applied, in Britain and the United States during the 1980's, the economic legacy has been horrific. If these maps promised to guide us on the golden road to Samarkhand, but led us instead off the end of a disintegrating pier into knee-deep mud, why are they allowed to persist?

At one epistemological level it is simply because core theories are extremely hard to displace. As Lakatos (1978) pointed out, core theories are protected by a series of peripheral qualifications, which can be modified or abandoned without threat to the underlying core theory. In the case of economics, the methodology is protected from contrary evidence by attributing the source of any perturbation to excluded variables. This may protect economics as an explanation of an abstract world; the consistency of the core theory is not affected; but however useful this may appear as a defence of economic methodology, it unfortunately implies that the core theory is inapplicable when it refers to circumstances in which the excluded variables cannot be excluded: in our lives, for example. It seems more than possible, then, that we need not follow economists in their evaluation of the role economics should play in determining social policy.

An interpretation based on Kuhn's analysis would reinforce these doubts. He argued that scientific communities are extremely resistant to fundamental changes of conception, in part because competing methodologies are incommensurable, so that, by every canon, a new metatheoretical proposition is utterly indigestible. This interpretation could certainly apply to economics. Practitioners may be right to consider it a deterministic science dealing with closed systems; but if it is concerned with closed systems and quantifiable criteria, can it reasonably be seen as a predictive mechanism for human systems which are open, and which involve values which are neither readily quantified nor easily expressed through the market?

The methodology of economics does not equip it to act as a reliable instrument for social planning. Data describing the objectives of economic activity and explaining some of the anomalies that arise are systematically excluded from the paradigm. If economics is to be used at all in a political role, it must almost certainly be made subordinate to other, more holistic approaches to map-making.

REFERENCES

Central Statistical Office, 1992, United Kingdom National Accounts, HMSO, London.

Friedman, M., 1979, The methodology of positive economics, *in:* "Philosophy and Economic Theory", F. Hahn and M. Hollis, ed., Oxford University Press, Oxford.

Friedman, M., 1984, "Market or Plan?", CRCE, London.

Keynes, J.M., 1964, Essays in persuasion, *in*: "Economic Philosophy", J. Robinson, ed., Penguin, Harmondsworth.

Kuhn, T., 1970, "The Structure of Scientific Revolutions", University of Chicago Press, Chicago.

Lakatos, I., 1978, "The Structure of Scientific Research Programmes", Cambridge University Press, Cambridge.

Max-Neef, M., 1986, Human scale economics, *in:* "The Living Economy", P. Ekins, ed., RKP, London.

Morgernstern, O., 1963, "On the Accuracy of Economic Observations", Princeton University Press, Princeton, N.J..

Meadows, D.H., and Robinson, J.M., 1985, "The Electronic Oracle", Wiley, New York, p51.

H.M. Treasury, 1985-1992, Financial Statement and Budget Reports, HMSO, London.

"AM I USING A SYSTEMS APPROACH?" - A FRAMEWORK FOR RESPONSE

Gordon Dyer

Open University
Cintra House
12 Hills Road
Cambridge CB2 lPF

INTRODUCTION

The difficulties of defining the systems approach are not surprising; systems is an eclectic discipline. Members of the contributory disciplines will, even within a systems approach, firstly want to incorporate their own visions and techniques to problems they encounter before they readily accept others. The mathematician-systemist will look for outlets for computer modelling; the psychologist-systemist will look for ways for human-behaviour centred exploration. Other points adding to our communication problem are that:

* we can assume change in the systems approach in the last 20/30 years
* each of us engaged with the systems movement at different starting lines
* each of us has undergone a process of personal development - a learning journey - as we have faced and interpreted issues in our "real worlds" and have then been exposed to, and experienced, the application of systems ideas. But, if the systems community is to achieve its full potential and collaborate and cooperate effectively to achieve systemic re-design of future social systems, the communication problem must be addressed. It is vital for us all to recognise the roots of our own system approach and to appreciate how our contribution, and that of colleagues, fits into the totality of systems practice.

The structure of the paper follows from a perspective that systems approaches have developed from two major themes ie to provide a better way of either, "managing the present", or, "re-designing the future". For a full version of this paper, see Dyer[1].

MANAGING THE PRESENT c.1975

The optimism in the 70s derived from the notion that we now had an approach which could be used for more effective management of human activity systems, which could help with the complexity of modern life and its technology. We had a new approach to understanding how and why current designed systems behaved the way they did. So when

Systems Science, Edited by F.A. Stowell *et al.*,
Plenum Press, New York, 1993

change was needed, what was important was to do a thorough systems analysis to identify the crucial inter-relationships between system components and to recognize the complexity of objectives present in the situation. Thus by following a systems-based methodology it should still be possible for managers/problem owners or consultants to choose a course of action which would lead to improvements in efficiency and effectiveness through improved control and coordination. The environment of the 70s was perceived as being relatively stable which allowed the luxury of this view. The roots of this management view are also reflected in the following pragmatic definition of a "system" used by the OU:

* a system is an assembly of components connected together in an organized way.
* the components are affected by being in the system and are changed if they leave
* the assembly of components does something
* the assembly has been identified by a human being as of interest

The main clue to the management focus in the definition is the phrase "identified by a human being as of interest". The emphasis is on the individual manager or owner of a problem, not on a community. I shall return to this point later.

SYSTEMS METHODOLOGIES c 1975

From the mid 70s a wide range of systems based methodologies were available. Some were characteristically top-down management approaches, as reflected, for example in the decision analysis of De Neuville and Keeney[2]. I shall characterise such a top-down-management-influenced approach as a FIRST generation methodology. Checkland's Soft System Methodology also appeared in this period. His definition of a system .. "a model of a whole entity; when applied to human activity, the model is characterised fundamentally in terms of hierarchical structure, emergent properties, communication and control"[3] - hinted at a top down vision but in practice the methodology has proved very flexible and allows for a wide range of consultation by the problem solver. Effective within organisations it has its limitations in dealing with community-based and social problems.

Methodologies differ considerably in their detail but typically all will have five iterative phases where evidence of a systems approach might be identified. Three phases

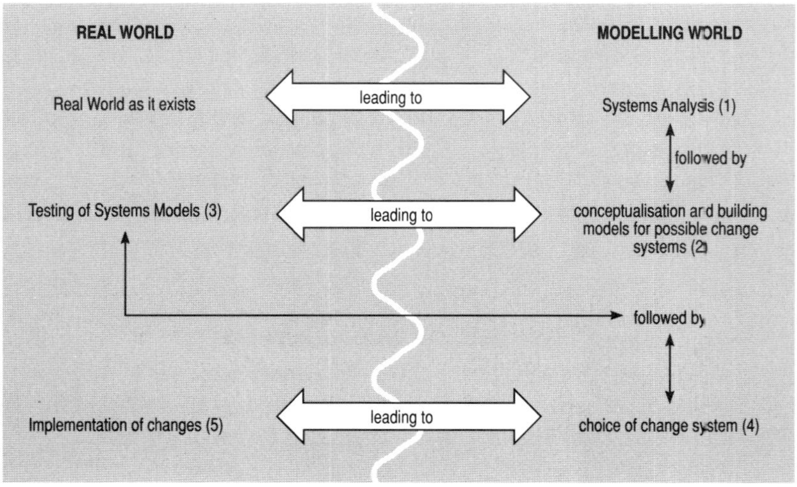

Figure 1. Typical Systems Methodology Structure.

involve the interface between the "real world"and the "modelling world" of the system thinker; two phases are within the modelling world, as shown in Figure 1.

The prevailing management philosophy will strongly influence actions at each stage although we can see considerable variety in the contribution of different systemists as they work to personal/professional weltanschauung. Hence in 1975:

(1) **SYSTEMS ANALYSIS**
- holistic approach reflected in eg systems maps, influence diagrams, numerical data collected on interconnected subsystems. But what maps are drawn, what data is collected, who is interviewed is conditioned by an underlying vision of how organizing in the real world has failed in terms of structures and processes. Hence in the 1975 view of an effective organization, the emphasis was on searching for lack of control and coordination, interviews were largely with those who manage, control and coordinate, or who had resource or position power

(2) **CONCEPTUALISATION AND MODELLING**
- concern with more effective forms of controlling and coordinating from a top-down approach
- consultants search for what might be "feasible and desirable" in that context
- mathematical simulations of various kinds aimed at clarifying and reducing the amount of uncertainty in the decision-making process. An important contribution but some danger in quantifying the unquantifiable as modelling boundaries pushed out

(3) **TESTING MODELS**
- the feasible and desirable options discussed with a limited number of stakeholders, those in control
- mathematical simulations may not be sufficiently verifiable in the real world

(4) **CHOICE**
- often made following recommendation, or by limited group of managers; many disenfranchised
- output from simulation may carry great weight

(5) **IMPLEMENTATION**
- left to problem owner/manager to implement. Emphasis in reorganized systems and their monitoring likely to reinforce bureaucracy and closed-loop control rather than open-loop control (delegation)

These then, are the roots of systems practice deriving from the mid-70s. You may recognize the organization/people or modelling themes in your own personal development and experience with systems. If not, it could be that your point of contact with the systems movement occurred later, say, 1988 by when the practice of systems was influenced by new management thinking, to which we now turn.

MANAGING THE PRESENT c.1988

New thinking from about 1988 came from a recognition of the environment as being extremely fast-changing (even chaotic), effected by a combination of new competition, rapid change in technology, and social upheavals affecting consumer choice[4]. This requires organizations to be flexible, responsive and adaptive to change in their of organizational structures, product/ service delivery, processes and attitudes. These requirements dictate that the experience and potential of all members of staff needs to released, in order to meet the challenges. Table 1, based on the work of Newman[5], contrasts the differentiating characteristics of the bureaucracy and "new managerialism", labelled by Newman as an organizational form better matched to current conditions:

Table I. Characteristics of Organisation Forms

BUREAUCRACY	NEW MANAGERIALISM
Position-centred	Person-centred
Repetition oriented	Creativity oriented
Efficiency seeking	Innovation and efficiency seeking
Rules oriented	Results oriented
Rewards adherence to rules	Rewards outcomes
Information tightly channelled through formal structure	Maximise communication flow
Internally focused	Externally focused - alliances and coalitions
Uniformity	Diversity
Perception of a workforce	Human Resource Management
Top down	Bottom up

Devolution, decentralisation and delegation of responsibility and power have become essential. Matrix- rather than bureaucratic-structures are now much more desirable. Control is more open-loop with semi-autonomous and self-managing teams common place. The role of the manager or leader is now to develop an inspiring vision: to manage by example rather than to exercise control.

SYSTEMS METHODOLOGIES c.1988

The ideas flowing from new managerialism are reflected in latest systems methodologies throughout all of the five stages, together with new mathematical modelling techniques now being explored. From 1988:

(1) **SYSTEMS ANALYSIS**

- emphasis now on collecting a wide range of views from all categories of staff eg through Nominal Group Techniques. Conflict not buried but allowed to surface and to be sifted and discussed for the best ideas. The investigation likely to be extended to the range of stakeholders associated with the system.
- emergence of critical systems theory[6]
- Group work emphasises democratic, participative, sharing processes. "Imaging" or "dreaming" techniques might be used.
- explorations with chaos and catastrophe theory to explain the behaviour of complex phenomena

(2) **CONCEPTUALISATION AND MODELLING**

- a new interpretation on what is feasible and desirable. Emphasis on inculcating a culture which will lead to preferred forms of participative, democratic styles of work and to encourage effective team-work. Typically this leads to options for matrix structures in which influence is more widely dispersed, eg ideas for quality circles.- users involved as designers, emancipatory systems[7]
- usually requires parallel investigation of alternatives for personal development of managers to encourage new styles of leadership. Emphasis in new models for structures and processes is on devolution of responsibility and power to teams.
- explorations with new forms of simulation with large-scale systems eg with Emergy models[8]

(3) **TESTING**

- options for change discussed and explored with a wide range of stakeholders. Options might be mapped onto the real world through imaging sessions

(4) **CHOICE**

- a group decision within arrangements for devolved power. The role of top-management is to provide strategic direction within which the choice can be made.

(5) **IMPLEMENTATION**

- team implementation
- the "guarantee of the design" is based on individual/collective team agreements to perform bolstered by mutual respect. Control is largely open-loop (delegated)

The differences - in terms of how the stages are conducted and the perception behind the techniques which are employed between the typical 1975 and 1988 version of methodology - is sufficient to call the latter SECOND generation.

The comparison and contrast between the methodologies should enable members of the systems community to see where their own contribution, and that of colleagues, fits into systems practice. There is no suggestion in this categorisation that SECOND generation approaches have replaced FIRST generation. Jackson confirms that with his argument for complementarism[9]. Both have a range of applicability. FIRST generation approaches are particularly useful for education in systems. As the predominant metaphors are derived from bio-systems and mechanisms, whose behaviour can be predicted and controlled, they provide a good starting point for understanding concepts. This then leads easily to critique of method when extended to the real world task of managing an organization or activity.

An individual whose work maps onto one or more of the 5 stages of either generation methodology has the right to claim that they are using a systems approach. Students of the systems approach and systems theorists are more likely to work with Stages 1 - 3 than with Stages 4 - 5. The latter are more likely to be in the realm of systems consultants/practitioners and enlightened problem owners.

REDESIGNING THE FUTURE

While systems approaches have been used successfully to find effective solutions to a wide range of problems, from within small to large organizations and have helped with some large scale public projects, they have been unsuccessful in terms of the other great hope: that they could also be used to seek a better future for mankind. One exception perhaps, is with the protection of world ecology. In the major social systems, including those for education and human development, however, we sense continuing failure. Transforming this situation is an enormous challenge; it now seems clear that improvements will only be possible if we can find higher human values and ideals to use for guidance. The systems movement is responding in a number of ways. In recent Annual Conferences of the ISSS, papers on the topics of wisdom and justice have been increasing; spirituality has emerged as a new topic for systems thinkers. Banathy[10] has provided a vision of a way forward, through his concept of evolutionary guidance and the possibility of transforming education through purposeful design. There are, as yet, no formal methodological structures to assist this work; in contrast there is some disenchantment with such methodologies in that their pragmatism has proved an obstacle in their use for designing a better future. We can see some characteristics emerging in this work:

* a search for a guiding vision and a set of core values
* based on community-led design with concern for community survival and growth
* the environment is conceived as part of the wider system (community) rather than a distinctly separate and potentially hostile entity. Thus the design considerations must extend to the environment
* SECOND generation approaches which stress democratic, participative and consultative approaches are tending to be explored
* design is seen as a continuous and evolutionary process rather than a one-off solution to a defined problem

Banathy has provided the vision of a direction, and some of the necessary milestones to plan this design journey. The challenges of this journey will require coordination of all the systems expertise we have to hand. As we develop tentative systems approaches to designing the future under this vision, we will be justified in calling them THIRD generation approaches.

CONCLUSION

Systems practice can be grouped under five activity headings: analysis; model building; model testing; selection of change system; and, change implementation. This list of activities can help us to identify where the techniques and methods we - and colleagues - use, fit into systems practice. Put together and applied iteratively these activities can constitute a systems-based methodology for "managing the present". Methodologies of the 70s reflect the dominant management thinking of that time. Bio-system and engineering metaphors influence models of desirable change, with resulting emphasis on top-down control, tendency for imposed change, and coordination via bureaucratic forms of organization. By the late 80s, with a new management vision that organizational survival depends on responsiveness and flexibility, systems approaches now reflect the search for new goals of participative change, delegation of responsibility and power, and open-loop control. While the same five activity headings are identifiable in the methodologies of the 70s and the late 80s, the techniques employed are so different that they justify categorization as FIRST and SECOND generation methodologies respectively. Both generation types remain important. They provide complementary elements for systems education and practice in managing change in organizations.

Systems practice can record little success in terms of designing a better future. But work on new approaches is now developing. This is not based on formal methodological structure, but some key characteristics in the design endeavours can be identified: there is emphasis on high human ideals a guiding vision of collective community concern; design is community-led and is seen as a continuous and evolving process; the design considerations extend to the environment. The challenges inherent in the task will justify an appellation of THIRD generation systems approaches, but it is likely that these will draw largely from SECOND generation approaches.

REFERENCES

1. G.C. Dyer, Am I using a Systems Approach?, *Systems Practice*, Vol. 6, 3, (forthcoming, June 1993).
2. R. de Neuville and R.L. Keeney, Systems Evaluation through Decision Analysis: Mexico City Airport, *Journal of Systems Engineering*, Vol 3, No 1. pp 34-50 (1972).
3. P.B. Checkland. "Systems Thinking, Systems Practice", Wiley, Chichester pp.317-318 (1981).
4. J. Krantz and T.N. Gilmore, The Splitting of Leadership and Management as a Social Defense, *Human Relations Journal*, Vol 43, No. 2 pp183-204 (1990).
5. J. Clarke and J. Newman, Managing to Survive - Dilemmas of Changing Evolutionary Forms in the Public Sector, in: "Social Policy Association Conference", University of Nottingham, July 1992.
6. R.L. Flood and M.C. Jackson. "Creative Problem Solving: Total Systems Intervention", Wiley, Chichester, (1991).
7. M.C. Jackson. "Systems Methodologies for the Management Sciences", Plenum, New York (1991).
8. H.T. Odum, Principles of Emergy Analysis for Public Policy, in: Emergy Analysis of Shrimp Mariculture in Ecuador. Working Paper prepared for Coastal Resources Center, University of Rhode Island, Narragansett, R.I. pp 89-110 (1991).
9. B.H. Banathy, Systems Design: A Creative Response to the Current Educational Predicament, A Paper presented at the Asilomar Systems Conference, California, USA, December 1991.
10. M.C. Jackson. ibid. pp 253-270.

EXPERIENCE OF TEACHING THE SYSTEMS-BASED METHODOLOGY OF INTERACTIVE MANAGEMENT

F.R.Janes[1], R.K.Ellis[2], K.Hammer[1]

[1] Department of Systems Science
City University
Northampton Square
London EC1V 0HB

[2] Centre for Systems Studies
University of Hull
Hull HU6 7RX

INTRODUCTION

The aim of this paper is to report on the authors' experience of teaching the methodology of interactive management at university level. The paper begins with a brief review of the framework of interactive management as an approach to making group work more efficient and effective. The way in which the methodology has been taught on various postgraduate and undergraduate courses is then illustrated through discussion of a case study on an ailing high-tech manufacturing company. The paper describes the sequence of methods used in the case and gives examples of the products obtained from the group work undertaken. The benefits in terms of learning and the difficulties of teaching using this approach are considered.

THE INTERACTIVE MANAGEMENT METHODOLOGY (IMM)

In using the expression Interactive Management, the term "interactive" refers to the way that Interactive Management both promotes communication and understanding among interacting participants, and enables the interactions among problem components to be examined. The term "management" refers to the way that Interactive Management is concerned both with the management of issues and problems, and with the management of the process of group work. The framework for Interactive Management involves a number of important aspects. In particular, the notion that work with groups can be made more productive through due attention to five components: the group, the facilitator, the methods, the computer support, and the decision support room. Also, the notion that appropriate

Systems Science, Edited by F.A. Stowell *et al.*,
Plenum Press, New York, 1993

methods may be selected for the intelligence, design and choice phases of management decision making. The framework owes its origins to the work of Warfield (1982, 1990).

The need to bring together individuals of diverse backgrounds when tackling complex issues such as planning often results in group work involving meetings. Handy (1985:154) suggests that, on average, 50% of the time of managers is spent working in one kind of group or another, whilst for senior managers this may be as high as 80%. Christakis (1984) reported that a study by the EXXON and XEROX Corporations in 1982 had shown that 17.5 million meetings per day took place in the U.S.A. This is approximately four billion meetings per year. However, all too often dissatisfaction with meetings is widespread. Wainwright (1987:5) lists 16 sources of dissatisfaction including: badly run meetings, avoidance of decisions, attendance by the wrong people, wandering from the point, and lack of participation. Poor preparation and poor chairmanship are two primary sources of such dissatisfaction.

In the context of meetings of groups for complex problem solving, the five components referred to above help to ensure that the resulting meetings are productive and overcome many of the common sources of dissatisfaction.The components may be defined as follows.

Participant Group. Individuals, including stakeholders, with content knowledge relevant to the issue.
Facilitator. To manage the work of the group and guide participants through the steps of the methods used.
Methods. To assist the group in intelligence, design and choice activities.
Computer Support. To run software associated with the methods and to provide display facilities and word processing capability.
Decision Support Room. To provide an environment equipped for efficient group work.

The methods form a core part of the framework. Each provides a process which, under the guidance of a facilitator, can be used to structure the participants' activities and focus their thinking and effort. Warfield (1982, 1990) describes the methods as "consensus" methods suitable for intelligence, design and choice activities. The term "consensus" is used to signify a number of characteristics. First, that the participative design of the methods is such that all participants have an opportunity to contribute, and to indicate that they have had that opportunity, e.g., by voting. Second, that the ideas of participants are acknowledged by the group so that individuals feel that their views have been listened to and understood. Third, that the products resulting from the use of the methods are generally agreed by the individuals involved and, that even if they are not in complete agreement with the products, they can "live with" the outcome . The term "method" is used here rather than "methodology" to describe the individual methods, in the sense that each method provides a way of doing something, or a process for attaining an objective. Interactive Management as a whole may then be considered as offering a process of inquiry constituting a 'methodology".

The seven methods selected by Warfield are as follows.

Ideawriting. A quick and efficient idea-generation method for use by groups.
Nominal Group Technique (NGT). An efficient idea-generation method for use by groups which allows for claTifying, editing and obtaining an initial ranking of the generated ideas.
Delphi. A method for systematically developing the views of a panel of individuals on an issue through written responses to successive questionnaires when it is not practical for the participants to meet together or anonymity is desired.
Interpretive Structural Modelling (ISM). A computer-assisted method for mapping the elements of a complex issue into an agreed diagrammatic structure by an informed group.
Options Field. A method for conceptual design by groups which shows in one diagram all the conceived dimensions of the prospective designs and all the conceived options within each dimension.

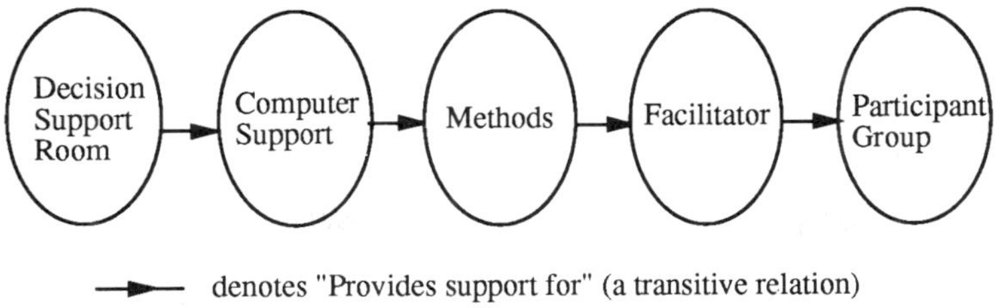

Figure 1. The five components shown as a transitive digraph.

Options Profile. A method for selecting specific alternative conceptual designs from an options field.
Tradeoff Analysis Method. A method for evaluating and choosing between alternative designs by comparing the differences between the altematives, and for clearly documenting the basis of decisions made.

Important features which these methods possess include:

- Capability of being used effectively by a group for idea generation, idea structuring, design or decision making activities,
- Sound behavioural and technical design,
- Promoting contributions from all participants,
- Encouraging focussed and open dialogue,
- Clear definition of the roles of participants and facilitator,
- Provision for iteration,
- Provision for documentation of results,
- Making efficient use of participants' time.

The five components for productive group work described ahove support one another as shown in Figure 1. The arrows in the figure represent the transitive relation "provides support for". Thus the facilitator provides support for the group whilst the methods support both the group and the facilitator; the decision support room supports all four other components.

EXPERIENCE OF TEACHING THE METHODOLOGY

A major difficulty which faces a group of people who are confronted by a complex issue is that of generating, controlling and then effectively using ideas relevant to the issue. The Interactive Management Methodology (IMM) encourages an "open system" attitude among group members and provides informed ownership of decisions based upon commitment to plans and solutions.

To be able to to use IMM in effective intervention it is necessary for potential users to be taught both the theoretical basis and practice in application. To that end, IMM has been taught - often as an integral part of courses dealing with problem management - on undergraduate and postgraduate degrees in Management and Systems, Business Studies and

Engineering Management. Students are provided with a series of lectures which establish the theoretical foundations of IMM. Based upon the premise that the best way to understand a methodology is to practice using it, the remainder of the teaching of IMM is case study driven.

One such case study is a derivative of a real-world intervention using IMM in a poorly performing organisation engaged in electronic equipment design and manufacture. From the real-world intervention an organisation history and culrent scenario were derived, as were executive roles for students to assume. For the purposes of this paper we shall use an example in which experienced engineers, following a postgraduate course in engineering management, undertook this particular case study. What was very interesting for us as teachers was the recognition that although the students possessed specialised working knowledge, they had no previous knowledge of IMM. This combination serves as a good example of the support offered to interactive decision making by IMM.

Students were asked to consider the fundamental question of what the organisation should be trying to achieve over the next three to five years. A basic learning requirement was that the students should use IMM to address the complex problem situation facing them in the roles of executives of the organisation. Various IMM methods were used in the sequence shown in Figure 2. As teachers we acted strictly in facilitation roles, having no part to play in the decision making or the outcomes. Eacn group of students proceeded through the case, building an Intent Structure, an Options Field and Options Profiles to produce an organisational mission and strategy. These were formally presented using audio/visual presentations and reports. Examples of outputs achieved are provided in Figures 3 and 4.

It is worth commenting that this group of engineering managers experienced difficulties, early in the case study, in achieving consensus through critical debate. This was - partly we feel - due to their individual experience and training which demands immediate and "decisive" action. Once they had learned to listen to each other, to respect and recognise the

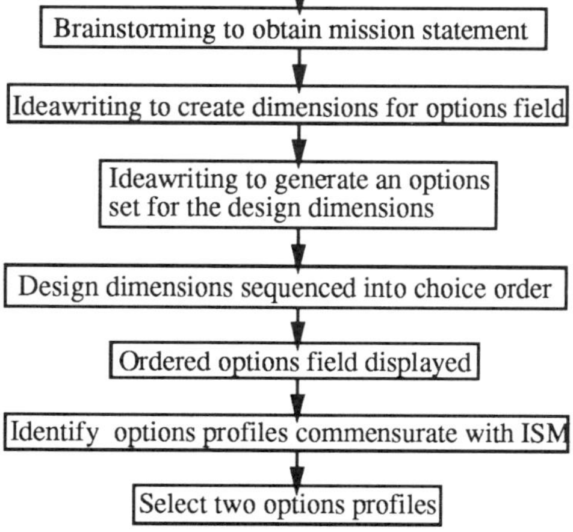

Figure 2. List of processes/techniques used.

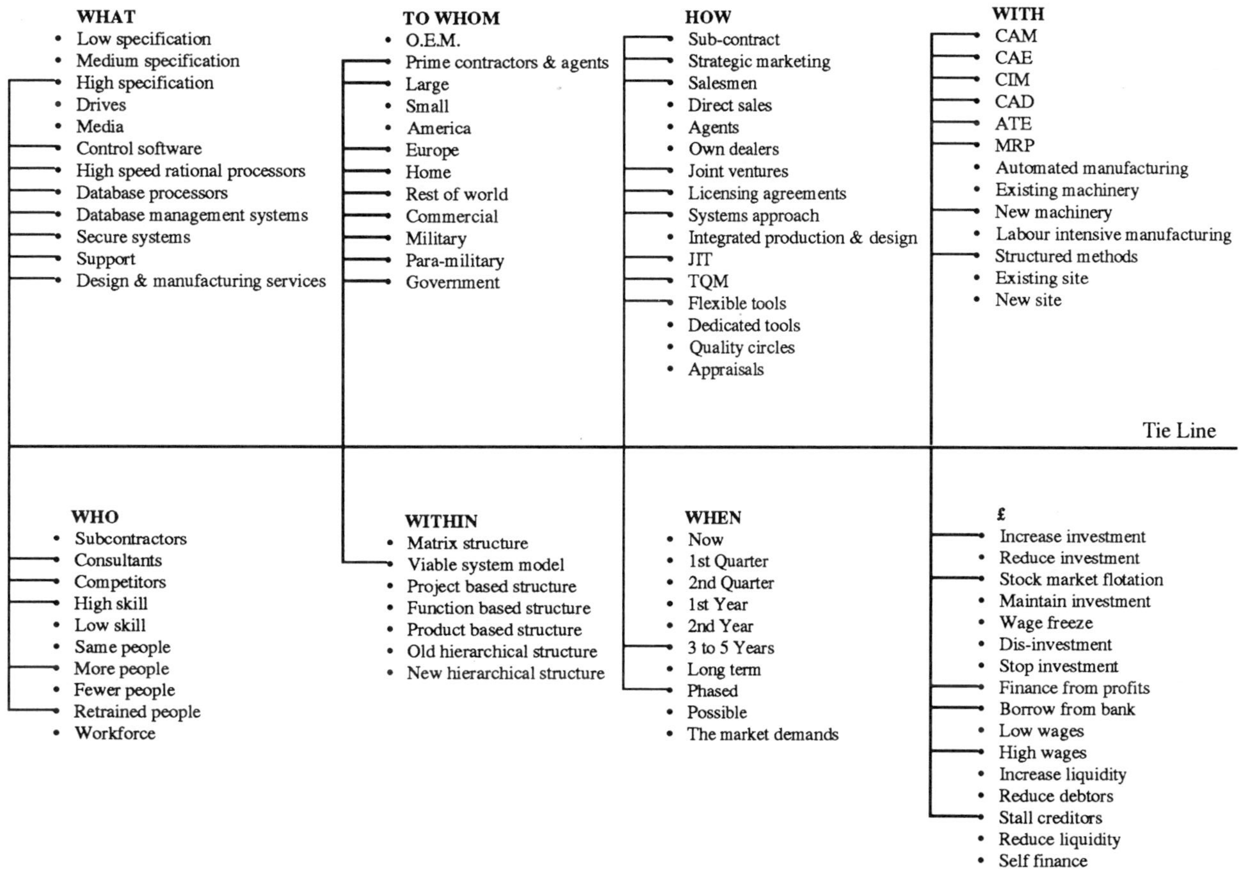

Figure 3. ISM intent structure.

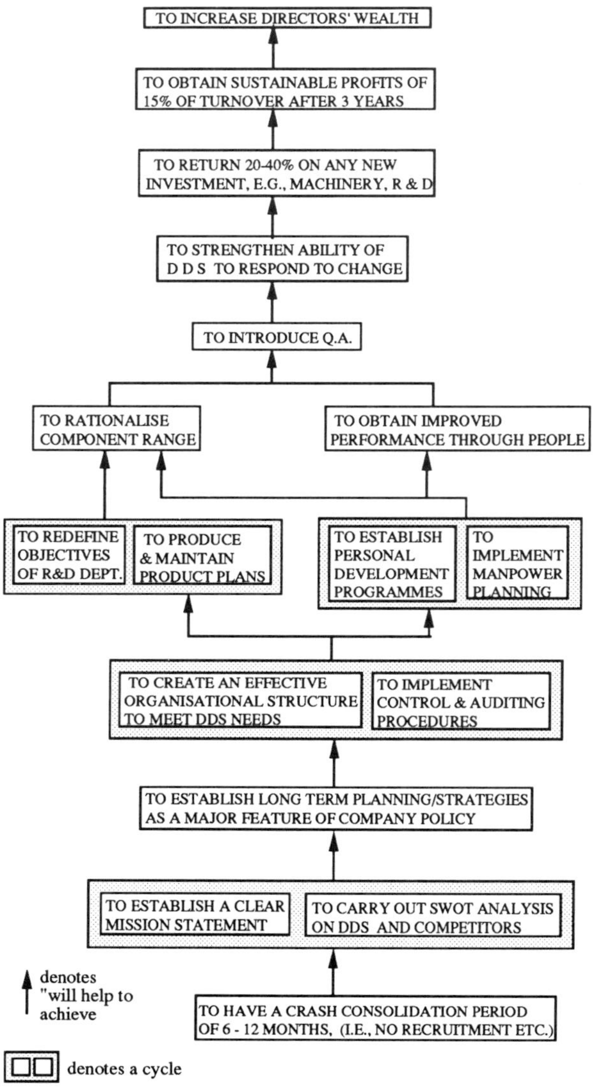

Figure 4. Options profile for a growth strategy for DDS.

views and participation of their colleagues, the process became smoother but not devoid of differences of opinion. This difficulty has been experienced with other groups to a greater or lesser degree, and we have concluded that this need to take immediate action is a common factor in our culture and is reinforced by our educational system.

Commenting briefly on undergraduate groups, it is worth noting that despite a lack of real world experience the methods were handled with surprising maturity. There is, however, a tendency to produce "ideal-type" strategies which do not always take full account of potential real-world consequences.

In all our teaching experiences associated with IMM we have found the reaction of student groups to be generally positive, with many individuals going on to use the processes in other aspects of their professional and/or academic work.

REFERENCES

Christakis, A.N., 1984, Management of complexity: the interactive approach, *Continuing the Process,* Newsletter, Graduate School, U.S. Department of Agriculture, Washington, D.C., 1(4):1-2.

Handy, C.B., 1985, Chap. 6, On the workings of groups, *in:* "Understanding Organisations", Penguin, London.

Wainwright, G.R., 1987, "Meetings and Committee Procedure", Hodder and Stoughton, London.

Warfield, J.N., 1982, Organisations and systems learning, *General Systems,* 27: 5-74.

Warfield, J.N., 1990, "A Science of Generic Design: Managing Complexity Through Systems Design", Intersystems Publications, Salinas, California.

A CONTINGENCY APPROACH FOR THE SELECTION OF A RESEARCH METHODOLOGY (RM)

Louis Jacques Filion

Universite du Quebec a Trois-Rivieres
C.P. 500
Trois-Rivieres, QC G9A 5H7
Canada

INTRODUCTION

Why do researchers select a particular RM? What are the reasons behind that choice? Before I began to do research, I was a practitioner for 15 years. It seemed to me then, generally speaking, that quantitative research helped influence how and when I would do something, whereas qualitative research helped me structure my thinking about the doing. When I first started doing research, the field I was researching - entrepreneurship - was not well-structured, and I thought a qualitative approach would be of most help in structuring what I needed to find for education purposes; in other words, to develop relevant material for training would-be entrepreneurs. This type of approach seemed to allow me to remain closer to the empirical world. My view was that using a human activity systems approach would help me understand emerging categories of entrepreneurs' activity systems.

I have now been researching in the field of entrepreneurship for a decade (Filion, 1991), and over the years have come to realize that there may also be many other reasons for choosing a particular research methodology. This paper presents and discusses a suggested selection process.

First, the rationale of the selection of a given research methodology (RM) is discussed. Seven elements are presented and explained: customer, researcher's W of what is scientific, researcher's training in the craft of a particular methodology, area of specialization, category, purpose and resources of the research. It is suggested that most research can be classified in one of two categories: content or structure research. Quantitative RMs seem to be used more frequently for content research and qualitative RMs for structure research. The paper concludes that there should be a contingency approach to support the selection of a RM which takes into account the elements of the proposed model.

Systems Science, Edited by F.A. Stowell *et al.*,
Plenum Press, New York, 1993

SELECTION OF A RESEARCH METHODOLOGY

There are several elements that may help explain the selection of a particular RM. The following sections will examine some such elements identified by this researcher as having a particular impact. The sequence in which they are introduced in no way reflects their level of importance.

Customer

Many elements can influence whatever happens in a human activity system. Of the six expressed by the CATWOE[1], three seem to be more important in this context: the W, the owner and the customer. There may be many different customers, depending on whether the research is to be published, and if so, where. If it is to be published in a journal, what kind of journal is it? If it is to be published in a special report, who will be reading it: managers of a specific enterprise, technocrats from funding bodies, etc.? Will it be read by students, practitioners, researchers, policy makers?

Often, the customers for academic research are other academics. This seems to be very much the case in my own field, entrepreneurship and small business, where most research results are of no practical use for owner-managers as such. Some research results are incorporated in course material. Some are used by other researchers. In fact, it often happens that researchers are their own best customers, since they keep building and adding to what they have already done.

Other customers are policy makers - especially in Europe, where integration of so many countries is taking place - who seem to use research results increasingly for policy development in complex environments. This may also be a result of higher education levels among policy makers, who now seem better able to understand research and its implications for policy making.

However research customers are identified, they can be broken down into four main categories: academics, students, practitioners and policy makers. The area of specialization of each will influence the category of research. The situation they are in will influence the purpose of the research. These three elements - area of specialization, category and purpose of research - will influence the selection of the RM. The selection process will also be influenced by the culture or habits of the customer's domain. For example, economists and psychologists seem to use essentially quantitative methodologies, while management and education specialists tend towards qualitative methodologies. The customer's values - especially if the customer is the funding body - may also considerably influence the final choice.

W of What is Scientific

Researchers may be influenced by what is recognized as scientific in their area of specialization, or conversely, they may become dissatisfied with the state of knowledge in their field and modify their W of what is generally considered in their area as being "scientific". Researchers seem to be looking increasingly for alternative ways of doing research. Whatever the situation, their W of what is scientific has an undeniable impact on

---

[1] CATWOE: the customers (beneficiaries or victims of a notional system), the system's actors, its transformation process, the Weltanschauung which gives it meaning, the owners who could destroy it and the environmental constraints to which it is subject. (Checkland, 1981: 18)

RM selection. What is considered scientific by one may not be by another (Feyerabend, 1975; Suppe, 1977; Chalmers, 1982).

The Researcher's Training

How researchers look at research - their W - is a key element here. It derives in part from their W of what is scientific, but essentially from the way they have been trained to do research. It is easy to say that one type of RM should be used for this research, and another type for that research, but in practice it takes so long to master the craft of one RM properly that many researchers, having got used to it, stay with the same one indefinitely. Very few manage to master more than one RM in their careers, and when they do, the mastered methodologies are usually in the same research paradigm. Clearly, epistemological considerations are important here, since they lead researchers to view research from a particular perspective (Burrell and Morgan, 1979). For instance "radical structuralists" and "functionalists" who look at the objective world tend to prefer quantitative methodologies, while "radical humanists" and "interpretive" researchers who look at the subjective world tend to prefer qualitative methodologies.

Most researchers prefer to stretch their RM to do what they want, rather than having to learn another to deal with research of a different type. For example, when building a house, we use different craftspeople, and even within the same craft, some things can be done more easily by some people than by others. The same applies to any human activity. The training and subsequent expertise acquired by people for doing things in a certain way sets a trend which they tend to follow throughout their working lives.

You may buy a car to meet your regular transportation needs, and if, on one occasion, you have a lot of goods or people to transport, you will perhaps do it with that car because you are familiar with it. You may take out the back seat to leave more room for your boxes, or you may ask your passengers to sit in certain ways to maximize the space available. But if the situation arises again, you may think of subcontracting to a truck or bus firm, or even buying a bus or a truck if it happens on a regular basis. And if you have small quantities of goods or people to carry on a frequent basis, you may consider exchanging your vehicle for an estate car.

In the same way, some RMs are more multifunctional than others. One thing is certain: we are moving towards a plurality of RMs[2] (Morgan,1983). But does this mean that researchers are now trained to use more than one RM? Increasingly, today, the answer is "yes", and in general it is accepted that many different RMs exist to generate research results that are useful for theory and practice (Filion, 1990). The fact remains, however, that the basic training and expertise gained by researchers in a given RM is one of the key elements determining RM selection.

Area of Specialization

Observation at academic conferences (Academy of Management (US), Administrative Science Association of Canada, International Council for Small Business, etc.) reveals that in some areas of specialization mainly quantitative RMs are used, while in others qualitative RMs (including systems approaches) are preferred. Almost invariably, research papers fall into one of two categories: content analysis and structure research (call it container or framework research). Research in the first category normally involves the use of quantitative RMs, and in the second, of qualitative RMs. Table 1 shows a division of areas of specialization.

---

[2] As an example, take a look at the number of books on research approaches and methodologies published by just one publisher (Sage) over the last ten years.

Table 1. Areas of Specialization and Categories of Research

Content Analysis	Structure Research
Industrial Psychology	Management
Accounting	Strategy
Human Resources Management	Education
Small Business	Entrepreneurship
Marketing	
Finance	
Economics	

Some areas could be listed in both columns, since their characteristics involve both content and structure research: information systems and operations management are two obvious examples. The Table 1 classification is derived strictly from publications in major academic journals and conferences. There seem to be cultures and ways of doing things in each area that almost naturally demand the use of a certain type of RM. Perhaps the state of knowledge in a given area may have something to do with the choice too: are we in a pre-paradigmatic state (Reason and Rowan, 1 9 8 1), an advanced paradigmatic situation or a well-established paradigm? For instance, economics and psychology, compared to other fields, seem to be in better-established paradigms, and this may be a reason why researchers in those areas tend to look at specific elements in a more precise way. They tend to use quantitative approaches. However, some areas are less well-established, and may require some structuring. For this reason qualitative approaches seem to be more appropriate.

Finally, it is worth noting that subjects classified under content are areas that affect part of the organization, while those classified under structure tend to include the organization as a whole.

Research Categories

As mentioned earlier, almost all research can be classified as either content or structure research. Table 2 presents the characteristics of each category.

Research category seems to be a key element in determining RM choice. Apart from the factors appearing in Figure 1, and apart from the culture of the domain and the paradigmatic stage, there also seems to be a logic, or rationale, that leads researchers doing content research to select quantitative RMs and those doing structure research to select qualitative RMs. Naturally, the two types are done from different perspectives. However, the main difference is that content research works on parts, while structure research takes into account inter-relations between the parts.

Purpose of Research

What is the immediate purpose of the research? Is it to solve a practical or a theoretical problem, to increase efficiency or effectiveness in a given organization, to support policy makers or to add to knowledge? The purpose, like the customer, may affect the resources and time available. It will also influence RM selection.

Table 2. Two Categories of Research

CONTENT	STRUCTURE
Part	Whole
An isolated element is studied	Inter-relationships between elements are studied
Vertical: deep research on a specific element	Horizontal: research on effects of one element on related ones
Functional	Systemic
Ontologic	Morphologic
Look at essence	Look at flux
Look for causal effects	Look for emerging patterns
Look for universal laws	Look for models
Involve rational analysis	Involve imaginative thinking
Offer knowledge that helps improve managerial activities	Offer thinking framework that supports activity process
Mechanistic: the research customer applies it as such when appropriate conditions are met	Organic: the research customer adapts it to meet activity needs
Absolute	Contingent to time, space, location, situation
Appeal to the specialist	Appeal to the generalist
Knowledge of domain means power	Power lies in equilibrium of whole

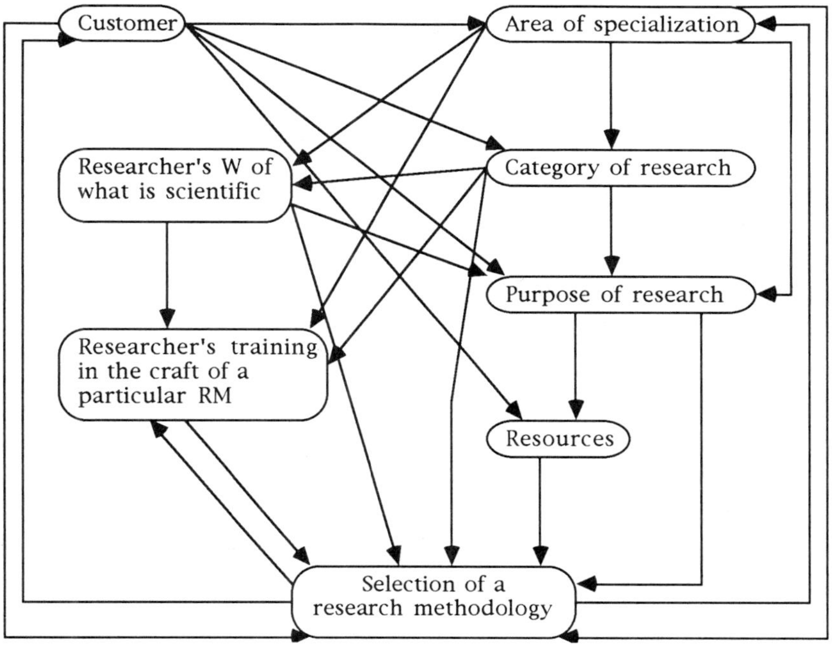

Figure 1. Selection of a research methodology.

CONTINGENCY APPROACH FOR THE SELECTION OF A RM

All the above elements influence RM selection to some degree, but some seem to influence it more than others. The main elements appear to be the researcher's W of what is scientific and the research training received. But the area of specialization, which includes the subject being researched and determines the research category, seems crucial. Everything else seems to flow from there. It determines the category of research and influences t h e researchers also seems to depend considerably on it. All these factors together have a major impact on RM selection.

Due to the shortage of space, not much has been said here about the different types of RM, which can almost all be classified under two main headings: quantitative and qualitative. As observed earlier, content research is usually done using quantitative methodologies, and structure research using qualitative methodologies. Clearly, the process does not stop here. However, the purpose of this paper is simply to show that there is a rationale in the selection of a given RM. It should not be absolute, it should be contingent: it all depends on the situation of each element presented in Figure 1 and how they are inter-related. It may well be that one way of doing innovative research is to select RMs that do not follow the usual pattern in the area of specialization and that focus on the customer's needs and the purpose of the research. If people keep using the same type of RM, they may well keep finding the same things!

REFERENCES

Burrell, G. and Morgan, G., 1979, "Sociological Paradigms and Organisational Analysis", Heinemann, London.

Chalmers, A.F., 1982, "What is This Thing Called Science?", The Open University Press, Milton Keynes.

Checkland, P.B., 1981, "Systems Thinking, Systems Practice", Wiley, Chichester.

Feyerabend, P., 1975, "Against Method", Verso, London.

Filion, L.J., 1988, "The Strategy of Successful Entrepreneurs in Small Business: Vision, Relationships and Anticipatory Learning", Ph.D. Thesis, University of Lancaster, Great Britain (UMI 8919064).

Filion, L.J., 1990, Research, epistemology and research methodologies in administrative science: ten reflective guidelines for the 1990s. Working Paper Series No. NC90-09. National Centre for Management Research and Development, The University of Western Ontario, Canada, January.

Filion, L.J., 1991, Vision and Relations: Elements for an Entrepreneurial Metamodel, *International Small Business Journal*, Vol. 9, No. 2, Jan., 2640.

Morgan, G., 1983, "Beyond Method", Sage, London.

Reason, P. and Rowan, J., 1981, "Human Inquiry: A Sourcebook of New Paradigm Research", Wiley, Chichester.

Suppe, F., 1977, "The Structure of Scientific Theories", Illini Book Edition, Illinois.

VULNERABLE SYSTEMS, FAILURE AND DISASTER

Tom Horlick-Jones, Joyce Fortune and Geoff Peters

Centre for Technology Strategy
Open University
Milton Keynes, UK

ABSTRACT

This paper reports on a contribution to the continuing debate about the influence of changes in political economy on the propensity for major accidents and disasters. Disasters are seen as resulting from systems made vulnerable by interaction with their socio-economic environment, together with interactions between these systems and with their environments.

INTRODUCTION

During the latter years of the 1980s much speculation took place about the possibility that UK disasters such as those at Kings Cross, Clapham and Bradford shared certain common underlying causes related to the state of the economy and Government policies (e.g. Cook, 1989). Attempts to make simple "cause-effect" linkages have received little support, but a wider debate has taken place about the influence of factors such as economic changes, public administration and the regulation of industry on major failures (Horlick-Jones, 1990; 1991; Handmer, 1992; Hood and Jackson, 1992). Such speculation raised important questions about the nature of disasters, the possibility that certain mechanisms may affect the likelihood of their occurrence across a broad range of unconnected failures, and the significance of perceived trends of such events.

Disasters rarely arise from technical failures alone, or from the results of simple human errors, and research suggests that it is informative to regard them as the failure of complex socio-technical, or human activity, systems (Bignell and Fortune, 1984; Peters and Fortune, 1991).

The work of Turner (1978) and Perrow (1984) has been very influential in creating conceptualizations of some of the processes at work in the evolution of failures and disasters. Turner introduced the notion of an "incubation" period, possibly extending over a long period, during which minor failures and other events accumulate to create an underlying causal chain. A "precipitating event" can then reveal the latent structure that has evolved during that period. Perrow categorises technologies displaying high risk by two parameters, interactive complexity and tight coupling. The former refers to the capacity of the socio-technical system associated with that technology to generate unforeseen, and unexpected interactions. The latter

Systems Science, Edited by F.A. Stowell *et al.*,
Plenum Press, New York, 1993

is a measure of the speed of interactions, and the extent to which parts of the system can be isolated.

Reason (1990) has described both Turner's and Perrow's models as being founded on a "resident pathogen metaphor", in which there is a suggestion of an analogy with the aetiology of multiple causation illness, involving the "triggering" of an underlying "system morbidity". He describes such "resident pathogen" theories as being "far from workable", claiming the necessity to and difficulty of demonstrating clear causal links between indicators of "morbidity" and accident likelihood.

Despite these criticisms, recent research by the authors (Horlick-Jones, 1990; 1991; Horlick-Jones, Fortune and Peters, 1991a; 1991b) questions how the influence of a socio-technical system's environment might contribute to the creation of "system morbidity". In the following section this research is reviewed, followed by an assessment set in the context of related work, and recent developments. Finally some possible implications for future technology policy will be considered.

SYSTEM VULNERABILITY

Figure 1 illustrates the flow of ideas comprising the recent work. The investigation was prompted by a widespread perception of a significant frequency of major disasters in a relatively short space of time.

In addition to the small data sets which present difficulty for an orthodox statistical analysis, research has identified a number of interpretative difficulties, including contextual factors that have an important bearing on the significance and reliability of statistics that do exist. Disasters, it was argued, take place, and are perceived, in specific socio-economic, historical, cultural and chronological contexts.

Measures of disaster magnitude, such as the so-called Bradford Disaster Scale, which seeks to define the magnitude of a disaster as a function of the logarithm of resulting deaths, may fail to represent the totality of harm generated by these events, and, it is argued, are not value-free. In addition, the psychological impact of disasters may generate very significant reflexive "learning" processes, so reducing the applicability of stochastic models to investigate trends.

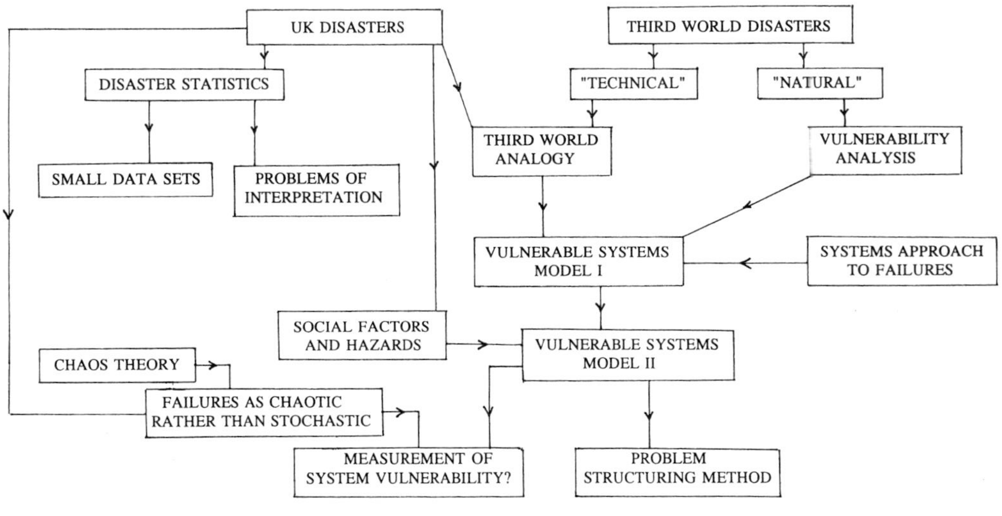

Figure 1. The structure of the ideas.

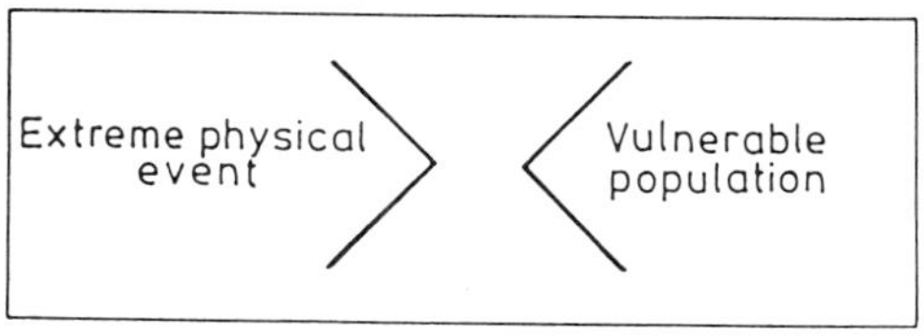

Figure 2. Vulnerability.

A further development in the understanding of the possible underlying processes which may lead to disastrous failures, was the identification of an apparent correspondence between a number of UK disasters and features commonly associated with "technological" failures in developing countries. According to this "The Third World Analogy", these features are:

Large numbers of people involved and overcrowding
Out-of-date and poorly maintained equipment
Poorly trained workers
Inadequate preparations for emergencies

Subsequent, examination has focussed on possible explanatory mechanisms. In particular, attention has focused on the powerful framework centred on the concept of "vulnerability", introduced by geographers and others almost twenty years ago, to explore the key role played by socio-economic factors in creating a weakness in responding to, and recovery from, extreme climatological or geophysical events.

According to this model, disasters are "the interface between an extreme physical event and a vulnerable population". This model is illustrated in figure 2. The use of this conceptual framework has recently been reviewed by Horlick-Jones (1993).

The vulnerability framework can be combined with approaches used to understand disasters in terms of systems failures. The result is illustrated in figure 3. According to this model, a complex socio-technical system is made vulnerable to failure by a dialectical interaction of numerous factors, both internal and environmental. The resilience of the system, i.e. its ability to absorb disturbance, is eroded to the extent that failure may be triggered by relatively minor events.

Recognition that in technologically advanced societies correlations exist between social factors and vulnerability to various hazards leads to a second "vulnerable systems" model. This is illustrated in figure 4. The concept of vulnerability is now used in a generalised way, referring to the erosion of a system's resilience resulting from an interaction between the perceived system, and its socio-economic environment.

Figure 4 provides a first general model of disasters, whether "natural" or "technological". In both, the resilience of social systems may be eroded by the interaction with the socio-economic environment. In "technological" disasters, a dialectical interaction between a socio-technical system's state variables and its environment yields a sensitivity, or metastability, that can be triggered by a relatively small disturbance to produce a catastrophic failure. That failure may in turn be instrumental in the generation of a hostile environment for an associated social system. In "natural" disasters, the hostile environment arises from some physical event. Hybrid disasters, such as famine or disease, may combine both processes.

SIGNIFICANCE OF THE WORK

Two important themes emerge from this work on vulnerable systems, the key role of organisational and operational factors such as maintenance and regulation in a socio-technical system's environment, and the prospects for diagnosing system vulnerability.

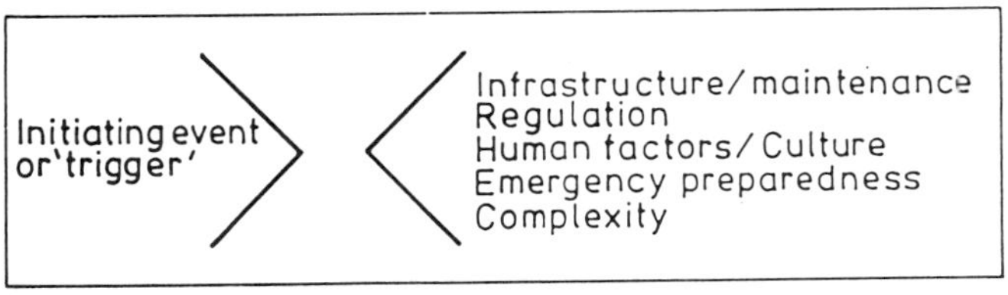

Figure 3. Vulnerable systems model stage I.

First, the importance of stressing that socio-technical systems are open, and therefore interact with their socio-economic environments. Perrow's model has been criticised for being too deterministic (Wynne, 1988), or for ignoring the influence of specific human factors and corporate cultures (Douglas, 1992; Horlick-Jones, 1992a). In addition it can be added that Perrow's categories of interactive complexity and tight or loose coupling are themselves both aspects of complexity. Models such as Perrow's provide important structural indicators of propensity for failure. However real-world interactions are highly significant in this context.

Secondly, whilst the vulnerable systems model has proved useful in structuring investigations into failures of specific technologies, such work has occurred after the occurrence of disasters. The great promise of systems approaches to understanding failures is the possibility of developing predictive methodologies offering the possibility of a preventative tool (e.g. Van Gigch, 1988; Fortune and Peters, 1990). Whilst, as noted above, Reason has criticised "resident pathogen" models, his more recent work (Reason, 1991) has stressed the importance of monitoring indicators of organisational weakness, an idea which is conceptually close to "system vulnerability". Reason's work focuses on empirical analysis of modes of failure, and its conclusions are reminiscent of Toft's (1991) claim that structurally similar failure modes occur in different disasters, a phenomenon he terms "isomorphism". This School of thought, namely that clear indicators of system vulnerability exist, and can be monitored prior to a failure, was implicit in the UK Health and Safety Commission's (1990) call for engineers to report "warnings of disaster". This statement was in part justified by the assertion that "inquiries after recent disasters in the UK have *on every occasion* identified people who were aware of shortcomings in the safety system before the event" (italics added). In contrast, researchers such as Wagenaar and Groeneweg (1987) point out that accidents are the consequences of highly complex coincidences, and that such accidents appear "impossible" beforehand. Wildavsky (1988) goes further in suggesting that failures may result from the behaviour of systems that is inherently unpredictable.

Furthermore, the possibility of measuring, or finding indicators of system vulnerability, relates to the work of Haastrup and Funtowicz (1992), who utilise chaos theory to analyse accident time series. It is important to stress that this approach seeks to capture

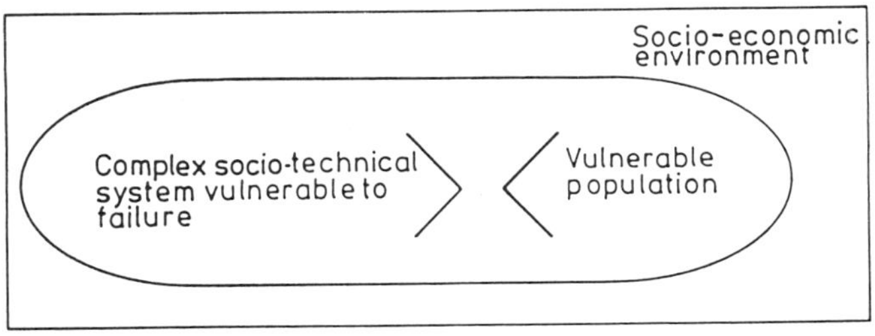

Figure 4. Vulnerable systems model Stage II.

systemic behaviour, rather than searching for discrete organisational or other artefacts that may have accumulated during an hypothesized incubation period. The debate between those who believe that diagnostic methodologies can be developed to identify aspects of system vulnerability and so reduce disastrous failures ("anticipationists"), and those who regard this task as unworkable ("resilientists") was recently identified by a Royal Society Study Group as one of the seven key areas of debate in risk management (Hood et al, 1992).

CONCLUSIONS

Evidence for the significance of disaster trends is difficult to find. This stems from the high consequence-low probability nature of disasters and resulting sparse data sets, and from difficulties in the interpretation of data that are incomplete, prone to error and influenced by subjective factors.

A model in which disasters arise from the exposure of vulnerable populations to hostile environments generated by the failure of complex systems has been reviewed. This invokes a mechanism in which the resilience of systems may be eroded by certain interactions with their socio-economic environment.

The interaction of complex socio-technical systems, such as those associated with safety-critical software and military technologies, with their environments, has been examined elsewhere (Bennett, 1984, Horlick-Jones 1992b). This work suggests the increasing resource implications of avoiding failure of these systems may have important implications for technology policy. Therefore in view of the anticipated future utilisation of information technology in a wide range of applications including nuclear reactors, various modes of transport, water supply systems and medicine, research into system vulnerability may take on increasing importance for the stability and safety of both economically developed and developing countries.

REFERENCES

Bennett, J. M., 1984, "Some Reflections of Large Systems - Computing and Others" Technical Report 196 Basser Department of Computer Science University of Sydney.

Bignell, V. and Fortune, J., 1984, "Understanding Systems Failures", Manchester University Press, Manchester.

Cook, J., 1989, "An Accident Waiting to Happen", Unwin Hyman, London.

Douglas, M., 1992, "Risk and Blame: Essays in Cultural Theory", Routledge, London.

Fortune, J. and Peters, G., 1990, "The formal system paradigm for studying failures" *Technology Analysis and Strategic Management*, 2, 4, 383-390.

Haastrup, P. and Funtowicz, S., 1992, Accident Generating Systems and Chaos: A Dynamic Study of Accident Time Series, *Reliability Engineering and System Safety*, 35, 31-37.

Handmer, J., 1992, Hazard management in Britain: another disastrous decade: *Area*, 24, 2113-122.

Health and Safety Commission, 1990, Responsibility of the Individual in the Prevention of Disasters, *News Release 6th September*, London.

Hood, C., Jones, D.K.C., Pidgeon, M.F., Turner, B.A. and Gibson R., et al, 1992, Risk Management *in:* Royal Society Study Group "Risk: Analysis, Perception and Management", Royal Society, London.

Hood, C. and Jackson, M., 1992, The New Public Management: A Recipe for Disaster, *in:* "Hazard and Disaster Management", D.J. Parker and J.W. Handmer, ed.,James and James, London.

Horlick-Jones, T., 1990, Acts of God? An Investigation into Disasters, EPICENTRE, London.

Horlick-Jones, T., 1991, The Nature of Disasters, *in:* "Emergency Planning in the 1990s", A.Z. Keller and H.C. Wilson, ed., British Library/Technical Communications, Letchworth.

Horlick-Jones, T., 1992a, Think Big and Stay Out of Trouble, *New Scientist*, 10th October, Vol. 136, pp42-43.

Horlick-Jones, T., 1992b, Complexity, Vulnerability and the Failure of Computer Systems, *Etudes et Dossiers Geneva Association*, No. 170, August.

Horlick-Jones, T., 1993, Patterns of Risk and Patterns of Vulnerability, *in:* "Report of a Workshop on Emergency Management Commission of the European Communities Joint Research Centre", A. Amendola and B. De Marchi, ed., Ispra.

Horlick-Jones, T. and Peters, G., 1991a, Measuring Disaster Trends Part One: Some Observations on the Bradford Fatality Scale, *Disaster Management,* Vol. 3, No. 3, pp144-148.

Horlick-Jones, T., Fortune J., and Peters, G., 1991b, Measuring Disaster Trends Part Two: Statistics and Underlying Processes, *Disaster Management,* Vol. 4, No.1, pp41-45.

Perrow, C., 1985, "Normal Accidents", Basic Books, New York.

Peters, G. and Fortune, J., 1991, Are all Failures Systems Failures *in:* "Systems Thinking in Europe", M. Jackson et al., ed., Plenum, New York.

Reason, J., 1990, "Human Error", Cambridge University Press, Cambridge.

Reason, J., 1991, Too Little and Too Late: A Commentary on Accident and Incident Reporting Systems, *in:* "Near-Miss Reporting as a Safety Tool", T. van der Schaff,D. Lucas and A. Hale, ed., Butterworth Heinemann, Oxford.

Toft, B., 1991, The Failure of Hindsight, Paper given to the *Changing Face of Europe: Disasters, Pollution and the Environment Conference*, University of Bradford September, 1991.

Turner, B., 1978, "Man-Made Disasters", Wykeham, London.

van Gigh, J.P., 1988, Diagnosis and Metamodelling of Systems Failures, *Systems Practice* Vol. 1, No. 1, pp31-45.

Wagenaar, W. and Groeneweg, J., 1987, Accidents at Sea: Multiple Causes and Impossible Consequences, *Int. J. Man-Machine Studies,* 27, 587-598.

Wildavsky, A., 1988, "Searching for Safety", Transaction Books, New Brunswick.

Wynne, B., 1988, Unruly Technology: Practical Rules, Impractical Discourses and Public Understanding, *Social Studies of Science,* Vol. 18, pp147-167.

THE SYSTEMS LAW OF ORDINARY PEOPLE

Peter Haine

Managing Director
Savant Associates
1 Bow Court
Fletchworth Gate
Coventry CV5 6SP

INTRODUCTION

In his excellent treatise on software engineering Lowell Arthur[1] makes the suggestion that:

Program complexity grows until it exceeds the capability of the programmer who must maintain it

What Arthur doesn't comment on but I would like to address in the paper is the effect which complex programs indeed complex systems have on their users. I hold the view, for which I believe there is ample empirical evidence that systems fail because they are at the same time too clever for the people who have to use them, and naively unaware of the real business needs.

I will give some example of such failure and will advance my Systems Law of Ordinary People which tries to account for such failure.

I will then discuss how in our work with clients we try to put business people rather than technology at the heart of the systems development process. In so doing, we believe we can avoid many of the problems commonly experienced with information systems:
- short term failure due to poor human factoring
- medium term problems due to lack of functionality
- longer term problems resulting from business change.

SYSTEMS FAILURE

I would like to give some simple everyday examples of the way in which systems fail. Some may seem trivial but they do illustrate an important aspect of the problem I am addressing which is that the perception of the system's effectiveness is all important. That is does it actually do the job for which it was designed <u>and</u> is it *perceived* to do so? Consider this example. In the early 70's I was involved in a stock control system for a medium sized manufacturing company. The finished goods warehouse manager would compete with the system on stock replenishment. When challenged as to why he was placing orders in the factory which the system would place in the normal course of events, he would argue that he was ahead of the computer because he had despatched goods against special orders which the computer did not know about. What we had here was a simple case where man and computer were not in tune! There was no way that that warehouse manager could ever get his mind around the way the computer handled factory ordering.

A more recent and undoubtably more serious example was the failure of the computer aided despatching system at the London Ambulance Service. I don't want to argue the rights and wrongs of the technological aspects of that system. What does distress me though is the failure of the users and management to come to terms with the impact of the problem.

Another example demonstrating the way in which systems fail to keep pace with events is a local authority finance system which reports monthly on spend against budgets to every one of the 200+ Council departments. To my certain knowledge the majority of those departments run their <u>own</u> system to keep track of expenditure. They simply don't believe the information produced by the central system: finding it to he too out of date to be useful. The business sponsor for this system is the Council Treasurer and he cannot understand why the departmental staff are so dissatisfied. This only goes to show that it's not just systems people who are out of touch with the ordinary user! Not only do those responsible for systems fail to deliver the systems needed by the business in the short term but they store up problems for themselves to the point where in many organisations their credibility is seriously questioned. The amount of "maintenance" necessary to keep systems even remotely in touch with the business is extreme. Management derive minimal benefit from operational systems and business opportunities fail to be addressed because necessary systems cannot be put in place in time. I cite an instance of this in which one of the banks came second to market with d new investment product because of a fifteen month delay in adapting an existing system. As a result **the product** was a failure in the marketplace since their rivals were already there!

DEFINING THE BUSINESS NEED

What are the underlying reasons tor these failures? Is the length of time it takes us to respond to systems requirements some kind of inevitable overhead? Are the development tools now available to us never going to have the kind of impact necessary to make the response time acceptable? Are we always going to find that by the time we have implemented anything the business needs have changed?

We have made strenuous efforts over the years to address these issues - both from a technological and methodological standpoint. On the technological front we have seen an ever increasing array of tools coming to market which promise to increase systems development productivity: 4gls, code and application generators, as well as query and reporting tools which enable existing data to be exploited more readily.

From the methodological standpoint there have been two strands of advance which arc in some ways mutually supportive hut in other ways in conflict.

The first of these in the emergence of total life cycle structured development methods. I mean by this those methods which identify systems needs from a top down business perspective and develop a coherent framework of systems to deliver those needs. There is still much controversy as to the practicality of achieving a stable framework using such approaches but certainly in our experience better fit systems are delivered. This may well be because people in the organisation have actually got together to discuss systems as a business issue.

To this extent these top down structured development methods are consistent with the other strand of improved practise I want to address. This is the area concerned with the human aspects of systems. Interestingly significant advances were made on this before structured development methods became fashionable. I'm thinking of Enid Mumford's work in the early 80's which resulted in the ETHICS Methodology[2] and IBM's use of JAD[3] in their Canadian facility at about the same time.

Regrettably, though, there seems to be something of a polarisation into two camps: in the one those who advocate the quick fix approach - using rapid development tools JAD sessions and heavy user involvement; in the other camp those who wish to adopt the more formal top down structured development approach using CASE tools heavily prescriptive methodology and quality management and long timescales.

Of course the latter camp have had to make some concessions to the former: and every structured development method and CASE toolset now has its RAD - Rapid Application Development - capability. But we have to ask whether this capability is really consistent with the method's objectives - or is it just a way of bypassing the method and delivering a quick fix?

PEOPLE CENTRED SYSTEMS DEVELOPMENT

In People Centred Systems Development we seek a synthesis between the benefits of a structured top-down approach to systems development and those addressing the human factors. The approach is founded on a pragmatic law: the Systems Law of Ordinary People which states quite simply that 'People make Systems Work'. This law is self-evident to anyone who has ever tried to implement a system in any environment where those affected by the system are not 100% behind it.

What is important is not so much the law itself, (which as I say is self evident) hut rather its implications for the systems development process. To understand this I believe we have to go back to first principles. Gwilym Jenkins published a very helpful paper in this regard[4] where he discussed the philosophy underlying a systems approach to solving problems. A point he stressed was the way in which new systems make fundamental changes in the way people go about their work and hence the need for an interdisciplinary approach to be taken in devising new systems and procedures.

Another issue Jenkins and others have addressed is the essential subdivision of higher order systems into subsystems. In the business situation we must necessarily re-design systems at the sub-system level - but often do so without considering the wider picture. Hence we arrive at naive solutions which address only part of the business requirement or interfere with the effective operation of related sub-systems; even worse they often run counter to overall business objectives. For example I have seen systems which seek to improve the efficiency of processes which are no longer relevant to the business.

So what then are the implications of the Systems Law of Ordinary People for information systems development? I believe that in simple terms they are that **systems** development has to establish a much higher profile for itself within business organisations and has to involve more senior people in those organisations much more fundamentally.

Several things have to happen of this is to be achieved. First we have to ensure that IT people have much greater business awareness. Even quite senior people in IT departments often feel much more comfortable with the technical issues surrounding systems than they do with the business issues. What is more they sometimes find it difficult to explain the technical issues in terms which business people can understand. This is in my view a very serious problem which is holding back IT departments and the businesses they are supposed to support to a considerable extent.

So, first, we have to develop the business skills of IT people and help them win the respect of other senior managers in the business. Second they must then use their new found respectability to involve senior managers in taking a more structured view of the business.

In this way they should become more involved in the business planning process and more able to plan the future systems infrastructure to support business strategy.

At the working level too there has to be a much greater level of interaction between technical people and business people. Again this will be facilitated if IT staff have a greater awareness of business issues. Another factor which militates against such interaction is the inability of the IT function to deliver results following such interaction. There is no point in a team coming together in facilitated JAD sessions to accelerate the process of requirements elicitation if the resulting requirements take eighteen months to deliver!

I suppose it could be argued that all I am really doing is supporting the argument for soft systems methodology in place of systems engineering approaches which try to treat business systems as purely deterministic. I don't really see it this way. The problem I think we really have is that there is too much of a gap between those who attempt to devise and implement systems and those who are the 'victims' or users of those systems. Before we are able to claim we have found an acceptable way of delivering systems with long term business value we have to close this gap and ensure that the people who will work with systems at every level are closely involved in the development of both the human and the technical aspects.

In our work we have attempted to achieve this with a People Centred Systems Development approach[5] which harnesses business executives, senior managers and other relevant staff in the business organisation to the systems development process.

SUMMARY

The key issue I have been addressing in this paper is the need for a modified role to he established for the systems development function of a business organisation. In my experience many IT departments have too much focus on technology and not enough on information and its role in the business. They are not close enough to their users and are not able to involve them sufficiently in their work. As a result there is a very real barrier to delivering the systems the business really needs. This militates against IT being regarded as of strategic value in the business and results in systems which support only much mechanistic aspects of the business process rather than management and policy making activities.

REFERENCES

1. Arthur, L., 1983, "Programmer Productivity", John Wiley, Chichester.
2. Mumford, E., 1985, Defining Systems Requirements to meet Business Needs, *Computer Journal*, Vol. 28, No. 2, pp97-104.
3. GUIDE, 1986, "Proceedings of GUIDE Session No Mp-746" GUIDE, White Plains, USA.
4. Jenkins, G.M., 1969, The Systems Approach, *Journal of Systems Engineering*, (1), 1, pp70-74.
5. Haine, P.M., 1992, "People Centred Systems Development", Seminar Documentation, Savant.

NEW DIRECTIONS IN THE USE OF STRUCTURED TOOLS IN REQUIREMENTS ENGINEERING AT GLASGOW POLYTECHNIC

Alan Hogarth and Gopal Rao

Department of Computer Studies
Glasgow Polytechnic
Cowcaddens Road
Glasgow G4 OBA

INTRODUCTION

Requirements analysis and requirements specification are arguably two of the most important aspects of systems development. Failure to elicit a correct and feasible set of requirements from the customer client at the outset of development can result in the user receiving a final system that bears little resemblance to the original system he envisaged. The economic and manpower costs of this failure to deliver the 'right ' system on time' is one of the major concerns of systems professionals and researchers at the present time. Many solutions to the problems of Requirements Engineering have been proposed with varying claims of success. In this paper the authors will briefly consider some of the approaches to requirements engineering and demonstrate in some detail how Structured Common Sense[1] and SRCA[2], are implemented at Glasgow Polytechnic. An analogy between "policy-setting" and requirements specification is drawn.

PROBLEMS WITH REQUIREMENTS ENGINEERING

Requirements are a collection of statements that should describe in a clear, concise and consistent an unambiguous manner what a systems behaviour is to be. This requirements statement will generally be in the form of some event-response observation of the system. They should contain enough information to enable the developer to model a systems behaviour that will create a common understanding between customers and analysts. In creating this we must consider analyst/user conceptualisation/communication and the conflicts that can ensue due to their differing viewpoints. The System Requirements statement can be seen as a "base model" of the real world constructed to represent the customer view. The requirements usually have to include much more detail such as that which will allow the following three views to be established; information flow, information content and information structure.

On reviewing the user's base model the analyst's role in Requirements analysis is concerned with eliciting, clarifying and documenting the desired functionality of some system. He will then derive a "lumped model" or developmental model. Conflicts of misconceptions can arise at this stage between the user and the analyst. The authors suggest that by considering a "modelling framework", SCS and the relevant tools can be utilised to produce a quality requirements specification that adheres to the desired objectives of languages with a single interpretation without ambiguities, completeness(all and only the relevant facts), size and complexity, clarity (by partitioning), consistency(by review and cross-reference), verification(checking by tracing), validity (mapping to real world and describing the right system), modifiability (change the requirements easily) and by reducing "noise" making statements more concise.

Many studies have shown that since the communication content between analysts and users is high the chances for misinterpretation and misinformation abound. It is clear that no tools or techniques exist at the moment which are capable of dealing satisfactorily with all areas of requirements. "Despite the introduction of development methods and CASE tools in the project life cycle, the process of requirements specification remains problematical". It been claimed that over 50% of system malfunctions have their origin in the process of requirements capture. This indicates a large gap between the user's world and the developer's world.

Loucopolous and Champion[3] maintain that a more appropriate approach to development would be one that has been recognised in almost all project development methods and that provided tools for the effective undertaking of these activities. This approach is increasingly receiving the attention of researchers in the field of systems development.

SOLUTIONS TO REQUIREMENTS PROBLEMS

General solutions include prototyping, animation and reviews. More specific methods and tools include CORE (Controlled Requirements Expression), Structured Common Sense(SCS), SRCA and Soft Systems Methodology. Some of these techniques are very restrictive ie. solutions looking for problems. However, of these newer techniques ,SCS,SRCA and SSM seem to offer methodologies that are not overtly "technical". SCS an SRCA are being implemented at Glasgow Polytechnic using various non-standard development tools such as Hypercard, Agenda and Toolbook. Consideration should also be given to the philosophical base of requirements specification a review of Sutherland's thoughts on *policy-setting*[4] could shed some light on present day problems of requirements.

THE RELEVANCE OF POLICY-SETTING TO REQUIREMENTS SPECIFICATION

Requirements specifications can be viewed philosophically as *policies*. All policies have assertions about desirable systems properties, and hopefully serve. This behaviour could readily apply to a requirements specification where desirable requirements are put forward for the new' system following elimination of the problem areas encountered during requirements analysis. In essence requirements analysis could be classed as "the customer requirements statement to design transition". To this extent policy-setting becomes the rational activity associated with the isolation of those means by which a transition from a less favourable to a more favourable state may be most efficiently effected. The problem of irrationality however can still exist between the customers view and the developers view of requirements and this is an area where SCS and SRCA can be utilised constructively. The aforementioned description of policy-setting underpins the concept of requirements specification.

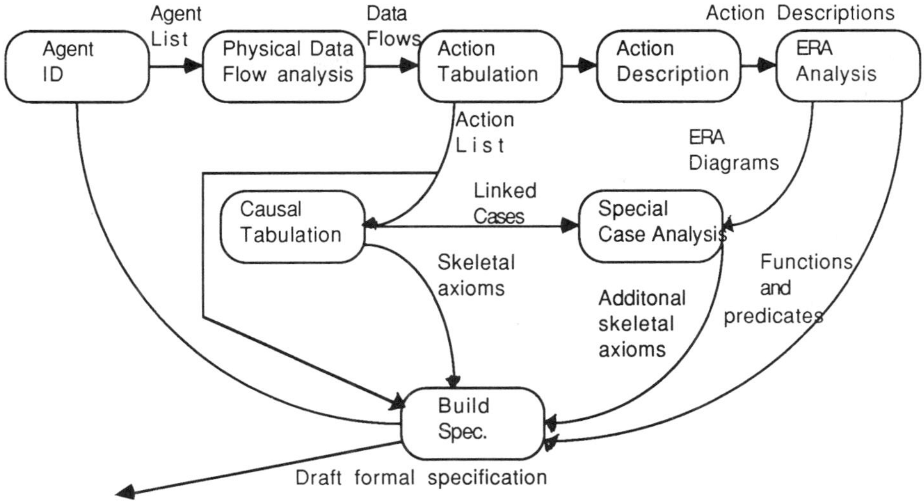

Figure 1. Steps in Structured Common Sense.

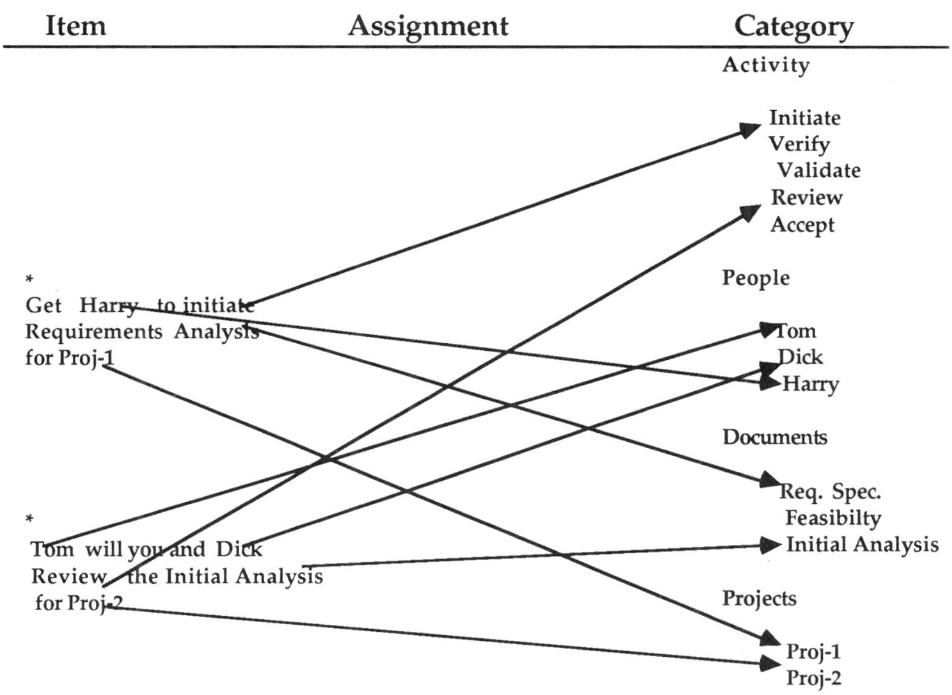

Figure 2. How Agenda stores and connects the data.

WHAT ARE STRUCTURED COMMON SENSE (SCS) AND STRUCTURED REQUIREMENTS CAPTURE AND ANALYSIS (SRCA)?

At Glasgow Polytechnic work is being carried out in area of requirements engineering, utilising SCS and SRCA for the elicitation and formalisation of systems requirements. SCS is goal-directed and specifically targets the method on the form of specification to be produced and utilises where possible concepts and techniques from existing methods. It can be viewed as the construction of a requirements engineering method. As an exercise where creativity, experience and common sense are minimally constrained, while at the same time structuring the process into a number of specific and cognitively manageable steps.

SCS consists of a number of distinct steps each of which has two aspects, a work plan or strategy and heuristics or tactics. The application of of SCS steps can be viewed as "filling in" slots in a skeletal format specification and where possible information from a step is used as a support in the completion of the proceeding steps. These steps in SCS are shown in Figure 1.

SRCA is based on CORE and is a result of work done to improve requirements specifications. SRCA attempts to address the problems associated with the more traditional approaches to requirements specification by using a model of requirements which can be tested and animated to ensure that the requirements expressed actually make sense and are complete. SRCA regards the required system as a part of a *system of systems,* taking the view that the required system "exists to fulfil a need in its environment". Requirements are expressed in terms of both what the system does and what the users of the system do (and need from it). The analyst will thus examine the behaviour of both the required system and the relevant parts of the environment.

SRCA consists of **seven** stages, **1) Identify the customer authority 2) Problem definition 3.) Viewpoint structuring 4) Tabular collection 5) Single viewpoint modelling 6) Combined viewpoint modelling 7) Constraints analysis.**

The benefits suggested for this approach include: 1) The requirements for new system are captured by ascertaining what different parts of its operating environment will require of it. 2) By understanding the behaviour of the system's surroundings, better decisions can be made about the requirements. 3) The requirements can be checked more completely by following threads of activity through the required system and its environment. There is also an iterative aspect SRCA. The analyst must know enough to understand what the user is talking about and he must continue to build up his knowledge of the system and its behaviour so that the next fact fits into place and supplies a step to the next more detailed question. SRCA considers that a system is made up of a number of primary objects which have their own view of a system, the analyst's task is to extract the view points of these particular object. This paper will demonstrate with an example on a PC how these techniques, utilising such automated tools as Agenda and Toolbook can be used to produce, review and validate a Requirements Specification.

THE IMPLEMENTATION OF SCS AT GLASGOW POLYTECHNIC

It is accepted by the authors that total understanding of requirements may never be achieved in the distant future due to the vagaries of natural language. However in utilising the conceptual underpinning developed earlier in the paper and using Hypertext/Hypermedia tools we believe that such an approach may go some way towards alleviating the problems of requirements engineering. Toolbook was utilised to represent the pertinent areas of SCS and SRCA and Agenda was used for "policy setting". In this way it was demonstrated that, via prototyping , users gained a clearer view of what they required. How does Agenda help in requirements analysis eg. elicitation, specification, verification and validation? Agenda can be used to organise random, "free-form text" and as such is less rigid than a database while still allowing the user to set relationships so that common data can be linked. In eliciting the user's requirements and entering them via Agenda, data may be grouped in the order supplied by the user. In considering verification and validation of requirements Agenda can store seemingly disparate, and unconnected data in a "cross-indexed" form. As the data is stored it can be automatically assigned to a preset *category* or item by scanning the text. It can scan free text and search for words, phrases, nouns or verbs that have been preset as our categories. An *item* represents any kind of information that is to be tracked, these items are *assigned* to *categories* which provide the structure for these items of data. Assignments are the connections between an item and a category and can be made for one item to many categories or many items to one category. See figure 2. on how Agenda can store and connect its data. Any text entered, as an item, would be stored under its appropriate category. Thus Agenda can help in requirements analysis verification and validation by allowing the storage of free-form text and then allocating it to preset "categories".

A model of the system can then be created by "moving or linking" several categories together using data flow diagrams and/or logical data structures in order to show various "views" for inspection and review. Reports can be printed from various views eg. one report could show people involved in a project another could show the uncompleted tasks associated with a project. Using Toolbook a model of SCS's Agent Identification Hierarchy was created.

Within this structure each *agent* has a textual description giving details of the role of that agent in the system. This is linked to each *"agent box"* on the model. An agent "Programme Committee" has been set up with complete cross-referencing facilities available eg. Descriptive text, data flow diagrams and Action Table, for verification and validation.

CONCLUSION

Many of the problems discussed arise from a single source. To read and understand the requirements , to validate and verify them, and to check them for consistency and completeness, demands that the language used to express them is interpreted. It is this interpretation that causes the problems. As such problems arise it is essential to distinguish between good and bad requirements specifications. The many methods, tools and techniques available go some way to alleviating problems with requirements as was demonstrated using SCS and Hypertext, but it is still perceived that in order to achieve a perfect requirements specification much work is still to be done. What then does the future hold for requirements analysis? To encourage the development of languages and methods that are practical for real systems requires better links between developers and users of these methods to be forged. At Glasgow Polytechnic we have attempted to address these issues by utilising the various tools referred to above as a means of improving the requirements analysis and requirements specification stages of development.

REFERENCES

1. A. Finkelstein and C. Potts, Structured common sense: the elicitation of systems requirements in: "FOREST Project", GEC Research, Chelmsford, (1990).
2. GEC Plessey, "SRCA (Structured Requirements Capture and Analysis", (1990).

3. P. Loucoupoulos and R.E.M. Champion, "Knowledge based support for requirements engineering" *in:* "Info. and Software Tech". vol. 31, No. 3, Apr., pp124-135, (1989).
4. J.W. Sutherland, "Systems Analysis, Administration and Architecture", pp 241-277, Van Nostrand Reinhold Co., New York, (1975).

TOWARDS THE CASE FOR MORE DELIBERATE STRATEGIC IS

J.G. Howell

Information Systems
Department of Computing Science
University of Paisley
Scotland

INTRODUCTION

"Systems thinking is traditionally practice driven where other disciplines are not"[1]. It is the purpose of this paper to deliver an account of major change driven by an IT department in one of Europe's largest local authorities and hence help bridge Systems Thinking and Practice.

This paper is based on the experience gained over several years of working for the authority in various IT capacities and focuses on the strategic issues that permeated this experience. The case is argued that this experience indicates that benefits are gained from the consideration of strategic Information Systems (IS) in light of recent claims that, "the track record of corporate strategies is dismal"[2] and further that, "in recent years, there has been an increasingly loud chorus of discontent about corporate strategic planning"[3]

There will be some critical evaluation of the route taken during the process of change from a fairly traditional centralised mainframe DP environment using, mainly, unstructured methods to a distributed environment using structured methods and CASE tools. The account will be embedded with references to current thinking and literature.

Strategic IS has been variously defined, indeed it has been suggested that,

> "almost no consensus exists about what corporate strategy is, much less about how a company should formulate it".[4]

In this paper strategic IS is the consideration of the policy and position of Information Systems organisation-wide for some years ahead.

Clear conclusions will be drawn to support the case for more considered and deliberate strategic IS planning.

BACKGROUND

The organisation referred to in this paper is one of the largest local authorities in Europe, employing approximately 105,000 people and servicing a region which is

Systems Science, Edited by F.A. Stowell *et al.*,
Plenum Press, New York, 1993

geographically one of the largest in Europe. The past IT environment reflected that of a typical central Data Processing department offering timesharing connections to mainframes, with a large percentage of the work being undertaken in batch mode. Little or no formal techniques were generally employed in analysis although some database design techniques were used for mainframe bound databases. A strategic IT (but not IS) policy had been partially formulated but there had been difficulty in concluding this exercise. The PC/LAN databases did not enjoy even database design techniques. Formal project management methods were seldom used nor were Strategic IT policies made explicit. There had been a degree of belief in improvements in systems production brought about by the increased capabilities of technology and languages, in particular 4GLs. However, the experience was generally disappointing and, in spite of increased machine power, the use of such "4GLs" often brought the machine "to its knees". This experience was not dissimilar to that of other organisations,

> ".....in the past, we thought we could save money by applying a new 4GL or applications generator; but that did not get to the heart of the problem."[5]

The heart of the problem was identified to be in the methods of analysis and design with much greater weight on analysis. It became the express desire of the department to "formulate and work in an environment where effective analysis and design techniques were employed methodically to produce (computer) systems of the required quality, on time and within budget"[6]. This, in turn, meant a new way of working with user departments and personnel. Over recent years the user population had altered significantly and their expectations of delivered systems had risen dramatically. For example, their desire to use simply mainframe terminals had reduced in favour of PCs with terminal emulation packages connected to Local Area Networks (LANs), they desired to own their own data, they wanted to play an *active part* in the development process, and some had gone as far as to create new posts for their own IT staff, for example, LAN manager.

PRESSURES TO CHANGE

The reasons for change within the DP/IT department were many and included the changing client base, the advance of technology, the desire to keep abreast of the marketplace in view of impending government decentralisation and privatisation legislation and the genuine desire to provide a better service, on time and within budget. It would be misleading to consider such a local authority as a properly homogeneous entity, rather it is more helpful to view the organisation heterogeneously in terms of its component parts as modelled in the diagram below.

It became necessary to reformulate the role of the IT department to match the changing needs of the *internal environment*, as different client departments perceived their changing future.

A dominant, but perhaps not consensus, view of the IT departments new role was as a leader and advisor to other departments, a standards bearer, and to act effectively as an interface to external IT providers. As suggested by Galliers[7], several views of the departments new role were expressed:

> "........ the likelihood that there will be a range of views expressed by senior management regarding an organisation's primary purpose and goals"

This recognised that an organisations primary purpose should be established and communicated to all levels as early as possible to encourage and support staff to pull in the same direction.

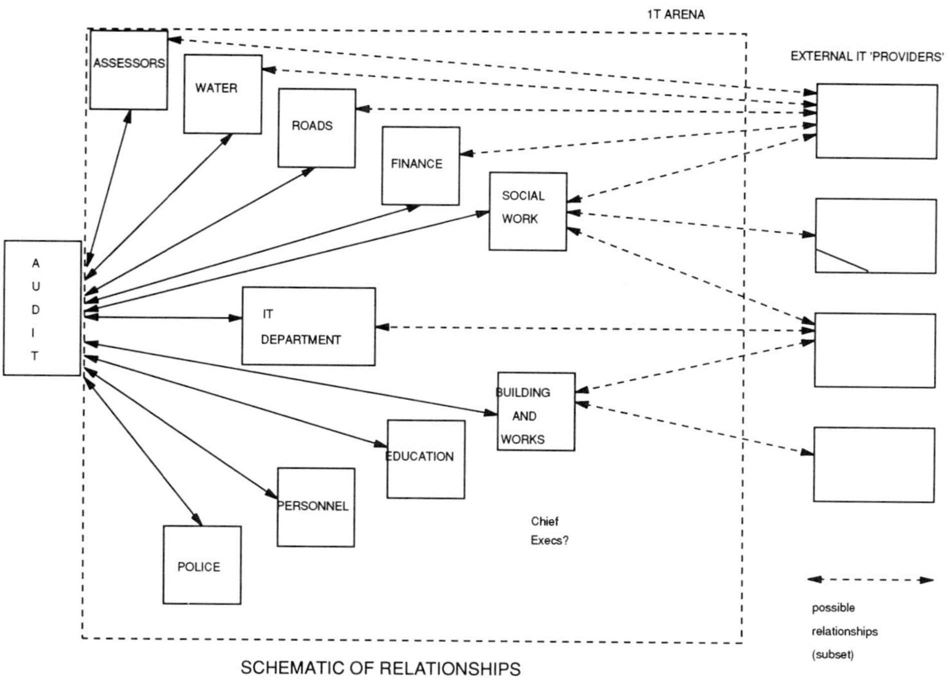

Figure 1. Relationships to/from the IT department.

THE NEEDS OF MANAGEMENT

have already been indicated above (Background) and are similar to those suggested by Avison[8],

> "we can identify three main categories of perceived need: a better end product, a better development process, and a standardized process."

THE STRATEGY

generally adopted was emergent and not deliberate except for technological considerations (such as electing for relational databases to be employed, moving towards "open systems", ...) and the aforementioned move towards structured methods. Some of the main emergent implementation issues will now be identified although it should be appreciated that within the scope of this paper this list is not exhaustive.

Building on Previous Success - Structured Programming

In the past success had been achieved in introducing Structured Programming which had resulted in improvements for analysts/programmers, users and management. This, however, brought limited benefits since it was recognised that,

> "Structured Programming is not a panacea. Your programmers are still capable of writing lousy code, especially if they try."[9]

This is doubly true for structured analysis and design, your analysts are still capable of (analyzing and) designing lousy systems.

By building on the success of structured programming, management was able to lay down a foundation for success in employing structured analysis and design.
This strategic approach of building on existing strengths proved beneficial.

PILOT PROJECT

Another tactic adopted was to pick a pilot project as a learning vehicle and a medium through which to grow experience and build standards. Identifying a pilot project with user management agreement proved to be a tortuous process ultimately failing for several reasons with one of the major reason beings the lack of consideration of the priority rating of the pilot project(s) in relation to other requirements for information systems to be developed. This problem was not simply monopolised by these pilot projects, but also existed for other "real" projects. A disparity of views of the relative importance of individual projects was evident. This theme of different user groups, different senior executives and different analysts holding different views consistently appeared at all levels.

One of the perceived reasons for this limited success was that other projects had become more strategically important and hence acquired the necessary resources, including personnel, leaving the pilot project struggling to survive.

Essentially, pilot projects which had begun life as strategically highly important had suffered a reduction in priority in favour of others through time.

It was recognised in management that it was beneficial for any future pilot project that to plan holistically and, if possible, for some years ahead. Such planning should include pilot projects in order to assure them of a reasonable life span and hence serve as a pilot project. Perhaps any such project can be considered not to be a suitable candidate as a pilot project but it is a *Catch Twenty-two* situation.

Suitable pilot projects were identified and gained success in many ways, however, the lack of explicit strategic planning culminating in action plans continued to affect the work.

THE CHAMPION

A senior and energetic member of staff was chosen to champion this major change, an element of successful *strategic change management* previously recognised,

> "Another finding of the study evidenced by outcomes and by comments in interviews, is the importance of sustained support in top management or some individual in top management"[10]

STRATEGIC MONITORING

To be able to deal with strategic events is part of the purpose of strategic management. In the same way that we build an Event Catalogue within systems we can build (or extend) to produce a Strategic Event Catalogue classifying events as, for example:

EVT#ID	Date Created	Use #ID	Description	Internal/External	Frequency	C Factor	Span
EVTnnn	aa/aa/aa	XXXnnn	XXX	External	D/W/M/Y/		

to be able to monitor the external environment and internal systems.

Attributes of EVENTS may change, for example, confidence factor.

This approach was considered and employed using the data dictionary of a CASE tool and strategic events were added to the dictionary from the experience within developing projects.

PROBLEMS ASSOCIATED WITH THE EMERGENT STRATEGY

It would be misleading to suggest that no problems existed in this major exercise of change, however, it was felt that some of the problems were associated with an emergent

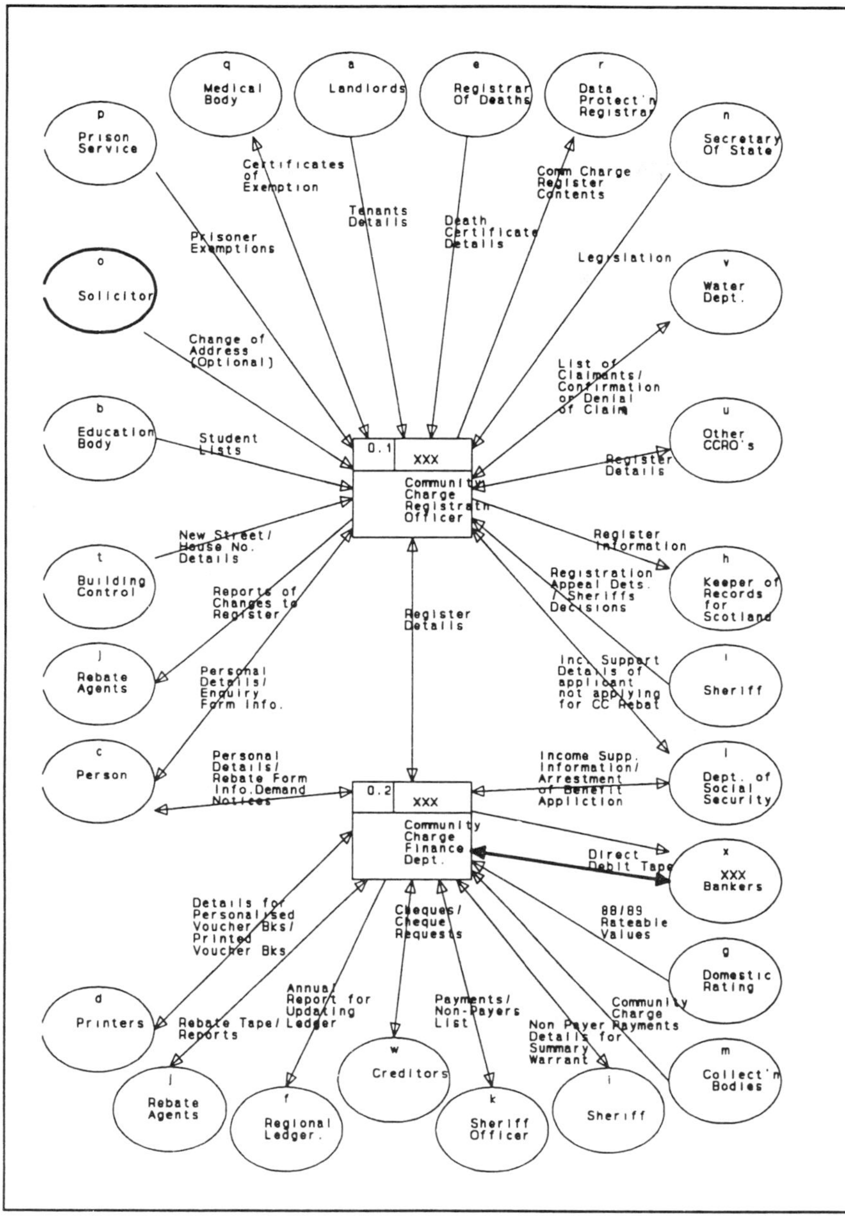

Figure 2. "Context Diagram" showing interfaces.

strategy which would have been better addressed by a more formal and deliberate strategy. One particular problem must be discussed, *Interfaces*.

> "Before a system can be discussed sensibly you have to establish its interfaces to other systems".[11]

In considering the example CONTEXT DIAGRAM shown above it is apparent that the system interfaces to many others, there are in fact 25 external entities.

External entities are "whatever or whoever donates information to or receives information from the system".[12]

Closer inspection of this diagram reveals that of the 25, 5 of these are internal systems within the organisation. It is important that these interfaces are analyzed and the necessary specifications result.

To give a project (eg. valuation) a reasonable chance for success it was imperative that other systems(e.g. domestic rating) be considered prior to the feasibility stage, in the strategic plan.

CHANGES IN THE ENVIRONMENT

Within local government it is a difficult and complex task to monitor changing legislature, both major and minor. This was a continuous exercise which benefited the department at many levels in planning the work ahead.

CONCLUSION

Strategic Management is not distinct from the rest of management, it does not "sit at a level above" but permeates the work at all levels and across levels. In order to reduce the risk of projects failing it is important that they are considered in relation to those existing and planned. This should be done as part of the strategic IS planning and probably across functional boundaries if necessary. More deliberate strategic IS planning is sought in order to better direct and protect systems development.

This change scenario has demonstrated some of the benefits of adopting a strategic IS approach. Although a complex exercise, this should be a continuous process if the future strategic position of the organisation is to be maintained and the threats of a rapidly changing environment overcome.

REFERENCES

1. M.C. Jackson, G.J. Mansell, R.L. Flood, R.B. Blackham and S.V.E. Probert, "Systems Thinking in Europe", Plenum, New York (1991).
2. M.E. Porter, From Competitive advantage to Corporate Strategy, *in*: "Readings in Strategic Management", D. Asch and C. Bowman, eds., p234, MacMillan, London (1989).
3. J.B. Quinn, "Managing Strategic Change: Readings in Strategic Management", p33, MacMillan, London, (1989).
4. M.E. Porter, From Competitive advantage to Corporate Strategy, *in*: "Readings in Strategic Management", D. Asch and C. Bowman, eds., p234, MacMillan, London (1989).
5. T. Hill, "Swissair: Flying High with CASE", SSADM Newsletter no. 8, Hyperion, Plymouth (1990).
6. J.G. Howell, Changing to an Integrated Structured Systems Development Environment (ISSDE) - An Empirical Study of The Determining Key Success Factors, *in*: Proceedings of the Third International Conference on Information Systems Developers Workbench, S. Wrycza, ed., Sopot, Poland (1992).
7. R.D. Galliers, A Scenario-based Approach to Strategic Information Systems Planning, *in*: "Systems Thinking in Europe", p77, Plenum, New York (1991).
8. D.E. Avison and G. Fitzgerald, "Information Systems Development, Methodologies Techniques and Tools", pp264-267, Blackwell, London, (1991).
9. E.Yourdon, "Managing the Structured Techniques", p120, Yourdon Press, New York, (1979).
10. J. Engledow and R.T. Lenz, Whatever happened to environment analysis, *in*: "Readings in Strategic Management", D. Asch and C. Bowman, eds., MacMillan, London (1989).
11. C.P. Allen, "Effective Structured Techniques, from Strategy to CASE", p62, Prentice Hall, (1991).
12. C. Ashworth and M. Goodland, "SSADM - A Practical Approach", p38, McGraw-Hill, London (1990).

MACRAME, A TOOL TO SUPPORT THE ANALYST IN ORGANIZATION

Maria Franca Norese

Dip. Sistemi di Produzione ed Economia dell'Azienda
Politecnico di Torino - Corso Duca degli Abruzzi, 24
10129 Torino - Italy

ABSTRACT

This paper is concerned with a tool, a multidimensional orienting schema, that helps the analyst to structure problems and models in multiple actorial contexts where differently structured, conflictual and equivocal representations which have arisen from the process of interaction between analyst and organization actors or those that the actors propose become essential objects of the analyst's action.

CONTEXT OF USE

In organizational contexts such as the management of public services, where a lot of actors are involved from other organizations at different levels, with specific roles, functions and competences, limited formal interactions between these various actors may make the coordination of a collective action difficult.

The analyst's intervention, inducing the participants' investment in the process of information gathering, evaluation, debate and presentation helps to produce consensus, or at least an accepted integration of the different points of view. When various stakeholders are affected, in the presence of complex problems, incompatible goals and multiple criteria of success, the analyst has to integrate organizational and political dimensions within a cognitive-operational paradigm (see for instance Moscarola, 1984; Norese and Ostanello, 1984, 1989; Kling, 1987; Rosenhead, 1989). It implies both an individual level of analysis and reconstruction of problem and context situation and an individual and/or collective action on the representations, developed by the interaction with different logics of action and representation of the actors involved and then explored and redefined (see for instance Bourgine and Espinasse, 1987; Smith, 1989) until an acceptable level of stability is obtained. In order to act on these representations the analyst needs supports which are pertinent to the context and which easily evolve with the action.

MACRAME (Multiple ACtor RepresentAtion ModElling) has been used in different organizational and problematic contexts, as a methodology of modelling in multiactorial contexts and as a tool to support the action in these contexts, and it seems to be sufficiently elaborate to deal with complex situations and free enough from structural limits. It was

Systems Science, Edited by F.A. Stowell *et al.*,
Plenum Press, New York, 1993

elaborated initially after a long and "heavy" intervention in a public sector (Norese and Ostanello, 1985; Norese, 1988) as an answer to requirements of documentation (to help model understanding, use and updating), coordination of teams of analysts and support to all their different activities.

MACRAME was then tested in action and made more general and flexible in two different studies. The first is related to an organizational change in the Departments of Emergency in Piemonte, a region of Italy (Mucci and Norese, 1990) and the other connected to the experimental insertion of a DSS in some public hospitals (Massa-Rolandino and Norese, 1992).

During the structuring process MACRAME helps the analyst to memorize, in a formalized structure, the elements which have arisen from investigation and observation and to explore and redefine these abstract representations

- to verify reliability (of sources and indications), relevance (at different levels of generality), consistency and completeness (both in relation to a single information-representation source and to a multiplicity of them);
- to distinguish between complementary and antithetical representations (Mason and Mitroff, 1973), collective opinions and subjective points of view, contextual solutions and more general ideas;
- to single out recurring concepts and to select information for operational model elaboration;
- to orient new investigation cycles, different additional representation development and redefinition of conceptual models.

More activities of Communication, Identification, Development, Control and Structuring (Norese and Ostanello, 1989) alternate and each cycle induces a new scheme, different in topic, content and structure of levels. Problem formulation, context representation and model structure change until a stability level becomes temporarily or definitively acceptable (see for instance Bowen, 1983; Smith, 1989).

The component parts of this scheme and some elements, which have arisen from applications of MACRAME are presented in the next sections in order to describe the use of the tool in some detail.

COMPONENT PARTS OF MACRAME

MACRAME is a multidimensional orienting schema, that is (Neisser, 1976) an active, information-seeking structure that orients a technical action. It helps to structure and select, in the operative phase, and then memorizes the meaning of each technical choice and becomes a communication tool in the explicative phase.

It assembles knowledge elements, from different sources, in different forms and with different meanings, in a scheme at more levels of aggregation.

For each level, from the most general to the most specific, the following are defined:

- a **statement of problem formulation,** declaring the nature of the definite level of structuring and making the adopted technical perspective explicit;
- a network that represents the actorial structure related to the level; each node denotes an actor and his role or principal function, the arcs qualify multiple relationships between the actors **(actor structure network);**
- a class of representation network in which nodes represent concepts and proposing sources, arcs denote relationships between concepts **(representation networks);**
- the dimensions of the problem, that is the disaggregation of the topic of the level into its terms of reference, so they can be deduced from the actorial representations explained in the networks **(problem dimensions);**
- the dimensions of the representation structuring model, that is the structure of transition from one scheme to another and one level to another, activated when a new level of

representation structuring becomes possible or necessary, consistently with the multi-dimensional exposition of the **problem (model dimensions).**

The scheme evolves with the process that it orients, documents logic and context of every technical choice, controls the coherence of the process and the use of models which have arisen from this structured and systematic reading of interactions between representations. At the end the scheme becomes a document structure in which the first level presents the global approach to the problem as the result of the problem formulation process. The other levels propose local perspectives generally related to sectorial or logical subsystems, where only some actors or users are involved, which become more and more specific and which are often connected to only one information source. The same framework of structuring may be read in relation to the model that is generally defined at the first level of the scheme and becomes more and more detailed and related to local perspectives and information sources.

COMPONENTS OF APPLICATION

In MACRAME a declarative statement introduces the elements that locally constitute the problems; the actor structure network presents the actors in some way involved in the local problem and the relationship structure that is peculiar to the specific context (communication or information; functional, hierarchical,...); representation networks connect local causes, goals, proposed solutions, definitions and remarks to the actorial sources and coordinate these concepts that are fragments of information and also problem elements.

The actorial structure and some representation networks related to the problem situation described in (Massa-Rolandino and Norese, 1992) are proposed in fig.1 and fig.2. A general statement declares that D.AN.T.E., a DSS elaborated by the Artificial Intelligence Laboratory of CSI (Regional Consortium for Information Systems) is now ready to be experimentally inserted in some Dialysis Centres. Purposes and ways of this insertion and the related operational validation (Landry et al., 1983) of the system have to be defined.

All the involved actors are presented in fig.1; they operate in different organizations and some informal groups have been created in relation to this DSS elaboration and testing. The relationship structure, at this general level, makes the different relations between organizations evident and defines the nature and articulation of the ad hoc groups.

In fig. 2 a class of network represents key concepts that the actors considered during the first phase of DSS development and validation. Some relationships between concepts have been proposed in this previous phase, both in terms of mutual learning and reinforcement and in a more conflictual and/or political approach (Newman and Noble, 1990). In the scheme, at the general level, all the key concepts are listed and placed in classes that express themes for discussion. A tentative qualification of the concepts is proposed; its validation and the explicitation of relationships are postponed to a collective analysis of this scheme.

The specific nature of relationships among concepts may generate causally-chained networks (see for instance Nakamura et al., 1982), relevance diagrams as in the knowledge maps proposed in (Howard, 1989) or a richer structure that includes consequences and explanations such as the cognitive mapping in (Eden, 1988). The few applications of MACRAME discovered a multiplicity of different situations, that have to be represented consistently, and need for specification that grows with the problem and model structuring.

In all cases and at all the levels of the scheme, the disaggregation of the topic of the level into its themes for discussion and terms of reference allows a multidimensional structuring of the perceived problem. The decomposition of a problem situation into sub-problems is common in prescriptive literature (Smith, 1989) "being needed for techniques to be properly targeted" and decompositional strategies and frameworks are proposed in the problem solving, decision analysis and problem formulation approaches.

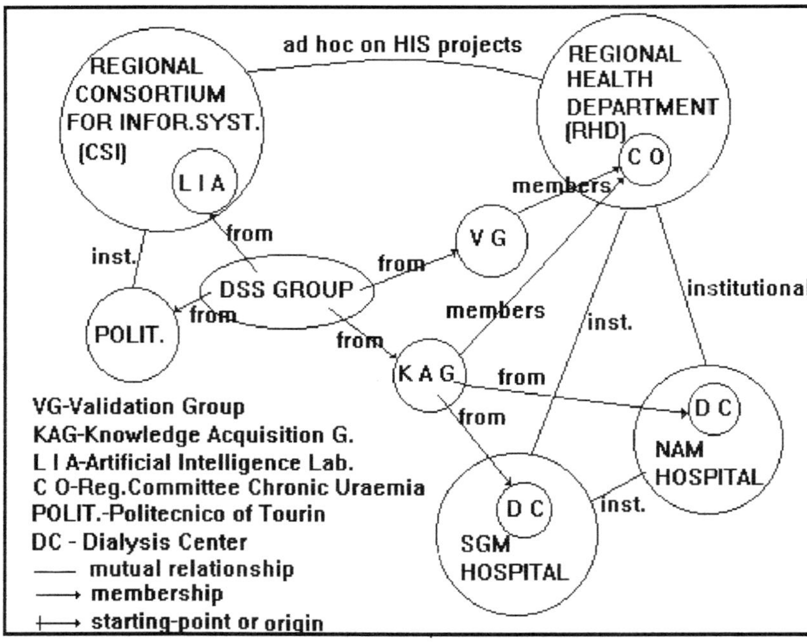

Figure 1. Actorial structure at the general level.

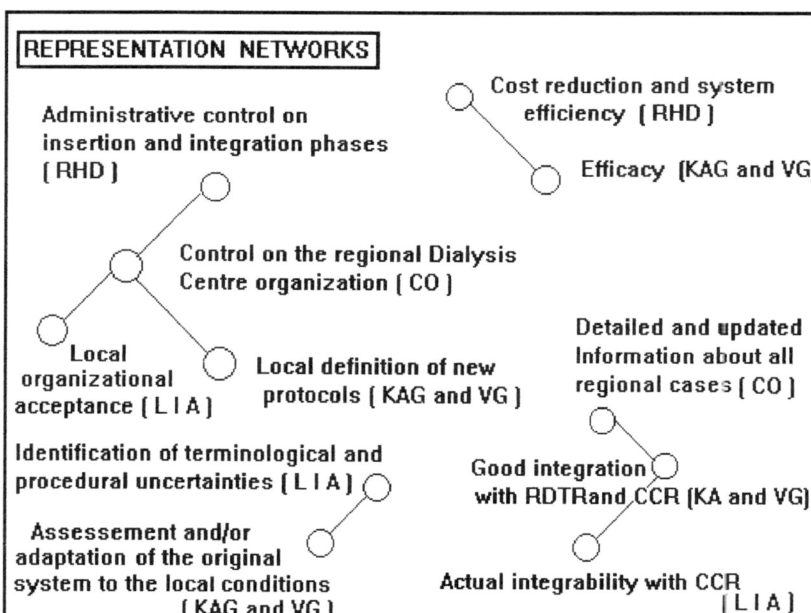

Figure 2. Actorial representations at the general level.

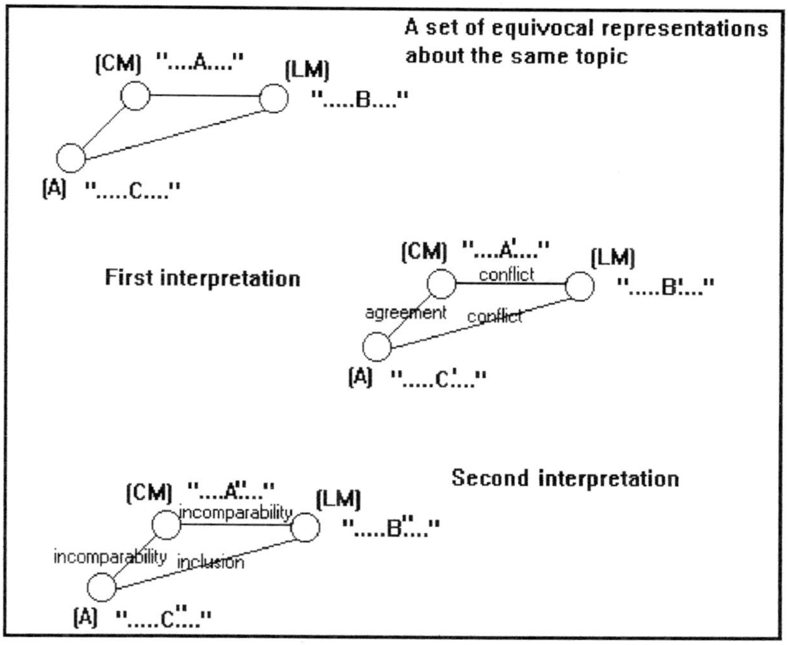

Figure 3. Multilevel structuring model

In this case problem dimensions are chiefly related to Multiple criteria approach (see for instance De Montgolfier and Bertier, 1978; Roy, 1985; Norese, 1991), its non-problematic disaggregation of the perceived problem and the possibility of aggregating points of view and problem dimensions at the appropriate moment and by one of the multicriteria methods.

Problem dimensions and model structuring dimensions are strictly connected.The analysis of the actorial representations, at the local level, gives rise to the consistent problem dimensions; the class of network in fig.2, for example, induces the dimensions "Efficiency and cost reduction", "Organizational acceptance", "Organizational control" and " Integration and development". Problem dimensions may induce a new structuring of the general perceived problem or of one upper level of problem formulation and then a revision of the general scheme, or they may stimulate the structuring of one or more sub-problems and then the transition from one level to another. The dimensions of the structuring model explain the governing process (Schneeweiss, 1987), that is the set of control and validation activities oriented to verification, evaluation and selection of different structuring opportunities, new exploration processes and actions on the developed representations, formalizations and structuring of operational model, applications and testing of these formalized models in relation to conceptual model and then to the reality.

In fig. 3 the problem of an experimental insertion of a DSS in Dialysis Centres is disaggregated, at the first level, in three different procedures of insertion that may be activated in sequence or contemporarily. Each of them is complicated enough to be analyzed alone, to require formalized validation protocols and a multicriteria evaluation in relation to the problem dimensions of the general level. Protocols and evaluations have then to be analyzed with key actors. New problem formulations are generally developed in this phase. MACRAME can help to explore and redefine them before the last stabilization in a formal project for the institutional decision maker, the Regional Health Department.

CONCLUSIONS

The interaction with the context and its key actors is essential to elaborate a model for the organization, especially when a formal Information System and a normative Frame of Reference are lacking. The analyst needs support in all the activities that characterize this interaction. His action which allows the problem to be structured and a consistent model to be formulated and validated must be aided at both an individual and a collective level. MACRAME is a proposal of methodological and operational support in these contexts.

REFERENCES

Bourgine, P., and Espinasse, B., 1987, Aide a la decision: une approche constructiviste, in: "Aide a la decision dans l'organisation", AFCET, Paris.

Bowen, K., 1983, An experiment in problem formulation, *J.Opl.Res.Soc.*, 34: 685.

De Montgolfier, J., and Bertier, P., 1978, "Approche multicritere des problemes de decision", Editions Hommes et Techniques, Paris.

Eden, C., 1988, Cognitive mapping, *Eur.J.Opl.Res.*, 36: 1.

Howard, R., 1989, Knowledge maps, *Management Sci.*, 35: 903.

Kling, R., 1987, Defining the boundaries of computing across complex organizations, in "Critical issues in: Information Systems Research", Boland, R. and R. Hirschheim, eds., Wiley, Chichester.

Landry, M., Malouin, J.L., and Oral, M., 1983, Model Validation in: Operations Research, *Eur.J.Opl. Res.*, 17:207.

Massa-Rolandino, R., and Norese, M.F., 1992, Modeling-validation process as decision support in organization, Working Paper PTdispea 92/11.1, in italian.

Mason, R. and Mikoff, I., 1973, A program for research on Management Information Systems, *Management Sci.*, 19: 475.

Moscarola, J., 1984, Organizational Decision Process and ORASA Intervention, in: "Rethinking the process of Operational Research and Systems Analysis", R.Tomlinson and I.Kiss, eds., Pergamon, Oxford.

Mucci, L. and Norese, M.F., 1990, A multidimensional model of the organizational answer to a change project, Working Paper PTdai 90/10.1.

Neisser, U., 1976, "Cognition and Reality. Principles and implications of cognitive psychology", Freeman, New York.

Nakamura, K., Iwai, S., and Sawaragi, T., 1982, Decision support using causation knowledge base, *IEEE Trans. Syst., Man, Cybern.*, SMC-12: 765.

Newman, M. and Noble, F., 1990, User involvement as an interaction process: a case study, *Inform.Syst. Res.*, 1: 89.

Norese, M.F., and Ostanello, A., 1985, A procedure to evaluate domand/supply of sport services and simulate intervention policies, Atti AIRO, Tecnoprint, Bologna.

Norese, M.F., 1988, A Multidimensional Model by a Multiactor System, in: "Compromise, Negotiation and Group Decision", B.R.Munier and M.F.Shakun, eds., Reidel, Dordrecht.

Norese, M.F., and Ostanello, A., 1989, Identification and Development of Alternatives: Introduction to the recognition of process typologies, in: "Improving decision making in organisations", Springer Verlag, Berlin.

Norese, M.F., 1991, A multiple criteria approach to complex situations, in: "Systems thinking in Europe", M.C. Jackson et al., eds., Plenum Press, New York.

Rosenhead, J., ed., 1989, "Rational analysis for a problematic world: problem structuring methods for complexity, uncertainty and conflict", Wiley, Chichester.

Roy, B., 1985, "Methodologie Multicritere d'Aide a la Decision", Economica, Paris.

Schneeweiss, Ch., 1987, On a formalization of the process of quantitative model building, *Eur.J.Opl.Res.*, 29:24.

Smith, G., 1989, Defining managerial problems: a framework for prescriptive theorizing, *Management Sci.*, 5:963.

BUSINESS AND ACADEMIC COLLABORATION IN A SYSTEMS PROJECT

Keith Sawyer[1], and Colin van Orton[2]

[1] University of the West of England
Coldharbour Lane
Bristol BS16 1QY
Avon, United Kingdom

[2] Digital Equipment Co. Limited
Basingstoke, RG22 4BS
Hants, United Kingdom

PURPOSE OF THE PAPER

This paper concerns a collaborative project undertaken by business consultants from Digital Equipment Corporation and the University of the West of England (formerly Bristol Polytechnic) with contributions from the 'client' organisation, Frenchay Hospital (Neurosurgery Department). The 'client' requirement was 'to develop a high quality information collecting, manipulating and dissemenation system for the Neurosurgery Department'. A second phase, driven by the successful outcome of this effort, was to produce a workstation to store and retrieve information for supporting decisions and research.

The central focus of the paper is to provide an account of the systems investigation which was conducted to establish the key issues and strategies which would needed to link into the second phase. In the process of investigating the information requirements, the similarities and differences between the two investigatory teams arose, and the paper shows how the backgrounds and skills knitted together to provide a comprehensive final report (Digital 1992).

BACKGROUND TO THE PAPER

The Neurosurgery Department at Frenchay Hospital, Bristol, has a formidable team of highly trained and powerfully motivated surgeons and supporting staff, and their enthusiasm ensures that a strong cooperative spirit is assured. Both the IT Department of the Univerity and Digital had separetly established links with the Frenchay Hospital, so this connection already provided a common concern and interest for the overall approach and requirements.

Systems Science, Edited by F.A. Stowell *et al.*,
Plenum Press, New York, 1993

DETERMINING THE APPROPRIATE METHOD OF APPROACH

Two early meetings were held with teams from Digital, Frenchay, and the University to establish the groundwork for investigating the situation, and from these meetings it was clear that a soft systems approach would be needed, due to the large number of stakeholders drawn from a variety of sections and the complex nature of the situation. This recognition leant itself to adopting a heuristic approach so that a clearer and more specific definition of the problem issues could be established, achieved through prototyping to uncover with increased clarity the strategy which would produce the successful operation of a workstation.

Digital have their own heuristic methodology to attend to complex social situations, and the University made use of models taken from OPIUM (Sawyer (1990a and 1990b) so an important part of the investigation became a process of deciding what would be the best methods and models to employ, and how these techniques could be woven together to provide a clear, common approach. From the beginning, it was agreed that the Digital consultants would lead the project and that the Digital methodology would be the backbone of the analysis. The OPIUM methodology is similar to the Digital style in that user-participation and the movement from a soft towards a hard, consensus established situation is sought. Workshops are of central importance especially in the Digital methodology, and a Future Requirements Workshop, to be discussed later, was run by Digital at the end of the investigation.

PRE-INTERVIEW INFORMATION SYSTEMS MODELS

Meetings took place with the Director of Neurosurgery and the IT Department at Frenchay Hospital to obtain an overall understanding of the current and desired situations. Three systems models were produced following the IT meetings, namely a Current Activities model (Sawyer (1992), Problem Situation Questions, also from the OPIUM methodology, and the Digital Top Mapping model, each of which is outlined below. The Problem Situation Questions model and the Digital Top Mapping model were consequently modified as a result of the interviews.

Current Activities model

A clear understanding of the main activities which surround a patients' entry, stay, and exit from the Neurosurgery section was of prime importance and a Current Activities model was developed for the following main reasons:

- the provision of an agreed overall systems boundary of the problem domain which would undergo changes following further negotiations and fresh learning derived mainly from building the systems models
- the identification of problem 'hot spots' which require specific attention
- an understanding of the technical and cultural jargon

The construction of the Current Activities model also proved useful as a consensus building method among the Frenchay IT staff since no such model had been produced which contained this information.

The Current Activity model consists of verb-fronted activity bubbles with (unspecified) input and output flows, decision diamonds with 'yes' and 'no' output flows, 'hot spot' areas with explanations, external source and receiving activities and a systems boundary, which may have connections to other Current Activity models. This was the case in this investigation, where it was found useful to clarify the activities undertaken by the neurosurgeons and the radiologists in two Current Activity models with interconnecting flows.

Digital Top Mapping model

A broader model was developed by both Digital and the University to outline the key entities and information flows for the Neurosciences Division at Frenchay Hospital, which encompases the Neurosurgery Department, the focus of the investigation. The Digital Top Mapping model thus showed the Neurosciences Department as the system, with 'islands' such as the departments of radiology, neurology, neurosurgery, and 'highways' and 'motorways' to portray the kind of information 'traffic', such as patient notes and scans, passing in both directions between the 'islands'. The Top Mapping model and the Current Activity models complimented each other well, since each was at a different resolution, with clear boundaries where the Current Activities models mapped into the Top Mapping model. The Top Mapping model was used at each interview; an account of this will be given later.

Problem Situation Questions

The final model produced before the interviews, (for later iteration with the other systems models, such as the Top Mapping model, and changes following the interviews,) was the Problem Situation Questions (from the OPIUM methodology). This model shows headings of key (problem) issues with questions raised by identified stakeholders. The final version of this model displays the problem questions in prioritised order, which allows a comparison to be made with the overall strategic plan, depicted in the Digital Future Objectives model and OPIUM's Strategy Tree, to ensure that the most important problems are addressed through well defined goals.

INTERVIEWS

Two typical problems encountered by systems investigations, namely identifying and then enticing the key stakeholders to particpate in interviews were overcome through the cooperation of the Director of Neurosurgery, who helped to arrange the interviews by sending out letters to those considered by Digital and himself as the most important to interview. Seventeen people were interviewed in total over a period of three weeks, longer than desired but a consequence of the pressure of work of the interviewees. Each interview was conducted by a Digital consultant with the assistance of a University analyst. The Digital Top Mapping model played an important role at each of the interviews, as it provided the opportunity to gain an overall appreciation of the inter-operations between and amomg departments, especially in terms of gaining a common perspective among the interviewees about the major elements which comprised the problem domain. The Top Mapping model was modified at each interview according to the comments made, and a consensus model was developed as a consequence. This consensus was to prove of vital importance in laying the foundation for an overall strategic position.

Each interviewee was given a set of cards, each card showing a problem which was originally identified during an interview with the Director of Neurosurgery. The interviewee was asked to arrange the cards in order of importance, and to include any other problems in the ranked order. The overall result of this problem prioritisation is given below.

POST-INTERVIEW INFORMATION SYSTEMS MODELLING

Apart from the iteration of the systems models already produced prior to the interviews, two systems models were developed from information provided at the interviews, namely the Multiview Diagram (Sawyer 1990a) and the Strategy Tree (Sawyer, 1990b), and

Digital's Future Objectives table, and Improved Information tables, developed as a consequence of the Future Requirements Workshop, set up to confer the results of the investigations and spearhead the future strategy. This section deals first with the models fuelled by the interviews, and then the Future Requirements Workshop.

The Multiview Diagram

The Multiview Diagram (Sawyer 1990a) forms part of the OPIUM methodology (PhD thesis at press) and depicts a focus statement or entity together with viewpoints expressed by named stakeholders, each with a separate segment. The diagram thus looks like a hub (the focus) with spokes, the spaces between the spokes representing the stakeholder segments where the viewpoints are displayed. Comparisons can be made both among and between stakeholders to identify conflicts and supporting attitudes and understandings. Several Multiview diagrams can be produced, one for each of the focuses deemed to warrant investigation. In this investigation, three Multiview Diagrams were produced, one with the focus 'improve patient care', the second 'improve information recording and retrieval' and the third 'administrative support'. These diagrams were discussed in the Future Requirements Workshop.

The Strategy Tree

This model forms part of the OPIUM methodology (see Sawyer, 1990b) and shows the overall mission at the top of the tree, followed by critical objectives, then one or two purpose levels and finally a goal level. Goals are defined as those activities which have a testable outcome for delivery within a specified period of time, whereas purposes are defined as desirable activities which need to be continually performed, such as 'improve patient treatment'and 'improve information support, both taken from the Strategy Tree developed during the investigation. A top/down 'how', and bottom/up 'why interrogation is performed to ensure that there is coherency in the hierarchical relationships, and that the systems boundaries of each activity are consistent. The mission of the Strategy Tree was shown as 'enhance Neurosurgery Performance' and consisted of four levels, with goals such as 'improve patient selection' and 'reduce time span'. These goals were iterated with the Problem Situation Questions and a direct correspondence between key problems and goal activities was achieved.

Digital's Future Objectives table

This table was presented at the Future Requirements Workshop for discussion and modification, and represents the impact of each of the prioritised objectives, shown vertically, against 'enterprise functions', shown horizontally and reading, from left to right, 'demand', 'supply', management and control' and 'customer/patient satisfaction'. A range of codes, from 'most significant impact' to 'no impact' indicates, for example, that 'reduced timespan' has 'little impact' on 'demand' but has 'major significant impact' on 'management and control'. This table presented interesting debate at the Workshop, and brought out many opinions heard for the first time.

Digital's Improved Information tables

These tables, also discussed and further developed at the Workshop, compare information requirements, such as the 'availability of the right information' at the 'right place', and at the 'right time', against the prioritised objectives, such as 'improve diagnosis', 'improve treatment and cure' and 'improve records/analysis'. These tables drew out the importance of matching information problems to prioritised objectives, leading the way to producing an overall strategy.

OUTCOMES AND CONCLUSIONS

The Digital methodology and the OPIUM methodology used by the University appoximated very well, each adopting a heuristic, contingent approach while incorporating models to aid in consensus building. The Multiview Diagrams and Strategy Tree from the OPIUM methodology complimented the Digital models, such as the Top Mapping model, the Future Objectives table, and the Improved Information tables. Because Digital led the analysis, the administrative burdons were well managed, with clear mandates and requirements. In preparation for the Future Requirements Workshop, a prioritised list of key problems, drawn from the card sorts performed by each interviewee, was prepared, by weighting each of the nine problems with 8 for the highest and 0 for the lowest. The list, which surprised the analysts, is shown below, with the highest ranking first:

- improved diagnosis
- improved treatment records and statistical analysis of outcomes
- improved treatment and care
- improved selection of better targeted patients to call forward
- reduced time-span from first to final contact
- improved utilisation of scarce resources
- improvement of the ratio of patients treated to staff employed
- reduced waiting lists
- other

A booklet outlining the work covered in this investigation has been produced by Digital (1992).

This investigation has paved the way for the next operation, which is the construction of the workstation, headed by Professor Sam Waters at the University of the West of England. Teams of researchers will be investigating areas such as distributed databases, multi-media, and end-user interfaces. One of the major problems highlighted in this investigation, namely the question of how patient notes and radiology scans and films can be always available together, permeates throughout all hospitals. For example, concurrent consultancy and research at the cardiology department of BRI (Bristol Royal Infirmery) by Sawyer and Gillies (paper in press) confirms the widespread requirement of holding all information about a patient in a workstation which is easily accessable to all staff. The Neuorosurgery investigation which this paper has outlined serves to contibute towards this essential requirement.

REFERENCES

Digital Equipment Corporation, "Frenchay Trust - an integrated Neurosurgery strategy and I.S. Strategy", June 1992.

Sawyer, K., "Multiview Diagrams - an introductory description", Systemist, Volume 12, Number 3, August 1990.

Sawyer, K., "Goals, Purposes, and the Strategy Tree", Systemist, Volume 12, Number 4, November 1990.

INTEGRATING SCIENTIFIC DISCIPLINES:
AN EVERGREEN CHALLENGE TO SYSTEMS SCIENCE

Olav Solem

The Norwegian Institute of Technology
The University of Trondheim
Trondheim, Norway

INTRODUCTION

The scale of global change exposes more clearly than ever before the critical shortcomings of discrete scientific disciplines. The single-minded pursuit of disciplinary research can well prove to be counterproductive to industrial productivity. This was one of the most important statements from the U.N. conference "Action for a Common Future" held in Bergen, Norway 1990 (Willums, 1990).

At the University of Trondheim,Norway, the Norwegian Institute of Technology, at the Department of Economics and Industrial Management, we have newly introduced a special programme in engineering management. The main objective of this programme is to educate engineers having the competence to solve problems requiring knowledge not only in technology, but also in economy and management. The underpinning idea is to integrate technical, economical and management disciplines in a type of combi-competence.

PURPOSE

As mentioned the vision of our department is to give an education with an interdisciplinary orientation. The purpose of the present paper is to describe a contribution to the realization of this challenging idea.

A dominating and characteristic feature of business problems are their cross-scientifical nature. Being able to solve such problems requires therefore knowledge from different scientific disciplines.

With the declared purpose in mind, a conceptual framework will be described. The idea behind the framework is to use it as a tool to raise challenging questions as to the underlying metatheoretical assumptions of individual different scientific disciplines. Followed by a comparative analysis, clarifying and reconsidering the underlying assumptions, further advances in knowledge and understanding should be achieved.Through the increased consciousness about metatheoretical underpinnings gained, an increased integration of different scientific disciplines and greater interdisciplinarity should be the result of this process.

Systems Science, Edited by F.A. Stowell *et al.*,
Plenum Press, New York, 1993

SCIENTIFIC INTEGRATION: TWO CHALLENGES TO EDUCATION

Having the intention to realize an extended interdisciplinarity in education, we are facing two major challenges which respectively will be referred to as the practical and the theoretical challenge.

The practical challenge is a question of the form of education and research that contributes to interdisciplinarity.

The traditional way of teaching is expert- or discipline oriented. What the students learn depend very much on the teachers who are responsible for the courses that they select. Further, a number of experts handle their specialities without any overall coordination.

The main concern of this paper is, however, more the second challenge, the theoretical challenge, which is a question of developing a perspective, or a frame of reference, together with methods and concepts to stimulate interdisciplinarity.

Who can help us to meet with this theoretical challenge? What about systems theory in this connection?

SYSTEMS THEORY: FROM UNITY OF SCIENCE TO COMMITMENT TO COMPLEMENTARISM

All from the beginning, one of the main aims of General Systems Theory (GST), was to promote interdisciplinarity by improving the communication between specialists (von Bertalanffy 1968). The idea was to develop principles crossing the individual sciences and which should bring us closer to a unity of science.

The wish for holistic thinking, synthetic thinking and a cross-conceptual thinking is therefore clearly expressed in the philosophical foundation of GST (Solem 1985)

The objective of the GST pioneers to develop a unity of science has proved to be an utopia. The dream is not, however, totally given up.

The critical systems movement has more or less given rebirth to that idea,but under a new conceptual umbrella: the commitment to complementarism/pluralism.

Critical systems thinkers are committed to a pluralist path which recognizes the value of different approaches for dealing with problematic situations. In order to develop a pluralist path for the complementary development of different systems approaches a taxonomy of approaches has to be created.

Jackson and Keys (1984), the originators to the commitment to complementarism, has proposed a "system of systems methodologies" to guide method selection and to provide insights about the values of different methods. This framework has later been extended by Jackson (1988).

Recently, Flood and Jackson (1991) have constructed a meta-methodology called "total systems intervention" describing procedures for translating complementarism into action in the real world.

Banathy's classification of "systems methodology" (Banathy 1984), Oliga's identification of "methodological foundations" (Oliga 1988) and Ulrich's description of "different levels of systems practice" Ulrich (1988) are other seminal contributions to complementarism in the spirit of critical systems thinking.

The conceptual framework developed and to be presented in the forthcoming is also a work committed to complementarism and is very much in the same road as the works mentioned above.

A CONCEPTUAL FRAMEWORK FOR ANALYZING METHODOLOGICAL ASSUMPTIONS

The conceptual framework to be presented is built on a metatheoretical framework for methodology in general, developed originally by Burrell and Morgan (1979), and later extended and clarified by Morgan and Smircich (1980) and Morgan (1983a).

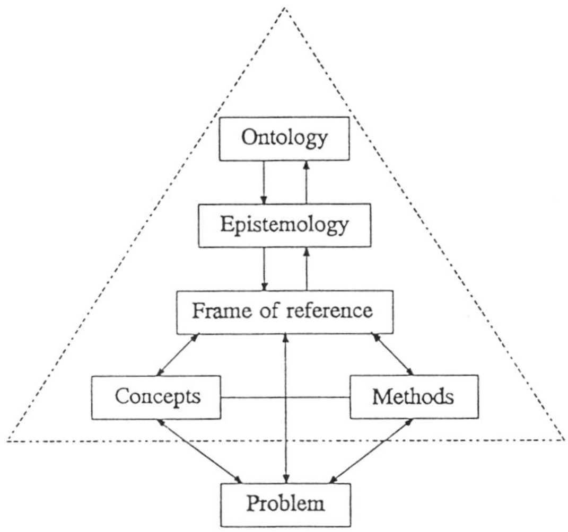

Figure 1. Conceptual framework for analyzing methodological suppositions.

Burrell and Morgan (op. cit.) present in terms of a two-dimensional perspective a general framework for the analysis of social theory.The framework reflects the basic philosophical presuppositions or metatheoretical assumptions underlying scientific inquiry. First, it reflects a philosophy of science in terms of four basic assumptions, related to ontology, epistemology, human nature, and methodology, along a subjective- objective dimension. Second, it reflects a theory of society in terms of a regulation- radical change dimension.

Overarching any social research, following Morgan (1983a), is the constitutive ontological assumptions regarding the researcher's view about the existential nature of the social world and human subjectivity. Such constitutive assumptions define the basic underlying paradigm for a particular inquiry. Concretization of these foundational assumptions in terms of favoured metaphors, or images, through which the assumptions become meaningful, represents a particular epistemological stance, or a particular view of the possibility of knowledge about the social world (Oliga 1988).

In the conceptual framework to be presented here, only one dimension, the subjective-objective, is to be treated. Ignoring modes of domination, conflict, and radical change, serious omissions are committed (Morgan and Smircich 1980). This simplification is, however, done to make the work of a manageable length. Due to the same reason, the nature of man is neither included.

Ontology and epistemology directly influence the frame of reference underlying a scientific discipline. Further, the frame of reference determines the corresponding methods and concepts accompanying a specific framework (Figure 1).

Morgan and Smircich (1980) have defined a typology for classifying different ontological assumptions with an objective approach, respectively a subjective approach as extremes (Figure 2). Between these two extremes, there is a continuum of intervening theories of reality.

A typology for the corresponding epistemological stances is presented in Figure 3. Positivism respectively anti-positivism represents here two extremes. Between these two extremes, there is a continuum of intervening stances of epistemology.

Epistemology deals with the groundwork of knowledge, i.e. with how we understand reality and communicate this to other people. The groundwork of knowledge will be different due to different fundamental conceptions about the world and social realities. It is therefore a close relationship between ontological suppositions and epistemological stances. Ontology is in this way clearly prior to epistemology as also illustrated in Figure 1.

Objectivist approaches to social science
Realism

- Reality as a concrete structure
- Reality as a concrete process
- Reality as a contectual field of information
- Reality as a realm of symbolic discourse
- Reality as a social construction
- Reality as a projection of human imagination

Subjectivist approaches to social science
Nominalism

Figure 2. Core ontological assumptions. (Morgan and Smircich 1980).

A central concept in epistemology is the metaphor concept. A metaphor is an image, or a description of one thing with another (Morgan, 1986). With the help of metaphors we try to generate knowledge about the world. Visions and images are often more easily to understand than words and therefore the best way in which to communicate.

USE OF THE DEVELOPED CONCEPTUAL FRAMEWORK: A PRACTICAL EXAMPLE

With the simplifications, as presented above, in mind, the two basic questions to be asked in a practical analysis of a specific scientific discipline, are as follows:

1) What are the core ontological suppositions underlying its frame of reference?
2) What are the basic epistemological stances for its frame of reference?

In other words:

1) How does the discipline look at and understand reality?
2) How does the discipline learn about reality?

Figure 2 gives a scale, or a standard, on which to place answers to question 1) above. Similarly Figure 3 gives a corresponding scale for answers to question 2). The use of the conceptual framework presented in Figure 1 in this way, results in an overview over core ontological assumptions as well as core epistemological stances. In turn this will give the

Objectivist approaches to social science
Positivism

- To construct a positivist science
- To study systems, processes, changes
- To map contexts
- To understand patterns of symbolic discource
- To understand how social reality is created
- To obtain phenomenological insight, revelation

Subjectivist approaches to social sciences
Anti-positivism

Figure 3. Basic epistemological stances. (Morgan and Smircich 1980).

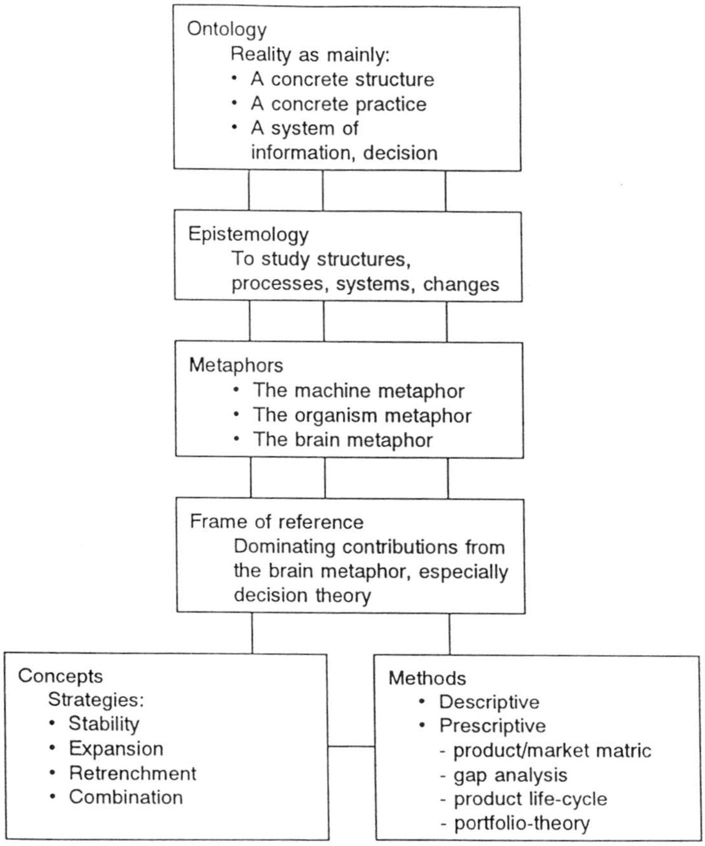

Figure 4. Methological suppositions for the rational analytical tradition within the strategic management area.

basis for comparative analysis between different scientific disciplines, i.e. to elaborate on their metatheoretical suppositions.

More concretely, we will demonstrate how to use our conceptual framework by analyzing the methological assumptions underlying the rational analytical school or tradition within the strategic management area. According to Bengtsson and Skärvad (1988), famous authors like Ansoff (1965) and Porter (1980) mainly belong to this school.

Figure 4 shows the result of an analysis of the methodological suppositions underlying the frame of reference of the rational analytical school as defined above.

As shown in Figure 4, the rational analytical school within the strategic management area is intensively influenced from realism regarding ontology and from positivism as to epistemology, i.e. it builds heavily on objectivist methodological suppositions. Decision theory represents consequently a dominating contribution in the reference framework of this school. Strategic management, from the rational analytical perspective, is therefore mainly a question of deciding strategy for a business which means deciding the content of the strategy. This will in turn determine the corresponding methods and concepts belonging to the actual frame of reference, confer Figure 4 and methods and concepts referred to there.

The rational analytical school represents only one school among many others within the strategic management area. A similar analysis, as demonstrated above, can be performed for other approaches, for instance, approaches looking at strategic management as a social process of change and learning (Mintzberg, 1987) or as a cybernetic process (Morgan 1983b).

Further, the strategic management area is only one managerial theory among several others working with changing organizations. Following the analytical scheme developed and used in this work a similar analysis can be performed to elaborate on their underlying metatheoretical assumptions.

As mentioned earlier some major simplifications have been done in this work by not taking into account, for instance, the ethical dimension or the "regulation-radical change" dichotomy (Burrell and Morgan, 1979) and the nature of man (Morgan and Smircich, 1980). However, this is only a question of extending the analysis. The excluded dimensions may be taken care of in the same manner as illustrated in the present work for the cognitive dimension or the subjective-objective dichotomy.

FURTHER RESEARCH

Separate analyses of the type demonstrated in the present work might be extended to include different scientific disciplines giving in turn the opportunity to perform comparative analysis. The intention is therefore to go further and to include economical as well as technical disciplines to elaborate on.

The purpose of this work has been to develop a conceptual analytical scheme to be used as a tool to ask challenging questions as to the core methological suppositions underlying different scientific disciplines. Elaborating metatheoretical underpinnings in this way should stimulate and promote interdisciplinarity.

REFERENCES

Ansoff, H.J., 1965, "Corporate Strategy", McGraw-Hill, New York.

Banathy, B.H., 1984, "Systems Design in the Context of Human Activity Systems", Internatinal Systems Institute, San Francisco.

Bengtsson, L. og Skärvad, P.H., 1988, "Företagsstrategiska perspektiv" (Corporate Strategic Perspectives). Studentlitteratur, Lund, p213.

Burrell, G. and Morgan, G., 1979, "Sociological Paradigms and Organizatoinal Analysis", Heineman Educational Books, London, p432.

Flood, R.L. and Jackson, M.C., 1991, "Creative Problem Solving. Total Systems Intervention", John Wiley & Sons, Chichester, p259.

Jackson, M.C., 1988, Some methodologies for community operations research, *Journal of the Operational Research Society*, 39 (8), pp715-724.

Jackson, M.C. & Keys, P., 1984, Towards a system of systems methodologies, *Journal of the Operational Research Society*, Vol. 35, 1984, pp473-486.

Mintzberg, H., 1987, Crafting strategy, *Harvard Business Review*, No. 4, July - August 1987, pp137-145.

Morgan, G., 1983a, "Beyond Method: Strategies for Social Research", G. Morgan, ed., Sage, London.

Morgan, G., 1983b, Rethinking corporate strategy: A cybernetic perspective, *Human Relations*, Vol. 36, No. 4, pp345-360.

Morgan, G., 1986, "Images of Organization". Sage Publications, Beverly Hills.

Morgan, G. and Smircich, L., 1980, The case for qualitative research, *Academy of Management Review*, Vol. 5, No. 4, pp491-500.

Oliga, J., 1988, Medthodological Foundation of Systems Methodologies, *Systems Practice*, Vol.1, pp87-112.

Porter, M.E., 1980, "Competitive Strategy", Free Press, New York.

Solem, O., 1985, "Systemteori for administrasjon og ledelse" (Systems Theory for Administration and Leadership). Tapir, Trondheim, p112.

Ulrich, W., 1988, Systems thinking, systems practice, and practical philosophy: A program for research, *Systems Practice*, Vol. 1, No. 2, pp137-163.

von Bertalanffy, L., 1968, "General Systems Theory, Foundations, Development, Applications", Georg Braziller, New York. p289.

Willums, J-O., ed., 1990, The Greening of Enterprise. International Chamber of Commerce. Papers presented at the Industry Forum on Environment, the "1990 UN Conference: Action for a common future", Bergen 1990.

A SURPRISING UNION? - SOFT SYSTEMS ANALYSIS AND SOCIOMETRY

Roger W. Stewart

School of Information Systems
Penrhyn Road
Kingston upon Thames
Surrey, KT1 2EE

ABSTRACT

Communication has long been regarded as an important factor in the performance and behaviour of organisations, existing in theory as depicted by formal organisation charts, and in practice by a mixture of formal and informal patterns. The perspective of organisations as social systems has similarly a well written history. The approach then of a social analysis of communication patterns to understand or predict behaviour of organisations is therefore a logical result.

Soft systems analysis is an approach that is well suited for the richness of social systems and sociometry is a family of techniques for displaying patterns of human behaviour in social groups.This paper examines how these two complementary approaches can be linked to provide new insights into organisational behaviour.

INTRODUCTION

The linking of these two approaches is part of a research programme by the author which examines a current trend in organisational change, that of creating flatter more flexible organisation structures. This type of change encompasses the reduction of the number of levels of management and the creation of task based teams set up to provide a continuing function/service to the whole organisation or a multi-disciplinary task based team with a high degree of autonomy set up for a specific project. A team in this sense could be a small group or a larger department. In parallel with these structural and operational changes from tall hierarchical to flat structures, there is a growing awareness of the need to understand the social processes involved in the changes and in the newly created teams.

It is for the above environment that a research approach has been developed which combines social systems and behavioural analysis with communication analysis. A basic proposition underpinning the approach is that communication patterns both reflect and affect the performance of these teams and that analysis of communication networks employed, both 'formal' and 'informal' will provide an explanation of behaviour of the team and its relationship to the organisation.

Systems Science, Edited by F.A. Stowell *et al.*,
Plenum Press, New York, 1993

In order to undertake a 'Systemic' analysis of the behaviour of teams in a flattened organisational environment, three linked and interactive streams of analysis have been developed as a single approach whilst retaining their own theoretical underpinnings (Figure 1). In this sense the notions of complementarism and social awareness in critical systems thinking have been adopted (Flood and Jackson, 1991).

The three analysis streams are described below in order to provide a background to the linking of Sociometry and Soft Systems analysis.

The **Process** stream gives the ability to measure and see the communication patterns of who talks to whom, how often, on what subject, and how important. This is a key representation of the reality of operation of the organisation. The Process stream of analysis utilises the Netmap System, a computer package based on a relational database that depicts a graphical representation (Sociogram) of communication links between people. It is shown in a network format that can be manipulated to show how and what is communicated (the informal network) and compared to how the organisation believes what should be communicated (the formal network). This enables the creation of a picture of how the organisation and teams and are working. The concept on which this is based is Sociometry orginally devised by J. Moreno in the 1930's.

The **Meta-process** stream of analysis is a systemic logic-based analysis, developed from Checkland's Soft Systems Analysis (1990), which provides the positioning of the communication network in the operating environment. It can be used to both design the operation of a new team or to analyse an existing team in terms of why it was created, how it operated and what activities with associated communication requirements have to be done to achieve its stated goals.

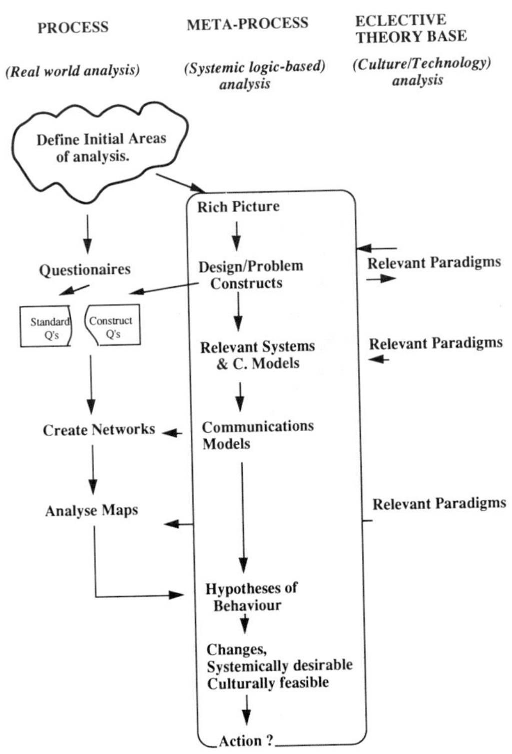

Figure 1. Linked Approach.

The **Elective Theory Base** stream.There has been much research in the fields of Organisational Behaviour, Organisational Psychology and Sociology which can help the understanding of the performance of an organisation. The eclective theory base is a set of paradigms or models which describe behavioural situations, their associated communication patterns, possible causes and possible theories that may be applicable to the area under analysis.

This paper will consider the first two streams based on Sociometry (Process analysis) and Soft Systems (Meta-Process analysis), the third stream that of the construction of paradigms can be seen in previous papers by the author.

SOCIOMETRY

Sociometry is the name given to the technique of displaying patterns of human relationships that exist within groups (Huczynski and Buchanan, 1985), and in its orginal setting was used to reveal the feelings and emotions of individuals towards each other. The visible output of a sociometric study is the use of a sociogram which depicts graphically the emotional feelings of individuals towards each other and deals with human social choice and judgement. Jacob Moreno created sociometry and produced the first sociograms in 1934 which proved to be significant in showing what was happening in a social system. Various uses of this were found in the commercial setting during the 1950's and early 1960's with work by Van Zelst and Hart, however these were still aligned to emotional networks amongst co-workers but nevertheless showed rudimentary communication networks when plotted as maps following the result of tests or questionaires. The work was time consuming and little use was made of computers, which is probably one of the main reasons why this technique apparently went into decline. Now with the availability of small powerful computers, relational databases and good graphics capability and the ability to handle large volumes of information and process them rapidly, the opportunity to use this technique is enhanced. In addition the scope of potential analysis has now been widened and is limited solely to what attributes are given to the nodes and what attributes are given to the links. For example intelligence maps and skills matrices. Here I will restrict the scope to people and communications between them.

The basis of a 'Sociogram' is, "simply a map of who communicates with whom in an organisation, without regard to formal channels" Mintzberg (1979) and is constructed from questionaires completed by the people in the systems under investigation. Although the concept is simple it can be a powerful tool, giving the ability to produce an analysis form showing a communication network map of not only the formal channels of an organigram but also the informal communications that occur . It provides a gestalt picture complete with emergent properties and enables many of the concepts of sociologists, anthropologists and behavioural psychologists to be seen reflected in actual communication patterns.

Sociogram interpretation

The interpretation of sociograms is obviously dependant on the purpose of the study, the selected group size and the data collected. It should also be remembered that each map is contextually dependant and interpretation of a particular pattern will vary. General examples given here are based on a selection of Netmap patterns showing actual interactions both formal and informal with the conclusions that might be drawn from them.

AGREED/DISAGREED COMMUNICATIONS - An important distinction is shown between interactions where both individuals agree that the communication is of a specific frequency and importance, and a disagreed communication is where one party differs in perception. The latter indicates that one of the parties does not believe that the communication happens as often or that it is of greater or lesser importance. There may even be a degree of conflict. This difference in perception can show that inappropriate information is being sent, important information is being disregarded, the format is incorrect ie. too much-too little or indeed that there is a difference in personal regard between the individuals.

INTERGROUP AND INTRAGROUP - The maps depict two types of communication, those where the interactions are within defined groups, eg. departments and those interactions between departments. At a broad level this will show how an individual group is operating, for example, many agreed communications between all members of the group generally means that there is a high degree of cohesion, conversely a high level of disagreed communications can indicate that the group has not fully formed its own norms and identity or there is major conflict between group members. Similarly between groups if there is a high volume of agreed communication links it can indicate that the group is well integrated within the organisation or that it does not know what it is doing and is desperately trying to find out its role. If there is a high volume of disagreed external communications, yet high agreed internal communications, this can be indicative of a group that is alienated to the organisation (Touraine, 1971), the organisation is fragmenting or typical of a "Newstream" group (Kanter, 1989).

We can now look at the relationship of an individual to the organisation as a whole or to specific groups, remembering that the following patterns can occur in any combination of the above.

ISOLATES - For a chosen frequency/importance/task a single isolate has no communication links to any other individual and is detached from the main network. The result is that the individual is not taking part in the activities of the formal group in which he/she belongs. The reasons for this vary, for example in a newly formed group it may be conflict with the group or non acceptance of the individual by the group eg. 'reactance' (Brehm and Brehm, 1981), or indeed non fulfillment of personal needs (Maslow, 1954). Isolates may also be in small groups of pairs and trios, this is particularly true in the case of group polarisation (Isozaki, 1984).

STARS AND BLEEDERS - Where a particular individual has many agreed communication links it is apparent that the individual is having an important effect on the behaviour of the organisation either good or bad. Many disagreed links could be a case of delusions of grandeur. Two aspects that may be considered are bottlenecks and gatekeepers. The first is straightforward in that this individual is affecting adversely the performance of the group or organisation by restricting the flow of communication, information or decisions. The second is more complex in that the individual is a gatekeeper of information and adding or detracting value to the to a particular process (Kanter, 1989) or of exercising personal power (Huczynski and Buchanan, 1991). Stars and bleeders are not necessarily the obvious or expected individuals.

EMERGENT GROUPS - Emergent groups are those where the members share more than half of their links irrespective of which formal group they belong to. This clustering can be in direct contrast to the formal organisation chart. There are many reasons that can explain this. It may be a function of their job or part of a matrix managed or team oriented organisation. However it may be that they have been rejected by, isolated from or never really joined their formal group. It could be a case of Ingroup/Outgroup (Tajfel, 1981) where their allegiance still lies with a previous group. Isolates, stars and bleeders can also be seen in emergent groups.

We can therefore see in the communication patterns shown by the sociogram the behaviour exhibited by individuals within their teams and the relationship of teams to the organisation. This social behaviour does not happen in isolation but within an environment. It is the understanding of this environment of the organisation that a Soft Systems approach is used to provide context and focus to the analysis.

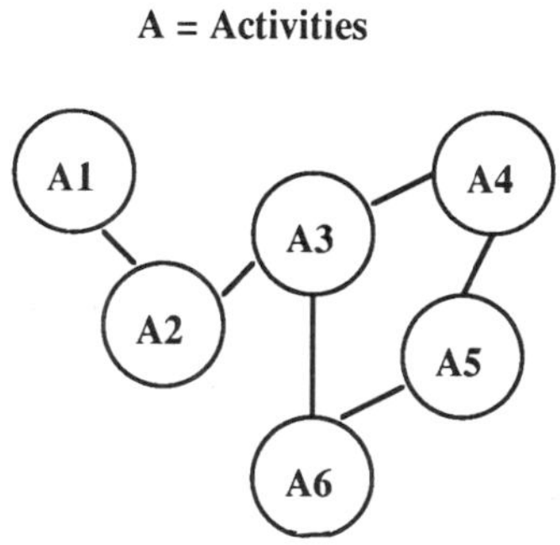

Figure 2. Conceptual Model.

SOFT SYSTEMS ANALYSIS

It is not proposed here to go into detail on Soft Systems but rather emphasise the links that can be used to guide the sociometric analysis. The case of investigating an existing team will be taken.

Rich Picture interpretation

Once the initial boundaries have been placed around the analysis, a variant of Soft Systems Analysis is used to understand or design the operating environment of the team. Members of the team and other related personnel are interviewed and a Rich Picture is created. This depicts, in one or more frames, the structure of the team, the relationship to other areas of the of the organisation, what processes and happenings occur together with "soft" information such as feelings and conflict. The Rich Picture thus forms the gestalt of the area of analysis as seen by the analyst and is the basis from which this stream of analysis proceeds. This gestalt allows an interpretation as to problems and successes shown in the behaviour of the team. This interpretation can be aided by using Kelly's Grids and construct theory to cluster tasks and issues, inter- and intra-relationships, and social considerations. For example this may be accomplished by using a pc package called Pegasus which enables the analyst to examine his or her perception of the situation to form groups of problems and constructs of team behaviour. These can be used to form specific focussed questions that can be added to the standard sociogram questionaire of to who talks to whom, how often and how important. This is the first link.

Conceptual Modelling

From the clustering of problems and the constructs of behaviour relevant systems are named to address these areas. The most relevant systems to the boundary of analysis are taken further with the construction of Root Definitions, a written statement of what each relevant system must do in order to achieve its purpose. From the activities stated or implied within the root definitions a Conceptual Model is developed as a linked activity diagram.

Each of these activities are extended to include the communication requirements between people necessary for the functioning of the activities contained in the conceptual model, thus forming a conceptual communication model. This extended conceptual model is thus the result of "systems thinking" in that it is the analysts perception of what the activities and communications are in order for the team to operate successfully.

In order to use this information as part of a sociometric analysis and construct sociograms, a variation on the Maltese Cross technique of B. Wilson can be utilised. The north part of the cross containing activities from the model, the east and west the input and output communications and the south the relevant pairs of people (Node information). The information from the Maltese Cross can then be linked back and inputted to Netmap to produce Conceptual Communications maps. This second link enables an important comparison to be made between what the organisation is structured to do (Formal organisation map), what actually happens in practice (Informal organisation maps) and what conceptually should happen (Ought to be maps). The results of these comparisons of communication maps and the interpreted individual and organisational behaviour are a description of what happens in reality and what should conceptually happen. This can be augmented by a comparison back to the rich picture. From these a list of potential changes that could provide changes to improve the situation can be constructed. However a discussion with the initiators of the analysis from the organisation is needed to establish those changes which are both practical and culturally feasible in the organisation in order to establish any intervention process.

CONCLUSIONS

The linking of Sociometry and Soft Systems analysis has been discussed in the context of providing a communication analysis approach to understand a particular type of organisational change. Although the two approaches have different theoretical bases they are complementary in that their respective strengths address different perspectives of social behaviour. It is hoped that this paper has provided some justification for using them in combination to provide new insights into organisational behaviour.

REFERENCES

Brehm, S., Brehm, J. W., 1981, "Psychological Reactance: A Theory of Freedom and Control", Academic Press, New York.

Checkland, P., 1981, "Systems Thinking, Systems Practice", Wiley, Chichester.

Checkland, P., Scholes, J.,1990, "Soft Systems Methodology in action", Wiley, Chichester.

Hart, J., Nath, R., 1979, Sociometry in business and industry, *Journal of Group Psychotherapy, Psychodrama & Sociometry,* 32, pp32-36.

Huczynski, A., Buchanan, D., 1991, "Organisational Behaviour", Prentice Hall, London.

Isozaki, M., 1984, The effect of discussion on polarization of judgements, *Japanese Psycnological Research,* 26, pp72-76.

Kanter, R. M., 1989, "When Giants Learn to Dance", Unwin, London.

Maslow, A., 1954, "Motivation and Personality", Harper and Row, New York.

Mintzberg, H., 1979, "The Structuring of Organisations", Prentice-Hall, Englewood Cliffs.

Moreno, J., et al, 1960, "The Sociometry Reader", The Free Press, Glencoe.

Stewart, R., 1991, The use of social paradigms in the analysis of team behaviour, *in*: "Systems Thinking in Europe", M. Jackson, G. Mansell, R. Flood, B. Blackham, and S. Probert, eds., Plenum, New York.

Tajfel, H., 1981, "Human groups and social categories: Studies in social psychology", Cambridge University Press, London.

Touraine, A., 1971, "The Post-Industlial Society", Random House, New York.

ORGANISATIONS WITHOUT "SYSTEM":
WORKING WITH HOLONS AGAINST UNIVERSALITY?

Julie Travis[1] and Duane Truex III[2]

[1] School of Information Systems
Curtin University of Technology
Bentley, Western Australia

[2] School of Management:MIS Group
Binghamton University
State University of New York at Binghamton
Binghamton, New York

INTRODUCTION

The social organisation just like the biological organism is in a constant process of evolution. In the social world all parties to the organisational change process come to the organisations with various lenses which shape the way they view and describe the change. Actors will have their own preconceived ideas about how organisational change *should* be implemented. The actor who also fulfils the role of interventionist enters the internal organisational environment and becomes part of the change process itself. Whatever methodology the external actor or interventionist brings into the internal environment is yet another framework to be incorporated into the social discourse embedded within the organisation.

This paper addresses the nature of change and emergence in organisations and questions the rigidity of inflexible methods or methodologies and the inapplicability of universal rules to the emergence process. In negotiating change, it is considered that the property of emergence arises from use of the language and this in turn will affect how systems are constructed. Finally, it suggests that when dealing with emergence, the interventionist consider approaches that are (w)holistic rather than approaches that always view the organisation as being systematic or systemic.

THE AUTOPOIESIS OF ORGANISATIONS

As suggested in the General Systems literature, social organisations are autopoietic and self-referential. They are always in the process of self-creation through the vehicle of discourse and must sustain themselves to survive and prosper. (Thyssen, 1992; Baskerville,

Travis and Truex, 1991; von Foerster, 1984; Luhmann, 1986; Aulin, 1986). As an open system, organic and yet in a state of flux, the organisation acknowledges its environment as a part of itself. In this way it maintains a self-consistent set of evolving relationships and self-making or self-referential only with respect to self-organisation.

Morgan (1986) states that:

> *The theory of autopoiesis thus recognizes that systems ...* have... *"environments", but insists that relations with any environment are internally determined. While there may be countless chains of interactions within and between systems... there is no independent pattern of causation... wholes evolve as complete fields of relations that are mutually determining and determined*

Open system autopoiesis suggests that that the systems pattern must be understood as a whole and as possessing a logic of its own yet the organisation need not be self-contained (*op cit*). Whereas in the theory of closed system autopoiesis, change is considered to be a result of random or chance internal events giving rise to new system relations. In favour of the closed system argument Thyssen (1992) states that an autopoietic system creates its own elements referring only to its environment/s if they become threatening in any way. We argue that the organisation is an open system that must also create itself in accordance with its environment, suggesting that it must be self-creating and yet aware of and also existing in alignment with external factors. Where boundaries exist may change depending on how the organisation wishes to see itself or project its own image within the *world* in which it exists.Perceived boundaries tend to be renegotiated as they are re-created. (Hedberg, 1991; Powell, 1987; Toffler, 1990). Therefore, how the organisation is viewed is largely determined on how it is socially constructed, and this will be affected by emergence, change, how we construct and understand world views, how we construct rules of negotiation and acceptance, and then how we apply them.

EMERGENT THEORY: PROCESS VS PROPERTIES

We view the organisation as being a network of social contracts where the context of the language is of great importance (Habermas, 1987), and as being organic and fluid. We would rather view behaviour, discourse and action from the perspective of emergence. That is a theory that states that there are no fixed organisational *structures* as the form is fluid and is always arising from a previous state of being.

Emergence as a theory is not equivalent to a theory of change. The term *emerging* may be better indicative of the change process. Change implies a transition between one period of time and another, from one state of being to another, or one structure to another (Baskerville, Travis and Truex, 1992; Truex and Klein, 1991). Emergent processes are different in that they are temporal agreements frozen and captured for observation, research, analysis or some kind of use in the recreation process, or in order to bring about action.

The concept of emergent processes is built upon the belief that they are products of constant social negotiation, shared understanding and awareness. There may be emergent regularities, but they are also shifting and changing as is the nature of sense-making and negotiation. As a result, the notion of organisational structure is replaced by the idea of organisational regularities. Agreements, decisions, ways of doing tasks and interpreting may move towards achieving a structure, but are never fully captured in a structure.

This theory considers emergence as a process, whereas Checkland views emergence as a *property* of the system (Checkland, 1981a, p3). Thus, *emergent properties* are described as entities which are natural and man made, existing as *holons* of the system, where entities are only meaningful in terms of the whole. For example, the properties of a bicycle relate and are meaningful only to the assembled whole, the bicycle itself, or the smell of ammonia as a concept having no meaning in terms of the properties of nitrogen and hydrogen being parts of the ammonia molecule. The term *holon* is adapted by Checkland and Scholes (*op cit*) from Koestler to mean ... *the abstract idea of a whole having emergent properties, a layered structure and processes of communication and control which in principle enable it to survive in a changing environment.* The use of holons as adapted by Checkland (1991) suggests perceived realities, or conceptual models of what the view of the world may *be*, and may *becoming*.

Our definition of a holon is further adapted from Checkland's adaptation (1991)to suggest that a holon may also be determined through use in a local setting, thus it could be ..th*e abstract idea of a whole, having emergence and change as a theme, the interpretations of which can only be meaningfully described in context.*

LOCAL CONTEXT VS UNIVERSALITY

From the perspective described above there are two major concerns of applying espoused information systems methods or methodologies to the organisational structure. The *first* is that the concept of structure is static, so the interventionalist is attempting to capture apparent organisational regularities given the organisational structure embedded within the method itself. So wherever methods or methodologies are linear and unyielding to emergent properties, it appears that the organisation must fit the rules, standards, norms, language and value system of the method. It is therefore argued that when the organisation is directed towards some technological design process through any *structured approach* of planning development or design, whether it be Business Systems Planning, SSADM, data modelling for example, the structure of that method enforces limitations. This in turn endangers growth in terms of improving the overall organisational, and hence endangers rather than enhances the social condition inhibiting the change process.

The *second* concern is that most development methods or methodologies have a rigid structure, and therefore have universal rules regardless of the type of organisation, the culture, the size of the organisation, the location, even the diversity of the collection of individuals in the same industry would conclude that no two organisations are alike, even though they may share commonalities. Metcalf (1990) demonstrates that even the use of technology within the organisation is dependent on the usage within the local context and the evolution of a particular organisation.

Identifying organisational emergence, either for the actor as interventionist, or the actor as the organisational *insider*, requires understanding the local organisational discourse, organisational *work languages* or other linguistic means by which organisational actors come to agreement about local knowledge, and hence the social constructs as defined by the local language.

LANGUAGE AND ORGANISATIONAL EMERGENCE

It may be theorized that organisational emergence occurs in the continuous discussion of established and new work situations. Therefore, how do we communicate what the new situation is and how is it to be described? Lyytinen and Klein (1985) suggest that in order to make sense of new situations one must create new meanings. We may understand from this that language is never used the same way twice and the use and construction of language itself is very much a part of the emergent process.

To illustrate this point Checkland and Scholes (1990, pp22-23) and Checkland (1991, p.26) suggest that even the use of the word *system* has been confused by many authors and has been used to describe parts of the real world perceived to be complex wholes. They argue that Koestler's neologism *holon* should be used for the abstract notion of a whole. They concede the word *system* to everyday language still, as ... *it is now too entrenched to be rescued*, therefore according to Checkland (*op cit*) greater clarity of the use of the word *holon* is needed and it should be used to describe the *abstract notion of a whole*. The question thus arises: "Is it really more accurate to describe the *systemic* process as being a *holonic* process"?

Consensus and sense making though requires that language rules be shared and applied in mutually understandable ways. These rules are based upon dialogue between individuals in a social setting until there is a shared understanding of the meaning behind the message. (Habermas, 1987). As the emergence of grammar alters the basis of meaning, language rules also allow the setting of frameworks and structures, and hence be considered as a power base for the enforcement of particular world views (Hopper, 1990).

The linguistic regularities of social *systems* that emerge as human activity are affected by politics, power relationships, legitimacy of leadership, line of authority, bargaining, negotiation and agreements. Legitimising and preserving the world view of a particular group of people in authority through the use of one activity system model over another may formalise the system and set a power game in motion. This raises questions about emergent inhibitors, unfair advantages, unequal power bases and distorted communication.

Sufficient power can freeze human organisational design, just as the manager, the business or information analyst, the corporate planner may impose their perspective upon other employees and hence on the system design. Language thus captured for the use of power play and regulated to impose or gain control will present idealistic or artificial inhibitors that distort meaning and the communicative process.

Transfer of power through language has been evident even without location of specific authority according to Boguslaw (1965) where...

A designer of systems, who has the defacto prerogative to specify the range of phenomena that "his" system will distinguish, clearly is in possession of enormous degrees of power ... computer programmers, the designers of computer equipment and the developers of computer languages possess power. To the extent that decisions are made by each of these participants in the design process serve to reduce, limit or totally eliminate action alternatives, they are applying force and wielding power in the precise sociological meaning of these terms.

What allows these regularities to become established? The lack of critical thinking must go beyond asking... "Why do things change?". We must also ask : "What forces inhibit change?" (Hopper, 1990).

ORGANISATIONS WITHOUT SYSTEM?

The basic concept of a systems *approach* differs throughout the literature. (Checkland, 1991; Gasparski, 1991). Both authors emphasise the confusion associated with the use of the word "system" and Gasparski promotes the idea that *style* is the real systems approach. He explains that because the systems approach is not independent from the authors of the approach, it is more accurate to consider the *approach* as a *style*. On the other hand, Gasparski (1991, p17) on quoting Tartarkiewicz (1988, p206) reminds us that style is also the style of the *epoch* and not just the individual style of the author. He suggests that styles change with the culture under the influence of social, economic and psychological factors, they also reflect the times. Is a systems approach rather a *style* and the term *holon* as described by Checkland (1991, p27), a more coherent epistemology for understanding emergence, change processes, behaviours and actions within the local context of the organisation?

CST, SSM AND EMERGENCE AS ORGANISATIONAL CONSTRUCTS

This paper discusses the nature of the organisation as a social construct. The dynamics of the self-making organisation and the coupling of the interventionist or change agent raises many issues and concerns of forces that inhibit change. These forces which may be impositions of perspectives and hence enforced rules or structures need some kind of technique to attain distortion-free communication or else be trapped by its own imposed structures and perspectives. Whether there is distortion or not may best be examined through Habermas' Critical Social Theory (CST). On the other hand Checkland's Soft Systems Methodology (SSM) may be adapted to accommodate *holonic* thinking techniques dealing with core concepts of emergent properties and to view emergent properties of the organisation in some kind of iterative or prototyping fashion. The *methodology* may so change itself through use in the local context as to look very little like the espoused methodology that may be found in Checkland's Systems Thinking Systems Practice text (1981b), that it may consider norms and values where it previously has not. Explicitly considering social norms and values is a strength of CST and perhaps a weakness of SSM.

Ethnographic techniques, prototyping techniques, techniques of group discussion and communication either towards consensus or shared understanding, may all be validated or guided by what are deemed relevant SSM activities, or a hybrid of CST and SSM. A hybrid may provide the necessary critique or self-evaluation currently not explicit within the SSM approach, or should we call it *style* ? (Gasparski, 1991; Mingers, 1980, p48). Of concern within SSM is the lack of external validation on the possible *Weltanschauungen*, which may serve to justify perspective s of top management rather than emancipate the human condition, or facilitate organisational change in the broadest sense of the organisation. Ethical concerns are related, and the evidence may be lacking as to whether all relevant view points are truly represented and dealt with fairly, where *fairly* means having equal veto of power.

Sense-making in the context of local norms and values are implicit within SSM, as are facilitator skills where the role of the interventionist is seen as being one to facilitate change. SSM could probably link with the strengths of Group Decision Support (GDSS) techniques in this case.

SSM and CST are both considered to be important and appropriate as they relate theory to practice and both deal with change processes and emergent processes and properties in their own ways. They both consider conceptual and applied techniques and purposeful human activity, though the techniques suggested differ significantly also. For example, where CST understands and brings to the fore the importance and power of social linguistic constructs, SSM does not. And where SSM devises relationships between activities and graphic modelling techniques of possible states of *becoming*, CST does not.

CONCLUSION

This paper does not profess to discuss the similarities and differences of CST and SSM, but rather hopes to raise enough questions about how current day organisations are viewed and to examine in brief some important issues concerning organisational awareness, survival , sense-making, growth and direction. It offers some alternative suggestions to espoused or traditional methods, methodologies and another perspective on systems thinking.

The truly global understanding of the organisation where global means multiverse rather than universe, must consider multiple truths forming perceived reality. For it is only then that we may enhance rather than inhibit organisational change and prosperity.

REFERENCES

Aulin, A., 1986, Notes on the Concept of Self-Steering, "Sociocybernetic Systems: Observation, Control and Evolution of Self-Steering Systems", F. Geyer and J. van der Zouwen, ed., Sage, Beverly Hills.

Baskerville, R., Travis, J., and Truex, D., 1992, Systems without method: The impact of new technologies on system development projects, "Proceedings of The IFIP WG 8.2 The Impact of Computer Supported Technologies on Information Systems Development", K.E. Kendall, K. Lyytinen, and J.I. DeGross, ed., Amsterdam: North Holland

Boguslaw, R., 1965, "The New Utopians: A Study of System Design and Social Change", Prentice Hall, Englewood Cliffs, NY.

Checkland, P.B., 1991, Towards the Coherent Expression of Systems Ideas, *Journal of Applied Systems Analysis*, Vol. 18, pp25-27.

Checkland, P.B. and Scholes, J., 1990, "Soft Systems Methodology in Action", Wiley, Chichester.

Checkland, P.B., 1981a Rethinking a Systems Approach, *Journal of Applied Systems Analysis*, Vol 8, pp3-13.

Checkland, P.B., 1981b, "Systems Thinking, Systems Practice", Wiley, Chichester.

Gasparski, W.W., 1991, Systems Approach as Style: A Hermeneutics of Systems, *in:* "Systems Thinking In Europe", M.C. Jackson et al., ed., Plenum Press, New York.

Habermas, J., 1987, "The Theory of Communicative Action", Vol. 2, Polity Press, Cambridge.

Hedberg, B., 1991, The Role of Information Systems in Imaginary Organisations - Collaborative Work, *in*: "Social Communications and Information Systems", R.K. Stamper et al., ed., North Holland, Amsterdam, pp1-8.

Hopper, P., 1990, Emergence of the Category "Proper Name" in Discourse, *in:* "Redefining Linguistics H. Davis and T. Taylor, ed., Routledge, London.

Luhmann, N., 1986, The Autopoiesis of Social Systems, *in:* "Sociocybernetic Paradoxes", F. Geyer and J. van der Zouwen, ed., Sage, London, pp172-179.

Lyytinen, K.J. and Klein, H.K., 1985, The Critical Theory of Jurgen Habermas as a Basis for a Theory of Information Systems, *in:* "Research Methods in Information Systems", E. Mumford et al., ed., North Holland, Elsevier Science, Amsterdam.

Metcalf, S., 1990, Introduction to Evolutionary Economics, from the lecture at Curtin University of Technology, Western Australia, presented to the Economics Society.

Mingers, J.C., 1980, Towards an Appropriate Social Theory for Applied Systems Thinking: Critical Theory and Soft Systems Methodology, *Journal of Applied Systems Analysis*, Vol. 7, pp41-49.

Morgan, G., 1986, "Images of Organisation", Sage, Beverly Hills.

Powell, W., 1987, Hybrid Organisational Arrangements: New Form or Transitional Development?, *California Management Review*, Fall, pp67-87.

Thyssen, O., 1992, The conceptual Scheme of Luhmann, Presentation notes at Binghamton University, State University of NY at Binghamton, Binghamton, NY, October 1992.

Toffler, A., 1990, "Powershift: knowledge, Wealth and Violence at the Edge of the 21st Century", Bantam, New York.

Truex, D. and Klein, H.K., 1991, A rejection of structure as a basis for Information Systems, *in:* "Collaborative Work, Social Communications and Information Systems", R.K. Stamper et al., ed., North Holland, Amsterdam, pp213-236.

von Foerster, H., 1984, Principles of Self-Organisation in a Socio-Managerial Context, *in:* "Self Organisation and Management of Social Systems", H. Ulrich and G.J.B. Probst, ed., Springer-Verlag, Berlin.

HERMENEUTIC APPROACHES TO LEARNING METHODOLOGY

Heather Watson[1] and Trevor Wood-Harper[2]

[1] Information Technology Institute
[2] Department of Mathematics and Computer Science
University of Salford
Salford, M5 4WT, ENGLAND

ABSTRACT

A methodology in theory differs from a methodology in practice. Consequently, texts describing methodologies contain no single prescriptive meaning. Learning to use a methodology is therefore a hermeneutic problem insofar as the practitioner's task is to appropriate meaning from the text by interpreting it according to contexts arising from particular situations. The interpretation of methodologies can therefore be understood in terms of the contingent application of practitioners' expertise in modelling information systems. The purpose of this paper is to argue that the application of expertise arises from practice and that a methodology should assist in elucidating this process. We describe this in terms of Multiview methodology. We conclude by proposing further research to develop explicit guidelines for expressing and representing methodology since these could assist practitioners in thinking about their expertise by providing theoretical frameworks for its expression. As such, resulting frameworks function as explicit links between theory and practice and so represent the potential for greater understanding of the relationship between methodological theory and its application.

INTRODUCTION

As shown in Figure 1 below, we suggest that an information system develops from a triangle of interpretation involving the practitioner, the situation, and the methodology (Avison and Wood-Harper, 1990). However, interpretative processes are surrounded by ambiguities. These include interactions within and between various sociotechnical elements. Such ambiguities make representation of interpretive processes problematic both for those who write and for those who apply methodologies. Difficulties begin with the realisation that methodological theory differs from practice. Consequently, texts describing theory are incomplete and so contain no single prescriptive meaning. Instead, meaning must be appropriated and applied by practitioners according to contexts arising from particular situations.

Systems Science, Edited by F.A. Stowell *et al.*,
Plenum Press, New York, 1993

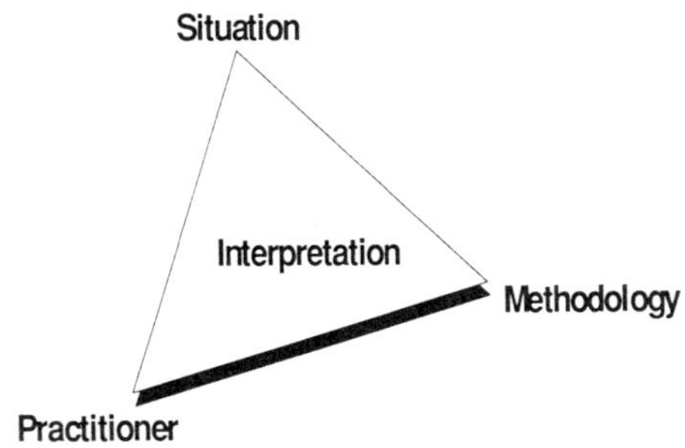

Figure 1. Information systems development arises from a triangle of interpretation.

Consequently, we propose that the interpretation of methodologies can be understood in terms of the contingent application of practitioners' expertise in modelling information systems. Because theory is inextricably linked with practice, we suggest hermeneutic approaches to learning methodology. We begin by outlining hermeneutic epistemology. We then discuss its application to learning methodology. We describe Multiview methodology (MVM) as one model consisting of a synthesis of multiple views, and we conclude by discussing ways of addressing limitations inherent in such models, and by proposing areas for further research.

HERMENEUTIC EPISTEMOLOGY AND THE INCOMPLETENESS OF REPRESENTATION

Hermeneutics is the study of interpretation, and in developing his phenomenological hermeneutics, Ricoeur (1991), extends epistemology to apply to situations and actions as well as to written documents. Moreover, he explains that texts are by necessity incomplete representations and that consequently in interpreting them we should not try to uncover an author's originally intended meaning. Instead, Ricoeur argues for interpretations in terms of the dialectic of distanciation and appropriation. We summarise this below.

Distanciation refers to the way that producing a text objectifies and thereby decontextualises any single meaning the authors may have intended. The text thus becomes open to actors' subsequent interpretations, resulting in meanings that surpass the author's original intentions (Atkinson, 1986). Distanciation therefore sets conditions for actors' appropriations of meaning. These are the processes through which they make the text their own, not by trying to understand what the author originally meant, but by trying to apply the text's potential meanings to the present situation (Ricoeur, 1991, p87).

Hermeneutic epistemology thus recognises that any representation is incomplete since it reduces complexity and depends on actors to respond to contexts of its subsequent interpretation. Instead of being prescriptive, hermeneutic approaches place the actor engaged in the interpretive process at the centre of the epistemology. As such, it requires theory to be reflective and self-critical and so suggests ways of learning that are open to disconfirmation and revision forced by the actor. Learning implies a continuing relationship between the actor and the action: the action is intelligible only insofar as intention and consequence appear as parts of the action to the actor.

Boland (1991) has proposed hermeneutics as a way of understanding information system use insofar as the system's outputs are made meaningful by users who then incorporate this understanding into subsequent decision and action. When applied to learning methodology, resulting action also becomes a reflection of the practitioner. It is thus distinguished not only from involuntary or chance action, but also from the larger class of actions regarded as independent of their origins. Consequently, the presuppositions of practice-- the continuous articulation of a self and the internal relation between the subject and the object-- openly dominate hermeneutic interpretations. In the following sections we suggest the suitability of this epistemology for learning methodologies.

HERMENEUTIC APPROACHES TO LEARNING METHODOLOGY

Texts describing methodological theory are inevitably incomplete in the sense that they cannot fully describe the contexts of subsequent applications. Yet, the polysemy of words requires sensitivity to the selective role of contexts for determining meaning (Ricoeur, 1991, pp54-55). Hermeneutic epistemology is relevant here because in learning to apply a methodology, the practitioner forces theory by hitting upon the openings which its representation necessarily leaves. Consequently, practice is founded on a principle of incompleteness since there is always a gap in a methodology, a requirement for an added presence in the activities of the practitioner (Checkland, 1983; Wood-Harper, 1985; Atkinson, 1986). Hence, in interpreting methodology, practitioners read themselves, not in the sense that perception inevitably bears the mark of the perceiver; not in the sense that texts are always directed at someone; but because texts in their composite expectations are always dependent upon the reader to appropriate meaning.

The practitioner in conceiving and then in applying theory looks for a ground to the methodology. The first step towards this end is in terms of coherence and consistency-- reconciling the families of possible meaning both within and outside the methodology -- and second in excluding alternatives. What arises from this activity is the reconstructed figure of the methodology in which, by that time, the practitioner himself appears. Having forced an entry in the opening lines of interpretation, the practitioner unavoidably looks out of the methodology thereafter. In this sense, complete descriptions of methodologies are impossible. We suggest therefore that theory should explicitly assist practitioners in structuring their expertise contingently with each situation.

MODELLING EXPERTISE IN MULTIVIEW METHODOLOGY

Hermeneutic approaches to learning methodology recognise the incompleteness of theory, and so they are not prescriptive. Consequently, learning the methodology is not confined to learning the texts describing the theory. Instead, hermeneutic approaches attempt to structure the course of inquiry such that theory assists practitioners in adapting their own expertise to meet the contexts of each situation. We discuss this in terms of MVM, which emphasises different perspectives in defining an information system.

Multiple Perspectives of Definition

MVM stresses the relative autonomy of differing dimensions of an organisation and of the roles information and information systems play for individuals and groups in that society (Wood-Harper and Avison, 1992). Therefore, it represents a broad range of perspectives. Multiple perspectives suggest that one's vantage point and resulting interpretive model never mirrors reality exactly, moreover, that it is always reductive and unavoidably mediated by one's given assumptions, theories, values, and interests (Atkinson, 1986; Checkland and Scholes, 1990). These represent continuity, but they also change over time. Multiple perspectives also imply that representations can never fully depict the complexity of any single phenomena, let alone the infinite aspects of a social reality (Bell and Wood-Harper, 1992). Since there exists no single continuously valid perspective in which one can ground theory, that theory must address its limitations by continuously adapting to new emerging perspectives (Wood-Harper, 1989).

Multiple Models of Inquiry

MVM structures different perspectives from other methodologies with the objective of defining an information system. For example, MVM differentiates information systems issues that may be mainly or entirely organisational from those that demand technical or computational approaches. This assists practitioners in adapting their expertise to address changes in understanding information and information systems within current social contexts. Stages of MVM can therefore be thought of as a progression from a fuzzy definition in a problematic situation to the specification of the information system's technological components of the solution. The structure of inquiry is not necessarily linear, but consists of the following possible stages: human activity exploration, information analysis, sociotechnical analysis and design, technical specification, and human or user/computer interface design. (More detailed descriptions of MVM are given in Antill and Wood-Harper, 1985; Episkopou and Wood-Harper, 1984, 1986; and Avison and Wood-Harper, 1990.)

MVM thus assists in structuring expertise in terms of five different perspectives. These delineate a range of existing models of inquiry available to information systems theory at a given moment in history. All of these can be emphasised, reduced in scale or even omitted according to the context of each situation. Since there is no single meta-level model structuring expertise, explicit multiple perspectives help to bring out the theory's inherent incompleteness. This assists practitioners in hitting upon the openings which theory necessarily leaves, resulting in a more comprehensive analysis and a process of inquiry that contingently adapts expertise. In the following section we discuss ways of addressing limitations inherent in such models.

ADDRESSING LIMITATIONS OF MODELLING ACTIVITIES

One advantage of representational epistemology is that resulting modelling activities are ways of socially structuring understanding in complex and changing situations. However, one disadvantage is that such models can become so effective at structuring inquiry that they are subsequently mistaken for reality. When this happens, modelling activities become unreflective formulae for response, and this can lead to a passion for fine-tuning existing methodological theory that atrophies our critical abilities to respond to present situations.

MVM attempts to address this limitation by assisting in challenging existing and in building new understandings of developing information systems. That is, in interpreting methodology that explicitly combines different perspectives, practitioners learn to question critically the assumptions upon which their expertise in modelling information systems is based. MVM therefore attempts to account not just for the structuring activities necessary for understanding but also for ambiguities surrounding limitations of understanding. Resulting theory is therefore explicitly open to change and is accordingly revised in light of subsequent developments.

However, to avoid mere eclectic pluralism, theory based on multiple perspectives must also attempt to address limitations inherent in its epistemology. Most obviously, it must avoid extreme relativism since some dimensions of interpretation are more important and useful than others in specific contexts. The challenge of developing such theory is therefore to combine its perspectives to decide in given contexts which aspects of competing theories are useful in representing an information system.

CONCLUSIONS

In this paper we suggested that the interpretation of methodologies can be understood in terms of the development and contingent application of practitioners' expertise in modelling information systems. Expertise thus arises and is adapted through structured inquiry in which theory is inextricably linked with practice. As such, we suggested hermeneutic approaches to learning methodology. We began by outlining hermeneutic epistemology. We then explained its application to learning methodology. We described MVM as one model for structuring expertise, and we discussed ways of addressing limitations inherent in such models. We conclude by proposing to research these areas further through a systematic analysis of a single methodology. From this, we hope to obtain a better understanding of how methodology is interpreted in practice and to develop a framework for expressing and representing methodology such that it assists in the contingent application of expertise.

REFERENCES

Antill, L., and Wood-Harper, A.T., 1984, The who's doing what for whom design stage, Soft, 2: 2.

Atkinson, C.J., 1986, Towards a plurality of soft systems methodology, *Journal of Applied Systems Analysis*, 13, 19-31.

Avison, D.E., and Wood-Harper, A.T., 1990, "Multiview: An Exploration in Information Systems Development," Blackwell Scientific, Oxford.

Bell, S., and Wood-Harper, A.T., 1992, "Rapid Information Systems Development," McGraw-Hill, London.

Boland, R.J., Jr., 1991, Information system use as a hermeneutic process, *in:* "Information Systems Research: Contemporary Approaches and Emergent Traditions," H.E. Nissen, H.K. Klein, and R. Hirschheim eds., Elsevier Science Publishers B.V. , North-Holland, 439-458.

Checkland, P.B., 1983, O.R. and the systems management movement: mappings and conflicts, *Journal of Operations Research Society*, 34: 8.

Episkopou, D.M., and Wood-Harper, A.T., 1984, The Multi-view Methodology: applications and implications, *in:* "Beyond Productivity: Information Systems Development for Organisational Effectiveness," T.M.A. Bemelmans, ed., North-Holland, Amsterdam, 1985.

Episkopou, D.M., and Wood-Harper, A.T., 1986, Towards a framework to choose appropriate IS approaches, *Computer Journal*, 29:3.

Ricoeur, P., 1991, "From Text to Action: essays in hermeneutics II", K. Blamey and J.B. Thompson , trans., The Athlone Press, London.

Scholes, J., and Checkland, P.B., 1989, "Soft Systems Methodology In Action," Wiley and Sons, New York.

Wood-Harper, A.T., Information Research Methods: using action research, *in:* "Research Methods in Information Systems," North Holland, Amsterdam.

Wood-Harper, A.T., 1989, "Comparison of Information Systems Definition Methodologies: An Action Research Multiview Perspective," Ph.D. Thesis, University of East Anglia, Norwich.

Wood-Harper, A.T., and Avison, D.E., 1992, Reflections from the experience of using Multiview: through the lens of Soft Systems Methodology, *Systemist*, 14:3, 136-144.

THE CHANGING ROLE OF SYSTEM DYNAMICS

Eric F. Wolstenholme

Department of Management Science
University of Stirling

ABSTRACT

The object of this paper is to contribute to a clarification of the role of System Dynamics as a framework for strategic debate. A revised framework for System Dynamics will be presented within a philosophy of 'modelling as learning', covering knowledge acquisition, qualitative and quantitative modelling supported by microworlds and archetypes to support dissemination of insights. The appropriateness of modelling in the strategic arena is discussed and it is suggested that System Dynamics models, when used interactively with different groups of participants in a problem situation, can assist in developing a shared understanding of how culture, power and politics combine to affect the behaviour of managerial processes.

INTRODUCTION

The purpose of this paper is to present an up to date description of System Dynamics and how it is being used to assist with the process of strategic debate in complex managerial situations.

The role of the subject is changing rapidly and these changes can be conceived of as a response to two causes. The first of these is due to a changing environment and the second is due to changes in the subject itself. Each of these sources of change will be discussed it turn.

THE NEED FOR NEW METHODS TO SUPPORT STRATEGIC DEBATE

The first factor which has contributed to a change in the role of System Dynamics in recent years is due to the increasing complexity and rate of environmental and organisational change. The development of new methods which can shed light on strategic alternatives and act as a catalyst to develop share understanding of multiple ownership issues is long overdue.

Traditional methods in this arena have tended to focus on learning from the past, largely using static case studies. As the rate of external change grows and tomorrows

Systems Science, Edited by F.A. Stowell *et al.*,
Plenum Press, New York, 1993

increasingly are made up of situations radically different from yesterdays, so the relevance of static and historic methods diminishes and the relevance of dynamics increases. There is a general movement in strategic analysis towards recognising the inadequacies of many current approaches and the need for more rigorous ideas for knowledge capture and communication of mental perceptions based on modelling methods. Additionally, it is recognised that models must be capable of addressing problems in future environments which differ significantly from those of the present (De Geus, 1988).

DEVELOPMENTS IN THE SUBJECT OF SYSTEM DYNAMICS

The second trend which has contributed to the changing role of System Dynamics is due to changes within the subject itself (Wolstenholme, 1990; Morecroft, 1992; Senge, 1990). These include:

* a recognition of the need to build models interactively with participants (managers, owners and associated actors) of problem situations.
* a recognition of the relevance and importance of both qualitative and quantitative modelling approaches to analysis.
* a move towards explicitly incorporating organisational boundaries into qualitative models as a means of highlighting the demarcation of responsibilities within management processes and for addressing aspects of culture, power and politics between participants.
* the development of more appropriate software to lubricate the ease of developing simulation models where these are necessary.
* the development of the concept of microworlds to help the dissemination of modelling insights.
* the use of simple archetypal structures by which to express insights from complex models.

KNOWLEDGE CAPTURE AND INTERACTIVE MODEL BUILDING

The problem of formulating appropriate models for strategic debate has been one of the major blockages in the use of modelling in this arena. All too often models have been created by analysts in isolation from participants in the problem situation or at best based on the biassed perceptions of a small number of participants. It is now recognised both from a point of view of relevance and implementation that models must be created interactively with and owned by participants. Further, it is recognised that the purpose of modelling should be to support reasoning and understanding. Only then is it possible for models to become successful change agents.

Obviously, the more unitary the situation (that is, the fewer participants there are and the more consistent their objectives), the easier it is to produce a model which satisfies the needs of participants. However, methods have emerged in recent years to assist the involvement of very diverse groups of participants in model building. For example, in a recent modelling study of Dutch Health care over 100 professionals participated in a modelling effort by means of Delphi questionnaires and structured workshops on specific sectors of the model (Vennix and Gubbels, 1992). Additionally, the use of electronic Delphi methods (using Hypermedia) has recently been investigated as a means of involving a geographically dispersed group of professionals in a model building effort (Wolstenholme and Corben, 1992).

QUALITATIVE MODELLING AND ORGANISATIONAL BOUNDARIES

The value and power of influence/causal loop diagrams as a metaphor by which to structure strategic analysis has been recognised for many years. The case for qualitative System Dynamics as a free standing subject was formally made in 1985 (Wolstenholme, 1985) and has recently been consolidated (Wolstenholme, 1990; Senge, 1990). This phase of the method is based on creating cause and effect diagrams or system maps (known as causal loop or influence diagrams) according to precise and rigorous rules and using these to explore and analyse the system. The diagrams must be developed with system actors to allow their mental models concerning system structure and strategies (and those of the environment of the system) to be made explicit and to consolidate their ownership of the model.

The incorporation of organisational boundaries into influence diagrams was also first introduced in the early 1980s (Wolstenholme, 1983) and described in detail in 1990 (Wolstenholme, 1990). This development is extremely important since it helps to overcome one of the areas where System Dynamics has been criticised in the past. That is, in its focus on process structure at the expense of organisational structure. These boundaries can be within organisations (for example between production and marketing) or between organisations.

The inclusion of boundaries on influence diagrams can often focus attention on mismatches between process and organisational structure. Boundaries may also dramatically highlight situations where no feedback exists within organisations and indicate the need to design control procedures.

As process chains become longer, the definition of precisely where responsibility for the process changes hands becomes increasingly important. One of the most common reasons for problems encountered in lengthy processes in large systems is the number of different organisations controlling different parts of the process. If organisations do not integrate their control strategies with adjacent organisations, the processes will not flow smoothly. The breakdown of diagrams into spheres of responsibility facilitates the incorporation of aspects of culture, power and conflict in the diagrams.

SOFTWARE AND MICROWORLDS FOR QUANTITATIVE MODELLING

The power of quantitative System Dynamics has been significantly enhanced in recent years by the development of the desk top computer and associated software. The creation of computer simulations of dynamic models has always been significant factor in improving systemic understanding. This is because there is a severe limit in the cognitive ability of the human brain to process multi-variate problems without such help.

However, never before has computer power been so readily accessible and the potential this creates for experiential learning through questioning is enormous (De Geus, 1988). Both DYNAMO (Richardson and Pugh, 1981) and DYSMAP (Dangerfield and Vapenikova, 1987; Dangerfield,1992) have been enhanced recently and other powerful purpose built programmes introduced, for example Microworlds Creator (Diehl, 1992) and Vensim (Eberlein and Peterson, 1992). In particular, the development of the STELLA/ithink software (Peterson, 1992) has greatly facilitated the development of quantitative simulation models directly with managers and amongst many other features allow the superimposition of organisational sectors on models.

Additionally, some of the recently developed software such as 'ithink' (when combined with Hypermedia) or 'Microworld Creator' can be used to formulate specific micoworlds by which models can be used in a 'hands on' gaming mode by other participants in the problem area and to recreate for themselves the learning process experienced by the team who developed the model. By providing an experiential learning environment, based on relevant and not hypothetical models, this approach reinforces model ownership and greatly improves the chances of implementation of change.

ARCHETYPES

The ease of development of models facilitated by some of the developments in System Dynamics presented in this paper, has quickly added to the general body of insights generated by models. This has promoted the idea of composing generic structures or archetypes of managerial problems and dilemmas which can be used both to simplify and reinforce the messages of change and to aid transfer of concepts between systems and form the basis of new modelling efforts in these new arenas (Senge, 1990). The use of such qualitative structures has prove immensely useful as a means of communicating managerial parables.

THE APPROPRIATENESS OF STRATEGY MODELS

A rich debate within the systems community at the present time concerns the appropriateness of models and problem solving methodologies for different purposes (Flood and Jackson, 1991). System Dynamics models are sometimes perceived as only appropriate to the consideration of simple and unitary situations and not relevant to complex, pluralistic or coercive situations where the organisation under study has a multitude of interactions and participants in the problem under debate perceive the organisation in different ways and have differing objectives.

It is suggested that the revised framework for System Dynamics described in this paper has enhanced the ability of the subject to address some elements of complex and pluristic situations and that System Dynamics modelling does have a role to play in such problems. This role is twofold.

Firstly, it can provide a basic map or systemic view containing some of the main variables recognised as existing in the world. For example, there really are waiting lists for hospital treatment and and council houses and people really do move into and our of the housing market. Making a shared vision of the 'physics' of an organisation explicit is a powerful starting point for developing alternative perspectives of how an organisation works.

Secondly, these basic maps can be used to develop a multitude of analyses of how problem situations develops over time as participants attempt to manipulate the evolution of the organisation in line with their own ends and objectives within their own spheres of responsibility.

In particular, the maps can help to trace out informal practices which occur in organisations when they are constrained. In general, when any organisation reaches the capacity of its formal processes, there is always an escalation in the emergence of informal routes and pathways to bypass bottlenecks and these routes often become more important determinants of organisational behaviour than the formal routes. Obvious examples of this phenomenon are the increases observed in efforts to persuade people to plead guilty as court capacity reaches capacity in the criminal justice system and the asking of people whether they really want to have operations as hospital congestion forces up the numbers of people on hospital waiting lists.

Whilst the core maps of the subject might be seen to take a simple and unitary perspective of problems, the analyses which develop from them involve a high degree of complex and pluralist characteristics. The purpose of the approach is to allow each group of participants in the problem to share understanding of how their actions will change the organisation. This learning may never result in an amicable compromise, but will always facilitate debate from a common and transparent basis rather than from opaque personal agendas.

CONCLUSIONS

This paper has suggested that there has been a mixture of internal and external trends which have combined together to create a revised role for the subject of System Dynamics. This is one of developing transparent models which, at the qualitative phase, facilitate knowledge capture and pluristic exploration of process, structure and strategy and at the quantitative phase are capable of being developed into computer based microworlds, games and archetypes by which insights can be disseminated in a 'hands on' framework.

REFERENCES

Dangerfield, B. and Vapenikova, O., 1987, DYSMAP2 User Manual, University of Salford.

Dangerfield, B., 1992, The system dynamics modelling process and DYSMAP2, in "Modelling for Learning", J.W.D. Morecroft and J.D. Sterman, ed., *European Journal of Operational Research.* 59:203.

De Geus, A.P., 1988, Planning as Learning, *Harvard Business Review*, Volume 66, Number 2, March-April 1988.

Diehl, E.W., 1992, Participatory simulation software for managers; the design philosophy behind Microworld Creator, in "Modelling for Learning", J.W.D.Morecroft and J.D.Sterman ed., *European Journal of Operational Research.* 59:210.

Eberlein,R.L. and Peterson, D.W., 1992, Understanding models with Vensim, in: "Modelling for Learning", J.W.D.Morecroft and J.D.Sterman ed., *European Journal of Operational Research.* 59:216.

Flood, R.L. and Jackson,M.C., 1991, "Creative Problem Solving - Total System Intervention", Wiley, Chicester.

Morecroft, J.W.D., 1992, Executive knowledge, models and learning, in: "Modelling for Learning", J.W.D. Morecroft and J.D. Sterman, ed., *European Journal of Operational Research.* 59:19.

Peterson, S., 1992, Software for model building and simulation: an illustration of design philosophy, in "Modelling for Learning", J.W.D.Morecroft and J.D.Sterman ed., European *Journal of Operational Research.* 59:197.

Richardson, G.P. and A.L. Pugh (1981), "Introduction to System Dynamics Modelling with DYNAMO", MIT Press, Cambridge, Mass..

Senge, P. 1990, "The Fifth Discipline", Doubleday, London.

Wolstenholme, E.F., 1990, "System Enquiry - a System Dynamics Approach", Wiley, Chichester.

Wolstenholme, E.F. and Corben, D.A., 1992, A computerised delphi tool for knowledge acquisition in model building, "Proc. 1992 International System Dynamics Society Conference", Utrecht, Netherlands.

Wolstenholme, E.F., 1985, A methodology for qualitative system dynamics, "Proc. International System Dynamics Society Conference", Denver, U.S.A..

Wolstenholme, E.F., 1983, Modelling national development programmes - an exercise in system description and qualitative analysis, *JORS*, 34:1131.

Vennix, J.A.M. and Gubbels, J.W., Knowledge elicitation in conceptual model building: A case study in modelling a regional Dutch health care system, in "Modelling for Learning", J.W.D.Morecroft and J.D.Sterman ed., *European Journal of Operational Research*, 59:17.

INFORMATION SYSTEMS IN SUPPORT OF RESOURCE MANAGEMENT IN THE COMMUNITY

Carys Y Parker[1], Prokash Chatterjee[1] and Bob Wood[2]

[1] Frenchay Health Care Trust
Bristol

[2] Department of Computer Science
University of Salford
Salford

INTRODUCTION

> "Now, if we all have a technical, a practical and an emancipatory interest in the functioning of organisations and society, then a management science which can support all these various interests has an important role to play in human well-being and emancipation." (Flood & Jackson, 1991)

Computer-Based Information Systems have long been seen as instruments for organisational improvement and there is a substantial tradition of both theory and practice which has sought to explain the conditions under which the successful application of Information Technology can be achieved. Over the last few years a considerable amount of attention has also been paid to the alleged 'failures' of this technology with particular focus being placed upon the 'dissatisfaction' of clients and users as they have seen their expensive investments in IT fail to live up to expectations (KPMG, 1990). It has become clear to many people involved in the running of complex human activity systems that the problems we face stubbornly refuse to yield to 'solution' by conventional means, even when such strategies are supported by powerful and sophisticated new technologies. There is increasing evidence that this disenchantment is already leading many managers to question the conventional wisdom that IT is fundamental to the success of any modern organisation (Strassmann, 85). Although such scepticism is probably healthier than the almost unquestioning acceptance of the last thirty years there is a danger that an anti-technology view may become the new norm which could have profoundly harmful effects both for the IT industry and for end-users generally.

Systems Science, Edited by F.A. Stowell *et al.*,
Plenum Press, New York, 1993

AUTOMATING AND INFORMATING

When computers were first introduced into industrial and commercial organisations they were used almost exclusively to replace existing manual, clerical operations. This led to a view of technology dominated by thinking about possible cost savings (mainly labour), increases in speed and improvements in efficiency. Systems development was seen as a process by which the most efficient solution to a problem could be engineered correctly. This mixture of rationalistic thought about organisational problem solving and technical concerns in the design process has dominated our thinking about the use of IT over the last thirty years. The working assumption has always been that we can somehow 'fix' the organization before we move on to 'automate' the required procedures. Information System Development Methods require, therefore, a clear and correct view of the problem situation before 'design' can take place and the preferred solution engineered in a cost-effective manner.

Recent research into IS Development, influenced strongly by writers such as Ackoff, Checkland and West Churchman, has paid much more attention to the issue of problem formulation. In particular, researchers have investigated the possibility of linkages being developed between an approach such as Soft Systems Methodology and more conventional development methods such as SSADM, or particular tools and techniques such as Data Flow Diagrams or Entity Attribute Relationship Models. For many of these researchers, however, and certainly for those practitioners working in this area, the objective of this approach seems to be the clarification of the problem situation in advance of design taking place, so that development may proceed with an increased chance of success.

Following the direction taken by Wilson much of the work based on the use of SSM has sought to develop a 'Consensus Primary Task Model' which can then be used as a firm basis for the establishment of user requirements, particularly with respect to data and information needs (Wilson, 1990). By implication, rather than by expressed intent, the systemic properties of such an agreed conceptual model also become normative statements about the desirability of such properties in the CBIS to be designed and implemented as part of the process of desirable and feasible change. Although undoubtedly an improvement on existing approaches to systems analysis, the direction being taken by such work does not address two fundamental issues of concern:

- where the problem situation refuses to yield up a clear and agreed 'structure' (i.e., those situations characterised by Ackoff as "messes") (Ackoff, 1978).

- the role of information systems themselves in the creation of the problem situation (i.e., the process Zuboff has called "informating") (Zuboff, 1988).

The research work described in this paper seeks to investigate how systems ideas might be applied at the very earliest stages of an intervention in order to help clarify the issues of concern, which in turn can then inform a process of emergent system design which is then fed back into the intervention and so on in a process of evolutionary development tied inexorably to the inherent change of the human activity system itself.

THE PROBLEM CONTEXT

The Frenchay Healthcare NHS Trust

The Frenchay Healthcare Trust acts as provider for a near complete range of health care services. All such services are either tendered for in competition with other trusts in the region (with the South West Regional Health Authority performing the role of purchaser) or are bought annually on a block contract basis by the Bristol and District Health Authorities.

The headquarters of the Trust are situated at Frenchay Hospital which is also the main centre for acute services. The district covered by the Trust is approximately sixteen miles square and is partly urban and partly rural in nature. It contains little heavy industry and in 1991 was estimated to be serving a population of around two hundred and twenty three thousand people. This figure is projected to increase by seven per cent, or some fifteen thousand people over the next decade.

Overall, the people of Frenchay enjoy a high socio-economic status and this is reflected in the fact that both the overall mortality rate, as well as that from the commonest forms of death such as cancer and heart disease, are amongst the lowest in the region. The district also contains, however, two older inner-city residential areas which are amongst the most underprivileged in the county .

The organisation and structure of Frenchay Community Health Services

Within the Trust each functional area, such as Pathology, Neurology, Obstetrics and Community, is organised as an independent unit with a Clinical Director and a General Manager at its head, and each such Directorate is responsible for its own budget. At the sub-unit level there are a number of multi-disciplinary clinical teams led by the respective Heads of Specialty Services.

Each Clinical Director has delegated responsibility for all services within a Directorate, and is ultimately responsible to the Trust Board. Day-to-day operational management is undertaken by the appropriate General Manager with the Heads of Specialty Services being managerially accountable to her for the delivery of the service contract. Each specialty head is considered to have a high degree of autonomy, albeit within broadly agreed parameters and a shared vision of service objectives.

The Community Services Directorate currently employs some seven hundred staff and has a funded establishment of 468.96 WTE. In the financial year 1992/93 Bristol and District Health Authorities purchased community care from the Directorate worth some ten and a half million pounds.

A definition of community-based care

The concept of 'care in the community' is a somewhat fragile and elusive one. It is also, as various studies have shown, not to be seen as a cheap alternative to hospital-based care. A simple definition of community care might be:

> "those activities sponsored by the National Health Service, and local authority social services departments, which relate to patient care outside hospital." (NAHA, 1987).

The local concept of community care is perhaps best captured in the current business plan for Frenchay, where the declared mission of the Community Services Directorate is given as:

> "a service which seeks to offer a range of health care to the residents of Frenchay which promotes health, prevents and detects illness, and provides and facilitates care to people in their homes and immediate locality. It also seeks to respond to the changing pattern of care in line with national and regional guidelines." (Frenchay, 1992).

Community care is also heavily dependent on inter-agency working and thus there is close collaboration, both at the formal and informal local level, between the other agencies involved, such as Social Services, GP Fundholders and a variety of voluntary bodies.

Resource Management in the Community

The Resource Management (RM) Initiative was announced formally in 1986 and was originally piloted in six Acute Units (hospitals which treat acutely-ill patients rather than chronic or emergency cases) with the declared purpose of improving patient care by:

> "helping clinicians and other managers to make better informed judgements about how the resources they control can be used to the maximum effect." (DHSS, 1986).

At the centre of the RM Initiative lies the idea of a collaborative management structure, in which clinicians and managers become jointly responsible for resource allocation and control, so that all available resources can be efficiently and effectively directed towards better patient care. The new management structures together with the timely provision of accurate Patient-related as well as cost-related data are intended to encourage carers to exercise responsibility and accept accountability, in return for a significant influence over the use of resources (RM Directorate, 1989). In 1989 the RM Initiative was extended to other Acute sites and has now become a major part of central government policy for the NHS.

The Frenchay RM Project

This project, instigated in December 1991, represents a concerted attempt by the Community Services Directorate to develop a strategy for the introduction of RM into all aspects of the provision of community care under their control. The proposal is to develop RM in such a way that both individual and total patient care can be planned, delivered and costed more effectively. It is recognised that clinicians, if given adequate and detailed information, are in the best position to influence the way in which resources are used, as well as to ensure that treatment is provided in a way which is cost-effective without being compromised by any loss of quality. The major objectives of the Frenchay project can be summarised as follows:

a) the integration of information systems throughout the range of agencies providing community-based health care,

b) the provision of access to information on 'whole' individual patients to community health workers as required,

c) the provision of the best information available to facilitate planning for the usage of finite resources.

CONCLUSION

From the evidence of the cursory description given above it should be very clear that the problem domain we are concerned with here exhibits all the characteristics of a 'messy' situation. Computer-Based Information Systems within the NHS are an order of magnitude greater in their complexity than those which are generally found in manufacturing or financial organisations. If this is true for CBIS *within* hospitals, then any envisaged 'system' in the Community with its intricate networks of clinicians, voluntary bodies, Regional and District management, GP's, the Social Services and so on, is likely to be even more difficult, if not impossible, to specify. To this complexity can be added a plethora of state generated legislation and standards coupled with massive structural change. Flood and Jackson have suggested the use of various systemic metaphors as a way into the initial analysis of a problem situation and we intend to explore this further in the context of the RM in the Community Initiative. Our aim is to move beyond the present concern with rational/consensus accounts of information requirements and to explore the potential for the use of CBIS in complex-coercive domains so that we may contribute to the development of a "management science which has an important role to play in human well-being and emancipation."

REFERENCES

Ackoff R.L., 1978, "The Art of Problem Solving," Wiley, New York.

DHSS, 1986, "Health Service Management - Resource Management (Management Budgeting) in Health Authorities," Health Notice, HN(86)34, DHSS, London.

Flood R.L., and Jackson M.C., 1991, "Creative Problem Solving," John Wiley, Chichester.

Frenchay, 1992, "Community Services Business Plan," Frenchay NHS Trust.

KPMG, Peat, Marwick, (1990) "Runaway Computer Systems," KPMG Peat Marwick, London.

NAHA, 1989, "NHS Handbook," Macmillan, London.

RM Directorate, 1989, "RM Management Initiative - Information Package for Acute Hospitals," Department of Health, London.

Strassmann, P.A., 1985, "Information Payoff," Free Press, New York.

Wilson, B., 1990, "Systems: Concepts, Methodologies and Applications," John Wiley, Chichester, 2nd Edn.

INTRODUCTION TO THEORY AND CONCEPTS STREAM

The globe is too small to contain systems thinking. Systems thinking may or may not contain the globe. The theories and concepts of systems science begin to provide a vocabulary for notating the larger concerns of realisation in our time, but this is a vocabulary perhaps best understood subjectively, although the eventual realisation may be objective enough. Meantime, debate and agreements among systemists can happily be conducted while the mutually-tokened concepts are abstract enough to accommodate multiple interpretations, so that one can confirm one's own reflection in the words of another, or find an identity by disputing them.

What's more: talk's cheap! Theories and concepts only registered in language seem to lack a quality of ultimacy, regardless of appeals to elegance and other distracting, but essentially aesthetic qualities; or appeals to street-cred by expedient references to mundane and (usually economically-governed) application instances. Pleas, whether for the overdue implementation of a theory or for the perfectionistic refinement and generalisation of constructs, can evidence dissatisfaction with words, but does that mean that those who don't plead thus believe it is enough to permute linguistic tokens (and make the world go round)? Such questions beg and beget the Kantian gobsmacker "What is to be done?".

Theories and concepts which are merely vehicles for promulgating some value based Utopia while disguising an entrenched political or personal dogmatism may be insidious, but who cares? As long as we are all free to do it, and can tacitly conspire that theories and concepts remain only in the field of language, with appropriate lip service to future vindication through enhancements to daily life, or justified by the nobility and purity of intellect in itself. The tacit values, materialist, pragmatic, transcendental, environmental, self-referential or what-you-like that inform theories and concepts in the (usually individual) systemist's thinking may often be more real than the theories themselves. "More real" may imply some inn-accurate view of reality, but this is the chance for my dogma, not debate to prevail. If you want that, the papers in this session give you the opportunity.

John Gammack
Information Systems
Department of Computing Science
University of Paisley

Systems Science, Edited by F.A. Stowell *et al.*,
Plenum Press, New York, 1993

SELF-REFERENTIAL SYSTEMS THEORY:
A DISCUSSION OF SOME THEORETICAL DEVELOPMENTS

Michael C. Charlton

Centre for Systems Studies
The University of Hull
Hull, HU6 7RX, U.K.

ABSTRACT

This paper is essentially one about reflection upon theoretical development in Systems thinking. It starts from the ideas popularised by Fenton Robb and John Mingers in a debate about social autopoiesis in 1989. With some regret this debate has lost its vigour and an attempt is made in this paper to revive some of the arguments by discussing the self-referential systems theory of Niklas Luhmann. With a basic outline of Luhmann's ideas in place, and a more general theory of social autopoiesis explained, I will use this framework to reflect upon some of the intellectual activity in contemporary systems thinking.

INTRODUCTION

In 1989 John Mingers and Fenton Robb made a notable effort to bring the theory of autopoiesis closer to the practical affairs of Management and Organization. They had recognised the contributions of Maturana and Varela who had devised and applied the theory of autopoiesis in the physical domain, and Mingers (1989) and Robb (1989) were beginning to speculate about the theory's application in the social domain.

The social world is rather different though, and Maturana and Varela have aired strong reservations about the theory's possible relocation from the physical to the social domain. This did not prevent Mingers and Robb asking some pertinent questions about the theory's potential in terms of social application. It now seems opportune to review that debate and the important ideas it generated in terms of the self-referential systems theory of Niklas Luhmann.

REFLECTIONS UPON THE SOCIAL AUTOPOIESIS DEBATE

If we were to map out a brief and simplified specification for a theory of autopoiesis we would probably make reference to some of the key concepts extracted from the theory as

Systems Science, Edited by F.A. Stowell *et al.*,
Plenum Press, New York, 1993

it relates to the physical domain (a good example is Zeleny, 1981). These are: organization, structure, structural coupling, and arguably the most important, the ability of a system to self-create or self-renew. The organization of such a system is invariant and it constitutes an identifiable unity. Its structure is a particular configuration of components existing spatio-temporally within the organization. The structure can change by determining what the environment may trigger. This is what is known as structural coupling. It is the structure of a system that determines what the environment can and cannot trigger, it is therefore unacceptable to assume that the environment is stimulating the system as we tend to assume in the more conventional case of open-sytems theory. The defining characteristic of autopoietic systems is that they self-create or self-renew. They do so by producing components which interact to generate the network that originally produced them. Hence self (auto) producing (poiesis) systems.

The problem with such a theoretical outlook becomes apparent when we compare it with social(human) organization. We undoubtedly recognise that human agency plays a fundamental role in our social systems, but to identify humans as components of social systems appears to be a pointless proposition in light of the above argument. If, on the other hand, the theory compels us to consider the idea that humans (concrete individuals) are part of the environment then we are faced with the possibility of autopoietic systems that we have little control over. So what is the significance of this for systems thinking?

Both Mingers and Robb made considerable ground with autopoietic theory, yet both settle for recognising the theory's metaphorical power in explaining certain social behaviour.

Robb believed that certain social phenomena can be explained by autopoietic theory, though he had rather hoped that this was not possible because it gives rise to the possibility of unmanageable social systems. The implication of this is that we are unable to "intervene" in such systems because this type of organization is of "a higher logical order" than our general notion of human organization. These are what Robb, rather awkwardly refers to as "supra-human autopoietic systems". He cites conspiracy and conspiratorial thinking as examples of social behaviour which escape our capacity to manage and intervene no matter how hard we try.

Mingers's approach is slightly different. His undisguised admiration for the original ideas in the Biological Sciences allowed him to consider a meticulous and "truthful" transfer of the theory of autopoiesis from the living system domain to the social domain. Mingers denies the possibility of autopoietic social organization, suggesting there are too many unanswered questions. For example what would be the boundaries and components of such social systems? Mingers's point is that it may be relatively easy to identify the components and boundaries of physical systems, but to define the equivalents in social systems is highly debatable. He concludes, "...the definition of autopoiesis as such is too strongly tied to physical production, but more general versions may be useful, at least metaphorically, in examining social institutions" (Mingers, 1989a, p178).

We are left feeling that the theory of autopoiesis invented to suit physical approaches does not sit comfortably with social ideas.

However, a more "general theory of autopoiesis" can be developed to meet these requirements. It will certainly mean that we must shift our analysis to a higher level of generality, and it will also entail that we temporally relinquish our tendency to make human individuals the focus of our attention. We can rekindle the debate about social autopoiesis only at the cost of focusing our energies at a more general level. This necessitates some radically different thinking about the social context that our organizations or preoccupations about management operate within. Creating a more general theory of autopoiesis which is more acceptable in the social domain is Niklas Luhmann's achievement.

LUHMANN'S SELF-REFERENTIAL SYSTEMS THEORY

Luhmann defines three types of self-referential autopoietic processes. One such process bases its autopoiesis on the grounds of life (eg. the cell) and we have heard much about this owing to Maturana and Varela's work. The other two ("psychic" and "social" systems) base their autopoiesis on the grounds of meaning (Luhmann, 1990). They are "meaning constituted systems" using consciousness (psychic systems) and communication (social systems) as their respective modes of meaning based reproduction (Luhmann, 1990). In this paper I am only interested in the autopoietic features of social systems and so greater emphasis will be placed upon the communicative form of reproduction. We will neglect psychic (conscious) systems.

These remarks should immediately attend to some of the problems identified in the Mingers-Robb exchange. A social system defines its boundaries based on the idea of meaning and its components are communications, and not as one might expect, human individuals. When we talk about social systems generally, and especially with regard to management, we are reluctant to relinquish ideas about the individual and his/her subjective preferences, but we are forced to do this if we are going to postulate a more general theory of autopoiesis. Luhmann is particularly clear about this

> Human beings, concrete individual persons, take part in all social systems. But they do not enter into any of these as determinate parts themselves nor into society itself. Society is not composed of human beings it is composed of communication among human beings. (Luhmann, 1990, p30)

Social systems then are a necessary reduction of complexity. They "acquire an identity through their characteristic means of reducing complexity" (Luhmann, 1982). Of course the most important social system is society itself. But society, according to Luhmann, is made up of functional sub-systems which define the unity of society through their own self-referential operations.

Now, if we are prepared to accept this thesis, we could focus our efforts on studying systems that have had success in "differentiating" themselves. This is a contentious point. It is one where I believe that Luhmann, much to his protestations, is arbitrary about selecting certain systems to study. His criteria is that the functional sub-systems of society have differentiated themselves, because the communications that they use to self-create are more successful than other possible communications. He cites as examples the economic system, the political system, education, science and religious systems. Despite the allegations that these selections may be arbitrary, it is difficult to deny - that for arguments sake - that the political or economic system exist; and in any case, as observers of autopoietic systemic phenomena, we are likely to demarcate an entity which we can hardly claim to be the "true" domain, but as an approximation it allows us to apply concepts such as autopoiesis.

How does a functional sub-system continually produce itself? Let us briefly consider the case of the economic system, bearing in mind that it is impossible to give a full account within the bounds of this paper. The interested reader should consult Luhmann (1989, 1990).

The economy is "differentiated through the monetary mechanism". Luhmann is not interested in who makes payments. This is not important, because to meet the requirements of the theory, the economy is closed and self-referential. Nevertheless for the economy to work we must assume that someone has a stock of money and can "effect payments". Payments are determined by "prices" and we all know a 'fair' price, but with or without us prices have a kind of autonomous existence. They are above all a "successful" communication and they can be communicated with indifference to circumstances (Luhmann, 1982, p203). That is, we can make quick decisions based upon this simple communications media. They may not be the right decisions, but the fact that the dominant economic system has differentiated itself, and the fact that it may be autopoietic (which I admit requires more argument) must be significant for the systems inquiring community.

Thus, thanks to Luhmann we have the beginnings of a conceptual framework that allows us to explain social phenomena in terms of an autopoietic system theory. It is highly contentious, but it is very significant and answers many of the nagging doubts that both Robb and Mingers brought to the systems community. To further the interest in the debate I proposes to compare some of the ideas to recent intellectual activity in contemporary systems thinking.

My choice of these trends is necessarily partial, though they have been chosen to show their divergence from the ramifications of a self-referential systems theory outlined above. They are deliberately chosen in an almost provocative manner, to consider whether recent theorists using systems ideas are as Luhmann would say "barking up the wrong tree".

RECENT DEVELOPMENTS WITHIN SYSTEMS THINKING

The work of both the Interpretive Systemologists and Critical System Thinkers are an exciting development within systems thinking. Despite their use of the same "systems" conceptual idiom as Luhmann however, their approaches could not be more different.

Interpretive Systemology is a welcome newcomer to the systems movement. It continues the theoretical development of soft (interpretive) systems thinking, only, in its new form it attempts to add a critical dimension. In contrast to the above brief outline of self-referential system theory it focuses its level of analysis on the subject and subjectivity. Let us consider one of the many concerns that stimulates the research agenda of Interpretive Systemology.

The Venezuelan researchers have found that the "social purposes" and "formal objectives" of the institutions and organizations that constitute Venezuela are often at odds with everyday social behaviour and "the idealised utopian state of affairs" that Venezuelans seek to bring about (Fuenmayor and Lopez-Garay, 1991). They define this as "institutional schizophrenia". It is felt that certain interpretations of society do not surface because official interpretations tend to dominate. Thus the research effort that the Interpretive Systemologists recommend is one of conducting interpretive investigation of all institutions. This is a huge project, it should require the wholesale redesign of "human activity systems" to allow maximum participation and unrestrained discussion of interpretive themes.

In a similar way Midgley (1992) is an example of someone who is using the systems idea and critical theory in a refreshing way. He is continuing the line of thought started by Ulrich (1981) and Critical System Thinking generally (Jackson, 1992), which attempts to promote critical reflection upon system boundaries. He recognises that when we define systems boundaries we necessarily marginalise other factors that could otherwise have been included in the definition. Aspects which become marginal or exhibit a "profane" status, to use Midgley's adopted language, do so through "cultural reactions". The process of marginalisation can bring about an "ethical tension" of various degrees, but Midgley does not consider how the marginal remains marginal other than in terms of "ritual", which has the effect of perpetuating power relationships, and consequently helps to sustain "sacred" and "profane" boundaries.

This explanation is not in dispute but there is a second possibility of course, and this would entail that systems theorists must use self-referential and autopoietic theory. For instance, consider what the effects of such a theoretical stance would have on the relation between the idea of a dominant autopoietic (economic) system and the process of marginalisation. For example, ecological issues might be considered as marginal. Luhmann's theoretical endeavour can explain why ecological issues remain marginal (Luhmann, 1989) because the "socially most consequential communication" occurs in functionally differentiated autopoietic sub-systems, and not in the 'lifeworld'. We made reference to the economic sub-system earlier, and if we recognise its dominance and its capacity to self-create then this takes us down a different avenue of inquiry. If the economic system is so successful, as Luhmann continually suggests in his arguments, then commercial organizations will exploit this socially "successful" communication and concerns like environmental protection will always be second best (marginal).

The Interpretive Systemologists have similarly followed an intellectual route which will be unable to explain certain ideas if they continue to put all their efforts in to interpretive thinking. They recognise that Venezuela has transplanted the organizations and institutions of Europe, but are not prepared to consider whether some of those societal structures exhibit autopoietic features.

CONCLUSION

This paper is about reflection upon theoretical development in systems thinking. It does not set out to accuse theorists of "barking up the wrong tree", though it seeks to remind theorists and practitioners that this possibility is always present. It is however felt that autopoietic system theory, especially the ideas of Niklas Luhmann, deserve more attention in the wider systems community. I hope this paper has gone some way to demonstrating that.

REFERENCES

Fuenmayor, R., and Lopez-Garay, H., 1991, The scene for Interpretive Systemology, *Systems Practice*, 5:401.

Jackson, M.C., 1991, "System Methodology for the Management Sciences", Plenum, London.

Luhmann, N., 1982, "The Differentiation of Society", S. Holmes and C. Larmore, trans., Columbia University Press, New York.

Luhmann, N., 1989, "Ecological Communication", J. Bednarz jr., trans., Polity, Cambridge.

Luhmann, N., 1990, "Essays on Self-Reference", Columbia University Press, New York.

Midgley, G., 1992, The Sacred and Profane in Critical Systems Thinking, *Systems Practice* 5:5

Mingers, J., 1989, An Introduction to Autopoiesis - Implications and Applications, *Systems Practice*, 2:159.

Robb, F., 1989, Cybernetics and Supra-human Autopoietic systems, *Systems Practice*, 2:47.

Ulrich, W., 1983, "Critical Heuristics of Social Planning: A new approach to Practical Philsophy", Haupt, Berne.

Zeleny, M., ed., 1981, "Autopoiesis: A Theory of Living Organization", North Holland, New York.

DESIRABILITY AND FEASIBILITY OF LINKING 'SOFT' SYSTEMS METHODOLOGY [SSM] WITH STRUCTURED METHODOLOGIES

Nimal Jayaratna[1] and Brian D'Arcy[2]

[1] Heriot-Watt University
Riccarton
Edinburgh EH14 4AS, UK

[2]Sheffield Hallam University
100 Napier Street
Sheffield S11 8HD, UK

ABSTRACT

There is considerable claims by methodology proponents and methodology users as to the desirability and their ability for linking SSM with structured methodologies. This paper is concerned with an evaluation of some of the philosophical, conceptual and pragmatic issues arising from such attempts. A brief clarification of the term 'Information Systems' will be attempted before discussing the linkages.

INFORMATION SYSTEMS DISCIPLINE AND MEANING

The origins of information systems as a discipline can be traced to the discipline of Computer Science. Computer Science is concerned with computer technology while information systems is concerned [as part of its domain interest] with the application of information technology. In essence the terms 'computer technology' and 'information technology' refer to the same technology. However, the term 'information technology' is used to emphasise the <u>use</u> of computers for information processing, storage and presentation with a clear need for satisfying some user needs, while the term 'computer technology' is used to emphasise the experimentation and explorations of the computational power of the technology whether or not that achieves any given end. In the former it is only a means to an end while in the latter it is an end in itself. Therefore in many instances what becomes a research issue in computer technology today, if successful, may end up as an application issue for tomorrow's information technology. Similarly, what emerges as a well defined problem issue in information technology may lead to a hypothesis for exploration within computer technology. However, information systems is concerned with a much wider domain than the mere application of information technology. It can be defined as a discipline that embraces information technology [means], information activities [role, tasks and functions] and organisational activities [context].

The dominant message is clear. Whenever we engage in the activity of information systems development we are inevitably involved with at least two systems:-

Systems Science, Edited by F.A. Stowell *et al.*,
Plenum Press, New York, 1993

[1] the information system itself

[2] the context system within which the information system will, or will not, deliver value.

Of equal clarity is the realisation that understanding the context system must precede any consideration of current or envisaged information systems. There is of course no assurance that using SSM as a context understanding approach will automatically give primacy to information systems development unless of course we deliberately insist on coercing it to do so.

These observations call into quite serious question the development of information systems from a purely technical perspective where the development is seen as a set of activities that are very much related to what is possible with the use of current technology. Conversely the approach to development of information systems from an organisational perspective tend to address the meaningful aspects of human activities but ignore the capabilities of the technology. This seems to invite what is generally termed as an interdisciplinary approach and this in turn is often taken to mean the linking of SSM with structured methodologies.

As part of an attempt to examine this proposition it is useful to clarify the term 'information system' in view of the many interpretation that are given to it. An information system can be defined as:

> "a system for the most efficient and effective means of identifying the 'real' needs of users and developing information processing systems for satisfying these needs; ensuring that the resulting information processing systems continue to satisfy changing user needs by the most efficient means of acquiring, storing, processing, disseminating & presenting information; providing facilities and a learning environment for users to enable them to improve the effectiveness of their decision models; supporting operations, control and strategic organisational objectives" (Jayaratna, 1991a).

According to this definition, the notion of an information system includes both technology and non-technology issues, information task activities and the organisational context. It can be considered to consist of five functions, namely: information processing and usability; educating and learning; information systems development; management and control and strategy and planning.

Given this wider role of information systems, the links between SSM and information systems can be considered at several levels.

LINKAGES TO INFORMATION SYSTEMS DEVELOPMENT

An obvious and a transparent link between SSM and information systems can be considered at the information systems development function level within the broad area of the information systems domain as discussed earlier. SSM is a methodology for structuring and bringing about transformation in ill-structured situations. The development function is concerned with the development of information systems. In order to undertake these developments, some form of methodologies need to be deployed. Therefore there is no conceptual or intellectual reason as to why SSM cannot be used for structuring situations where client groups cannot agree or are uncertain about the nature of information systems to be developed. Indeed, some authors have confirmed their use of SSM for developing information systems (Rodriguez-Ulloa, 1988; Wood-Harper et al., 1985; Avison & Wood-Harper,1990). It is also possible through this function to use SSM for structuring strategy and planning activities of the organisation (Galliers, 1991) or management control systems.

NATURE AND FORM OF LINKAGES [GRAFTING OR EMBEDDING]

This issue arises from those who have attempted to make specific links between SSM and structured methodologies i.e. combining two well known methodology types. Miles (1988) was the first to raise the conceptual & practical issues of embedding and grafting these two methodology types. This can be considered as the linkages of a methodology approaching from IT or related activities [structured] and a methodology approaching from organisational context activities [SSM]. These links falls within the domain of information systems development function.

In the context of structured and SSM methodologies, there is no reason why the data flow diagrams of structured methodologies cannot form part of the 'rich' picture construction in SSM in order to help improve the understanding of a 'situation of concern' any more than graphs, statistical tables, sketches or animation. In fact any model that help to develop the richest picture possible need to be encouraged. Data flow diagrams can be also used as part of the conceptual models [though it may be a costly exercise and may not be advisable to do so until the notions of 'root' definitions are discussed, debated, modified, reconstructed and agreed upon as the basis to proceed] within SSM. According to Miles (1988) this notion of embedding makes the use of 'hard' methods sub-ordinate to the meta-level iterations of SSM.

Equally, there is no reason why the conceptualisation processes and the 'weltanschauung' notion used in SSM cannot be employed in the development of not one but many data flow diagrams representing part of each conceptual model. In 'action research' case studies we have employed both forms illustrating the conceptual nature of boundary construction, domain definitions. Lewis (1992) illustrates a radical attempt to bring the notion of 'weltanschauung' to data modelling.

PHENOMENOLOGY VS POSITIVISM [NATURE OF REALITY]

SSM with its attendant philosophical underpinning of promoting many 'world views' [phenomenological] is concerned with pursuing many client concerns and their desirable states. Jayaratna (1986; 1988) defined the philosophical notions of SSM to be a particular form of phenomenology [i.e. hermeneutics] because of the attempts made to understand the individual world images and their historicity through which phenomena are interpreted. Checkland (1981) discusses the heroic mood of Singarian inquiring system being central to SSM which is used for analysing organisations from a hermeneutic stance. Oliga (1988) too interprets the philosophical essence of SSM to be hermeneutics.

Structured methodologies on the other hand subscribe to a particular 'world view'. In many situations the dominant 'world view' is taken-as-given from the client for directing the design activities. Structured methodologies do not alert the designer to this dimension.

The effects of these philosophical differences become evident at many levels in systems development. For example at a diagnosis level, the representation of 'real world' understanding within SSM is taken to mean an expression ['rich picture'] for the exploration of the 'world views' of the client set. In structured methodologies this expression [data flow diagrams] is taken to be the correct representation of observer independent 'real world' systems. De Marco is explicit about this widely held view. "Working closely with the users, you learn and document the way things currently work. Rather than do this from the point of view of *any one user or set of users, you* attempt to assess operations from the viewpoint of the data" (De Marco, 1979, my italics). De Marco fails to realise that data has no viewpoints, only humans do. Checkland contrast the differences between data and information. " We can obtain the data which is the position of the hands on the clock and convert it to information that we are late for an appointment or that we have time for a cup of tea " (Checkland, 1988).

Structured methodologies with their underlying scientific paradigms are seeking to develop an observer independent 'single truth' value while SSM with its underlying hermeneutic paradigms is attempting to make users search for 'many truth' values. The way in which these differences are resolved in grafting are not explained explicitly in the current literature. Lewis's (1992) efforts to develop object oriented databases from different 'world views' is a heroic attempt to address this issue. However, this is an example of using notions of SSM within structured methodologies and not about grafting methodologies. Lewis (1992) does not claim this to be grafting either.

EPISTEMOLOGICAL VS ONTOLOGICAL [NATURE OF THE TRANSFORMATION]

SSM prepares the analyst to use 'systems' as an epistemological notion for creating radically different 'root' definitions and associated conceptual models. In structured methodologies the use of 'systems' is in an ontological sense of the term. This means that the same term when used within the two methodologies lead to quite different conclusions.

The different interpretation in the use of the term 'system' has also quite different impact on the notions of transformation they attempt to bring about in practice. In SSM, the use of 'systems' in the epistemological sense of the term help it to use the methodology as an instrument for learning about the situation while the ontological use of the term within structured methodologies help its users to identify an existing system i.e. an independent system out there in the 'real world' to be explored, identified, documented, analysed and changed. This latter view is confirmed by De Marco (1979) when he states his belief that " the analyst and user ought to communicate across logical/physical boundary that exists in any computer project". The notions of learning in SSM leads it to establish the **relevance** of systems for the situation while the notions in structured methodologies lead them to establish the **correctness** of the system.

Once again current literature on methodology grafting is not explicit about the way they have resolved these philosophical differences.

However it is possible to use the stages, phases, models or techniques from one methodology within the operation of another (embedding). In order to do this the philosophical and conceptual notions underlying the elements of one methodology have to be sacrificed, altered or adapted to fit within those of the other methodology. This is what would effectively take place if data flow diagrams are used as part of 'rich picture' construction or the 'weltanschauung' idea from SSM is used in the construction of data flow diagrams or data models as has been attempted by Lewis (1992).

SYSTEMIC VS SCIENTIFIC [DOMAIN OF INTEREST]

SSM with its use of the term 'system' as an epistemological device makes the users open themselves to any organisational dimension. In this sense the methodology can be used for focusing on different issues (Stowell, 1987; Galliers, 1991; West, 1992). In contrast, structured methodologies are about structuring data and related activities. They manage this by partitioning the problem domain into decomposable models. The systemic notions advocated in SSM makes the methodology user focus on the exploration of <u>context</u> issues while the use of scientific notions advocated in structured methodologies focus attention on the organisation of <u>content</u> issues albeit data dimension.

Problems for grafting arise in the transition from SSM to structured methodologies when the designer lose sight of the systemic issues and become engrossed in the structuring of formal data. Thus what gets captured with techniques of structured methodologies are observable but formally identifiable data and data related activities. Even at the data level they exclude data relationships that have no pattern, stability or regularity. The domain issues of what changes need to be undertaken to tasks, roles, functions, structures etc. are excluded from the design. The restaurant example in Jayaratna (1991) illustrate these exclusions.

Those who attempt to graft the methodologies do need to abandon the use of SSM at its final stage and switch to the use of scientific based structured methodologies for the development of agreed Notional Systems as is the case with Multiview. Stowell observes that " ... benefits gained from the development of information systems through the use of SSM are essentially lost " (Stowell, 1991). Miles (1992) also makes similar conclusions and warns of this stage not to be " .... a signal for soft systems practice to grind to a halt". However under grafting it is very difficult to maintain the philosophical nature of SSM because of the switch over. The other approach is to continue to use structured techniques within the SSM thus increasing the case for embedding (not grafting) models and techniques within the philosophy and structure of SSM.

PROBLEM FORMULATION VS SOLUTION DESIGN

This offers the best possible area for integration provided the questions raised earlier in this paper can be addressed.

SSM is problem formulation oriented while structured methodologies are design oriented. This offers the best opportunity for grafting the two methodologies. However, SSM is not claimed to be a problem solving methodology. Indeed, according to Checkland in some situations the methodology may lead only to a better understanding of the situation. In contrast, structured methodologies can be defined as solution driven opportunity seeking methodologies.

Problems for grafting comes from this orientation of SSM. If the exploration of a problem situation leads to the abandonment of the solution design phase then indeed those who attempt to graft the methodologies must be prepared to abandon the structured methodology phase. The examples illustrated by those attempting grafting shows that they have used SSM as a front-end extension of a structured methodology thus using the strength of SSM's problem formulation abilities to structure only the data related activities. In so doing they have had to sacrifice the systemic nature of the situation and SSM.

CONCLUSION

Serious reservations seem to exist when attempts are made to embed or graft the totality of methodologies that are epistemologically and ontologically diverse. This, given the preceding debate should not come as any surprise, indeed it could be anticipated. There is no a-priori analytic justification for assuming that the application of SSM will lead to a natural continuation into structured information systems development. Nor is there any a-priori analytic justification for assuming that it will not. Only if we distort the essence of soft systems by insisting that it will serve the predetermined intention of IS development may a 'natural' transition to structured methodology occur.

In contrast, if we retain the essential openness of the SSM approach there may well be occasions when, having run its course, our understanding now contains to a greater or lesser extent, the recognition of an IS development opportunity. Such opportunities will, by virtue of SSM contribution, be clearly resident in a well understood context, one which we may well have conceptualised as a relevant context system for discussion and understanding. The reality appears to be that these very different philosophical/methodological propositions:-

[1] are not naturally connected on any a-priori analytic grounds

[2] address very different domains and interests, and serve different purposes

[3] may well, on occasions, have nothing to say to each other [see 5b]

[4] may well, on occasions, have a great deal to say to each other, and when they do, it is a dialogue between two highly relevant systems; the context system and the information systems [see 5a]

[5a] naturally emergent dialogues are likely to improve the quality and value of information systems that we develop

[5b] enforced, predetermined, contrived dialogues are unlikely to provide anything more than enforced, predetermined, contrived justifications for making the outcome what we always anticipated, an information systems of some sort, and finally:

[6] confronted with the obvious weakness of structured methodologies to address 'context systems' there are powerful incentives to address this issue. Intuitively SSM seems to be a promising means of progress, explicitly that progress must be debated, researched and justified in terms of the conclusions of this paper.

REFERENCES

Avison, D. E. and Wood-Harper, A. T., 1990, "Multiview", Blackwell Scientific Publications, Oxford.

Checkland, P. B., 1981, "Systems Thinking, Systems Practice", Wiley, Chichester.

Checkland, P. B., 1988, Information Systems and Systems Thinking: Time to Unite?, *International Journal of Information Management*, Vol. 8, No. 4, pp239-246.

Checkland, P. B. and Scholes, J., 1990, "Soft Systems Methodology in Action", Wiley, Chichester.

Computer Journal, The : Special Issue in Information Systems, 1991, Cambridge University Press, Vol. 34 No. 2, pp160-169.

De Marco, T., 1979, "Structured Analysis and Systems Specification", Yourdon Press, New York.

Galliers, R. D., 1991, Strategic Information Systems Planning: Myths, Reality and Guidlines for Successful Implementation, *European Journal of Information Systems*, Vol. 1, No. 1, pp55-54.

Jayaratna, N., 1986, Normative Information Model-based Information Systems Analysis and Design (NIMSAD): A Framework for Understanding and Evaluating Methodologies, *Journal of Applied Systems Analysis*, Vol. 13, pp73-87.

Jayaratna, N., 1988, Guide to Methodology Understanding and Information Systems Practice, *International Journal of Information Management*, Vol. 8, No. 1, pp43-53.

Jayaratna, N., 1991, Systemic Analysis: The Weak Link in the Systems Development Process!, *Journal of Applied Systems Analysis*, Vol. 18, pp61-67.

Jayaratna, N., 1991a, "A Conceptual Framework for understanding and Evaluating Infomnation Systems Methodologies", Unpublished PhD Thesis.

Lewis, P., 1992, "The Feasibility and Desirability of a Closer Linking of SSM with Data-Focused Information Systems Development", *Systemist*, Vol. 14, No. 3, pp168-179.

Miles, R.K., 1988, Combining 'Soft' and 'Hard' Systems Practice: Grafting or Embedding, *Journal of Applied Systems Analysis*, Vol. 15, pp55-60.

Miles, R.K., 1992, Combining 'Hard' and 'Soft' Systems in Practice: Grafting and Embedding Revisited, *Systemist*, Vol. 14, No. 2, pp62-66.

Oliga, J., 1988, Methodological Foundations of Systems Methodologies, *Systems Practice*. Vol. 1, No. 1, pp87-112.

Rodriguez-Ulloa, R., 1988, The Problem Solving System: Another Problem Content System, *Systems Practice*, Vol. 1, No. 3, pp211-217.

Stowell, F., 1987, Cooperation, Power and the Impact of Information Systems, *Systems Practice*, Vol. 1, No.2, pp181-192.

Stowell, F., 1991, Towards Client-led Development of Information Systems, *Journal of Information Systems*, Vol. 1, No. 3, pp173-189.

West, D., 1992, Knowledge Elicitation as an Enquirying System: Towards a Subjective Knowledge Elicitation Methodology, *Journal of Information Systems*, Vol. 2, No. 1, pp31-44.

Wood-Harper, A. T., Antill, L. and Avison, D.E., 1985, "Information Systems Definition: The Multiview Approach", Blackwell Scientific Publications, Oxford.

SYSTEM SCIENCE: A NEW IMPERIALISM?

David G. Jenkins

Department of Computing Science
University of Paisley
Scotland, UK, PA1 2BE

INTRODUCTION

The author's interest in the impact of cultural and social norms on the conduct of the systems' approach arises from his experience on being appointed as a Principal Systems Technologist in a British manufacturing enterprise with a global business. This was in 1970, at the very outset of the pioneering work of Jenkins and Checkland at Lancaster University. Great difficulty was experienced in making any use of the approach within the enterprise. In the author's judgement, this was due to the highly traditional management style and organisational structure, and the intellectual ethos, Taylorism, that underlay it. To control information was to exercise power, with inescapable consequences for anyone trying to take a systems approach or to do systems engineering, which in turn, was poorly understood, being understood then, as Checkland (1979) indicates, in terms of *hard* systems engineering.

THE PROMISE OF SOFT SYSTEMS METHODOLOGY

Since the pioneering work noted above, systems science has moved on from the approach of *hard* systems engineering, in which the addressing of any, and hence global issues also, requires the construction of a benefit function over the state variables of the system being modelled, to the approaches of *Soft Systems Methodology (SSM)* with its focus on *Human Activity Systems* rather than on engineering systems. Here Checkland *op. cit.* can be quoted to advantage:

> The major impression which the work has made on those doing it is that *soft* system studies are markedly different from the *hard* systems engineering which was the intellectual starting point. Engineering proper aims to provide a solution to a defined problem which is optimally efficient in the terms in which the problem is posed. ... The best chemical engineering is that which defines the mcst efficient process, the one yielding most product at least cost, and all of the work can be measured against the achievement of this single value, efficiency. This kind of procedure turns out not to be a useful model for studying soft systems. The special nature of human activity systems means that system studies concerned with them are always multi-valued, with

Systems Science, Edited by F.A. Stowell *et al.*,
Plenum Press, New York, 1993

> many relevant and often conflicting values to be explored. The outcome is never an optimal solution to a problem, it is rather a learning which leads to a decision to take certain actions, in the knowledge that this will lead in general not to the problem becoming solved but to a new situation in which the whole process can begin again.

He also writes:

> The thinking embodied in *systems engineering* and *systems analysis* is essentially the same. Analysis of many different accounts of these activities shows that they all assume that problems can be formulated as the making of choices between alternative means of achieving a known end. The belief that real-world problems can be formulated in this way distinguishes all *hard* systems thinking from *soft* systems thinking which in dealing with soft problems finds the designation of objectives and goals itself problematic.

Although Checkland sharply distinguishes between the activity of *hard* systems engineering approach and his open ended *SSM*, the appearance of *fuzzy logic* and the recognition that often design engineers, and certainly architects (Heath 1984), do examine the systems context at the early stages of design, blurs the sharp distinction between the approaches which he makes. But it is undoubtedly true that early acceptance that there is a well-defined system to optimize, and the assumption that an artifact, building or computer-based information system is the required solution moves the problem solver into a phase of the work in which the *optimisation of a known system* paradigm works, and is used.

BARRIERS TO CLIENT-LED DESIGN

As remarked in (Stowell 1991): *The whole process of developing and designing a computer-aided information system necessitates an intervention into the working practices and customs of an organisation's culture.* He rightly avers that: *The modern information systems professional, therefore, must be a competent technical advisor and possess many of the skills of a social scientist.*

The range of skills implied here may be gauged from Schutz (1978) who described the task of the social scientist as *to investigate to what extent the different forms of systems of relevancy in the life-world, motivational, thematical and, most of all, interpretational systems, are socially and culturally conditioned.*

Checkland discusses *SSM* in the light of the need to change *Human Activity Systems* and as a method by which the problem solver, together with the problem owners, move to the point when they understand what has to be done to improve the *Human Activity Systems* in ways that are seen as improvements within the *Weltanschauungen* of all participants. This is very close to the concept of client-led design advocated by Stowell, and to the method advocated by Heath (1984) that seeks to involve clients in this way.

Realistically, Heath points out that the architect/designer must often of necessity construct a mental image of a composite user which he or she interacts with to guide the design decision making process when access to the real clint population is impossible by reason of distance, &c. It is this construction of a mythical (in the strict sense) customer that leads technology orientated designers of application generators to build generators that can only build applications which mirror his/her view as a technician of what the client needs, and not what the client actually needs and the sort of interface they could use, that is, they unreflectively impose their *Weltanschauung*. This might well be described as *technocratic imperialism.*

However, as we shall see, *SSM* and *client-led design* do not automatically avoid the imposition of one *Weltanschauung*, that of the problem-solver, upon the *solution seeking*, or, more correctly, *understanding building*, process. This process is advanced by the decisions as to matters of fact and of judgement made by the group. *SSM* avoids the trap of presenting information in technical jargon to users which they cannot understand with the consequent paralysis of decision making and abandonment of involvement and responsibility seen in many IS projects in which (pseudo) progress has been made by the analyst imposing his/her solution. It seeks to do this by requiring the analyst to reflect back to the client/user group a conceptual model of an ideal understanding of the problem situation in the light of a selected and explicit *Weltanschauung*, and to do this for each *Weltanschauung* identified. But this conceptual model is expressed (for Checkland) in English, and, as Jenkins (1975) notes, in the English of *distinctive client subcultures.*

THE ROLE OF LANGUAGE IN SYSTEMS PRACTICE

By stating that the language *may use all the verbs of English*, Checkland identifies the concern of this paper: *Does the selection of a language constrain the possible outcome in a way that would result in the redesigned Human Activity System, using, perhaps, some computer-based information system, being distorted by the cultural baggage of the chosen language in which the analyst is working?*
As Checkland *op. cit.* himself says:

> In SSM, *cultural feasibility* is one of the criteria which potential changes in the real world must meet if they are to be implemented. The implication is that the culture of a particular problem situation, with its unique norms, roles, and values, will be able to accept as meaningful and possible, a certain range of changes.

As any visitor to Scotland knows, Winston Churchill's *dictum* about the English and the North Americans as *two peoples separated by a common language* conveys a good insight into this problem. Indeed, even at Independence, it was possible to convince Congress that there was no need to adopt a native North American language as the official language of the United States since American English was sufficiently different to constitute a separate language.

Even within one organisation (Jenkins *op. cit.*), the analyst must observe these *Musts in user relationships*:

1. Realise that the users collectively know more about the problem than you ever will.
2. Be aware that this knowledge is fragmented across departments and up and down the hierarchies within the enterprise.
3. Realise that different sorts of decision must be made at each level.
4. Each sort of decision requires communication at the corresponding level both in language, type of model, and in level of detail, strategic, tactical, technical.

Checkland *op. cit.* observes that *A designed system is a man-made entity which an observer elects to treat as a whole having emergent properties, it may be concrete or abstract.*
Lewis (1992), writing about methods such as SSADM, endorses the power of this election:

> If entities are taken to have a real world existence then the analyst must be choosing, from the infinitely large number of entities that exist, and might be included in a data model, only some entities as being *relevant* and suitable for inclusion. Over twenty years worth of literature provides no suggestion that this is done other than subjectively, as a choice of the analyst. This means that, however well researched and rigorous are the techniques for manipulation and refinement of the conceptual data model, they ultimately rely upon a subjective and interpretive identification of entities and of objects.

> Consequently, the process of creating such a model is not value-free and the resulting data model is not *neutral*; it will be affected very much by the values, beliefs and expectations of those creating it.

This is the point where the choice of language is all: To quote Popper (in Schlip 1974): *At any moment of our time we and our values are products of existing institutions and past traditions. Admittedly this imposes some limitations on our creative freedom and on our powers of rational criticism...*
And in (Lakatos and Musgrave 1970): ... *at any moment we are prisoners caught in the framework of our theories; our past experience; our language.* More recent thinking on *Information Systems Design* reinforces the role of language. Gregory (1992) advocates and uses the *logico-linguistic model* as a bridge from SSM's *Conceptual Models* towards the more detailed view need to construct either process (SADT) or data (SSADM) oriented analytical and synthetic models from which the elementary processes and data flows of an IS solution can be derived.
Frayn (1974) has written: *The complexity of the universe is beyond expression in any possible notation. Our notations are by their very nature simpler than what we denote. Our reading of the world and our mastery of notations are intimately linked. We read the world in the way that we read a notation - we make sense of it, we place constructions upon it.*

The range of expression, the range of constructions, available in a language must then constrain our thinking about the world, for surely if something cannot be said then it cannot be rationally considered.

LANGUAGE AS A POWER STRUCTURE

In developing his critique of Western society, Habermas (1970) contrasts ***purposive-rational action*** *governed by technical rule based on empirical knowledge* and ***communicative action***, or ***symbolic interaction***, *governed by binding consensual norms which define reciprocal expectations about behaviour.* In science based capitalist societies purposive-rational action has become dominant through its institutionalisation; and positivist natural science, supposedly value-free, has supplied an apparent legitimation for this process. We have, then, a society in which behaving in accordance with technical recommendations is the only rational way to behave; our culturally-defined self-understanding of a social life-world (as in traditional pre-capitalist societies) is replaced by *the self-reification of man under categories of purposive-rational action and adaptive behaviour.* For Habermas (as summarised in Checkland *op. cit.*) RAND Corporation Systems' Analysis supplies a **sinister means by which cultural control can be exercised** (author's emphasis). To avoid the negative Utopia which will result otherwise, Habermas (1970) contends:

> it will be necessary to remove cultural restrictions on communication, to achieve public unrestricted discussion, free from domination, of the suitability and desirability of action-orienteering principles and norms, which requires, in turn, both equal participation in discussion, undistorted by power relationships, and unlimited scope for a radical questioning of societal structures and procedures.

Thus while Habermas is aware of the cultural dominance of the Western *Weltanschauung* it is Marcuse that Checkland *op. cit.* quotes as expressing the role of language by quoting in turn Adorno: *language itself has been so much permeated by the Establishment that clear intelligible writing itself expresses ... the control and manipulation of the individual by the power structure*, as a measure of the rejection of the science-dominated culture of Western societies by the Frankfurt school, to the point that even Marx's analysis is seen as needing further distancing from positivism and extension to cover more than economic factors.

We conclude, then, that language will be a powerful determiner of the outcomes of system practice when faced with global issues.

LEARNING FROM HISTORY

Western systems practice is in much the same situation as the Christian missionary movement (Neill 1964) at the start of the Victorian era, a previous example of a behaviour changing movement. The content of the message and the type of change are, for the purpose of this paper, not relevant. Discussing the growth of the indigenous church during 25 years persecution in Madagascar, he notes the role played by the Malagasy New Testament and indicates that this justified the conviction of protestant missionaries that *the first duty of the missionary, after he had once learned the language, is to provide the people with the Christian Scriptures without note or comment in their own language.*

The strategy was to present, as it were, the *model* in the mother tongue so that an unbiased by the cultural overtones &c of English, German, French, Dutch, ... But systems practitioners must learn from the fact that this was not enough to prevent new Christians absorbing the (denominational) church culture of the missionary, an absorption reinforced by the inability of many, but not all (Jowett Murray 1919), missionaries to even consider that leadership could be non-Western.

Here then the role of language in introducing change is central. But the structures, and personal attitudes of the systems practitioner and his or her cultural sensitivity, can be seen to be crucial to success. Thus Stowell's inclusion of Social Science skills is as necessary as the linguistic skills needed to translate into the receptor culture. The attitude of militant Catalan nationalists to Spanish is notable, they will not speak[1] it for they see it as a language of oppression. The task of the systems analyst is to select and use language, the native language of the problem stake-holders, in such a way that he or she does not act in an *imperial* fashion.

Thus the choice of English, even international English, as the language for systems science, and the SSM, et al., risks imposing a Western, capitalist, scientific *Weltanschauung*[2] of the analyst upon the other participants in the situation in which they are not discussing the problem situation in their mother tongue. It may inhibit the problem owners or stake holders from expressing their viewpoint and concerns in a way that enables the consensual problem clarification central to *SSM* to proceed.

Government to government aid will inevitably be channelled into the existing structures, which have a vested interest in inhibiting change (Heath op. cit.). But this is a recognised problem at the political level which can be addressed within *SSM* while our concern here is with the challenge of the necessarily *built in* problems of the method. The problems are not, of course, those of *SSM* alone, all methods share it, but the nature of the method means that its promise is blighted by this problem if unrecognised and dealt with by the practitioner.

---

[1]Report in The Guardian, 19 November 1992

[2]This analysis also applies to the gulf between the middle-class academically-trained systems worker and the inhabitants of any large urban conurbation.

CONCLUDING REMARKS

SSM, with its recognition of multiple *Weltanschauungen* and the provisional nature of the analyst's *Conceptual Model* can form the basis of a non-imperialist non-Western systems science but only if the skills of the Social Scientist, but also those of the linguistically equipped translator are available to the systems worker when the problem is owned by, or the customers are, native speakers of a language other than English[3].

This in turn means that serious attention needs to be given to *curriculum development* for courses in *Business Information Systems* where the aim is to prepare the graduate to work outside his or her native culture.

The indicated skills and the attendant methodologies of both the social scientist and the linguistically equipped translator must be acquired to a high degree. An institution wishing to mount such a course must have both the appropriate range of professional skills and the institutional culture which permits their integration in this way. In the author's experience, while the required breadth may be found in both traditional and newer universities, the later are more likely to be free of inter-disciplinary rivalry, and it is to these that we should look for successful courses of this type.

ACKNOWLEDGEMENTS

The author wishes to thank his colleagues Dr. Marek Machura and Arshad Mahmood, for many stimulating discussions about the problems of translation and cultural dominance.

REFERENCES

Checkland, P.B., 1979, "Systems Thinking, Systems Practice", Wiley, Chichester.

Frayn, M., 1974, "Constructions", Collins, London.

Gregory, F., 1992, "SSM to Information Systems: A Logical Account", *Systemist*, Vol. 14, No. 3, pp180-189.

Habermas, J., 1970, On systematically distorted communication, and, Towards a theory of communicative competence, *Inquiry*, Vol. 13, pp205,360.

Heath, T., 1984, "Method in Architecture", Wiley, Chichester.

Jenkins, D.G, 1975, "Professional programming' Lecture Notes for Junior Honours and Advanced Ordinary Classes in Computing Science, Department of Computing Science, University of Glasgow, Glasgow.

Jowatt Murray, A.H., 1919, The Upbuilding of the Church *in:* "The Chinese Recorder", March 1919, pp153-4; quoted in (Neill, 1964), p534.

Lakatos, I. and Musgrave, A., ed., 1970, "Criticism and the Growth of Knowledge", Cambridge University Press, London.

Lewis, P.J., 1992, "The Feasibility and Desirability of a Closer Linking of SSM with Data-focused Information Systems Development", *Systemist*, Vol. 14, No. 3, pp168-179.

Neill, S., 1964, "A History of Christian Missions", Penguin Books, Harmondsworth.

Schlip, P.A., ed., 1974, "The Philosophy of Karl Popper", Open Court, La Salle.

Schutz, A., 1978, Some Structures of the Life-World, reprinted in Luckman, T., ed., "Phenomenology and Sociology", Penguin Books, Harmondsworth.

Stowell, F.A., 1991, Client-Led Design, *in:* "BA in Business Information Technology, IS Workshop notes", A.J. Quin, ed., University of Paisley, Paisley.

---

[3] or another Western language, Spanish, French, German, ...

METAPHORS AS SEEN BY A KNOWLEDGE ELICITOR

Dr Ali Behrooz[1] & Dr Garry Marshall[2]

[1] Director of Postgraduate Research Studies
Department of Management Systems
Bournemouth University
Bournemouth BH12 5BB

[2] GEC Hirst Research Centre
Wembley
Middlesex

INTRODUCTION

A methodology for the design of knowledge-based systems (KBSs) to operate in complex domains has been described by Behrooz (1986). The basic stages of the methodology are the elicitation of the domain knowledge, the determination of the structure of this knowledge, in terms of its elements and their interrelationships, and the use of the structure as the structure of a KBS to operate in the domain. When domain knowledge is elicited from a human expert, it is the structure of this person's knowledge that determines the form of the KBS. This ensures not only that the KBS possesses the expert's knowledge but also that the KBS adopts the same approach and displays the same behaviour as the human expert. This ensures that people interacting with the KBS will find themselves on familiar ground in that they can expect to be treated, and, indeed, will be treated, in the same way as they would when dealing with a human expert.

Clearly, the success of this enterprise depends, in the first place, on the successful elicitation of knowledge from the expert. This paper suggests that an important part of this process, and one which has not so far received sufficient attention, is the realisation that experts often use metaphors to codify their expertise. Consequently, a knowledge elicitor may usefully seek to recognise the metaphors used by an expert and to understand their significance by, for example, clarifying the correspondence between the components of the metaphor and the things that they stand for; finding how far the metaphor may be pursued before the correspondences break down, and why the breakdown occurs; and detecting whether there are multiple levels of reference within the metaphor.

Systems Science, Edited by F.A. Stowell *et al.*,
Plenum Press, New York, 1993

OVERALL PROCESS OF KNOWLEDGE ELICITATION

Knowledge Elicitation is an interaction between two contributors, one acting as the observer (the elicitor) and the other being observed (the expert). The knowledge elicitation process is a complex structured interactive activity between these two contributors.

The elicitor-expert interaction can be modelled by regarding it as a super-activity (Behrooz, 1992a). This super-activity structurally and functionally is formed by five components, each being a complex activity on its own (Fig. 1).

When the elicitor is conversant with the basic concepts and vocabulary of the domain to be explored, he then begins continuous sessions with the expert, while studying textbooks and cases, etc. which enable him to create his questions.

This process is known as the identification of components and concepts, during which the elicitor would establish the very basic logic of the expert's reasoning.

As the human expert, who has taken his implicit reasoning knowledge process for granted, begins to verbalise his thinking, his explanation is usually unstructured. He wants to explain almost everything in general terms within the first session. The expert, naturally, feels that whatever he does is reasonably straightforward, but in reality it is not so. To the elicitor who is not familiar with the domain, the process is neither very simple nor straightforward. This shortcoming of the elicitor's understanding of the expert's specific reasoning will eventually be rectified, although this is depends to a large extent upon the elicitor's successful familiarisation with the terminology and metaphors used by the expert.

It is the elicitor's function to map the metaphors of the expert to their actual processes and conceptualisations.

DEFINING METAPHORS

Weaver (1967) and Garber (1976) and many others who are deeply involved with conceptualisation and the implication of metaphors in socio-political aspects of life have deduced that metaphor is an essential element of a natural language, in that it defines certain aspects of an object, yet distracts observers from other portions of that object (Madsen, 1989). Here, we would not attempt either to review these authors' contributions or to examine the soundness of their definitions. A metaphor, so far as the elicitor is concerned, is the terminology or concept that could lead the elicitor to a new environment.

These motivating concepts, as defined by Behrooz (1986), trigger the elicitor to associate certain elements of the sub-domain to some other conceptualisation.

This may be demonstrated in the following example extracted from the minutes of a session between an elicitor and a medical expert:

> "....The syndrome is something that is a pathway on the formation of a specific diagnosis . . . . it is a cotlection of symptoms and signs, or symptoms and test results which runs you down towards a particular set of diseases.... "

Here, the expert is employing the figure of speech 'running down a pathway' as a metaphor for the procedure by which a diagnosis is made. Regardless of ambiguities in the definitions of syndrome, symptom, and signs, the elicitor can deduce that:

1- If there is a *"specific"* diagnosis, it can be assumed that there must exist some nonspecific (perhaps general) diagnosis too. But what it is, and how it could be distinguished is another matter.

2- *"Syndrome"*, this indirectly indicates that the domain (of medicine) possesses knowledge with classifications.

3-*"Runs you down"*, emphasises that the diagnostic procedure has some levels. This phrase also suggests that the structure of diagnostic procedures is of a top-down hierarchical nature.

Now consider another example taken from telecommunications where, in the sub-domain of digital communications, system designers share a considerable amount of expertise expressed in the form of metaphors. The figures of speech incorporated in these metaphors include 'transparency', 'crossing a threshold' and, as illustrated in the following example which is a quotation from a system designer, the opening and closing of an eye.

> *"The eye diagram shows whether the signal has sufficient margin against the noise. If the eye is closed, the signal-shaping filters are too narrow and there is no margin against noise. If the eye is wide open, the noise margin is satisfactory, although the signal-shaping filters may be wider than is necessary or desirable. "*

Here, the degree of openness of the eye corresponds to the amount of signal distortion. In the designer's world, the more the signal is distorted, the more vulnerable it is to noise or, in the designer's words, the lower its margin against the noise becomes.

If the extent to which the metaphor may be pursued is examined, it emerges that there is no meaningful correspondebnce in the domain to a blinking, or winking, eye. The designer's problem is to find the appropriate fixed level at which the eye should be open.

The metaphor can also be seen to possess an intermediate level because when the designer talks of an *eye,* he may on some occasions be referring to a metaphorical eye and on others to a display on the screen of an oscilloscope (derived from the communications signal) which looks like an eye. Thus, the use of the word *eye* may be metaphorical while referring to an eye on the screen of the oscilloscope which, in turn, corresponds to a degree of signal distortion.

It has been shown in many cases (Behrooz, 1987) that the concept of metaphors is not domain dependent, that is, the means of communication in any system somehow accommodates metaphors which act as a short-cut facility during the process of communication. Lanzara (1983) and Schon (1979) have discussed the influence of natural language on systems analysis and systems research. Metaphors in a everyday language could be understood by any layman, where the metaphors in a knowledge domain as discovered by an elicitor are specifically domain dependent, and may well be due to the expert's own understanding of the knowledge domain.

If we want to apply one of Clancy's (1989) types of metaphors, such as family, journey, organism, war, social, etc., the nearest would be the organism and family metaphors, as knowledge elicitation is a goal-oriented interaction between at least two participants with certain characteristics and relationships. It is also a time dependent activity that involves at least two human beings. Each participant is expected to have his own views and approaches to a given problem, his own personal vocabulary, and his personal way of seeing and analysing things.

CONCLUSION

By realising that experts use metaphors, and by taking advantage of it, a knowledge elicitor can improve the success of knowledge elicitation. Since metaphors tend to encapsulate 'chunks' of knowledge, their detection and decoding also helps with the structuring of an expert's knowledge. Thus, an awareness of the use of metaphors by experts and an ability to take advantage of them are valuable additions to the repertoire of the knowledge elicitor. Once the procedures are fully developed, they may be used in the course of automatic knowledge elicitation and embedded in, for instance, knowledge-acquisition machines (Behrooz, 1992b), to help overcome the knowledge-elicitation bottleneck.

If machines are to be used to perform tasks that have otherwise been performed by people, it seems reasonable that they should carry them out in the same way as people. If they do not, there is a distinct danger that, even if they can perform their tasks successfully, they will be shunned by their intended users. One way to develop such a machine is to employ a development methodology based on the expertise of the person it is intended to replace, whether it is intended as a partial replacement (an assistant) or a total replacement. A methodology for the development of KBSs which is driven by the elicitation of an expert's knowledge has been described elsewhere. The elicitation is followed by a determination of the structure of that knowledge, and the re-use of this structure as the structure of the KBS itself. Such systems will not fail because they do not behave in a sympathetic fashion. On the other hand, accurate and efficient elicitation of knowledge is a goal that is still waiting to be achieved. A recognition of the potential value of metaphors in knowledge elicitation should help toward the achievement of this goal.

REFERENCES

Behrooz, A., 1986, "Meta Description of Experimental Identification of Medical Knowledge", Ph.D. Thesis, Department of Computer Science, Brunel University, U.K.

Behrooz, A. and Kohout, L.J., 1987, Experimental Identification and Formalisation of Knowledge Structure representing Uncertainty, *in*: "Proc. 7th Congress on Cybernetics and Systems", Imperial College, London, U.K.

Behrooz, A., 1992a, A New Knowledge Elicitation Method, *in*: "KBSs for Multiple Environments", L. Kohout, J. Anderson & W. Bandler, eds., Ashgate Publishing Ltd, Gower House, Aldershot, U.K. 105-124.

Behrooz, A., Marshall, G. and Fatmi, H., 1992b, Knowledge Acquisition MAchine: A Constructive Element of Co-operative Expert Systems (COOPES), *in*: "13th Int. Cong. on Cybernetics", Namur, Belgium.

Clancy, J.J., 1989, "The Invisible Powers; The Language Of Business", Lexington Books, Lexington, MA.

Garber, D.A., 1976, "Verbal Behaviour And Politics", University Of Illinois Press, Chicago.

Lanzara, G.F., 1983, The Design Process: Frames, Metaphors, And Games, *in*: "SystemsDesign For, With And By The Users", U. Briefs, C. Ciborra and L. Schneider, eds., North Holland, New York.

Madsen, K.H., 1989, Breakthrough By Breakdown: Metaphors And Structured Domains, *in*: "Systems Development For Human Progress", H.K. Klein and K. Kumar, eds., North Holland, New York.

Schon, D., 1979, Generative Metaphor: A Perspective On Problem Setting In Social Policy, *in*: "Metaphor And Thought", A. Ortony, ed., Cambridge University, Cambridge, UK.

Weaver, R.M., 1967, "A Rhetoric And Handbook", Holt, Rinehart, and Winston, New York.

AUTHOR INDEX

INDEX